ANNALS OF THE NEW YORK ACADEMY OF SCIENCES

Volume 934

EDITORIAL STAFF

Executive Editor
BARBARA M. GOLDMAN

Managing Editor
JUSTINE CULLINAN

Associate Editors
STEFAN MALMOLI

The New York Academy of Sciences
2 East 63rd Street
New York, New York 10021

THE NEW YORK ACADEMY OF SCIENCES
(Founded in 1817)

BOARD OF GOVERNORS, September 2000 – September 2001

BILL GREEN, *Chairman of the Board*
TORSTEN WIESEL, *Vice Chairman of the Board*
RODNEY W. NICHOLS, *President and CEO* [ex officio]

Honorary Life Governors
WILLIAM T. GOLDEN   JOSHUA LEDERBERG

JOHN T. MORGAN, *Treasurer*

Governors

ELEANOR BAUM	D. ALLAN BROMLEY	KAREN BURKE
LAWRENCE B. BUTTENWIESER		PRAVEEN CHAUDHARI
JOHN H. GIBBONS	MICHAEL GOLDEN	RONALD L. GRAHAM
ROBERT G. LAHITA	JACQUELINE LEO	WILLIAM J. McDONOUGH
JOHN F. NIBLACK	SANDRA PANEM	RICHARD RAVITCH
RICHARD A. RIFKIND	SARA LEE SCHUPF	JAMES H. SIMONS

HELENE L. KAPLAN, *Counsel* [ex officio]   PETER H. KOHN, *V.P. & Secretary* [ex officio]

HEAT TRANSFER IN GAS TURBINE SYSTEMS

ANNALS OF THE NEW YORK ACADEMY OF SCIENCES
Volume 934

HEAT TRANSFER IN GAS TURBINE SYSTEMS

Edited by Richard J. Goldstein

The New York Academy of Sciences
New York, New York
2001

Copyright © 2001 by the New York Academy of Sciences. All rights reserved. Under the provisions of the United States Copyright Act of 1976, individual readers of the Annals *are permitted to make fair use of the material in them for teaching or research. Permission is granted to quote from the* Annals *provided that the customary acknowledgment is made of the source. Material in the* Annals *may be republished only by permission of the Academy. Address inquiries to the Permissions Department (editorial@nyas.org) at the New York Academy of Sciences.*

Copying fees: *For each copy of an article made beyond the free copying permitted under Section 107 or 108 of the 1976 Copyright Act, a fee should be paid through the Copyright Clearance Center, Inc., 222 Rosewood Drive, Danvers, MA 01923 (www.copyright.com).*

⊗ *The paper used in this publication meets the minimum requirements of the American National Standard for Information Sciences—Permanence of Paper for Printed Library Materials, ANSI Z39.48-1984.*

Library of Congress Cataloging-in-Publication Data

Heat transfer in gas turbine systems / editor, Richard J. Goldstein.
   p. cm. — (Annals of the New York Academy of Sciences; v. 934).
   Includes bibliographical references and index.
   ISBN 1-57331-328-9 (cloth: alk. paper) — ISBN 1-57331-329-7 (pbk.: alk. paper)
   1. Gas-turbines. 2. Heat—Transmission. I. Goldstein, Richard J. II. Series.

Q11.N5 vol. 934
[TJ778]
500 s—dc21
[621.43′3]   2001030670

PCP
Printed in the United States of America
ISBN 1-57331-328-9 (cloth)
ISBN 1-57331-329-7 (paper)
ISSN 0077-8923

ANNALS OF THE NEW YORK ACADEMY OF SCIENCES

Volume 934
May 2001

HEAT TRANSFER IN GAS TURBINE SYSTEMS

Editor
RICHARD J. GOLDSTEIN

[This volume is the result of a conference entitled *Turbine-2000: International Symposium on Heat Transfer in Gas Turbine Systems* sponsored by the International Centre for Heat and Mass Transfer (ICHMT) on August 13–18, 2000, in Çesme, Izmir, Turkey.]

CONTENTS

Preface. *By* R. J. GOLDSTEIN	xi
Synopsis. *By* V. SRINIVASAN, T. W. SIMON, AND R. J. GOLDSTEIN	1

Invited Papers

Secondary Flows in Axial Turbines—A Review. *By* L. S. LANGSTON	11
Heat Transfer near Turbine Nozzle Endwall. *By* M. K. CHYU	27
Transition to Turbulence under Low-Pressure Turbine Conditions. *By* T. W. SIMON AND R. W. KASZETA	37
Turbulence Modeling in Simulation of Gas-Turbine Flow and Heat Transfer. *By* G. BRERETON AND T. I-P. SHIH	52
A Review of Turbine Blade Tip Heat Transfer. *By* R. S. BUNKER	64
Unsteady Flow Modelling in Turbine Stage. *By* F. MARTELLI	80
The Detailed Structure and Behavior of Discrete Cooling Jets in a Turbine. *By* F. LEBOEUF AND O. SGARZI	95
Flow and Heat Transfer Predictions for Film Cooling. *By* S. ACHARYA, M. TYAGI, AND A. HODA	110
Film Cooling: What Did We Learn from Our Measurements? *By* T. ARTS	126

Combustor Liner Cooling Technology in Scope of Reduced Pollutant
    Formation and Rising Thermal Efficiencies. *By* A. SCHULZ 135

Jet-Impingement Heat Transfer in Gas Turbine Systems. *By* B. HAN AND
    R. J. GOLDSTEIN 147

Recent Developments in Turbine Blade Internal Cooling. *By* J-C. HAN AND
    S. DUTTA 162

Heat Transfer Technology for Internal Passages of Air-Cooled Blades for
    Heavy-Duty Gas Turbines. *By* B. WEIGAND, K. SEMMLER, AND
    J. VON WOLFERSDORF 179

Cooling Systems for Ultra-High Temperature Turbines. *By* T. YOSHIDA 194

Some Current Research in Rotating-Disc Systems. *By* J. M. OWEN AND
    M. WILSON 206

Selection of a Turbine Cooling System Applying Multidisciplinary Design
    Considerations. *By* B. GLEZER 222

Contributed Papers

Analysis of Particle-Laden Flow and Heat Transfer in Cascade and Rocket
    Nozzle. *By* H. H. CHO, W. S. KIM, M. S. YU, AND J. C. BAE 233

Convective Heat Transfer on an Inlet Guide Vane. *By* M-L. HOLMER,
    L-E. ERIKSSON, AND B. SUNDEN 241

Effect of Reynolds Number, Turbulence Level, and Periodic Wake Flow on
    Heat Transfer on Low-Pressure Turbine Blades. *By* D. SUSLOV,
    A. SCHULZ, AND S. WITTIG 249

Surface Temperature Mapping of Gas Turbine Blading by Means of
    High-Resolution Pyrometry. *By* S. L. F. FRANK 257

Studies on Free Stream Turbulence as Related to Gas Turbine Heat Transfer:
    A Review of Authors' Past Work and Future Implications. *By*
    S. YAVUZKURT AND G. R. IYER 265

A Conjugate Heat Transfer Procedure for Gas Turbine Blades. *By* G. CROCE . . . 273

Heat/Mass Transfer Characteristics on Turbine Shroud with Blade Tip
    Clearance. *By* H. H. CHO, D. H. RHEE, AND J. H. CHOI 281

Heat Transfer and Flow Characteristics on a Gas Turbine Shroud.
    By M. OBATA, M. KUMADA, AND N. IJICHI 289

A Novel Digital Image Processing System for the Transient Liquid Crystal
    Technique Applied for Heat Transfer and Film Cooling Measurements.
    By G. VOGEL AND A. BOELCS 297

Contribution of Heat Transfer to Turbine Blades and Vanes for High-Temperature Industrial Gas Turbines: Part 1. Film Cooling. *By* K-I. TAKEISHI AND S. AOKI 305

Experimental Investigation of Film Cooling Flow Induced by Shaped Holes on a Turbine Blade. *By* S. BARTHET AND F. BARIO 313

Film Cooling from Rows of Holes—Effect of Cooling Hole Shape and Row Arrangement on Adiabatic Effectiveness. *By* J. DITTMAR, I. S. JUNG, A. SCHULZ, S. WITTIG, AND J. S. LEE 321

Effects of Bulk Flow Pulsations on Film Cooling with Shaped Holes. *By* H-W. LEE AND J. S. LEE 329

Film Cooling: Case of Double Rows of Staggered Jets. *By* E. DORIGNAC, J. J. VULLIERME, P. NOIRAULT, E. FOUCAULT, AND J. L. BOUSGARBIÈS ... 337

Characteristics of Various Film Cooling Jets Injected in a Conduit. *By* H. TAKAHASHI, C. NUNTADUSIT, H. KIMOTO, H. ISHIDA, T. UKAI, AND K-I. TAKEISHI 345

Film Cooling Performance on Curved Walls with Compound Angle Hole Configuration. *By* P-H. CHEN, M-S. HUNG, AND P-P. DING 353

The Variation of Heat Transfer Coefficient, Adiabatic Effectiveness, and Aerodynamic Loss with Film Cooling Hole Shape. *By* J. E. SARGISON, S. M. GUO, M. L. G. OLDFIELD, AND A. J. RAWLINSON 361

Numerical Investigation of Film Cooling Flow Induced by Cylindrical and Shaped Holes. *By* S. BARTHET AND P. KULISA 369

Numerical Investigation of Heat Transfer on Film-Cooled Turbine Blades. *By* P. GINIBRE, M. LEFEBVRE, AND N. LIAMIS 377

Comparison between Two Models of Cooling Surfaces Using Blowing. *By* L. MATHELIN, F. BATAILLE, AND A. LALLEMAND 385

Finite Element Analysis of Flow Field in the Single Hole Film Cooling Technique. *By* F. BAZDIDI-TEHRANI AND A. A. MAHMOODI 393

Effects of Entrance Cross-Flow Directions to Film Cooling Holes. *By* C. SAUMWEBER, A. SCHULZ, S. WITTIG, AND M. GRITSCH 401

Mach Number Effect on Jet Impingement Heat Transfer. *By* P. BREVET, E. DORIGNAC, AND J. J. VULLIERME 409

Numerical Investigation of Combined Impingement and Convection Heat Transfer. *By* A. ABDON AND B. SUNDÉN 417

Mass/Heat Transfer in Dimpled Two-Pass Coolant Passages with Rotation. *By* F. ZHOU AND S. ACHARYA 424

Detailed Heat/Mass Transfer Distributions in a Rotating Two-Pass Coolant Channel with Engine-Near Cross Section and Smooth Walls. *By* L. RATHJEN, D. K. HENNECKE, S. BOCK, AND R. KLEINSTÜCK 432

Analyses of Heat Transfer in Stationary and Rotating Ribbed Blade Cooling Passages Using Computational Fluid Dynamics. *By* R. A. BREWSTER AND S. JONNAVITHULA 440

Prediction of Pressure Loss and Heat Transfer in Internal Cooling Passages. *By* K. HERMANSON, S. PARNEIX, J. VON WOLFERSDORF, AND K. SEMMLER 448

Numerical Simulation of Local Heat Transfer in Rotating Two-Pass Square Channels. *By* A. I. KIRILLOV, V. V. RIS, E. M. SMIRNOV, AND D. K. ZAITSEV 456

Experimental Determination of Average Turbulent Heat Transfer and Friction Factor in Stator Internal Rib-Roughened Cooling Channels. *By* L. BATTISTI AND P. BAGGIO 464

Contribution of Heat Transfer to Turbine Blades and Vanes for High-Temperature Industrial Gas Turbines: Part 2. Heat Transfer on Serpentine Flow Passage. *By* K-I. TAKEISHI AND S. AOKI 473

Secondary Flow Effect to Heat Transfer of a Duct with Discrete Rib Turbulators. *By* K. TATSUMI, H. IWAI, K. INAOKA, AND K. SUZUKI ... 481

Development of Nondestructive Inspection Method for the Performance of Thermal Barrier Coating. *By* M. MORINAGA AND T. TAKAHASHI 489

Numerical Modelling of Flow and Heat Transfer in the Rotating Disc Cavities of a Turboprop Engine. *By* J. FARAGHER AND A. OOI 497

Verifying Heat Transfer Analysis of High-Pressure Cooled Turbine Blades and Disk. *By* S. YAMAWAKI 505

Computation of Flow and Heat Transfer in Rotating Cavities with Peripheral Flow of Cooling Air. *By* M. KILIÇ 513

Index of Contributors 521

Financial assistance was received from:

- AMERICAN SOCIETY OF MECHANICAL ENGINEERS (ASME)
- MIDDLE EAST TECHNICAL UNIVERSITY (METU), ANKARA, TURKEY
- SCIENTIFIC AND TECHNICAL RESEARCH COUNCIL OF TURKEY (TÜBITAK)
- UNIVERSITY OF MINNESOTA, MINNEAPOLIS, MINNESOTA

The New York Academy of Sciences believes it has a responsibility to provide an open forum for discussion of scientific questions. The positions taken by the participants in the reported conferences are their own and not necessarily those of the Academy. The Academy has no intent to influence legislation by providing such forums.

Preface

Turbine 2000 is the second symposium related to heat transfer in high-performance gas turbines sponsored by the International Centre for Heat and Mass Transfer (ICHMT). The first, held in Marathon, Greece, in August 1992, resulted in the book *Heat Transfer in Turbomachinery*. However, the general nature of the present volume and the meeting on which it reports is quite different from the earlier one.

Turbine 2000, conducted in Çesme, Turkey, 13–18 August 2000, under the auspices of ICHMT, is a summation of the present state of knowledge of heat transfer in high-performance gas turbines. Highlighted by 16 invited keynote lectures by some of the world's best-known authorities on gas turbine heat transfer, the present book contains a wealth of information in one place. The topics cover most of the major issues in turbine heat transfer: from internal cooling to blade heat transfer to film cooling; from design issues to detailed experimental techniques to the latest numerical issues in predicting gas turbine heat transfer phenomena. Material comes from key industrial, academic, and nonprofit laboratories.

The editor owes thanks to many individuals: Terry Simon, Vice Chair of the Symposium, for his continued support in recommending key issues to be covered; Vinod Srinivasan, Symposium Secretary, for help in many details of planning and conducting the meeting, and for collecting the papers for publication; and, finally, Faruk Arinc, Secretary General of ICHMT, and his staff, who were very helpful in organizing and leading the actual details and conduct of the meeting, and for choosing an excellent meeting site.

Also to be acknowledged are the Scientific Committee: T. Arts (Von Karman Institute for Fluid Dynamics, Belgium), R. Bunker (General Electric, USA), P-H. Chen (National Taiwan University, Taiwan), M. Chyu (University of Pittsburgh), A. W. Date (Indian Institute of Technology, Bombay, India), B. Glezer (Optimized Turbine Solutions, USA), R. J. Goldstein [Symposium Chair] (University of Minnesota), J-C. Han (Texas A&M University), D. Hennecke (University of Darmstadt, Germany), H. Hodson (Cambridge University, UK), T. Jones (Oxford University, UK), M. Kumada (Gifu University, Japan), L. Langston (University of Connecticut), F. Leboeuf (Ecole Centrale de Lyon, France), J-S. Lee (Seoul National University, Korea), A. Leontiev (Institute for High Temperature, Russia), F. Martelli (University of Florence, Italy), E. North (Siemens Westinghouse, USA), M. Owen (University of Bath, UK), Y. V. Polezhaev (Institute for High Temperature, Russia), J. Seume (Siemens AG, Germany), T. Simon [Symposium Vice Chair] (University of Minnesota), V. Srinivasan [Symposium Secretary] (University of Minnesota), K. Takeishi (Mitsubishi Heavy Industries, Japan), Z. Tao (Beijing University of Aeronautics and Astronautics, China), and T. Yoshida (National Aerospace Laboratory, Japan).

—R. J. GOLDSTEIN

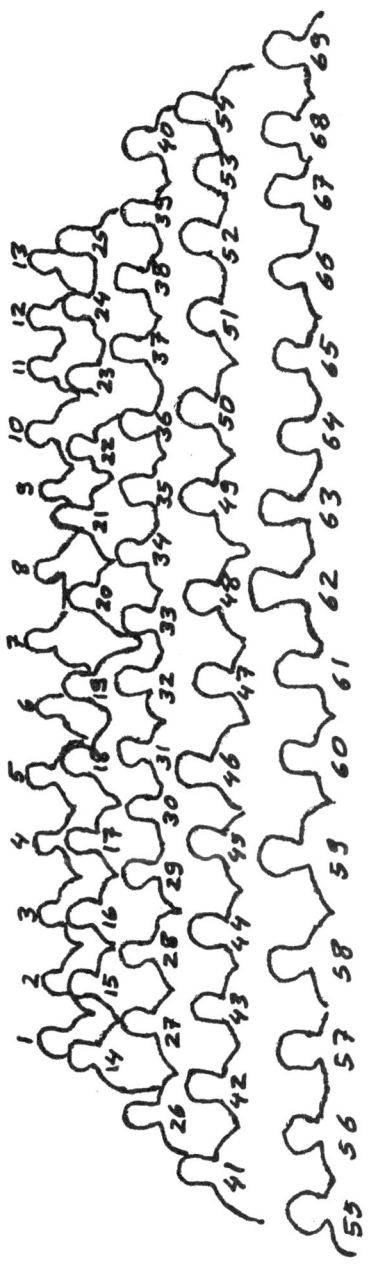

Participants at the conference, August 2000: (1) K. Tatsumi, (2) A. Toker, (3) B. Sunden, (4) K. Asato, (5) S. L. F. Frank, (6) F. Leboeuf, (7) C. J. Saumweber, (8) D. Suslov, (9) S. Küçüka, (10) H. Kimoto, (11) H. Takahashi, (12) S. Yamawaki, (13) H-W. Lee, (14) J. Faragher, (15) R. S. Bunker, (16) J. M. Owen, (17) M. Kiliç, (18) A. Boelcs, (19) B. Weigand, (20) K. Vogeler, (21) A. Schulz, (22) M. Obata, (23) G. Descombes, (24) T. Arts, (25) B. Han, (26) A. I. Kirillov, (27) G. Vogel, (28) L. Mathelin, (29) K. Takeishi, (30) J. E. Sargison, (31) T. Yoshida, (32) T. W. Simon, (33) J-C. Han, (34) B. Leger, (35) S. Kakaç, (36) M-K. Chyu, (37) F. Bazdidi-Tehrani, (38) L. S. Langston, (39) I. Ayranci, (40) F. Arinç, (41) L. Battisti, (42) T. I-P. Shih, (43) P. Ginibre, (44) V. Bregman, (45) A. I. Leontiev, (46) R. J. Goldstein, (47) V. Srinivasan, (48) P-H. Chen, (49) E. Paykoç, (50) A. S. Üçer, (51) S. Yavuzkurt, (52) S. Parneix, (53) T. Okutucu, (54) A. M. Gürün, (55) N. Egrican, (56) F. Martelli, (57) D. K. Hennecke, (58) J. Dittmar, (59) L. Rathjen, (60) I. Koç, (61) G. Croce, (62) J. S. Lee, (63) P. Brevet, (64) E. Dorignac, (65) S. Duranti, (66) R. A. Brewster, (67) M. Morinaga, (68) S. Alaybeyi, and (69) S. Barthet.

Synopsis

V. SRINIVASAN, T. W. SIMON, AND R. J. GOLDSTEIN
Department of Mechanical Engineering, University of Minnesota, Minneapolis, Minnesota 55455, USA

TECHNICAL ISSUES COVERED IN TURBINE 2000

A number of presentations, invited and contributed, are included in the present volume. There is valuable material for the gas turbine heat transfer researcher, student, and designer. The invited papers clearly summarize the state of knowledge and much of the current effort related to key heat transfer phenomena important in modern-day gas turbine systems, generally serving as an excellent review of activities in the field. The contributed papers bring current research design and analysis to a number of these areas.

Lee Langston's keynote paper begins with a brief summary of the pre-1985 work on secondary flows in blade passages and then presents a post-1985 survey of additions and improvements to models of the secondary flow and passage vortex. These include data that indicate that the suction side of the horseshoe vortex formed at the leading edge of turbine blades wraps itself around the passage vortex instead of getting entrained by it, and other work that shows the existence of a small, intense vortex at the leading edge and counter-rotating vortices at the blade-endwall junction. He points out a recently discovered parameter that appears to correlate well with the formation of swirl in the flow approaching the leading edge, and therefore affects the saddle-point flow behavior. He reviews the various methods that have been tried to reduce the losses due to secondary flow, such as contouring, fences, and grooves on endwall and blade surfaces. The technique that shows the most promise is endwall and blade contouring. Modifications of the blade leading edge profile at the endwall junction have been shown to result in a significant reduction in secondary flow losses. He also notes that shortcomings in turbulence modeling still preclude use of CFD in predicting secondary flow losses in turbine cascades.

Minking Chyu notes that the flatter radial temperature profile in today's combustors imposes higher endwall cooling requirements. Studies have shown that the heat transfer distribution on the endwall is very complex and can be closely correlated with the vortex system set up by the secondary flow. Introduction of film cooling sets up a complex interaction between the cooling jets and the secondary flow system. He points to the need for careful design of endwall film cooling configurations since a poor design may actually strengthen the secondary flow in some cases. High blowing rates are needed to energize the endwall boundary layer and reduce entrainment into the passage vortex. Endwall contouring has been shown to be promising in reducing heat transfer, but needs to be studied further. He also points to the lack of data on the film cooling effect of leakage from the disc cavity into the hot gas path.

Other papers on external heat transfer include contributed papers 1 through 6.

The keynote paper by Terry Simon and Richard Kaszeta discusses boundary layer transition and its importance to low-pressure turbine performance and heat transfer in the hot sections. The authors begin with the background of transition in the low-disturbance environment, including such concepts as Tollmien-Schlichting waves and turbulent spots. They then discuss the differences that one finds in turbines where background disturbance levels are high, called "bypass transition". Next, they review recent work on transition in turbines, both computational and experimental. Computational contributions are by Direct Numerical Simulation (DNS) and by refinements of correlations. Finally, they review recent work from the University of Minnesota in which a wind tunnel facility is used to simulate the flow through a low-pressure turbine, including wakes that pass through the approach flow.

Giles Brereton and Tom Shih concentrate on popular k-ε models and analyze the reasons for their poor performance for predicting gas turbine flows. These models are often calibrated towards prediction of simple shear flows. However, the flow in a gas turbine is characterized by shear, rotation, and streamline curvature. Using several benchmark problems as illustrations, they point out the situations where such models can be used with reasonable accuracy, in terms of parameters that describe the magnitude of curvature, rotation, and shear, in gas turbines, and other situations where one can a priori expect poor performance of k-ε models.

In his review of turbine blade tip heat transfer, Ronald Bunker explains the difficulties involved in designing efficient cooling systems for blade tips and summarizes work on understanding the complex flow field caused by the tip clearance. He examines the early studies, both numerical and experimental, that investigated the flow and heat transfer near the tip using simple 2-D models. He points out that, so far, film cooling of blade tips has been studied only for such idealized models. He also emphasizes that recent studies in linear cascades have shown the limited applicability of 2-D models to prediction of heat transfer. The heat transfer distribution is shown to be sensitive to blade shape and changes significantly with other changes in geometry, such as the introduction of squealers, winglets, etc.

He notes the various studies in progress that seek to examine the effects of relative motion, film cooling, and other parameters. CFD predictions are noted to be at best within 20% of experimental data. Also, he points to the lack of data that model real engine conditions, such as the variation in heat transfer distribution with time, as the blade profile changes due to erosion.

Other papers on blade tip heat transfer are contributed papers 7 and 8.

Francesco Martelli summarizes three broad approaches to modeling the effects of unsteadiness, especially rotor-stator interactions through the velocity and pressure fields, shock waves, wakes, and hot streaks, on turbine performance. The passage-to-passage averaging procedure of Adamczyk uses steady modeling, but seeks to retain the effects of unsteadiness through new deterministic stress terms, body forces, etc. The quasi-steady approach involves linearization of the Euler equations, with the unsteadiness considered to be a small perturbation. Alternatively, results for a single blade row are computed with unsteady boundary conditions in order to model rotor-stator interaction.

Fully unsteady modeling involves overcoming several problems such as transfer of data across the rotor-stator boundary, management of the periodic boundary condition between blades in a row, and modeling unsteady viscous effects. Martelli discusses different methods that have been used to tackle the above problems and concludes that, while current methods can predict unsteady blade loading reasonably well, there is a need to develop better models in order to predict unsteady losses and unsteady heat transfer.

The Symposium featured 4 keynote lectures and 14 contributed papers (papers 9 through 22) on film cooling, attesting to the renewed interest in this area.

Francis Leboeuf and Olivier Sgarzi give detailed numerical results for the case of a single laminar jet issuing normally into a laminar cross-flow. They note that near-hole physics determine the overall effectiveness of the cooling configuration, and thus fine meshes need to be used in order to capture the relevant scales. Using vorticity fields, they provide a detailed description of the effects of the deceleration and entrainment of the upstream boundary layer, and the various vortices formed, such as the horseshoe vortex, the counter-rotating kidney vortices, etc. They also explain, in terms of these vortices, why shaped film cooling holes tend to perform better than round holes. They conclude with a brief description of the unsteady behavior of the jet, with the upstream boundary being dominated by a set of Kelvin-Helmholtz-type vortices.

In their paper on numerical predictions for film cooling, Sumanta Acharya, Asif Hoda, and Mayank Tyagi review various turbulence models that have been used for design of film cooling geometries. They note that, at blowing ratios used for blade cooling, the cross-flow significantly distorts the boundary layer in the duct and therefore, for realistic simulation, the hole and plenum should be included in the computational domain, instead of assuming a velocity profile at the exit. They compare the performance of algebraic models, the traditionally used k-ε models, the Reynolds stress transport models, and the relatively new approaches such as Large Eddy Simulation (LES) and DNS. While algebraic models such as the Baldwin-Lomax model have been used with reasonable success for a variety of flow geometries, they (along with the two-equation models such as the k-ε, κ-ω, and q-ω models) suffer from the assumption of an isotropic eddy viscosity, which is usually tuned so that it matches the streamwise turbulence parameters. Reynolds stress transport models, which do not make this assumption, fail to deliver any significant advancement in the accuracy of their predictions. This is because they fail to capture the large-scale coherent structures found in film cooling flows.

LES has the advantage that it makes no assumptions on the large-scale structure of the flow and thus captures the large scales that contain most of the kinetic energy, while the small scales that are responsible for dissipation are modeled as isotropic. LES has been shown to be much more accurate than the two-equation models used earlier. DNS involves solving for all temporal and spatial scales without any modeling and therefore gives very accurate solutions; however, the computing resources required make DNS prohibitive as of today, except for low Reynolds number flows.

Tony Arts' keynote paper presents measurements of film cooling performance from the Von Karman Institute. Effects studied are blade geometry, coolant emission geometry, blade loading, airfoil Reynolds number, free-stream turbulence, blowing ratio, and density ratio. Measurements of effectiveness and heat transfer coefficient are presented for cases representative of engine environments. The overall reduction of heat transfer is presented to show total performance. Measurements are used to develop calculational methods or correlations and to provide test cases for code validation. Aerodynamic studies of film cooling include detailed flow-field measurements from a VKI, low-turbulence facility, which document effects of density ratio, blowing rate, and Mach number. A strong correlation is noted between jet spreading and momentum flux ratio. When the momentum flux ratio is greater than unity, the jet breaks through the boundary layer and interacts less with the near-wall flow than when the momentum flux ratio is less than unity. Performance with rows of jets is similar to that with a single jet until the lateral hole spacing becomes three diameters or less; then, interaction of the neighboring jets is observed.

In his overview of combustor liner cooling, Achmed Schulz first notes the shift from long can-type combustors operating at near-stoichiometric conditions to short annular combustors operating on lean mixtures due to NOx emission requirements. He then discusses various methods for cooling combustor liners, such as impingement, film cooling, and convection heat transfer. The effects of slot film cooling are achieved by creating wall parallel films using overlapping walls. Combustion chambers of aero-engines run closer to stoichiometric conditions with temperatures significantly hotter than turbine inlet temperatures, and therefore require a combination of cooling techniques. The current trend is to use impingement cooling on the outer surface of the liner and then use the spent air for film cooling, preferably full-coverage film cooling with ejection at very shallow angles. He discusses features of film cooling that are peculiar to combustor liner cooling, such as the interaction of a mixing jet with the film.

A review of impingement heat transfer is presented by Bumsoo Han and Richard Goldstein. The different regions of the flow, such as the free jet region, the stagnation flow region, and the wall jet region, are discussed. Heat transfer distributions on the impingement plate are presented. Artifacts that may arise due to improper definition of the heat transfer coefficient are pointed out. Other areas that need to be studied for extension to gas turbine systems, such as the effects of curvature, deflection of the jet in a cross-flow, angle of impingement, and interaction between multiple jets, are presented and discussed.

Other aspects of impingement heat transfer are discussed in papers 23 and 24.

In a review of techniques for blade internal cooling, Je-Chin Han and Sandip Dutta cover rib turbulated cooling for internal channels, jet impingement cooling, and pin-fin cooling. Various parameters that affect the performance of cooling channels are described, such as the effect of rib angle, channel aspect ratio, 180° sharp and rounded bends, and coolant bleed through film cooling holes. They also discuss the different effects of the Coriolis and buoyancy forces on the velocity profile in the channel, depending on the radially inward/outward direction of the flow.

In the case of jet impingement cooling, they point out that very few studies concentrate on rotational effects, such as jet deflection, on the flow and heat transfer distribution on the impingement surface. Pin-fins, which are used in the trailing edge region of blades, also have been studied for the stationary case only. The effects of various fin array arrangements, flow acceleration, pin shape and orientation, and trailing edge ejection are discussed. They also cite examples where the effects of using different cooling techniques in conjunction are studied, such as impingement on ribbed rotating walls, dimples, and new concepts such as heat pipes and mist cooling.

Bernhard Weigand, Klaus Semmler, and Jens von Wolfersdorf discuss the special problems in designing internal cooling systems for land-based power generation gas turbines, such as long-life requirement, variety of fuels used, and the difficulties posed by large blades. These lead to cooling designs that are somewhat different from those for aero-engines. They illustrate some of these changes for the cases of rib design in internal cooling channels, rotation effects, and impingement cooling.

Contributed papers 25 to 33 also discuss the performance of internal cooling channels.

In his article on recent trends in cooling systems for high-temperature turbines, Toyoaki Yoshida reviews the advances in closed-loop steam cooling for vanes and blades of stationary turbines. Steam cooling offers significant benefits since it eliminates the need for film cooling (except at the leading edge) and reduces the difference between the combustor exit temperature and the turbine inlet temperature. New cooling systems involving heat pipes and evaporatively cooled rotors are introduced. Currently, tests are being conducted to explore the idea that coolant mass flow can be reduced when the coolant is precooled through a heat exchanger. For this, the blade needs to be made in three layers: a TBC coating, a high-temperature alloy, and copper.

In their keynote paper on the internal cooling and sealing systems of gas turbine engines, Mike Owen and Michael Wilson summarize recent work done on rotor-stator systems with ingress/egress of hot gas into the wheel-space cavity. They also describe the effect of using preswirled air in conjunction with a cover plate to reduce the total temperature of the cooling air. Work done on disc systems with a cavity rotating with respect to a stationary casing is described. This models the peripheral flow of cooling air used in platform cooling systems. The authors show that the effect of cavity geometry on heat transfer is not very significant and that the radial distribution of the tangential velocity in the cavity is similar to that of a Rankine vortex.

The axial throughflow of cooling air in the center of rotating compressor discs is modeled using the geometry of two corotating discs, with buoyancy effects. The paper describes the structure of the resulting stratified flow, and its change in behavior depending on the disc that is heated (inner vs. outer) and the state of the flow (laminar vs. turbulent), and gives heat transfer correlations for all cases studied.

Contributed papers 34 to 36 also discuss rotating discs and cavities.

Traditional methods of designing gas turbines involve a loop of thermodynamic cycle analysis, aerodynamic design, mechanical design, and then thermal design before returning to the thermodynamic analysis. Uncertainties involved in the design process are accounted for within each discipline by a factor of safety, resulting in conservative design. In his keynote paper, Boris Glezer argues that, instead, probabilistic risk analysis should be used in order to share the factor of safety among various disciplines, thereby improving efficiencies without sacrificing turbine life. This requires a system of concurrent design that involves close interaction between each discipline at every stage of the design process, instead of having a sequential process.

This keynote paper illustrates the concept with specific examples. In the case of the decision on distribution of the workload among the turbine stages, it is shown that the mechanical constraints of rotational speed and hub diameter determine the blade aspect ratio and gas path divergence angle. Similarly, combustor design has to be performed in conjunction with the turbine first-stage nozzle design, using appropriate endwall contouring and employing thermal barrier coatings in order to meet the cooling air constraints. A number of recently developed turbine cooling techniques, which are applicable to critical turbine components and combustors, are discussed to illustrate their potential benefits for an integrated multidisciplinary engine design approach.

Other examples where cross-disciplinary design enhances performance are active/passive systems for controlling tip clearance and decisions on the number of blades, hub-to-tip radius ratio, sealing, etc., which need to be included in design process to satisfy structural integrity of the disc (in designing against failure of the disc).

PANEL DISCUSSION

In the last session of the Symposium, a panel was assembled to discuss the future needs in gas turbine heat transfer. Members were Ron Bunker of General Electric Corporate Research and Development, Tony Arts of the Von Karman Institute for Fluid Dynamics, Dietmar Hennecke of the University of Darmstadt and formerly from MTU, and Ken-Ichiro Takeishi of Mitsubishi Heavy Industries. The text below summarizes their presentations and the discussion that followed:

Ron Bunker:

Requirements for future work, as seen by industry: We want to get to the point where we are fully predictive. We need to continue development of CFD techniques and experimental work in support of CFD development.

In the category of secondary flows, more studies are required on airfoil/endwall region heat transfer. Manufacturing processes today allow production of 3-D, exotic shapes that may reduce losses. We need to take advantage of the opportunities they create. Aero-efficiency, or High Cycle Fatigue (HCF), is the primary goal. Studies in these areas that show promise will also require research in other, nontraditional areas, such as shaped impingement arrays, microtechnology/nanotechnology, etc.

Further information is required on the effects of endwall leakage and film cooling flows as they interact with the main gas path flow. The effects of coolant migration and the combustor-to-vane gap flow need to be quantified. Also, the effects of endwall contouring on the leakage and film coolant interaction with the main flow need description.

Researchers need to concentrate less on "simple" flat-plate situations to examine film cooling and need to work with more realistic geometries that model the effects of curvature, passing wakes, streamwise acceleration, and secondary flows. Recent improvements in RANS, LES, and DES (Detached Eddy Simulation) turbulence modeling can bridge the gap between the basic flat-plate geometry and the more complicated engine geometries while helping to explain the physics involved.

While a large number of papers presented at the Symposium discuss shaped holes, it is important to remember that not all of them may be manufactured economically. Research on shaped holes must take into account real turbine effects, for example, the deposition of particulates with time. Also, some of these shapes may be ruled out due to the need to maintain acceptable stress levels. Many papers discuss relatively low frequency effects, such as those associated with combustors, but there have been very few studies at higher, engine-representative frequencies, such as the wake shedding frequency.

Industry needs data on unsteady heat transfer effects at high frequencies in order to apply a correction to data obtained under steady conditions. The study of unsteady effects on leading edge heat transfer has had some attention. Areas of interest today are airfoil blade tips and shrouds.

Transition to turbulence is an ongoing problem. There are no reliable and robust models that predict the start, length, and path of transition.

The connection between heat transfer data and life of the blade is still very "soft". Industry needs life models of heat transfer, that is, heat transfer = f (time, conditions). These models should take into account the changing operating conditions with time.

In the immediate future, blade cooling research should work on the following areas:

- Consider the effects of impingement at an angle on nonflat surface regions.
- Consider various blade platform cooling geometries.
- Consider the fillet region at the endwall-to-airfoil junction, either impingement or convective cooling.
- Consider manufacturing innovations that may enhance heat transfer, such as dimples, etc.
- Find methods to reduce sensitivity of heat transfer to rotation and buoyancy.
- Investigate nonuniform, nonsymmetric geometries in cooling, such as variable impingement jet diameters, or nonuniform distribution of impingement array jets or pin-fins.

Blade tip heat transfer will be dealt with by a couple of new and ongoing research projects, but more data are welcome. The goal is a generalized tip heat transfer model that can take into account the effects of film cooling, squealer tips, and relative motion between blade and shroud.

Techniques for nonintrusive hot section measurements are needed. The applicability of lab data to engine design still remains weak because too many parallel effects come together in the real engine. Experience remains the best designer. Laboratories need to have more realistic inlet conditions when studying blade or vane heat transfer.

Combustor cooling schemes with a lower delta-P are needed, but high heat transfer and low NOx emissions must be maintained.

Real engine effects need to be modeled: combustor turbulence characteristics, surface roughness characteristics, as-manufactured film cooling holes and supply plenum geometries, plugging, surface erosion, oxidation, TBC spalling, interface steps and leakages, turbulator core wear, manufacturing defects and tolerance effects, etc.

Tony Arts:

More work needs to be done on transition modeling to determine the onset, length, and path of transition. Models for transition can be improved only by generating more high-quality data. Modeling attention is also needed for the turbulent heat flux and the turbulent Prandtl number.

In the area of film cooling, more flow-field measurements are needed at real engine conditions. Besides heat transfer and effectiveness measurements, it is important to take measurements quantifying aerodynamic losses in film-cooled situations; thus, both pressure and temperature measurements are needed. Film cooling geometries tested need to be closer to reality, such as proper surface roughness and effects of manufacturing tolerances.

Internal cooling now has an extensive database of heat transfer measurements, but data are lacking on the aerodynamic side. Time-resolved velocity and temperature measurements are needed so that modeling of the turbulent heat flux term, $v't'$, can be improved. The effects of rotation must also be included. Again, models should take into account real manufacturing constraints, such as those on turbulator ribs. While many measurements report area-averaged heat transfer, there is a lack of data for heat transfer and temperature over the rib surface itself.

Laboratory experiments also need to model the effect of unsteady interaction phenomena, such as stator/rotor interaction effects on aerodynamic quantities such as blade lift and on heat transfer and film cooling. Losses need to be quantified in terms of pressure and temperature, for both one- and two-stage configurations.

For low-pressure turbines, unsteady, wake-induced transition is in need of study. Fully quantitative data are needed for transition model development. Surface-mounted hot-film anemometry is a possible tool for getting such data.

Dietmar Hennecke:

Most papers presented at the Symposium were on blades and vanes and a few on disks. Virtually no papers were on combustors, so how are they going to evolve with the introduction of RQL and LPP designs. This is especially important since RQL combustors cannot use traditional cooling techniques. Heat transfer issues in other systems of the gas turbine, such as labyrinth seals, air systems, heat exchangers, bearings and shafts, etc., have largely been ignored. Although they operate at cooler temperatures, it is important to consider them in order to reduce stresses and minimize clearances.

There is a lack of vision at this Symposium regarding novel cooling schemes. We must be thinking 10 to 50 years into the future.

The present industrial design process follows an integrated approach in that all groups involved in the process work simultaneously instead of following a sequential approach. This has important consequences for the heat transfer community and should affect the way researchers approach their work. Researchers need to recognize that the best solution is not the one with the most or least heat transfer, but one that represents an optimum among various factors, such as cooling, aerodynamics, stresses, life, manufacturability, materials, and cost. For example, aerodynamic losses need to be quantified when examining any blade film cooling configuration. Similarly, internal cooling schemes should be designed keeping the pressure drop penalty in mind. Conjugate heat transfer including both internal and external convection must be considered as heat transfer design configurations are chosen.

Areas that need further work are as follows:

- Consider unsteady effects (possibly using them to one's benefit).
- Consider the effects of shaped entry geometry of film cooling holes.

- Consider designing cooling passages with novel 3-D turbulators or with swirling flow with/without the effect of rotation.
- Consider the inlet guide vane to be an integral part of the flame tube assembly.
- Consider the use of transpiration cooling of turbine blades. (There was some work more than 25 years ago, but now, with advances in manufacturing processes, it should be reexamined.)
- Consider novel cooling configurations with and without film cooling for RQL combustors.
- Consider new heat transfer problems due to the introduction of active controls into future engines.
- Move from RANS to LES for design in industry.

About Turbine 2000: Because of the focus it had, the Conference was very useful and beneficial to the research community. There was a good mixture of invited and contributed papers. However, more representation from industry is needed. The next symposium on this topic is recommended in 6 years.

Ken-Ichiro Takeishi:

We must produce a more environmentally benign engine with low NOx and a capability of using LNG fuel. Important considerations during the design of industrial gas turbines are high reliability, high efficiency, and minimal environmental impact. Turbine inlet temperatures have risen at a rate of 200 K every 10 years, reaching about 1800 K today. Mitsubishi engines today use premix, low-NOx, multiburner combustors that operate on lean mixtures. There is very little cooling air available. As a consequence, a steam-cooled transition piece was designed and placed before the first-stage vane. The temperature profile of the gas entering the first stage is flat as a second consequence.

Researchers need to devote their attention to means of reducing aerodynamic losses while reducing or keeping constant heat transfer coefficients on the airfoil. Film cooling effectiveness has to be improved. CFD techniques need to progress to the point where they can be used as predictive design tools. Recently, Mitsubishi altered the aerodynamic design of turbine and compressor airfoils and eliminated separation solely by using a 3-D viscous solver for design. However, the complexity of heat transfer phenomena makes extension to estimation of turbine external heat transfer difficult.

Experimental research work in the near future will be focused on complex heat transfer phenomena, such as film cooling on blade tips and shrouds, the effect of unsteady flow on heat transfer from airfoils, heat transfer in the 180° bend of serpentine flow passages, and other complex configurations in internal cooling passages. These experiments must be designed such that they add to the database for comparison with and improvement of CFD analyses.

Work in CFD should be focused on development of turbulence models that accurately predict heat transfer. LES shows promise as a design tool for the future. Computing constraints will limit the near-future use of DNS to evaluating LES and RANS models and as an analysis tool for giving data rather than as a predictive design tool.

New cooling schemes need to be devised for industrial gas turbines: steam cooling, mist cooling, or more innovative methods such as heat pipes. Another possible area of innovation is the use of wafered or stacked turbine vanes and blades manufactured using nanotechnology. These could employ microheat transfer technology to significantly enhance their thermal performance.

Mitsubishi already uses 3-D viscous codes in the aerodynamic design of turbine and compressor airfoils. However, it is difficult to apply these codes to estimate external heat transfer coefficients since a very fine mesh is needed and imposes a large CPU requirement. Therefore, these codes are not useful as heat transfer design tools.

General Discussion Session:

CFD researchers today can accurately resolve the large-scale vortices present in film cooling flows by using crude RANS models. Such models can predict Kelvin-Helmholtz instabilities that lead to transition and vortex formation, provided we do not force steadiness on the computed solution. What is needed to improve modeling is development of quasi-steady turbulence models near to the wall that mimic the unsteady behavior of the boundary layer.

Acharya: The major problem hindering faster growth of LES as a predictive tool is that it breaks down near the wall, where the length scales are very small. LES can satisfactorily compute the flow in free-stream situations, such as flow structures in film cooling jets, and mixing in combustors where the flow scales are large. Subgrid-scale modeling must be used for smaller scales. LES is competitive with RANS solvers in terms of computing effort in such situations. Currently, hybrid LES/RANS looks more promising for solving wall-bounded flows. More research needs to be done in terms of improving LES solvers on unstructured meshes, body-fitted coordinates, and the use of immersed boundary methods with LES.

Simon: Regarding combustor turbine interaction on turbine heat transfer, with strong acceleration and the use of catalytic convertors, is there a possibility of low free-stream turbulence in the entry flow and a need for new framing of the transition problem?

Bunker: The DOE AGTSR program now has projects to address the turbulence effect on heat transfer where this is being considered.

Simon: Often, transition section, endwall, impingement, and film cooling configurations need to be tailored to the specific blade, vane endwall, or transition section profile, but the profile data are not available because they are proprietary. How then do we, in academia, appropriately analyze such situations?

Bunker: Industrial representatives are willing to help frame the studies. Give us a call.

Secondary Flows in Axial Turbines – A Review

L.S. LANGSTON

University of Connecticut
Mechanical Engineering Department
191 Auditorium Road, U-3139
Storrs, CT 06269 USA

ABSTRACT: An important problem that arises in the design and the performance of axial flow turbines is the understanding, analysis, prediction and control of secondary flows. Sieverding[1] has given a review of secondary flow literature, covering up to 1985. In this paper a brief review of pre-1985 work is given, and then a survey of open literature secondary flow investigations since the Sieverding review is presented. Most of the studies reviewed deal with plane or annular cascade flows. Tip clearance effects are not covered. The basic secondary flow picture for a turbine cascade, as measured and verified by a number of investigators is described. Recent work that shows refined secondary flow vortex structures is examined. A flow parameter based on inlet boundary layer properties used to predict horseshoe vortex swirl is presented. Work on secondary flow loss reduction, involving airfoil geometry, endwall fences and endwall contouring is briefly reviewed. A new leading edge bulb geometry that has demonstrated impressive loss reduction is considered. It is concluded that accurate routine prediction of secondary flow losses has not yet been achieved, and must await either a better turbulence model or more experiments to reveal new endwall loss production mechanisms. Lastly, loss is examined from the standpoint of entropy generation.

INTRODUCTION

Bradshaw[2] has aphoristically observed: "Of all the fluid-dynamic devices invented by the human race, axial-flow turbomachines are probably the most complicated." One important contributor to this complexity is the fluid flow and heat transfer brought about by the existence of endwalls in axial flow turbomachine gas path passages. Due to viscous effects, endwalls divert primary flow produced by blades and vanes, to give rise to what has come to be called secondary flow, or endwall flow.

The secondary or endwall flow in a cascade of turbine blades or vanes (such as depicted in Fig. 1) constitutes one of the most commonplace and widespread three-dimensional flows that arise in the generation of electrical and motive power. Such fluid flows occur in all axial flow turbines (gas, steam and water) used to generate most of the world's electricity. They occur in all of the jet and turboprop engines (30,000 in the inventory (1993) of the U.S. Air Force, alone) which power most of the aircraft of the world.

By turbine designer conventions, the effects of the highly interactive flow picture in Fig. 1 is artificially broken down to those caused by the blade or vane "profile" surface and those caused by the endwall. (A third category of tip clearance effects will not be treated in this paper.) The aerodynamic losses so attributed to the endwall – usually termed secondary flow losses or secondary losses – can be as high as 30-50% of the total aerodynamic losses in a blade or stator row, according to Sharma and Butler[3]. Inlet guide vanes, with lower total turning and higher convergence ratios, will have smaller secondary losses, amounting to as much as 20% of total loss for an inlet stator row.

Because of the essential importance of secondary flow in all kinds of power turbines, research in the last 25 years has been quite extensive. Sieverding[1] has reviewed secondary flow literature up to 1985. The goal in this paper is to build upon this excellent 1985 review and to examine some of the secondary flow research since its publication. Text length considerations here allow for either a shallow listing of a large number of secondary flow studies, or a focus on a smaller number. The author has chosen the latter path (the focus) and asks the indulgence of those precluded.

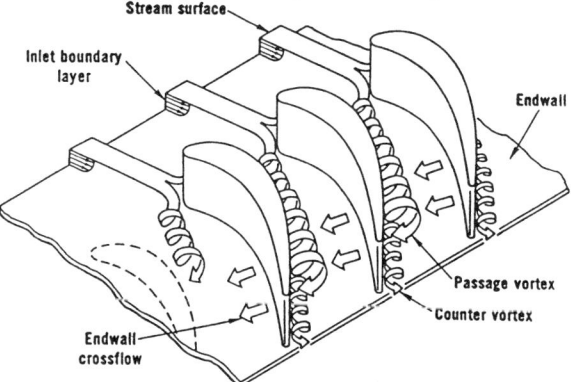

Fig. 1 The three-dimensional separation of a boundary layer entering a turbine cascade. The saddle point occurs where the vortex is formed.

THE BASIC FLOW PICTURE

The hardware sketched in Fig. 1 represents a plane (or linear) cascade, depicting the airfoils and endwalls in a turbomachine with a very large (infinite) radius. For many years now, experimenters studying these intriguing, but complex three-dimensional flows in axial turbines, have made use of planar cascades to sort out and measure fluid flow and heat transfer features. Numerical calculators modeling these flows have also relied on simple plane cascade geometries to attempt to "postdict" existing cascade data, or to separate out the effect of various analytical techniques (such as turbulence models).

The basic flow through a plane turbine cascade has been fairly well studied, documented and verified by many experimenters (See Sieverding[1]). One of the earliest complete studies was done by Langston, Nice and Hooper[4] who used a very large scale, low aspect ratio plane cascade of four turbine airfoils. In their work, detailed measurements of subsonic flow were made at axial planes upstream, within and downstream of the plane cascade.

The three-dimensional flow they measured is that shown schematically in Fig. 1. This figure, taken from Langston[5], shows that at the endwall of the cascade, the inlet boundary layer separates at a saddle point and forms a horseshoe vortex. One leg of this vortex (sometimes called the "pressure" leg), drawn into a cascade passage, is "fed" by the passage pressure-to-suction endwall flow and becomes the passage vortex. The other leg (called the "suction" leg) is drawn into an adjacent passage and has an opposite sense of rotation to the larger passage vortex. This smaller vortex is labeled as a counter vortex in Fig. 1 and can be thought of as a "planet" possibly rotating about the axis of the passage vortex (the "sun"). Thus the position of the counter vortex relative to the passage vortex may be different than that shown in Fig. 1. The ribbon arrows in the figure have been drawn to exaggerate vortex motion. The actual rotation of the vortices is much less than shown (about two rotations for the passage vortex).

Sieverding and Van den Bosche[6] showed the development and interaction of the passage and counter vortices by marking with smoke, two stream surfaces initially parallel to an endwall in low speed air flow entering a cascade. As shown in the sketch of their results in Fig. 2, one can see the evolution and movement of each vortex (as evidence by stream surface curling), using the planet and sun analogy of the last paragraph.

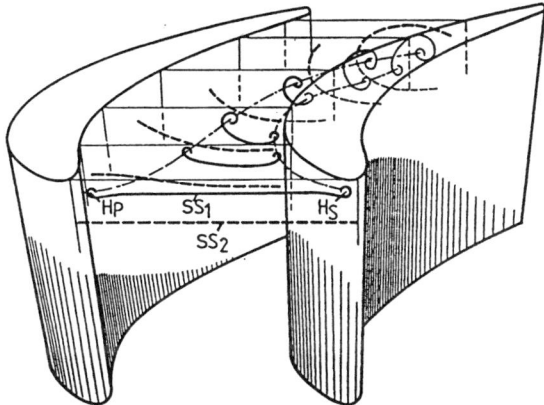

Fig. 2 Laminar flow stream surfaces SS_1 (within inlet boundary layer) and SS_2, from Sieverding and Van den Bosche[6], where H_p – pressure side horseshoe vortex leg, and H_s – suction side horseshoe vortex leg.

From the important standpoint of the turbine designer, this endwall vortex secondary flow is responsible for a loss of lift (i.e. loss of turbine work) and an increase in aerodynamic loss (i.e. a decrease in turbine efficiency), when compared to a hypothetical two-dimensional cascade flow.

The loss in lift is shown by static pressure data plotted in Fig. 3, from Langston, Nice and Hooper[4]. Values of the static pressure coefficient, c_p, taken with pressure taps mounted on the suction and pressure airfoil surfaces at 2.3, 12.15, 25 and 50 percent of span (measured from one endwall) are shown in Fig. 3, as a function of nondimensional axial distance, x. The two curves plotted in Fig. 3, the result of a two-dimensional incompressible potential flow calculation, are shown for purposes of comparison.

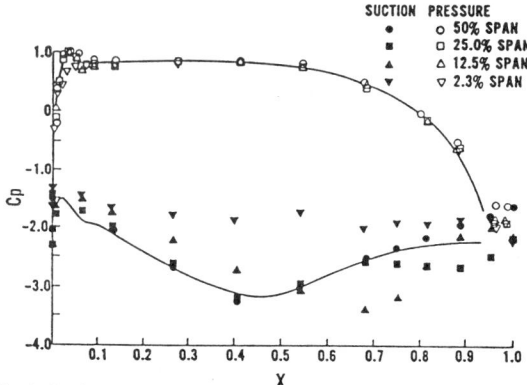

Fig. 3 Static pressure distribution on airfoil surfaces compared with potential flow (Langston[4]).

The pressure-side data shows two-dimensional behavior for all spanwise locations. On the suction-side, the data at midpsan (50 percent) exhibits two-dimensional behavior, except near the trailing edge, where viscous effects (trailing edge separation) influence the flow. As discussed in more detail in Langston[4] the suction-side pressures at other spanwise locations (25, 12.5 and 2.3 percent) show the effect of the passage vortex formed on the endwall of the cascade. The net effect of this three-dimensional "secondary flow" is to decrease the area between the suction and pressure side pressure distribution data points, for a given spanwise position. This area is a measure of the net lift (work output), so that the Fig. 3 data clearly shows a decrease of lift on the airfoil, as the endwall is approached from midspan.

The growth of mass-averaged aerodynamic loss coefficient, c_{pt}, a function as nondimensional axial distance x through the cascade is shown in Fig. 4, taken from Langston[4]. Also shown is the measured mass coefficient, η, (the inverted triangular data points) which show that mass was conserved (i.e. all values are close to $\eta = 1$). The measured mass-averaged loss values in Fig. 4 show that near the inlet of the cascade there is only a slight increase in loss as the inlet boundary layer (the value near $x = -0.2$) becomes the horseshoe vortex (up to about $x = 0.3$). Then, beyond a value of $x = 0.5$, the mass-average loss values rise rapidly as the passage vortex dramatically increases in size in the region of decelerated suction-side flow ("uncovered" turning, where the pressure rises rapidly in the downstream direction). Then, as the flow leaves the cascade ($x > 1.0$) the mass-averaged loss values rise rapidly again as passage vortex flow and the skewed endwall boundary layer are mixed-out and turned to a nominal exit flow angle, by the mainstream flow.

SECONDARY FLOW RESEARCH SINCE 1985

The basic secondary flow picture described in the last section was arrived at from examination of detailed velocity, pressure, and flow visualization measurements made upstream, within and downstream of turbine cascade flow passages. Pre-1985 investigations were reported by Langston, Nice and Hooper[4], and Marchal and Sieverding[7]. Post-1985 works that report on complete flow field measurements include

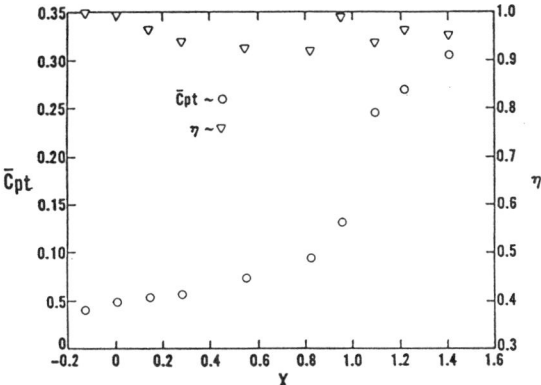

Fig. 4 Mass flow coefficient and mass averaged loss coefficient as a function of axial distance through the cascade (Langston[4]).

Yamamoto[8,9], Gregory-Smith, Graves and Walsh[10], and Harrison[11]. One can now build upon this firm foundation of endwall flow physics to examine some of the secondary flow research that has gone on since Sieverding's[1] 1985 review.

Other Endwall Vortex Flow Patterns

Sharma and Butler[3] proposed a slightly different version of the vortex pattern of Fig. 1. Their modified flow pattern is shown in Fig. 5, with the counter vortex wrapping around the passage vortex. (This wrapping wasn't evident in the velocity and pressure flow field measurements of Langston[4].) They deduced the flow pattern of Fig. 5 from the work of others and their own experimental observations. They used this approach to formulate a semi-empirical model for estimating losses in a turbine. Secondary flow loss correlations based on experimental data have long been a basic design tool, especially in the early design stages of any new axial flow turbine. Usually, it is not until later in the design cycle that CFD (computational fluid dynamics) codes are used (and then, sometimes with limited success in predicting losses, as will be discussed later). One of their key findings was that inlet boundary layer losses are convected through the cascade without causing any additional losses, and is independent (counter to classical secondary flow theory) of the amount of turning. In effect, the saddle point separation of the inlet boundary layer decouples it from the downstream effects of turning.

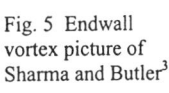

Fig. 5 Endwall vortex picture of Sharma and Butler[3].

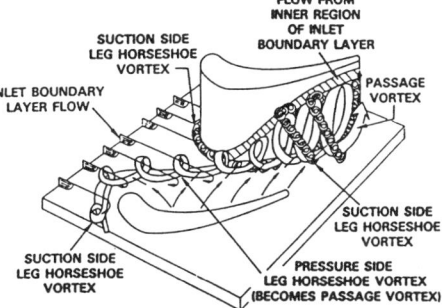

Another endwall vortex pattern revision to Fig. 1 has been put forth by Goldstein and Spores[12] of the University of Minnesota. The Minnesota group has a long history of reporting detailed mass transfer measurements (naphthalene sublimating into wind tunnel air) to infer heat transfer coefficient distributions on endwalls and airfoils in a cascade. Their measurements of mass transfer show a very high concentrated mass transfer rate (5 times flat plate values) right at the leading edge-endwall junction of their high pressure turbine blades, at a Reynolds number in the range of engine operating conditions. This led them to postulate the existence of another very small and intense vortex at the junction, which Goldstein and Spores called a leading edge vortex.

In a later, very detailed flow visualization study of a similar cascade flow, Wang, Olson, Goldstein and Eckert[13] documented the endwall flow using laser light and multiple smoke wires (similar to Sieverding and Van den Bosche[6]). Their multivortex flow pattern is shown in Fig. 6. Because they ran at a low Reynolds number (for flow visualization with smoke) some of the features they found seem to be similar to the complex periodically varying horseshoe vortex pattern formed in front of an endwall mounted cylinder in laminar flow (e.g. see Baker[14]).

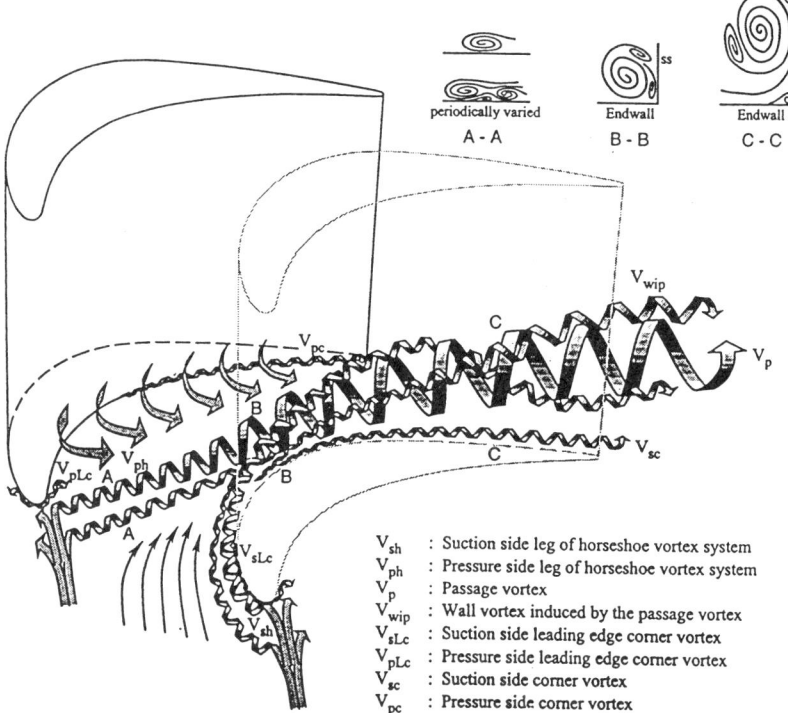

V_{sh} : Suction side leg of horseshoe vortex system
V_{ph} : Pressure side leg of horseshoe vortex system
V_p : Passage vortex
V_{wip} : Wall vortex induced by the passage vortex
V_{sLc} : Suction side leading edge corner vortex
V_{pLc} : Pressure side leading edge corner vortex
V_{sc} : Suction side corner vortex
V_{pc} : Pressure side corner vortex

Fig. 6 Endwall Vortex Picture of Wang, Olson, Goldstein and Eckert[13].

The modifications to the secondary flow in Fig. 1 proposed by Sharma[3] (Fig. 5) and by Goldstein and Spores[12] and Wang, et al[13] (Fig. 6) add to the complexity of the vortex picture. Further work is necessary to verify that these inferred modifications exist, through the use of flow field measurements of pressures, velocities and vorticity. A good example of such verification was the low-pressure turbine cascade study by Hodson and Dominy[15]. However, their attempt to make a complete endwall vortex tracing was hampered somewhat by the relative small size of their cascade relative to the size of their instrumentation.

As Green[16] points out, "vortex" is an inherently fuzzy term. He offers the definition of a vortex "......as the feature of a flow field that a majority of fluid dynamicists would label as such." This definition is not of much help to a turbine designer who might be trying to account for the effects of secondary flows on endwall heat transfer or aerodynamic losses. Other definitions are based on concentrated or distributed vorticity, and on circulation. For turbine design purposes the author defines a vortex as a flow feature in the turbine gas path, composed of a collection of throughflow streamlines spiraling around a center of low static pressure to produce simple closed curve isobars (of static pressure) in successive planes normal to the throughflow. (An example is shown in Figs. 8 and 9 of Langston[4] and in Figs. 6 and 9 of Eckerle and Langston[17].)

Inlet Boundary Layer Separation

Some progress has been made on understanding the flow conditions that occur when the inlet endwall boundary layer separates at the saddle point on the endwall and rolls up into the horseshoe vortex (see Figs. 1, 5 and 6). Thus far there is no quantitative criterion for determining the saddle point location. Data presented by Langston[4] shows that its location occurs in a region of stream wise adverse pressure gradient, and that it is most strongly dependent on the incidence angle of the cascade flow. The work of Moore and Ransmayer[18] which is reviewed in Sieverding[1], involved the modification of a cascade blade cylindrical leading edge with a sharper, wedge-shaped extension of the leading edge. Moore and Ransmayer saw no change in their airfoil pressure distribution and no appreciable change in endwall loss production, with the addition of the sharp leading edge to the cascade airfoils. They assumed they had eliminated the horseshoe vortex with the sharp leading edge, and, since they saw no change in exit loss, concluded the horseshoe vortex has little effect on loss. However the author draws the opposite conclusion from their results; that is, their sharp leading edge had no effect on the formation of the horseshoe vortex, and it was not eliminated in their tests. The separation at the saddle point is not dependent on the radius of curvature of the airfoil edge, but is dependent on the pressure field generated by the overall shape and size of the pressure and suction surfaces of the airfoils. The axial chord of the airfoil would be the significant length since it would characterize the airfoil pressure field, rather than a leading edge radius.

The symmetrical separation of boundary layer flow normal to a circular cylinder mounted on an endwall is perhaps the simplest of geometries that demonstrate a horseshoe vortex system where the cylinder diameter is the important dimension. One such study was carried out by Eckerle and Langston[17]. They provided a detailed description of the vortex system around the base of a single cylinder mounted in a wind tunnel. The side walls of the tunnel acted as planes of symmetry so that their actual single cylinder was one element of a cascade of cylinders at zero incidence angle. The Eckerle-Langston single cylinder experiment yielded a symmetrical saddle point (asymmetric for a turbine cascade) whose position occurred in a region of an adverse pressure gradient (as in the turbine cascade), upstream of the cylinder. Recently, research on the characteristics of horseshoe vortices has been extensively reviewed by Ballio, Bettoni and Franzetti[19].

Eckerle and Awad[20] carried out more detailed studies of the horseshoe vortex formation in front of a cylinder and varied the free stream velocity, making extensive velocity measurements with a laser doppler anemometer system. Their most important finding was that they could correlate their data and the data of others with the nondimensional parameter E, where

$$E = (\text{Re}_D)^{\frac{1}{3}} \left(\frac{D}{\delta^*} \right) \qquad (1)$$

Re_D is the cylinder Reynolds number, D is the cylinder diameter and δ^* is the displacement thickness of the boundary layer at the position of the leading edge of the cylinder (without the cylinder).

Their results showed that for E > 1000, no swirling motion was present in the plane of symmetry upstream of the cylinder. The vortex motion of the separating boundary layer developed at an angular distance from the plane of symmetry. For E < 1000 the swirling motion of the vortex was initiated in the plane of symmetry. As the authors point out, the identification of these two regimes gives a turbine designer a better way to predict effects of the separating flow on the heat transfer and effectiveness of a coolant film in the saddle point region of the endwall. The Eckerle-Awad parameter given by Equ. (1) allows one to predict saddle point flow behavior in a turbine. Because of endwall clearance gaps between blade and stator rows in a turbine one can argue that certainly there are instances when the upstream developing endwall boundary layer is very small. This would make E large, and if its value is greater than 1000 as given by Equ. (1), the turbine heat transfer designer can expect the gas path flow in the region of the saddle point to be swirl-free.

Recently, Kang, Kohli and Thole[21] measured heat transfer and the flow field in the leading edge region of a first stage stator cascade, at two Reynolds numbers of 6×10^5 and 1.2×10^6, based on inlet velocity and stator true chord. Fig. 7 shows the inplane velocity vectors for the two Reynolds numbers they measured in what they term a stagnation plane. This is a plane extending radially outward from the stator leading edge and passing through the saddle point. For their smaller Reynolds number case (0.6×10^6) the value of the Eckerle-Awad parameter was E=990 (based on stator axial chord). As can be seen in Fig. 7, the inplane velocity vectors do show a swirling flow as the horseshoe vortex forms, in agreement with the Eckerle-Awad limit. For the larger Reynolds number case (1.2×10^6) E=1640, and the velocity vectors in Fig. 7 exhibit no swirl, for E>1000.

Thus the parameter E and the Eckerle-Awad limit of 1000 seem to accurately characterize turbine cascade saddle point flow on the endwall for swirl or no-swirl, if axial chord is used as an effective cylinder diameter. Certainly more work needs to be done to validate this conclusion, but Equ. (1) holds the promise of providing one of the first parameters to quantify the flow resulting from the separation of a turbine cascade inlet boundary layer.

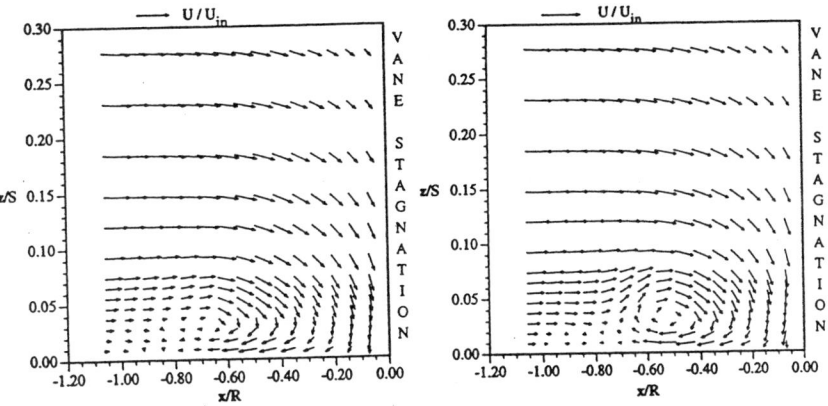

Fig. 7 Kang et al[26] mean velocity vectors on stagnation plane. On the left: E=1640, Re = 1.2X10^6. On the right: E=990, Re = 0.6X10^6.

Secondary Flow Loss Reduction

Because secondary or endwall losses can be so high (30-50% of total loss, as stated earlier) there have been many attempts and studies to reduce them. Various "bowed" and "leaned" airfoils have been experimented with and are in use on many newer jet engines designs, to manage endwall and profile losses. As far as the author is aware, there have been no dramatic loss reductions by either the use of bowing or leaning. Harrison[22] presents a systematic study of both, and more recently Duden, Raab and Fottner[23] show the results of an experimental and design program for three-dimensional cascade airfoils and endwall contouring.

The use of fences and grooves on endwall and suction side airfoil surfaces was extensively investigated by Prumper[24] to reduce secondary flow losses. Prumper tested upwards of 400 different inlet guide vane configurations on a low aspect ratio annular cascade. His optimum configuration involving shallow grooves cut on the suction side vane surface at the endwall, improved the efficiency (by one percentage point) and off-design performance of an existing commercial single stage turbine (see Gallus and Kummel[25]). More recent studies on endwall fences have been carried out by Chung and Simon[26] and Aunapu, et al[27].

Endwall contouring, especially in inlet guide vane passages, has been a major focus for many years. Among the first to test an "S" shaped endwall was Deich, et al[28] who were able in increase the efficiency of an existing turbine by profiling the hub endwall of turbine inlet guide vanes into a sigmoidally curved surface converging in the streamwise direction. Later, Morris and Hoare[29] did detailed planar cascade S-wall studies, showing that endwall losses were significantly reduced at the non-contoured endwall. Kopper and Milano[30] carried out S-wall testing on inlet guide vanes in linear cascades and measured a 17% reduction of full passage mass-averaged loss relative to a full planar endwall configuration. More recent studies involving testing and CFD calculation on S-walls by Dossena, Perdichizzi and Savini[31] yielded results similar to Kopper and Milano[30].

Preliminary research on one unusual, but promising approach to endwall contouring has been reported by LaFleur, Whitten and Araujo[32]. This technique makes use of the shape of ice that has been formed on the refrigerated endwalls of turbine cascade passages, as water is run through them. Where water circulation or wall shear forces are low ice grew in thickness, and where either was high, less ice formed. By applying this technique to a cylinder mounted on an endwall, LaFleur and Langston[33] were able to show through wind tunnel testing that the total drag (or aerodynamic loss) for the ice-shaped endwall surface was reduced by 18% when compared to smooth, non-contoured endwall and cylinder combination.

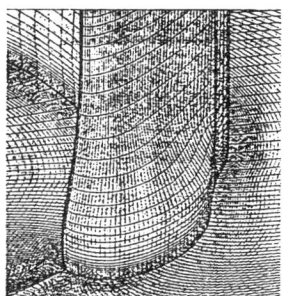 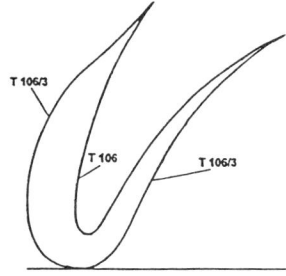

Fig. 8 Turbine endwall modification by Sauer, Müller and Vogeler[34]. The leading endwall bulb (T106/3) reduced endwall losses (of T.106) by 47%.

One of the most promising results on reducing secondary flow losses by contouring has been presented by Sauer, Muller and Vogeler[34]. They positioned a leading edge "bulb" at the endwall-airfoil junction of high turning turbine airfoils in a plane cascade. Their bulb geometry (presumably analogous to the bulb on a ship's prow used to reduce bow wave drag) fairs into the leading edge in a spanwise direction and has its largest thickness at the suction side-endwall junction, as shown in Fig. 8. The authors report an experimentally determined 47% reduction in net secondary loss using the bulb in a low pressure turbine airfoil cascade with an aspect ratio of 3.0 at a Reynolds number of 4×10^5. In an earlier work using similar bulb geometry lower turning, inlet guide vane-like turbine cascades, Sauer and Wolf[35] reported reductions of net endwall loss of up to 25%. Sauer, et al[34,35] reason that these remarkable reductions in loss are brought about by the bulb increasing the strength of the counter vortex (see Fig. 1), which reduces the strength of the passage vortex with its opposite sense of rotation. But one can also argue that such a proposed mechanism would actually increase losses, as measured downstream of the cascade trailing edge plane. (Sauer, et al[34] were unable to provide answers through the use of a CFD model.) Clearly some experiments involving endwall measurements within cascade passages are needed to give more information on how these leading edge bulbs give such promising secondary loss reductions.

Secondary Flow Prediction

There are no closed form analytical solutions to the secondary flow shown in Fig. 1. Since the early 1970's there has been a great deal of effort to model this complex flow using a variety of CFD codes and associated turbulence models. Only a few representative examples will be briefly examined here, of the many reported in the open literature. Much progress has been made and it would be safe to say that most turbine manufacturers use 3-D CFD codes routinely in the mid to later stages of the design process for a new machine. Generally loading curves (i.e. airfoil pressure distributions as shown in Fig. 3) can be predicted accurately even when secondary effects are quite large. However, the ability to *routinely* predict aerodynamic losses with strong secondary flows (e.g. Fig. 4) has been more limited.

Hah[36] in 1983 was one of the first investigators to develop a numerical code that captured most of the secondary flow features (Fig. 1). Comparing his calculations with the data of Langston, Nice and Hooper[4], he used a modified algebraic Reynolds stress model for the turbulence closure problem. An examination of his results show very good agreement with static pressure data on the airfoil (Fig. 3) and fairly good agreement with the endwall static pressure distributions. His calculations of aerodynamic loss through the cascade passage showed the overall trends of the data in Fig. 4 (e.g. the slow growth of loss in the first part of the passage and then the sudden jump near the trailing edge), but the calculated mass averaged loss coefficients, c_{pt}, were 20-30% higher near to and downstream of the trailing edge.

Nearly 10 years later, Dorney and Davis[37] calculated fluid flow and heat transfer for the Langston et al[4] cascade, using the thin inlet boundary layer heat transfer and aerodynamic data of Graziani, Blair, Taylor and Mayle[38], with much better accuracy. The authors' results using a Baldwin-Lomax two-layer algebraic turbulence model show excellent agreement for static pressures on airfoil surfaces and the endwall when compared to data. A grid refinement study was done, and for their finest grid, they predicted the measured overall area-averaged loss just downstream of the cascade within 2%. (It should be noted that the experimental loss values were not given in the original Graziani et al[38] paper.) The Dorney and Davis work is not representative of typical CFD design simulations used for engine design. They used a CRAY-2 supercomputer and their computational times were relatively large. They used very fine grids (up to 900,000 grid points for full span) whose construction was influenced by knowledge of existing heat transfer and fluid flow experimental data.

The recent work of Hartland, Gregory-Smith and Rose[39] might be more typical of CFD calculations used in a turbine design system. They report on mass averaged total pressure loss measurements made in the large scale turbine cascade of Gregory-Smith and Cleak[40]. Their measurements were made first in a planar endwall cascade and then in a non-axisymmetric endwall cascade, contoured to reduce the strength of the passage vortex. The net mass-averaged total pressure loss, c_{pt}, for each of their configurations was calculated using a Rolls-Royce standard turbomachinery CFD code, described by Harvey, et al[41]. Their net mass-averaged losses across the cascade, measured and predicted, are given in Table 1.

From the values given in Table 1, it can be seen that the CFD code over-predicted full losses by as much as 75%, overpredicted mid-span losses by as much as

173% and underpredicted the secondary losses by as much as 47%. Also, the CFD calculations predicted the planar endwall to be the smaller loss producer, whereas the experimental values showed just the opposite, with a reduction in "full" loss of 20% provided by the contoured endwall. Thus the CFD results were not correct either from a quantitative or a qualitative standpoint. Hartland et al[39] attribute this limited success in loss prediction to the lack of an adequate turbulence model. The authors' state: "The CFD predicts the secondary kinetic energy well, but the mixing length turbulence model is not adequate to translate this into accurate losses. The modelling required to improve on this is not clear, however. Moore and Gregory-Smith[42] found that for the code used in this design, a k-ε turbulence model performed rather worse than a mixing length model, especially if laminar and transitional regions are allowed for in the latter. More work in this area is needed."

Table 1. Net losses, c_{pt}, for the Durham cascade as given in Table 3 of Hartland, Gregory-Smith and Rose[39].

Endwall Configuration Measured/CFD	c_{pt}		
	Full	Mid-Span	Secondary
Contoured	0.1108	0.0557	0.0551
Planar	0.1377	0.0598	0.0780
CFD Contoured	0.1937	0.1518	0.0419
CFD Planar	0.1926	0.1512	0.0414

Bradshaw[2] has observed that turbulence modeling for complex flow (e.g. turbomachinery) represents an "unfinished business". He further concludes: "The present state is that even the most sophisticated turbulence models are based on brutal simplifications of the N-S equations and cannot be relied on to predict a large range of flows accurately. For example – a particular embarrassment to the modeling community – a model that is adjusted empirically to give good results for plane jets from "two-dimensional" nozzles often gives poor results for jets from circular nozzles."

In addition to the possible shortcomings of turbulence modeling, the author is of the opinion that at least one loss producing mechanisms may not be being modeled correctly, namely the highly skewed boundary layer on the endwall. This boundary layer is very thin and much of it is laminar as seen early on by Senoo[43] and Langston[4], and later by Harrison[11,22]. Strongly accelerated laminar flows (e.g. Falkner-Skan flows) can give local friction coefficient and heat transfer coefficient values many times greater than nonaccelerated cases. Skewing such flows (as occurs on an endwall) further increases heat transfer and wall shear forces from collateral boundary layer values. More experimental work is needed to correctly model the laminar portion of the endwall boundary layer.

Secondary Flow Loss Reconsidered

Thus far in this paper secondary flow "loss" has been treated without regard to a rigorous definition, rather in the same way that Oliver Heaviside, the English physicist, defended some of his important (but unproven) mathematical results: "Shall I refuse my dinner because I do not fully understand the process of digestion?"

A clear and simple explanation and example of aerodynamic loss – or simply, loss – has been given by Taylor[44]. Consider the adiabatic flow of an ideal gas through a stationary screen or vane row. If the upstream (station 1) and downstream (station 2) flow is steady and uniform, the thermodynamic state can be completely specified by the stagnation temperature, T_0, and the stagnation pressure, P_0, at each station. Since no work is done and no heat is transferred, $T_{01} = T_{02} = T_0$, but due to the friction forces caused by the interaction of the gas and the screen (or vanes), the stagnation pressure, P_o, will change. Taylor shows that the specific work, w, that would be required to restore the fluid (the ideal gas) to its initial (station 1) state would be

$$w = \frac{1}{\rho_{01}}(P_{01} - P_{02}) \qquad (2)$$

where ρ_{01} is the upstream stagnation density and $P_{01} - P_{02}$ is the change in total or stagnation pressure. From this idealized example one can see that what is "lost" in the irreversible frictional flow through the screen or vanes is the ability to do work, as given by Equ. (2), and that the change in stagnation or total pressure ($P_{01} - P_{02}$) is a direct measure of the lost work. This pressure difference can be nondimensionalized with the station 1 velocity head to form the basis of the mass-averaged total pressure loss coefficient, c_{pt}, used in Fig. 4 and in Table 1. Following Taylor[44], Equ. (2) can be rewritten as

$$s_2 - s_1 = \frac{R}{P_{01}}(P_{01} - P_{02}) \qquad (3)$$

where R is the ideal gas constant and $s_2 - s_1$ is the entropy increase in the fluid after it has passed through the screen or vane row. Thus the change in total pressure is also a direct measure of entropy generation.

Denton[45], in the 1993 IGTI Scholar Lecture, has proposed that the understanding of loss in turbomachines will be improved by thinking of it in terms of entropy generation, such as given by the simple example that resulted in Equ. (3). In this insightful paper he treats some of the sources of entropy that cause loss, such as viscous effects in boundary layers, mixing processes, shock waves and heat transfer across temperature differences. In addition to a fresh approach to aerodynamic loss in turbomachinery, the paper provides an expert's review of work in many areas of gas path fluid mechanics for both axial and radial machines.

Using the Denton entropy generation approach to loss, O'Donnell and Davies[46] at the University of Limerick report the results of very detailed measurements of entropy generation in the suction side midspan boundary layer of a turbine blade mounted in a subsonic linear cascade at a Reynolds number of 1.85 x 10^5. They found that 75% of the entropy generated by the boundary layer occurred within the laminar region (50% of the suction surface length). Conventional wisdom, based on considerations of aerodynamic drag, holds that turbulent boundary layer losses are significantly higher than laminar losses. The Limerick work shows that the reverse is true, with the greater portion of entropy being generated by the high shear forces found in the very thin, laminar midspan suction side boundary layer. One is led to speculate that such detailed measurements of entropy made in an endwall boundary layer might yield similar results, concerning laminar flow endwall contributions to secondary flow losses in turbine cascades.

CONCLUSIONS

Over the last 25 years, through the research efforts of many investigators, a sharper and more quantified picture of secondary flow in turbine airfoil cascades has emerged. In his review of the subject, Sieverding[1] called for the need to evaluate the significance of various aspects of the complex flow. Some progress in that evaluation has been reviewed here, from which one may conclude the following:

1) The basic secondary flow picture (e.g. Fig. 1) put forth by various teams of investigators in the 1970's has since been verified by a number of other studies.

2) More details of this basic picture have emerged or been proposed (e.g. Figs. 5 and 6). Detailed measurements of velocity and pressure are needed to quantify their significance.

3) A more complete understanding of the separation process of the inlet boundary layer (or more generally, the redistribution of the inlet vorticity) is needed. The flow parameter given in Equ. (1) is a first step in the direction of predicting properties of the separated flow.

4) Research into and development of geometries and devices to reduce or control secondary flow has made some progress. One recent investigation (the leading edge bulb in Fig. 8) is particularly promising. The apparently strong effect it has on losses points out the need for basic endwall flow measurements.

5) In the last twenty years the overall features of secondary flow have been successfully modeled by investigators using a variety of CFD codes. However, the lack of an adequate turbulence model seems to be one major cause of an inability to routinely predict endwall loss accurately. Denton[45] maintains that even for the straightforward problem of calculating the loss of a two-dimensional cascade, an *a priori* prediction using the best available methods, is unlikely to be accurate to better than about ± 20%. Endwall loss *a priori* CFD predictions are probably no more accurate than Denton's estimate.

6) The underlying mechanisms causing secondary flow losses are not yet fully understood. The approach of thinking of loss in terms of entropy generation proposed by Denton[45] may provide new ways of dealing with these important problems.

Lastly, the author would make a strong case for more basic experiments to study secondary flows, either with cascades or with experiments such as endwall-cylinder flows that capture some aspects of the endwall flow physics. It has been the author's experience that CFD follows the experimenter, not the other way around. It is with experiments that new flow features are found out. These new features can then be modeled by numerical codes or other analytical models.

ACKNOWLEDGEMENTS

The author thanks Laurie Hockla of the University of Connecticut and Sandor Becz of Pratt & Whitney, United Technologies Corporation for their help in preparing this paper.

REFERENCES

1. Sieverding, C.H. 1985. Recent Progress in the Understanding of Basic Aspects of Secondary Flows in Turbine Blade Passages. ASME Jour. of Turbomachinery. **107**: 248-257.
2. Bradshaw, P. 1996. Turbulence Modeling with Application to Turbomachinery. Prog. Aerospace Sci. **32**: 575-624.
3. Sharma, O.P., and Butler, T.L. 1987. Predictions of Endwall Losses and Secondary Flows in Axial Flow Turbine Cascades. ASME Jour. of Turbomachinery. **109**: 229-236.
4. Langston, L.S., Nice, M.L. and Hooper, R.M. 1977. Three-Dimensional Flow Within a Turbine Cascade Passage. ASME Jour. of Engineering for Power. **99**: 21-28.
5. Langston, L.S. 1980. Crossflows in a Turbine Cascade Passage. ASME Journal of Engineering for Power. **102**: 866-874.
6. Sieverding, C.H. and Van den Bosche. 1983. The Use of Colored Smoke to Visualize Secondary Flows in a Turbine-Blade Cascade. Jour. of Fluid Mechanics. **134**: 85-89.
7. Marchal, P.H. and Sieverding, C.H. 1977. Secondary Flows Within Turbomachinery Bladings. Secondary Flows in Turbomachines. AGARD CP No. 214. Paper No. 11.
8. Yamamoto, A. 1987. Production and Development of Secondary Flows and Losses in Two Types of Straight Turbine Cascades: Part I – A Stator Case. ASME J. of Turbo. **109**: 186-193.
9. Yamamoto, A. 1987. Production and Development of Secondary Flows and Losses in Two Types of Straight Turbine Cascades: Part 2 – A Rotor Case. ASME J. of Turbo. **109**: 194-200.
10. Gregory-Smith, D.G., Graves, C.P., and Walsh, J.A. 1988. Growth of Secondary Losses and Vorticity in an Axial Turbine Cascade. ASME Jour. of Turbomachinery. **110**: 1-8.
11. Harrison, S. 1990. Secondary Loss Generation in a Linear Cascade of High-Turning Turbine Blades. ASME Jour. of Turbomachinery. **112**: 618-624.
12. Goldstein, R.J. and Spores, R.A. 1988. Turbulent Transport on the Endwall in the Region Between Adjacent Turbine Blades. ASME Jour. of Heat Transfer. **110**: 862-869.
13. Wang, H.P., Olson, S.J., Goldstein, R.J. and Eckert, E.R.G. 1997. Flow Visualization in a Linear Turbine Cascade of High Performance Turbine Blades. ASME J. of Turbo. **119**: 1-8.
14. Baker, C.J. 1979. The Laminar Horseshoe Vortex. Jour. of Fluid Mechanics. **95**: 347-367.
15. Hodson, H.P. and Dominy, R.G. 1989. Three-Dimensional Flow in a Low-Pressure Turbine Cascade at Its Design Condition. ASME Jour. of Turbomachinery. **109**: 177-185.
16. Green, Sheldon I. 1995. *Fluid Vortices*. 1-9. Kluwer Academic Publishers.
17. Eckerle, W.A. and Langston, L.S. 1987. Horseshoe Vortex Formation Around a Cylinder. ASME Jour. of Turbomachinery. **109**: 278-285.
18. Moore, J. and Ransmayr, A. 1984. Flow in a Turbine Cascade: Part 1 – Losses and Leading Edge Effects. ASME Jour. of Engineering for Gas Turbines and Power. **106**: 400-408.
19. Ballio, F., Bettoni, C. and Franzetti, S. 1998. A Survey of Time-Averaged Characteristics of Laminar and Turbulent Horseshoe Vortices. ASME Jour. of Fluids Engineering. **120**: 233-242.
20. Eckerle, W.A. and Awad, J.K. 1991. Effect of Freestream Velocity on the Three-Dimensional Separated Flow Region in Front of a Cylinder. ASME Jour. of Fluids Engineering. **113**: 37-44.

21. Kang, M.B., Kohli, A. and Thole, K.A. 1999. Heat Transfer and Flowfield Measurements in the Leading Edge Region of a Stator Vane Endwall. ASME J. of Turbo. **121:** 558-568.
22. Harrison, S. 1992. The Influence of Blade Lean on Turbine Losses. ASME Jour. of Turbomachinery. **114:** 184-190.
23. Duden, A., Raab, I., Fottner, L. 1999. Controlling the Secondary Flow in a Turbine Cascade by Three-Dimensional Airfoil Design and Endwall Contouring. ASME J. of Turbo. **121:** 191-199.
24. Prümper, H. 1972. Application of Boundary Fences in Turbomachinery. AGARD AG No. 164. Paper No. II-3: 315-331.
25. Gallus, H.E. and Kummel, W. 1977. Secondary Flows and Annulus Wall Boundary Layers in Axial-Flow Compressor and Turbine Stages. AGARD CPP No. 214. Paper No. 4: 1-15.
26. Chung, J.T. and Simon, T.W. 1993. Effectiveness of the Gas Turbine Endwall Fences in Secondary Flow Control at Elevated Freestream Turbulence Levels. ASME Paper No. 93-GT-51.
27. Aunapu, Nicole V., Volino, Ralph J., Flack, Karen A. and Stoddard, Ryan M. 2000. Secondary Flow Measurements in a Turbine Passage with Endwall Modification. ASME No. 2000-GT-212.
28. Deich, M.E., Zaryanskin, A.E., Fillipov, G.A. and Zatsepin, M. 1960. Method of Increasing the Efficiency of Turbine Stages with Short Blades. *Teploenergetikia*. **2:** 240-254.
29. Morris, A.W.H. and Hoare, R.G. 1975. Secondary Loss Measurements in a Cascade of Turbine Blades with Meridional Wall Profiling. ASME Paper No. 75/GT-13.
30. Kopper, F.C. and Milano, R. 1981. Experimental Investigation of Endwall Profiling in a Turbine Vane Cascade. AIAA Journal. **19:** 1033-1040.
31. Dossena, V., Perdichizzi, A. and Savini, M. 1999. The Influence of Endwall Contouring on the Performance of a Turbine Nozzle Guide Vane. ASME Jour. of Turbomachinery. **121:** 200-208.
32. LaFleur, Ronald S., Whitten, Timothy S. and Aranjo, Juan A. 1999. Second Vane Endwall Heat Transfer Reduction by Iceform Contouring. ASME Paper No. 99-GT-422.
33. LaFleur, R.S. and Langston, L.S. 1993. Drag Reduction of a Cylinder/Endwall Junction Using the Iceformation Method. ASME Jour. of Fluids Engineering. **115:** 26-32.
34. Sauer, H., Müller, R. and Vogeler, K. 2000. Reduction of Secondary Flow Losses in Turbine Cascades by Leading Edge Modifications at the Endwall. ASME paper No. 2000-GT-473.
35. Sauer, H. and Wolf, H. 1997. Influencing the Secondary Flow in Turbine Cascades by the Modification of the Blade Leading Edge. 2. European Conference of Turbomachinery. Antwerpen.
36. Hah, C. 1984. A Navier-Stokes Analysis of Three-Dimensional Turbulent Flows Inside Turbine Blade Rows at Design and Off-Design Conditions. ASME Jour. of Engineering for Gas Turbines and Power. **106:** 421-429.
37. Dorney, D.J. and Davis, R.L. 1992. Navier-Stokes Analysis of Turbine Blade Heat Transfer and Performance. ASME Jour. of Turbomachinery. **114:** 795-806.
38. Graziani, R.A., Blair, M.F., Taylor, J.R. and Mayle, R.E. 1980. An Experimental Study of Endwall and Airfoil Surface Heat Transfer in a Large Scale Turbine Blade Cascade. ASME Jour. of Engineering for Power. **102:** 257-267.
39. Hartland, J.C., Gregory-Smith, P.G., Harvey, N.W. and Rose, M.G. 2000. Nonaxisymmetric Turbine End Wall Design: Part II – Experimental Validation. ASME J. of Turbo. **122:** 286-293.
40. Gregory-Smith, D.G. and Cleak, J.G.E. 1992. Secondary Flow Measurements in a Turbine Cascade with High Inlet Turbulence. ASME Jour. of Turbomachinery. **114:** 173-183.
41. Harvey, Neil W., Rose, Martin G., Taylor, Mark D., Shahrokh, Hartland, Jonathan and Gregory-Smith, David G. 2000. Nonaxisymmetric Turbine End Wall Design: Part I – Three-Dimensional Linear Design System. ASME Jour. of Turbomachinery. **122:** 278-285.
42. Moore, H. and Gregory-Smith, D.G. 1996. Transition Effects on Secondary Flows in a Turbine Cascade. ASME Paper No. 96-GT-100.
43. Senoo, Y. 1958. The Boundary Layer on the End Wall of a Turbine Nozzle Cascade. Transactions of the ASME. **80:** 1711-1720.
44. Taylor, E.S. 1971. Boundary Layers, Wakes and Losses in Turbomachines. MIT Gas Turbine Report No. **105:** 1-3.
45. Denton, J.D. 1993. Loss Mechanisms in Turbomachines. ASME J. of Turbo. **115:** 621-650.
46. O'Donnell, F.K. and Davies, M.R.D. 2000. Turbine Blade Entropy Generation Rate Part II: The Measured Loss. ASME Paper No. 2000-GT-266.

Heat Transfer near Turbine Nozzle Endwall

MINKING K. CHYU

Department of Mechanical Engineering
University of Pittsburgh
Pittsburgh, PA 15261

ABSTRACT: This paper gives an overview and reviews recent findings concerning turbine endwall cooling in the literature. The text below begins with a brief discussion of the secondary flows and heat transfer around cascade endwall. This will be followed by a review of recent developments in cooling concepts and related heat transfer results. Key topics include: film cooling, upstream bleeding, endwall contouring, and leakage through component interfaces.

INTRODUCTION

Endwall heat transfer is an important design consideration for modern gas turbines. The first stage vanes of the turbine are exposed to severe convective loads from the high temperature gases exiting the combustor. In an effort to increase the efficiency of the turbine cycle, these temperatures are as high as the materials will allow. Thus, cooling is required for protection of the hot turbine components. Previously, combustor cooling and dilution air reduced the temperature of the flow adjacent to the endwall, thus reducing the endwall thermal load. As a result, the main focus of turbine thermal protection was directed primarily on airfoils. In recent years, combustor re-design has flattened the temperature distribution at the combustor exit, thus raising endwall temperatures. This has increased thermal loading on the endwall and renewed interest in endwall cooling.

Cooling of turbine endwall poses considerable challenges due mainly to the complex secondary flow patterns near the endwall. The secondary flow tends to lift the coolant off the endwall surface and entrain to the core flow in the passage. Although this phenomenon has been known for some time, current understanding toward the detailed transport features in the region remains to be insufficient. Because of such uncertainty, coolant supplies for the endwall are often excessive and distributed in a less than optimal fashion. Identifying more effective coolant management strategies as well as new cooling concepts for the region are major turbine design issues. They are the driving force of the recent research efforts in turbine endwall cooling.

The main intent of this paper is to give an overview concerning the nature of endwall cooling and review recent findings in the literature. Since the surface heat transfer characteristic is closely related to the adjacent flowfields, the text below begins with a brief discussion of the secondary flows in turbine passage. This will be followed by a review of recent developments in cooling concepts and related heat transfer results. Key topics for the latter include: endwall contouring and coolant leakage or slot bleeding.

SECONDARY FLOWS IN TURBINE CASCADE PASSAGE

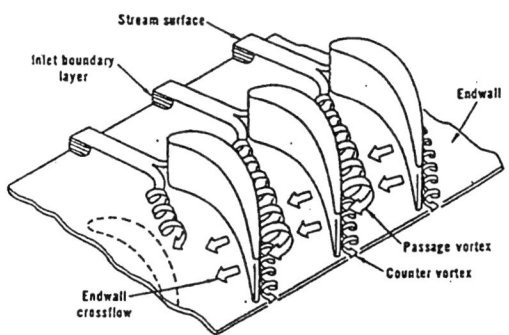

Fig. 1   Vortex pattern described by Langston (1980).

The subject of secondary flow in a turbine passage as well as its relevant control strategies has been the center of many studies concerning turbine flows, aerodynamic losses and component cooling. The nature of such flow is highly complex as it consists of several vortex systems in the vicinity of endwall, particularly near an airfoil-endwall junction. Although the detailed characteristics of secondary flows vary with stage geometry and blading, the general flow features, as shown in Fig. 1, can be revealed as a three-dimensional cascade flow model developed by Langston et al. [1,2]. The figure shows that the inlet boundary layer separates and forms a horseshoe vortex with two divided legs wrapping around the leading edge of an airfoil. The leg merge on the pressure side of the airfoil eventually becomes the passage vortex. The other leg, labeled as counter vortex, remains along the junction of suction-surface and endwall. The counter vortex has a sense of rotation opposite to the passage vortex and is much smaller in size compared to the passage vortex.

As a result of further studies on convection transport near turbine endwall, additional findings pertaining to passage

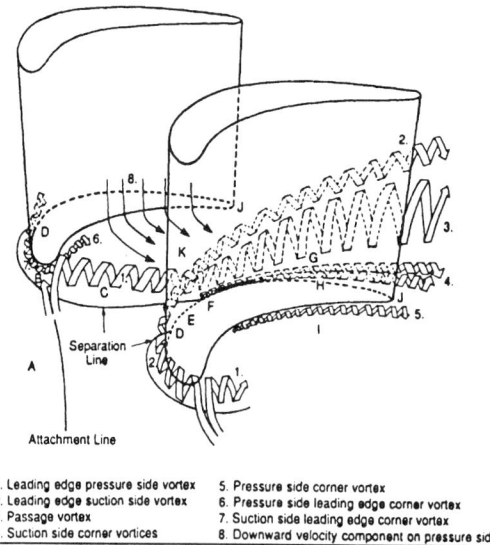

1. Leading edge pressure side vortex
2. Leading edge suction side vortex
3. Passage vortex
4. Suction side corner vortices
5. Pressure side corner vortex
6. Pressure side leading edge corner vortex
7. Suction side leading edge corner vortex
8. Downward velocity component on pressure side

Fig. 2 Passage Vortices (Goldstein and Spores 1988)

flow were discovered. Figure 2 exhibits a schematic of important features summarized by Goldstein and Spores [3]. Labeled as "C" and " A" in Fig. 2 are two important reference lines. The line C is the separation line of the endwall boundary layer as it approaches the turbine airfoils. The line A is the attachment line that extends from the incoming flow to the stagnation point at the leading edge of the turbine airfoil. This attachment line divides the incoming boundary layer flow entering a blade passage from the flow entering the adjacent passage. The intersection of these two lines is a saddle point.

Also revealed in Fig. 2 is the existence of several corner vortices. The counter vortex on the suction side, in fact, consists of two corner vortices. These vortices originate just downstream of where the passage vortices lift off from the endwall (location F). On the other hand, there is a single vortex existing in the corner of the pressure side, which prevails from about one-third of the chord length from the airfoil leading edge. Additional small but intense corner vortices also found embedded underneath the horseshoe vortex just ahead of an airfoil.

HEAT TRANSFER AND FILM COOLING ON CASCADE ENDWALL

Convective heat transfer on the turbine endwall is strongly affected by the complex features of secondary flow aforementioned. Hylton et al. [4] performed heat transfer measurements on an uncooled cascade endwall and reported that Stanton number near the airfoil leading edge is approximately three times higher than that of the mid-pitch value at the passage entrance. Their finding was later verified by Gaugler and Russell [5]. Using an analogous mass transfer technique based on naphthalene sublimation, Goldstein and Spores [3] reported probably the most detailed heat transfer distribution on an uncooled cascade endwall. Figure 3 shows their results in terms of local heat transfer enhancement relative to its flat plate counterpart. Attributed to the leading edge corner vortex, very high heat transfer prevails in the region immediately ahead of the airfoil. The corner vortex on the pressure side continues carrying such a high heat transfer trend toward downstream. The passage vortex modestly increases the heat transfer over the central section of the endwall extending from the leading edge of one airfoil to the suction surface of the adjacent airfoil. Near where the passage vortex encountered the suction surface, elevated heat transfer exists on the upstream side and low heat transfer on the downstream side. Due mainly to strong mixing of two flow streams from both sides of airfoil, significant increase in heat transfer exists near the airfoil trailing edge. Recently, Kang and Thole [6] and Radomsky and Thole [7] have examined the effects of free stream turbulence on the heat transfer on the uncooled endwall.

One method to cool the turbine endwall is the use of film cooling, where coolant air is discharged typically through discrete holes on the endwalls of a turbine passage. The characteristics of film cooling is expected to be affected by the secondary flows present in the vicinity. Less significant is that the coolant injection may conversely alter the secondary flows in the passage. These features combined mark the major distinction of film protection over an endwall compared to similar design concern over airfoil surfaces. Studies related to discrete-hole injection over a cascade endwall include Goldman and McLallin [8], Sieverding and Wilputte [9], Bario et al. [10], Harasgama and Burton [11], Freidrichs et al. [12-14]. The general conclusion drawn from these studies is that secondary flows have shown a strong impact on film cooling on the endwall, especially near the suction side corner. The excessive mixing inherited in a number of vortices

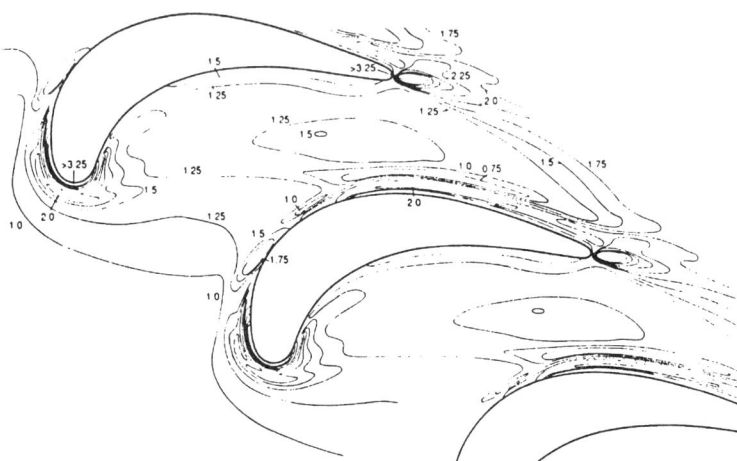

Fig. 3 Heat Transfer Enhancement on Cascade Endwall (Goldstein and Spores, 1988)

dissipates the overall cooling effectiveness in these regions. The migration of passage flows from the pressure surface toward the suction surface depletes the coolant near the pressure side of the cascade.

In an attempt to optimize the hole placement for nearly full coverage over the endwall surface, Freidrichs et al. [12-14] performed a series of studies measuring the distributions of adiabatic film effectiveness and aerodynamic losses in a large-scale, low-speed linear cascade. They concluded that coolant injection downstream of the separation line on the endwall does not alter the secondary flow structure. In addition, placement of film holes in regions of high static pressure benefits the reduction of aerodynamic losses incurred by coolant injection. They also suggested a division of endwall into four distinct regions for individual cooling consideration and hole placements. The region between the separation line of the leading edge horseshoe vortex and the leading edge itself is the most difficult to cool.

As an auxiliary focus to the primary cooling objective, coolant injection is often used to weaken or control the secondary flows in the passage. This is achieved typically by slot injection near the inlet of a cascade. Blair [15] was probably the first to report reduced heat transfer with film cooling injected through the endwall upstream of the throat and suggested a possible impact of the coolant on the secondary flows. To simulate the leakage from an endwall interface, Biesinger and Gregory-Smith [16] used upstream slot injection exclusively for control of secondary flows. Their results implied that the effects of low-velocity bleeding is to thicken the boundary layer, strengthen secondary flows, and induce more losses. High velocity injection, on the other hand, re-energizes the inlet boundary layer and weakens the passage vortex. Recently, Burd et al. [17,18] performed a series of studies examining both the heat transfer and aerodynamic effects of upstream slot injection over a realistic cascade with a contoured endwall. Oke et al. [19] studied the similar effects but with two rows of discrete hole injection, instead of slot injection. Based on the same cascade geometry, Shih et al. [20] performed a numerical simulation with slot injection to evaluate the effects of endwall contouring.

EFFECTS OF ENDWALL CONTOURING

An improved turbine design generally implies lower losses in aerodynamics and reduced coolant amount for cooling of turbine stages. To a great extent, these two design objectives are closely related and non-conflicting. As mentioned, secondary flows in turbine passage and deleterious pressure gradients within blading are the major causes for excessive aerodynamic losses and coolant requirements near the endwall region. It is often the case that the blading of first stage vanes is characterized by low aspect ratios and, due to cooling requirements, high-pitch-to-chord ratios. Both features enhance the strength of secondary flows and increase the secondary losses, to the extent comparable or even surpassing the profile losses. A viable approach to control the secondary flows and lessen the losses is endwall contouring.

The concept of endwall contouring was first introduced by Deich et al. [21] as a reduction of the airfoil span between the leading edge and trailing edge at the tip. This results in a reduction of velocity in the region of highest turning and a shift of the diffusion part of pressure distribution on the suction surface toward downstream. In addition, the more favorable pressure gradient induced by the axial contraction impedes the growth of the boundary layers on both the airfoil surface and the endwall. This latter feature marks the major benefit of endwall contouring and offers an effective means for controlling the secondary flow and reducing the aerodynamic losses. Ewen et al. [22] later reported an increase of turbine efficiency up to 3.5% with a contoured endwall. Morris and Hoare [23] documented reduction of secondary losses with various contoured endwall profiles. They observed that reduction of losses for non-axisymmetric profiling exists near the flat endwall, but an increase in losses prevails near the contoured endwall. Kopper et al. [24] reported similar findings with a total loss reduction up to 17%. Several recent studies [26, 27] also tested non-axisymmetric endwalls and cited improved secondary losses with reduced cross-flow pressure gradients.

While the thermal transport near the endwall is expected to be closely related to the secondary flows in the passage, the effect of contouring on the endwall heat transfer is little studied. Kopper et al. [24] observed a significant change in heat transfer distribution on a contoured endwall relative to its flat counterpart. As the energy of fluid migrating across from the pressure side of passage is reduced by the endwall contouring, the values of heat transfer coefficient are evidently lower close to the downstream portion of the suction side of passage. Recently, a series of flow and heat transfer studies [17-20, 25]. The main focus of these studies was to investigate the combined effects of upstream film injection and endwall contouring. Further studies on optimizing contouring shape with innovative cooling arrangements are recommended.

GAP LEAKAGE AND ITS INTERACTION WITH FILM COOLING

Turbine endwall heat transfer research has recently focused on one class of aerothermal problems that has been much disregarded. This problem however, occurs in many locations within the turbine hot flow path and concerns the flow and heat transfer over component-to-component interfaces. Interfaces are the necessary result of turbine mechanical requirements, including casting limitations, manufacturing, assembly, wear, load carrying capability, and maintenance. The mechanical design must also accommodate other imposed constraints of the system design, such as non-uniform

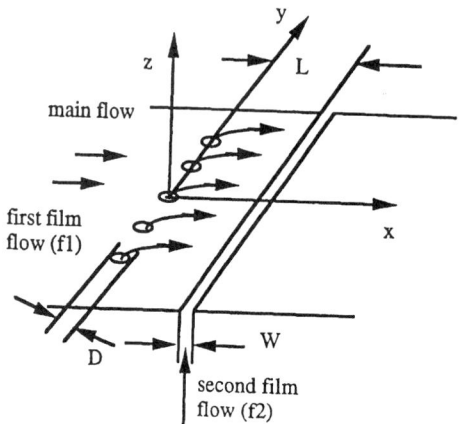

Fig. 4 Gap Leakage Downstream of Fim Cooling Holes

circumferential and/or radial pressure and thermal loads, and thermal growth mismatches during transients, for which component interfaces are indispensable. The most severe interface locations are found around the turbine inlet guide vanes. Such vanes are usually made as singlets or doublets, i.e., one or two vanes per segment. Free faces that must be sealed against leakage of valuable air represent the perimeter of the segment. At the same time, a certain minimal amount of leakage -- either passive or active -- is required to keep these faces cool and to prevent hot gas ingestion below the flow path. Ideally, the interfaces are perfectly smooth transitions, but in reality they are subject to misalignment due to manufacturing tolerances and assembly. Thus the interfaces may present steps together with a gap in the flow path having undesirable aero and thermal consequence.

For most cases, the transport phenomena concerning interface leakage is fundamentally similar to that of film cooling with slot injection. From the standpoint of heat convection, the problem as such is known as the "three-temperature problem" since the heat transfer from a film cooled wall is collectively determined by temperatures of the mainstream, the injectant, and the wall. If film cooling holes are present in the vicinity of a leakage gap, the two streams of injectants with different temperatures may interact. This situation is a "four-temperature" problem.

Vedula and Metzger [28] probably is the first resolve the three-temperature, film cooling problem using the transient liquid crystal (TLC) technique. Their technique is capable of revealing the local film effectiveness and heat transfer coefficient by solving two coupled transient conduction equations obtained from two separate tests. This method was extended by Yu and Chyu [29] to resolve a four-temerature problem with a leakage gap located downstream of film injections holes, as shown in Fig. 4. Their method is capable of revealing three major heat transfer parameters; i.e. heat transfer coefficient, the first effectiveness (η_1), and the second effectiveness (η_2). The local heat transfer on a film-protected location is a function of these three parameters. Chyu and Hsing [30] extended the transient principle to resolve the three-temperature problem using the laser-induced theremographic phosphor (TGP) fluorescence imaging [31]. One of the major advantages of TGP is its capability of resolving multiple-temperature problems with a single test. In addition, TGP can survive under a realistic turbine environment with high temperature and high pressure. With this approach, Chyu et al. [32] made a detailed measurement of film effectiveness and heat transfer coefficient in the region downstream to a leaking gap.

Figures 5 to 8 show sample results of a four-temperature situation similar to that shown in Fig. 4, but with misaligned interface. The measurements were performed using the TGP imaging. All the coolant jets in a low-blowing strength status i.e., M_1=0.32.

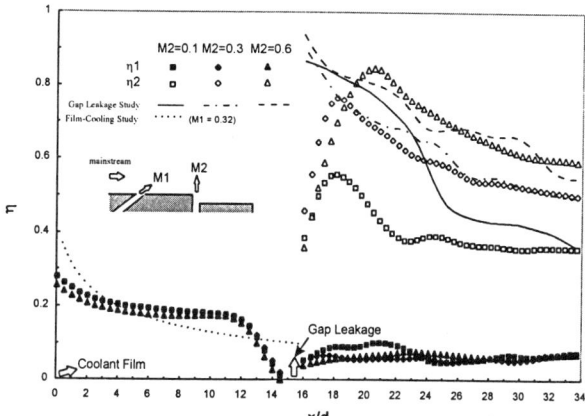

Fig. 5 Effectiveness with Backward-Facing Misalignment

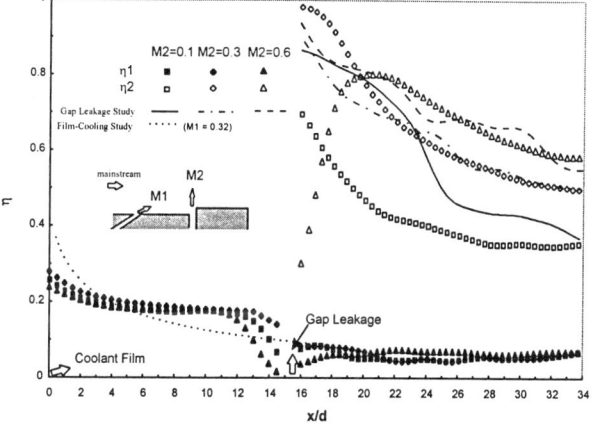

Fig. 6 Effectiveness with Forward-Facing Misalignment

Figures 5 and 6 show the effectiveness of the distribution of coolant jets over a leaking gap where the misalignment forms a backward-facing step or a forward-facing step, respectively. In order to envision how the cooling effectiveness is disrupted by the presence of another flow, the results of previous film cooling study [30] and gap leakage [32] are also labeled in these figures. Since the gap leakage forms a hydrodynamic obstacle that behaves like a solid wall, the flow field of the present situation is similar to an aligned interface situation. Hence, similar distributions of film effectiveness are found for both cases. Overall the leakage offers very effective protection over the surface downstream of the gap. For the case with a forward-facing step, the liftoff phenomenon appears to occur when the leakage has a strong injection at $M_2 = 0.6$.

Figures 7 and 8 show the corresponding results of heat transfer coefficient. As a result of strong flow interaction between the jets and the leakage, a local peak of heat transfer coefficient exists in the region upstream to the gap for both misaligned cases. An

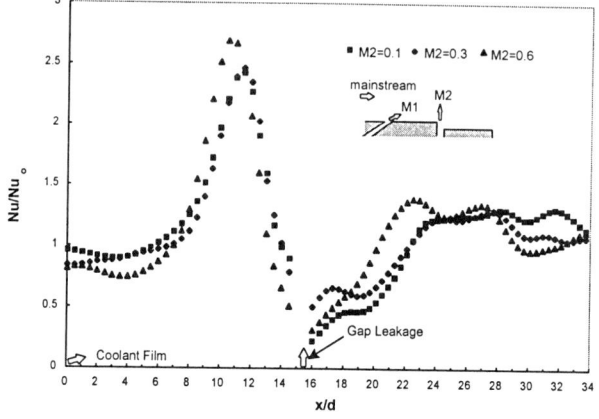

Fig. 7 Heat Transfer Coefficient with Backward-Facing Misalignment

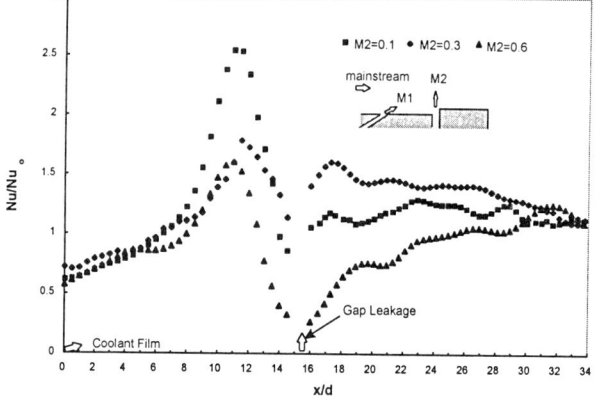

Fig. 8 Heat Transfer Coefficient with Forward-Facing Misalignment

interesting observation is that the magnitude of the peak decreases with an increase in leakage blowing ratio. The heat transfer coefficients downstream to the leaking gap show significantly different distributions between the cases. For the case of backward-facing misalignment, the heat transfer coefficient for any injection strength exhibits the characteristic of recirculation zone. The corresponding result for the forward-facing misalignment shows the trend of a newly developed boundary layer over a sharp corner.

CONCLUDING REMARKS

As a result of combustor redesign with a flattened outlet temperature profile, cooling of first stage nozzle endwall has emerged as a critical turbine design issue. Previous studies based on turbine cascades have shown that secondary flows are complex, three-dimensional and consist of a number of vortices. Surface heat transfer distributions on uncooled cascade endwalls reveal pattern highly influenced by the adjacent secondary flows. For the cases of film cooled endwall, recent research strives for better film coverage and reduced coolant requirements. Film injection with proper hole-placements may be a viable means to control the secondary flows. Endwall contouring also shows great potential for improved cooling with reduced aerodynamic losses. A series of recent studies have been devoted to exploring the combined effects of endwall contouring and upstream film injection. Heat transfer pertaining to gap leakage and its interaction with the active cooling schemes nearby is a new class of endwall cooling problem. Quantifying the leakage aero-thermal characteristics can be highly complex as it often involves multiple temperatures as the driving potential for the transport processes in the region. New experimentation and measurement techniques have been developed for tackling this issue. Similar approaches may be applicable for characterizing various cooling systems involving multiple coolant injections. Overall, there remain many unresolved issues concerning turbine endwall cooling. Continuing research in this area is recommended and expected.

REFERENCES

1. Langston, L.S., Nice, L.M., and Hooper, R.M., 1976, "Three-Dimensional Flow within a Turbine Cascade Passage," ASME Paper 76-GT-50.
2. Langston, L.S., 1980, "Crossflows in a Turbine Cascade Passage," ASME J. of Engineering for Power, Vol. 102, pp. 866-874.
3. Golstein, R.J. and Spores, R.A., 1988, "Turbulent Transport on the Endwall in the Region Between Adjacent Turbine Blades," ASME J. of Heat Transfer, Vol. 110, pp. 862-869.
4. Hoylton, L.P., Mihelc, M.S., Turner, E.R. and York, R.E., 1981, "Experimental Investigation of Turbine Endwall Heat Transfer," AFWAL-TR-81-2077, Vols. 1-3.
5. Gaugler, R.E., and Rusell, L.M., 1984, "Comparison of Visualized Turbine Endwall Secondary Flows and Measured Heat Transfer Patterns," ASME J.of Engineering for Gas Turbines and Power, Vol. 106, pp. 168-172.
6. Kang, M.B. and Thole, K.A., 1999, "Flowfield Measurements In The Endwall Region of A Stator Vane," ASME Paper 99-GT-188.
7. Rodamsky, R.W. and Thole, K.A. 2000, "High Freestream Turbulence Effects on Endwall Heat Transfer for a Gas Turbine Stator Vane," ASME Paper 2000-GT-0201.
8. Goldman, L.J. and McLallin, K.L., 1977, "Effect of Endwall Cooling on Secondary Flows in Turbine Stator Vanes," AGARD CP-214.
9. Sieverding, C.H. and Wilputte, Ph., 1981, "Influence of mach Number and End Wall Cooling on Secondary Flows in a Straight Nozzle Cascade," ASME J. Engineering for Power, Vol. 103, pp.257-264.
10. Bario, F. Leboeuf, F., Onvani, A. and Seddini, A., 1990, "Aerodynamics of Cooling Jets Introduced in the Secondary Flow of a Low Speed Turbine cascade," ASME J. of Turbomachinery, Vol. 112, pp. 539-546.
11. Harasgama, S.P. and Burthon, C.D., 1992, "Film Cooling Research on the Endwall of a Turbine Nozzle Guide Vane in a Short Duration Annular Cascade Part 1: Experimental Technique and Results," ASME J. of Turbomachinery, Vol. 114, pp. 741-746.
12. Friedrichs, S., Hodson, H.P. and Daws, W.N., 1996, "Distribution of Film-Cooling Effectiveness on a Turbine Endwall Measured Using Ammonia and Diazo Technique," ASME J. of Turbomachinery, Vol. 118, pp. 613-621.
13. Friedrichs, S., Hodson, H.P. and Daws, W.N., 1997, "Aerodynamic Aspects of Endwall Film-Cooling," ASME J. of Turbomachinery, Vol. 119, pp. 786-793.

14. Friedrichs, S., Hodson, H.P. and Daws, W.N., 1996, "The Design of an Improved Endwall Film-Cooling Configuration," 98-GT-483.
15. Blair, M.F., 1974, "An Experimental Study of Heat Transfer and Film Cooling on Large-Scale Turbine Endwalls," ASME J. of Turbomachinery, Vol. 96, pp. 524-529.
16. Biesinger, T.E. and Gregory-Smith, D.G., 1993, "Reduction of Secondary Flows and Losses in a Turbine Cascade by Upstream Boundary Layer Blowing," ASME Paper 93-GT-114.
17. Burd, S.W. and Simon, T.W., 2000a, "Flow Measurements in a Nozzle Guide Vane Passage with Low Aspect Ratio and Endwall Contouring," ASME Paper 2000-GT-0213.
18. Burd, S.W. and Simon, T.W., 2000b, "Effects of Slot Bleed Injection over a Contoured Endwall on Nozzle Guide Vane Cooling Performance: Part II – Thermal Measurements," ASME Paper 2000-GT-0200.
19. Oke, R.A., Simon, T.W., Burd, S.W. and Vahlberg, R., 2000, "Measurements in a Turbine cascade over a Contoured Endwall: Discrete Hole Injection of Bleed Flow," ASME 2000-GT-214.
20. Shih, T. I-P., Lin., Y.-L. and Simon, T.W., 2000, "Control of Secondary Flows in a Turbine Nozzle Guide Vane by Endwall Contouring," ASME Paper 2000-GT-0556.
21. Deich, M.E., Zaryankin, A.E. Fillipov, G.A. and Zatsepin, M.F., 1960, "Method of Increasing the Efficiency of Turbine Stages and Short Blades," Teploenergetika, No. 2, February, Translation No. 2816, Associated Electrical Industries (Manchester) Ltd., April, 1960.
22. Ewen, J.S., Huber, F.W. and Mitchell, J.P., 1973, "Investigation of the Aerodynamic Performance of Small Axial Turbines," ASME Paper 73-GT-3.25.
23. Morris, A.W.H. and Hoare, R.G., 1975, "Secondary Loss Measurements in a Cascade of Turbine Blades with Meridional Wall Profiling," ASME Paper 75-WA/GT-13.
24. Kopper, F.C., Milano, R. and Vanco, M., 1981, "Experimental Investigation of Endwall Profiling in a Turbine Vane Cascade," AIAA J., Vol. 19, No. 8, pp. 1020-1040.
25. Burd, S.W. Satterness, C.J. and Simon, T.W., 2000c, "Effects of Slot Bleed Injection over a Contoured Endwall on Nozzle Guide Vane Cooling Performance: Part I – Flow Field Measurements," ASME Paper 2000-GT-0199.
26. Hartland, J.C., Gregory-Smith, D.G., and Rose, M.G., 1998, "Non-Axisymmetric Endwall Profiling in a Turbine Rotor Blade," ASME Paper 98-GT-525.
27. Yan, J., Gregory-Smith, D.G., and Qalker, P.J., 1999, "Secondary Flow Reduction in a Nozzle Vane Cascade by Non-Axisymmetric Endwall Profiling," ASME Paper 99-GT-339.
28. Vedula, R.P. and Metzger, D.E., 1991, "A Method for the Simultaneous Determination of Local Effectiveness and Heat Transfer Distributions in Three-Temperature Convection Situations," ASME Paper 91-GT-345.
29. Yu, Y. and Chyu, M.K., 1996, "Influence of a Leaking Gap Downstream of the Injection Holes on Film Cooling Performance," ASME Paper 96-GT-175.
30. Chyu, M.K. and Hsing Y.C., 1996, "Use of a Thermographic Fluorescence Imaging System for Simultaneous Measurement of Film Cooling Effectiveness and Heat Transfer Coefficient," ASME Paper 96-GT-430.
31. Bizzak, D.J. and Chyu, M.K., 1994, "A Rare-Earth Phosphor Laser-Induced Fluorescence Thermal Imaging System," Rev. Scientific Instruments, Vol. 65, No. 1, pp. 102-107.
32. Chyu, M.K., Hsing, Y.C. and Bunker, R.S., 1998, "Measurements of Heat Transfer Characterisitics of Gap Leakage Around a Misaligned Component Interface," ASME Paper 98-GT-182.

Transition to Turbulence Under Low-Pressure Turbine Conditions

Terrence W. Simon[a] and Richard W. Kaszeta

Heat Transfer Laboratory, Mechanical Engineering Department, University of Minnesota, Minneapolis, Minnesota 55455

ABSTRACT: In this paper, the topic of laminar to turbulent flow transition, as applied to the design of gas turbines, is discussed. Transition comes about when a flow becomes sufficiently unstable that the orderly vorticity structure of the laminar layer becomes randomly oriented. Vorticity with a streamwise component leads to rapid growth of eddies of a wide range of sizes and eventually to turbulent flow. Under "natural" transition, infinitesimal disturbances of selected frequencies grow. "Bypass transition" is a term coined to describe a similar process, but one driven by strong external disturbances. Transition proceeds so rapidly that the processes associated with "natural" transition seem to be "bypassed." Because the flow environment in the turbine is disturbed by wakes from upstream airfoils, eddies from combustor flows, jets from film cooling, separation zones on upstream airfoils and steps in the duct walls, transition is of the bypass mode. In this paper, we discuss work that has been done to characterize and model bypass transition, as applied to the turbine environment.

INTRODUCTION

Boundary Layer Laminar-to-Turbulent Transition

A considerable body of literature is dedicated to transition to turbulence of a laminar boundary layer subjected to infinitesimal disturbances. The first signs of this growth are Tollmien-Schlichting, two-dimensional waves which contain vorticity that is still perpendicular to the flow. Such flow is amenable to analysis by perturbation methods. As time passes, higher order modes grow and randomly-oriented vorticity appears. A complete presentation of this is given by Schlichting[1]. For a boundary layer to pass through transition to turbulence under this mode, the background disturbance level must be low, perhaps less than 0.1% turbulence intensity, although some signs of some of the instability modes identified with this "natural" transition have been viewed in flows with turbulence levels of ten times this value. Because the turbulence levels in the low-pressure turbine range from 2 to 10%, natural transition analysis and the above discussion cannot be directly applied. For this reason, natural transition is not discussed further. It is worthy of note, though,

[a]Corresponding Author

that linear stability theory continues to have value. As noted by Simon & Ashpis[2], bypass transition can occur near the critical Reynolds number of linear instability even though it proceeds so rapidly that the instability modes associated with linear instability are not seen.

Bypass Transition

To separate the description of transition on the low-pressure turbine blade from "natural" transition, the term "bypass" has been coined. Though the instability processes that lead to natural transition may be present to some degree in the low-pressure turbine flow, the higher disturbance levels characteristic of bypass transition disturb the boundary layer with a higher strength and excite 3D instabilities that apparently grow much more rapidly than the instabilities associated with natural transition, leading to transition early in the growth of the boundary layer. Hence the name "bypass transition" is used to imply that this rapid growth leads to a bypassing of the "natural" instability modes. Bypass transition is the topic of this paper; natural transition would not be expected in turbine flows.

Why transition is important to low-pressure turbine design:

Design analyses of turbine blade losses and heat transfer often treat the flow through the turbine as a steady, attached, turbulent flow. This may be justified in the high-pressure turbine in lieu of (1) the high turbulence of the flow entering the turbine stage from the upstream components, and (2) the disturbances inside the turbine passage, such as film cooling, endwall cooling, and surface steps and roughness. However, in the low-pressure turbine, the pressures are considerably lower while the temperatures remain rather high, making the Reynolds numbers relatively low by virtue of property effects. One can show that for an attached flow over an airfoil of a particular shape, as the Reynolds number decreases, the pressure gradient effects increase. With low Reynolds numbers, the suction surface flow is stabilized tending to hold the boundary layer in the laminar state. If transition were to begin in this zone, it would be extended by this stabilizing acceleration. It is not uncommon for an airfoil to have laminar or early transitional flow for the first one-third of the suction surface under low-Re conditions.

Similarly, the deceleration portion of the airfoil flow is destabilized, leading either to quicker transition or to separation. If transition occurs first, the enhanced mixing of the turbulent flow may prevent or delay separation. It is well known that heat transfer rates, skin friction and aerodynamic losses are strongly affected by separation and transition so it is important to the designer that transition and separation be predicted well. For instance, it is known that the performance of the low-pressure turbine at altitude is several points below its performance at take-off due to separation on the suction surface. This is a result of delayed transition on the upstream portion of the suction surface and separation on the downstream portion. Proper design to reduce such losses will require better prediction of transition to turbulence. Turbine designers must account more carefully for viscous effects, including rapid boundary layer growth, laminar-to-turbulent transition and boundary layer separation. If transition effects were properly incorporated, efficiencies could be improved or stages with fewer, more highly loaded airfoils may be possible[3].

In recent years, a number of trends have led to lower Re designs. Combined cycle gas turbine plants have become cost competitive in the 0.5–170 MW range and development of microturbines in the 25–100 kW range is proceeding well, now at $300–500 per kW[4]. The military is looking into small, gas-turbine-powered UAV aircraft. Newer noise regulations are leading to engines with higher bypass ratios and reduced core flows. Also, small aircraft are being considered by Internet companies, where stratospheric flying aircraft are proposed to fly slowly and continuously over major metropolitan areas to provide relay services for increased intracity Internet links[5]. Market forces are leading to small fan jet engines in commuter aircraft[6].

Furthermore, in recent years there has been a strong tendency to reduce the number of blades and stages within turbomachinery, which leads to stronger pressure gradients due to increased blade loading. Howell *et al.*[7] showed how a better understanding of transition, and particularly the effects of wakes, can be exploited to reduce the number of blades in the LP turbine. They did their study with rod-generated wakes to augment transition in the highly loaded airfoil stage.

The modern low pressure turbine sees a lower turbulence of the approach flow than experienced by upstream turbine stages. Transitional flows in lower freestream turbulence environments are more sensitive to pressure gradients and changes in freestream turbulence. Such enhanced sensitivity, in turn, leads to strong unsteadiness, resulting from a greater influence of wakes convected from upstream airfoils. This unsteadiness is seen as strong temporal and spatial variations in transition length and separation bubble strength, both known to be important to aerodynamic and thermodynamic performance of the engine.

The combination of all these various effects has led to intense interest in the prediction of boundary layer transition and separation on the low-pressure turbine suction surface, particularly as influenced by pressure gradients and wakes from upstream airfoils.

Why is it so difficult?

The difficulty in predicting the state of the flow on the airfoil surface lies in the prediction of transition, whether it be of an attached boundary layer or within a separated flow zone. Transition is complex for it is a process whereby certain frequencies in the disturbance signal are selectively amplified by the unstable shear flow giving rise to very rapid changes in state, behaving differently from case to case according to the distributions of energy in the disturbance. Transition depends also on the stability of the flow, as influenced by streamwise pressure gradients, roughness and maturity, in ways that are not yet completely described. A further complication is that such instabilities are not uniform. Thus, transition proceeds in spots (transitional spots similar to those described by Emmons[8]) which grow and merge as they are convected downstream. Also, the scales represented in the flow ran from very long wavelengths of the first instability waves to the fine dissipation scales of the fully mature turbulent flow. Modeling of transition must capture this wide range of scales. Steady progress is being made with a concerted effort between analysis, experimentation and computation.

This paper will provide a review of recent results, as applied to the low-pressure turbine, and will discuss some recent contributions by the authors.

CLASSICAL TRANSITION RESEARCH AND THE LOW-PRESSURE TURBINE:

Laminar-to-turbulent transition has been a well-studied topic. However, most of the literature has focused on "natural transition." Transition in the engine environment, with elevated disturbance levels and periodic unsteadiness, has been less well documented until recently. High levels of freestream turbulence cause earlier transition than with lower turbulence levels and such transition can prevent separation in the adverse pressure gradient region on the trailing portion of the suction surface of a turbine airfoil. Thus, blades in such an environment can be designed for higher loading if the effects of bypass transition are properly included in the design. This concept is just beginning to be exploited. In a seminal paper on this topic, Mayle[9] observed in 1991 that the majority of the experimental work focused upon laminar-to-turbulent transition under lower turbulence and steady flow conditions. However, the actual flow present in turbine engines has turbulence levels of 2–10%, with significant unsteadiness due to wakes. Since that time, considerable progress on the topic has ensued, as will be discussed. Mayle further suggested that investigations should be conducted to document the effects of wakes on transition over turbine airfoil surfaces. This also has been intensely addressed in recent years.

Spalart & Strelets[10] used DNS to study the mechanisms responsible for separation bubble transition. They noted a large negative surge in skin friction and a strong violation of Reynolds analogy at the onset of transition. The mechanism involves the wavering of the shear layer, then Kelvin-Helmholtz vortices, which become three-dimensional.

Hughes & Walker[11] looked for evidence of instability waves in the Tollmien-Schlichting frequency range in a compressor airfoil boundary layer. They found extensive regions of amplifying instability waves in nearly all the cases studied.

Andersson et al.[12] used a steady boundary layer approximation to calculate the upstream disturbances that experience maximum spatial energy growth. The optimal disturbances were streamwise vortices developing streamwise streaks. This led to a transition model.

Dietz[13] determined the receptivity of a Blasius boundary layer to harmonic disturbances created by a vibrating ribbon. Hot wire anemometry was used to confirm the generation of Tollmien-Schlichting waves. An agreement between experiments and theory was obtained over a range of frequencies and Reynolds numbers.

Davies et al.[14] compared computational results from an Renormalization Group (RNG) model of turbulence with experiments, discussing the results in terms of the entropy generation rate. Details of the boundary layer measurements were given in O'Donnell & Davies[15].

Modeling

Bypass transition modeling on a fundamental level has seen numerous contributions in the last few years. Contributions are from Steelant & Dick[16–18], Goldstein & Wundrow[19], Volino[20], Beck et al.[21], Hatman & Wang[22], Mayle[23], Mayle & Schulz[24], Hu & Fransson[25], Johnson[26], Dietz[13], Leib et al.[27,28], Johnson & Dris[29], Johnson[30], and Walker et al.[31], Solomon et al.[32]. Modeling of the effects of wakes convected from upstream was presented by Schulte & Hodson[3] and Wu et al.[33].

Dorney et al.[34] applied an unsteady Navier-Stokes solver, a two-layer algebraic turbulence model and transition models and compared the results to data for transition in turbines. They were most successful with the Abu-Ghannam & Shaw[35] transition model.

Roach & Brierley[36] developed a bypass transition model which accounts for turbulence level and scale. They noted that the model shows a sensitivity to turbulence at the leading edge of the test surface not to the turbulence locally outside of the boundary layer.

Experiments

Steady Flow Experiments

Much of the early work concerning transition in turbomachinery flows focused primarily upon compressor blade passages. Evans[37] is representative of this early work, presenting hot-wire boundary layer measurements on a singular compressor stator blade. Similar investigations were conducted in compressor passages by Hodson[38] (hot-wire) and Deutsch & Zierke[39] (single-component LDV).

In order to take a fundamental look at turbomachinery transition, recent experiments have been conducted in somewhat simplified geometries which represent certain aspects of the low-pressure turbine flows. These include the works of Blair[40], Baughn et al.[41], Qiu & Simon[42], Simon et al.[43], Sohn et al.[44], Murawski et al.[45], Jonas[46], Funazaki & Kitazawa[47], Matsubara et al.[48], Boyle et al.[49], Volino & Simon[50–52], Boyle & Simon[53], Volino & Hultgren[54], Hatman & Wang[55,56], Chakka & Schobeiri[57], Funazaki et al.[58,59], and Schobeiri et al.[60]. The Volino & Simon study was on a curved surface with various pressure gradients imposed. The Boyle & Simon study argued that compressibility was unimportant to the turbulent spot formation rate below Mach 2, after which the estimate of this rate ought to be reduced. Blair showed that the response of heat transfer to transition was slower than that of momentum transfer. Recently, Alfredsson & Matsubara[61] used simultaneous PIV and visualization to show the penetration of elevated free stream turbulence in the transitioning boundary layer on a flat plate. Bons et al.[62] documented the effects vortex generators on the stability of transitional boundary layers on a turbine surface. They noted a significant reduction in losses and were able to reduce mass flow by pulsing the jets at a particular frequency. Lake[63] and Lake et al.[64] looked at the effects of adding dimples, V-grooves and trips to the surface. The dimples are successful in reducing losses over the Reynolds number range investigated, including cases in which the flow on the smooth surface was not separated.

Unsteadiness due to passing wakes

Like the steady-state investigations, much of the early work involving the effects of unsteadiness in turbomachinery flows started with compressor studies. In one such study, Dong & Compsty[65,66], hot-wire data from a compressor cascade allowed comparing the incidence of wakes to increased levels of turbulence, both resulting in early transition to turbulence. IT was suggested that the incorporation of wakes be a special case to existing bypass transition models.

It is important to note that in unsteady turbine passage flows, there are several wake phenomena: (1) oscillating freestream velocity component due to wake, (2) oscillating angle of attack and (3) wake turbulence. All of these have separate effects upon transition and separation in the turbine blade boundary layer, as discussed by Mayle & Schulz[24] and Lou & Hourmouziadis[67].

In the last five years, detailed measurements have included wakes. Some used wake simulators and others were done within rotating rigs. The works of Halstead et al.[68–71], Funazaki & Kitazawa[47], Kim & Crawford[72], Dorney et al.[73,74], Wu et al.[33], Chakka & Schobeiri[57,75], Funazaki et al.[58,59], Schobeiri et al.[60], Walker et al.[31] and Chernobrovikin & Lakshminarayana[76] include wake effects. Of these, the works of Halstead et al.[68–71], Funazaki & Kitazawa[47], Chakka & Schobeiri[57,75], and Dorney et al.[74] show detailed inflow data, with wakes. The experiments of Kittichaikarn et al.[77] showed with liquid crystals that "puffs" could be observed as precursors to wake-induced transition. Experiments in actual rotating turbomachinery include those by Halstead et al.[68–71], Tiedemann & Kost[78], and Kost et al.[79].

Brunner et al.[80] noted a loss reduction of 34% and 28% in two airfoil profiles due to the unsteady inlet flow associated with passing wakes. Stadtmüller et al.[81] showed that the wake effect on the reduction of losses depends on the frequency and strength of the wake. Figure 1. shows a comparison between their experimental and numerical results. The setting is that of bar-generated wakes passing over a LP turbine airfoil suction surface. The colors indicate the power supplied to small heated patches on the surface (called quasi shear stress, by analogy) versus time and meridianal position ($M/M_{ges} \sim x/L_{ax}$). The effects of passing bars moving at 20 m/s are shown.

In results similar to those presented by Dong & Compsty[66], Teusch et al.[82] documented a 20% reduction in losses under low Reynolds number conditions due to wakes, but a 20% rise in losses due to wakes at higher Reynolds numbers where the separation bubble is small.

Solomon[83] looked at the effects of wake passing and elevated freestream turbulence for two blade spacings in a low-speed rotating rig. He noted that the effects of turbulence on the transition onset location were smaller when the solidity was reduced. He also noted an optimum clocking of the rotor, or most effective wake passage timing, for effecting transition before separation.

Computation of wake passing data with a more accurate blade approximation was shown by Dorney et al.[73] to reduce the computational unsteadiness.

Lou & Hourmouziadis[67] documented separation bubble transition, contouring the opposite wall to create the appropriate pressure gradient on the test wall. By using a downstream rotating valve, they were able to separate the effects of the wake velocity change from the effects of boundary layer turbulence in the wake and identify characteristic instability frequencies in the shear layer over the bubble.

Funazaki & Aoyama[84] showed the effects of wake-generated turbulence on the separation bubble by using a stationary wake generator. They were able to identify a critical location of the bar for minimal loss and discussed the results in terms of the optimum clocking of the stage.

Schreiber et al.[85] showed the effects of Reynolds number and free-stream turbulence on transition for a controlled-diffusion compressor airfoil. They noted that for increased turbulence, the transition location moved upstream and the thin pre-transitional boundary layer flow became more sensitive to surface roughness.

Kim & Crawford[72] computed heat transfer rates with wake-induced transition using an unsteady Navier-Stokes solver and an unsteady transition model. They noted that the prediction of time-averaged heat transfer is improved when unsteady modeling is used.

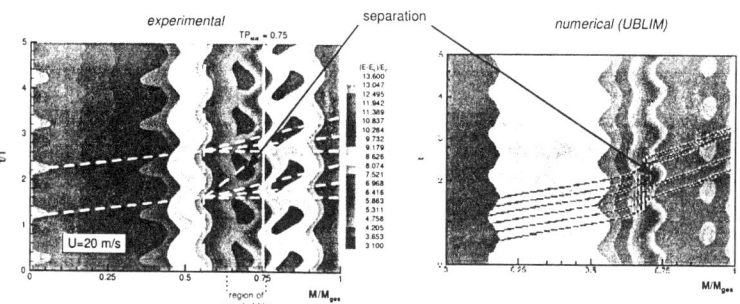

Figure 1: Comparison of quasi wall shear stress from experimental and numerical results from Stadtmüller et al.[81]. Figure indicates that rod velocity is 20 m/s

Analysis and Computation

Numerous contributions to transition modeling by way of CFD analysis have been presented in recent years. Some solve the Reynolds-Averaged Navier-Stokes (RANS) equations, others employed Large Eddy Simulation (LES) and others Direct Numerical Simulation (DNS). Those specific to bypass transition have come from Chernobrovikin & Lakshminarayana[76,86], Suzen & Huang[87], Dorney et al.[73,74], Kang & Lakshminarayana[88], Chakka & Schobeiri[75] and Kim & Crawford[72]. Large eddy and direct numerical simulations of transitional flows have been made by Madavan & Rai[89], Rai & Moin[90], Berlin & Henningson[91], Wu et al.[33], and Alam & Sandham[92]. The Wu et al. study includes DNS simulation of the effects of passing wakes. Rai & Moin initiated their computations with random inlet values and allowed sufficient streamwise distance to let them develop into proper turbulence. This was shown to require considerable computation time and some compromise in grid resolution.

Figure 2 shows slices through the flow results of a DNS simulation by Jacobs[93]. The three planes are (1) free-stream, (2) top of the boundary layer, and (3) inside the boundary-layer. Frame (1) shows decaying grid turbulence and in frame (3) streaks are observed in the upstream boundary layer, similar to those shown in Emmons[8].

In McDaniel & Hassan[94], transition was treated with turbulence modeling. Correlations taken from the literature were used for the transition length and the intermittency distribution through transition. The turbulent diffusivity was determined by the turbulence model.

Alam & Sandham[92] showed by DNS the development of a separating flow where transition is via oblique modes and Λ-vortex-induced breakdown.

Müller et al.[95] investigated the modeling of separated flow transition to find that a combined onset model of Mayle[23] and production rate model of Walker et al.[96] was best. It performed well on the pressure side but was calculated to be far too downstream on the suction surface.

Enomoto et al.[97] showed by computation and experimentation that the process of laminar flow separation, reattachment and subsequent flow transition at low Reynolds numbers

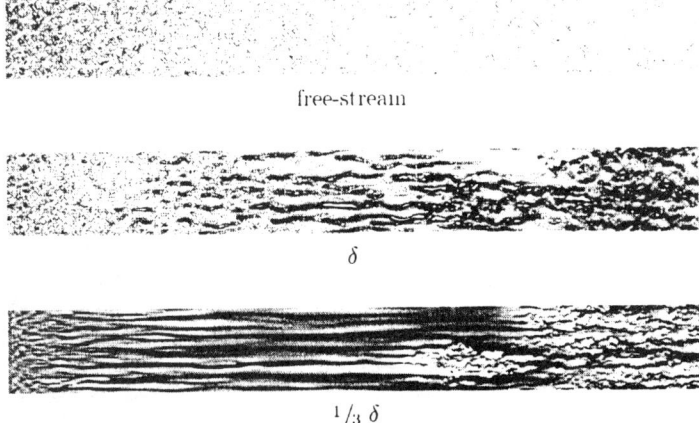

Figure 2: Planes through a simulated transition provoked by grid turbulence. From Jacobs[93]

is dominated by relatively large eddies near the wall which can be simulated with unsteady numerical codes. Hobson & Weber[98] described the application of Navier-Stokes solvers to the computation of controlled-diffusion compressor blades.

Calculations were made for the PAK-B turbine blade to show the effects of turbulence and boundary layer trips[74]. The Baldwin-Lomax turbulence model, with a transition model was shown to produce satisfactory results for low Reynolds number, as compared with data. An intermittency transport equation formulated by Suzen & Huang[87,99] from the works of Steelant & Dick[100] and Cho & Chung[101] was shown to be successful against the transition data assembled by Savill[102,103].

RECENT WORK AT THE UNIVERSITY OF MINNESOTA

The University of Minnesota research addressed the need for detailed experimental data which document transition in boundary layers and separated flows over highly-loaded airfoils, including the effects of passing wakes. The program objectives are accomplished with the following steps:

1. The effects of freestream turbulence and varying Reynolds numbers were documented in a facility which simulates the flow through a modern, highly-loaded, low-pressure turbine, but without wakes.
2. The same flow was documented, but with the influence of simple, rod-generated wakes.

   By comparing these data with the steady-state data, one can identify the effects of these simple-geometry wakes. These data will be unsteady as a result of the passing wakes so they must be correlated with position within the wake-passing period. These simple-geometry wakes are amenable to computer simulation.

Part 1 is documented by Qiu & Simon[42] and Simon et al.[43]. The results show cases with strong separation at low Reynolds numbers and low turbulence levels and cases with much smaller separation bubbles as the Reynolds number or freestream turbulence is raised. They show also that an algebraic correlation in the literature for the streamwise distance from separation to the start of transition by Davis et al.[104] is quite accurate. This correlation is based upon the effective turbulence intensity at the point of separation. The experimental results show also that a model for the intermittency path by Dhawan & Narasimha[105] is remarkably good, in spite of its derivation from attached boundary layer flow transition data. A need for better prediction of the transition length is indicated, however.

Part 2 is in progress. The facility (Fig. 3) is built and qualified, wakes are generated by sliding a rack of rods through the approach flow. Rods can be used to simulate stator wakes since they produce wakes similar to airfoils[106]. The wakes are convected the low-pressure turbine passage simulator used in part 1. The airfoil geometry is the PAK B shape offered for research by Pratt and Whitney. On the slider is a photogate which records the rods' positions and allows ensemble averaging on location within the wake passing period. Ensemble averages from 150 traverses of the wake generator (900 rod passings) are made. Mean and rms values of the approach flow velocity are presented in Figs. 4. Examining, we see that the minimum velocity of the wake is approximately 87.5% of the average value, which matches the work of Halstead et al.[68] in which a rotating airfoil stage (simulating a rotating turbine stage) was used to create wake profiles. Examining the turbulence intensity, however, we see that it peaks at 17.5%, more than twice that reported by Halstead[107]. This may be consistent with Halstead's assertion that rods seem to produce more turbulence than airfoils of the same loss coefficient. It should be noted, however, that the flows over the

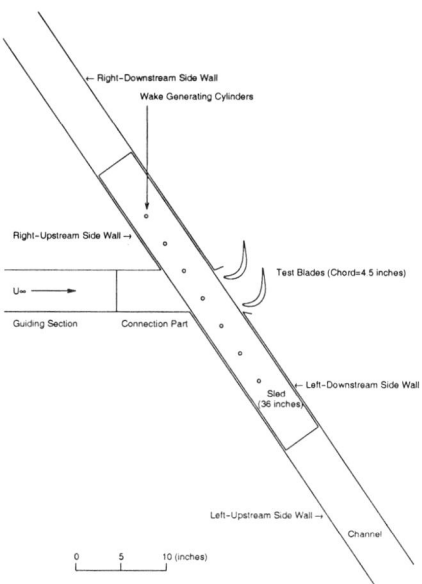

Figure 3: Schematic of the wake generator with rods inserted as generation devices.

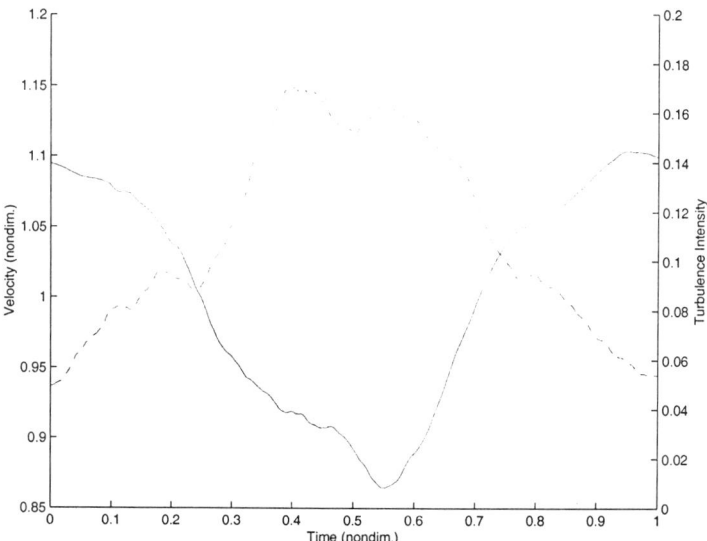

Figure 4: Ensemble-average velocity and TI of 900 wakes, taken 16 rod diameters downstream of the rods' centerlines.

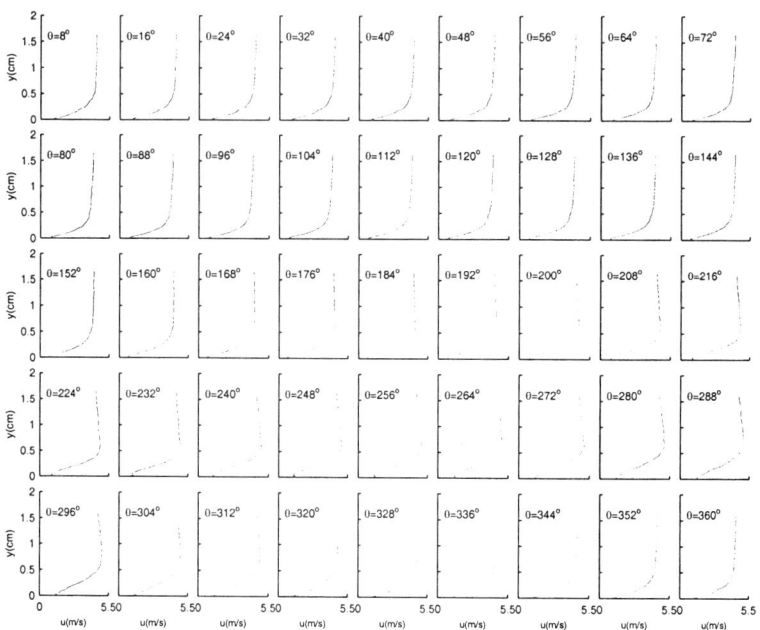

Figure 5: Phase-Averaged Velocity $\bar{u}(y,\theta)$ at 68% of the suction surface length with 2.5% TI at various θ values

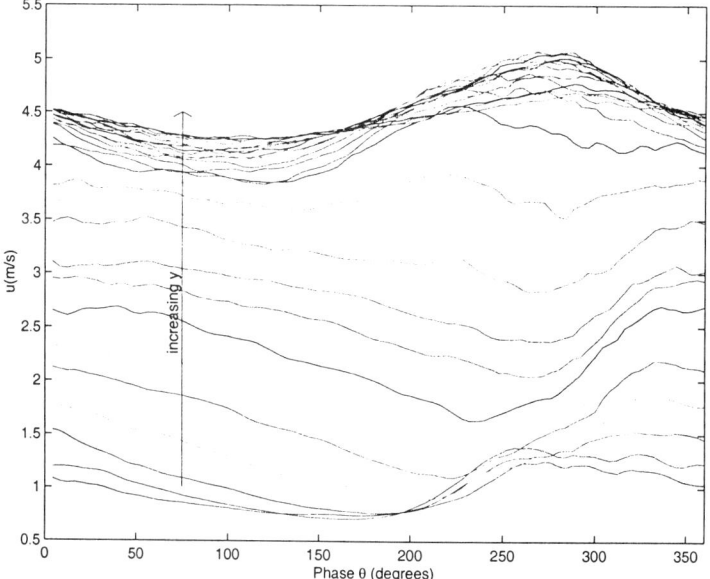

Figure 6: Phase-Averaged Velocity $\bar{u}(y,t)$ at 68% of the suction surface length with 2.5% TI at various y values

airfoils of the Halstead study were not strongly separated and a highly loaded airfoil, such as that of the Minnesota study, would be inclined to separate more strongly at the lower Reynolds numbers and, thus, the wakes would contain higher levels and larger scales of turbulence.

The unsteady boundary layer measurements include the ensemble-averaged, period-resolved profiles of velocity, rms velocity fluctuation and intermittency over the surface. The velocities are measured by a single-sensor, hot-wire anemometer. Sample data at 68% of the suction length from this study are shown in Fig. 5, in which the ensemble average velocity $\tilde{u}(y,\theta)$ is shown for a complete wake period from $\theta = 0$ to $\theta = 360°$. Examining the figure, we see normal turbulent boundary layer profiles, initially ($0 < \theta < 100°$). As the wake travels down the passage, deceleration destabilizes the flow to the point that the profile becomes inflectional and the flow separates from the wall (starting at $\theta = 168°$). The wake passes this station and the flow begins to accelerate and restabilize until it reattaches (at $\theta = 312°$). These observations are consistent with Lou & Hourmouziadis[67], who noted similar oscillations in the location and length of separation as a function of phase angle.

Figure 6 shows the same data presented differently, with $\tilde{u}(y,\theta)$ shown as a series of complete wake periods at different y values. From these data, we can see that the location of the wake center differs from one y-value to another. We can conclude that the wake has different propagation speeds at different elevations in the boundary layer. This is also consistent with Lou & Hourmouziadis[67], who observed that the temporal location of their wakes were phase-shifted with respect to the free-stream flow.

Additional measurements are planned in this flow. A rapid-response flow direction sensor is being developed to determine location of separation and re-attachment points.

PROJECTION TO THE FUTURE

The importance of raising the low pressure turbine stage loading, removing blades and vanes, while increasing stage efficiency is clear. These incentives will lead to a continued emphasis on this topic of bypass transition in unsteady flows. Now that experience has been gained in the fundamental experiments, more complex experiments which will allow detailed flow measurements in rotating rigs, will be brought on line. The industry will develop a better feel for the limits of their design methodology which uses algebraic correlations and simple models for implementation into RANS codes and they will be used with more confidence in design. It is clear that LES and DNS will take a more important role in the prediction of transition behavior in design models. Thus, such modeling will become more commonplace in the design methodology. Also, as computer speed grows, DNS will take on increasing importance as a means of generating "data" for further development of simpler transition models.

REFERENCES

1. Schlichting, H., 1979. Boundary Layer Theory. McGraw-Hill, New York, 7th edn.
2. Simon, F. F. & Ashpis, D. E., 1996. Progress in modeling of laminar to turbulent transition on turbine vanes and blades. NASA TM-107180.
3. Schulte, V. & Hodson, H. P., 1996. Unsteady wake-induced boundary layer transition in high lift low pressure turbines. In ASME J. of Turbomachinery **120**: 28–35.
4. Langston, L. S., 2000. Gas turbine industry overview. In Global Gas Turbine News **40**(1).
5. Ianotta, B., 2000. Stratospheric hopes for wireless internet link. In Aerospace America, April 2000.
6. New York Times, 2000. Twilight of the turboprop? Passengers go out of their way to catch jets. In The New York Times Business Page, Feb. 18.

7. Howell, R. J., Ramesh, O. N., Hodson, H. P. et al., 2000. High lift and aft loaded profiles for low pressure turbines. ASME 2000-GT-261.
8. Emmons, H. W., 1951. The laminar-turbulent transition in a boundary later—Part I. In J. Aero. Science 490–498.
9. Mayle, R. E., 1991. The role of laminar-turbulent transition in gas turbine engines. In ASME J. of Turbomachinery **113**: 509–537. ASME 91-GT-261.
10. Spalart, P. R. & Strelets, M. K., 2000. Mechanisms of transition and heat transfer in a separation bubble. In J. of Fluid Mechanics **403**: 329–349.
11. Hughes, J. D. & Walker, G. J., 2000. Natural transition phenomena on an axial compressor blade. ASME 2000-GT-264.
12. Andersson, P., Berggren, M. & Henningson, D. S., 1999. Optimal disturbances and bypass transition in boundary layers. In Physics of Fluids **11**: 134–150.
13. Dietz, A. J., 1999. Local boundary-layer receptivity to a convected free-stream disturbance. In J. of Fluid Mechanics **378**: 291–317.
14. Davies, M. R. D., O'Donnell, F. K. & Niven, A. J., 2000. Turbine blade entropy generation rate, part 1: The boundary layer defined. ASME 2000-GT-265.
15. O'Donnell, F. K. & Davies, M. R. D., 2000. Turbine blade entropy generation rate, part 2: The measured loss. ASME 2000-GT-266.
16. Steelant, J. E. & Dick, E., 1996. Calculation of transition in adverse pressure gradient flow by conditioned equations. ASME 96-GT-160.
17. Steelant, J. E. & Dick, E., 1999. Prediction of by-pass transition by means of a turbulence weighting factor, part ii. ASME 99-GT-30.
18. Steelant, J. E. & Dick, E., 1999. Prediction of by-pass transition by means of a turbulence weighting factor, part i. ASME 99-GT-29.
19. Goldstein, M. E. & Wundrow, D. W., 1998. On the environmental realizability of algebraicly growing disturbances and their relation to klebanoff modes. In Theoretical Computational Fluid Dynamics **10**: 171–186.
20. Volino, R. J., 1998. Wavelet analysis of transitional flow data under high free-stream turbulence conditions. ASME 98-GT-194.
21. Beck, K. H., Henningson, D. S. & Henkes, R. A. W. M., 1998. Linear and nonlinear development of localized disturbances in zero and adverse pressure gradient boundary layers. In Physics of Fluids **10**(6): 1405–1418.
22. Hatman, A. & Wang, T., 1998. A prediction model for separated flow transition. ASME 98-GT-237.
23. Mayle, R. E., 1998. A theory for predicting the turbulent-spot production rate. ASME 98-GT-256.
24. Mayle, R. E. & Schulz, A., 1997. The path to predicting bypass transition. In ASME J. of Turbomachinery **119**: 405–411.
25. Hu, J. & Fransson, T. H., 1998. On the application of transition correlations in turbomachinery flow calculation. ASME 98-GT-460.
26. Johnson, M. W., 1998. Turbulent spot characteristics in boundary layers subjected to streamwise pressure gradient. ASME 98-GT-124.
27. Leib, S. J., Wundrow, D. W. & Goldstein, M. E., 1999. Effect of free-stream turbulence and other vortical disturbances on a laminar boundary layer. In J. of Fluid Mechanics **380**. 169–203.
28. Leib, S. J., Wundrow, D. W. & Goldstein, M. E., 1999. Generation and growth of boundary layer disturbances due to free-stream turbulence. AIAA 99-0408.
29. Johnson, M. W. & Dris, A., 1999. The origin of turbulent sports. ASME 99-GT-32.
30. Johnson, M. W., 1999. Prediction of turbulent spot growth rates. ASME 99-GT-31.
31. Walker, G. J., Hughes, J. D. & Solomon, W. J., 1998. Periodic transition on an axial compressor stator—incidence and clocking effects, part i. ASME 98-GT-363.
32. Solomon, W. J., Walker, G. J. & Hughes, J. D., 1998. Periodic transition on an axial compressor stator—incidence and clocking effects, part ii. ASME 98-GT-364.
33. Wu, X. H., Jacobs, R. G., Hunt, J. C. R. et al., 1999. Simulation of boundary layer transition induced by periodically passing wakes. In J. of Fluid Mechanics **398**: 109–153.
34. Dorney, D. J., Ashpis, D. E., Halstead, D. E. et al., 1999. Study of boundary layer development in two-stage low-pressure turbine. AIAA 99-0742, also NASA TM-1999-208913.

35. Abu-Ghannam, B. J. & Shaw, R., 1980. Natural transition of boundary layers—the effects of pressure gradient and flow history. *In* J. of Eng. Science **22**(5): 213–228.
36. Roach, P. E. & Brierley, D. H., 2000. Bypass transition modeling: A new method with accounts for free-stream turbulence intensity and length scale. ASME 2000-GT-278.
37. Evans, R. L., 1978. Boundary-layer development on an axial-flow compressor stator blade. *In* Journal of Engineering for Power **100**: 287–293.
38. Hodson, H. P., 1984. Boundary layer and loss measurements on the rotor of an axial flow turbine. *In* ASME Journal of Engineering for Gas Turbines and Power **106**.
39. Deutsch, S. & Zierke, W. C., 1987. The measurement of boundary layers on a compressor blade in cascade: Part 2: Suction surface boundary layers. ASME 87-GT-249.
40. Blair, M. F., 1982. Influence of free-stream turbulence on boundary layer transition in favorable pressure gradients. *In* ASME J. of Heat Transfer **105**: 33–40.
41. Baughn, J. W., Butler, R. J., Byerley, A. R. *et al.*, 1995. An experimental investigation of heat transfer, transition, and separation on turbine blades at low reynolds number and high turbulence intensity. ASME 95-WA/HT-25.
42. Qiu, S. & Simon, T. W., 1997. An experimental investigation of transition as applied to low pressure turbine suction surface flows. ASME 97-GT-455.
43. Simon, T. W., Qiu, S. & Yuan, K., 2000. Measurements in a transitional boundary layer under low-pressure turbine airfoil conditions. NASA Contractor Report NASA/CR-2000-209957.
44. Sohn, K. H., Shyne, R. J. & Dewitt, K. J., 1998. Experimental investigation of boundary layer behavior in a simulated low pressure turbine. ASME 98-GT-34, also NASA/TM-1999-207921.
45. Murawski, C. G., Vafai, K., Simon, T. W. *et al.*, 1997. Experimental study of the unsteady aerodynamics in a linear cascade with low reynolds number low pressure turbine blades. ASME 97-GT-95.
46. Jonas, P., 1997. Experimental investigation of by-pass transition in the institute of thermomechanics in prague. Minnowbrook Workshop Proceedings.
47. Funazaki, K. & Kitazawa, T., 1997. Boundary layer transition induced by periodic wake passage: Measurements of the boundary layer by hot wire anemometry. Bulletin of the GTSJ.
48. Matsubara, M., Alfredsson, P. H. & Westin, K. J. A., 1998. Boundary layer transition at high levels of free stream turbulence. ASME 98-GT-248.
49. Boyle, R. J., Lucci, B. L., Verhoff, V. G. *et al.*, 1998. Aerodynamics of a transitioning turbine stator over a range of reynolds numbers. ASME 98-GT-285, also NASA/TM-1998-208408.
50. Volino, R. J. & Simon, T. W., 1997. Boundary layer transition under high free-stream turbulence and strong acceleration conditions: Part 1, mean flow results. *In* ASME J. of Heat Transfer **119**(3): 420–426.
51. Volino, R. J. & Simon, T. W., 1997. Boundary layer transition under high free-stream turbulence and strong acceleration conditions: Part 2, turbulent transport results. *In* ASME J. of Heat Transfer **119**(3): 427–432.
52. Volino, R. J. & Simon, T. W., 1997. Measurements in a transitional boundary layer with Görtler vortices. *In* J. of Fluids Eng. **119**: 562–568.
53. Boyle, R. J. & Simon, F. F., 1998. Mach number effects on turbine blade transition length prediction. ASME 98-GT-367, also NASA/TM-1998-208404.
54. Volino, R. J. & Hultgren, L. S., 2000. Measurements in separated and transitional boundary layers under low-pressure turbine airfoil conditions. ASME 2000-GT-0260.
55. Hatman, A. & Wang, T., 1998. Separated flow transition, Part I. ASME 98-GT-461.
56. Hatman, A. & Wang, T., 1998. Separated flow transition, Part II. ASME 98-GT-462.
57. Chakka, P. & Schobeiri, M. T., 1999. Scales of turbulence during boundary layer transition under steady and unsteady flow conditions. ASME 99-GT-221.
58. Funazaki, K., Tetsuka, N. & Tanuma, T., 1999. Effects of periodic wake passing upon aerodynamic loss of a turbine cascade, Part I. ASME 99-GT-93.
59. Funazaki, K., Tetsuka, N. & Tanuma, T., 1999. Effects of periodic wake passing upon aerodynamic loss of a turbine cascade, Part II. ASME 99-GT-94.
60. Schobeiri, M. T., Chakka, P. & Pappu, K., 1998. Unsteady wake effects on boundary layer transition and heat transfer characteristics of a turbine blade. ASME 98-GT-291.
61. Alfredsson, R. H. & Matsubara, M., 2000. Free-stream turbulence, streaky structures and transition in boundary layer flows. AIAA 2000-2534.
62. Bons, J. P., Sondergaard, R. & Rivir, R. B., 2000. Turbine separation control using pulse

vortex generator jets. ASME 2000-GT-262.
63. Lake, J. P., 1999. Flow Separation Prevention on a Turbine Blade in Cascade at Low Reynolds Number. Ph.D. thesis, Air Force Institute of Technology.
64. Lake, J. P., King, P. I. & Rivir, R. B., 2000. Low reynolds number loss reduction on turbine blades with dimples and v-grooves. AIAA 2000-0738.
65. Dong, Y. & Compsty, N. A., 1990. Compressor blade boundary layers: Part 1—test facility and measurements with no incident wakes. In ASME J. of Turbomachinery **112**: 221–230.
66. Dong, Y. & Compsty, N. A., 1990. Compressor blade boundary layers: Part 2—measurements with incident wakes. In ASME J. of Turbomachinery **112**: 231–240.
67. Lou, W. & Hourmouziadis, J., 2000. Separation bubbled under steady and periodic-unsteady main flow conditions. ASME 2000-GT-270.
68. Halstead, D. E., Wisler, D. C., Okiishi, T. H. et al., 1995. Boundary layer development in axial compressors and turbines: Part 1 of 4: Composite picture. ASME 95-GT-461.
69. Halstead, D. E., Wisler, D. C., Okiishi, T. H. et al., 1995. Boundary layer development in axial compressors and turbines: Part 2 of 4: Compressors. ASME 95-GT-462.
70. Halstead, D. E., Wisler, D. C., Okiishi, T. H. et al., 1995. Boundary layer development in axial compressors and turbines: Part 3 of 4: Low pressure turbines. ASME 95-GT-463.
71. Halstead, D. E., Wisler, D. C., Okiishi, T. H. et al., 1995. Boundary layer development in axial compressors and turbines: Part 4 of 4: Computations and analyses. ASME 95-GT-464.
72. Kim, K. & Crawford, M. E., 2000. Prediction of transitional heat transfer characteristics of wake-affected boundary layers. In ASME J. of Turbomachinery **122**: 78–87.
73. Dorney, D. J., Flitan, H. C., Ashpis, D. E. et al., 2000. Effects of blade count on boundary layer development in a low-pressure turbine. AIAA 2000-0742.
74. Dorney, D. J., Lake, J. P., King, P. I. et al., 2000. Experimental and numerical investigation of losses in low-pressure turbine blade rows. AIAA 2000-0737.
75. Chakka, P. & Schobeiri, M. T., 1999. Modeling unsteady boundary layer transition on a curved plate under periodic unsteady flow conditions: Aerodynamic and heat transfer investigations. In ASME J. of Turbomachinery **121**: 88–97.
76. Chernobrovikin, A. & Lakshminarayana, B., 1999. Turbulence modeling and computation of viscous transitional flow for low pressure turbines. Proceedings of the 4th International Symposium on Engineering Turbulence Modeling and Measurements, Corsica, France.
77. Kittichaikarn, C., Ireland, P. T., Zhong, S. et al., 1999. An investigation of the onset of wake-induced transition and turbulent spot production using thermochromic liquid crystals. ASME 99-GT-126.
78. Tiedemann, M. & Kost, F., 1999. Unsteady boundary layer transition on a high pressure turbine rotor blade. ASME 99-GT-194.
79. Kost, F., Hummel, F. & Tiedemann, M., 2000. Investigation of the unsteady rotor flow field in a single HP turbine stage. ASME 2000-GT-432.
80. Brunner, S., Fottner, L. & Schiffer, H.-P., 2000. Comparison of two highly loaded low pressure turbine cascades under the influence of wake-induced transition. ASME 2000-GT-268.
81. Stadtmüller, P., Fottner, L. & Fiala, A., 2000. Experimental and numerical investigation of wake-induced transition on a highly loaded low pressure turbine at low reynolds numbers. ASME 2000-GT-269.
82. Teusch, R., Brunner, S., Fottner, L. et al., 2000. The influence of multimode transition initiated by periodic wakes on the profile loss of a linear compressor cascade. ASME 2000-GT-271.
83. Solomon, W. J., 2000. Effects of turbulence and solidity on the boundary layer development in a low pressure turbine. ASME 2000-GT-273.
84. Funazaki, K. & Aoyama, Y., 2000. Studies on turbulence structure of boundary layers disturbed by moving wakes. ASME 2000-GT-272.
85. Schreiber, H.-A., Steinert, W. & Küsters, B., 2000. Effects of reynolds number and free-stream turbulence on boundary layer transition in a compressor cascade. ASME 2000-GT-0263.
86. Chernobrovikin, A. & Lakshminarayana, B., 2000. Turbulence modeling and computation of viscous transitional flows for low pressure turbines. Presented at the 4th International Symposium on Engr. Turbulence Modeling and Measurements, Corsica, France.
87. Suzen, Y. B. & Huang, P. G., 2000. An intermittency transport equations for modeling flow transition. AIAA 2000-0287.
88. Kang, D. J. & Lakshminarayana, B., 1997. Numerical prediction of unsteady transitional

boundary layer flow due to rotor-stator interaction. AIAA 97-2752.
89. Madavan, N. K. & Rai, M. M., 1995. Direct numerical simulation of boundary layer transition on a heated flat plate with elevated freestream turbulence. AIAA 95-0771.
90. Rai, M. M. & Moin, P., 1993. Direct numerical simulation of transition and turbulence in a spatially evolving boundary layer. *In* Journal of Computational Physics **109**(2): 169–192.
91. Berlin, S. & Henningson, D. S., 1999. A nonlinear mechanism for receptivity of free-stream disturbances. *In* Physics of Fluids **11**(12): 3749–3760.
92. Alam, M. & Sandham, N. D., 2000. Direct numerical simulation of 'short' laminar separation bubbles with turbulent reattachment. *In* J. of Fluid Mechanics **403**: 223–250.
93. Jacobs, R. G., 1999. Transition Phenomena Studied by Numerical Simulation. Ph.D. thesis, Stanford University.
94. McDaniel, R. D. & Hassan, H. A., 2000. Study of bypass transition using the k-zeta framework. AIAA 2000-2310.
95. Müller, M., Gallus, H. E. & Niehuis, R., 2000. A study on models to simulate boundary layer transition in turbomachinery flows. ASME 2000-GT-274.
96. Walker, G. J., Subroto, P. H. & Platzer, M. F., 1988. Transition modeling effects on viscous/inviscid interaction analysis of low reynolds number airfoil flows involving laminar separation bubbles. ASME 88-GT-32.
97. Enomoto, S., Hah, C. & Hobson, G. V., 2000. Numerical and experimental investigation of low reynolds number effects on laminar flow separation and transition in a cascade of compressor blades. ASME 2000-GT-276.
98. Hobson, G. V. & Weber, S., 2000. Prediction of a laminar separation bubble over a controlled-diffusion compressor blade. ASME 2000-GT-277.
99. Suzen, Y. B. & Huang, P. G., 2000. Modeling of flow transition using an intermittency transport equaton. *In* J. of Fluids Eng. **123**(2).
100. Steelant, J. E. & Dick, E., 1996. Modeling of bypass transition with conditioned navier-stokes equations coupled to an intermittency transport equation. *In* Intl. Journal for Numerical Methods in Fluids **26**: 193–220.
101. Cho, J. R. & Chung, M. K., 1992. A k-epsilon-gamma equation turbulence model. *In* J. of Fluid Mechanics **237**: 301–322.
102. Savill, A. M., 1993. Some recent progress in the turbulence modeling of by-pass transition. In R. M. C. So, C. G. Speziale & B. E. Launder, eds., Near-Wall Turbulent Flows, 829–848. Elsevier Science Pub.
103. Savill, A. M., 1993. Further progress in the turbulence modeling of by-pass transition. In R. M. C. So, C. G. Speziale & B. E. Launder, eds., Near-Wall Turbulent Flows, 583–592. Elsevier Science Pub.
104. Davis, R. L., Carter, J. E. & Reshotko, E., 1985. Analysis of transitional separation bubbles on infinite swept wings. AIAA 85-1685.
105. Dhawan, S. & Narasimha, R., 1958. Some properties of boundary layer flow during the transition from laminar to turbulent motion. *In* J. of Fluid Mechanics **3**: 418–436.
106. Pfeil, H. & Eifler, J., 1979. Turbulencezeverhältnisse hinterroteirenden zylindergittern. *In* Forschung im Ingenieurwesen **42**: 27–32.
107. Halstead, D. E., 1996. Boundary Layer Development in Multi-Stage Low Pressure Turbines. Ph.D. thesis, Iowa State University, Ames, IA.

Turbulence Modeling in Simulation of Gas-Turbine Flow and Heat Transfer

G. BRERETON and T. I-P. SHIH[*]

*Department of Mechanical Engineering, Michigan State University
East Lansing, MI 48824-1226, U.S.A.*

ABSTRACT: The popular k-ε type two-equation turbulence models, which are calibrated by experimental data from simple shear flows, are analyzed for their ability to predict flows involving shear and an extra strain – flow with shear and rotation and flow with shear and streamline curvature. The analysis is based on comparisons between model predictions and those from measurements and large-eddy simulations of homogenous flows involving shear and an extra strain, either from rotation or from streamline curvature. Parameters are identified, which show the conditions under which performance of k-ε type models can be expected to be poor.

NOMENCLATURE

F	curvature factor defined by Eq. (9)
k	turbulent kinetic energy
u', p'	velocity and pressure fluctuation about the mean
S, S_{ij}, S′	mean rate of shear (S′ is defined by Eq. (8))
t, t^*	time coordinate ($t^* = t$ S)
$-\overline{u'_i u'_j}$	Reynolds stress divided by density
U, U_i	mean velocity component
x_i	Cartesian coordinate (x, y, or z), inertial or rotating at constant speed
ε	dissipation rate of k
ν_τ	turbulent kinematic viscosity
Ω, Ω_{ij}	rate of rotation

---

[*] Corresponding author.

INTRODUCTION

Considerable progress has been made in the application of computational fluid dynamics (CFD) to predict the flow and heat transfer in gas turbines.[1,2] However, our ability to make accurate and reliable predictions of these extremely complicated problems is constrained by the number of grid points or cells that can be afforded in the calculations and by the trustworthiness of the mathematical models used to describe the turbulence.

With today's computing capabilities, turbulent flows in gas turbines are typically modeled by the ensemble- or Reynolds-averaged Navier-Stokes (RANS) equations in which the Reynolds stresses are closed by a two-equation eddy-diffusivity model such as k-ε, k-ω, or shear-stress transport (SST). Nearly all of these models are calibrated for simple shear flows. Thus, they predict well when shear is the dominating flow feature. But, gas-turbine flows are characterized by streamline curvature, rotation, and complex "freestream" turbulence structures. As a result, it is difficult to know the conditions under which these models might be of questionable applicability, prior to carrying out a complete simulation.

In this paper, we first summarize the basis of popular two-equation turbulence models. Afterwards, their adequacy for computing flows with shear and rotation and flows with shear and streamline curvature is analyzed by comparing their predictions with results from large-eddy simulations and experimental measurements of homogeneous turbulent flows subjected to the same strain rates.

BASIS OF TWO-EQUATION TURBULENCE MODELS

For many turbulent flows, the most important characterizing quantity is the mean rate of the strain, S_{ij}, given by

$$S_{ij} = \frac{1}{2}\left(\frac{\partial U_i}{\partial x_j} + \frac{\partial U_j}{\partial x_i}\right) \tag{1}$$

All popular two-equation models invoke the Boussinesq hypothesis, which assumes the Reynolds stresses, $-\overline{u'_i u'_j}$, to be aligned with the mean rate of strain; i.e.,

$$-\overline{u'_i u'_j} = \nu_t S_{ij} - \frac{2}{3}\delta_{ij}k \tag{2}$$

where the eddy viscosity, ν_t, is an isotropic scalar. For the k-ε model, ν_t is modeled by

$$\nu_t = C_\mu \frac{k^2}{\varepsilon} \tag{3}$$

with the turbulent kinetic energy, k, and its dissipation rate, ε, computed by two transport equations. The proportionality constant, C_μ, in Eq. (3) is chosen so that

$$-\frac{\overline{u_i' u_j'}}{k} = 0.3 \qquad (4)$$

Equation (4) is a calibration to match some regions of equilibrium homogeneous shear-layer experiments in stationary frames, where production of turbulent kinetic energy equals its dissipation rate. Though the closure given by Eqs. (1) to (4) along with the transport equations for k and ε was designed to perform well for steady, two-dimensional, boundary-layer flows over a flat plate, it also performs well for more complex flows if the term, $\partial U/\partial y$ (i.e., the strain normal to the wall; henceforth referred to as shear), is the dominant one in S_{ij}.

Since it is difficult to know when k-ε type turbulence models will fail in the presence of shear and an additional "extra" strain, it is useful to characterize candidate flows to be modeled according to the kinds of S_{ij} imposed on the turbulence, and to test the proposed turbulence models against data from measurements, direct numerical simulations (DNS), and large-eddy simulation (LES) of flows that are dominated by those same strain-rate tensors. For gas turbines, two of the most important attributes that complement shear are rotation and streamline curvature. Since there exist measurement, DNS, and LES results in which shear with rotation and shear with streamline curvature have been studied, it seems rational to assess the performance of turbulence models by comparing model predictions with those results. This exercise can be particularly useful in identifying regions of a flow, where a given turbulence model might be of questionable applicability, prior to carrying out a complete computation.

PERFORMANCE UNDER SHEAR AND ROTATION

Benchmark Test Problem for Rotation

A useful benchmark test problem is homogeneous turbulent shear flow in a rotating frame. When the shear is in the plane of rotation, as shown in Fig. 1, the combination of shear and rotation is known to have some particularly interesting effects on turbulence. There are LES[3] and DNS[4] solutions for this problem over a range of different values of Ω. Although truly homogeneous rotating shear flows in gas turbines do not exist, there are certainly many regions, which undergo shear in the plane of rotation and so this case is certainly relevant, if somewhat simplistic. It is, therefore, useful to examine some solutions to this problem to gain a qualitative understanding of when turbulence closures like the k-ε, k-ω, and SST models might perform well, and when they might be inadequate.

The problem, shown in Fig. 1, is characterized by the following three dimensionless parameters[5]:

$$\frac{S k_o}{\varepsilon_o}, \; \frac{\Omega}{S}, \; \frac{k_o^2}{\nu \varepsilon_o} \qquad (5)$$

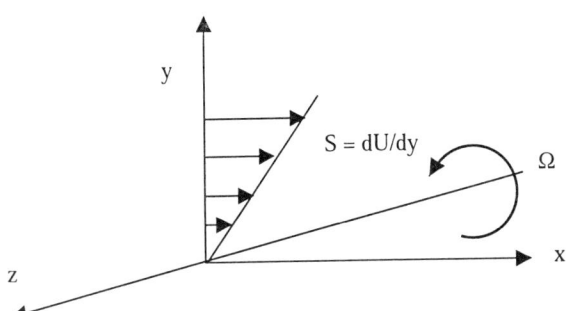

Fig. 1. Turbulent homogeneous shear flow in a rotating frame.

where S is the homogeneous shear, and k_o and ε_o are respectively the initial values of the turbulent kinetic energy and its dissipation rate, and Ω is the rate of rotation. The first parameter, Sk_o/ε_o, represents the initial turbulent-to-shear time scale ratio. It is a useful indicator of when flows are likely to be in equilibrium ($(\Delta S)k_o/\varepsilon_o \leq 1$) and when non-equilibrium behavior such as rapid distortion might be expected ($(\Delta S)k_o/\varepsilon_o \gg 1$). The second parameter, Ω/S, is the ratio of rotation-to-shear rate (related to the Richardson number), and its magnitude and sign have a strong influence on the turbulence of this problem. Finally, $k_o^2/\nu\varepsilon_o$ is the equivalent of a turbulent Reynolds number and, given the usual Reynolds-number invariance of turbulent flows, is not expected to influence the turbulence significantly.

For this homogeneous shear-flow problem, the k-ε model, in both stationary and rotating reference frames, can be written as

$$\frac{dk}{dt} = C_\mu \frac{k^2}{\varepsilon} S^2 - \varepsilon \tag{6}$$

$$\frac{d\varepsilon}{dt} = C_{\varepsilon 1} C_\mu k S^2 - C_{\varepsilon 2} \frac{\varepsilon^2}{k} \tag{7}$$

where $-\overline{u'v'}$ was modelled by $-\overline{u'v'} = C_\mu (k^2/\varepsilon) S$ with the constants, C_μ, $C_{\varepsilon 1}$, and $C_{\varepsilon 2}$, set to their usual values.

From Eq. (6), it is clear that the strain rate, which is the symmetric part of the deformation tensor, produces turbulent kinetic energy by interactions with $-\overline{u'v'}$, whereas, the rotation rate, which is the antisymmetric part of the deformation tensor, does not appear. The role of rotation is to rotate or redistribute the turbulence structure, though it is know to have some effects on the dissipation rate of k. Thus, strain and rotation affect turbulence in very different ways.

The k-ε model obviously lacks any dependence on Ω/S, so its solutions are independent of the rate of rotation, Ω. The k-ω model, on the other hand, does include a rotation-rate sensitive closure coefficient in its ω equation though it is not included for the purposes of making the model applicable to rotating shear flows. These shortcomings of the k-ε and k-ω models and hence the SST model are an implicit consequence of their eddy viscosity models (i.e., Eqs. (2) and (3)) and how they are calibrated (Eq. (4)). In contrast, the full transport equations for each of the Reynolds stresses do include a kinematic rotation term and a pressure strain term[6], both of which can depend on Ω and whose solutions can be quite different from k-ε type models.

Comparison between LES and the K-ε Model for Flows with Shear and Rotation

It is straightforward to solve the coupled ordinary differential equations of the k-ε model given by Eqs. (6) and (7) and to compare them with results from LES carried out for different values of Ω from Bardina.[3] These results are shown in Fig. 2. They are for a flow, which is an initially decaying isotropic turbulence, suddenly subjected to shear S and rotation Ω at t = 0.

For the case of no rotation, Fig. 2(a) shows the k-ε model to describe the evolution of k to within about 30% accuracy up to dimensionless times of about $t^* = 5$, where $t^* = t\,S$. The reason for this relatively large error, even without rotation, arises because, as an equilibrium model, k-ε generates increased k levels on immediate application of S instead of allowing time for intercomponent transfer before growth of k, as the LES correctly does.

The ratio Ω/S = 0.25 is the most destabilizing and results in much more rapid growth in k, which the k-ε model significantly under predicts at long times. An axis transformation reveals that setting Ω/S equal to 0.25 corresponds to pure strain – since shear decomposes into the sum of strain plus rotation, adding/subtracting rotation leaves us with pure strain. Any rotation relative to this case reduces production since shear is reduced. It does so by rotating the principal axis of the Reynolds stress away from that of the strain field. This misalignment means that it takes longer for intercomponent transfer to the principal axis of the strain field, reducing the growth rate of turbulent kinetic energy.

At the higher value of Ω/S = 0.5, the effect of rotation is less pronounced and k grows more slowly than in the case without rotation – again the k-ε model wildly over predicts k at long times. Results for k-ω model calculations are very similar.

From these comparisons, it is clear that k-ε type closures are only appropriate for regions of turbulent flow when Ω/S is very close to zero; i.e.,

$$\frac{\Omega}{S} \ll 0.25 \tag{8}$$

and in flows close to equilibrium in which times of flight through shear regions are less than a few $t^* = t\,S$ units.

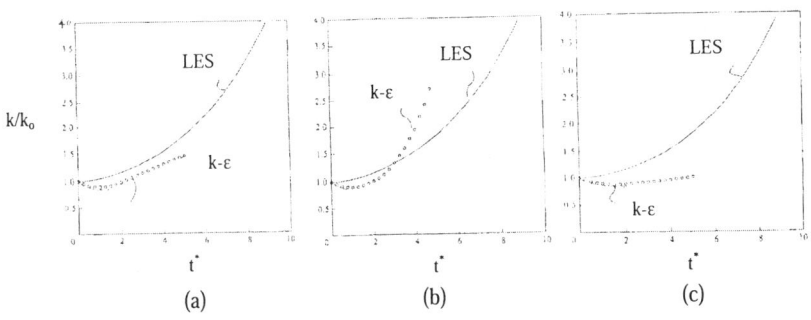

Fig. 2. LES and k-ε predictions of $k^* = k/k_0 = f(t^*)$ for homogeneous turbulent shear flow in a rotating frame at different Richardson numbers (Bardina, et al.[3]).
(a) $\Omega/S = 0$. (b) $\Omega/S = 0.25$. (c) $\Omega/S = 0.5$.

Magnitudes of Dimensionless Parameters for Flows between Blade Passages

Typical values of the dimensionless parameters relevant to turbulent flow and heat transfer in gas turbines are now estimated to the nearest order of magnitude. For illustrative purposes, we consider a rate of rotation of 3,600 rpm or roughly 400 radians/sec; a radial distance, R, from the axis of rotation of 1 m, and a kinematic viscosity, ν, of 2×10^{-5} m²/s². For external flow over the airfoils, we assume axial velocities through blade passages to be up to a few hundred m/s, and the spacing between the blades in the azimuthal direction to be about 2 cm apart with chord lengths of about 5 cm.

From these dimensions, we can estimate most of the relevant dimensionless quantities as follows. Reynolds numbers based on transverse dimension would to be around 3×10^5, and Reynolds numbers based on chord length would approach 10^6. By using the Prandtl steady boundary-layer thickness approximation, $\delta/z = 0.375/Re_z^{1/5}$, for order-of-magnitude purposes, we would anticipate boundary layers of the order of 1 mm in attached regions around the mid-chord region. Therefore, in shear layers along blade surfaces, when the axial component of velocity determines the shear rate, we might anticipate $S = \partial U/\partial y$ to be roughly U/δ or a few hundred thousand sec^{-1}. The blade curvature causes a small component of this shear to be in the plane of rotation. It is important to note that boundary layers are not homogeneous shear layers but, instead, feature progressively increasing shear as the wall is approached. Thus, S is practically zero at boundary-layer edges, but approaches our estimated value in the vicinity of the wall. While one might be tempted to characterize the low-shear outer regions of boundary layers as strongly susceptible to effects of rotation, since Ω/S might approach values above unity there, production and dissipation are essentially zero in the outer layer. Therefore, for the purposes of our approximate analysis, we need only consider the near-surface regions of high turbulence levels as the one for which Ω/S is relevant. For these shear layers, Ω/S would take values of about 10^{-3}. Thus, the effects of rotation on predictions of turbulence with k-ε, k-ω, and SST models would be negligible.

On the other hand, about stagnation regions near the root and tip at the leading edge of the blade, where $\partial U/\partial z$ leads to $\partial v_r/\partial r$ through continuity, boundary layers will be an order of magnitude thinner than at mid-chord and so shear rates an order of magnitude higher. Therefore, the absence of rotational effects is likely to affect turbulence prediction in these regions. However, typical times of flight of fluid through these regions correspond to values of $t^* = t\,S$ around 10^{-6} so that concerns about rotation effects on the long-time growth of k are not relevant to this application. Furthermore, the dominance of the strain-rate tensor by normal stresses like $\partial U/\partial z$ at the leading edges of vanes and blades is inconsistent with the calibration of eddy viscosity models to reproduce correctly only the $-\overline{u'v'}$ component of the stress tensor (see Moore & Moore[7] for further discussions). Thus, standard two-equation closures are likely to be untrustworthy at leading edges, on account of their calibration to shear (rather than normal) strain flows, and at the root and tip junctures, on account of the alignment of shear and rotation there. The corner flows in these junctures are also regions of secondary flow which two-equation closures are not capable of modeling. It follows that near surface quantities like heat transfer in these regions are also likely to be difficult to predict accurately with two-equation models.

Magnitudes of Dimensionless Parameters for Internal Blade-Cooling Flows

We consider, as a representative internal flow of cooling air along a passage within a turbine blade, the flow studied by Stephens and Shih[8]. This flow takes place in a square-cross-section duct, with hydraulic diameter $D_h = 1.27$ cm, which comprises a straight section of $L = 18$ cm, a 180°-bend with inner and outer radii of 0.28 cm and 1.8 cm, followed by an identical second straight return section. Both straight sections are in a plane perpendicular to that of rotation. The flow is studied without rotation and with rotation at 3133 and 6265 rpm, or $\Omega = 328$ and $\Omega = 656$ radians/sec. For a Reynolds number of 25,000 based on hydraulic diameter, the average velocity at the inlet is about 17 m/s.

For this flow, the momentum boundary layer is fully developed at the duct inlet with a shear S of about 6×10^4 sec^{-1}. Therefore, the value of Ω/S at entrance to the U-tube is 0.0 for the non-rotating case, and approximately 0.0055 and 0.0109 for the two cases with rotation. The particular boundary layers in which shear interacts with rotation are on the leading and trailing faces of the duct. The other two faces are unaffected. Thus, effects of rotation are potentially much more significant in internal blade-cooling flows than in flows over rotating blades if the Reynolds number is low and the rotation number is high.

PERFORMANCE UNDER SHEAR AND STREAMLINE CURVATURE

The k-ε, k-ω, and SST models do not change on account of streamline curvature within the flow, nor do their constants. However, the stabilizing and destabilizing effects of convex and concave flows tend to decrease and increase turbulence levels respectively, relative to their flat-wall values. Thus, one would not expect two-equation closures of this kind to be of sufficiently general validity to describe effects observed in flows with curvature.

The reason that k-ε, k-ω, and SST models cannot account for the effects of streamline curvature is that these models have just enough undetermined coefficients to allow them to be tuned to perform well in one kind of flow – the steady, two-dimensional, thin-shear flow like a boundary layer, jet or wake. The single value of the coefficient C_μ in the eddy-diffusivity model, chosen to provide adequate predictions of the Reynolds stress $-\overline{u'v'}$ in two-dimensional thin shear layers is simply not appropriate for flows with streamline curvature. Accurate predictions of other components of the Reynolds stress tensor like $-\overline{u'w'}$, and $-\overline{v'w'}$ are required too, and are generally beyond the capabilities of simple eddy-diffusivity models. In principle, two-equation models could be tuned to provide accurate prediction of a curved homogeneous turbulence at some prescribed value of F (see, e.g., Ref. 10). However, in wall-bounded flows, F is not single-valued, but varies across the flow. Furthermore, tuning a two-equation model in this way could diminish its ability to provide reasonable predictions of flat-wall-bounded flows. It follows that the performance of two-equation closures in flows with streamline curvature is likely to depend primarily on the value of curvature factor and, in particular, its proximity to zero.

More sophisticated closures, which have been used to study flows with streamline curvature include algebraic and differential-equation Reynolds-stress models. By moving beyond the Boussinesq approximation (which essentially calibrates eddy-diffusivity primarily for the purpose of predicting $-\overline{u'v'}$, without regard for the accuracy of other components of turbulent shear stress), these closures can, in principle, provide reliable estimates of all components of turbulent shear stress. This capability would appear to be essential, given the potential importance of all components of the Reynolds stress tensor in influencing momentum of the mean flow, though it must be included at the cost of having to solve more equations.

Algebraic models, which represent a low-cost alternate to solving the Reynolds stress transport equations, invoke an assumption that the anisotropy of turbulence has reached an equilibrium state when measured in an appropriate reference frame. It is noteworthy that this equilibrium algebraic approach leads to explicit representations of the Reynolds-stress tensor with C_μ not as a constant but as a variable, which typically depends on the magnitude of the local strain[9,10]. Thus, one can think of the eddy-diffusivity models in k-ε, k-ω, and SST closures with $C_\mu = 0.09$ as a special case of more general C_μ models, which happens to be appropriate for the turbulence producing regions of two-dimensional, steady, thin shear layers with almost parallel streamlines. Even if the complete differential Reynolds-stress transport equations were to be solved, there remain many questions about how to model faithfully the pressure-strain terms, and if quantities like mean shear, rotation, k, ε, and components of the Reynolds stress tensor even provide an adequate basis for general modeling of pressure-strain terms.[6]

Benchmark Test Problem for Streamline Curvature

The benchmark test problems that we consider for flows with streamline curvature are the curved homogeneous turbulent shear flow (the homogeneous idealization of a flow with circular streamlines) by Holloway and Tavoularis[11] and the wall-bounded inhomogeneous flow in a curved channel by Moser and Moin[12].

Shear flows of this kind, which, in the mean, follow curved streamlines, may be characterized by a shear-rate, S′, and a curvature-factor parameter, F, given by

$$S' = \frac{\partial U}{\partial r} \tag{9}$$

$$F = \frac{U}{r} \Big/ \frac{\partial U}{\partial r} \tag{10}$$

In the above equations, U is the mean tangential velocity, and r is the radial coordinate. The homogeneous curved shear flow corresponds to both F and S′ taking constant values. Positive values of the curvature factor correspond to flow over convex surfaces. In such flows, the angular momentum of the flow increases with increasing radius about its center of curvature, and Tollmien-Schlichting waves present in the flow are attenuated, resulting in decreased turbulence levels. Such flows are often referred to as simply convex or stabilizing flows. Negative F values describe flows found adjacent to concave surfaces. In this case, the angular momentum decreases with increasing radius which gives rise to Taylor-Gortler instability. This destabilizing, concave effect causes increased turbulent fluctuations in velocity and generally enhances turbulence levels.

The experiments of Holloway and Tavoularis[11] in almost homogeneous curved shear flow, over a range of both positive and negative values of F, provide an important set of target data against which to compare the predictions of turbulence models proposed for use in flows with streamline curvature, such as those in gas turbines. In these experiments, for each value of F, there is a corresponding value of S′. The DNS results from Moser and Moin[12], in a wall-bounded, mildly curved fully-developed turbulent channel flow, provide essentially exact results for a flow with streamline curvature with the additional complexity of inhomogeneity, imposed though a no-slip condition at each bounding wall. While homogeneous-flow experiments are invaluable for basic model development purposes, the DNS provides further information for model refinement and a means of gauging model effectiveness in a flow, which has similarities to many practical engineering ones.

Comparisons between Measurements and the K-ε Model for Flows with Shear and Streamline Curvature

As for the case of rotating homogeneous shear flows, solution of the coupled ordinary differential equations of two-equation turbulence models in curved homogeneous turbulent flow given by Eqs. (6) and (7) is straightforward. However, instead of yielding solutions in which k grows exponentially with time, as is observed for parallel homogeneous flows, curved homogeneous flows over a wide range of curvature factors evolve to equilibrium states at long times. Since two-equation models are typically designed for equilibrium flows, the most useful comparisons between experiment and data can be made if these solutions are allowed to evolve and reach steady states. Transient solutions for homogeneous curved flows, which evolve from isotropic turbulence using k-ε and other closures, can be found in the paper of Girimaji[9]. For the

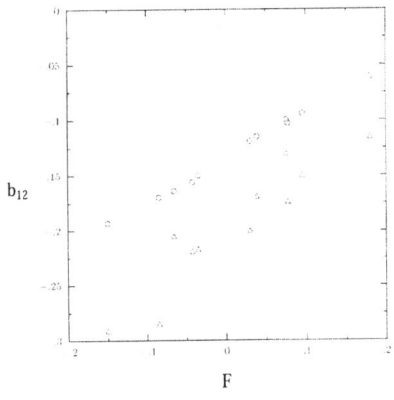
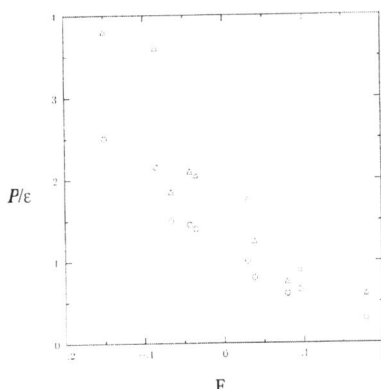

Fig. 3. Comparison of anisotropy of shear stress b_{12} between the k-ε model ($\triangle$) and measurents[11] ($\circ$).

Fig. 4. Comparison of production-to-dissipation ratio between the k-ε model ($\triangle$) and measurements[11] ($\circ$).

purposes of this paper, we compare equilibrium predictions of the k-ε model with the experimental data of Holloway & Tavoularis[11] in Figs. 3 and 4. With the k-ε model, the predicted values of ε have been replaced by those deduced in the experimental study to avoid additional ambiguities in interpretation, which would arise through the imperfections of the ε model. This use of experimental values of ε, however, diminishes the smoothness of the k-ε results.

In Fig. 3, the equilibrium shear-stress anisotropy component b_{12} ($\overline{u'v'}/2k$) predicted by the k-ε model, and its measured values are shown over a range of positive and negative curvature factors. The eddy diffusivity in the k-ε, k-ω, and SST models is calibrated to provide accurate prediction of this quantity in steady, two-dimensional flows with almost parallel streamlines (i.e., F = 0). It is, therefore, interesting to evaluate its predictions at positive and negative values of F. From Fig. 3, it is clear that the b_{12} predictions of the K-ε model are more strongly negative than those measured in experiments by between 50 and 100% over the entire range of curvature factors, though the trend is similar. The shear stress $-\overline{u'v'}$ is a larger proportion of turbulent kinetic energy at negative (concave / destabilizing / pressure side) curvatures and a smaller one at positive (convex / stabilizing / suction side) ones, consistent with many observations in wall-bounded homogeneous flows and the simulation of Moser & Moin[12]. These results indicate that, only for flows with |F| << 1, would one expect two-equation models to perform adequately when streamline curvature is present.

Figure 4 shows the ratio of production of turbulent kinetic energy, P, to its dissipation rate, ε, in the same experiments and model computations. The trends in both experiment and k-ε model prediction are very similar. In the case of negative curvature, the k-ε

closure significantly overpredicts production, consistent with its overly negative calculation of b_{12} in concave / destabilizing / pressure-side flows. When curvature is positive, the production of new turbulence can drop below the rate of dissipation of turbulent kinetic energy, which can be interpreted as the onset of relaminarization. Thus, it can be of great importance that turbulence models are in good agreement with experimental data at the value of F at which $P/\varepsilon = 1$, if they are to be used in convex flows. In this case, the k-ε model predictions are quite close to experimental measurements for large positive curvatures, and even appear to predict the onset of relaminarization.

Magnitudes of Dimensionless Parameters for Flows between Blade Passages

Order of magnitude estimates are now made of typical values of the curvature factor, F, in turbulent flows over rotor and stator blades in gas turbine. For illustrative purposes, we assume that the radius of curvature, r, over most of the suction side of a typical blade will be close to 5 cms, and the radius of curvature of the pressure side to be roughly twice as large. In order to estimate F, we use the previous estimate of attached boundary-layer thicknesses, δ, of about 1 mm. Since $\partial U/\partial r \approx U/\delta$, we can approximate F given by Eq. (9) as δ/r. At a mid-chord location on a typical blade of the dimensions chosen earlier, we might expect $-0.01 < F < 0.02$. The outer bounds of this range approach those F values at which serious discrepancies can appear when using the k-ε, k-ω, and SST models tuned to the case of $F = 0$. One might anticipate from Figs. 3 and 4, that these models would overpredict significantly $-\overline{u'v'}$ and production of k. At the leading edge, on both the pressure and suction sides, both the radii of curvature and boundary-layer thicknesses will be smaller and, if larger magnitudes of F were encountered, the fidelity of two-equation-model predictions would be compromised further.

Magnitudes of Dimensionless Parameters for Internal Blade-Cooling Flows

In the case of the internal-cooling flow along a passage within a turbine blade, studied by Stephens and Shih[8], the inner and outer radii of curvature are only 0.28 and 1.8 cm respectively, and are significantly smaller than those encountered on surfaces of typical compressor and turbine blades. Moreover, the greater boundary-layer thicknesses (≈ 5 mm) expected in this lower-speed (≈ 17 m/s) flow lead to estimates of $F \approx -0.3$ on the outer concave surface, and $F \approx 0.2$ on the inner convex surface. At the concave surface, one might expect the two-equation model to over predict $-\overline{u'v'}$ and turbulence production by a factor of two. At the convex inner surface, the curvature is too severe for any estimate of two-equation-closure adequacy to be made. It is quite possible that relaminarization takes place. Based on this analysis, it is difficult to make an assessment of the fidelity of turbulence predictions in this flow because of the lack of benchmark data at these high curvature factors. A differential Reynolds-stress closure would probably be preferred for prediction of this very challenging flow (see Chen, et al.[12]).

CONCLUSIONS

Comparisons between k-ε predictions with measurements and LES results for homogeneous turbulent flows with shear and an extra strain show the following. When there are shear and rotation, k-ε type models may be adequate if $\Omega/S \ll 0.25$. If Ω/S is not much less than 0.25, then the time of flight of fluid through the region with rotation, $t^* = t\,S$, must be small. When there is shear and streamline curvature, k-ε type models can be used in flows over convex surfaces, if they are calibrated to be in good agreement with the experimental data at the value of F at which $P/\varepsilon = 1$.

REFERENCES

1. Simoneau, R.J. and Simon, F.F., "Progress Towards Understanding and Predicting Heat Transfer in the Turbine Gas Path," *International J. of Heat and Fluid Flow*, Vol. 14, No. 2, 1993, pp. 106-128.
2. Shih, T.I-P. and Sultanian, B., "Computations of Internal and Film Cooling," *Heat Transfer in Gas Turbine Systems*, Editors: B. Sunden and M. Faghri, WIT Press, Ashurst, Southhampton, to appear.
3. Bardina, J., Ferziger, J. H., and Reynolds, W. C., "Improved Turbulence Models Based on Large Eddy Simulation of Homogeneous, Incompressible Turbulent Flows," Dept. of Mechanical Engineering, Report TF-19, Stanford University, 1983.
4. Salhi, A. and Cambon, C., "An Analysis of Rotating Shear Flow Using Linear Theory and DNS and LES Results," *J. of Fluid Mechanics*, Vol. 347, pp. 171-95, 1997.
5. Speziale, C. G. and Mac Giolla Mhuiris, N., "On the Prediction of Equilibrium States in Homogeneous Turbulence," *J. Fluid Mechanics*, Vol. 209, pp. 591-615, 1989.
6. Hanjalic, K., Advanced Turbulence Closure Models: A View of Current Status and Future Prospects, *International J. of Heat and Fluid Flow*, Vol. 15, No. 3, 1994, pp. 178-200.
7. Moore, J. G. and Moore, J., "Realizability in Turbulence Modelling for Turbomachinery CFD," ASME Paper 99-GT-24, 1999.
8. Stephens, M.A. and Shih, T. I-P., "Flow and Heat Transfer in a Smooth U-Duct with and without Rotation," *AIAA J. of Propulsion and Power*, Vol. 15, No. 2, 1999, pp. 272-279.
9. Girimaji, S. S., "A Galilean Invariant Explicit Algebraic Reynolds Stress Model for Turbulent Curved Flows," Physics of Fluids, Vol. 9, No. 4, 1997, pp. 1067-1077.
10. Rumsey, C.L., Gatski, T.B., and Morrison, J.H., "Turbulence Model Predictions of Strongly Curved Flow in a U-Duct," *AIAA J.*, Vol. 38, No. 8, 2000, pp. 1394-1402.
11. Holloway, A. G. L. and Tavoularis, S., "The Effects of Turbulence on Sheared Turbulence," J. of Fluid Mechanics, Vol. 237, 1992, pp. 569-585.
12. Moser, R. D. and Moin, P., "The Effect of Curvature on Wall-Bounded Turbulent Flows," J. of Fluid Mechanics, Vol. 175, 1987, pp. 479-510.
13. Chen, H.-C., Jang, Y.-J., and Han, J.-C., "Computation of Flow and Heat Transfer in Rotating Two-Pass Square Channels by a Reynolds Stress Model," ASME Paper 99-GT-174, June 1999.

A Review of Turbine Blade Tip Heat Transfer

RONALD S. BUNKER

General Electric Corporate Research & Development Center
Niskayuna, NY 12309 USA

ABSTRACT: This paper presents a review of the publicly available knowledge base concerning turbine blade tip heat transfer, from the early fundamental research which laid the foundations of our knowledge, to current experimental and numerical studies utilizing engine-scaled blade cascades and turbine rigs. Focus is placed on high-pressure, high-temperature axial-turbine blade tips, which are prevalent in the majority of today's aircraft engines and power generating turbines. The state of our current understanding of turbine blade tip heat transfer is in the transitional phase between fundamentals supported by engine-based experience, and the ability to *a priori* correctly predict and efficiently design blade tips for engine service.

INTRODUCTION

The design of high efficiency, highly cooled gas turbines is achieved through the orchestrated combination of aerodynamics, heat transfer, mechanical strength and durability, and material capabilities into a balanced operating unit. While decades of research have been dedicated to the study and development of efficient aerodynamics and cooling techniques for turbine airfoils, there remain regions, which retain a somewhat more uncertain design aspect, requiring more frequent inspection and repair. One such region particular to high-pressure turbines is the blade tip area. Blade tips are comprised of extended surfaces at the furthest radial position of the rotating blade, which are exposed to hot gases on all sides, typically difficult to cool, and subjected to the potential for wear against the outer shroud flow path. The blade tip operates in the transitional environment between the rotating airfoil and the stationary flow path casing, which experiences the extremes in most fluid-thermal conditions within the turbine.

What Is At Stake?

It has long been recognized that the effectiveness of the blade tip design and its effect on the subsequent tip hot gas path leakage flows is a major contributor to the aerodynamic efficiency of turbines. The derivative of turbine efficiency with blade tip clearance can be significant, driving turbine designers to improve efficiency by decreasing tip-to-shroud operating clearances, or by implementing more effective tip leakage sealing mechanisms. Figure 1a depicts the general effect of tip clearance on turbine efficiency, which in an operational measure is manifested as an decrease in the turbine exhaust gas temperature margin (ΔEGT). This sensitivity to tip clearance is very high for smaller engines as the clearance represents a larger portion of the overall working flow path annulus height. As any turbine engine accumulates steady running time at temperature, the blade tip region is oxidized and eroded away to some degree. This can be countered by over-cooling, or simply by running at low firing temperatures, both of which severely impact overall efficiency. Certain off-design point conditions for the engine may also subject the blade tips to short durations of even higher temperatures. During transient operation, blade tips can be subjected to rubs against the shroud, which may remove substantial amounts of tip material, resulting in larger clearances for subsequent steady operation.

As a consequence of these operational realities, blade tips must be designed to balance material type and operating temperature with oxidation rate, cooling flow, and other effects, such that an acceptable or economic mean time between repairs (MTBR), or maintenance cost per hour (MCPH), is achieved. Clearly, knowledge of the heat transfer on all surfaces and under all conditions is essential. Figure 1b depicts this design choice in terms of the material oxidation resistance. Considerations of material ductility, fabrication, repair-ability, and cost must enter into the design of blade tips. There are several blade tip designs in current use within the industry, which emphasize various aspects of the total problem.

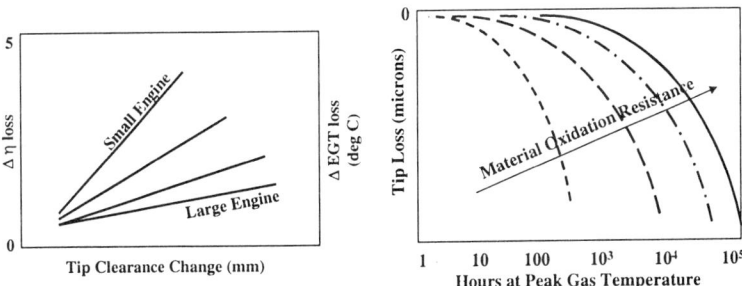

Figure 1. (a) Loss of Turbine Performance/Margin as Tip Clearance Opens in Service; (b) Loss of Tip by Oxidation in Service

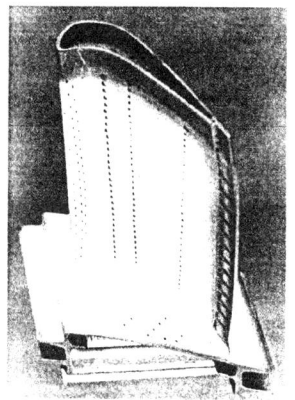 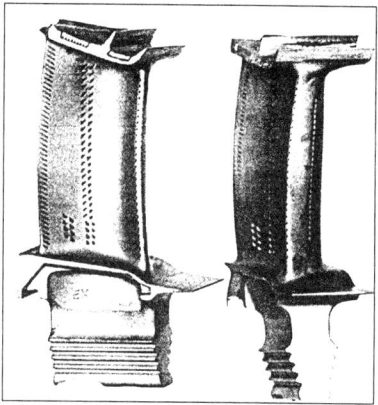

Figure 2. (a) GE Aircraft Engines HP Turbine Blade with Squealer Tip Design, (b) Rolls-Royce Trent 800 HP Turbine Blade After 8299 hrs (cleaned)

These designs include: (1) flat unshrouded blade tips, (2) unshrouded tips with various forms of squealer rims, such as shown in Figure 2a, to reduce hot gas leakage while providing protection against shroud rubs, and (3) shrouded blade tips, such as shown in Figure 2b, which seek to establish high aerodynamic efficiency with better sealing at the tip, but also incur greater tip weight. High performance, high-temperature versions of such blade tip designs usually also utilize some amount of film cooling to reduce regional heat loads and thermal stress, and to bring metal temperatures down to minimum acceptable oxidation rates.

Why Is It So Difficult?

The cooled turbine blades found in today's modern gas turbine engines represent very complex heat exchangers of specific aerodynamic shape and strict structural integrity. The blade tip region is arguably the most three-dimensional portion of the blade in terms of hot gas flow interaction, coolant delivery, and geometry. Issues which challenge the best turbine designers include:
- The blade tip is subjected to some of the highest convective heat transfer loads on the entire blade, as well as some of the lowest.
- The flow field is extremely complex, including periodic unsteadiness from upstream airfoils, and shroud region leakage flows.
- The combustion systems impose sometimes severe steady-state radial gas temperature profiles in the tip area.
- The tip can be subject to extreme variations in both gas temperature and pressure as operating conditions change.
- Internal-to-external surface thermal gradients can result in high thermal stresses and cracking.

- The loss of material due to oxidation or rubs effectively alters the flow field and thermal conditions over time.
- Aerodynamic and thermal boundary conditions change during short transients as the tip clearance varies.
- The blade tip weight directly impacts the blade root stresses, LCF life, and blade creep, and can also effect aero-mechanical responses.
- Any increase in blade tip cooling represents a chargeable flow penalty on the engine cycle.
- The addition of film cooling adds further complexity or constraints in many factors.

THE TIP FLOW FIELD WHICH DRIVES HEAT TRANSFER

No matter the design choice selected for any particular turbine blade tip, a detailed knowledge of the tip flow field is required to achieve the proper balance of elements for efficiency with durability. The flow in and around turbine blade tips has been under investigation much longer than the heat transfer aspects, spurred by the great impact on efficiency for both turbines and compressors in axial and radial turbomachinery. An early study by Allen and Kofskey[1], examined the secondary flows in the blade tip region of a low-speed rotating turbine rig through smoke visualization. Lakshminarayana[2] developed predictive models for stage efficiency and compared these to existing data for several classes of turbomachinery. A comprehensive study by Booth et al.[3] and Wadia and Booth[4] measured overall and local blade tip losses for many configurations of tip geometries, and developed predictive methods based on discharge coefficients. Later work of Moore et al.[5] examined flat tip region flows from laminar to transonic conditions and compared their predictions with available experimental data. Detailed measurements of velocity and pressure fields have been obtained within an idealized tip gap by Yaras et al.[6] and Sjolander and Cao[7]. Bindon[8] and also Yamamoto[9] performed detailed blade cascade studies of the tip clearance loss mechanisms. The effects of tip clearance, tip geometry, and multiple stages on turbine stage efficiency have lately been quantified by Kaiser and Bindon[10] within a rotating turbine rig environment. Many other works, too numerous to list here, have studied the effects of tip clearances in axial turbines with the primary emphasis on total stage leakage and efficiency loss prediction. Figure 3 depicts the general time-averaged flow features that are present on simple blade tips without film cooling. All of the various features of separated flows, vortices, and reattached flows have been observed for stationary test cases without film coolant. Rotational flow field data within the blade tip region is essentially non-existent. Rotational information is limited to that obtained by the acquisition of discrete local surface pressures or surface flow visualization, or by flow field traverses downstream of the blade row. Even this simple schematic, underscores the complexity of the associated blade tip heat transfer.

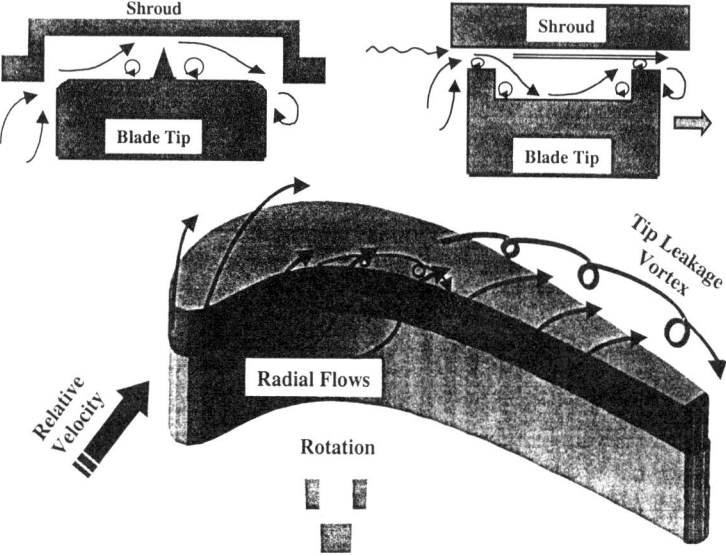

Figure 3. Schematic Representation of Time-Averaged Blade Tip Flows

BLADE TIP HEAT TRANSFER

Fundamental Research Related to Turbine Blade Tip Heat Transfer

The geometries of idealized blade tip regions, as shown in Figure 3, resemble either a sudden contraction between two parallel plates, or a shrouded shallow transverse cavity, or a labyrinth seal. Common aspects of these simplified cases include sudden contractions, backward facing steps, forward facing steps, and the associated separation regions. A comprehensive review of heat transfer in compressible separated and reattached flows can be found in Merzkirch et al.[11]. Studies with particular relevance to blade tip heat transfer are noted here. Boelter et al.[12] performed a comprehensive series of tests, which determined the heat transfer downstream of sudden contractions of various inlet geometries and thermal boundary conditions. Heat transfer in the separation region, reattachment location, and developed flow downstream of a backward facing step have been reported by Seban[13], Vogel et al.[14], Orlov et al.[15], Scherer and Wittig[16], and Tsou et al.[17]. All have come to the same general conclusions regarding the effect on local heat transfer, showing enhancement factors of as much as 2 to 3 occurring within the entry contraction region, as well as at the reattachment region some six step heights downstream of a step. Heat transfer in the separated flow zone ahead of a forward facing step has been reported by

Luzhanskiy and Solntsev[18]. In general, the various studies have used very thin incoming boundary layers prior to the step, so that the results are indicative of step heights many times the size of the momentum thickness, which is the case within blade tip regions. Cavity heat transfer has been studied by a number of researchers with the focus being primarily on the cavity internal walls or floor. Seban[19], Fox[20], and Yamamoto et al.[21], all measured heat transfer on the cavity surfaces for cavity depths typical of blade tips, but in open channel flow conditions. In most cases, the downstream surface immediately after the forward facing step had heat transfer enhancement factors of 1.5 to 2 times the fully developed level.

Early Studies on Modeled Blade Tips

Work directly aimed at blade tip heat transfer began nearly 20 years ago with the study of Mayle and Metzger[22] in which tip average heat transfer coefficients were measured for nominally flat tip models with various flow Reynolds number and rotational speeds. For the parameter ranges tested, they found that the average tip heat transfer was only a weak function of the rotational speed; ie. the average heat transfer was mainly determined by the pressure driven flow through the tip gap. A subsequent study of Metzger et al.[23] examined the local details of tip heat transfer coefficients for both flat and grooved, stationary rectangular tip models as a function of geometry and Reynolds number. In the case of a flat blade tip model, heat transfer was shown to be very similar to that for flow between parallel plates with an abrupt entry region. The addition of a transverse cavity adds flow resistance, resulting in lower averaged and local heat transfer levels at least in the cavity region. Chyu et al.[24] then carried this study one step further by introducing a moving shroud surface over the rectangular cavity. Here again it was determined that the relative motion had a minor influence on the average tip heat transfer, though some local effects were observed. Figure 4 shows a typical set of detailed local mass transfer coefficients[24] for the cavity rims (X/W<0 and X/W>1) and floor (0<X/W<1) as a function of relative tip clearance. The features of entry region enhancement, cavity floor reattachment, and separation are all apparent. Entry region enhancement is highest, though very localized, at the tighter clearance gap C, while the cavity floor heat transfer level consistently increases as tip clearance increases. All heat transfer magnitudes increase with the effective Re number.

Additional heat transfer experimental studies have focused on other aspects of blade tips, which are equally important to the design of turbines. Metzger and Rued[25] and Rued and Metzger[26] performed fundamental studies showing both the flow field and heat transfer characteristics of the blade pressure side sink flow region as leakage enters the tip gap, and the blade suction side source flow region as leakage exits the tip gap, respectively. The sink flow region, idealized as a channel flow with streamwise extraction of fluid from one corner, showed that leakage generates increases in the heat transfer near the tip gap of up to 200%. The source flow region, idealized as a channel flow with streamwise

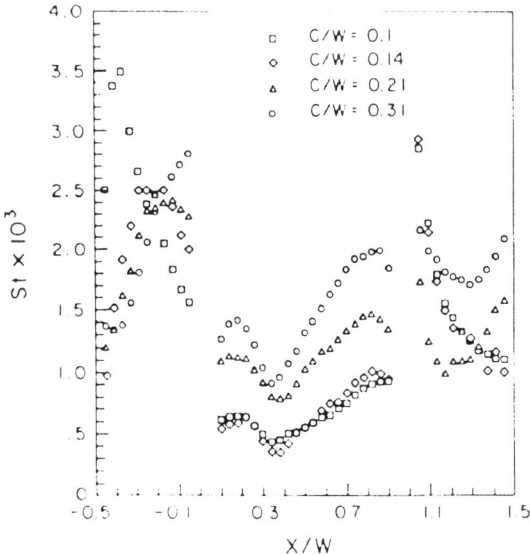

Figure 4. Gap Clearance C Effect on Mass Transfer for Cavity Depth-to-Width Ratio 1.0, Re = 20,000, and No Relative Shroud Motion (Reproduced with permission from the Transactions of the ASME).

transverse injection of fluid from one corner, showed three distinct heat transfer behavioral characteristics, depending upon the relative strengths of the leakage and mainstream flows. The typical characteristic heat transfer is shown in Figure 5. Again, the heat transfer enhancement near the tip gap can be as high as 200%, with results clearly showing the leakage vortex influence. The effects of such enhanced heat transfer for both sink and source flow can be seen in the distress typical of engine blade tips.

The effects of film injection on blade tip local heat transfer and adiabatic film effectiveness distributions, again using idealized rectangular tip models, is summarized in Kim et al.[27]. Several arrangements of cooling holes and slots were investigated in this simple format, showing a wide range of resulting film mixing and heat transfer. To date, no other fundamental blade tip film cooling research has been reported, though film ejection is widely used in practice in the form of tip dust/cooling holes, tip pressure-side film cooling holes, and other design-specific orientations (see Figure 2 for examples). This lack of information highlights two points, that blade tip cooling is highly complex and experience based, and that increases in turbine firing temperatures have out-paced our research for understanding the blade tip heat transfer.

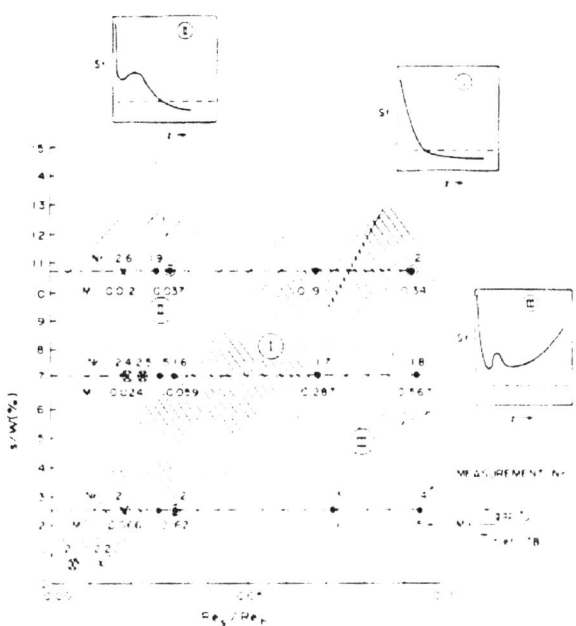

Figure 5. Blade Suction Side Near-Tip Mass Transfer Characteristics Expressed in Terms of Relative Strength of Gap Re_s to Mainstream Re_h, and Relative Gap Clearance S/W (Reproduced with permission from the Transactions of the ASME).

Detailed Tip Heat Transfer in Cascade Flow Environments

Beyond the information available from fundamental heat transfer investigations on simplified models, the next step is to obtain tip heat transfer characteristics within environments which model the actual tip aerodynamics, and therefore contain the principal three-dimensional pressure fields driving the tip leakage flows. In this respect, stationary airfoil cascades offer the opportunity to obtain both discrete and detailed local heat transfer data. While several such studies have been performed for flow field data over many years, heat transfer experiments have only been conducted within the last five years. Yang and Diller[28] modeled a turbine blade tip with recessed cavity in a stationary linear cascade, and deduced a local heat transfer coefficient from a high-response frequency heat flux gage placed at a single discrete location within the cavity at aft-midchord. This study found similar heat transfer behavior with changing gap clearance to that of Metzger et al.[23]. In addition, the unsteady component of this cavity floor heat flux was found to be at least 25% of the total heat flux, for an otherwise nominally steady flow cascade. The first study to measure detailed blade tip heat transfer coefficient distributions was that of Bunker et al.[29], in which full-surface measurements were made in a high-speed linear blade cascade. Figure 6 presents a sample of these measurements for a flat blade tip case with a clearance gap equal to about 1% of the blade height. Such data, when combined with detailed pressure distribution measurements in the same facility, clearly indicate the complex features of tip heat transfer as they change in various locations. The previous fundamental model study results really only apply to a limited portion of a real blade tip, primarily the narrow trailing edge region which experiences a large tip pressure differential. Other regions of the tip experience highly complex three-dimensional flows. These varying flows may include a prominent tip entry vortex region near midchord, and a sizable "sweet spot" of low heat transfer in the forward tip region. Effects due to the variation of tip clearance height and freestream turbulence intensity, were found to be reasonably linear in this study. The further study of Bunker and Bailey[30] presented tip heat transfer distributions for even more complex cases involving the addition of various tip treatments in the form of simple azimuthal and chordwise sealing strips intended to decrease hot gas leakage over the tip. Bunker and Bailey[31] also investigated a second blade tip cascade using a different aerodynamic design, which showed similar general heat transfer effects, but with varying local effects due to the specific aero definition.

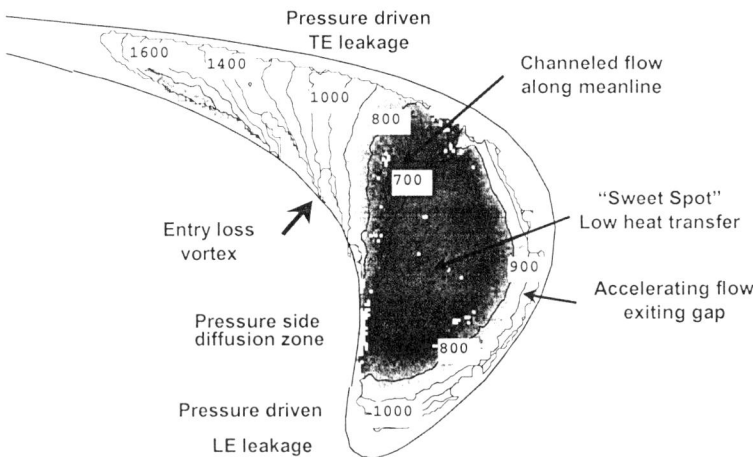

Figure 6. Detailed Heat Transfer Coefficient Distribution for Smooth, Flat Blade Tip Cascade Model, W/m2/K (Reproduced with permission from the Transactions of the ASME).

Additional studies are now being reported for detailed blade tip heat transfer distributions in stationary cascades. Teng and Han[32] report detailed tip heat transfer measurements within a low-speed cascade. Two current studies of Azad et al.[33,34] report tip heat transfer for the GE E^3 design in a high-speed cascade, using flat and squealer tip geometries, respectively. Flat tip heat transfer is very similar to that of Bunker and Bailey[31], with a similar aerodynamic shape, clearly showing a strong pressure side entry vortex and enhanced heat transfer across the midchord region. The addition of a squealer tip geometry served to decrease the midchord heat transfer by as much as 25%, but also increased the forward region levels by this amount due to a redistribution of leakage flows. Reattachment of leakage flows within the tip cavity was evident in the forward areas where the cavity width-to-depth ratio is lowest, causing the enhanced heat transfer. The entire squealer rim around the tip perimeter experienced the highest heat transfer levels, remaining essentially unchanged with varying clearance gap. It is the intention of these studies to further incorporate the effects of tip film cooling discharge within the cascade setting[36]. Another cascade investigation by Urban and Vortmeyer[35] looked at the thermal loading for both flat and squealer tip models, including the effects of clearance gap, mainstream Re and Mach number, without film cooling. This study reported general tip surface cooling effectiveness distributions present with a constant level of internal tip cooling flow, and compared the flat tip to the squealer tip. A blade tip section of lower total flow turning than the other noted studies was utilized. Flat tip heat transfer features were very similar to those found by Bunker et al[29]. Squealer tip heat transfer was found to be nearly constant over the cavity surface at a given clearance and flow condition, and also more sensitive to changes in flow conditions than the flat tip distribution. The variety of these results reinforce the very three-dimensional nature of blade tip flow and heat transfer, where seemingly slight changes to geometry and flow field conditions can have major influence on the tip heat loading.

Research is also in progress at this time[37,38] to measure detailed blade tip and near-tip mass transfer coefficient distributions for other geometries, including film cooling and the use of innovative design features. Much of this research will also include CFD modeling, some of which will begin to incorporate improvements from DNS or LES methods.

Moving to Full Turbine Rig Testing

The best information is always what the turbine engine tells us directly as a result of field experience under various operating conditions. But this information is costly, after-the-fact, and not always good news. It is also very difficult, if not impossible at this time, to discern individual effects or functional relationships from such data given the myriad of turbine parameters involved. Rotating turbine rigs offer the next best alternative to field data, though they tell us little about blade tip heat transfer in engine service exposure cases. Since most turbine rigs intended for the acquisition of heat transfer data operate without combustion, the considerable challenges of instrumentation are at least somewhat manageable. Metzger et al.[39] measured several local, time-resolved tip heat fluxes in the blade forward region on the flat tips within a rotating, single-stage turbine (Garrett TFE-731-2) at two differing tip clearances. A computational method was used to predict the local tip leakage flow rates, while a simple model of tip gap heat transfer was employed based upon Metzger's earlier works. The comparison of time-averaged data and model showed fairly good agreement. Such agreement would be expected since the blade tip was a thin flat section with leakage flows driven over in a transverse manner at each location. Differences between the data and the model might be attributed to the unsteady effects present in the flow due to wake passing. Blair[40] measured some near-tip time-averaged heat transfer coefficients for a similar geometry blade tip in a low-speed rotating turbine rig. In this case, the blade was of constant cross-section and had a tip clearance gap of 1% blade height. Blair found that the near-tip suction surface heat transfer was about 100% higher than that at midspan, at least over the latter portion of the surface which is affected strongly by the tip leakage vortex. The pressure side near-tip region heat transfer was only slightly higher than the midspan by about 10%. Flow visualization in this rig confirmed the transverse tip leakage paths, and showed the extent of the tip leakage vortex. In a recent study by Dunn and Haldeman[41] using the Allison VBI stage, the time-resolved effect of unsteadiness on blade tip heat transfer has been demonstrated for heat flux gages located both in the cavity of a squealer tip and on the squealer rim surfaces. Time-averaged heat transfer magnitudes were found to be very dependent upon the location, and ranged from values similar to those in the lowest heat flux regions of the blade midspan, to values exceeding even the leading edge stagnation heat flux. The frequency response of the heat flux was also very location dependent, with some areas dominated by the blade passing frequency and others appearing quite disperse in response.

Additional research using rotating turbine rigs is currently underway in various locations. An important set of research results by Dunn and co-workers[42] is being examined at this time. This blade tip data concerns both flat tip and squealer tip geometries on a common turbine stage representing the GE YF120 turbine. Dey and Camci[43] have begun a series of rig experiments to investigate tip heat transfer and improved sealing with tip blowing. Other work sponsored by the US Dept. of Energy[44] with the aid of the Air Force Research Lab[45], is aimed at investigating very detailed tip heat transfer in the Turbine Research Facility at Wright-Patterson AFB, perhaps with the use of pressure and temperature sensitive surface coatings.

Outside of such research, actual engine testing can be used to discern the local temperatures that blade tips experience, for example through the use of temperature indicating paints or metallurgical analysis, but this information must still be calibrated against models describing the full conjugate heat transfer conditions of the blade. While such testing and analysis is common within turbine companies, very few examples have been reported in the open literature.

One recent example of such an analysis and test for a turbine blade tip is that of Gegg et al.[46], in which a 3D CFD analysis was performed including the blade tip gap, with and without tip pressure-side film holes of varying orientation. The CFD gas-side predictions were combined with a finite element model of the blade and correlations for internal heat transfer to determine blade tip metal temperatures. The heat transfer prediction for a flat blade tip with angled pressure-side film holes was compared to thermal paint results from an engine test. The analysis over-predicted the benefit of the tip cooling flows compared to the engine test, highlighting again the very complex nature of the problem. Another study of Takahashi et al.[47] performed similar computational predictions for the conjugate heat transfer of a simple radially cooled turbine blade with tip ejection of coolant. Comparisons were made to the metallurgical aging indications of a service engine blade showing approximate agreement in the suction side downwash region at the tip. This study also attempted to provide some analysis of the sensitivity of regional blade temperatures to variations in the gas temperature profile and cooling flow rate.

Computational Progress in Blade Tip Heat Transfer

Numerical investigations are playing an increasing important role in the study and design of turbine blade tips for both flow and heat transfer considerations. An early work of Chyu et al.[48] used a two-dimensional finite difference solver to predict the flow and heat transfer in rectangular grooves modeling blade tips, with and without the effects of rotation. This work was then extended by Chyu and Schwarz[49] to include the effects of tip cavity injection. More recently, three-dimensional CFD analyses have been performed by Ameri and Steinthorsson[50,51] and Ameri et al.[52,53] showing the predicted effects of tip clearance, tip geometry, and shroud casing for several blade tip designs. The CFD predictions of Ameri et al.[52,53] were performed using the GE E^3 geometry. Comparing these predicted squealer tip and flat tip heat transfer distributions to the very recent experimental data obtained by Azad et al.[33,34] for the same overall tip geometry shows mixed success. While the CFD model does not capture the total cascade geometry in this case, certain heat transfer features are reproduced by the analysis, while others such as the midchord cavity reattachment location are not well predicted. Figure 7a shows one example of such a CFD prediction. While the details of CFD analyses are astounding, there is still very little detailed validation data available for comparison. Moreover, such analyses have yet to incorporate unsteady or multi-stage effects. The only study which has compared steady CFD predictions to detailed experimental data is that of Ameri and Bunker[54]. In this work, the entire stationary cascade of Bunker et al.[29] was modeled numerically. Figure 7b shows a sample of the CFD heat transfer results, which can be compared against the data shown in Figure 6. As this case shows, the CFD predictions are generally very good in form, and perhaps within 15-20% magnitude agreement at best with the experimental data.

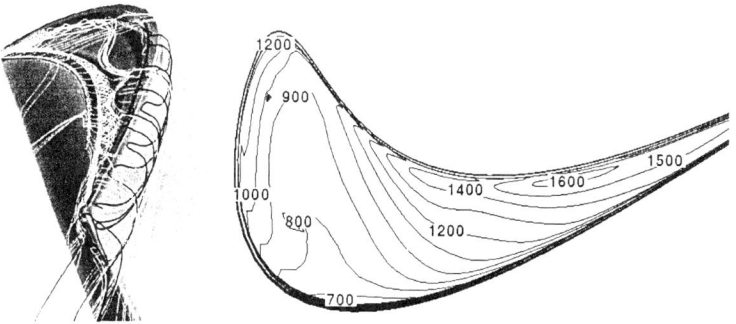

Figure 7a. Computed Flow Traces for E3 Squealer Blade Tip.
Figure 7b. Predicted Blade Tip Heat Transfer Coefficients for Geometry and Flow Conditions of Figure 6, W/m2/K (Reproduced with permission from the Transactions of the ASME).

CONCLUSIONS

Our understanding of turbine blade tip heat transfer today is certainly far beyond the fundamental level of knowledge initiated nearly twenty years ago. In truth however, our ability to correctly perform pre-service blade tip designs rests primarily on our engine experience. The pace of technology advancement in high-temperature turbines has outstripped our investigation of the complex thermal conditions in blade tip regions. While we have concentrated more efforts on the fundamentals of airfoil heat transfer and film cooling, blade tips have advanced through invention and testing. We are at a point now when CFD heat transfer predictions are becoming very attractive and cost effective means for designing blade tips, but this still requires closure of the ever present turbulence modeling issue. Though methods such as Large Eddy Simulation may help to bridge the CFD needs, there still remain many areas requiring research and development to allow a full understanding of turbine blade tips. To date, we have only engine experience to describe the effects of tip film coolant mixing fundamentals, both on blade tips as well as on the airfoil surfaces near the tip. Our knowledge of the effects of unsteadiness on tip heat loads is very limited at this time, covering only a few discrete locations in some uncooled rig tests. There are as yet no reported studies on heat transfer for blade tips utilizing attached shrouds. The ability to predict blade tip heat transfer also rests on a better understanding of the hot gas migration and temperature profiles present in the many types of combustion-turbine systems. The understanding of how heat transfer coefficients and film effectiveness change as the blade tip are altered in service is a wholly unexplored area. As such information becomes more plentiful, innovative solutions will be required to extend blade tip life.

REFERENCES

1. Allen, H.W. & Kofskey, M.G. 1955. Visualization Study of Secondary Flows in Turbine Rotor Tip Regions. NACA Technical Note 3519.
2. Lakshminarayana, B. 1970. Methods of Predicting the Tip Clearance Effects in Axial Flow Turbomachinery. J. of Basic Engineering. **92**: 467-482.
3. Booth, T.C. et al. 1982. Rotor-Tip Leakage: Part I - Basic Methodology. J. of Engineering for Power. **104**: 154-161.
4. Wadia, A.R. & T.C. Booth. 1982. Rotor-Tip Leakage: Part II - Design Optimization Through Viscous Analysis and Experiment. J. of Engr for Power. **104**: 162-169.
5. Moore, J. et al. 1989. Flow and Heat Transfer in Turbine Tip Gaps. J. of Turbomachinery. **111**: 301-309.
6. Yaras, M. et al. 1989. Flow Field in the Tip Gap of a Planar Cascade of Turbine Blades. Journal of Turbomachinery. **111**: 276-283.
7. Sjolander, S.A. & D. Cao. 1995. Measurements of the Flow in an Idealized Turbine Tip Gap. J. of Turbomachinery. **117**: 578-584.
8. Bindon, J.P. 1989. The Measurement and Formation of Tip Clearance Loss. Journal of Turbomachinery. **111**: 257-263.
9. Yamamoto, A. 1989. Endwall Flow / Loss Mechanisms in a Linear Turbine Cascade with Blade Tip Clearance. Journal of Turbomachinery. **111**: 264-275.
10. Kaiser, I. & J.P. Bindon. 1997. The Effect of Tip Clearance on the Development of Loss Behind a Rotor and a Subsequent Nozzle. Paper No. 97-GT-53, IGTI Turbo Expo, Orlando, USA.
11. Merzkirch, W. et al. 1988. A Survey of Heat Transfer in Compressible Separated and Reattached Flows. AIAA Journal. **26**: 144-150.
12. Boelter, L.M. et al. 1948. An Investigation of Aircraft Heaters XXVII - Distribution of Heat Transfer Rate in the Entrance Region of a Tube. NACA TN 1451.
13. Seban, R.A. 1964. Heat Transfer to the Turbulent Separated Flow of Air Downstream of a Step in the Surface of a Plate. Journal of Heat Transfer, Transactions of the ASME: 259-264.
14. Vogel, J.C. et al. 1983. Combined Heat Transfer and Fluid Dynamic Measurements Behind a Backward Facing Step. Paper No. 83-WA/HT-11. ASME Winter Meeting.
15. Orlov, V.V. et al. 1984. Investigation of Turbulence and Heat Transfer in a Separating Flow Behind a Step. Heat Transfer - Soviet Research. **16**: 58-68.
16. Scherer, V. & S. Wittig. 1991. The Influence of the Recirculation Region: A Comparison of the Convective Heat Transfer Downstream of a Backward Facing Step and Behind a Jet in a Crossflow. Journal of Engineering for Gas Turbines and Power. **113**: 126-134.
17. Tsou, F.K. et al. 1991. Starting Flow and Heat Transfer Downstream of a Backward Facing Step. Journal of Heat Transfer. **113**: 583-589.

18. Luzhanskiy, B.Ye. & V.P. Solntsev. 1971. Experimental Study of Heat Transfer in the Zone of Turbulent Boundary Layer Separation Ahead of a Step. Heat Transfer - Soviet Research, **3**: 200-206.
19. Seban, R.A. 1965. Heat Transfer and Flow in a Shallow Rectangular Cavity with Subsonic Turbulent Air Flow. Int. Journal of Heat and Mass Transfer. **8**: 1353-1368.
20. Fox, J. 1965. Heat Transfer and Air Flow in a Transverse Rectangular Notch. Int. Journal of Heat and Mass Transfer. **8**: 269-279.
21. Yamamoto, H. et al. 1979. Forced Convection Heat Transfer on Heated Bottom Surface of a Cavity. Journal of Heat Transfer. **101**: 475-479.
22. Mayle, R.E. & D.E. Metzger. 1982. Heat Transfer at the Tip of an Unshrouded Turbine Blade. In Proc. Seventh Int. Heat Transfer Conf.: 87-92. Hemisphere Pub.
23. Metzger, D.E. et al. 1989. Cavity Heat Transfer on a Transverse Grooved Wall in a Narrow Flow Channel. J. of Heat Transfer. **111**: 73-79.
24. Chyu, M.K. et al. 1989. Heat Transfer in the Tip Region of Grooved Turbine Blades. J. of Turbomachinery. **111**: 131-138.
25. Metzger, D.E. & K. Rued. 1989. The Influence of Turbine Clearance Gap Leakage on Passage Velocity and Heat Transfer Near Blade Tips: Part I - Sink Flow Effects on Blade Pressure Side. J. of Turbomachinery. **111**: 284-292.
26. Rued, K. & D.E. Metzger. 1989. The Influence of Turbine Clearance Gap Leakage on Passage Velocity and Heat Transfer Near Blade Tips: Part II - Source Flow Effects on Blade Suction Sides. J. of Turbomachinery. **111**: 293-300.
27. Kim, Y.W. et al. 1995. A Summary of the Cooled Turbine Blade Tip Heat Transfer and Film Effectiveness Investigations Performed by Dr. D.E. Metzger. J. of Turbomachinery. **117**: 1-11.
28. Yang, T.T. & T.E. Diller. 1995. Heat Transfer and Flow for a Grooved Turbine Blade Tip in a Transonic Cascade. Paper No. 95-WA/HT-29, Int. Mech.E. Congress.
29. Bunker, R.S. et al.. 1999. Heat Transfer and Flow on the First Stage Blade Tip of a Power Generation Gas Turbine Part 1: Experimental Results. J. of Turbomachinery. **122**: 263-271.
30. Bunker, R.S. & J.C. Bailey,. 2000. An Experimental Study of Heat Transfer and Flow on a Gas Turbine Blade Tip with Various Tip Leakage Sealing Methods, In 4th ISHMT / ASME Heat and Mass Transfer Conference, India.
31. Bunker, R.S. & J.C. Bailey. 2000. Blade Tip Heat Transfer and Flow with Chordwise Sealing Strips. In Proceedings 8th ISROMAC Conference, Honolulu, Hawaii.
32. Teng, S. & J.C. Han. 2000. Detailed Heat Transfer Coefficient Distributions on a Large-Scale Gas Turbine Blade Tip. In Proceedings 8th ISROMAC Conference, Honolulu, Hawaii.
33. Azad, G.M.S. et al. 2000. Heat Transfer and Pressure Distributions on a Gas Turbine Blade Tip. Paper No. 2000-GT-194. IGTI Turbo Expo, Munich.

34. Azad, G.M.S. et al. 2000. Heat Transfer and Flow on the Squealer Tip of a Gas Turbine Blade. Paper No. 2000-GT-195. IGTI Turbo Expo, Munich.
35. Urban, M.F. & N. Vortmeyer. 2000. Experimental Investigations on the Thermal Load and Leakage Flow of a Turbine Blade Tip Section with Different Tip Section Geometries. Paper No. 2000-GT-196. IGTI Turbo Expo, Munich.
36. Han, J.C. 1999. Personal communication.
37. Goldstein, R. 1999. Personal communication.
38. Ekkad, S. 2000. Personal communication.
39. Metzger, D.E., M.G. Dunn & C. Hah. 1991. Turbine Tip and Shroud Heat Transfer. J. of Turbomachinery. **113**: 502-507.
40. Blair, M.F. 1994. An Experimental Study of Heat Transfer in a Large-Scale Turbine Rotor Passage. Journal of Turbomachinery. **116**: 1-13.
41. Dunn, M.G. & C.W. Haldeman. 2000. Time-Averaged Heat Flux for a Recessed Tip, Lip, and Platform of a Transonic Turbine Blade. Paper No. GT-0197. IGTI Turbo Expo, Munich.
42. Dunn, M.G. 2000. Personal communication.
43. Dey, D. & C. Camci. 2000. Development of Tip Clearance Flow Downstream of a Rotor Blade with Coolant Injection from a Tip Trench. *In* Proceedings 8th ISROMAC Conference, Honolulu, Hawaii.
44. Kapat, J. 1999. Personal communication.
45. Stevens, C. 2000. Personal communication.
46. Gegg, S.G. et al. 1999. Computational Modeling and Thermal Paint Verification of Film-Cooling Designs for an Unshrouded High-Pressure Turbine Blade. Paper No. 99-GT-330. IGTI Turbo Expo, Indianapolis, USA.
47. Takahashi, T. et al. 2000. Thermal Conjugate Analysis of a First Stage Blade in a Gas Turbine. Paper No. 2000-GT-251. IGTI Turbo Expo, Munich.
48. Chyu, M.K. et al. 1987. Heat Transfer in Shrouded Rectangular Cavities. J. of Thermophysics. **1**: 247-252.
49. Chyu, M.K. & S.G. Schwarz. 1990. Effects of Bottom Injection on Heat Transfer and Fluid Flow in Rectangular Cavities. J. of Thermophysics. **4**.
50. Ameri, A.A. & E. Steinthorsson. 1995. Prediction of Unshrouded Rotor Blade Tip Heat Transfer. Paper No. 95-GT-142, IGTI Turbo Expo, Houston.
51. Ameri, A.A. & E. Steinthorsson. 1996. Analysis of Gas Turbine Rotor Blade Tip and Shroud Heat Transfer. Paper No. 96-GT-189, IGTI Turbo Expo, Birmingham, UK.
52. Ameri, A.A. et al. 1997. Effect of Squealer Tip on Rotor Heat Transfer and Efficiency. J. of Turbomachinery. **120**: 753-759.
53. Ameri, A.A. et al. 1998. Effects of Tip Clearance and Casing Recess on Heat Transfer and Stage Efficiency in Axial Turbines. Paper No. 98-GT-369, IGTI Turbo Expo, Stockholm, Sweden.
54. Ameri, A.A. & R.S. Bunker. 1999. Heat Transfer and Flow on the First Stage Blade Tip of a Power Generation Gas Turbine Part 2: Simulation Results. J. of Turbomachinery. **122**: 272-277.

Unsteady Flow Modelling in Turbine Stage

FRANCESCO MARTELLI

Energetic Department, University of Florence Italy

ABSTRACT:The paper deals with the problems of unsteady flow modelling in turbine stage. The impact of unsteadiness in the design procedure as well as in the real flow phenomena are briefly addressed. After a discussion of the physical aspects of the real flow in turbine stage which differ from the unsteady flow in a single row environment, the paper reports a survey of the actual and more recent approaches to the numerical modelling on the unsteady flow in multiple rows environment. The classical steady models, used for long time in the design procedure, are addressed and their limits are focused. The more complex procedures ranging from quasi-unsteady methods to the fully unsteady methods are reported and investigated in details. The discussion on the specific features of the different approaches is pointed to assess the accuracy, the feasibility, the robustness and the usability (i.e. the computer time and storage requirements). Sample results from the author research activity are reported and presented. Discussion of crucial and open questions to improve the prediction capability of the methods are finally reported as well as some comments on the possibility to joint classical (steady methods) and advanced approaches (fully unsteady methods) to improve the actual design procedure are addressed.

INTRODUCTION

The real flow in the stage of any kind of turbomachinery is highly three dimensional and unsteady for several reasons; the first and more relevant phenomena is related, of course, to the relative motion of rotor and stator airfoils rows, as well as the effect of movement of circumferential and span-wise gradients in total pressure and temperature induced by the upstream airfoils rows, burners etc.; other unsteady phenomena occur which can have significant effect especially on the stage behaviour in off—design conditions , i.e. the vortex shedding at trailing edge, the tip leakage vortex which can be unstable interacting with other unsteady and unstable situation in the flow passages. The effect of unsteadiness has significant effect on the performance of the stage, and actually it is demonstrated that the behaviour of a single row differs from its when located in a multiple row environments. In the turbine blades of the firsts stages of Gas Turbine, where metal temperature are very important, the effect in not only on performance but on the blade heat transfer as well and therefore on the possible life of the bucket itself. Further the unsteady forces can interact with blade elasticity producing flow-induced vibration (flutter) with strong effect on structural survival of the rows.

For long time the stage have been computed, analysed and designed with steady assumption, based on CFD codes that relay on numerical solution of three dimensional steady flow equations through turbine airfoils rows. The complexity of unsteady flow, and unsteady viscous effect modelling have been accounted by some empirical correlations and "experience factors" deduced in each company by the on field results. The meaning of these approach is not always clear and do not reflect properly the real physics

More recently the improvements in computing capacity and the need to increase the efficiency in the last few percentage points, the maximum operating temperature and life of turbine stage has stressed the research activity in the area of unsteady modelling in order to develop more physics-based prediction systems for multi stage turbines.

Several factors have contributed to the rise of such activity both on the experimental and numerical sides as well. The development of short duration test facility (MIT[1], VKI[2], SHULTZ[3] ..) and the related instrumentation, data acquisition and data reduction systems on one side has provide deep and detailed investigations on the unsteady flow phenomena and offering suitable reference test for numerical modelling. On the other side the improvements of computing capability both for the availability of new powerful computer in terms of storage, computing speed, new architecture and for the improved solution algorithms and computer graphics and animation techniques which allow a real time visualisation and significant interpretations of the results itself have provided new powerful modelling tools for predicting unsteady flow behaviour in multi rows configurations.

Physical Unsteady Flow Features

The physical behaviour of unsteady flow in the multi rows environments is characterised by the non uniform outlet flow field produced by the each row which is seen as an unsteady periodic condition from the following blade row. The gap between the adjacent blade rows play an other relevant effect as the blockage of the downstream blades affects the flow in the upstream blade passages which , of course, not constant in time and space . All these periodic distortions can be classified in two main different categories potential and viscous and in more detail as follows:

- Potential interactions: they are related to the velocity and pressure field produced by one row on the adjacent one, and are strongly affected by the axial gap and/or the high flow mach number;
- Shock wave interaction : the shock system produced by one row , which is intrinsically unsteady can extents to the upstream, down stream row appearing as strong unsteady discontinuity ;
- Vortical and entropy wakes: large scale organised flow structures (streamwise vortices) and random moving small-scale flow structure(wakes) , the induced unsteady flow depends on the scale of the up stream distortion;
- Hot streaks: they are produced by non uniform temperature distribution at the combustor outlet and it is typical of gas turbine first stage, the non uniformity is convected downstream from the stator to the rotor and can produce preferred heating of pressure side with respect to suction side; the behaviour of these temperature distortion can be grouped with the previous one according to their scale.

The effects of the last two categories can be well represent by the Kerrenbock[4] and others [5] and it is widely addressed in the review paper of Sharma[6] where the migration effect of the low-high energy particle is reported. The migration of fluid particle can affect the boundary layer characteristic of the following blade through the effect on transition process, the secondary flow generation in the downstream passages and can produce a redistribution of total enthalpy. In the paper[6] various and deep interpretations of the different phenomena are based on experimental test both on streamwise vortices as well as on the wakes and hot streaks behaviour.

Some of the reported phenomena can be treated with unsteady Euler codes once the proper inlet boundary conditions are set up other as wake, transition and heat transfer induced by temperature distortion can only be investigated with full Navier-Stokes code with appropriate turbulence modelling.

NUMERICAL SIMULATION : TYPE OF APPROACHES

Turbine design engineers have utilised many three-dimensional steady CFD codes with significant success over the past twenty years. Both Euler and Navier Stokes codes have been used to improve turbine performance allowing to control loading and exit flow angles distribution along the blades. Loss prediction was and is not jet completely assessed because of relevant lack of proper physical modells in Turbulence and transition. The evaluations of performance in a multistage environment has started with Euler codes, using appropriate inflow boundary conditions and loss evolution models like that proposed by Horlock[7] and further used by Denton[8]. Further improvements in the performance evaluation of multistage turbine require on one side improvements of physical models for loss and heat load (Turbulence and transition) and on the other hand account for unsteady effect which can not properly be accounted for in the steady code.

The research activity in the area of unsteady effect has been focused over the last fifteen years on three different approaches of increasing complexity and simulation capability:
- Steady models based on the Adamczyk averaging procedure (determinist stress);
- Quasi unsteady methods where the unsteadiness is computed in hybrid way ranging from linear decomposition forms to other numerical approaches based only on the unsteady Boundary Conditions.
- Fully unsteady calculation with different features of the numerical solvers and the physical models implemented in the code.

At present time we are more interested in the study of turbine stage aerodynamics and heat transfer; these items require a fully unsteady simulation in order to catch crucial points of the heat transfer phenomena and dissipation mechanism in the unsteady boundary layer and the losses related to the vortex shedding as well. We will briefly report the first two approaches as well to give a complete picture of the unsteady modelling.

Steady Modelling

This approach is based on the fundamental work of Adamczyk[9], where he introduces three different averaging procedure: "Ensemble Averaging", "Time Averaging" and "Passage to Passage Averaging" , and using these procedure in sequence with the help of special "gate functions" leads to the definition of models hierarchy[10] of decreasing complexity as reported in the fig. 1. The final results of the procedure is the a set of "AVERAGE PASSAGE EQUATION SYSTEM" which can be applied to each single blade row and retains the effect of unsteadiness through new terms: Deterministic Stress (DST), Body Forces, etc. which have to be modelled some way. Several approaches have been suggested to model the new terms by different authors: Adamczyk and other[10] ,on the basis of a previous work of Adkins[12], propose a simple procedure based on the use of through flow similar correlations, in further works[11] a more precise approach is developed through an iterative procedure to compute the DST accounting the non uniform velocity distribution in circumferential direction. Many other authors[15,16,17,18,19] have proposed different attempt to closure the Average Equations and a wide discussion of it is reported in Adamczyk[20]. A special case of this approach is the intensively used method of "Mixing Plane" which disregards all the unsteady effect an treat each row, single passage, with boundary conditions from the other row with a mixing process of the transported quantities. The mixing process can be slightly different among authors[21,22] but our experience is that it has no relevant effect on the results. Interesting comparison are reported in Adamczyk[11] and in Turner[23]

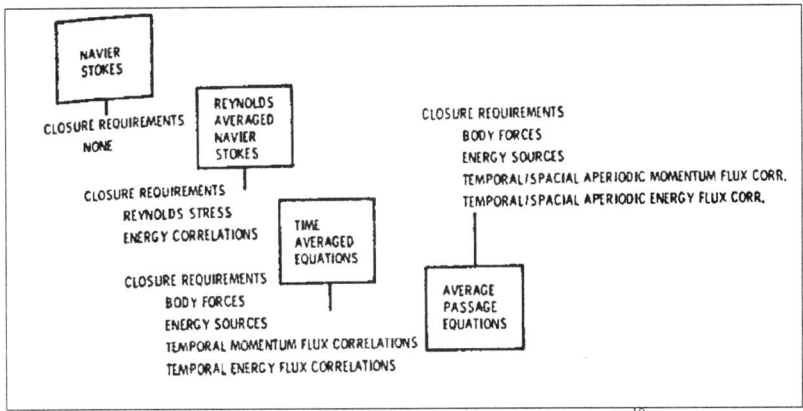

Fig. 1 models hierarchy (by courtesy of Adamczyk [10])

Quasi Unsteady Modelling

The aim of this kind of approaches is the attempt to get information on the unsteadiness but at the same time to try to reduce computational effort. Part of them have been developed to study aeroelasticity and are based on a process of linearisation of the Euler equations; the unsteadiness is considered as a small perturbation of the steady flow and therefore strong simplification can be introduced in the code, leading to a linear system of equation in the parameters of the spinning mode selected. Whitehead, Crawley and Hall[24,25] have work intensively in this direction mainly for aeroelasticity and noise studies. Anyway Orkwiis suggests to use this approach to evaluate the DST needed in the Average Passage Equation and despite the fact of the limitation introduced by the linear hypothesis i can be a good first approximation.

A different approach is based on the use of unsteady boundary condition for the single blade row, deduced by the early theory of Sears (1941) which read the unsteadiness as the results of an "incident gust" (wake) and the aerodynamic response of the blade; in other terms the flow is splitted in a part , Time averaged, and in unsteady parts whose excitation is computed by the steady solution.. In the work of Chen[26] an unsteady code working for the single row is developed and the unsteady boundary conditions are simulated through the periodic passing wake. The use of the Riemman invariant to manage the boundary complete the method which results in a very fast code.

A further improvements in the direction of economical calculations with improved accuracy is the " Loosely Coupled Blade Row" approach, Dorney and others[27] establish an unsteady code by the assembling of a single row unsteady code applied to the different rows of the turbine. The interface between the single row is treated with special care to allow the transmission of the information The results is a fast code with reasonable accuracy

Fully Unsteady Modelling

The fully unsteady modelling of turbomachinery stage has been approached at the beginning of the 90' by some researchers, first of them Giles[28,29,30] with an Euler code and Rai [31,32] with a Navier-Stokes solver. The main problems which researchers have to face with were the followings:
- How to treat the boundary between stator and rotor, i.e how to transfer data, and flux across a moving boundary?
- How to manage the periodic boundary in space and time; in other words when the pitches of the rotor and stator are different ,which always happens in practice, how can be organised the number of passages to be computed and the transfer of data from on periodic boundary to the other ?
- How can the far boundaries be treated in order to avoid the introduction of spurious perturbations in the tine periodic solution we are looking for ?
- How the unsteady physical viscous effects can properly be modelled .

Part of these problems have been approached with success from the first authors and mainly the last one is the most crucial and up-to day relevant. Further problems arise from the need to have reasonable computing time, time accuracy and space resolutions, data storage, data reduction, but these have found different answers according to the different authors and to the numerical algorithm used.

Giles proposed, in his previous works, a solutions to the first three problems:
For the moving boundary, in 2D, he proposed the use of special shearing cells which fill up the space between stator and rotor and modify their forms because of the relative movements of one domain respect to the other. Fig.2 shows an example of how it works in 2D, of course its extension to 3d is not straight forward.

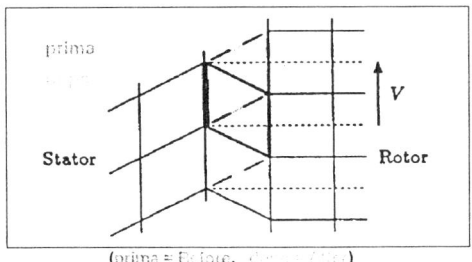

Fig.2 Sharing cells at interface between stator/rotor

For the problem related to the different pitches of rotor and stator Giles introduces the interesting concept of inclined computational grid, so that what happens on a periodic boundary at time t corresponds to the same at the other periodic boundary at time t+ΔT where $\Delta T = (Ps - Pr)/V$ and Ps, Pr, are stator, rotor pitches and V the rotor speed. In other words, if u is an unknown, it can be stated that : $u(x,y,t) = u(x, y + Ps, t + \Delta T)$ for the rotor (see Fig.3) .By the use of these approach is possible to work in new computational plane where a new time is introduced t' = t-λy (y is the tangential coordinate) and λ = ΔT /P . The new set of equations are solved in these computational plane where periodic conditions are easily fulfilled and only one stator and rotor passages grids are used. The approach becomes more complex in 3D and with viscous terms and few authors [33] have used it. On the other hands it is quite elegant and reduces the number of passages to be computed to the minimum and avoid any approximation on the geometry as it can be needed to maintain the computations reasonable, see Rai [31,32]

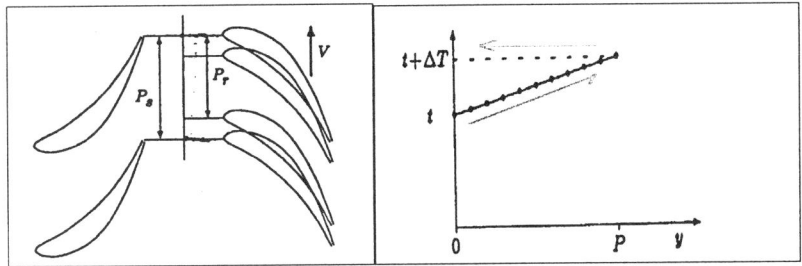

Fig. 3 Shetch of the inclined plane features

The third problem on the far boundary condition is addressed by Giles with the use of non reflecting B.C. based on the Riemman characteristic theory and which are more deeply treated by Lakshminarayana,[34] who present a deep investigation of an unsteady solver applied to an 3D Euler simulation of the wake effect on a rotor.

One of the first approaches to the viscous simulation is due to Rai[31] (thin layer approximation) who used multi zone grid approach to fulfil the B.L. resolution requirement and at the same time used overlapping grids in the interface boundary (Fig.4). This results, on one side, in a better accuracy in the gap area, especially when it is narrow, but on the other hand increases the number of operation for data transfer. The pitches difference (22 stators → 28 rotors) was treated by changing the rotor geometry in order to have the same pitch as the stator one and to use only one passage for each row. The rotor enlargement leads to inaccurate results, but main feature of the viscous effect could be retained despite the fact the turbulence models came from steady formulations. It was the beginning of the 90' and the computer capability did not allowed treatment of very large grids.

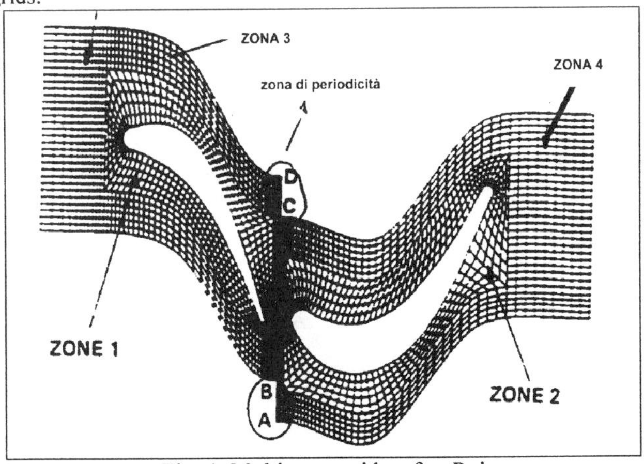

Fig. 4 Multi-zone grids after Rai

Successively several formulations - among which we can report someone[33-38,40,41,43-45] - have been presented for the simulation of fully unsteady flow in turbine stage. They differ each other for several features, most of them related to the numeric and algorithms used, but the main characteristic are just provided in the above mentioned papers. Few authors[33,45] use the Giles inclined plane for computing single passages and most of them try to get good approximation on the pitches differences using reduced number of passages. In the interface boundary the procedure to transfer data depend on the type of grid and accurate interpolation are used. Of course this is easier in 2D with structured grids and become more complex in 3D and more and more with unstructured one. The numerical features that distinguish the different formulation can be stated as follows:

- Physical time integration and accuracy : direct integration, fix time step and runge-kutta or other explicit, implicit scheme, or dual time steps, classical Time marching procedure applied to unsteady physical equations; this takes advantages of the well known speeding up procedures, it is easy to be implemented in existing code, and can select the more appropriate physical time discretization (2nd, 3rd order,...)
- Type of discretization & grid: Finite Volume/Difference/Elements; structured/ unstructured, multi domain approach, cell centered, vertex centered,...etc. stretching ratio, aspect ratio etc. Many of these features are not always addressed in the papers
- Numerical solvers and scheme used: Time marching explicit/implicit schemes, pressure correction schemes (real few), flux discretization (up-wind, centered Jameson, Roe ,etc..); matrix solvers, ADI, Approximate Factorisation, LU, GMRES ,etc.
- Boundary, interface treatment and inizialisation : non reflecting B.C. or other; linear interpolation, overlapping flux interpolation etc; steady solution or approximate unsteady guess.

The real critical points depend on the target of the simulation, which can be, classified as follows:
1. Unsteady load prediction and span wise secondary flow behaviour; accurate Euler solver with appropriate B.C.[29,33] can properly approach the problem with relatively coarse meshes.
2. Unsteady loss prediction; this requires Navier-Stokes Solver with appropriate physical models and relevant number of mesh points to account for viscous effect especially if separated flow exists in off-design operations; Further to the appropriate turbulence/transition model, which is still an open topics, accurate numerical resolution methods are required. An interesting comparison on the different capability of different approaches is reported in Dorney[17], where losses and cpu time are reported for various code:

	FCBR	SCBR	SSBR	LCBR
η_{tt}	0.894	0.902	0.915	0.904
CPU TIME(min)	FCBR	SCBR	SSBR	LCBR
IGV	-	-	1704	2454
ROTOR	-	-	2401	3391
Total	33744	7582	4105	5845

FCBR= Full Unsteady, SCBR= Mixing plane-steady, SSBR= Unsteady Single Row, LCBR[27]

3. Unsteady heat transfer, leakage, and internal flow prediction; This requires N-S solvers with very fine meshes ($y+ < 1.0$ for good heat transfer prediction), physical models with special care for transition onset and development (still open point) and appropriate strategy to maintain mesh point and cpu time within reasonable value[39,42].

A wide comparison of the results of different authors is quite heavy and out of the scope of this work that is focused on the features, up-to-day capabilities and future improvements of the unsteady modelling Therefore before drawing some conclusion and suggestion on future we would like to present some results from our research group which are well positioned in present scenario of calculations.

SOME EXPERIENCES AND RESULTS

The activity of the CFD group at the "Energetic Department in Florence" have been characterised by several directions in the Computational Fluid Dynamics area, N-S solvers development, turbulence and transition modelling and implementation in analysis/design procedures, etc.. The main interest, related to the Gas Turbine, is since about ten years, the aerodynamics and heat transfer in unsteady flow. Two Navier-Stokes codes, in house developed, have been used and are still under continuos improvements:

XFLOS[40] ; Implicit Time marching, structured grid (I type),with FDM for spatial discretization (artificial dissipation, second order accuracy) and various iterative scheme available based on the approximate factorisation technique (ADI ,   scalar factorisation after Pulliam, Diagonal Dominant ADI, etc.). The code have several turbulence models available , always two equations models   with Kato-launder correction and Durbin' realizability constrain, k-ω model is the most used. Low mach regimes are treated by a preconditioning technique.

HYBFLOW[55,56] ; Implicit Time marching, unstructured hybrid-grid (multi-blocks type),with FVM for spatial discretization (up-wind,Roe and linear reconstruction) and Newton iterative scheme plus relaxation and linear solvers GMRES, approximate LU factorisation and other techniques. Same turbulence models as the previous one with some further algebraic and non linear k-ε models. Low mach regimes are treated by a preconditioning as well the transonic one with the TVD technique.

Both the codes use the implicit dual time stepping to treat unsteady flow with second order accuracy in time. If U in the unknown vector, t the physical time (τ the iterative time) it sounds as follows:

$$\frac{\partial U}{\partial t} + R(U) = 0 \; ; \quad \frac{\partial U}{\partial \tau} + \left[\frac{\partial U}{\partial t} + R(U)\right] = 0 \quad R^*(U^{n+1}) = \frac{\partial U}{\partial t} + R(U^{n+1})$$

and therefore:

$$\frac{\partial U}{\partial \tau} + R^*(U) = 0$$

the time derivative is compute with a second order accurate formula:

$$\frac{\partial U}{\partial t} \approx \frac{1}{\Delta t} \cdot \left[\frac{3}{2} U^n - 2U^{n-1} + \frac{1}{2} U^{n-2}\right]$$

The XFLOW code has been successfully applied to Q3D rotor/stator interaction[37,40]

The computer simulations of the stator-rotor interaction required a slide deformation of the rotor to model three rotor and two stator vanes instead of the correct number i.e. : 64/43 vanes. The physical time step is of crucial importance for the success of the simulation and must be selected on the basis of the following expected frequencies encountered in the flow field:

i)   The rotor passing frequency (7 kHz, according to RPM))

ii)   The stator passing frequency (4.6 kHz, according to RPM)

iii) The stator trailing edge vortex shedding

iv) The rotor trailing edge vortex shedding

According to the expected Strouhal number after Sieverding and Heinemann[58], these resulting vortex shedding frequencies for the stator and the rotor are approximately $f_{st}=101$kHz and $f_{rot}=40$kHz. Preliminary inviscid tests have confirmed the ability of the algorithm to capture the unsteadiness of the flow field and compare correctly with experiments. The frequency analysis show an excellent agreement on the first two value i) and ii). The physical time step used in most of the calculations is Δt_p of 4.29×10^{-6} [s], which requires 100 physical time steps for a periodic block passage and larger than that one (order of $\Delta t=9\times 10^{-7}$ [s]) required to capture the vortex shedding, but still, the flow angle and pressure unsteadiness are in excellent agreement with the expected experimental value. In Figure 5 they compare the computed unsteady pressure profiles with the measurements by Sieverding et al.[57] and the agreement is real good. Viscous flow calculations have been performed and used to study the unsteady and averaged heat transfer parameters and losses as well. In fig. 6 the average values of Nusselt number Nu are reported against the experimental one and the agreement is still good , but some further work has to done to improve the prediction capability. The effect of the passing wakes on the rotor heat transfer can be traced in Figure 7 which

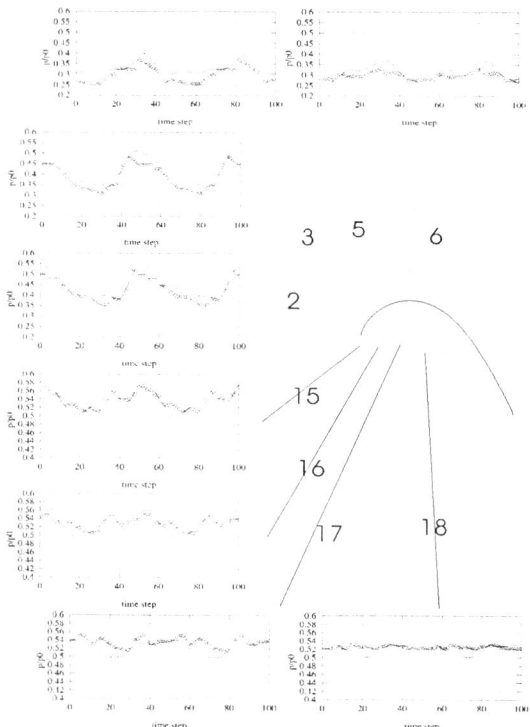

Figure 5. Unsteady pressure profiles on the rotor blade

shows how the peak of Nu is moving downstream. It is now easy to follow the development of the transition point in time. The triggering effect of the wake is not very large since the transition point on the suction side ranges between 0.2<S/Smax<0.38. This is understandable on account of the large turbulence level which is mainly responsible for the onset of transition, at least in the computer simulation code.

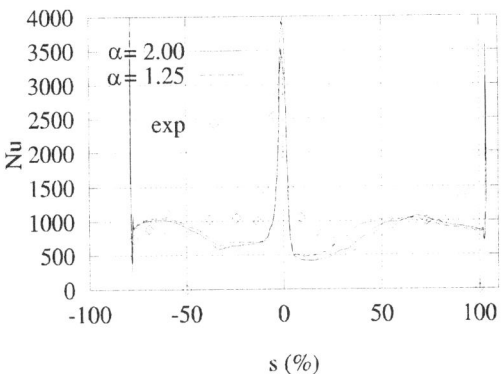

Fig. 6 Avearge heat transfer on the rotor (α = Turb. Model parameter)

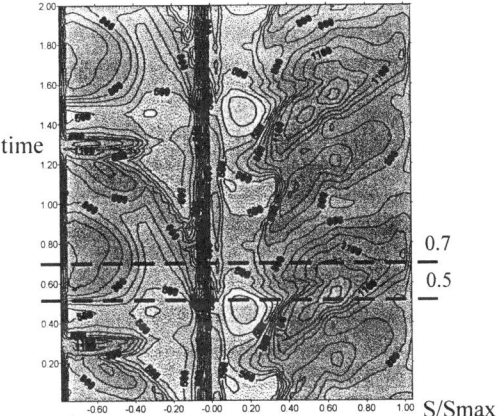

Fig. 7 Unsteady Nusselt number

The unsteady flow field in a transonic turbine stage was computed and the results were compared favorably with experiments in terms of blade load, rotor blade pressure and Nusselt number distribution. The rotor blade load fluctuations were reproduced with a good degree of accuracy. While in terms of heat transfer, the accuracy of the predictions deteriorated, but still, the computations seem to capture the essential features of the flow. Although the code is able to capture the essential features of the flow, more work is needed in the field of unsteady transition. An other interesting comparison is performed with the HYBFLOW code in comparing the steady and unsteady calculations on the same stator of previous work, that of the a BRITE project, where the use of unstructured code allows a better description of inside passages (Cooling passages of the stator) and a reduction of mesh points with increasing accuracy at the wall, Fig. 8 shows part of the 3D grid. In Fig. 9 the convergence behavior of the code is reported, each pick corresponds to a physical time step. In Fig.10 we compare the steady calculation (black) and unsteady ones (symbols) with experiments and it appears that the problem in capturing the first recompression, that all the steady codes (the other involved in the project show the same discrepancy) exhibit, disappears in the unsteady calculation which of course being transonic does not change significantly in the sonic region.

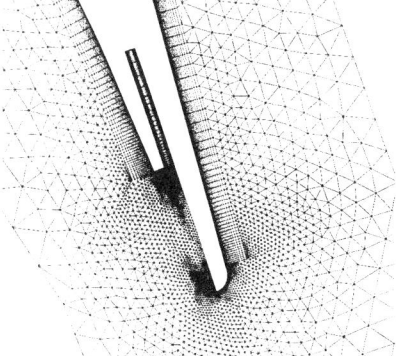 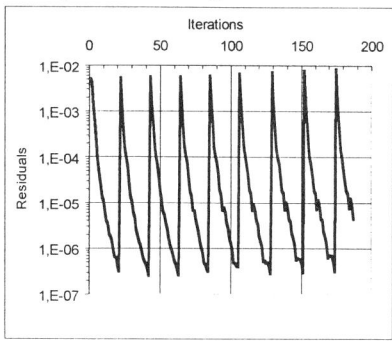

Fig. 8 2D view of 3D unstructured grid   Fig. 9 Convergence History

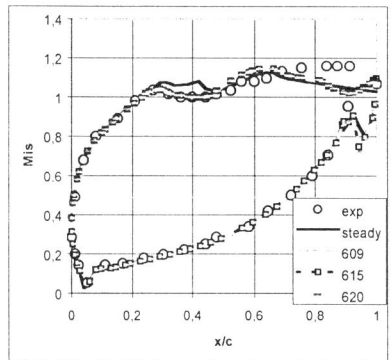 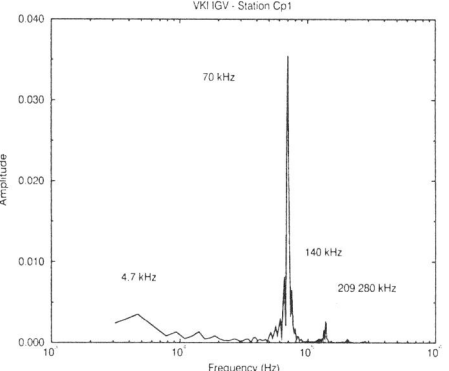

Fig. 10 Comparison steady/unsteady calculations   Fig.11 Pressure frequency spectrum

The calculation reports a good agreement on the various frequency expected (fig 11), the rotor passing and the vortex shedding as well. New Calculation are in progress on the fully 3D stator/rotor interaction, and preliminary investigations on the single rotor with the passing wake have shown the needs in terms of computational effort in order to capture viscous effect and heat transfer as well. Structured code requires more than 1 million mesh points per passage to get $y^+ \approx 2$, but is faster, Unstructured one requires less than half points, $y^+ <0.2$ but is slower ! Appropriate strategy and movements towards parallel computing is in progress.

COMMENTS AND FUTURE WORK

It has been shown, from previous examples and from literature references[37-45], that at present time exists the capability to run 3D unsteady stage calculations to

capture the main feature of the flow field. Referring to the above mentioned targets it can be stated that:
- Unsteady load and potential effect can be predicted with good accuracy;
- Unsteady loss and heat transfer can be predicted qualitatively (trends and main features) but still work has to be done to improve results and to result in better reliability for the use in design procedure.

To reach a higher usage capability the methods requires work on the numeric, speed up of computing and grid arrangements, but the most relevant and crucial topics for future research to improve accuracy of unsteady calculations is the unsteady modelling of transition and turbulence. As we have seen in stage calculations, the interaction between stator and rotor has a deep impact on the boundary layer state. Although the stator experiences a marginally unsteady behaviour (Michelassi et al)[42], it is the rotor where the boundary layer is highly perturbed. In LP turbine stages the perturbation is mostly given by wakes and mild pressure gradients as shown by Addison and Hodson[46]. Hodson[47,48] analysed the becalmed region present in LP turbines and proposed a simple simulation method to be cast into a 2D model. The influence of wakes on the boundary layer state has been studied in details, among many others, by Funazaki and Koyabu[49], who analysed the effect of pressure gradients on transition. Fan and Lakshminarayana[50] presented a model able to compute unsteady boundary layers subject to wakes in turbulent and transitional regime. A two-equation turbulence model is able to give a fair description of transition. The boundary layer state in axial compressor is of crucial importance to control stall. In this respect Solomon et al.[51] propose to use a correlation based on integral parameters, derived from that of Abu-Ghannam and Shaw[52], which proves appropriate for unsteady flows. In HP turbines flow field complexity is generally increased by the presence of shocks. Denos et al[53] presented a detailed investigation on the boundary layer state on the rotor subject to the unsteady wakes and shocks. The analysis proved that, at least under the test rig operating condition, the boundary layer experiences sharp and/or deferred transition depending on the complex interaction between stator and rotor. Martelli et al[42] and Denos et al[54] modelled the same geometry. The simulation of the flow proved that a two-equation turbulence model together with a transition criterion based on integral parameters is able to reproduce the heat transfer on the rotor blade with acceptable accuracy.

In conclusion, the prediction of unsteady transition in LP and HP turbomachines is feasible with turbulence and transition models generally developed and tested for steady flows, but for improved and more general simulation work has to be done to assess the models presently available and to improve their capability against the experiments which now become more and more detailed and available. Prospective exists in the use of intermittency functions transport or in the direction of LES once improved computational capability will help in CPU time reduction.

ACKNOWLEDGEMENTS

The author want grateful acknowledge the support of EEC, in the frame of the BRITE-EURAM Program, IMT Area 3 Turbine Project, the Italian Ministry of Scientific Research MURST and Italian Space Agency (ASI), to the researches which some of the results of has been presented here. He thanks his colleagues Prof. Michelassi and Dr. Adami for the support in preparing, discussing the material collected with the help of Mr. Zaccaria.

REFERENCES

1 Epstein, A.H., Guernette, G.R., and Norton, R.J.G., 1984, "The MIT Blowdown Turbine Facility, "ASME Paper No. 84-GT-116
2 C.H. Sieverding, T. Arts, R. Dénos, J.-F. Brouckaert, 2000 "Measurement techinques for unsteady flows in turbomachines". Experiments in Fluids 28 PP 285-321
3 Schultz, D. L., Jones, T.V., Oldfield, M.L. G., and Daniels, L.C., 1977 "A New Transient Facility for the Measurement of Heat Transfer Rates". Conference Proceedings No. 220, High Temperature Problems in Gas Turbine Engines, PP 33-1 to 33-27
4 Greitzer, E. M., and Tan, C.S., 1987. "Unsteady Flow in Turbomachinery: Basic Phenomena and Practical Aspects, "Updated Version of the Special Lecture Given at the International Gas Turbine Congress, Tokyo
5 Kerrebrock, J. L., and Mikolajczak, A. A., 1970 "Intra Stator Transport of Rotor Wakes and Its Effect on Compressor Performance",ASME J.OF ENGINEERING FOR POWER, Vol.92, PP. 359-370
6 O.P.Sharma, G.F.Pickett, R.H.Ni, 1992, "Assesment of unsteady flows in turbines," ASME JOURNAL OF TURBOMACHINERY, VOL.114, pp. 79-90.
7. J.H. Horlock 1971 "On the Entropy Production in Adiabatic Flow in Turbomachines" ASME Trans. Journ. Of Basic Eng. 93D,587.
8 J.D.Denton, 1986, "The use of distributed body force to simulate viscous effects in 3D flow calculations", ASME paper No. 86-GT-144
9 J.J.Adamczyk, 1985, "Model equation for simulating flows in multistage turbomachinery," ASME paper n.85-GT-226
10 J.J.Adamczyk, R.A.Mulac, M.L.Celestina, 1986 "a model for closing the inviscid form of the average –passage equation system," ASME J. OF TURBOMACHINERY, VOL.108, PP. 180-186
11 J.J.Adamczyk, 1991 "A mathematical constraint placed upon inter-blade row boundary conditions used in the simulation of multistage turbomachinery flows," AGARD Conf. Proc. 510 CFD Technique for Propulsion
12 G.G.Adkins, Jr., I.H.Smith, Jr., 1982, "Spanwise mixing in axial-flow turbomachines", ASME JOURNAL OF TURBOMACHINERY, VOL.104, PP. 97-109
13 J.J.Adamczyk, M.L.Celestina, T.A. Beach, M. Barnett, 1990, "Simulation of three- dimensional flow within a multistage turbine", ASME JOURNAL OF TURBOMACHINERY, VOL.112, PP. 370-376
14 J.J.Adamczyk, R.W.Johnson, "Kinetic energy equations for the average-passage equation system", AIAA, JOURNAL OF PROPULSION AND POWER, VOL.5, No.2, PP. 252-254
15 A.P.Saxer, M. B.Giles, 1994, "Prediction of three-dimensional steady and unsteady inviscid transonic stator|rotor interaction with inlet radial temperature nonuniformity",ASME JOURNAL OF TURBOMACHINERY, VOL.116, PP. 347-357
16 G.Fritsch, M.B. Giles, 1995, "An asymptotic analysis of mixing loss", ASME JOURNAL OF TURBOMACHINERY, VOL.117, PP. 367-374
17 D.J.Dorney, O.P.Sharma, 1997, "Evaluation of flow field approximations for transonic compressor stages," ASME JOURNAL OF TURBOMACHINERY, VOL.119, PP. 445-451
18 J.J.Adamczyk, I.K.Jennions, 1997, "Evaluation of the interaction losses in a transonic turbine HP rotor/LP vane configuration", ASME JOURNAL OF TURBOMACHINERY, VOL.119, PP. 68-76

19 C.M.Rhie, A.J.Gleixner, D.A.Spear, C.J.Fischeberg, R.M.Zacharias, 1997, "Development and application of amultistage Navier Stokes solver –Part 1: multistage modeling using bodyforces and deterministic stresses;", ASME JOURNAL OF TURBOMACHINERY, VOL.119, PP. 445-451

20 J.J.Adamczyk, 2000 "Aerodynamic Analysis of Multistage Turbomachinery Flows in Support of Aerodynamic Design ," ASME JOURNAL OF TURBOMACHINERY, VOL.122, APRIL 2000 PP. 189-217

21 J.D.Denton, 1992, "The calculation of three-dimansional flow through multistage turbomachines," ASME JOURNAL OF TURBOMACHINERY, VOL.114, PP. 18-26

22 W.N.Dawes, 1992, "Towards improved throughflow capability : the use of 3D viscous flow solvers in a multistage environment," ASME paper No. 90-GT-18

23 G.Turner, 1996, "Multistage turbine simulations with vortex-blade interaction," ASME JOURNAL OF TURBOMACHINERY, VOL.118, PP. 643-651

24 K.C.Hall, W.S. Clarck, 1993, "Tlinearized Euler Prediction of Unsteady small Disturbance Flows in Cascades," AIAA JOURNAL, VOL.31,NO.5 PP.890-900

25 K.C.Hall, P.D.Silkowski, 1997, "The influence of neighbouring blade rows on the unsteady aerodynamic response of cascades," ASME JOURNAL OF TURBOMACHINERY, VOL.119, PP. 85-93

26 J.P.Chen,M.L.Celestina,J.J.Adamczyk, 1994, "A new procedure for simulating unsteady flows through turbomachinery blade passages," ASME paper n. 94-GT-151

27 D.J.Dorney, R.L.Davis, O.P.Sharma, 1996, "Unsteady multistage analisys using a loosely coupled blade row approach," AIAA, J. OF PROPULSION AND POWER, VOL.12, NO.2, PP. 274-282

28 M.B.Giles, 1988, "UNSFLO : A Numerical Method for Unsteady Flow in Turbomachinery" MIT Gas Turbine Lab. Cambridge MA TR 195

29 M.B.Giles, 1988, "Calculation of Unsteady Wake /Rotor Interaction" AIAA, JOURNAL OF PROPULSION AND POWER, VOL.4, NO.4, PP. 356-362

30 M.B.Giles, 1990, "Stator/rotor interaction in a transonic turbine," AIAA, JOURNAL OF PROPULSION AND POWER, VOL.6, NO.5, PP. 621-627

31 M. M. Rai, R.P. Dring, 1987, "Navier-Stokes Analysis of the Redistribuition of Inlet Temperature Distorsion in a Turbine" AIAA Paper, N. 87-2146

32 M.M.Rai,1989,"Three-dimensional Navier Stokes simulations of turbine rotor-stator interaction, Part1-methodology" AIAA, J. OF PROPULSION AND POWER,VOL.5,NO.3,PP. 305-311

33 R. E. Walraevens, H.E. Gallus, A.R.Jung, J.F. Mayer, H. Stetter, 1998, "Experimental and computational study of the unsteady flow in a 1,5 stage axial turbine with enphasis on the secondary flow in the second stator", ASME paper No. 98-GT-254

34 S.Fan, B.Lakshminarayana, 1996, "Time accurate euler simulation of interaction of nozzle wake and secondary flow with rotor blade in an axial turbine stage using nonreflecting boundary conditions," ASME JOURNAL OF TURBOMACHINERY, VOL.118, PP. 663-678

35 D. J. Dourney, R.L. Edwards, N. K. Madavan, 1990, "Unsteady Analysis of Hot Streak Migration in a Turbine Stage" AIAA Paper N. 90-2354

36 J. Zeschky, H.E. Gallus, 1993, "Effects of stator wakes and spanwise nonuniform inlet conditions on the rotor flow of an axial turbine stage",ASME J. OF TURBOMACHINERY, VOL.115, PP.128-136

37 V. Michelassi, F. Martelli, P. Adami P., 1996 - "An Implicit Algorithm for stator-rotor interaction Analysis" - presented ASME Turbo Expo June 10-13, Birmingham,UK

38 F. Martelli, V. Michelassi, (invited lecture) 1998 "Numerical Simulation of Stator-Rotor Interaction in BRITE Turbine stage", Lecture Notes on "Blade Row Interference Effects in Axial Turbomachinery Stages", Von Karman Institute for Fluid Dynamics, 9-12 February

39 V. Michelassi, F. Martelli, 1998 **"Modelling of unsteady heat transfer in a transonic turbine stage"** 29th AIAA Fluid Dynamics Conference. Albuquerque Convention Center - June15-18

40 F. Martelli, V. Michelassi, 1998 - " Unsteady Flow Simulation in Turbine Cascade " ECCOMASS 98 Athens September 7-11

41 R.Emunds, I.K.Jennions, D.Bohn, J.Gier, 1999, "the computation of adiacent blade-row effects ina 1,5-stage axial flow turbine", ASME JOURNAL OF TURBOMACHINERY, VOL.121, PP. 1-10

42 V. Michelassi, F. Martelli R. Denos, T. Arts, C.H. Sieverding, 1999 - "Unsteady Heat Transfer in Stator-Rotor Interaction by Two Equation Turbulence Model", Transaction of the ASME – Journal of Turbomachinery, Vol. 121, pp. 436-447.

43 L. He, 1999 "Three dimensional unsteady Navier-Stokes analysis of stator-rotor interaction in axial-flow turbines", Third European Conference on Turbomachinery Fluid Dynamics and Thermodynamics" London 2-5 March, 1999, PP. 289-306

44 M.Von Hoynengen-Huene, J. Hermeler, 1999, "Comparison of three approaches to model stator-rotor interaction in the turbine front stage of an industrial gas turbine" Third European Conference on Turbomachinery Fluid Dynamics and Thermodynamics" London 2-5 March, PP. 307-322

45 J.E. Krysinsky, A. Smonly, J.R.Blaszczak, H.E.Gallus, 1999, "Stator wake clocking effects on three-dimensional unsteady flow in a two-stage-low-pressure turbine" , Third European Conference on Turbomachinery Fluid Dynamics and Thermodynamics" London 2-5 March, PP 323-332

46 J. S. Addison, H.P. Hodson, 1989 "Unsteady Transition in axial flow turbine", ASME GT-289/290.

47 H.P. Hodson, 1990, "Modelling Unsteady Transition and Its Effect on Profile Loss" ASME J. Turbomach, 112 PP. 691-701

48 V. Shulte, H.P. Hodson, 1997 "Prediction of the becalmed region for LP turbine profile design", ASME 97-GT-398.

49 K. Funazaki, E. Koyabu, 1998, "Effects of periodic wake passing upon flat-plate boundary layers experiencing favourable and adverse pressure gradient", ASME GT-114.

50 S. Fan, B. Lakshminarayna, 1994, "Computation and simulation of wake-generated unsteady pressure and boundary layers in cascades, part 1: description of the approach and validation", ASME GT-140.

51 W.J. Solomon, G.J. Walker, J.D. Hughes, 1998, "Periodic transition on an axial compressor stator – incidence and clocking effects. Part II – Transition onset predictions", ASME GT-364.

52 B.J. Abu-Ghannam, R. Shaw, "Natural transition of boundary layers – the effects of pressure gradient and flow history", J. of Mechanical Engineering Science, Vol. 22 No. 5, pp. 213-228.

53 R. Denos, C.H. Sieverding, T. Arts, J.F. Brouckaert, G. Paniagua, V. Michelassi, 1999, "Experimental Investigation of the Unsteady Rotor Aerodynamics of a Transonic Turbine Stage", Third European Conference on Turbomachinery Fluid Dynamics and Thermodynamics" London 2-5 March IMECHE Journal.

54 R. Dénos, T. Arts, G. Paniagua, V. Michelassi, F. Martelli, 2000, " Investigation of the Unsteady Rotor Aerodynamics in a Transonic Turbine stage". ASME, Journal of Turbomachinery.

55 P. Adami, V. Michelassi, F. Martelli,1998 "Performances of a Newton-Krylov scheme against implicit and multigrid solvers for inviscid flows" 29th AIAA Fluid Dynamics Conference. Albuquerque Convention Center - June15-18, 1998

56 P. Adami, F. Martelli, V..Michelassi ,2000 "Three-Dimensional Investigations for Axial Turbines by an Implicit Unstructured Multi-block Flow Solver"" ASME TURBO 2000 8-11 May 2000 Munich , Germany

57 C.H. Sieverding, R. Dénos, T. Arts, J.F. Brouckaert, G. Paniagua,1998 "Experimental Investigation of the Unsteady Rotor Aerodynamics and Heat Transfer of a Transonic Turbine Stage, VKI Lecture Series on "Blade Row Interference Effects in Axial Turbomachinery Stages", February 1998,

58 C.H. Sieverding, H. Heinemann,1989 "The influence of Boundary Layer State on Vortex Shedding from Flat Plates and Turbine Cascades", 89-GT-296, 1989.

The detailed structure and behavior of discrete cooling jets in a turbine

PROFESSOR FRANCIS LEBOEUF

Ecole Centrale de Lyon, Laboratory of Fluid Mechanics and Acoustic, UMR CNRS 5509, 36 avenue Guy de Collongue, 69131 ECULLY Cedex, France

Dr. OLIVIER SGARZI

Snecma, Centre de Villaroche, 77556 - MOISSY CRAMAYEL, France

ABSTRACT: Three-dimensional jets are an efficient way of cooling the walls of modern high-pressure turbines. Introduced in the external flow that develops around the turbine blade, they are associated with a set of vortex structures. The purpose of this paper is to underline the respective origin and importance of these structures, with reference to both experimental and numerical results. Steady and unsteady vortices will be analyzed. Recommandations for numerical simulations will be proposed from these observations.

INTRODUCTION

The introduction of three-dimensional jets in a viscous cross flow involves a very intricate shearing process. Complicated flow structures appear near to the wall, which imply local three-dimensional flow separation, and a lot of vortex structures. The interest for a proper understanding of the local flow behavior comes from the difficulty to accurately predict the 3D jet in a cross flow by means of Navier-Stokes equations coupled with a turbulence model. It is now recognized that turbulence is highly anisotropic around the jet (Ardey et al.[1]). For instance, Hoda et al.[2] show that a Reynolds stress model does not improve the prediction of the flow in the vicinity of the jet, as compared to results from the standard k-ε model. However, the predictions by large eddy simulation show correct levels as observed experimentally. The improved LES predictions are linked to the ability of LES to resolve the large scales.

The first part of this paper deals with the analysis of the flow topology around the jet orifice, and particularly with the description of the origin and transfer of vorticity around a jet introduced in a cross-flow. The vorticity provides a powerful alternative for the discussion of three-dimensional flows. Because it implies the knowledge of the velocity at a spatial point, and also at the neighboring points, the dynamic mechanism of the shearing process is more finely described by the vorticity than by the velocity. This is particularly true in the neighborhood of a solid wall. Since the pioneer work of Lighthill[3], it is well recognized that the wall is at the origin of the flow vorticity. For instance, consider the force $\vec{F}_S$ that acts upon a surface element $\vec{dS} = dS\,\vec{n}$:

$$\vec{F}_s = -\left(p + \frac{2}{3}\rho k - \frac{4}{3}\mu_{tm} \text{div}\vec{V}\right)\vec{n} + \mu_{tm}\vec{\Omega} \wedge \vec{n} = -\Pi\vec{n} + \mu_{tm}\vec{\Omega} \wedge \vec{n} \qquad (1)$$

In equation (1), $\vec{V}$ is the velocity, $\vec{\Omega}$ is the vorticity vector, p is the pressure, ρ is the density, μ_{tm} is the sum of the viscous and turbulent viscosity coefficients, k is the turbulent kinetic energy. The shear force is then directly related to the vorticity vector. This expression of the force $\vec{F}_s$ is obtained for a turbulence model based on a turbulent viscosity, and for a compressible flow.

Morton[4] and more recently Wu and Wu[5] have shown that the establishment of the wall vorticity involves both the wall pressure gradient and the viscosity. This is a key point, as the pressure gradient has a weak effect on the dynamics of the flow vorticity in a turbine, but it is important to recognize its effect on the wall vorticity and the normal flux of vorticity. In this paper, we will often refer to the vorticity flux; it is then useful to give its expression:

$$\frac{\partial \mu_{tm}\vec{\Omega}}{\partial n} = \overrightarrow{\text{grad}}(\mu_{tm}\Omega_n) + \vec{n} \wedge \rho\frac{d\vec{V}}{dt} + \vec{n} \wedge \overrightarrow{\text{grad}}\Pi - 2\left(\overrightarrow{\text{grad}}\mu_{tm} \wedge \vec{n}\right)\text{div}\vec{V} \qquad (2)$$

This expression (2) differs from those given by Wu and Wu[5], through the first term of the right hand side and the use of the viscosity μ_{tm}. We see that a wall vorticity flux exists in the $\vec{n}$ direction if pressure and temperature gradients exist along the wall. A non-stationary process is also a source of vorticity flux through the wall acceleration. Finally, the normal flux of Ω_n is provided by the kinematic condition $\text{div}(\vec{\Omega}) = 0$, whose expression on the wall is:

$$\text{div}(\vec{\Omega}) = \text{div}\left(\frac{\vec{n}}{\mu_{tm}} \wedge \vec{\tau}\right) = \vec{\tau}.\overrightarrow{\text{rot}}\left(\frac{\vec{n}}{\mu_{tm}}\right) - \frac{\vec{n}}{\mu_{tm}}.\overrightarrow{\text{rot}}\vec{\tau} = 0 \qquad (3)$$

The first term of (3) is related to the wall curvature, while the second term is connected to the spiraling of the limiting streamlines on the wall.

It is also of interest to give the gradient of Π in the direction of $\vec{n}$ in order to underline the coupling with the value of the wall vorticity:

$$\frac{\partial \Pi}{\partial n} = -\vec{n}\rho\frac{d\vec{V}}{dt} - \vec{n}.\overrightarrow{\text{rot}}(\mu_{tm}\vec{\Omega}) - 2(\vec{n} \wedge \vec{\Omega}).\overrightarrow{\text{grad}}\mu_{tm} \qquad (4)$$

The term $\vec{n}.\overrightarrow{\text{rot}}(\mu_{tm}\vec{\Omega})$ of (4) corresponds to the spiraling of the vortex lines on the wall. The last term $2(\vec{n} \wedge \vec{\Omega}).\overrightarrow{\text{grad}}\mu_{tm}$ underlines the role of the temperature gradient on the wall.

The interest for the vorticity is also related to the losses: the more vorticity is included in a flow, the more kinetic energy the flow contains (according to Kelvin's minimum energy theorem), and so the more loss capacity it has also. Of course, the amount of losses in a turbine is not only dependant on the amount of vorticity, but also of other aerodynamic phenomena, such as the localization of the transition from laminar to turbulent in the viscous wall layers, and their capacity to avoid separation. But, it is well recognize that the introduction of cooling jets at the wall of turbine is responsible of a decrease of the aerodynamic efficiency (LeGrives[6]).

For the stationary case, we base our analysis on detailed numerical results obtained by Sgarzi[7], Sgarzi and Leboeuf[8]. The studies on the unsteady behavior of the jet in a cross-flow are mainly experimental. We will, in particular, analyze the behavior of the jet boundary.

STEADY BEHAVIOUR OF 3D JETS

The single jet in a cross-flow

Sgarzi[7] has performed a detailed numerical study for a laminar jet introduced in a laminar cross-flow. This work is based on the code CANARI, developed by ONERA[9]. This code is also widely used for the numerical simulation of turbine cooling (Fougère, Heider[10], Liamis et al.[11]).

We consider a laminar jet introduced in a direction normal to a flat wall into a laminar two-dimensional boundary layer. Brizzi[12] performed the related experiments with water. The temperature is kept constant in the experiment; the ratio of the averaged jet velocity to the external cross-flow velocity ($R_{inj}=V_j/U_e$) is 2.5. The external Reynolds number based on the orifice diameter ($R_{eD}=U_eD/\nu$) is 251. The ratio of the upstream boundary layer thickness to the orifice diameter (δ/D) is 4 at $X/D=-1.5$. For these values of R_{inj} and R_{eD}, the real flow is far from steady. This unsteadiness has been ignored in the numerical study. We will treat the aspect of unsteadiness in the second part of this paper.

Comparisons between the numerical and experimental results have already been presented in Sgarzi and Leboeuf[8] and have been shown to be of excellent quality. It is sufficient here to remember two main results. The original expressions of the artificial viscosity of Jameson[13], corrected by Eriksson[14] have been found to prevent the separation, which occurs upstream of the jet, in the region of the horseshoe vortex Ω_{HS}. A modification of the artificial viscosity has been introduced, which enables the capture of these fine structures, simultaneously avoiding numerical problems. In the wake region of the jet, the simulation tends to overpredict the separation area. Three reasons may be put forth: firstly, the turbulence has been neglected in the simulation; secondly, the real flow is unsteady in the wake, and thirdly the static pressure gradient is not properly predicted in the wake. It is difficult to conclude at this level on the respective importance of each of these phenomena. Chiu and al.[15] have shown in their simulation that the turbulence model was unable to improve the prediction of the static pressure field by comparison to the laminar case for a similar jet but with a higher injection ratio R_{inj}. This difficulty may be related to the mesh topology and size. We shall return to this discussion at the end of this paper.

The description of the vortex structures in the neighborhood of the 3D jet is widely based in this paper on particle trajectories that are introduced at specific locations of the flow domain.

Consider first the effect of the upstream boundary layer. Very close to the jet orifice, the upstream boundary layer suffers a strong deceleration that induces a flow separation; this leads to a well known hors-shoe vortex Ω_{HS} observed in Figure 1 by the traces that are ejected on both sides of the jet very close to the wall. The material that gathers into the horseshoe vortex is limited between the wall and the streamline that ends at P, the stagnation point. A saddle point S exists above the focus point F (not shown on the figure). All the material of the upstream boundary layer located between the stagnation line ending in P and the line that cross the saddle point S converges into a focus point F that concentrates strongly all the vorticity of that part of the boundary layer. The focus F collects also a part of the boundary layer that develops on the wall of the injection tube; a part of it is shown at the bottom of the figure. All the vorticity concentrated into the focus F is finally ejected into a vortex tube, the so-called lip vortex Ω_{LIP}. Perry et al[17] have proposed a similar pattern.

Figure 1: Flow trajectories in the meridian section, upstream of the jet (Sgarzi[7])

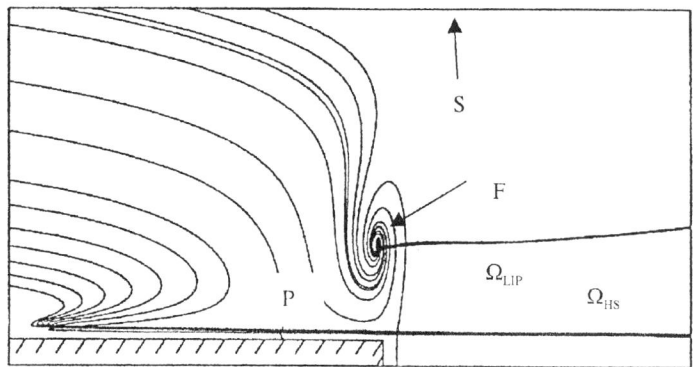

Figure 2: Balance of the transport equation of the Ω_Y component for the lip vortex for a particle of the wall boundary layer. $((S-S_0)/D=0$ corresponds to the ejection of the vortex from the meridian section).

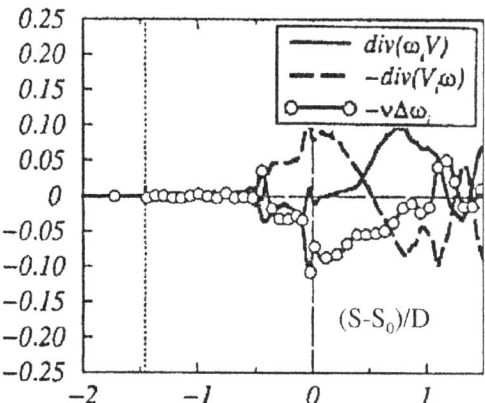

In order to appreciate the real mechanism that occurs in the lip vortex, a balance of the equation of transport of the vorticity is presented in Figure 2. It is necessary to take special care during the computation of the spatial derivatives of the vorticity from a set of numerical results whose primitive variables are based on the momentum. For that purpose, we use the following form of the transport equation of the vorticity, valid for an incompressible flow:

$$\text{div}\left(\overline{\overline{\Omega V}}\right) - \text{div}\left(\overline{\overline{V\Omega}}\right) - \nu\Delta\vec{\Omega} = 0 \tag{5}$$

The first term of equation (5) gathers the convective and stretching terms; the second is the reorientation term; the last is the viscous diffusion term. The results presented in Figure 2 show that up to the ejection from the meridian section, there is an almost exact balance between the stretching and the viscous terms for the Y component. The slight growth of the viscous term is linked to a transfer by viscosity from particles associated with the jet boundary that strongly mix as they concentrate into the focus point F (the corresponding balance for the jet particle is not shown here but is presented in Sgarzi[7]). Right after the ejection, the convection grows slowly with a parallel decrease of the viscous term. The change of sign of the reorientation term is due to a transfer into the other component Ω_X. The particular role of the viscous term underlines the necessity of a very fine mesh in order to capture this phenomena for a real cooling problem.

Figure 3: Trajectories of the flow particles introduced in the boundary layer of the injection tube (from Sgarzi[7])

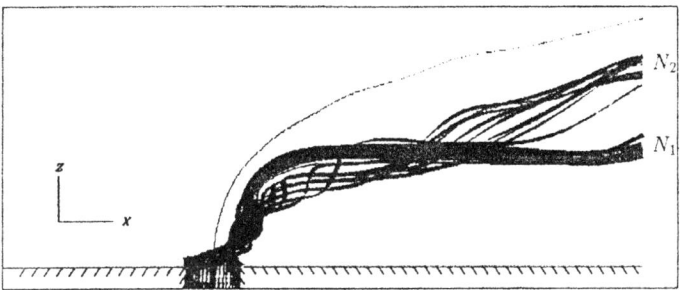

Figure 3 gives a meridional view of the particle trajectories introduced inside the boundary layer of the injection tube. The particles are clearly collected in a strong vortical movement, known as the kidney or counter-rotating vortices Ω_1. These vortices give a specific shape for the jet cross-section. They have a sense of rotation that promote the jet lift-off, and the entrainment of the cross flow towards the wall. Their origins are linked to the redistribution of the hole boundary layer under the influence of the curvature of the jet trajectory (Leboeuf and Sgarzi[16]). The particles of one of the Ω_1 vortices are collected in two groups. The group N1 corresponds to the particles injected in front of the axial location X/D=0.32 (X=0 corresponds to the middle of the orifice); they are quickly aligned with the cross-flow direction. The group of particles N2 gathers more vorticity and is associated with the minimum of static pressure. They tend to move at a higher elevation from the wall. The reason for this separation into two groups was proposed in Sgarzi and Leboeuf[8]: a strong concentration exists of fluid trajectories for the location X/D=0.32 on the jet boundary near to the solid wall. This is a direct consequence of the local minimum of static pressure on the wall surface that occurs slightly at the same location. This strong under-pressure is induced by the strong curvature of the streamlines induced by the two kidney vortices Ω_1. The jet flow is then transferred towards this pressure minimum. As a consequence, there exists an abrupt change of the sign the circumferential shear force for the same location (see Sgarzi and Leboeuf[8]). Perry et al.[17] have reported similar observations of this separation of the jet boundary layer in two groups. These authors perform detailed experiments for a jet introduced in a normal direction to the cross flow. For an injection ratio of 4 in water, they report a possible separation line on the inside wall of the pipe that may explain this separation in two groups.

Figure 4: Close view near to the orifice with particles showing the Ω_1, Ω_{HS} and Ω_{LIP} structures.

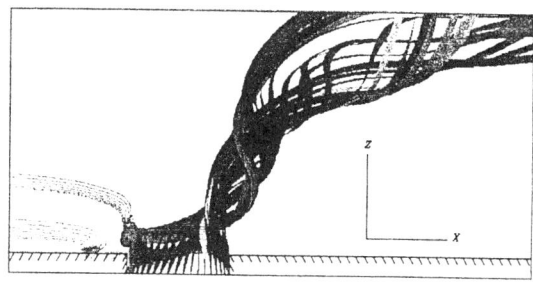

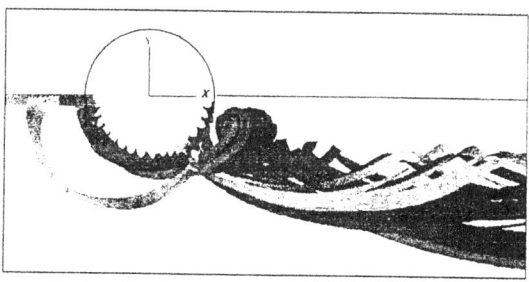

Figure 4 shows the local behavior near to the orifice of the lip and horseshoe vortices as they turn around the jet. The lip vortex closely follows the jet boundary. Along its path, it transports a part of the jet material inducing a sort of overflow of the jet on the sidewall. Although distinct of the Ω_1 vortex, it is also the most upstream occurrence of this vortex. The horseshoe vortex, on the contrary, is first transported at a constant radius from the jet center. Characterized by a low dynamic, it is then abruptly attracted towards the jet boundary, owing to the local static pressure decrease. At this location, it is ejected from the wall, and immersed in the jet wake. At the point of interception with the jet boundary, an abrupt exchange exists among the vorticity component that Sgarzi[7] has shown to be related to the viscous diffusion effect. Again, this phenomenon can only be captured locally with a sufficiently fine mesh.

Figure 5 shows the limiting streamlines associated with so-called Ω_5 vortex. The wall boundary layer streamlines are sucked into the wake of the jet. If the depression is sufficiently high, they will converge from both sides into the meridional section. This convergence up to the meridian plan seems to occur only when the injection ratio is greater than or of the order of one (Werlé[18], Beral[19]). A stagnation point appears here in X/D=1.9, and Y=0. As a consequence, the flow will behave in a similar way as in Figure 1. A horseshoe-like vortex Ω_5 appears, with two branches: one is ejected in the upstream direction and will be sucked into the jet lee-side (see Figure 6). The other branch is ejected downstream.

Figure 5: Limiting streamlines defining the Ω_5 vortex

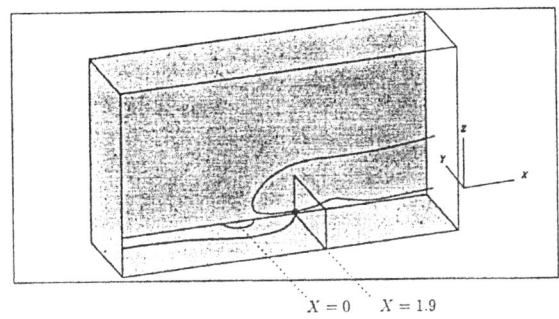

Figure 6: Trajectories of the particles associated with the lip vortex, the horseshoe vortex and the Ω_5 vortex.

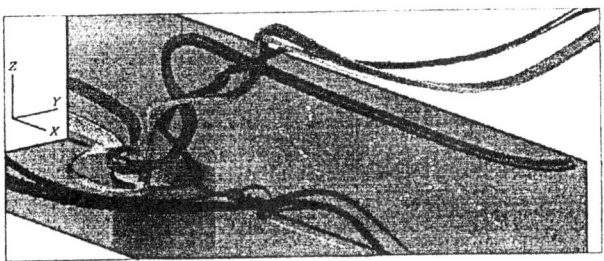

Figure 6 presents the trajectories of horseshoe and lip vortices, and of Ω_5. It is clearly observed how the horseshoe and Ω_5 merge in the jet wake, and are transported in the N_2 group of particles shown in Figure 3. The lip vortex, on the contrary, closely follows the path of the upper part of the Ω_1 vortex, and finally rises in the N_1 group.

Figure 7: Numerical simulation of Ersoy and Walker[20]

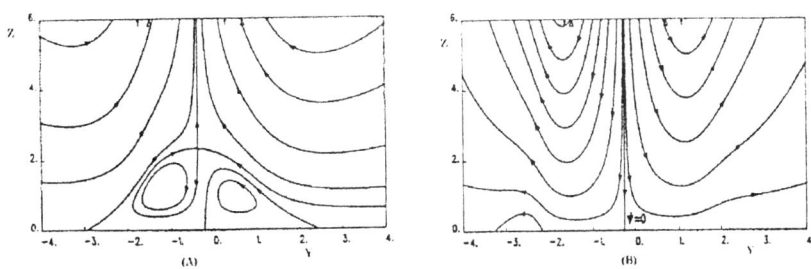

Ersoy and Walker[20] have obtained interesting results that complement the previous analysis. They have performed numerical simulations of a 3D-boundary layer subjected to the flow induced by two counter-rotating vortices whose axis of rotation are

aligned with the cross flow. In their case, the two vortices were slightly inclined with respect to the transverse Y direction. This inclination induces a non-symmetrical flow with respect to the meridian plan. Figure 7(A) presents a situation for the Ω_1 vortices typical of a single 3D jet; only one quarter of these vortices is shown on the upper part of the figure. A set of Ω_5 vortices is observed near to the wall. The non-symmetric flow path is a consequence of the inclination of the Ω_1 vortices with respect to Y.

Figure 8: Superposition of critical lines and singular points on a flow visualization for R_{inj}=1.17 (from Petit[21]).

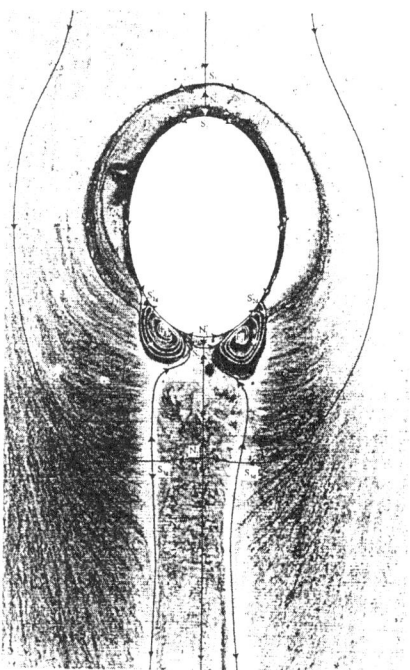

Detailed surface visualization has been performed by Petit[21] for a row of jets introduced at 45° from the sidewall with various injection ratios. We present here a typical result obtained for an injection ratio of 1.17 that is typical for the high injection rate according to Petit. Figure 8 presents the superposition of critical lines and singular points on a surface flow visualization. The number of critical points verifies the law given by equation (6) proposed by Hunt et al.[22]:

$$\left(\sum N + \frac{1}{2}\sum N'\right) - \left(\sum S + \frac{1}{2}\sum S'\right) = 1 - n \qquad (6)$$

N is the number of nodes and N' the number of semi-nodes, S the number of saddle points, and S' the number of semi-saddle points. Two surfaces are treated here: the sidewall and the jet boundary, so n=2.

The remarkable result is the existence of two small tornados on the jet lee side; they are located around the impact of the horseshoe vortex on the jet boundary. This

explains why the horseshoe vortex moves so abruptly in the Z direction just after the impact with the jet boundary (Figure 6). On this Figure 8, the set of singular points (N_2, S_{3g}, S_{3d}) are the starting location of the Ω_5 vortex in
Figure 5 and Figure 6. The node N_2 is slightly displaced above the surface, as can be inferred from Figure 7-A that could represent an orthogonal view in the same location.

Figure 9: Upstream flow structures for a row of jet (Sgarzi[7])

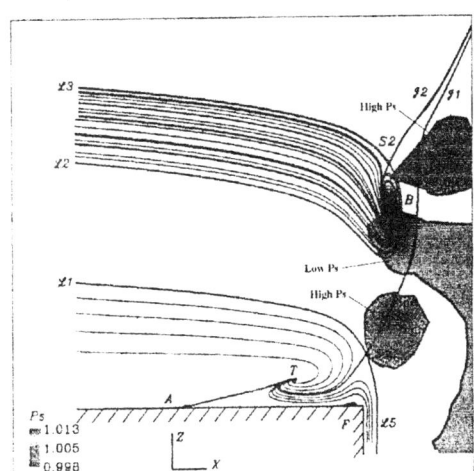

Figure 10: Overall flow structures for a row of jets (Sgarzi[7])

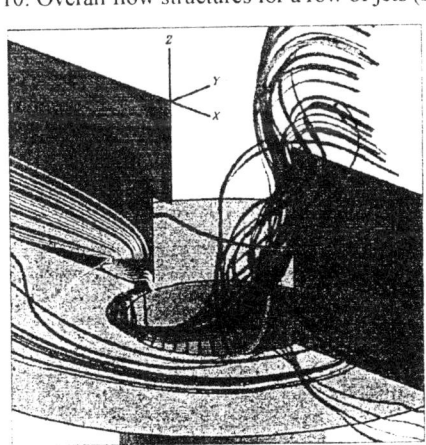

The row of jets

The effect of a row of jets with the same injection ratio of 2.5 as Brizzi[12] does not significantly change the overall structure of the vortices previously observed for a single jet. However, the intensity of their effect is strongly modified. Figure 9 presents the upstream structures in the meridian section for the upper part of the jet. The focus point B is now completely separated from the jet flow. All the flow located between the wall and the streamline that crosses the lower saddle point S_1 ends into the horseshoe vortex. A part of the jet flow comes along the same process. The direct consequence is that a significant part of the jet material overflows over the wall as can be seen in Figure9. The lip vortex that rises from the focus point B does not merge with the jet for that configuration, as can be seen also from Figure 10. The Ω_5 vortex is also shown for this case, although only the upstream branch is shown in Figure 10.

In the case of a row of jets, Lopez Pena and Arts[23] have shown a supplementary structure approximately located at one and a half diameter from the meridian plan for a case with low injection ratio (R_{inj}=0.5) (Figure 11). They associated this area with a larger region of turbulence. According to the simulations of Ersoy and Walker[20], this could be a result of a separation of the wall boundary layer because of a non-symmetry of the experimental vortices Ω_1 from two neighboring jets (see Figure 7(B)). It is interesting to notice that the completely symmetrical simulation of Sgarzi[7] did not produce any separation of this type.

Figure 11: Experience from Lopez Pena and Arts[23]: isocontours of turbulence intensity.

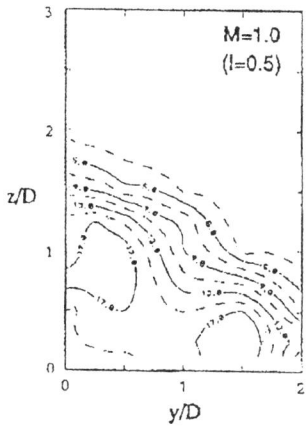

It is even more spectacular to notice that for this injection case of low injection ratio, the simulation did not generate any of the previous vortex structures, except for the main contra-rotating vortices Ω_1. The cross-flow boundary layer is transported above the jet without any of the upstream structures analyzed in Figure 1 and Figure9. The horseshoe vortex has disappeared, because the aerodynamic blockage induced by the jet is too small. The absolute minimum of static pressure is then located on the wall itself, on the corner of the jet orifice. A reverse flow still exists however close to the wall in the wake, but the sucking under the jet is much smaller. The cross-flow boundary layer is still attracted in the wake region, but this does lead to a Ω_5 vortex because the stagnation point is too close from the wall in order to enable a flow separation.

The interest for shaped holes

The interest of shaped hole is conventionally attributed to the reduction in coolant velocity caused by the increased hole exit area (Goldstein et al.[24]). Haven and Kurosaka[25] and Haven et al.[26] have performed a very detailed experimental study of the effect of the hole exit geometry on the near-field characteristic of the jet introduced in a cross flow. They investigate different hole shapes (round, elliptical, square and rectangular) with a fixed value of the cross-sectional area. By tracking the vorticity around the circumference of the jets, they evaluate its relative contribution to the downstream vorticity. They observe a pair of vortices above the kidney vortices Ω_1. They find their origin in the hole leading-edge boundary layer. The sense of rotation of these upper vortices depends on the hole shape. Considering the ratio D/L of the width D to the length L of the hole, they observe that for D/L greater that one, the jet is pushed deeper towards the jet lee-side in the meridional plane (Y=0) than on the jet extremities $Y = \pm L/2$. The cross flow then warps the upstream jet boundary near the section Y=0, and creates a realignment of the leading-edge vorticity (Figure 12). This upper pair of vortices rotates in an opposite sense compare to the kidney vortices. Provided the jet is kept close to the wall, they show that the jet lift-off may be controlled by the width L of holes, which determines the lateral separation between vortices and the existence of the anti-kidney pair of vortices. As a consequence, the cooling efficiency may be kept constant when the blowing ratio increases, because the anti-kidney pair reinforces the jet adherence to the wall.

Figure 12: Jet vortex tube for a hole with a large aspect ratio

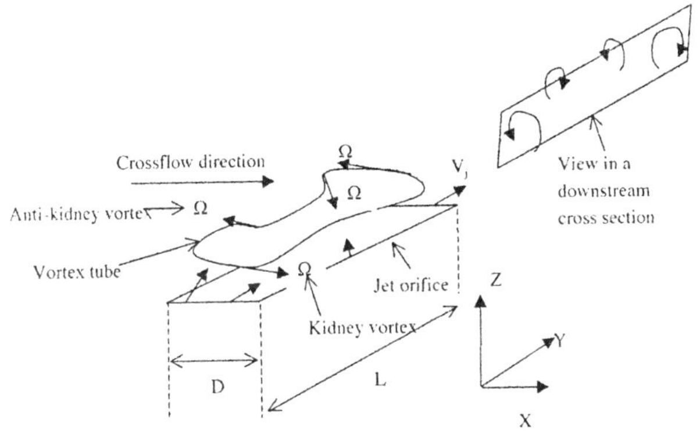

THE UNSTEADY BEHAVIOR OF THE 3D JET IN A CROSS FLOW

The steady description was presented in the first part of this paper is a strong simplification. Unsteadiness appears particularly in two specific regions of the 3D jet in a cross flow: on the windward jet boundary and in the jet wake.

Brizzi, Foucault and Bousgarbiès[27] proposed a good picture of the unsteady behavior of the three-dimensional jet boundary. A vortex ring instability of the Kelvin-Helmoltz type is observed on the windward boundary of the 3D jet. These vortex rings on the upstream half of the jet appear to remain coherent very far downstream (see figure 2 of Perry et al.[17]). Perry et al[17] have also observed a flow oscillation on the upstream hole corner; the flow oscillates between a cross flow reinjection into the hole and an overflow of the jet over the sidewall. These vortex rings ride intermittently over the top of the steady kidney vortices (Haven, Kurosaka[25]). They may exhibit a rotational direction that depends on the shape of the hole (see the corresponding paragraph above on the shaped hole); this rotational direction depends also on the injection ratio.

Figure 13: Unsteady behavior of the jet in a cross flow (Brizzi, Foucault and Bousgarbies[27])). (A) Injection ratio R_{inj} =0.54; (B) R_{inj} =1.73.

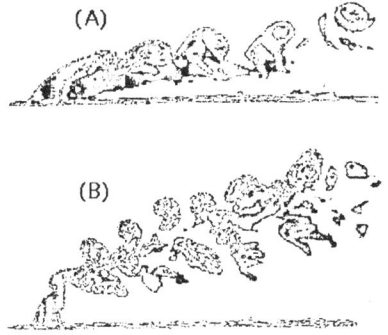

Figure 14: Scheme of the dynamics of the stretched vortex rings for a low injection ratio (A: view from above the wall, B: side view) (from Brizzi[12])

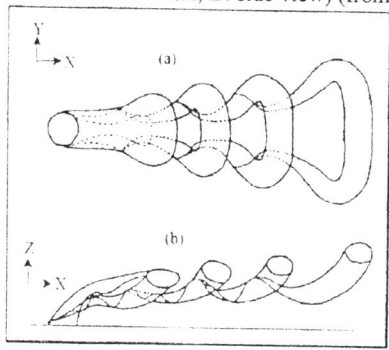

Considering the jet wind-side, we observe that the vortex structures change their sign for Rinj =0.54 and for Rinj =1.73. (Figure 13, Brizzi et al.[27]). These authors have given Rinj =1.5 as a limit value between the two regimes. One explanation of this behavior is linked to the flow path in the jet wind-side, near to the wall. Consider Figure 1 for a typical view of the upstream region of the jet that may help the explanation. We have already mentioned that the cross flow further to the side wall is reintroduced into a very small lip vortex F. This focus collects also that part of the jet boundary layer that transports the greatest part of the jet vorticity. The observations of Brizzi et al[27] show that this lip vortex is steady for the low values of Rinj, as the saddle point S above it blocks. According to Brizzi[12], the vortex rings are then also blocked on the windward side of the jet. On the jet lee-side on the contrary, they are ejected from the wall under the influence of the various effects described before in this paper. These vortex rings appear then strongly elongated downstream (Figure 14) with a clock-wise rotation

(Figure 13-A). For a very high stretching, the vortex rings will finally collapse in concentrated vortices (Neu[28]).

When the injection ratio Rinj is increased above 1.5, Brizzi et al.[27] have observed that the focus F may merge with the saddle point S located above it. As a consequence, the vorticity from the tube wind-side is no more blocked by the focus-saddle points. This will introduce a change in the sign of the vorticity on the jet boundary that will rotate counter clock-wise; this phenomenon was already observed by Fric and Roshko[29].

Unsteady flow ejections have also been reported from the wall boundary layer into the jet wake (Wu, Vakili and Yu[30], Fric and Roshko[29]). These ejections look like small tornados that exist also for high value of the Reynolds number (Wu, Vakili, Yu[30]). As mentioned by Perry et al.[17] and Fric and Roshko[29], all this shed vorticity is generated at the solid boundary, and are a way of reintroducing the side wall boundary layer into the jet (see equation (2)). According to Wu, Vakili and Yu, the tornado like vortices have a common root in the downstream branch of the Ω_5 vortex (Figure 15). They only appear in their experiments for the asymmetric jet and for injection ratio greater than 2.5. Their strength increases with the injection ratio. They may be related to the unsteady behavior of the tornadoes which are observed on the downstream side of the jet in Figure 8. It is also interesting to notice the similarity of the N_1 and N_2 trajectories in Figure 3 and the two upper traces of Figure 15.

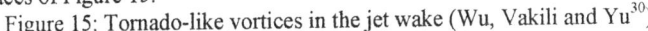
Figure 15: Tornado-like vortices in the jet wake (Wu, Vakili and Yu[30])

CONCLUSION

We have emphasized the complexity of the flow structures of a 3D jet introduced in a cross flow, using results from a Navier-Stokes simulation. We have observed that the familiar kidney vortices Ω_1 are strongly influenced by the flow behavior in the orifice or very close to it. A 3D separation occurs on the walls of the injection tube under the influence of the static pressure field; this pressure field is itself induced by the kidney vortices. The fluid particles are then collected in two groups separated by their localization with respect to this separation line; the two groups have very different trajectories.

The boundary layer of the cross flow generates two vortices in the upstream region of the jet. They belong to a complex set of flow singularities that depends on the injection ratio and the neighboring jets. They may collect both a part of the jet material. The trajectory of the lip vortex follows closely the jet boundary; thereby it is the most upstream manifestation of the jet. The horseshoe vortex intercepts the jet boundary in the region of the minimum of static pressure. This is also the localization of tornados, which eject the wall material in the jet wake. The downstream wake is dominated close to the wall by a vortex Ω_5 that mimics the horseshoe vortex. It is induced in the crossflow boundary layer by the secondary flow imposed by the kidney vortices. One branch is reintroduced in the jet lee-side, while the second branch develops close to the wall in the downstream direction.

The unsteady behavior is an important characteristic of the 3D jet. While the upstream boundary is dominated by a set of Kelvin-Helmoltz-like vortices, the jet wake transports small tornadoes that seem to connect the Ω_5 vortex to the jet wake.

All these flow characteristics require a lot of caution in the numerical simulation of the 3D jets. The mesh must be fine enough in order to capture all the local flow structures. In other case, the solution will still verify the topological law. This means that a completely different solution could be obtained. Think, for instance, of an unsteady simulation that would miss the lip vortex, and the direction of rotation of the vortices could be wrong.

A good numerical simulation could not be performed without a set of local experimental results. We know for instance that detailed field measurements are necessary in order to validate the choice of the turbulence model, particularly for the wake region. We have shown also that the surface visualizations are invaluable for a proper understanding of the flow behavior. If they are used with a good knowledge of the topological laws, and with 3D numerical simulations, they allow an access to a 3D understanding of the flow behavior.

Finally, we have given the formula of the vorticity flux. This equation emphasizes the strong coupling with the pressure field, and the role of surface topological singularities. This is a key point that underlines the necessity of an accurate pressure prediction on a wall in order to expect a good simulation of the 3D flow structures in the 3D jet environment.

REFERENCES

1. Ardey, S., Wolff, S., Fottner, L.,2000, Turbulence Structures of Leading Edge Film Cooling Jets, , ASME paper 2000-GT-255, May 8-11, 2000, Munich Germany
2. Asif Hoda, Sumanta Acharya and Mayank Tyagi, 2000, Reynolds Stress Transport Model Predictions and Large Eddy Simulations for Film Coolant Jet in Crossflow, ASME paper 2000-GT-249, May 8-11, 2000, Munich Germany
3. Lighthill, M.J., 1963, Introduction. Boundary Layer Theory. *In* Laminar Boundary Layers, Rosenhead, L., Ed. 46-113. Oxford University Press.
4. Morton, B.R., 1984, The Generation and Decay of Vorticity, Geophys. Astrpphys. Fluid Dynamics, vol 28, 277-308.
5. Wu, J.Z., Wu, J.M., 1993, Interactions between a solid Surface and a viscous compressible Flow, J.Fluid Mech., vol. 254, 183-211.
6. Le Grivès, E., 1986, Cooling Techniques for modern Gas Turbine, in Advanced Topic in Turbomachinery Technology, Ed. D.Japikse (Concepts ETI).
7. Sgarzi, O., 1996, Simulation d'un jet défléchi par une couche limite; Analyse des structures tourbillonnaires, PhD thesis, Ecole Centrale de Lyon, n°96-44.
8. Sgarzi, O., Leboeuf, F. , 1997, Analysis of Vortices in three-dimensional Jets introduced in a Cross-flow Boundary Layer, 97-GT-517, ASME Gas Turbine and Aeroengine Congress, Orlando, USA, June 2-5.
9. Cambier, L., Escande, B., 1989,Navier-Sokes simulation of shock-wave-turbulent boundary layer interaction in a three-dimensional channel, AIAA-paper 89-1851, Buffalo (see also ONERA TP n° 1989-82)

10. Fougère, J.M., Heider, R., 1994, Three-dimensional Navier-Stokes prediction of heat transfer with film cooling, ASME Paper 94-GT-14, The Hague.
11. Liamis, N, Lefebvre, M, Duboue, J.-M., 1999, CFD analysis of technological effects in high pressure turbines, ISABE Paper No.: 99-7106, Sept.
12. Brizzi, L., 1994, Contribution à l'étude de l'instabilité générée par un jet cylindrique débouchant perpendiculairement à un écoulement transversal, thèse de doctorat de l'Université de Poitiers.
13. Jameson, A., Schmidt, W., 1985, Some recent Developments in numerical Methods for transonic Flows, Computer Methods in Applied Mechanics and Engineering, vol. 51, 467-493. North Holland.
14. Eriksson, L.E., 1984, Boundary Conditions for artificial Dissipation Operator, Technical report, FFA TN 1984-53.
15. Chiu, S.H., Roth, K.R., Margason, R.J., Tso, J., 1993, A numerical Investigation of a subsonic jet in a Cross flow, Computational and Experimental Assessment of Jets in Cross Flow, AGARD-CP-534, n°22, 22-1, 22-14.
16. Leboeuf, F., Sgarzi, O., 1997, " Film cooling in turbine-A review of the behaviour of three-dimensional jets", in Turbomachinery Fluid Dynamics and Heat transfer, ed. Chunill Hah, Marcel Deckker Inc.
17. Perry, A.E., Kelso, R.M., Lim, T.T., 1993, Topological structure of a jet in a cross flow, Winchester, in Computational and experimental assessment of jets in cross flow, FDP Symposium, Agard-CP-534, 12, 12-1, 12-9, April.
18. Werlé, H., 1977, Ecoulement le long d'une paroi plane comportant des jets normaux ou inclinés de section circulaire, RT ONERA n°65/7106 ENA.
19. Beral, C., 1996, Etude expérimentale des écoulements se developpant sur un aubage de turbine en absence et en présence d'injections pariétales, Thèse de doctorat, Ecole Centrale de Lyon.
20. Ersoy, S., Walker, D.A., 1986, Flow induced at a wall by a vortex pair, AIAA Journal, vol.25, n°10, 1597-1605, October.
21. Petit, P., 1996, Ecoulement produit par une rangée de jets inclinés débouchant dans une couche limite turbulente ; application au refroidissement, PhD thesis, university of Poitiers, France
22. Hunt, J.C.R., Abell, C.J., Peterka, J.A., Woo, H., 1976, Kinematical studies of the flows around free or surface-mounted obstacles; applying topology to flow visualization., J.Flui Mech., col. 86, part 1, 179-200.
23. Lopez Pena, F., Arts, T., 1993, On the development of a film cooling layer, Heat transfer and cooling in gas turbines, 36, AGARD-CP-527, 36-1, 36-12, February.
24. Goldstein, R. J.,Eckert, E. R. G.,and Burgraff, F.,1974, "Effects of Hole Geometry and Density on Three-Dimensional Cooling," International Journal of Heat and Mass Transfer, Vol. 17, pp.595-607.
25. Haven, B.A., Kurosaka, 1997, M., Kidney and anti-kidney vortices in cross flow jets, J.Fluid Mech., vol. 352, 27-64.
26. Haven, B. A. ,Yamagata, D. K., Kurosaka, M.,Yamawaki, S., Maya, T., 1997, Anti-Kidney Pair of Vortices in Shaped Holes and their Influence on Film Cooling Effectiveness, ASME paper 97-GT-45.
27. Brizzi, L.E., Foucault, E., Bousgarbies, J.L., 1995, Sur les structures tourbillonnaires générées à la frontière d"un jet circulaire débouchant perpendiculairement dans une couche limite, C.R. Acad. Sci. Paris, t.321, Série II b, pp.217-223.
28. Neu, J.C., 1984, The dynamics of stretched vortices, J.Fluid Mech., vol. 143, 253-276.
29. Fric, T.F., Roshko, A., 1989, Structure in the near field of the transverse jet, 7th symposium on turbulent shear flows, Standford, 6p, August.
30. Wu, J.M., Vakili, A.D., Yu, F.M., 1988, Investigation of the interacting flow of nonsymmetric jets in cross flow, AIAA J., vol. 26, n°8, 940-947, August.

Flow and Heat Transfer Predictions for Film Cooling

SUMANTA ACHARYA, MAYANK TYAGI AND ASIF HODA

Mechanical Engineering Department
Louisiana State University
Baton Rouge, LA 70803

ABSTRACT: Film cooling flows are characterized by a row of jets injected at an angle from the blade surface or endwalls into the heated crossflow. The resulting flowfield is quite complex, and accurate predictions of the flow and heat transfer have been difficult to obtain, particularly in the near field of the injected jet. The flowfield is characterized by a spectrum of vortical structures including the dominant kidney vortex, the horse-shoe vortex, the wake vortices and the shear layer vortices. These anisotropic and unsteady structures are not well represented by empirical or ad-hoc turbulence models, and lead to inaccurate predictions in the near field of the jet. In this paper, a variety of modeling approaches have been reviewed, and the limitations of these approaches are identified. Recent emergence of Direct Numerical Simulation (DNS) and Large Eddy Simulation (LES) tools allow the resolution of the coherent structure dynamics, and it is shown in this paper, that such approaches provide improved predictions over that obtained with turbulence models.

INTRODUCTION

The flow of turbulent jets in crossflow are encountered in a variety of applications including pollutant discharges, VSTOL, combustion chamber design, and film cooing of turbine blades. These flows are difficult to predict accurately due to the inherent complexity of the jet-crossflow interaction. However, with advances in computing hardware and solution algorithms, our ability to better predict these flows have improved from the empirical and integral methods of the 1960's to DNS and LES methods of today. In this paper, a brief review of representative studies with different approaches have been provided, and the corresponding limitations have been identified. While considerable emphasis is given in this paper to two equation turbulence models (since these are most widely used in industry), discussion is also included on recent emerging DNS/LES approaches since they appear to remedy some of the inherent inadequacies of turbulence models.

The majority of the studies reported in the 1970's and 1980's were motivated by VSTOL-related applications and several flow visualization and experimental studies were conducted to understand the characteristics of the jet-crossflow interactions. Figure 1 shows a cartoon from Fric and Roshko[1] illustrating the various structures generated when a jet is injected normally into an unbounded crossflow. Unlike a rigid cylinder in crossflow, the boundaries of the jet are compliant and entraining, causing the jet to bend over. Periodic shedding of wake vortices have been observed particularly when the jet blowing ratio (V_{jet}/V_∞) is greater than 1. Considerable effort has gone into determining the origin of the wake vortices, and there is now experimental[1-3] and computational evidence[4] that the wake vortices are initiated by the entrainment of the crossflow boundary layer into the wake, and the upward reorientation of the entrained flow into the wake structures. The jet structure itself is dominated by a pair of kidney shaped counter-rotating vortex pair (CVP), and both the shearing between the jet and the crossflow and the vorticity issuing from the jet exit has been attributed to be the source of the CVP[3]. There are however different mechanisms proposed on the reorientation of the jet-hole vorticity into the CVP structure. Upstream of the jet, due to the adverse pressure gradients, a horse-shoe vortex system is formed, which wraps around the base of the jet travelling downstream with vorticity counter to the CVP[2,3]. Shear layer vortices on the leeward and windward edges of the jet have also been observed, and have been attributed to Kelvin-Helmholtz type instabilities[2,5]. These and other studies provide unambiguous evidence of the importance of the coherent structures and their dynamics in the near field of the injected jet. Clearly, any predictive model must embody the physics of the coherent structures to accurately predict the near-field jet behavior.

The application of interest in the present paper is the film cooling of turbine blades where a row of coolant jets are injected at an angle into the crossflow (see Goldstein[6] for a review of film cooling till 1971). This

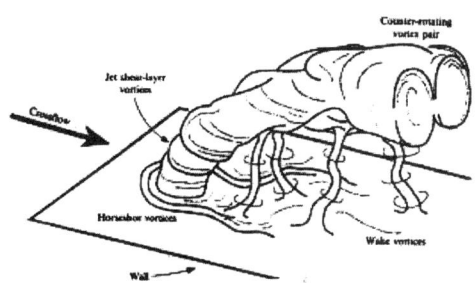

Figure 1: Schematic of the flow field of a jet in crossflow[1]

problem differs from the single-jet-in-crossflow studies, in that, the spanwise boundaries are no longer freestream boundaries. They are periodic or symmetry (for time averaged calculations with simple angle injection) boundary conditions. Further the blowing ratios of interest in turbine blade cooling are usually low (ranging from 0.5 to 2), and for low blowing ratios the flow development in the injection hole is affected by the crossflow leading to a highly non-uniform jet-exit profile. This is in contrast to the more commonly studied single-jet studies at high blowing ratios where the hole exit profile is more or less symmetrical and is not strongly influenced by the crossflow. Despite these differences, it is anticipated that coherent structures will play an important role in the film cooling flow and heat transfer behavior.

The present paper attempts to examine the available predictive models reported on film cooling, and the ability of these models to reproduce the measured trends. Thus the paper does not attempt to provide a comprehensive review of every film cooling computation reported; rather, it explores computational and modeling issues of film cooling calculations of different orders of sophistication. The focus of the paper is on the ability of these different predictive procedures to capture the correct hydrodynamic behavior; therefore, heat transfer issues are given less priority. This focus is based on the argument that the flow physics has to be correctly represented before modeling of the scalar flux can be addressed.

INTEGRAL MODELS

The first elaborate calculation procedures applied to predict the behavior of jets in cross flow were integral models in which integral equations were derived either by considering a balance of forces acting over an elementary control volume of the jet or by integrating in two spatial dimensions, the three-dimensional partial-differential equations governing the flow. In either case, a set of ordinary differential equations were obtained, which were solved analytically or numerically. The effect of pressure drag, entrainment of cross-stream fluid, and spreading rates were simulated by way of empirical relations. Analytical solutions were obtained by making ad-hoc simplifying assumptions. Abramovich[7] developed such a model for predicting the trajectory of a jet in a cross flow by assuming that the momentum of the jet in the direction normal to the cross flow was preserved and that the pressure difference across the jet was balanced by the centrifugal force due to its curvature. The model involved approximating the kidney-shaped cross-section of the jet with an ellipse in order to derive an equation for the trajectory of the jet. Crowe and Riesebieter[8] (1967) proposed a somewhat similar model Fan[9] (1967) developed an integral model incorporating the effects of drag forces and entrainment for buoyant jets in cross-flow and obtained fairly good agreement with his experimental data. Chien and Schetz[10] (1973) proposed a more elaborate model, which took into consideration the effects of drag forces, entrainment, buoyancy, axial pressure gradient, turbulent shear stress between jet and cross-stream fluid, and heat transfer due to forced convection from jet to cross-stream fluid. Model predictions showed relatively good agreement with experimental data. Changes of increasing complexity were made by several researchers in order to remove the ad-hoc simplifying assumptions, but the lack of generality of these models was a severe drawback. It quickly became clear that the next generation of models would have to be based on the solution of the complete, three dimensional Navier Stokes equations and energy equation leading to a transition from the integral approach to the finite-difference/finite-volume methodology.

ALGEBRAIC MODELS

In the 1980's, with the rapid advances in computational resources and with the development of better and faster algorithms, finite difference methods began to replace integral methods as the analysis tools of choice. Most numerical investigations of the jet in cross-flow after 1975 involved the solution of the Reynolds Averaged Navier Stokes (RANS) equations and the energy equation on a finite difference grid, with closure for turbulent quantities obtained through a turbulence model. The simplest form of the turbulence model employed is based on the Boussinesq eddy viscosity approximation where the turbulent stresses are represented as:

$$-\rho \overline{u'_i u'_j} = -\frac{2}{3}\rho k \delta_{ij} + 2\mu_t \overline{S}_{ij} \qquad (1)$$

where k is the turbulent kinetic energy, S_{ij} is the mean rate of strain, and μ_t is the eddy diffusivity. In algebraic models, μ_t is expressed simply by an algebraic expression. The most commonly used algebraic model is the Baldwin-Lomax model[11], where the eddy diffusivity is expressed in terms of mixing length in both the inner and outer layers. Fougeres and Heider[12] used a mixing length model to solve the unsteady three-dimensional Navier-Stokes equations, and obtained predictions for a film cooled flat plate and a nozzle guide vane. Comparison of the spanwise-averaged heat transfer coefficient on the vane with experimental data shows qualitative agreement in certain regions, and the disagreement is attributed to mesh characteristics. In a series of papers, Garg and Gaugler[13-15] used the Baldwin-Lomax model to obtain predictions of flow and heat transfer over a film-cooled C3X vane (data provided by Hylton et al.[16]) and a VKI rotor (for which measurements are reported by Camci and Arts[17]). The computational model consisted of a series of holes in the spanwise directions (covering, for example, 20% of the span), and the calculations were reasonably well resolved (nearly 0.5 million grid points). Their choice of the Baldwin-Lomax model was based, in part, on uncooled airfoil predictions reported by Ameri and Arnone[18], who observed that the Baldwin-Lomax model compared better with the experimental data of Graziani et al.[19] than the q-w model. Garg and Gaugler[13-15] obtained qualitatively good agreement with experimental data, but significant differences were noted in the vicinity of the film cooling injection rows and in the leading edge regions.

Although the Baldwin-Lomax model has been used extensively, and with reasonable success, the applicability of algebraic models is likely to be quite limited for film cooling flows with strong pressure gradients and separation in the immediate vicinity of injection.

TWO-EQUATION TURBULENCE MODELS

The majority of the RANS simulations for jet in a cross flow have employed a variant of the k-ϵ model (originally proposed by Launder and Spalding[20]) to obtain the distribution of eddy viscosity. Patankar et al.[21] were among the early researchers to use this model to perform a detailed study of the jet in a cross flow, and even with a relatively coarse (15x15x10) grid, obtained reasonable agreement with experimental data for the jet trajectory and streamwise velocity. Jones and McGuirk[22] (1980) used a grid containing 20x15x15 nodes and obtained only qualitative agreement with measured data due to the inadequate grid resolution. Grid resolution requirements were investigated by Demuren[23] in his computations for a row of jets in a crossflow. Results on a 37x70x14 (stream wise, vertical and spanwise directions) were shown to be grid independent and captured experimental trends fairly well. Demuren[24] also published a detailed analysis on modeling turbulent jets in cross flow, and presented a systematic review of the various models reported till 1985.

Claus and Vanka[25] used a refined grid (256x96x96) and the k-ϵ model and found that they could not capture the horseshoe vortex. Kim and Benson[26] (1992) employed a multiple-time-scale turbulence model to perform a detailed analysis of the flowfield of a row of jets in a confined cross flow. The horseshoe structure was predicted correctly using a non-uniform 165x59x80 grid and the good agreement was attributed partly to the multiple-time-scale model used for this study. An analysis of cooling jets near the leading edge of turbine blades was performed by Benz et al.[27] (1993). The RANS equations coupled with the standard k-ϵ model were solved using the SIMPLEC algorithm for compressible flows. A multi-block grid was used to simulate an actual blade geometry along with the coolant supply hole. Good agreement with experimental results was obtained due to the inclusion of the coolant delivery tube along with the main flow.

Garg and coworkers have systematically studied the effects of turbulence models[28] and the hole physics[29,30,13,14]. In Garg[28], an ACE rotor with five rows containing 93 film cooling holes were simulated. Three different turbulence models were explored (Wilcox's k-w, Coakley's q-w and the Baldwin-Lomax model). Results were compared with the experimental data of Abhari[31], and typical comparisons are shown in Fig. 2. Overall, the k-w model appears to provide the best agreement with the measurements, and particularly on the pressure side. Garg and Rigby[29] used Wilcox's k-ω turbulence model, and found that the coolant velocity and temperature profiles at the hole exit did not conform to the commonly used parabolic or 1/7-th power law distribution. The exit velocity profile appeared to significantly impact the heat transfer coefficient distribution on the suction side (Fig. 3), and to obtain reasonable predictions, it was shown that the flow development in the coolant delivery tube must be accounted for. In another application of Wilcox's k-ω turbulence model, Garg[30] computed heat transfer coefficient on the blade, hub and shroud for a rotating high-pressure turbine blade with 172 film-cooling holes in eight rows.

Leylek and Zerkel[33] included the coolant supply hole and the plenum in their calculations and used the standard k-ϵ model employing generalized wall functions prescribed by Launder and Spalding[19]. York and Leylek[34] presented predictions for mainstream pressure gradient effects in film cooling. A realizable k-ϵ model was used and the computations demonstrated the ability of the applied computational methodology to accurately model film cooling in the presence of mainstream pressure gradients. A detailed analysis of film-cooling physics, in a four part series, has been presented by Walters and Leylek[35], McGovern and Leylek[36], Hyams and Leylek[37], and Brittingham and Leylek[38], each dealing with different aspects of the film cooling problem. The standard k-ϵ model employing wall functions and a two layer model were used to obtain results for streamwise injection with cylindrical holes (part 1), compound injection with cylindrical holes (part 2), streamwise injection with shaped holes (part 3) and compound-angle injection with shaped holes (part 4).

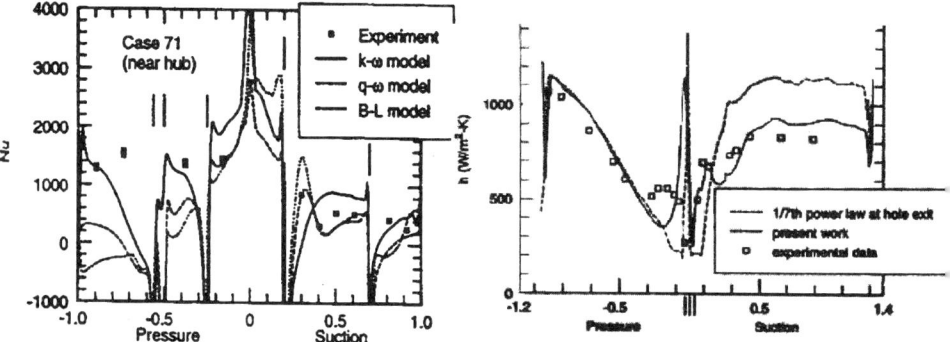

Fig. 2: Predicted[28] and measured Nusselt numbers[31] on the bade Surface near the hub

Fig. 3: Span-averaged heat transfer coefficient on the blade surface: Predictions[28] and

An aerodynamic and heat transfer analysis of discrete site film-cooled turbine airfoils was conducted by Edwards et al.[39] (1994). Ajersch et al.[40] (1995) made detailed measurements of multiple square jets injected normally into a cross flow and carried out an accompanying numerical simulation using a multi-grid, segmented, k-ϵ CFD code. Predictions and measurements did not compare well for velocities and stresses on the jet centerline, while values off the centerline matched those of the experiments much more closely. A similar study for an inclined jet was performed by Findlay et al.[41]. A numerical study of discrete-hole film cooling was conducted by Berhe and Patankar[42] on a three dimensional film cooling geometry that included the main flow, injection hole and the plenum. The effect of various variables like blowing ratio, density ratio, hole length, plenum height, plenum flow direction and turbulence level at inlet were discussed in detail. Berhe and Patankar[43,44] extended their flat plate studies and included the effect of curvature using a Richardson type correction and a two-equation model. The standard k-ϵ and the two-layer k-ϵ turbulence models were used by Lakehal et al.[45] for investigating film cooling effectiveness of a flat plate by a row of laterally injected jets. In order to match the measured lateral spreading, they employed an anisotropic correction for eddy viscosity proposed by Bergeles et al.[46]. Hoda and Acharya[47] compared seven different turbulence models for film cooling flows and concluded that the Lam-Bremhorst formulation provided the best comparison with the measurements[40].

More complex configurations have also been studied. A transonic film cooling investigation of the effect of hole shapes and orientations was carried out by Wittig et al.[48]. Bohn et al.[49] made detailed 3-D conjugate flow and heat transfer calculations of a film-cooled turbine guide vane at different operational conditions. More recently, Heidmann et al.[50] have reported a fully coupled calculations of an Allied-Signal film cooled vane with shaped holes. Their calculations included both the internal cooling channels, the coolant delivery tubes, and the external flow.

In general, while two equation models provide improved predictions with respect to algebraic models, they are unable to predict the near field accurately. The lateral spread and mixing of the film cooling jet is under-predicted by the two-equation models (Hoda and Acharya[47], Lakehal et al.[45], York and Leylek[34], Berhe and Patankar[42]) while the vertical penetration is over-predicted[42, 45, 47, 51]. Evidence of this is presented by Hoda and Acharya[47] who show that the u'w' stress component is significantly underpredicted (Fig. 9 shown later) and in Fig. 4 from Berhe and Patankar[42]. The lower lateral spreading has been linked to under-estimation

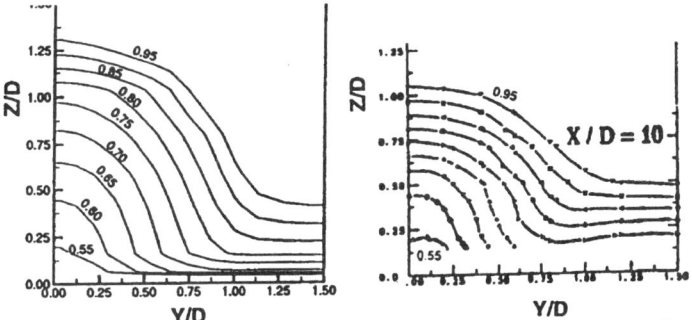

Fig. 4: Comparison of streamwise velocity predictions[42](left) with measurements[52](right). Note the greater vertical penetration and the smaller lateral spreading in the predictions

of the eddy viscosity in the lateral direction. Since the two-equation model assumes the same value of eddy-viscosity in each direction, anisotropic corrections for the eddy viscosity have been proposed (Bergeles et al. [45]). However, it has been shown through LES simulation and measurements (see discussion in the last section pertaining to Fig. 11) that the actual anisotropies have a far more complex distribution than that proposed by Bergeles et al[45].

It should be mentioned that the unresolved calculations can add numerical diffusion and inadvertently give you greater lateral spreading. Jones and McGuirk[22] performed computations with a 20x15x15 grid, and observed lateral spreading that over-predicted those reported by the measurements. These observations are in contrast with those noted in the preceeding paragraph, and are linked to the under-resolved calculations.

The importance of providing the appropriate jet-exit boundary conditions should be emphasized as one of the primary requirements for film cooling calculations. As noted in Fig. 3 earlier, and in several calculations[27, 28, 49, 50] that incorporate the plenum and the coolant delivery tube, the jet exit profile is not well represented by a $1/7^{th}$ power law or a parabolic profile (commonly made assumptions). Rather, they are a function of the blowing ratio, the local pressure gradient, the relative direction of the plenum flow with respect to the crossflow, and the coolant delivery tube length to diameter ratio.

REYNOLDS-STRESS TRANSPORT MODELS (RSTM)

In RSTM, each component of the Reynolds stresses is solved for, and therefore, turbulence anisotropy can be represented by these models. Only a few numerical studies for film cooling flows using second-moment closures have been reported in the published literature. Inze and Leschziner[52,53] carried out an investigation using a high-Re RST model employing wall functions in order to avoid solving the Reynolds stresses all the way to the wall. Demuren[54] also reported predictions with a high-Re model using a multigrid method and obtained fairly good prediction of mean flow trends. Alvarez et al.[55] compared the k-ε model predictions with the RSTM predicitons for blowing ratios of 2.35 and 5, and concluded that both models result in a similar level of agreement. Jansson and Davidson[56] applied near-wall corrections to the basic linear model and solved a low-Re RST model to predict effusion cooling in a double-row discrete-hole configuration and reported better predictions than a two layer k-ε model. Hale et al.[57] used a commercial code with a Reynolds-Stress model using non-equilibrium wall functions or a two-layer zonal approach. The two-layer zonal model performed better, although significant differences in quantitative predictions were noted in the velocity profiles. Chen et al.[58] compared the predictions of the RSTM with those obtained using a two layer k-ε model, and found improved predictions with the RSTM. Hoda and Acharya[59] used two different formulations of the RSTM model (one due to Chen[60], and the other due to Launder and Tselepidakis reported by Randriamampianina et al.[61]) and did not find any significant improvements with the RSTM predictions. These results are presented and discussed in greater detail in the next section (Figs. 8 and 9). Thus, despite the better representation of turbulence anisotropy, the RSTM predictions did not provide the expected improvements in the predictions.

DIRECT NUMERICAL SIMULATIONS AND LARGE EDDY SIMULATIONS

Numerical capabilities have increased by several folds during the last decade, and it is now becoming possible to perform DNS and LES studies in relatively simple geometries. As seen so far, the issues of modeling errors in RANS and the non-universality of such turbulence models have rendered the use of such approximations unreliable in the case of complex turbulent flows (particularly for the cases for which the turbulence models are not calibrated). A specific drawback of turbulence models is their inability to resolve large-scale coherent dynamics, and in problems where such dynamics are important (such as film cooling flows), turbulence models do not perform well. In DNS, higher-order schemes have to be used, and all temporal and spatial scales are resolved; therefore, no modeling is required. Due to the resolution requirements, the computational effort required is severe, and since the resolution requirements increase with Re, DNS is limited to only low Re values. In LES, one simulates the large scales of the flows that are dependent on the boundary conditions and contain most of the kinetic energy of the flow. The small scales or subgrid scales (SGS) are expected to be more universal and isotropic in nature, and these are modeled. Since, the small scales are problem independent, and are generally Gaussian in nature, eddy viscosity based models generally do well in describing the behavior of the small scales. Since only the large scales are resolved, resolution requirements are more modest, and high-Re LES calculations are feasible. To achieve decomposition in terms of resolved fields and subgrid fields, one generally applies a spatial filtering operation.

The current CFD practitioners in industry continue to use RANS modeling in view of the more modest computational requirements. However, current supercomputers and Beowulf clusters have provided enough computational capability to attempt DNS and/or LES of complex turbulent flows at moderate Reynolds numbers.

Hahn and Choi[62] (1997) presented unsteady simulations of circular jets in crossflow using second order spatial schemes with 14.6×10^6 grid points at a Reynolds number of 1750 (based on the jet velocity and the duct width) for a momentum ratio of 0.5. Sharma and Acharya[63] (1998) presented the direct numerical simulation of a rectangular coolant jet injected normally into a periodic crossflow using high order spatial discretization on a $128 \times 64 \times 64$ grid for a channel Reynolds number of 5600 and momentum ratio of 0.25. Jones and Willie[64,65] presented the results of LES of round and plane jets in crossflow on a $87 \times 30 \times 30$ mesh. The Reynolds number based on jet velocity and the nozzle width was 5,815, and the blowing ratio was 7.34. Yuan and Street[66], examined the trajectory and entrainment characteristics of a round jet in crossflow using LES at

Reynolds numbers of 1,050 and 2,100 based on jet diameter and freestream velocity. The simulations were performed for jet to crossflow velocity ratios of 2.0 and 3.3 on grids using 1.34×10^6 grid points (also Yuan[67], Yuan and Street[68], and Yuan et al [5]). Muldoon and Acharya[69] used higher order finite difference schemes to study the film cooling problem and included the flow development in the delivery tube. They presented the unsteady interactions of the upstream crossflow and horseshoe vortex system with flow development in the delivery tube, and showed that this unsteady interaction led to a periodic pulsing of the jet at a Strouhal number of 0.44. Takata et al.[70] presented the hybrid LES-RANS computations of film cooling flow on gas turbine blade. In a series of papers, Tyagi and Acharya[71-74,4] studied the influence of various parameters on the flow physics of the square coolant jets issuing normally in crossflow corresponding to the experimental setup of Ajersch et al.[40]

A CASE STUDY

In order to provide an assessment of the performance of different models, a film cooling configuration that was geometrically simple, but which incorporated all the essential physics of the film cooling problem was chosen. The configuration chosen corresponds to the experimental study of Ajersch et al[40] where measurements are presented for normal injection through square holes. The physical domain in Fig. 5 shows a single row of six square jets on a flat plate which represents the turbine blade surface. The computational domain is chosen to be a periodic module and is shown in both Figs. 5a and Fig. 5b

The Reynolds number of the jet based on the jet dimension is 4,700 which matches that

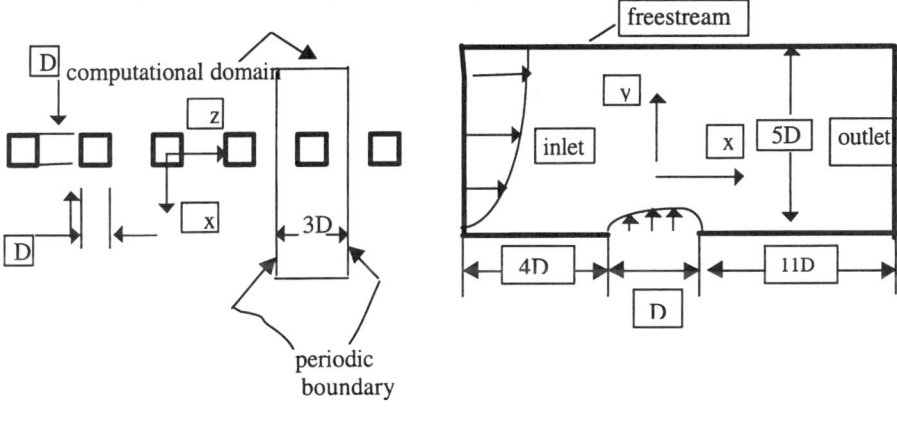

(a) (b)
Fig. 5: (a) Schematic of the physical problem-top view and (b) Side view of the computational domain

of the experimental study. The experimental investigation was carried out for velocity ratios R=0.5, 1.0 and 1.5. However, the computations have only been carried out for the lowest velocity ratio of R=0.5, and the general flow characteristics for this case as predicted by DNS (Muldoon and Acharya[69]), LES (Tyagi and Acharya[71-74]) and RANS with several different turbulence models (Hoda and Acharya,[47] 1999, Hoda et al.[59]) are evaluated by comparison with the measurements.

Results from RANS calculations will be presented first to determine the ability of the various turbulence models to capture the flow physics accurately, and to highlight the specific deficiencies in the models. These will be followed by predictions from DNS and LES, where the computed statistics will be compared with the measurements.

Comparison of mean velocity and turbulence statistics

Figure 6 shows the mean axial velocity and the turbulence kinetic energy along the jet centerplane (z/D=0) at several axial locations measured from the center of the jet-hole exit (Hoda and Acharya[47]). Predictions are shown for seven different turbulence models. These include the high-Re version of the k-ε model (called HRE), the Launder-Sharma (LS), and the Lam-Bremhorst low-Re models (LB), the k-ω model (KW), the Rodi-Mansour DNS-corrected low-Re k-ε model (MR), and the non-linear k-ε models of Speziale (SP), and Myong-Kasagi (MK). In the near field of the jet, and in the wake region, where the large scales play an important role in the mixing process, the streamwise velocity is over-predicted and the turbulence kinetic energy is underpredicted. The best prediction is shown by the Lam-Bremhorst k-ε model. Even the non-linear models which incorporate anisotropy are unable to correctly reproduce the measured trends. The underprediction of the turbulence kinetic energy with the different turbulence models is due to the inability of these models to capture the energy production and transport associated with the coherent scales.

Figure 7 shows the spanwise velocity, the turbulence kinetic energy and the spanwise shear stress $\overline{u'w'}$ along the spanwise-edge of the jet hole (z/D=-0.5). Again the model predictions exhibit substantial differences from the data with overprediction of the peak spanwise velocities and underprediction of the kinetic energy and the spanwise stress. In fact, the spanwise stress is significantly lower than the measurements indicating that the predictions underestimate the mixing and jet growth in the spanwise direction. This is typical of the two-equation models, and, as noted earlier, several investigators[42,45,47,71] have reported underprediction of the lateral spreading.

With the intent of improving the results, predictions were next obtained with RSTM and LES, and these results are displayed in Figs. 8 and 9. The predictions shown are from the Lam-Bremhorst k-ε model (which showed the best performance of all the two-equation models tested), predictions from two Reynolds-Stress-transport models (the L-T and Chen models noted earlier), and predictions from LES. The RSTM predictions do not show any significant improvement over the Lam-Bremhorst model, and in fact show poorer agreement in the kinetic energy predicitions. This points to the fact that the anisotropy in the flow turbulence is not the major contributor to the lack of agreement (since RSTM incorporates turbulence anisotropy), and the discrepancy comes from the inability of these models to capture the effects of large scale unsteadiness in the near field. The best agreement with the data comes from the LES predictions which is able to capture the dynamical behavior of the coherent structures. The LES predictions for the mean velocity (both streamwise and spanwise) are in good agreement with the data. Further, the LES predictions for the turbulence kinetic energy and the spanwise stresses are larger than the RANS predictions and are in much better agreement with the measurements. This is because the energy and transport associated with the fluctuations of the larger scales are resolved directly by the LES, and the correct representation of the energy associated with these scales results in an increase in k and spanwise stress.

The turbulence kinetic energy along the spanwise edge of the jet exit (z/D=-0.5) as computed from DNS statistics (Muldoon and Acharya[69]) is shown in Fig. 10. It is clear that the turbulence kinetic energy is correctly predicted, and that the DNS is able to reproduce the true physics of the near field accurately.

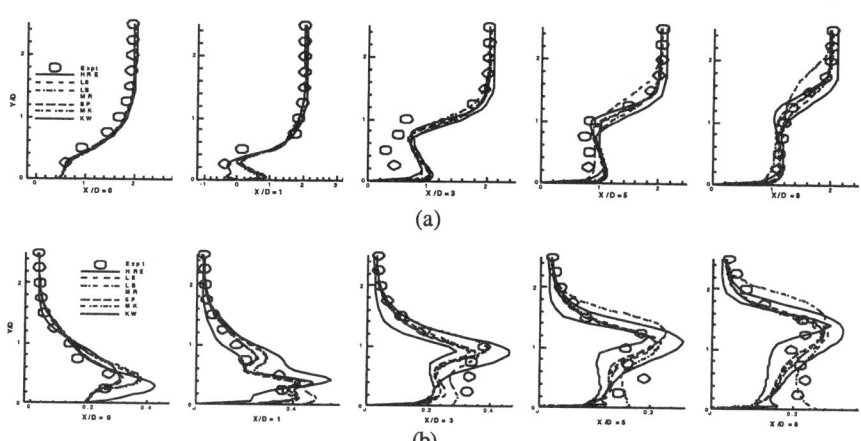

Figure 6: (a) Streamwise velocity, U/V_{jet} (top figure) and (b) turbulence kinetic energy, $\sqrt{k}/V_{jet}$ (bottom figure) along jet centerplane ($z/D=0$). Two-equation model predicitions.

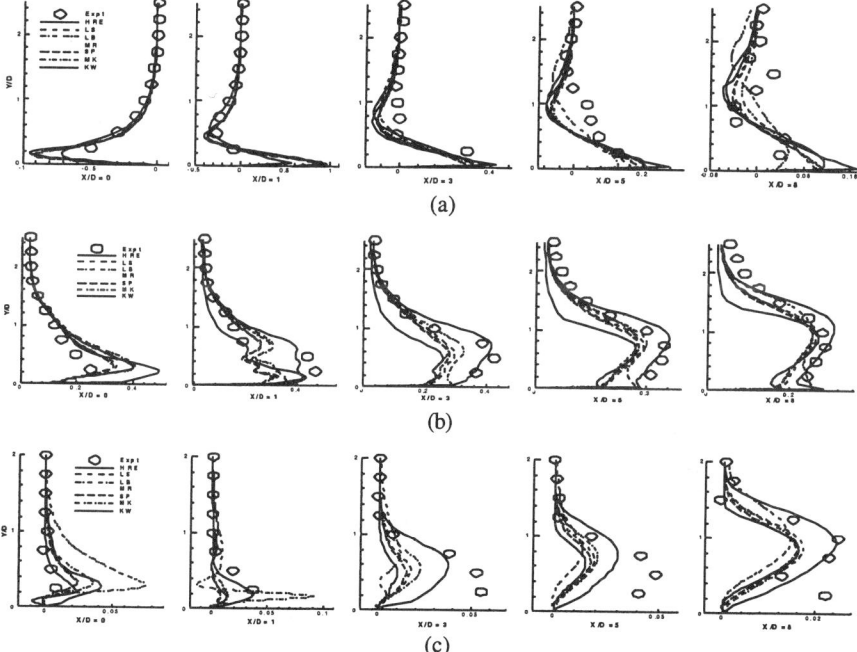

Figure 7: (a) Spanwise velocity, W/V_{jet} (top figure) (b) turbulence kinetic energy, $\sqrt{k}/V_{jet}$ (middle figure) and (c) transverse shear stress ($u'w'$) along $z/D=-0.5$. Two-equation model predicitions.

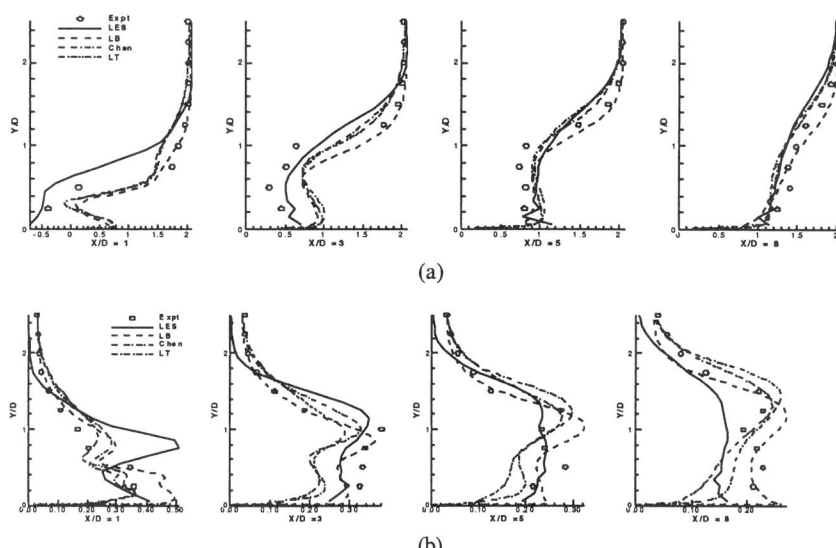

Figure 8: (a) Streamwise velocity, U/V_{jet} (top figure) and (b) turbulence kinetic energy, $\sqrt{k}/V_{jet}$ (bottom figure) along jet centerplane (z/D=0). RSTM model and LES predicitions.

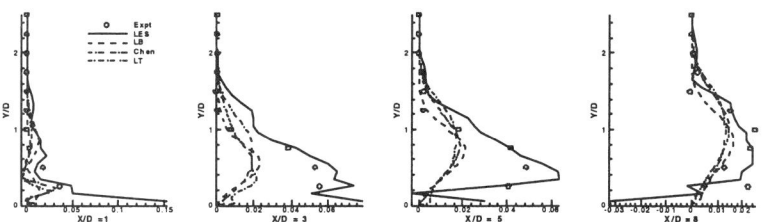

Figure 9: Transverse shear stress (u'w') along z/D=-0.5. RSTM model and LES predictions

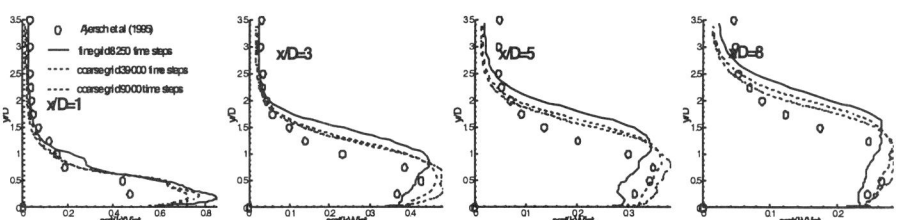

Figure 10: Turbulence kinetic energy predictions from DNS (Muldoon and Acharya, 1999)

The inadequacies of the isotropic eddy viscosity assumption are shown in Fig. 11, where LES results are used to compute the eddy viscosity in the xy, xz, and yz planes. These eddy viscosities are simply obtained by dividing the stresses with the corresponding rate of strain. In turbulence models, the primary focus is on tuning the model to compute the correct xy stress component. Such tuning has been extensively done for boundary layer flows, jets, flow over backsteps, channel flows, etc. Thus the turbulence model is expected to predict the xy eddy viscosity component correctly. However in a turbulence model, the same eddy viscosity is used to represent the diffusion of the stress components in all three directions. Fig. 11 clearly shows that this assumption is grossly incorrect, and that the lateral eddy viscosity (xz component) near the surface is significantly greater than the vertical eddy viscosity (xy component). Thus the use of the xy eddy viscosity component to represent the lateral stress $\overline{(u'w')}$ transport would be incorrect, and an anisotropy correction needs to be used. Such corrections have been proposed, for example by Bergeles et al.[46,] but the present LES calculations indicate that the anisotropy corrections are more complicated than those proposed so far. The anisotropy maps determined by the LES computations are quite similar to the measurements reported by Kaszeta and Simon[75].

Having demonstrated that the LES/DNS calculation schemes provide predictions that agree well with measurements, and that these simulations do capture the true flow physics, these simulations can be utilized to explore a variety of issues with reasonable confidence. We have explored the effect of freestream turbulence intensity levels (Tyagi and Acharya[71]), the effect of length scales in the freestream (Tyagi and Acharya[72]), the effect of jet-inclination angle (Tyagi and Acharya[73]), and the effect of jet-hole geometry (Tyagi and Acharya[74]). As an example, Figure 12 shows the effect of turbulence length scales in the freestream on the normal stress levels in the jet and the wake regions. Two cases were run both with 15% freestream turbulence intensity levels. In one case (termed small scale), all the energy is in the small scales (high frequency), while in the second case (termed large scale), a von-Karman energy spectrum is specified in the freestream with the peak energy at a length scale corresponding to 4D. The choice of 4 hole diameters is based on typical values reported in experiments.

Figure 12 shows that there is a dramatic effect of the freestream length scale on the streamwise and wall-normal stress levels, and that these stress levels are considerably higher for the case corresponding to the larger freestream length scales. This observation is true in the jet region, in the wake region, and also in the regions representing the horse-shoe vortex system (just upstream of the jet injection). For the large scale case, the freestream energy is concentrated at a length scale corresponding to 4D, which is larger than the length scales of the coherent structures (such as the horseshoe, the shear layer vortices on the windward side of the jet, and the wake vortices), and therefore, energy is transferred directly from the larger freestream scales to these coherent structures. This cascade of energy does not occur for the small scale case since the length scales associated with the freestream are smaller than those associated with the coherent structures, and no significant backscatter occurs. Thus the turbulent stresses for the large scale case are observed to be considerably greater in the near field of the jet injection than the corresponding stresses for the small scale case.

CONCLUDING REMARKS

A brief review of the capabilities of different predictive methods in correctly calculating the measured statistics of a film-cooling jet in a crossflow is presented. The focus of the paper is on the hydrodynamic predictions. A case study is presented where the predictions using several different turbulence models (k-ε models and the RSTM) are compared with measurements and the results obtained with LES and DNS. It is shown that the turbulence models do not accurately predict the near-field statistics, while considerable improvements are obtained with DNS and

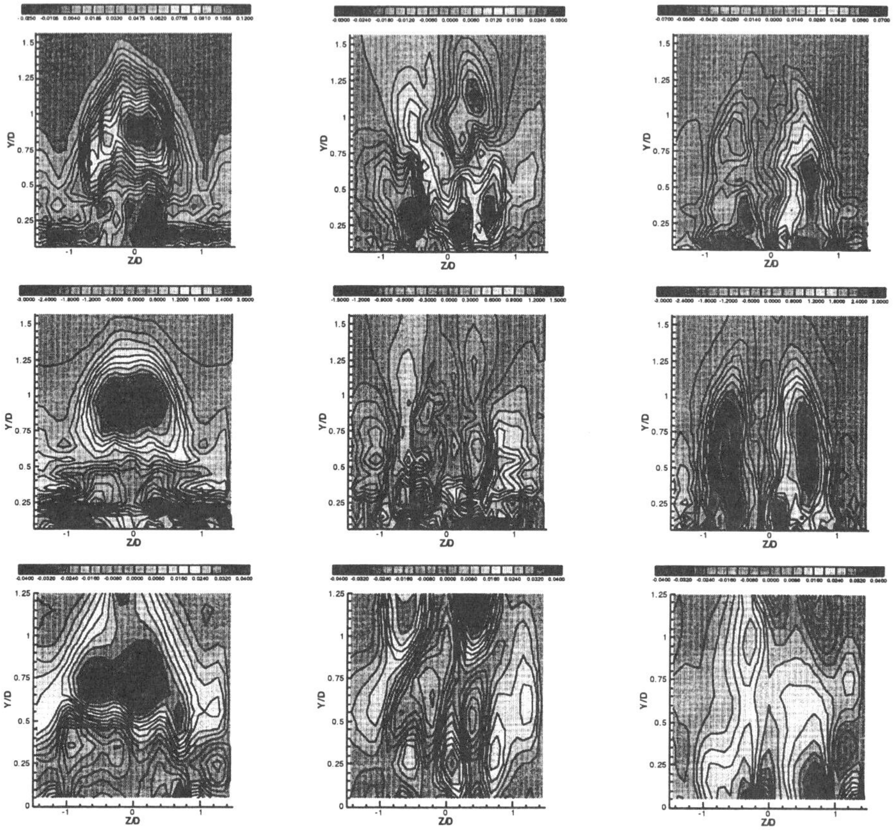

Figure 11: LES predictions of shear stress (top row), rate of strain (middle row), and eddy viscosity (bottom row) for the xy component (left column), yz component (middle column) and xz component (right column)

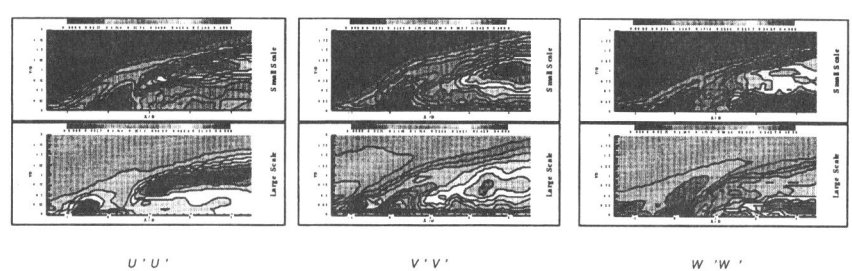

Figure 12: LES predictions showing the effect of length scales in the freestream with 15% freestream turbulence intensity (small scales in the top row, von-Karman freestream spectra in the bottom row with dominant energy at a length scale corresponding to 4D)

LES. The following issues should be kept in mind in performing the numerical calculations for film cooling.
1. Boundary conditions at the jet-exit hole play an important role in the predictions, and must either be directly provided by companion experiments or the coolant delivery tube and the plenum must be incorporated in the calculation procedure.
2. Two-equation model usually underpredict the lateral spreading of the film cooling jet and overpredict its vertical penetration. This is due to the incorrect computation of the isotropic eddy viscosity in two-equation models. Anisotropy corrections are required, but these are considerably more complex than those proposed so far in the literature.
3. The RSTM predictions were not found to be substantially better than the two-equation model predictions.
4. The LES and DNS predictions were able to better predict the mean velocities, and the turbulent stresses. These are presumably due to the resolution of the energy carrying large scale structures in the near-field of the jet.

ACKNOWLEDGEMENTS

The author's work on film cooling, referenced in this paper, have been supported by grants from NASA. This support is gratefully acknowledged. Discussions and help provided by Raymond Jones and Frank Muldoon is also appreciated.

REFERENCES

1. Fric, T. F., and Roshko, A., 1994, J. Fluid Mech., Vol. 279, pp. 1-47
2. Kelso, R. M., Lim, T. T., and Perry, A. E., 1996, J. Fluid Mech., Vol. 306, pp. 111-114
3. Andreopoulos, J., 1985, J. Fluid Mech., Vol. 157, pp. 163-197
4. Tyagi, M., and Acharya, S., 2000a, Phys. Of Fluids, in review
5. Yuan, L. L., Street, R. L., and Ferziger, J. H., 1999, J. Fluid Mech., Vol. 379, pp. 71-104
6. Goldstein, R. J., 1971, Film Cooling., In Advances in Heat transfer, Vol. 7, pp. 321-379
7. Abramovich, G. N., 1963, The Theory of Turbulent Jets, Cambridge, MA, MIT Press.
8. Crowe, C. T., and Riesebieter, H., 1967, AGARD Reprints.
9. Fan, L. N., 1967, California Institute of Technology, Report No. KH-R-15.
10. Chien, C. J. and Schetz, J. A., 1975,Trans. ASME: J. Appl. Mech., Vol. 42, pp. 575-579
11. Baldwin, B. S. and Lomax, H., 1978, AIAA Paper No. 78-257.
12. Fougeros, J. M., and Heider, R., 1994, ASME Paper 94-GT-14
13. Garg, V. K., and Gaugler, R. E., 1997, J. of Turbomachinery, Vol. 119, pp. 343-351
14. Garg, V. K., and Gaugler, R. E., 1997, Int. J. Heat and Mass Transfer, Vol. 40, pp. 435-445
15. Garg, V. K., and Gaugler, R. E., 1995, ASME 95-GT-275
16. Hylton, L. D., Nirmalan, V., Sutanian, B. K., and Kaufman, R. M., 1988, NASA CR 182133
17. Camci, C., and Arts, T., 1990, J. Turbomachinery, Vol. 112, pp. 497-503
18. Ameri, A., and Arnone, A., 1996, J. of Turbomachinery, Vol. 118, pp. 307-314

19. Garziani, R. A., Blair, M. F., Taylor, J. R., and Mayle, R. E., 1980, J. of Engng. Power, Vol. 102, pp. 257-267
20. Launder, B., E., and Spalding, D., B., 1974, Comp. Methods Appl. Mech. Eng., vol. 3, pp. 269.
21. Patankar, S. V., Basu, D. K., and Alpay, S. A., 1977, J. Fluids Engng., vol. 99, no. 4, pp. 758-762.
22. Jones, W. P. and Mc Guirk, J. J., 1980, Turbulent Shear Flows 2, Springer, pp. 233-245.
23. Demuren, A. O., 1983, Comp. Meth. App. Mech. Engr., vol. 37, pp. 309-328.
24. Demuren, A. O., 1985, Encyclopedia of Fluid Mechanics, vol. 2, Gulf Publishing Co., pp. 430-465.
25. Claus, R. W. and Vanka, S. P., 1990, AIAA Paper No. 90-0444.
26. Kim, S. W., and Benson, T. J., 1993, *AIAA Journal*, Vol. 31, No. 5, pp. 806-811
27. Benz, E., Wittig, S., Beeck, A., and Fottner, L., 1993, AGARD-CP-534.
28. Garg, V. K., 1999, Intl. J. of Heat and Mass tr., Vol. 42, pp. 789-802
29. Garg, V. K., and Rigby, D. L., 1999, Intl. J. of Heat and Fluid Flow, Vol. 20, pp. 10-25
30. Garg, V. K., 1999, ASME 99-GT-44, IGTI Congress and Exhibition, Indianapolis.
31. Abhari, R. S., 1991, Ph. D Thesis, MIT, Cambridge
32. Camci, T., and Arts, T., 1985, J. Eng. Gas Turbines and Power, Vol. 107, pp. 1016-1021
33. Leylek, J. H., and Zerkle, R. D., 1993, ASME paper 93-GT-207.
34. York, W. D., and Leylek, J. H., 1999, ASME 99-GT-166, IGTI, Indianpolis
35. Walters, D. K., and Leylek, J. H., 1997a, ASME-97-GT-269, IGTI, Orlando
36. McGovern, K. T., and Leylek, J. H., 1997b, ASME-97-GT-270, IGTI, Orlando
37. Hyams, D. G., and Leylek, J. H., 2000, J. of Turbomachinery, Vol. 122, pp. 122-132
38. Brittingham, R. T. and Leylek, J. H., 2000, J. of Turbomachinery, Vol. 122, pp. 133-145
39. Edward, J. H., et al., 1994, AIAA paper 94-3070.
40. Ajersch, P. et al., 1995, ASME paper 95-GT-9.
41. Findlay, M. J., He, P., Salcudean, M., and Gartshore, I. S., 1996, ASME-96-GT-167, IGTI, Birmingham
42. Berhe, M. K., and Patankar, S., 1996, ASME paper 96-WA/HT-8.
43. Berhe, M. K., and Patankar, S. V., 1999, J. of Turbomachinery, Vol. 121, pp. 792-803
44. Berhe, M. K., and Patankar, S. V., 1999, J. of Turbomachinery, Vol. 121, pp. 781-791
45. Lakehal, D., Theodoridis, G. S., and Rodi, W., 1998, International Journal of Heat and Fluid Flow, pp. 418-430
46. Bergeles, G., Gosman, A. D., and Launder, B. E., 1978, Num. Heat Transfer, Vol. 1, pp. 217-242
47. Hoda, A., and Acharya, S., 1999, ASME-IGTI, Indianapolis
48. Wittig, S. et al., 1996, ASME paper 96-GT-222.
49. Bohn, D. E., Becker, V. J., and Kusterer, K. A., 1997, ASME paper 97-GT-23, 1997.
50. Heidmann, J. D., Rigby, D. L., and Ameri, A. A., 1999, ASME Paper No. 99-GT-186, IGTI, Indianapolis
51. Pietrzyk, J. R., 1989, Ph. D thesis, Univ. of Texas at Austin
52. Inze, N. Z., and Leschziner, M. A., 1990, Engng. Turb. Modeling and Expts., Eds. Rodi and Ganic, Elseiver, pp. 155-164
53. Inze, N. Z., and Leschziner, M. A., 1993, AGARD-CP-534.
54. Demuren, A. O., 1993, Intl. J. Engng. Sc. , Vol. 31, pp. 899-913
55. Alvarez, J., Jones, W. P., and Seoud, R., 1993, AGARD-CP-534.
56. Jansson, L. S., and Davidson, L., 1996, Engng. Turb. Modeling and Expts., Eds. Rodi and Bergeles, Elseiver, pp. 731-740
57. Hale, C. A., Ramadhyani, S., and Plesniak, M. W., 1999, ASME-99-GT-162, IGTI, Indianapolis

58. Chen, H. C., Wei, G., and Han, J. C., 1999, ASME IMECE, Nashville
59. Hoda, A., Acharya. S., and Tyagi, M., 2000, ASME-IGTI, Munich
60. Chen, H. -C., 1995, Journal of Engineering Mechanics, Vol. 121, No. 10, pp.1136-1146.
61. A., Elena, L., Fontaine, J.P., and Schiestel, R., 1997, Phys. Fluids, Vol. 9, No. 6, pp. 1696-1713.
62. Hahn, S. and Choi, H., 1997, J. Comp. Phys., Vol. 134, pp. 342.
63. Sharma, C. and Acharya, S., 1998, NASA/CR-1998-208674
64. Jones, W. P. and Wille, M., 1996a, Engineering Turbulence Modeling and Experiments 3. Ed. Rodi, W. and Bergeles, G. pp.199-209.
65. Jones, W. P. and Wille, M., 1996b, Int. J. Heat and Fluid Flow, Vol. 17, pp.296-306.
66. Yuan, L.L. and Street, R. L., 1996, *ASME Fluids Engineering Division* Vol. 242, pp.253-260
67. Yuan, L.L., 1997, Ph.D. Dissertation, Stanford University.
68. Yuan, L.L. and Street, R. L., 1998, Phys. Fluids, Vol. 10, pp. 2323-2335
69. Muldoon, F. and Acharya, S., 1999, ASME-IGTI99
70. Takata, T., Takeishi, K., Kawata, Y. and Tsuge, A., 1999, AJTE99-6458.
71. Tyagi, M. and Acharya, S., 1999a, FEDSM 99-7799, 3rd ASME/JSME Joint fluids engineering conference.
72. Tyagi, M. and Acharya, S., 1999b, IMECE conference, Nashville, TN.
73. Tyagi, M. and Acharya, S., 2000, 4th ISHMT/ASME heat and mass transfer conference.
74. Tyagi, M. and Acharya, S., 1999c, Adv. in DNS/LES, proceedings of SAICDL, Eds. Sakell and Knight.
75. Kaszeta, R.W. and Simon, T.W., 2000, J. of Turbomachinery, Vol. 122, pp. 178-183.

Film Cooling:
What did we learn from our measurements?

Tony ARTS

von Karman Institute for Fluid Dynamics
Turbomachinery Department
72, chaussée de Waterloo – B1640 Rhode Saint Genèse - Belgium

Foreword

"Film cooling" is definitely an area of research that has received a considerable amount of interest from the scientific community over the three last decades, both from an industrial and an academic point of view. Any literature survey on this subject will end up with a number of references filling, only by themselves, more than the allowed page quota of a conventional publication. A choice had therefore to be made; it resulted in the presentation of the experience gained, via measurements, on this subject during a number of years at the von Karman Institute, both from a thermal and an aerodynamic point of view. The list of references is therefore, by far, not complete and the author apologizes for not giving credit to all scientists who contributed, by their research and publications, to the advances on film cooling.

Introduction

A continuous improvement in the performance of modern aero-engines requires more and more detailed optimization of each of their components. Especially in the field of high pressure turbine cooling, methods allowing a very accurate prediction of the airfoil temperature are essential to guarantee the lifetime of that component. A large number of research projects have therefore been addressed over the three last decades to various film cooling techniques. The numerous parameters to be investigated in this field concern the main (or freestream) and the secondary (or coolant) flow. They can be listed in a non-exhaustive way as follows:
- Airfoil geometry: curvature distribution, coolant emission location
- Coolant emission geometry: hole shape, diameter and spacing, inclination and/or sweep angle of the hole, number of rows of holes
- Blade loading: transition location, boundary layer status, shock/boundary layer interaction
- Freestream Reynolds number
- Freestream turbulence intensity
- Blowing ratio or coolant to freestream mass weight ratio
- Coolant to freestream temperature, or density, ratio

In the area of film cooling, a large number of basic experimental investigations have been presented, with the objective to identify and understand the thermal and aerodynamic behavior of the coolant film developing along a flat plate, e.g. Goldstein (1971), Forth and Jones (1976) and Pietrzyk et al (1988). The influence of coolant to mainstream temperature, or density, ratio was addressed by a number of authors, e.g. Loftus and Jones (1982). The effect of surface curvature also received some attention, e.g. Ko et al (1986) or, more recently, in a number of Brite-Euram programmes. A large fraction of these experiments were however performed at low speed and results were mostly presented in terms of adiabatic wall temperature and adiabatic effectiveness. As a matter of fact, in the severe environment of a film cooled turbine blade, the large temperature differences existing between the mainstream and the blade surface induce a wall temperature pattern quite different from an adiabatic distribution. Considering moreover the important spatial temperature variations due to internal cooling passages and the strongly varying heat flux distributions downstream of a film cooling hole or slot, the most representative heat transfer parameter seems to be the convective heat transfer coefficient h. Either an experimental or a numerical determination of h is essential to perform any detailed heat conduction or thermal stress analysis. The recent development of conjugate heat transfer codes (e.g. Bohn et al, 1997 or Montenay et al, 2000) becomes a serious alternative. Measurements on film cooled turbine cascade models were presented, among other researchers, by Lander et al (1972), Nicolas and Le Meur (1974), Ito et al (1978), Daniels (1979), Dring et al (1980), Horton et al (1985), Camci (1985), Béral (1996).

Two directions of research were followed at the VKI in the area of heat transfer investigations.

On one side, a systematic research program on leading edge, suction side and pressure side film cooled turbine vanes and blades was conducted in close collaboration with Snecma-Moteurs (e.g. Arts et al, 1990, 1992). The global objectives were to develop accurate and reliable calculation methods and to possibly identify relatively simple correlations to be used during the design phase. The selection of the airfoil geometries and cooling configurations were generally considered as general demonstrator test cases, representative of modern aerodynamic designs. The influence of various mainstream (Mach, Reynolds, turbulence, incidence) and coolant (location, mass flow, density) parameters was systematically investigated.

On the other side, efforts were made in the development of in-house, reliable and accurate test cases for the validation of numerical solvers used for turbomachinery viscous flow computations.

Heat transfer with film cooling – test case

The experimental effort developed at VKI in this area is a natural continuation of earlier work performed by Arts and Lambert de Rouvroit (1992) on the aero-thermal characteristics of a highly loaded transonic turbine uncooled guide vane mounted in a linear cascade arrangement. This study was complemented by adding the effect of film cooling to the already considered mainstream flow parameters. Two independent film cooling configurations were implemented, one located on the suction side and one on the pressure side. Both configurations consisted of two staggered rows of circular holes. A detailed description of the available results was presented by Arts (1995).

Typical examples of this investigation are presented, along the suction side, for nominal flow conditions ($Re_2=10^6$, $M_{2,is}=0.9$, $Tu=1\%$) in fig 1 for 4 different blowing ratios ranging between 0.252 and 0.831. From a general point of view, the heat transfer coefficient seems to decrease regularly with increasing blowing rate. These results were also plotted in fig. 2 in the form of a heat transfer reduction ($h/h_{zero\ cooling}$) as a function of blowing ratio at 5 different locations. Just downstream of the second row of holes (1.5 film cooling hole diameters), the reduction is quite modest and the optimum blowing ratio is of the order of 0.6. For larger values of this blowing ratio, the aerodynamic disturbing effect of the coolant film (local augmentation of turbulence) counteracts its thermal protection effect. A little farther (8 diameters) the reduction is much more important and the optimal blowing ratio value is shifted towards 0.7. Farther downstream (15, 63 and 131 film cooling hole diameters) the heat transfer reduction is monotonic. As opposed to the suction side results, the aerodynamic effect of the film is of prime importance along the pressure side just downstream of the film cooling holes, where important (up to 100 %) heat transfer coefficient enhancement factor are observed. This is demonstrated in fig. 3 that presents the evolution of this enhancement factor ($h/h_{zero\ cooling}$) as a function of blowing ratio at five different locations (1, 9, 16, 62 and 100 film cooling hole diameters). In the first measurement point, this enhancement factor increases monotically and this effect is maintained for quite a distance before some beneficial effects are observed. The local turbulence augmentation due to the aerodynamic effect of the film, together with the concave nature of the wall, is responsible for this behavior. On the contrary, far downstream, the heat transfer reduction is monotonic and effective.

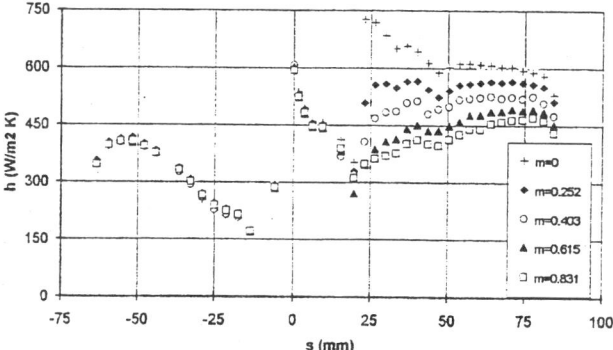

Fig. 1 – Suction side heat transfer coefficient (Film cooled VKI airfoil)

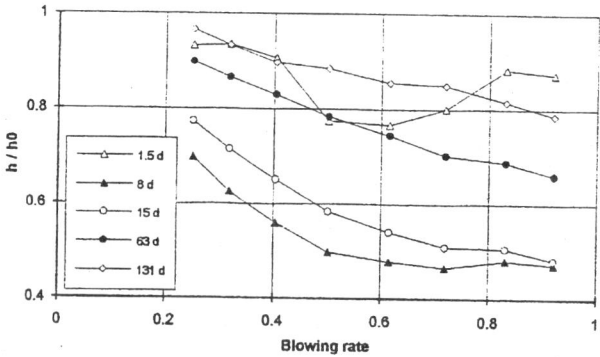

Fig. 2 – Suction side heat transfer reduction (Film cooled VKI airfoil)

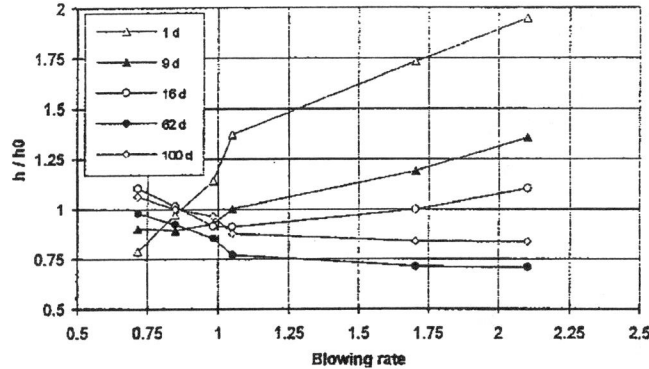

Fig. 3 – Pressure side heat transfer reduction (Film cooled VKI airfoil)

The VKI configuration only considered circular film cooling holes. Many efforts are also made to characterize different film cooling holes geometries, namely fan shaped or layed back fan shaped holes. Typical examples of this work have been presented, among other authors, by Gritsch et al (1998 a, b) (discharge coefficient, adiabatic wall effectiveness), demonstrating the advantage of this geometry. Film cooled leading edges (showerhead cooling) were recently addressed by, among others, Ekkad et al (1997), Hoffs et al (1997) and Ou et al (2000).

Aerodynamics of film cooling

Most of the experimental studies related with film cooling are dealing with a heat transfer analysis. Aerodynamic investigations appear more rarely; flow visualizations and, sometimes, partial aero-measurements are most of the time used to complement these heat transfer investigations.

The published experimental work on the aerodynamics of jets emerging from a wall into a cross flow is mainly devoted to the case of injection normal to the latter. An early contribution was published by Keffer and Baines (1963), who already reported the presence of contra-rotating vortices. Kamotani and Greber (1972) not only confirmed the existence of the vortex pair, but also report that these vortices can dominate the flowfield far downstream. Andreopoulos and Rodi (1984) made a detailed turbulence study and an aerodynamic description of a jet normal to a wall, using a triple hot wire probe to collect their data. Andreopoulos (1985) studied the structure and mixing of normal jets by means of spectral analysis, flow visualization and conditional sampling techniques.

Aerodynamic studies on inclined jets appear less often, and few of them attempt to measure turbulence characteristics. Yoshida and Goldstein (1984), as well as Kadotani and Goldstein (1979) measured the mean and RMS longitudinal and transverse velocity components at one streamwise position. Jubran and Brown (1985) also performed some measurements of turbulence intensity within this type of jet. Pietrzyk et al (1988) conducted a detailed aerodynamic measurement program on inclined jets. This study was made on a flat plate with a single row of holes inclined at 35 deg.

The large scale VKI Low Turbulence Wind Tunnel was used to conduct a detailed measurement program on the interaction between a single or a row of turbulent jets, inclined at 35 deg, with a laminar uniform mainstream along a flat plate. The jets were produced by injecting the secondary flow through holes made in an insert placed in the central part of the flat plate. Three different inserts were available. One was equipped with a single inclined film cooling hole, whereas

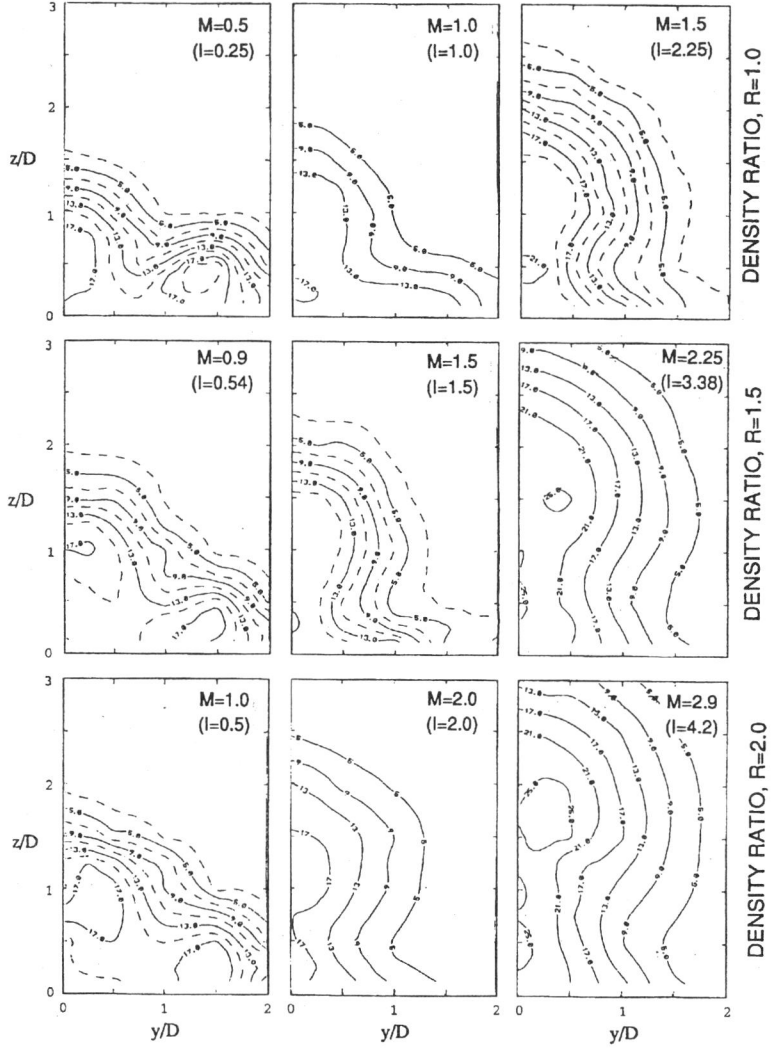

Fig. 4 – Density and blowing ratio effects on turbulence intensity distribution

the two others had a row of seven ejection holes with two different pitch to diameter ratios. To reproduce the real jet to mainstream density ratios, not only jets of air were used, but also mixture of air with a heavier gas, namely SF6.

The main characteristics of the experiment are summarized as follows:
- Freestream velocity : 5 m/s
- Injection angle : 35 deg
- Injection hole diameter : 12 mm
- Incoming boundary layer thickness : 8 mm
- Pitch to diameter ratio : ∞ (single hole), 3, 5
- Blowing ratio (M) : 0.5 ... 2.4
- Momentum ratio (I) : 0.25 ... 4.2
- Density ratio (R) : 1, 1.5, 2

A two-component fiber optic Laser Doppler anemometer system, provided by Dantec, was used to quantify the velocity field. The jets and the mainstream were seeded by two separate systems and measurements were taken in a series of vertical planes perpendicular to the freestream direction; the rotation of the receiving optics by 90 deg allowed the quantification of the third velocity component and the associated statistics. Lopez-Pena et al (1992) provided a detailed description of this setup and of the associated measurement techniques.

One of the striking features, observed in this data base, was the definite correlation between the lateral spreading of turbulence and the momentum flux ratio (I). This is shown in fig. 4, representing isolines of turbulence intensity for a single jet at 10 diameters downstream of the film cooling hole. Nine cases are displayed, three for each of the density ratios considered (1.0, 1.5 and 2.0). The corresponding blowing and momentum ratios are indicated on each graph. All cases having a momentum flux ratio less than unity present a turbulent spot growing besides the jet itself. This lateral spreading of turbulence disappears for momentum flux ratios greater than unity. This is an indication that the high momentum jets break through the boundary layer and produce a large interaction with the freestream, while low momentum jets get confined in the low levels producing a wide interaction with the boundary layer.

The growth of lateral turbulent spots has also been observed for the case of a row of jets. Fig. 5 represents the turbulence level in a plane at the same position as in fig.4, but in this case, to rows of holes of pitch to diameter ratios of 3 and 5 were considered. For simplicity, only the case of air injection is analyzed, and only a low and a high blowing rate are considered. The row with pitch to diameter ratio equal to 5 does not present major differences with the single jet case presented before. The interaction between the jets therefore eventually appears more downstream. On the contrary, the two cases of the row with pitch to diameter ratio of 3 appear to be different than the equivalent single jet cases. The first difference is that the turbulence intensity inside the jet appears to have lower values, and does not extend as high as the single jet case. Additionally, the lateral turbulent spot characteristic of the low momentum case has a higher turbulence level, and appears to be shared by two consecutive jets.

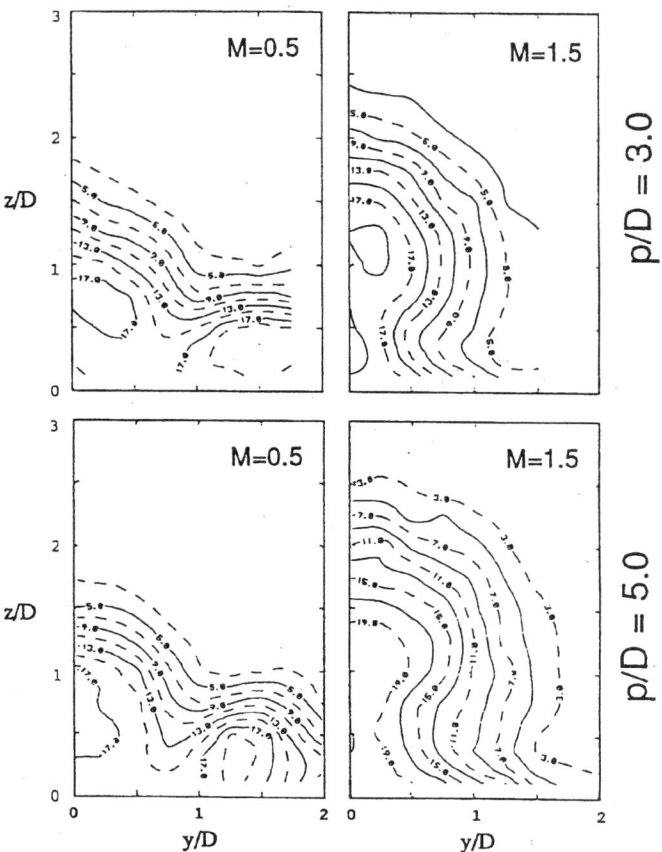

Fig. 5 – Density and blowing ratio effects on turbulence intensity distribution, Influence of pitch-to-hole diameter ratio

Typical examples on aerodynamic measurements performed on film cooling configurations were also reported by, among others, Wilfert and Fottner (1994), Osnaghi et al (1997) and Walters and Leylek (1999)

References

Andreopoulos, J. & Rodi, W. (1984): Experimental investigation of jets in a cross flow. Journal of Fluid Mechanics, Vol. 138

Andreopoulos, J. (1985): On the structure of jets in a cross flow. Journal of Fluid Mechanics, Vol. 157, pp 163-197.

Arts, T. & Bourguignon, A.E. (1990): Behavior of a two rows of holes coolant film along the pressure side of a high pressure nozzle guide vane. *J. of Turbomachinery*, Vol. 112, No 3

Arts, T. & Lapidus, I. (1992): Thermal effects of a coolant film along the suction side of a high pressure turbine nozzle guide vane. AGARD 80th PEP Symposium on Heat Transfer and Cooling in Gas Turbines, Antalya, Turkey.

Arts, T. & Lambert de Rouvroit, M. (1992): Aero-thermal performance of a two-dimensional highly loaded transonic turbine nozzle guide vane – A test case for inviscid and viscous flow computations. J. of Turbomachinery, ASME, Vol. 114, No 1

Arts, T. (1995): Thermal investigation of a highly loaded transonic turbine film cooled guide vane. 1st European Conference on Turbomachinery-Fluid Dynamic and Thermodynamic Aspects, Erlangen, Germany

Béral, C. (1996): Etude expérimentale des écoulements se développant sur un aubage de turbine en absence et en présence d'injections pariétales de refroidissement. Ph. D. Thesis, Ecole Centrale de Lyon, France

Bohn, D.E.; Becker, V.J.; Kusterer, K.A. (1997): 3-D conjugate flow and heat transfer calculations of a film cooled turbine guide vane at different operating conditions. ASME Paper 97-GT-23

Camci, C. (1985): An experimental and theoretical heat transfer investigation of film cooling on a high pressure gas turbine blade. Ph.D. Thesis, Katholieke Universiteit Leuven, Belgium

Daniels, L.C. (1979): Film cooling of gas turbine blades. Ph.D. Thesis, University of Oxford, United Kingdom

Dring, R.P.; Blair, M.F.; Joslyn, H.D. (1980): An experimental investigation of film cooling on a turbine rotor blade. *J. of Engineering for Power*, Vol. 102, No 1

Ekkad, S.V.; Han, J.C.; Dui, H. (1997): Detailed film cooling measurements on a cylindrical leading edge model: effect of freestream turbulence and coolant density. ASME Paper 97-GT-181

Forth, C.J.P. & Jones, T.V. (1986): Scaling parameters in film cooling. 8th Int. Heat Transfer conference, Hemisphere Publishing Corp., New-York.

Goldstein, R.J. (1971): Film cooling. Advances in Heat Transfer, Academic Press, Vol. 7, pp 321-379

Gritsch, M.; Schulz, A.; Wittig, S. (1998 a): Discharge coefficient measurements of film-cooling holes with expanded exits. J. of Turbomachinery, ASME, Vol. 120, No 3

Gritsch, M.; Schulz, A.; Wittig, S. (1998 b): Adiabatic wall effectiveness measurements of film-cooling holes with expanded exits. J. of Turbomachinery, ASME, Vol. 120, No 3

Hoffs, A.; Drost, U.; Bölcs, A. (1997): An investigation of effectiveness and heat transfer on a showerhead-cooled cylinder. ASME Paper 97-GT-69

Horton, F.G.; Schultz, D.L.; Forest, A.E. (1985): Heat transfer measurements with film cooling on a turbine blade profile in cascade. ASME Paper 85-GT-117

Ito, S.; Goldstein, R.J.; Eckert, E.R.G. (1978): Film cooling of gas turbine blade. *J. Engineering for Power*, Vol. 100

Jubran, B. & Brown, A. (1985): Film cooling from two rows of holes inclined in the streamwise and spanwise directions. Journal of Engineering for Gas Turbines and Power, Vol. 107

Kadotani, K. & Goldstein, R.J. (1979): On the nature of jets entering a turbulent flow. *J. Engineering for Gas Turbines and Power*, Vol. 101

Kamotani, Y. & Greber, I. (1972): Experiments on a turbulent jet in a cross flow. *AIAA Journal*, Vol. 10, No 11, pp 1425-1429

Keffer, J.F. & Baines, W.D. (1963): The round turbulent jet in a cross wind. *Journal of Fluid Mechanics*, Vol. 15, No 4.

Ko,S.Y.; Yao, Y.Q.; Xia, B.; Tsou, F.K. (1986): Discrete hole film cooling characteristics over concave and convex surfaces. 8th Int. Heat Transfer conference, Hemisphere Publishing Corp., New-York.

Lander, R.D.; Fish, R.W.; Suo, M. (1972): External heat transfer distribution on film cooled turbine vanes. *J. Aircraft*, Vol. 9, No 10.

Loftus, P.J. & Jones, T.V. (1982): Effect of temperature ratios on the film cooling process. ASME Paper 82-GT-305

López Peña, F. & Arts, T. (1992): On the development of a film cooling layer. AGARD 80th PEP Symposium on Heat Transfer and Cooling in Gas Turbines, Antalya, Turkey.

López Peña, F. (1992): Aerodynamic Aspects of Dilm Cooling. Ph.D. Thesis, Université Catholique de Louvain, Belgium.

Montenay, A.; Paté, L.; Duboué, J.M. (2000): Conjugate heat transfer analysis for an engine internal cavity. ASME Paper 2000-GT-0282

Nicolas, J. & Le Meur, A. (1974): Curvature effects on a turbine blade cooling film. ASME Paper 74-GT-156

Osnaghi, C.; Perdichizzi, A.; Savini, M.; Harasgama, P.; Lutum, E. (1997): The influence of film cooling on the aerodynamic performance of a turbine nozzle guide vane. ASME Paper 97-GT-522

Ou, S.; Rivir, R.; Meininger, M.; Soechting, F.; Tabbita, M. (2000): Transient liquid crystal measurement of leading edge film cooling effectiveness and heat transfer with high free stream turbulence. ASME Paper 2000-GT-0245

Pietrzyk, J.R.; Bogard, D.G.; Crawford, M.E. (1988): Hydrodynamic measurements of jets in cross flow for gas turbine application. ASME Paper 88-GT-174

Walters, D.K. and Leylek, J.H. (1999): Impact of film cooling jets on turbine aerodynamic losses. ASME Paper 99-GT-421

Wilfert, G. and Fottner, L. (1994): The aerodynamic mixing effect of discrete cooling jets with mainstream flow on a highly loaded turbine blade. ASME Paper 94-GT-235

Yoshida, T. & Goldstein, R.J. (1984): On the nature of jets issuing from a row of holes into a low Reynolds number mainstream flow. *J. Engineering for Gas Turbines and Power*, Vol. 106

Combustor Liner Cooling Technology in Scope of Reduced Pollutant Formation and Rising Thermal Efficiencies

A. SCHULZ

Lehrstuhl und Institut für Thermische Strömungsmaschinen
o. Prof. Dr.-Ing. S. Wittig
Universität Karlsruhe (TH)
D-76128 Karlsruhe, Germany

INTRODUCTION

Efficiency and power density of gas turbines have increased drastically during recent years. Today's gas turbines for power generation reach thermal efficiencies up to 38 % at net power outputs beyond 250 MW. Using gas turbines for aeroengines, thermal efficiencies clear beyond 50 % can be realized. The highest aeroengine thrusts today are at about 450 kN. The improvements are primarily a result of increased thermodynamic parameters such as pressure ratio and turbine inlet temperature (combustor exit temperature). Both parameters have direct impact on the cooling of hot gas ducting gas turbine components. The highest combustor exit temperatures of 2000 K and above are approximately 700 K beyond the highest allowable material temperatures while compressor exit temperatures (coolant temperatures up to 950 K) approach continually the maximum material temperatures due to increased pressure ratios. Besides increased temperature differences on the hot gas side and the decreased temperature differences available on the coolant side, the introduction of new low pollutant combustion concepts with their higher need of primary air represents a new challenge for efficient combustor liner cooling. Combustor liner cooling, therefore, is in a conflict between the following parameters:

- hot gas temperature (thermodynamic cycle)
- flame temperature (combustion concept)
- radiation (hot gas temp., combustion concept, pressure ratio, max. material temp.)
- coolant temperature (thermodynamic cycle, pressure ratio)
- available coolant mass flux (hot gas temperature, combustion concept)
- maximum allowable material temperature (service life)

The assessment of the various parameters differs between gas turbines for power generation and aeroengines according to the different demands of the combustion chambers.

IMPACT OF COMBUSTION PROCESS ON FLAME TUBE COOLING DESIGN

Besides efficiency and power especially the reduction of pollutant emissions is in focus of **gas turbines for power generation** development. During the last ten years the emission of nitrogen oxid could be reduced by one order of magnitude. This is primarily achieved by modifying the combustion process. In using only one or a few burners very long and almost stoichometric flames with high flame temperatures are produced in old silo combustion chambers. Short lean pre-mixed flames with considerably lower temperatures are generated in today's combustion chambers. Together with the combustion process the geometry of the combustor changed from big silo combustors with low power density to small highly loaded annular combustors with many burners along the circumference. The thermal load to the combustor walls in case of older combustion chambers is determined by the different combustion zones. In the primary zone the wall heat load is caused by a very hot flame with high radiation and by convection of the combustion gases. The radiative heat exchange with the walls, however, is reduced due to the relative long distance between flame and wall and the absorption by the colder combustion gases in between. Nevertheless, these flame tube areas have to be cooled intensively by establishing thick cooling films parallel to the wall or by use of heat resistant mineral bricks forming the liner. The high temperatures of the primary zone are reduced to turbine entry level by adding cold compressor air. In the mixing zone and in the transition zone to the turbine, the flame tube temperatures can be kept at an acceptable level by applying pure convection cooling on the outer surface (Fig. 1 left).

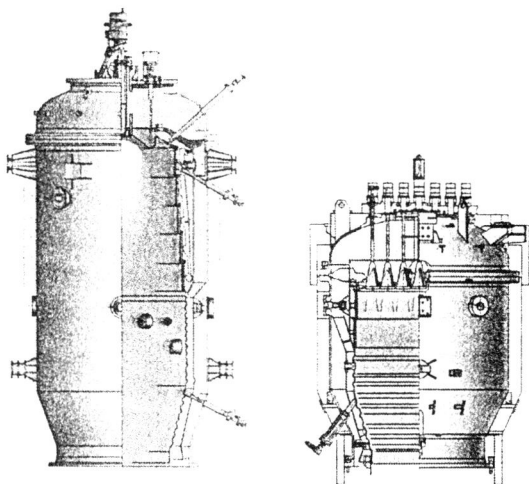

Figure 1: Silo combustors of the ABB GT13D gas turbine. Left: Big silo combustor with one diffusion burner. Right: Compact silo combustor with 36 EV burners. Film Cooled surface is reduced by 2/3 compared to the big silo combustor

Introducing pre-mixed flames in silo combustors and higher turbine inlet temperatures cause a strongly increased demand of primary air. Simultaneously, the amount of coolant for properly cooling the transition area and the turbine increases. Due to the excess air, resulting in low flame temperatures, mixing air injection can be avoided. In addition, the burner technology, which uses a higher number of burners, allows for shorter flames and hence compact combustors. The cooled flame tube surface is considerably reduced. Although the flame temperatures are low, the flame tube has to be film cooled in the flame zone since the flames are close to the wall with accordingly high radiative heat transfer. Here too, the remaining flame tube walls are convectively cooled by a cold counterflow in the double casing (Fig. 1 right).

The most recent versions of ground based gas turbines are equipped with annular combustors which have a three times higher power density compared to silo combustors operated in the pre-mixed mode. Such annular combustors penetrate into power density areas typical for jet engine combustors. The pre-mixed flame takes up almost the entire flame tube. The flame tube cooling has to be adjusted to the high thermal load. To reduce the expenditure of cooling, in many cases an impingement cooling is applied to the outer surface of the flame tube. Before blown through slots between flame tube tiles or ejected through holes in the wall, the coolant absorbs the heat transferred through the wall. On the inner surface the coolant forms a protective wall film (Fig. 2).

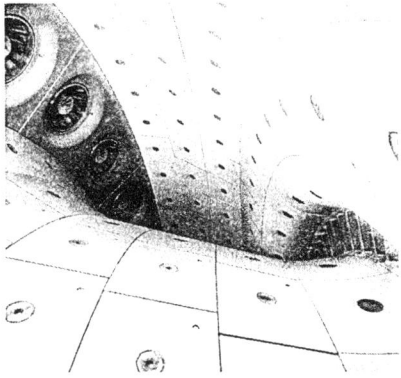

Figure 2: Compact annular cobustor of the SIEMENS V84.3a gas turbine

Whereas combustion chambers of ground based gas turbines can be operated under very lean conditions (the difference between flame temperature and turbine inlet temperature is not much higher than 100 K), the combustion chambers of **aeroengines** have to be run closer to stoichiometric conditions. The reasons on the one hand are the considerably higher turbine inlet temperatures despite simultaneously increased combustor inlet temperatures and the higher requirements on flame stability. On the other hand, it is necessary to add compressor air to the hot gases for establishing an adequate radial temperature profile at the exit of the combustor. The latter reason demands a compromise. Increasing the mean turbine inlet temperature at constant maximum wall temperatures necessitates a disproportionate rise of hot gas temperatures and more intensive wall cooling, i.e. the peak temperature in the radial temperature profile strongly increases hence requiring a higher cooling effort for the turbine blading. If wall temperature could be increased at constant mean turbine inlet temperature, the cooling effort of the flame tube as well as the turbine could be reduced, since the temperature profile at combustor exit would be more uniform. Flame temperatures could be reduced as well, which would result in lower nitrogen oxid formation. At ground based gas turbines, which have no mixing air injection, the temperature profile at combustor exit is almost constant at a high temperature level. This in turn requires an adequate blade cooling design. Since the turbine inlet temperatures of aeroengines increased much faster than the maximum allowable material temperatueres during recent years, the cooling methods had to be improved considerably. The major goal is a sufficient cooling of the flame tube at the highest allowable temperature level. This requires the application and combination of different cooling methods at the various locations of the flame tube. Most common is a combination of convective cooling on the outer surface of the flame tube and film cooling on the hot surface.

COMBUSTOR LINER COOLING METHODS

There are various possibilities to generate wall cooling films. With respect to film cooling effectiveness, an isokinetic two-dimensional film performs best. In practice, however, two-dimensional wall parallel slots can not be realized. In earlier designs, wall parallel films are formed by overlapping sheets kept on a constant distance by a corrugated strip. More recent designs consist of a machined ring with discrete cooling holes. The ejected air is then evenly distributed on the wall through a lip. Actual designs like the „Z"-ring (Rolls-Royce), rolled ring (General Electric) or double pass ring (Pratt&Whitney) are all lipless configurations since lips are subjected to pronounced temperature gradients resulting in high thermal stresses and hence reduced service life. Combined cooling methods often use impingement cooling on the outer liner wall for locally intensifying the convective heat transfer. Especially in wall areas where film effectiveness is poor, intensified convective heat transfer reduces the wall temperatures. The combined cooling methods usually have a double wall. An example, among others, is the float wall concept of Pratt&Whitney. Here, metal shingles are screwed to the outer casing in a way that allows thermal expansion of the shingles. The shingles are provided with cooling air from the back surface. The coolant then is ejected through slots between the staggered shingles.

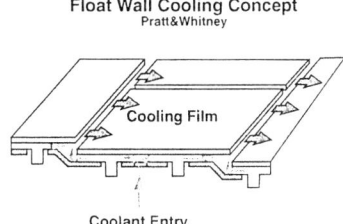

Figure 3: Example of a double wall design of modern combustor flame tubes combining different cooling methods

Latest designs use full-coverage film cooling for the combustor liner. This cooling method requires a large number of holes drilled into the liner. The major advantage of full-coverage film cooling is the uniform temperature distribution of the wall, the high overall efficiency and the low coolant consumption. Today´s combustion chambers of aeroengines use full-coverage film cooling for great portions of the flame tube. As with film cooling, full-coverage film cooling can be used either as single cooling method or in combination with an back side impingement cooling. The latter is subject of present research and development efforts.

GENERIC STUDIES OF FLAME TUBE COOLING CONFIGURATIONS
Slot Ejection

Since most flame tubes utilize discrete cooling films so far, numerous studies are performed with parallel ejection of cooling air onto a flat plate to model the mixing process of coolant and hot gas. Pioneering work regarding film ejection into a free-stream was performed by Wieghardt (1946), who first developed a correlation for the temperature profile far downstream of the ejection location and small blowing ratios. In these early studies (see e.g. Scesa 1954, Hartnett et al. 1961) it became obvious that the distribution of the adiabatic wall temperature can be correlated with the ratio M of mass flux density of coolant and hot gas, respectively.

$$M = \frac{\rho_C \cdot u_C}{\rho_{HG} \cdot u_{HG}} \quad \text{(Blowing Ratio)} \tag{1}$$

The adiabatic wall temperature is expressed in terms of an dimensionless adiabatic cooling effectiveness η_{aw}.

$$\eta_{aw} = \frac{T_{HG} - T_{aw}}{T_{HG} - T_C} \tag{2}$$

Further studies (Chin et al. 1961, Samuel and Joubert 1965) provide improved correlations for film cooling effectiveness and heat transfer in the near field of the ejection. Especially with high blowing ratios, a significant increase in heat transfer compared to flat plate results was found. Bittlinger (1994) derived a new correlation for film cooling effectiveness and heat transfer for tangential coolant ejection from own experiments. The correlation covers a wide range of geometric and thermodynamic parameters. Moreover, he gives a comprehensive survey on correlations developed for tangential coolant ejection so far (Bittlinger 1995).

Combining film cooling with convective cooling schemes

Although substantial work has been put in the improvement of film cooling schemes, purely film cooling the combustor liner is not an efficient use of coolant. It is rather necessary to use the cooling potential of the cooling air before it is ejected into the hot gas flow. Therefore, new cooling concepts include effective cooling schemes for the backside of the combustor liner such as ribbed channels, pin fins, or impingement cooling. A qualitative assessment of the different heat transfer enhancing measures is given in Figure 4. The total cooling effectiveness of a film cooled surface is plotted versus coolant mass flux. As reference, the adiabatic cooling effectiveness which can be achieved with the same amount of coolant, i.e. without convectively cooling the back surface, is plotted into the top left diagramm. It can be clearly seen that the amount of coolant can be more than halfed by applying forced convection on the outer surface of the flame tube. The effect of ribs, pin fins, and impinging jets is shown in the top right and bottom diagrams of Figure 4. Here, the total cooling effectiveness obtained with forced convection at a slot hight of 3mm is plotted as reference. The heat transfer augmentation of ribs and pin fins is very similar and offers the potential of a further reduction of the amount of coolant by about 50%. Compared to ribs and pin fins, impingement cooling depends more on the actual geometry. The cooling effectiveness increases with decreasing hole diameter. The impingement schemes with large hole diameters cannot compete with the ribs or pin fins. For the small holes, however, the best cooling performance is found.

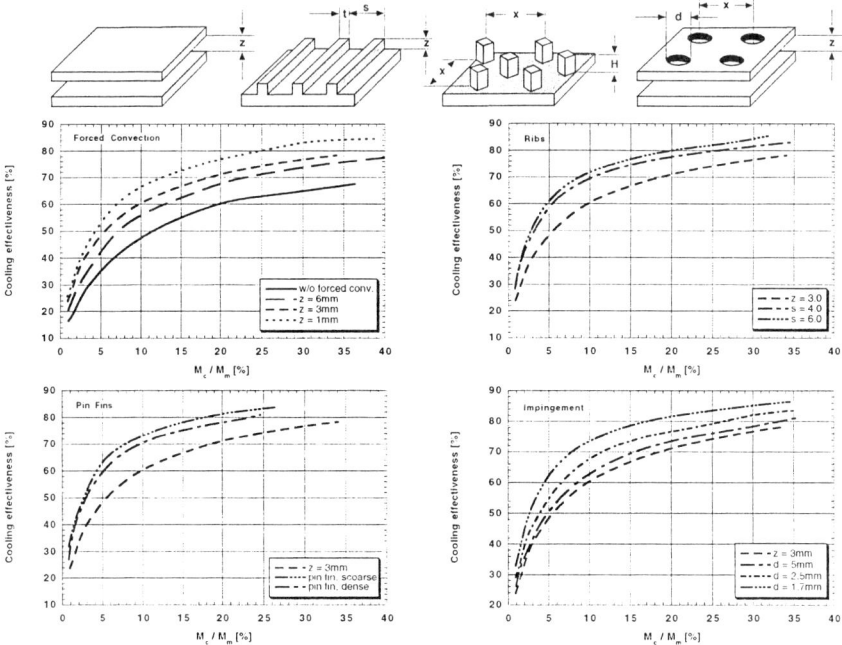

Figure 4: *Effect of heat transfer augmentation on the outer surface of flame tubes*

Interaction of a cooling film with a mixing jet

In real combustors the interaction of high momentum mixing jets and wall parallel cooling films has a strong impact on the combustor liner temperatures. Experimental and theoretical investigations of such configurations were performed by Best (1979). He found a considerable reduction in film cooling effectiveness downstream of a normal mixing jet and addressed it to the augmented turbulence in the wake of the jet. In a detailed study Martiny et al. (1997) demonstrated the effect of a mixing jet on film cooling effectiveness and heat transfer of a film cooled surface. On a flat plate model with slot injection and a normal mixing jet the complex flow was investigated qualitatively by use of a laser light sheet, and quantitatively by applying laser-Doppler-velocimetry.

Furthermore, detailed surface temperature mappings were performed by utilizing an infrared camera. Combining local surface temperatures with the known wall heat flux, local film cooling effectiness as well as heat transfer coefficients could be determined. As an example, the distribution of local isothermal heat transfer coefficients and a velocity vector plot is given in Figure 5 for a film blowing ratio of $M=1$ and a jet momentum flux ratio of $I=7$. In lateral direction the effect of the mixing jet on the heat transfer coefficient is noticeable up to 3 hole diameters in both directions. Beyond this area, no influence of the mixing jet is found, the heat transfer coefficients show the typical behaviour of a two dimensional cooling film ejection. Downstream of the mixing air injection, a pronounced increase in heat transfer can be observed for all test conditions investigated. For the example presented, the heat transfer coefficient is almost twice as high as in the undisturbed region. The high values decrease with downstream distance, approaching undisturbed conditions at a downstream distance of 15 mixing hole diameters. In contrast to the heat transfer, the adiabatic film cooling effectiveness is reduced by 15-25% downstream of the mixing air injection, depending on the combination of blowing ratio and momentum flux ratio of the mixing jet. Both effects lead to a drastically increased thermal load to the combustor wall downstream of a normal mixing air injection. Therefore, in most cases aeroengine combustor liners are equipped with additional film cooling holes downstream of a mixing jet injection.

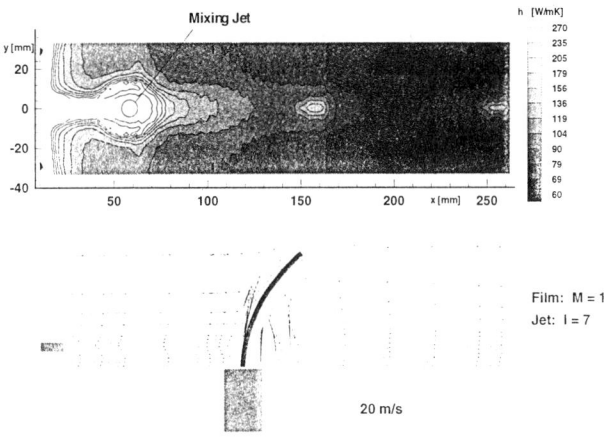

Figure 5: Heat transfer distribution of 2D slot ejection of coolant with a normal high momentum mixing jet (top), volocity vectors in the center plane of the mixing jet (bottom)

Ejection from cooling holes

Although the ejection of coolant through discrete cooling holes or rows of cooling holes is predominantly utilized in turbine bladings, it is worthwile to be considered since it provides an indepth insight into the complex interaction of coolant and hot gas flow. Furthermore, the ejection of a row of cooling holes represents a base case for the the full coverage film cooling of combustor liners to be discussed later.

Highly resolved flowfield measurements for typical cases of inclined coolant jets in crossflow were performed by Lee et al. (1994), illustrating the jet and main flow interaction. Detailed flowfield measurements and additional data on turbulence development of coolant ejection are provided by Pietrzyk et al. (1990), Burd et al. (1998), and Thole et al. (1998) for similar ejection situations. Experimental data on the temperature field connected to this type of flows are given by Ryndholm (1996) and Kohli and Bogard (1997). Regarding film cooling effectiveness downstream of a row of cooling holes, most research work concentrates on overall or laterally averaged effectiveness. Measurements of local effectiveness values were provided e.g. by Ekkad et al. (1997a), Goldstein et al. (1998), or Lutum and Johnson (1998). Investigations of local heat transfer coefficients were performed in the vicinity of the ejection holes by Kumada et al. (1981), Goldstein and Taylor (1982), and Cho and Goldstein (1995), using the mass transfer analogy at unity density ratio. They provide insight into the near ejection mixing phenomena. Investigations extending downstream into the film region have been undertaken by Ekkad et al. (1997b) and Goldstein et al. (1998).

The majority of the investigations concentrate on ejection angles of 30°-35° and only few heat transfer investigations were conducted at turbine like density ratios with high local resolution. Baldauf et al. (1999a,b) determined both, film cooling effectiveness and heat transfer coefficients for high density ratio film cooling on a flat plate downstream of a row of cylindrical cooling holes. They varied geometric and aerodynamic parameters in a wide range. Characteristic patterns of film cooling effectiveness and heat transfer distributions could be identified. Detailed observations of jet lift off effects and mixing phenomena confirm previous models (L'Ecuyer and Soechting 1985) and support the classification into flow regimes. The data documents significant and systematic changes in the flow behavior with the change from shallow ejection angle to normal ejection as well as low pitch ejection. As an example of the comprehensive study, local heat transfer coefficients downstream of a row of cylindrical cooling holes are shown in Figure 6.

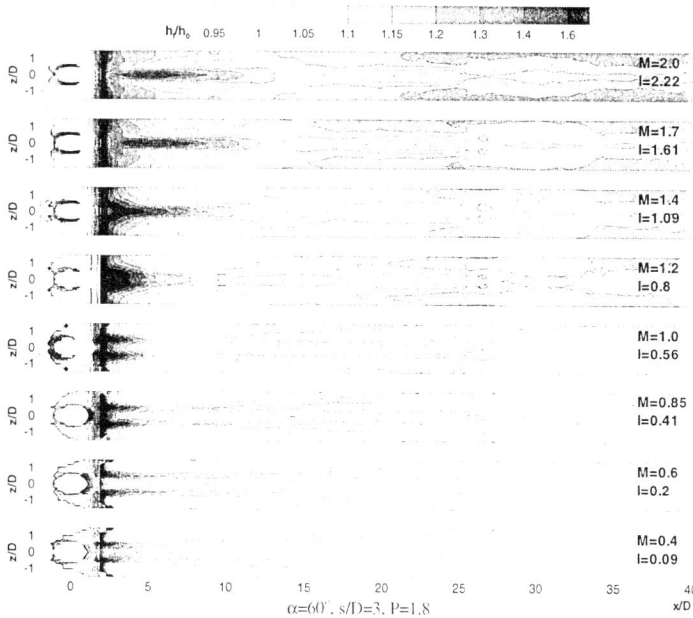

Figure 6: Local heat transfer coefficients downstream of 60° inclined row of cooling holes, at a density ratio of 1.8 and lateral spacing of 3 hole diameters

At low blowing rates, typical flow conditions of a fully attached coolant jet are present. The counter rotating vortices of the coolant jet flow are lying on the surface and cause significant traces of enhanced heat transfer. These are extending from the lateral edges of the ejection hole up to 20D downstream the ejection. Up to M=0.85 only minor changes in the vortex behavior are visible. At a blowing ratio of M=1.0, a significant change in the flow situation can be observed. The vortex surface contact is lost at around 7D downstream of the ejection, where the vortex traces are expected to be nearest to the centerline. Coolant flow and vortices are driven apart from the surface by the increased normal momentum of the coolant. For blowing rates of M=1.7 and beyond, a second significant change in the flow pattern takes place. Vortex interaction with the surface is suspended as the coolant jet completely detaches from the surface and penetrates into the hot gas flow.

Combustor flame tubes in general are provided with multiple cooling films to guaratee a sufficient protection of the whole surface. The interaction of such discrete cooling films and their optimal combination is of major importance. Figure 7 shows a typical result of a comprehensive study on the superposition of cooling films formed by discrete rows of cylindrical cooling holes. The lower contour plot shows the characteristic distribution of adiabatic film cooling effectiveness. At given blowing rates, extensively cooled regions form downstream of the respective cooling holes with pronounced maxima on the centerlines. Particularely downstream of the first row of holes, the laterally averaged film cooling effectiveness is low due to the fact that the coolant must spread in lateral direction to reach locations between the holes. In the staggered cooling hole arrangement shown, coolant ejected from the first row of holes hits the unprotected area between the holes of the second row. The fluid entrained by the counter rotating vorteces of the second coolant jets is cooled air ejected by the first row of holes. This favourable interaction of the cooling films allows an efficient use of higher amounts of coolant at the second row of cooling holes. The upper contour plot of Figure 7 shows the normalized heat transfer distributions. The interaction of vorteces generated by the first ejection with the coolant jets of the second row of holes leads to a stabilizing effect of downstream coolant ejection. The tendency of jet detachment is shifted towards higher blowing ratios.

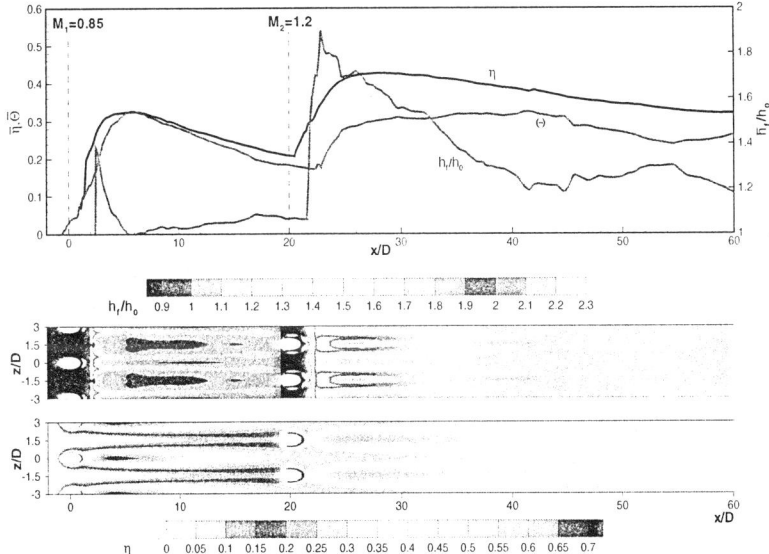

Figure 7: Superposition of film cooling effectiveness, heat transfer coefficient, and heat flux reduction for two consecutive rows of cylindrical cooling holes (ejection angle 30°)

The highest laterally averaged film cooling effectiveness is found in the area of the second ejection, although local values are less than those of the first ejection. The interaction of the two ejections leads to a high and steady film cooling effectiveness over extended surface areas. However, the high film cooling effectiveness goes hand in hand with considerably increased heat transfer coefficients. The cooling effect of two rows of cooling holes, therefore, is characterized by the heat flux reduction Θ. A comparison of the effectiveness and heat flux reduction downstream of the first injection reveals that the surface temperature reduction is almost entirely converted into a thermal relief of the wall. Downstream of the second ejection this conversion takes place in part only, since high heat transfer coefficients counteract the temperature reduction by the cooling film.

Full coverage film cooling

The literature available concerning full coverage film cooling is by far not as extensive as in the case of pure film cooling. First experimental studies regarding boundary layer development and adiabatic effectiveness at varying blowing angles and hole arrangements were performed by LeBrocq et al. (1973). Utilizing a transient measuring technique Metzger et al. (1973) investigated heat transfer and adiabatic effectiveness for inline and staggered hole arrays. The effect of hole density on the heat transfer was subject to the investigations of Mayle et al. (1975). Further fundamental experiments on heat transfer and cooling effectiveness at different ejection angles were performed by Crawford et al. (1980). Kumada et al. (1981) and Kasagi et al. (1981) applied the heat and mass transfer analogy in their experiments and demonstrated the strong impact of convection on the total cooling effectiveness. All these investigations contribute to a better understanding of the physical phenomena. For the design of effusion cooling, however, the results of these generic experiments have to be transferred to real application conditions. Experimentally demanding but crucial for the transferability, is the correct temperature and density ratio between coolant and hot gas. The previously mentioned investigations were all performed at small temperature differences between coolant and hot gas. Ammari et al. (1990) studied the influence of densitiy ratio on heat transfer coefficient. Their results show that heat transfer depends on density ratio especially at inclined ejection. The effect of density ratio on adiabatic cooling effectiveness was investigated by Foster et al. (1975). From their results it became obvious that density ratio has a much stronger influence on cooling effectiveness than on heat transfer.

As indicated earlier, the primary target in applying full-coverage film cooling to combustor walls is the reduction of cooling air since more air is needed for the combustion process itself and for properly cooling turbine vanes, blades, and disks. To prevent cooling jet separation and to increase the operational range of full-coverage film cooling towards higher blowing rates, cooling hole configurations with extremely shallow ejection angles are aimed on. Martiny et al. (1995,1997), therefore, concentrate on full-coverage film cooling at very shallow ejection angles (17°) under combustor typical conditions. The shallow ejection not only improves the formation of a protective cooling film but also increases the internal convective heat exchange due to an increased hole surface area.

Figure 8 shows the surface temperature distributions in terms of cooling effectiveness patterns and laterally averaged effectiveness for three blowing ratios. Attached cooling films are formed downstream of every ejection hole for small blowing rates (M=0.5). The coolant streaks superpose and merge with downstream distance. Downstream of the last row of holes cooling effectiveness drops rapidly due to the little amount of coolant introduced into the boundary layer. At a blowing rate of M=1.2 the coolant starts to detach from the wall, leading to a comparatively low effectiveness within the first rows of holes. Further downstream the coolant is brought back to the surface by diffusion. The cooling effectiveness increases considerably. The drop in efficiency downstream of the effusion hole array is not as pronounced as in the low blowing rate case. At the highest blowing rate (M=3.0) the ejection holes act as strong heat sinks. There are steep temperature gradients around the holes and almost no cooling film formation within the first rows of holes. Hot spots detected just downstream of the cooling holes are caused by downwash of hot gas under the detached coolant jet. The formation of a closed cooling film is shifted further downstream.

Cooling effectiveness stays on a high level even downstream of the hole array due to the high amount of coolant blown into the hot gas. A detailed description of the study, including the effect of an additional impingement cooling of the cold side of the full-coverage film cooling test specimen is given by Schulz et al. (2000).

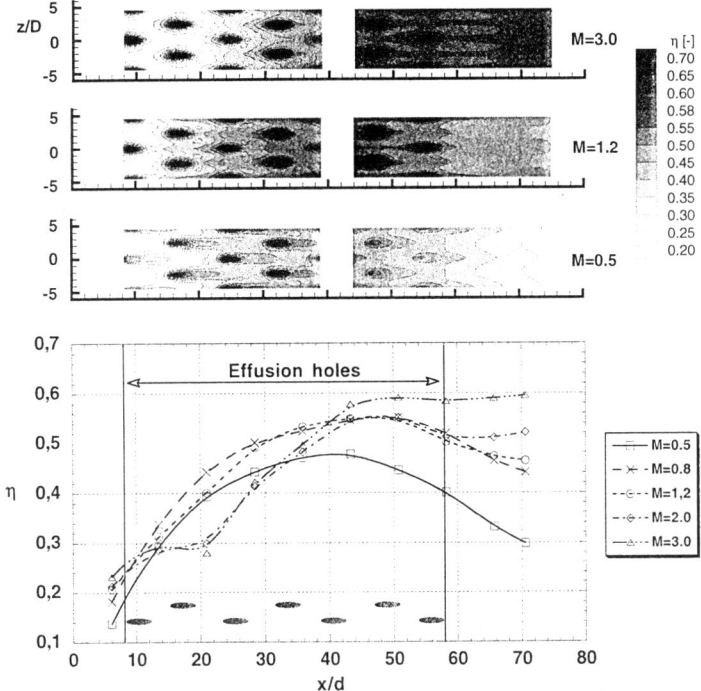

Figure 8: *Local and laterally averaged adiabatic full-coverage film cooling effectiveness*

CLOSURE

The preceeding discussion, although far from complete, should serve to introduce into combustor liner heat transfer. The emphasis was put on convective problems which are connected to highly complex, three-dimensional, and turbulent flowfields. As indicated right at the beginning, radiation is another important mode of heat transfer and must be taken into account since it can considerably contribute to the thermal loading of combustor walls. However, for reasons of brevity, radiation could not be included into the discussion. The same applies to numerical modelling of combustor liner cooling. A huge number of different approaches have been developed lately to provide engineers with more reliable theoretical design tools and to reduce experimental efforts.

REFERENCES

R. Best, (1979). "Investigation on the Cooling Effectiveness of Tangentially and Normally Injected Coolant Flows", Ger. Chem. Eng., Vol. 2, pp. 343-351

S. Burd, R. Kazeta, T. Simon, (1996). "Measurements in Film Cooling Flows: Hole L/D and Turbulence

Intensity Effects", ASME Paper No 96-WA/HT-7

J. Chin, S. Skirvin, L. Hayes, F. Burggraf, 1961, "Film Cooling with Multiple Slots and Louvers - Part I: Multiple Continuous Slots", Journal of Heat Transfer, Vol. 83, pp. 281-286

H. Cho, R. Goldstein, (1995), "Heat (Mass) Transfer and Film Cooling Effectiveness With Injection Through Discrete Holes: Part II - On the Exposed Surface", J. of Turbomachinery, Vol. 117, pp. 451-460

M. Crawford, W. Kays, R. Moffat, (1980), "Full-Coverage Film Cooling Part I: Comparison of Heat Transfer Data for Three Injection Angles", J. of Engineering for Power, Vol. 102, pp. 1000-1005

M. L'Ecuyer, F. Soechting, (1985), "A Model for Correlating Flat Plate Film Cooling Effectiveness for Rows of Round Holes", Heat Transfer and Cooling in Gas Turbines, AGARD-CP-390, Paper 19

S. Ekkad, D. Zapata, J. Han, (1997a), "Film Effectiveness Over a Flat Surface With Air CO_2 Injection Through Compound Angle Holes Using a Transient Liquid Crystal Image Method", J. of Turbomachinery, Vol. 119, pp. 587-593

S. Ekkad, D. Zapata, J. Han, (1997b), "Heat Transfer Coefficients Over a Flat Surface With Air CO_2 Injection Through Compound Angle Holes Using a Transient Liquid Crystal Image Method", J. of Turbomachinery, Vol. 119, pp. 580-586

R. Goldstein, P. Jin, R. Olson, (1998), "Film Cooling Effectiveness and Mass/Heat Transfer Downstream of One Row of Discrete Holes", ASME Paper No 98-GT-174

R. Goldstein, J. Taylor, (1982), "Mass Transfer in the Neighborhood of Jets Entering a Crossflow", J. of Heat Transfer, Vol. 104, pp. 715-721

J. Hartnett, R. Birkebak, E. Eckert, 1961, "Velocity Distribution, Temperature Distribution, Effectiveness and Heat Transfer in Cooling of a Surface with Pressure Gradient", International Developments in Heat Transfer, Part IV, pp. 682-689

N. Kasagi, M. Hirata, M. Kumada, (1981), "Studies of Full-Coverage Film Cooling Part 1: Cooling Effectiveness of Thermally Conductive Walls", ASME Paper No 81-GT-37

A. Kohli, D. Bogard, (1997), "Adiabatic Effectiveness, Thermal Fields, and Velocity Fields for Film Cooling With Large Angle Ejection", J. of Turbomachinery, Vol. 119, pp. 352-358

M. Kumada, M. Hirata, N. Kasagi, (1981), "Studies of a Full-Coverage Film Cooling Part 2: Measurement of Local Heat Transfer Coefficient", ASME Paper No 81-GT-38

P. Le Brocq, B. Launder, C. Priddin, (1973), "Experiments on Transpiration Cooling: Discrete Hole Injection as a Means of Transpiration Cooling: An Experimental Study", Proc. of the Institution of Mechanical Engineers, 187, pp. 149-157

S. Lee, J. Lee, S. Ro, (1994), "Experimental Study on the Flow Characteristic of Streamwise Inclined Jets in Crossflow on Flat Plate", J. of Turbomachinery, Vol. 116, pp. 97-116

E. Lutum, B. Johnson, (1998), "Influence of the Hole Length To Diameter Ratio on Film Cooling with Cylindrical Holes", ASME Paper No 98-GT-10

M. Martiny, A. Schulz, S. Wittig, M. Dilzer, (1997), "Influence of a Mixing Jet on Film Cooling", ASME Paper No 97-GT-247

M. Martiny, A. Schulz, S. Wittig, (1995), "Full-Coverage Film Cooling Investigations: Adiabatic Wall Temperatures and Flow Visualization", ASME Paper No 95-WA/HT-4

M. Martiny, A. Schulz, S. Wittig, (1997), Mathematical Model Describing the Coupled Heat Transfer in Effusion Cooled Combustor Walls", ASME Paper No 97-GT-329

D. Metzger, D. Tekeuchi, P. Kuenstler, (1973), "Effectiveness and Heat Transfer with Full-Coverage Film Cooling", J. of Engineering for Power, 95, pp. 180-184

R. Mayle, F. Camarata, (1975), "Multihole Cooling Film Effectiveness and Heat Transfer", J. of Heat Transfer 97, pp. 534-538

J. Pietrzyk, D. Bogard, M. Crawford, (1989), "Hydrodynamic Measurements of Jets in Crossflow for Gas Turbine Film Cooling Applications", J. of Turbomachinery, Vol. 111, pp.139-145

H. Ryndholm, (1996), "An Experimental Investigation of the Velocity and Temperature Fields of Cold Jets Injected Into a Hot Crossflow", ASME Paper No 96-GT-491

A. Samuel, P. Joubert, 1965, "Film Cooling of an Adiabatic Flat Plate in Zero Pressure Gradient in the Presence of a Hot Mainstream and a Cold Tangential Secondary Injection", Journal of Heat Transfer, Vol. 87, pp. 409-418

S. Scesa, 1954, "Effect of Local Normal Injection on Flat-Plate Heat Transfer", Dissertation, University of California, Berkeley, USA

A. Schulz, S. Wittig, M. Martiny, (2000), "Effusion cooled combustor liners of gas turbines - an assessment of the contributions of convective, impingement, and film cooling", Proc. Symp. on Energy Engineering (SEE2000), Ed. Ping Cheng, Hong Kong University of Science and Technology, 9-13 January 2000, Vol. 1, Paper B-10

K. Thole, M. Gritsch, A. Schulz, S. Wittig, (1998), "Flow Field Measurements for Film Cooling Holes with Expanded Exits", J. of Turbomachinery, Vol. 120, pp. 327-336

K. Wieghardt, 1946, "Hot air discharge for De-Icing", AAF Translation F-Ts 919-Re, August 1946, Wright Field

Jet-Impingement Heat Transfer in Gas Turbine Systems

B. HAN AND R. J. GOLDSTEIN

Heat Transfer Laboratory
Department of Mechanical Engineering
University of Minnesota, Minneapolis, MN55455, U.S.A.

ABSTRACT : A review of jet-impingement heat transfer in gas turbine systems is presented. Characteristics of the different flow regions for submerged jets - free jet, stagnation flow, and wall jet - are reviewed. Heat transfer characteristics of both single and multiple jets are discussed with consideration of the effects of important parameters relevant to gas turbine systems including curvature of surfaces, crossflow, angle of impact, and rotation.

INTRODUCTION

Impinging jets are used in many applications due to the high heat/mass transfer rates they can produce. Applications in engineering fields include electronic-chip cooling, paper and fabric drying, cooling of hot metal sheets, cryogenic tissue freezing and others. In addition to their important applications, impinging jets have many interesting fluid dynamics and heat transfer features.

In gas turbine systems, jet-impingment cooling is finding increased use as higher turbine inlet temperatures are utilized. For turbine components including turbine guide vanes, rotor blades, rotor disks and combustor walls, impingement cooling provides one of the useful ways to prevent overheating. However, the complex geometry of turbine systems, the high turbulence, and the rotation of the systems make the understanding of flow and heat transfer characteristics of impinging jets a challenging subject.

There have been numerous experimental and numerical investigations on flow and heat transfer characteristics of impinging jets. Since Martin[1] published his review on heat and mass transfer of impinging jets, several additional literature reviews[2-6] have appeared. At this time, a further review of impinging jet flow and heat transfer can provide a summary of current understanding and suggest directions for future research.

In this paper, research on impingement heat transfer as applied to gas turbine systems is reviewed. Therefore, the focus is laid on circular or slot jet impingement

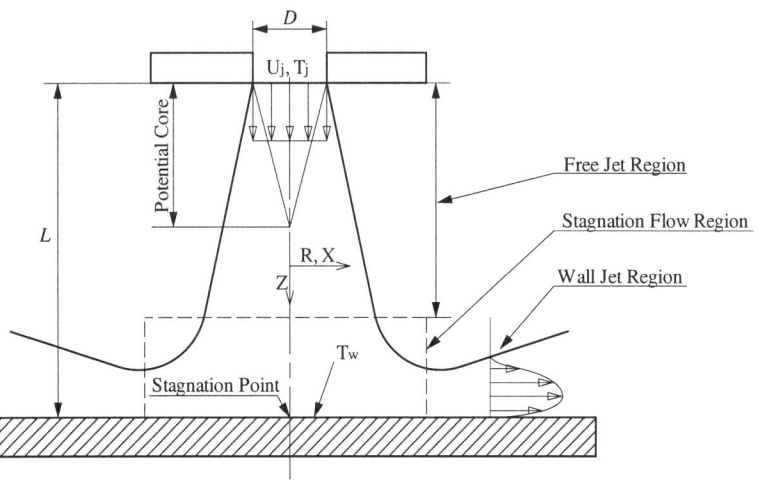

Figure 1: Schematic diagram of flow region around an impinging jet

and impingement of jet arrays. The flow fields of an impinging jet consists of a free jet region, a stagnation flow region, and a wall jet region. Flow characteristics of each regions are reviewed. The coherent structure of ring vortices and its role in turbulence and heat transfer are discussed. Local and averaged heat transfer characteristics for both a single jet and an array of jets are considered including effects of the Reynolds number and nozzle-to-plate spacing. Effects of other parameters to simulate gas turbine systems including impingement on a curved surface, effect of crossflow, effect of rotation and angle of impact are also considered.

FLUID DYNAMICS OF AN IMPINGING JET

The flow field of an impinging jet is illustrated in Figure 1. It is categorized by three regions - the free jet region, the stagnation flow region and the wall jet region. Characteristics of each region are briefly discussed in this paper. A detailed review of the flow field is beyond the scope of the present paper and can be found in many books[7-9] and reviews.[10]

The free jet region can be subdivided into a flow development region and a fully developed region. In the flow development region, the potential core can be observed until $L/D \approx 4$ to 6. The potential core is diminishing in width as the shear layer around the jet grows. After the jet is fully developed, axial velocity profile is approximated Gaussian. The free jet is considered to be turbulent[10] when the Reynolds number is larger than 3×10^3.

In the stagnation region, flow is affected by the presence of the impingement surface. Mean velocity and turbulent intensity were measured by Nishino et al.[11] They showed that the Gaussian distribution in axial velocity profile is valid even in the near vicinity of the surface. Radial velocity increases rapidly as the surface is approached.

The wall jet region is characterized by a flow in the outward radial or spanwise direction. In this region, the development of boundary layer from the stagnation point has a strong impact on the local and averaged heat transfer rate. Detailed flow field measurements can be founded in previous studies.[8,12]

Numerical studies[13-15] sought a suitable turbulence model for an impinging jet. Recently Olsson and Fuchs[16] performed Large Eddy Simulations(LES) of an impinging jet with a Reynolds number of 10^4.

Shear-driven entrainment of ambient fluid establishes a coherent structure of ring vortices around an impinging jet, which has been observed in many flow visualization studies.[17,18] Since the coherent structure of vortices was found, interest on the role of the structure in the development of turbulence and heat transfer enhancement has increased. Recently several studies[19-21] suggested the connection between the motion of ring vortices and "energy separation" in the impinging jet. There have also been some attempts to enhance the heat transfer rate by controlling the motion of coherent structure using acoustic excitation.[21-24]

HEAT TRANSFER OF IMPINGING JETS

Primary Variables

The local heat transfer coefficient, h, can be defined by;

$$h = \frac{q_w}{T_w - T_{ref}} \quad (1)$$

where q_w is the wall heat flux, T_w is the wall temperature, and T_{ref} is a reference temperature - usually either the jet total temperature T_j or the adiabatic wall temperature T_{aw}. The effect on h of the choice of reference temperature will be discussed later.

The local heat transfer coefficient, h, can be non-dimensionalized to the Nusselt number.

$$\text{Nu} = \frac{hD}{k} \quad (2)$$

where D is the nozzle diameter, and k is the thermal conductivity of the fluid.

The adiabatic wall temperature, T_{aw}, can be non-dimensionalized using a recovery factor (r)

$$r = \frac{T_{aw} - T_j}{U_j^2/2C_p} \quad (3)$$

and effectiveness (η)

$$\eta = \frac{T_{aw} - T_r}{T_j - T_\infty} \quad (4)$$

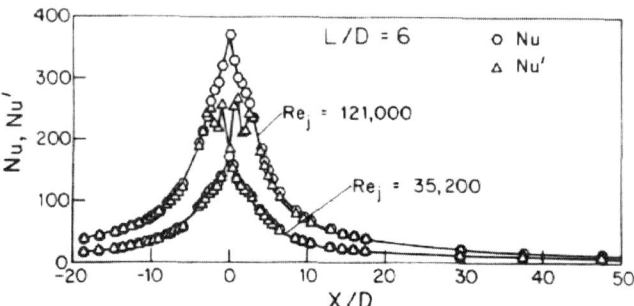

Figure 2: Variation of local Nusselt number for a single jet using different reference temperature[28]

where U_j is the jet velocity, C_p is the specific heat of the fluid at constant pressure, T_r is the adiabatic wall temperature when $T_j = T_\infty$, and T_∞ is the ambient fluid temperature.

One objective of research on an impinging jet is to find a reasonable empirical correlation for the local, stagnation point and averaged Nusselt number after assumed to be a power law relationship.

$$\mathrm{Nu} = C\mathrm{Re_D}^m \mathrm{Pr}^n \qquad (5)$$

where C is a constant, $\mathrm{Re_D}$ is the Reynolds number based on jet velocity and nozzle diameter, and Pr is the Prandtl number of the fluid.

Heat Transfer Characteristics of a Single Jet

The local Nusselt number distribution for a circular impinging jet is shown in Figure 2. Nu is calculated from a heat transfer coefficient based on the temperature difference between T_w and T_{aw}, but Nu′ is evaluated with T_j as the reference temperature. In the Nu distribution, there is a maximum heat transfer rate point at the stagnation point and Nu monotonously decreases regardless of the Reynolds number. For low Reynolds number, the variation of Nu′ is the same as that of Nu. However, secondary maximum points are observed in the Nu′ profile for high Re. These secondary peaks are artifacts and dependent on the wall temperature. they disappear when the heat transfer coefficient is defined properly. Secondary maximum points observed at small nozzle-to-plate spacing[25–27] have been attributed to energy separation within the shear layer around the jet and, by others, to transition in the wall boundary layer.

The recovery factor distribution at the stagnation point and 2 diameter away from the stagnation point are shown in Figure 3. At the stagnation point, for all Reynolds number, the recovery factor is close to unity at $L/D = 2$, increases until $L/D \approx 8$, and then weakly decreases at large spacing. For $R/D = 2$, the recovery factor is less than unity at small L/D due to the energy separation. The recovery

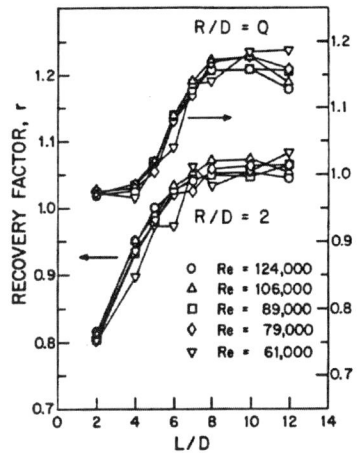
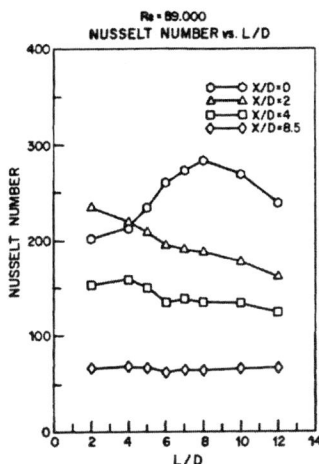

Figure 3: Variation of the recovery factor with nozzle-to-plate spacing[19]

Figure 4: Effects of nozzle-to-plate spacing on the local Nusselt number

factor is independent of the Reynolds number at both locations.

Figure 4 shows the variation of the Nusselt number with nozzle-to-plate spacing at different distances from the stagnation point. The Nusselt number at the stagnation point has its maximum value around $L/D \approx 8$. At small L/D, the stagnation Nusselt number is lower than the value of $R/D = 2$.

The influence of the Reynolds number on the Nusselt number can be formulated with a power law relationship such as in Equation (5). The exponent of Re_D is dependent on many parameters, but its value varies from approximately 0.5 for laminar flow to 0.8 for turbulent flow. A summary of selected previous investigations is presented in Table 1. Most of these correlated the local, stagnation or averaged Nusselt number in terms of Re_D and Pr. Geometric parameters such as the nozzle-to-plate spacing are often included in the heat transfer correlation.

Saad et al.[36] performed a numerical study on heat transfer and skin friction distribution with both uniform and parabolic velocity profile at the nozzle exit. For a parabolic velocity profile, early development of the shear layer induces higher heat transfer near the stagnation region than with a uniform velocity profile. A detailed reveiw of numerical studies on flow and heat transfer with impinging jets was performed by Polat et al.[37] More recently, Cziesla et al.[38] used LES to study heat transfer of impinging slot jets.

Heat Transfer Characteristics of Multiple Jets

With multiple-jet impingement, interactions among the jets affect the heat transfer characteristics. The local Nusselt number distribution with three impinging jets in a row is shown in Figure 5. The stagnation points of the two outside jets move

Table 1: Comparison of previous studies on a single jet

Author	D (mm)	Re_D	L/D	Comments
Perry[29]	16.5, 21.6	$7 \times 10^3 - 3 \times 10^4$	≥ 8	
Smirnov et al.[30]	2.5 - 36.6	$50 - 3.1 \times 10^4$	0.5 - 10	circular liquid jet
Gardon and Cobonpue[31]	2.3 - 9	$7 \times 10^3 - 1.12 \times 10^5$	≥ 0.5	single circular jet and array of jets
Gardon and Akfirat[32,33]	1.59 - 6.35	$2.8 \times 10^3 - 2.2 \times 10^4$	2 - 80	slot, circular jet and array of jets
Bouchez and Goldstein[34]	12.7	$3.5 \times 10^4 - 1.25 \times 10^5$	6, 12	air impingement in a well-defined cross flow
Popiel et al.[35]	13.8	$1 \times 10^3 - 1.9 \times 10^3$	2 - 20	combustion products in all directions
Goldstein and Behbabhani[28]	12.7	$2.5 \times 10^4 - 1.24 \times 10^5$	6, 12	air with or without crossflow
Goldstein et al.[19]	12.7	$6.1 \times 10^4 - 1.2 \times 10^5$	2 -12	recovery factor and Nu with varing L/D
Lytle and Webb[25,26]	7.8, 10.6	$3.6 \times 10^3 - 2.8 \times 10^4$	0.1 - 6	secondary maxima with small L/D

outward due to the crossflow so that the maxima in the Nusselt number occur away from the geometrical centers of the jet nozzles. When $L/D = 2$, local maximum points between the jets can be observed, which come from upwash stagnation formed by the spent air. A flow visualization study on the interaction between two impinging jets was carried out by Elbanna and Sabbagh.[39]

As in single jet impingement, the effect of the Reynolds number with multiple jets can be formulated by a power-law dependence. Goldstein and Seol[40] formulated $Nu/Re^{0.7}$ with geometric parameters including jet spacing and L/D. Selected previous studies on multiple jets are summarized in Table 2.

EFFECTS OF OTHER PARAMETERS

Impingement on a Curved Surface

In cooling of the leading edge of turbine blades and guide vanes, coolant jets impinge on the inner concave surface of the blades and vanes. The effect of curvature should be taken into account to simulate the leading edge cooling.

Metzger et al.[50] investigated heat transfer of impingement of a row of circular jets on a semi-circular surface. Dyban and Mazur[51] measured heat transfer coefficients

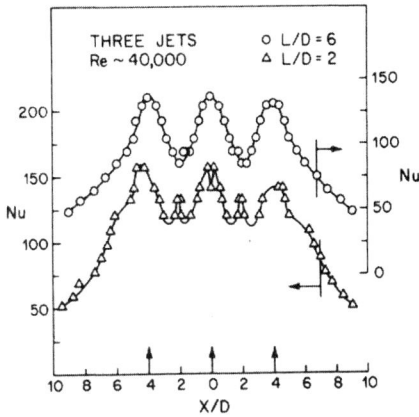

Figure 5: Local Nusselt number of a row of three jets[41]

Table 2: Comparison of previous stuides on multiple jets

Author	D (mm)	Re_D	L/D	Comments
Metzger and Korstad[42]	2.54	$2 \times 10^3 - 6 \times 10^3$	2 - 6.7	row of circular jets; effect of crossflow
Koopman and Sparrow[43]	6.35	$2.5 \times 10^2 - 1 \times 10^4$	2 - 10	row of circular jets; mass transfer analogy
Hollworth and Berry[44]	2, 5.5	$3 \times 10^2 - 3.5 \times 10^4$	1 - 20	square arrays of jets
Metzger et al.[45]	0.76 - 1.5	$1 \times 10^3 - 5.5 \times 10^3$	7 - 22.6	row and staggered array of jets
Hrycak[46]	9.5, 12.7	$2.5 \times 10^3 - 3.5 \times 10^4$	2 - 8	row of jets impinging on a concave surface
Goldstein and Timmer[41]	10	4×10^4	2, 6	3 jets in a row; a jet surrounded by 6 jets hexagonally
Behbahani and Goldstein[47]	5, 10	$5 \times 10^3 - 1.5 \times 10^4$	2 - 5	staggered array; spent air in one direction
Obot and Trabold[48]	3.175	$1 \times 10^3 - 2.1 \times 10^4$	2 - 16	square array; effect of spent air configuration
Goldstein and Seol[40]	6.35	$1 \times 10^4 - 4 \times 10^4$	≥ 6	row of circular jets
Son et al.[49]		$2 \times 10^4 - 2.0 \times 10^4$		staggered array; spent air in one direction

on a parabolic concave surface where a slot jet impinged. Several other investigators[27,46,52,53] studied the effect of curvature on the impingement heat transfer of a round jet, a slot jet, and a row of round jets.

Gau and Chung[54] perfromed flow visualization and heat transfer measurement of slot jet impingement on both concave and covex surfaces. Round jet impingement was investaged by Kornblum and Goldstein.[55] The results indicate that strong entrainment of spent air is induced with concave surfaces. This entrainment causes attenuation of the heat transfer rate and is stronger as L/D increases. Recently, Taslim et al.[56] examined the heat transfer enhancement on a concave surface with roughness elements, to achieve higher heat transfer rate near the leading edge region.

A typical leading edge of turbine blade is cooled by both impingement and film cooling. An experimental study of impingement cooling with film cooling holes on a concave impingement surface was performed by Metzger and Bunker.[57] The results show that the relative location of the film cooling holes to the impinging jets can cause significant variations in leading edge heat transfer.

Effect of Crossflow

Crossflow with impingement cooling can be characterized by a well-defined free stream type crossflow or a crossflow formed by the spent air from the jets. Metzger and Korstad[42] investigated the effect of well-defined crossflow on the impingement of a row of round jets. Flow visualization and effectiveness measurements with a well-defined crossflow were carried out by Bouchez and Goldstein[34] at various blowing ratios. Goldstein and Behbahani[28] measured heat transfer coefficients of a circular impiging jet with or without crossflow.

Obot and Trabold[48] investaged the effect of crossflow from spent air with different crossflow configurations. Three different configurations are illustrated in Figure 6. More research of the crossflow effect can be found in the literature.[43,58–60] Akella and Han[61] suggested a conceptual model of the effect of crossflow in a rotating impinging channel to simulate cooling of turbine rotors.

Effect of Angle of Impact

Perry[29] made quasi-local heat transfer measurements for an oblique impinging jet "using a calorimeter". Sparrow and Lovell[62] performed mass transfer experiments of an oblique impinging jet. Goldstein and Franchett[63] measured the local Nusselt number using a liquid crystal technique. Figure 7 shows the local Nusselt number variation with different impact angle. The local Nusselt number near the stagnation region is nearly independent of the impact angle until the angle becomes less than 30°. When the impact angle is 30°, average of the Nu near the stagnation point is 23% less than with a normal jet.

Stapountzis[64] investigated the combined effect of crossflow and angle of impact, when an oblique jet impinges on a flat plate. Ichimiya[65] considered an oblique circular jet with confined walls which are parallel and closely located to the impingement plate. The heat transfer characteristics of an inclined slot jet was stuided by Beitelmal et al.[66]

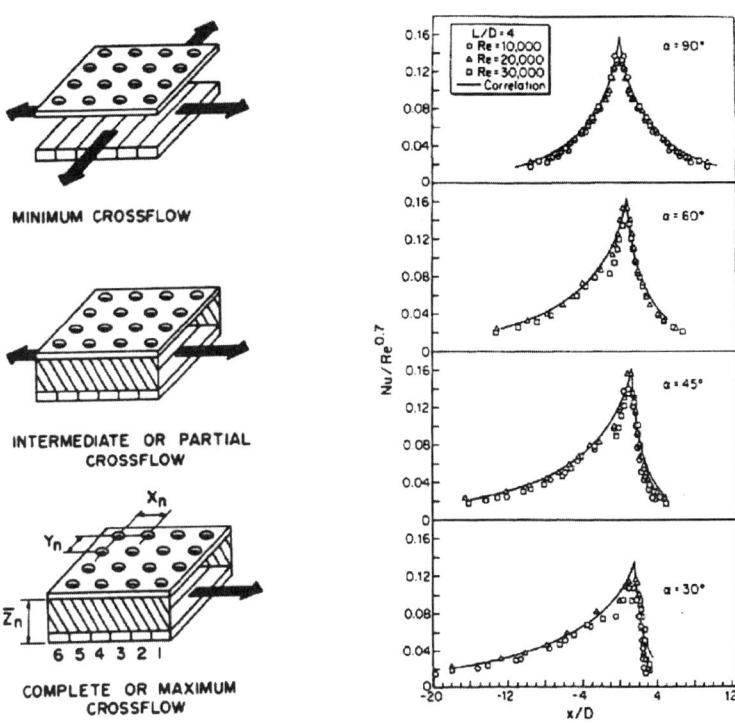

Figure 6: Schematics of crossflow configuration from spent air[48]

Figure 7: Local Nusselt number correlation including the effect of impact angle[63]

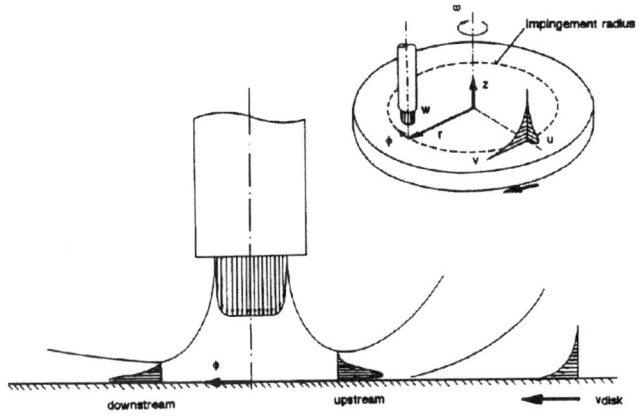

Figure 8: Schematic of flow structure of an impinging jet on a rotating disk[67]

Impinging on a Rotating Disk

Impingement on a rotating disk relates to cooling of a turbine rotor disk. A schematic of the flow situation is illustrated in Figure 8. The flow is different from the impinging jet with crossflow due to the thin wall boundary layer, established by the disk rotation, whose maximum velocity occurs at the wall.

Popiel and Boguslawski[68] investigated the local heat transfer characteristics of a round impinging jet on a rotating disk. Using a rotational Reynolds number, based on the rotational velocity and the distance between the nozzle and the center of the disk, they mapped three different heat transfer regiems - 1) impingement dominant regime; 2) transition regime; and 3) rotation dominant regime. A two-dimensional axisymmetric computation of flow and heat transfer fields was perfromed by Nakata et al.[69] A mass transfer experiment by Chen et al.[70] showed good agreement with heat transfer results.

Miscellaneous Parameters

Many other parameters can affect the heat transfer of impinging jets; for example, turbulence, entrainment effects, nozzle geometry and others. As a method to enhence the heat transfer, various types of turbulence generators have been suggested.[76,77] In that research, the heat transfer was correlated with the turbulence intensity.

If the fluid surrounding the jet is at a different temperature than the jet, three temperatures - the jet temperature (T_j), the wall temperature (T_w), and the ambient fluid temperature (T_∞) - must be considered in impingement heat transfer. The effects of T_∞ and the entrianment of the ambient fluid were studied to correlate heat transfer results for the three temperature problem.[20,78]

Obot et al.[71] experimentaly investigated the effect of nozzle geometry and found that it affects the heat transfer coefficient near the stagnation region for $L/D <$ 6. In addition to circular and slot orifice type nozzles, various types of nozzles such as radial jet reattachment nozzle, in-line jet nozzle,[72] and annular nozzle[73] have been suggested to enhance the heat transfer rate. However, the complexity in manufacturing may limit the usage of these nozzles over the orifice type nozzles in gas turbine related applications. Meola et al.[74] measured T_{aw} distirbution with different nozzle geometries. The effect of inlet geometery of orifice type nozzles was investigated by Brignoni and Garimella.[75]

SUMMARY

Impinging jets have important applications to local and area cooling in the hot section of high performance gas turbines. They have found wide applications for internal cooling of the stagnation region of blades and vanes.

The flow and heat transfer characteristics of impinging jets in gas turbine systems are reviewed. Characteristics of the major flow region - free jet, stagnation flow, and wall jet region are briefly described. Increasing attention on the role of the coherent structure on flow and heat transfer development is discussed.

Heat transfer characteristics of a single jet and multiple jets are considered. The variation of local Nusselt number with Reynolds number and nozzel-to-plate spacing is discussed. Interactions among multiple jets are also covered.

Other parameters which have effects on the heat transfer are discussed. Futher studies are still required for the effect of parameters, and there are few investigations on the combined effect of several parameters.

ACKNOWLEDGMENT

The authors would like to express their appreciation for the support by the Engineering Research Program of the Office of Basic Energy Sciences at the U. S. Department of Energy.

References

[1] Martin, H. 1977. Heat and mass transfer between impinging gas jets and solid surfaces. Adv. in Heat Transfer **13**: 1–60.

[2] Button, B. L. & D. Wilcook. 1978. Impinging heat transfer : A bibliography 1890-1975. Prev. Heat Mass Transfer **4**: 83.

[3] Hrycak, P. 1981. Heat transfer impinging jets : A literature review. AWAL Tech. Rep. AWAL-TR-81-3054.

[4] Button, B. L. & K. Jambunathan. 1989. Impinging heat transfer : A bibliography 1975-1985. Prev. Heat Mass Transfer **15**: 149.

[5] Viskanta, R. 1993. Heat transfer to impinging isothermal gas and flame jets. Experimental Thermail and Fluid Science **6**: 111–134.

[6] Jambunathan, K. & B. L. Button. 1994. Impinging heat transfer : A bibliography 1986-1991. Prev. Heat Mass Transfer **20**: 385.

[7] Abramovich, G. N. 1963. The Theory of Turbulent Jets. MIT Press, Cambridge, Mass.

[8] Rajaratnam, N. 1976. Turbulent Jets. Elsevier, New York.

[9] Schlichting, H. 1979. Boundary Layer Theory. 7th edn. McGraw-Hill, New York.

[10] Gautner, J. W., J. N. B. Livingwood & P. Hrycak. 1970. Survey of literature of flow characteristics of a single turbulent jet impinging on a flat surface. NASA TN D-5652.

[11] Nishino, K., M. Samada, K. Kasuya & K. Torii. 1996. Turbulence statistics in the stagnation region of an axisymmetric impinging jet flow. Int. J. Heat and Fluid Flow **17**: 193–201.

[12] Schneider, M. E. & R. J. Goldstein. 1994. Laser doppler measurement of turbulence parameters in a two-dimensional plane wall jet. Phys. Fluids **6**: 3116–3129.

[13] Cooper, D., D. C. Jackson, B. E. Launder & G.X.Liao. 1993. Impinging jet studies for turbulence model assessment - i. flow-field experiments. Int. J. Heat Mass Transfer **36**: 2675–2684.

[14] Craft, T. J., L. J. W. Graham & B. E. Launder. 1993. Impinging jet studies for turbulence model assessment - ii. an examination of the performance of four turbulence models. Int. J. Heat Mass Transfer **36**: 2685–2697.

[15] Hosseinalipour, S. M. & A. S. Mujumdar. 1995. Comprative evaluation of different turbulence models for confined impinging and opposing jet flows. Numerical Heat Transfer - Part A **28**: 647–666.

[16] Olsson, M. & L. Fuchs. 1998. Large eddy simulations of a forced semiconfined circular impinging jet. Phys. Fluid **10**: 476–486.

[17] Popiel, C. O. & O. Trass. 1991. Visualization of free and impinging round jets. Exp. Thermal Fluid Sci. **4**: 253–261.

[18] Cornaro, C., A. S. Fleischer & R. J. Goldstein. 1999. A visualization study of jet impingement on cylindrical surfaces. In Applied Optical Measurements, M. Lehner, ed., 307–317. Springer, Berlin.

[19] Goldstein, R. J., A. I. Behbahani & K. K. Heppelmann. 1986. Streamwise distribution of the recovery factor and the local heat transfer coefficient to an impinging circular air jet. Int. J. Heat Mass Transfer **29**: 1227–1235.

[20] Goldstein, R. J., K. A. Sobolik & W. S. Seol. 1990. Effect of entrainment on the heat trnasfer to a heated circular air jet impinging on a flat surface. J. Heat Transfer **112**: 608–611.

[21] Fox, M. D., M. Kurosaka, L. Hedges & K. Hirano. 1993. The influence of vortical structure on thermal fields of jets. J. Fluid Mech. **255**: 447–472.

[22] Kataoka, K., R. Sahara, H. Ase & T. Harada. 1987. Role of large coherent structures in impinging jet heat transfer. J. Chem. Eng. Japan **20**: 71–76.

[23] Kataoka, K., H. Ase & N. Sako. 1988. Unsteady aspects of large-scale coherent structures and impingement heat transfer in round air jets with and without controlled excitation. Int. J. Eng. Fluid Mech. **1**: 365–382.

[24] Liu, T. & P. J. Sullivan. 1996. Heat transfer and flow structures in an excited circular impinging jet. Int. J. Heat Mass Transfer **39**: 3695–3706.

[25] Lytle, D. & B. W. Webb. 1991. Secondary heat transfer maxima for air jet impingement at low nozzle-to-plate spacing. In Experimental Heat Transfer, Fluid Mechanics, and Thermodyanmics 1991, J. F. Keffer, R. K. Shah & E. N. Ganic, eds., 776–783. Elsvier, New York.

[26] Lytle, D. & B. W. Webb. 1994. Air jet impingement heat transfer at low nozzle-plate spacings. Int. J. Heat Mass Transfer **37**: 1687–1697.

[27] Choi, M., H. S. Yoo, G. Yang, J. S. Lee & D. K. Sohn. 2000. Measurements of impinging jet flow and heat transfer on a semi-circular concave surface. Int. J. Heat Mass Transfer **43**: 1811–1822.

[28] Goldstein, R. J. & A. I. Behbahani. 1982. Impingement of a circular jet with and without cross flow. Int. J Heat Mass Transfer **25**: 1377–1382.

[29] Perry, K. P. 1954. Heat transfer by convection from a hot gas jet to a plane surface. Proc. Inst. Mech. Eng. **168**: 775–784.

[30] Smirnov, V. A., G. E. Verevochkin & P. M. Badlick. 1961. Heat transfer between a jet and a held plate normal to the flow. Int. J. Heat Mass Transfer **2**: 1–7.

[31] Gardon, R. & J. Cobonpue. 1962. Heat transfer between a flat plate and jets of air impinging on it. In International Developments in Heat Transfer, 454–460. ASME, New York.

[32] Gardon, R. & J. C. Akfirat. 1965. The role of turbulence in determining the heat transfer characteristics of impinging jets. Int. J. Heat Mass Transfer **8**: 1261–1272.

[33] Gardon, R. & J. C. Akfirat. 1966. Heat transfer characteristics of impinging two-dimensional air jets. J. Heat Transfer **88**: 101–108.

[34] Bouchez, J. P. & R. J. Goldstein. 1975. Impingement cooling from a circular jet in a cross flow. Int. J. Heat Mass Transfer **18**: 719–730.

[35] Popiel, C. O., T. H. van der Meer & C. J. Hoogendoorn. 1980. Convective heat transfer on a plate in an impinging round hot gas jet of low reynolds number. Int. J. Heat Mass Transfer **23**: 1055–1068.

[36] Saad, N. R., W. J. M. Douglas & A. S. Majumdar. 1977. Prediction of heat transfer under an axisymmetric laminar impinging jet. Ind. Eng. Chem. Fundam. **16**: 148–154.

[37] Polat, S., B. Huang, A. S. Majumdar & W. J. M. Douglas. 1989. Numerical flow and heat transfer under impinging jets : A review. Ann. Rev. Num. Fluid Mech. Heat Transfer **2**: 157–197.

[38] Cziesla, T., E. Tandogan & N. K. Mitra. 1997. Large-eddy simulation of heat transfer from impinging slot jets. Numerical Heat Transfer - Part A **32**: 1–17.

[39] Elbanna, H. & J. A. Sabbagh. 1988. Flow visualization and measurements in a two-dimensional two-impinging jet flow. AIAA Journal **27**: 420–426.

[40] Goldstein, R. J. & W. Seol. 1991. Heat transfer to a row of impinging circular air jets including the effect of entrainment. Int. J. Heat Mass Transfer **34**: 2133–2147.

[41] Goldstein, R. J. & J. F. Timmers. 1982. Visualization of heat transfer from arrays of impinging jets. Int. J. Heat Mass Transfer **25**: 1857–1868.

[42] Metzger, D. E. & R. J. Korstad. 1972. Effects of crossflow on impingement heat transfer. J. Eng. Power **94**: 35–42.

[43] Koopman, R. N. & E. M. Sparrow. 1976. Local and average heat transfer coefficients due to an impinging row of jets. Int. J. Heat Mass Transfer **19**: 673–683.

[44] Hollworth, B. R. & R. D. Berry. 1978. Heat transfer from arrays of impinging jets with large jet-to-jet spacing. J. Heat Transfer **100**: 352–357.

[45] Metzger, D. E., L. W. Florscheutz, D. I. Takeuchi, R. D. Behee & R. A. Berry. 1979. Heat transfer characteristics for inline and staggered arrays or circular jets with cross-flow of spent air. J. Heat Transfer **101**: 587–593.

[46] Hrycak, P. 1981. Heat transfer from a row of impinging jets to concave cylinderical surfaces. Int. J. Heat Mass Transfer **24**: 407–418.

[47] Behbahani, A. I. & R. J. Goldstein. 1982. Local heat transfer to staggered arrays of impinging circular air jets. ASME Paper No. 82-GT-211.

[48] Obot, N. T. & T. A. Trabold. 1987. Impingement heat transfer within arrays of circular jets - part i. effects of minimum, intermediate, and complete crossflow for small and large spacings. J. Heat Transfer **107**: 872–879.

[49] Son, C., D. Gillespie, P. Ireland & G. M. Dailey. 2000. Heat transfer and flow characteristics of an engine representative impingement cooling system. ASME Paper No. 2000-GT-219.

[50] Metzger, D. E., T. Yamashita & C. W. Jenkins. 1969. Improvement cooling of concave surfaces with lines of circular air jets. J. Eng. Power **91**: 149–158.

[51] Dyban, Y. P. & A. I. Mazur. 1970. Heat transfer from a flat air jet flowing into a concave surface. Heat Transfer - Soviet Research **2**: 15–20.

[52] Metzger, D. E., R. T. Baltzer & C. W. Jenkins. 1972. Impingement cooling performance in gas turbine airfoils including effects of leading edge sharpness. J. Eng. Power **94**: 219–225.

[53] Lee, D. H., Y. S. Chung & D. S. Kim. 1996. Surface curvature effects on flow and heat transfer from a round impinging jet. In National Heat Transfer Conference, vol. 324 of *ASME HTD*, 73–83.

[54] Gau, C. & C. M. Chung. 1991. Surface curvature effect on slot air-jet impingement cooling flow and heat transfer process. J. Heat Transfer **113**: 854–858.

[55] Kornblum, Y. & R. J. Goldstein. 1997. Jet impingement on semicircular concave and convex surfaces, part two : Heat transfer. In Proc. Int. Symposium on the Physics of Heat Transfer in Boiling Condensation, 603–608. Moscow, Russia.

[56] Taslim, M. E., L. Setayeshgar & S. D. Spring. 2000. An experimental evaluation of advanced leading edge impingement cooling concepts. ASME Paper No. 2000-GT-222.

[57] Metzger, D. E. & R. S. Bunker. 1990. Local heat transfer in internally cooled turbine airfoil leading edge regions : Part ll - impingement cooling with film coolant extraction. J. Turbomachinery **112**: 459–466.

[58] Kercher, D. M. & W. Tabakoff. 1970. Heat transfer by a square array of round air jets impinging perpendicular to a flat surface including the effect of spent air. J. Eng. Power **92**: 1970.

[59] Andrew, G. E. & C. I. Hussein. 1986. Full coverage impingement heat transfer : Influence of channel height. Heat Transfer 1986 **3**: 1205–1211.

[60] Cho, H. H. & J. K. Ham. 2000. Influence of injection type and feed arragement on flow and heat transfer in an injection slot. ASME Paper No. 2000-GT-238.

[61] Akella, K. V. & J.-C. Han. 1998. Impingement cooling in rotating two-pass rectangular channels. J. Thermophysics Heat Transfer **12**: 582–588.

[62] Sparrow, E. M. & B. J. Lovell. 1980. Heat transfer characteristics of an obliquely impinging circular jet. J. Heat Transfer **102**: 202–209.

[63] Goldstein, R. J. & M. E. Franchett. 1988. Heat transfer from a flat surface to an oblique impinging jet. J. Heat Transfer **110**: 84–90.

[64] Stapountzis, H. 1993. Oblique impingement of a circular jet in a cross flow. Applied Scientific Research **51**: 231–235.

[65] Ichimiya, K. 1995. Heat transfer and flow characteristics of an oblique turbulent impinging jet within confined walls. J. Heat Transfer **117**: 316–322.

[66] Beitelmal, A. H., M. A. Saad & C. D. Patel. 2000. The effect of inclination on the heat transfer between a flat surface and an impinging two-dimensional air jet. Int. J. Heat Fluid Flow **21**: 156–163.

[67] Brodersen, S., D. E. Metzger & H. J. S. Fernando. 1996. Flows generated by the impingemetn of a jet on a rotating surface: Part i - basic flow pattern. J. Fluids Engineering **118**: 61–67.

[68] Popiel, C. O. & L. Boguslawski. 1986. Local heat transfer from a rotating disk in and impinging round jet. J. Heat Transfer **108**: 357–364.

[69] Nakata, Y., J. Y. Murthy & D. E. Metzger. 1991. Computation of laminar flow and heat transfer over an enclosed rotating disk with and without jet impingement. In Heat Transfer in Gas Turbine Engines, vol. 188 of *ASME HTD*, 15–27.

[70] Chen, Y.-M., W.-T. Lee & S.-J. Wu. 1998. Heat(mass) transfer between an impinging jet and a rotating disk. Wärme-und Stoffübertranung **34**: 195–201.

[71] Obot, N. T., A. S. Majumdar & W. J. M. Douglas. 1979. The effect of nozzle geometry on impinging heat transfer under round turbulent jet. ASME Paper No. 79-WA/HT-53.

[72] Seyed-Yagoobi, J., v. Narayanan & R. H. Page. 1998. Comparison of heat transfer characteristics of radial jet reattachment nozzle to in-line impinging jet nozzle. J. Heat Transfer **120**: 335–341.

[73] Maki, H. & A. Yabe. 1989. Heat transfer by the annual impinging jet. Experimental Heat Transfer **2**: 1–12.

[74] Meola, C., L. de Luca & G. M. Carlomagno. 1995. Azimuthal instability in an impinging jet : Adiabatic wall temperature distribution. Experiments in Fluids **18**: 303–310.

[75] Brignoni, L. A. & S. V. Garimella. 2000. Effects of nozzle-inlet chamfering on pressure drop and heat transfer in confined air jet impingement. Int. J. Heat Mass Transfer **43**: 1133–1139.

[76] Hoogendoorn, C. J. 1977. The effect of turbulence on heat transfer at a stagnation point. Int. J. Heat Mass Transfer **20**: 1333–1338.

[77] Wolf, D. H., R. Viskanta & F. P. Incropera. 1995. Turbulence dissipation in a freesurface jet of water and its effect on local impingement heat transfer from a heated surface: Part ii-local heat transfer. J. Heat Transfer **117**: 95–103.

[78] Striegl, S. A. & T. E. Diller. 1984. Analysis of the effect of entrainment temperature on jet impingement heat trasnfer. J. Heat Transfer **106**: 804–810.

[79] Nomoto, H., A. Koga, S. Ito, Y. Fukuyama, F. Otomo, S. Shybuya, M. Sato, Y. Kobayashi & H. Matsuzaki. 1997. The advanced cooling technology for the $1500^\circ c$ class gas turbines: Steam-cooled vanes and air-cooled blades. J. Eng. Gas Turbines Power **119**: 624–632.

[80] Li, X., J. L. Gaddis & T. Wang. 2000. Mist/steam heat transfer in confined slot jet impingement. ASME Paper No. 2000-Gt-221.

Recent Developments in Turbine Blade Internal Cooling

JE-CHIN HAN[a] and SANDIP DUTTA[b]

[a]Department of Mechanical Engineering
Texas A&M University, College Station, Texas 77843-3123, USA
[b]Department of Mechanical Engineering
University of South Carolina, Columbia, South Carolina 29208, USA

ABSTRACT: This paper focuses on turbine blade internal cooling. Internal cooling is achieved by passing the coolant through several rib-enhanced serpentine passages inside the blade and extracting the heat from the outside of the blades. Both jet impingement and pin-fin-cooling are also used as a method of internal cooling. In the past number of years there has been considerable progress in turbine blade internal cooling research and this paper is limited to reviewing a few selected publications to reflect recent developments in turbine blade internal cooling.

INTRODUCTION

Advanced gas turbine engines operate at high temperatures (1200-1400°C) to improve thermal efficiency and power output. As the turbine inlet temperature increases, the heat transferred to the turbine blades also increases. The level and variation in the temperature within the blade material (which causes thermal stresses) must be limited to achieve reasonable durability goals. The operating temperatures are far above the permissible metal temperatures. Therefore, there is a need to cool the blades for safe operation. The blades are cooled by extracted air from the compressor of the engine. Since this extraction incurs a penalty to the thermal efficiency, it is necessary to understand and optimize the cooling technique, operating conditions, and turbine blade configuration. Figure 1 shows the common cooling technique with three major internal cooling zones in a turbine blade. The leading edge is cooled by jet impingement, the trailing edge is cooled by pin-fins, and the middle portion is cooled by serpentine rib-roughened passages. This paper is limited to reviewing a few selected publications that dealt with the common cooling techniques. The compound and new suggested cooling techniques are briefly mentioned. In particular, this paper focuses on the effect of rotation on the rotor coolant passages heat transfer.

RIB TURBULATED COOLING

In advanced gas turbine blades, repeated rib turbulence promoters are cast on two opposite walls of internal cooling passages to enhance heat transfer. Thermal energy conducts from the external pressure and suction surfaces of turbine blades to the inner zones and that heat is removed by internal cooling. The internal cooling passages are mostly modeled as short, square or rectangular channels with different aspect ratios. The heat transfer performance in a stationary ribbed channel primarily depends on the channel aspect ratio, the rib configuration, and the flow Reynolds number. There have been many fundamental studies (Han and his co-workers[1-4]) to understand the heat transfer enhancement phenomena by the flow separation caused by ribs. These flow separations reattach the boundary layer to the heat transfer surface thus increasing the heat transfer coefficient. Moreover, the separated boundary layer enhances turbulent mixing; and therefore, the heat from the near-surface fluid can more effectively be dissipated to the main flow thus increasing the heat transfer coefficient. Ribs mostly disturb only the near-wall flow for heat transfer enhancement, consequently the pressure drop penalty by ribs is affordable for blade cooling designs. In general, ribs used for experimental studies are square in cross-section with a typical relative rib height of 5-10% of channel hydraulic diameter, and a rib spacing-to-height ratio varying from 5 to 15. However, today's airfoils have more complicated rib shapes and angles, and smaller gas turbines have high blockage ribs at closer spacing (Taslim and Lengkong[5]).

Effect of Rib Angle and Channel Aspect Ratio: Due to the curved asymmetric shape of a turbine blade, cooling channels near the trailing edge have broad aspect ratios and those near the leading edge have narrow aspect ratios. Normally, the suction and pressure sides are ribbed. Angle of attach is 90° for orthogonal or transverse ribs, and an angle of attack other than 90° is called skewed or angled ribs that develop secondary flows in the cross-stream direction to further enhance heat transfer. Park et al.[6] compared the heat transfer performance with different rib angles in different aspect ratio channels with aspect ratio varying from ¼ to 4 for Reynolds numbers between 15,000 and 60,000. The results show that the ribbed side heat transfer enhancements are about three times and the pressure drop penalties are about four to eight times the values for 45° and 60° ribs compared to a smooth channel. The pressure drop penalties are only two to four times for the 45° and 60° angled ribs with the same level of heat transfer enhancement for the narrow aspect ratio channels (W/H = ½ and ¼). However, for the same level of heat transfer enhancement in a broad aspect ratio channel (W/H = 4), the pressure drop penalties are as high as 8-16 times the friction factor in a smooth channel for angled ribs. It is concluded by Park et al.,[6] that the narrow aspect ratio channel performs better than a broad aspect ratio channel with angled ribs.

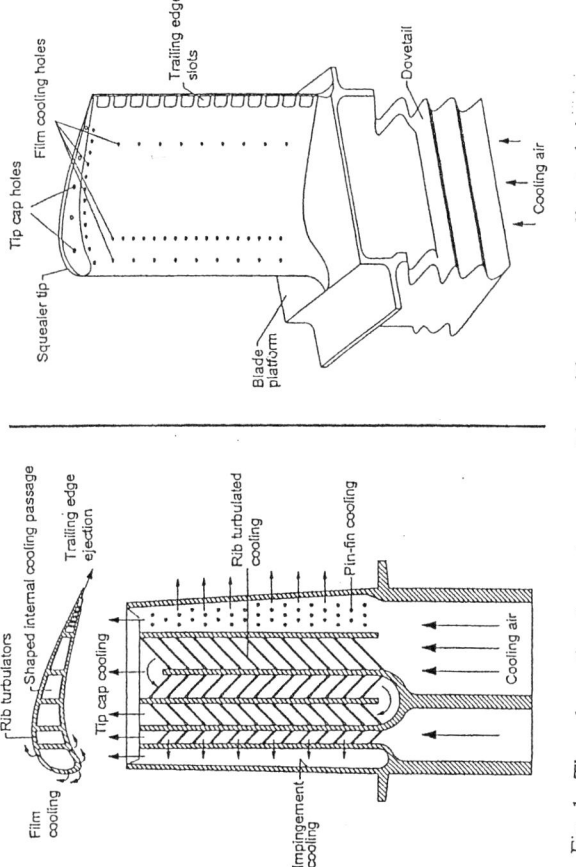

Fig. 1. The schematic of a modern gas turbine with common cooling techniques.

High Performance V-Shaped and Delta-Shaped Ribs: Han and Zhang[7] studied the high performance broken parallel and V-shaped ribs in a square channel. They presented the Nusselt number ratio of ribbed side versus the friction factor ratio (heat transfer Performance curve) for Reynolds numbers between 15,000 and 80,000. The ribbed-side Nusselt number ratios for 60° and 45° parallel broken ribs or V-shaped broken ribs are much higher than the corresponding 60° and 45° parallel continuous ribs or V-shaped continuous ribs. However, the corresponding friction factor ratios are comparable with each other for broken and continuous rib configurations. The high performance delta-shaped ribs were also reported. These ribs can be forward or backward aligned and Han et al.[8] has studied these special ribs. These heat transfer enhancement mechanisms combine the benefits of ribbed channel and the pin-finned channel. The isolated three-dimensional projections called broken ribs disturb the boundary layer and, like pins protruding in the flow, create a wake. Highest Nusselt number ratios are obtained with backward aligned delta-shaped ribs; whereas, backward offset delta ribs show a lower Nusselt number ratio but the friction factor ratio is significantly lower than other high performance ribs. For delta-shaped ribs, the backward alignment shows better performance than the forward direction. The backward delta-shaped ribs, the overall best performer in the group, produced 3 to 4 times heat transfer augmentation over a smooth surface result and the pressure drop was 7 to 9 times higher.

Effect of Rib Angle and 180° Sharp Turn: Han and Zhang[9] studied the effect of rib angle orientation on local mass transfer distribution in a three-pass rib-roughened channel. It was observed that the rib angle, rib orientation, and the sharp 180° turn significantly affected the local mass transfer distributions. The combined effects of these parameters increased or decreased the mass transfer coefficients after the sharp 180° turns. The angled ribs, in general, provided higher mass transfer coefficients than the transverse ribs, and parallel ribs gave higher mass transfer than the crossed ribs. However, the 180° turn direction and angled rib orientation can cause a reduction in mass transfer. This study shows that care needs to be taken in rib alignment in the turn regions and guidance in that respect is provided in the discussed results.

Effect of Film Cooling Hole: Most modern turbine airfoils have ribs in the internal coolant channel and film cooling for the outside surface. Therefore, some of the cooling air is bled through the film cooling holes. The presence of periodic ribs and bleed holes creates strong axial and spanwise variations in the heat transfer distributions on the passage surface. Shen et al.[10] studied the heat transfer enhancement by ribs in the presence of coolant extraction. They showed that with increasing discharge through the film-cooling holes, the heat transfer initially enhances (up to suction ratio = 4.4) and then decreases with further increase in the coolant extraction. Ekkad et al.[11] studied the detailed heat transfer coefficient distributions with different rib orientations in a two-pass channel with film-cooling bleed holes. The heat transfer coefficient distributions with the ribs show that the bleed holes increase heat transfer coefficient in the near-hole regions, but no broader impact by these holes is noticeable in these results. They showed that the regional-averaged Nusselt number ratios for different rib orientations are almost identical with and without bleed hole extraction. This indicates that 20 to 25% reduction of the main flow can be used for film cooling without significantly affecting the ribbed channel cooling performance.

ROTATIONAL EFFECT ON COOLING

Heat transfer in rotating coolant passages is very different from that in stationary coolant passages. Both Coriolis and rotating buoyancy forces can alter the flow and temperature profiles in the rotor coolant passages and affect their surface heat transfer coefficient distributions (Wagner et al.[12] and Dutta and Han[13]). It is very important to determine the local heat transfer distributions in the rotor coolant passages with impingement cooling, rib turbulated cooling, or pinned cooling under typical engine cooling flow, coolant-to-blade temperature difference (buoyancy effect), and rotating conditions. Effects of coolant passage cross-section and orientation on rotating heat transfer are also important. Figure 2 shows the schematic secondary flow and axial flow distribution in a rotating two-pass channel. Secondary flows in a two-pass channel are different for radial outflow and radial inflow passes. Since the direction of Coriolis force is dependent on the direction of rotation and flow, the Coriolis force has different direction in the two-passes. Figure 2 also shows the combined effects of Coriolis and rotational buoyancy on flow distribution (Han et al.[14]). For radial outward flow in the first channel, Coriolis force shifts the core flow towards the trailing wall. If both trailing and leading walls are symmetrically heated, then faster moving coolant near trailing wall would be cooler (therefore heat transfer would be enhanced) than the slow moving coolant near the leading wall (i.e., heat transfer would be decreased). Rotational buoyancy is caused by a strong centrifugal force that pushes cooler heavier fluid away from the center of rotation. In the first channel rotational buoyancy affects the flow in a similar fashion as the Coriolis force and causes a further increase in flow and heat transfer near the trailing wall of the first channel; whereas, Coriolis force favors the leading side of the second channel. The rotational buoyancy in the second channel tries to make the flow distribution more uniform in the duct.

Heat Transfer in Rib Turbulated Rotating Coolant Passages: Johnson et al.[15] compared 45° angled ribs with orthogonal 90° ribs (Wagner et al.[16]). Results show that the effect of rotation is more visible in the first-pass, and the following passes do not show much change in the presence of rotation. The rotation and buoyancy effects are in general less for the ribbed channel compared to that in a smooth channel. Results show that, like a stationary channel, 90° ribbed walls have higher heat transfer coefficients than smooth walled case. Results also indicate that the 45° ribs perform better than 90° ribs; the highest heat transfer coefficients on low-pressure surfaces in all three passes are obtained with 45° angled ribs. However, the heat transfer coefficients on high-pressure surfaces by 45° angled ribs are not so significantly better than 90° orthogonal ribs.

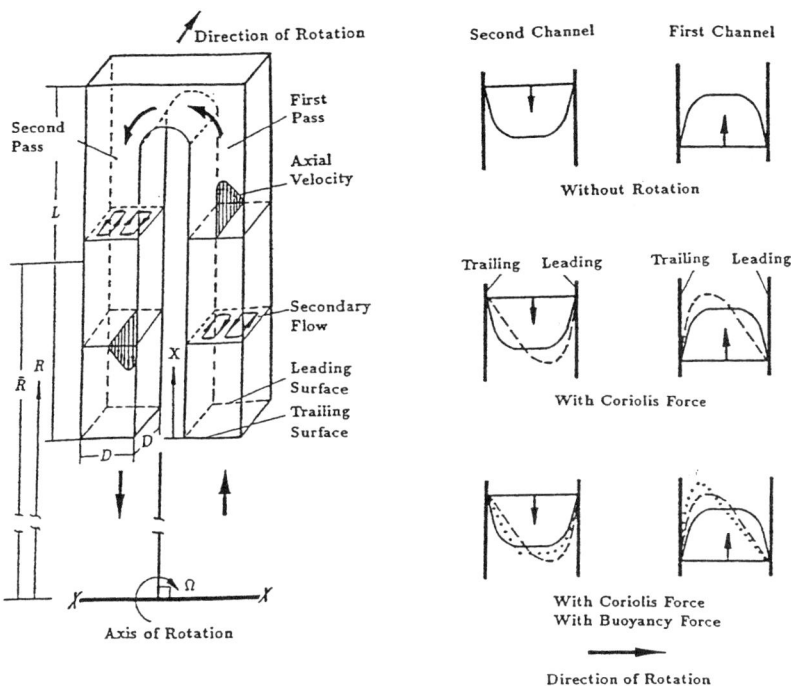

Fig. 2. Conceptual view of a two-pass rotating coolant flow distribution (Han et al.[14]).

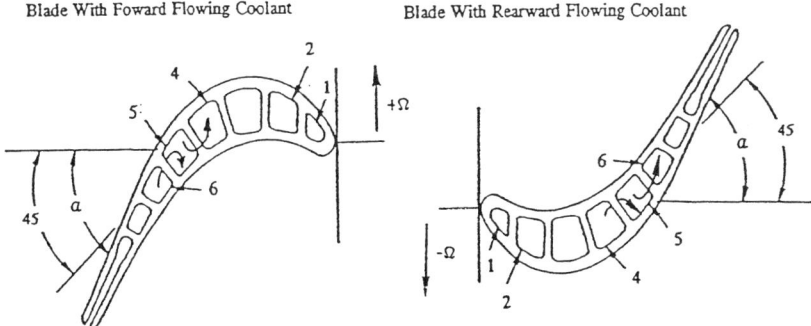

Fig. 3. Cooling channel orientation and rotation directions (Johnson et al.[17]).

Effect of Channel Cross Section and Orientation With Respect to the Rotation Direction on Both Smooth and Ribbed Channels: Besides the effects of rotation on ribbed surfaces, effects of channel orientation on heat transfer distribution in rotating ducts are also important in turbomachinery applications. Figure 3 shows the cooling channel orientation with respect to the rotation direction. Since the turbine blade is curved, the rotor blade cooling passage can have different channel orientation with respect to the rotating plane. Johnson et al.[17] experimented to determine the effects of model orientation as well as buoyancy and Coriolis forces on heat transfer. The results from channel orientations of 0° and 45° to the axis of rotation were compared. They presented the effects of rotation number on the heat transfer ratio for smooth and 45° ribbed channels. Results show that at a typical flow condition, the heat transfer on the leading surfaces for outward flow in the first-pass with smooth walls is twice as much for the channel at 45° compared to the channel at 0°. The heat transfer in the turn regions and immediately downstream of the turns in the second-pass with flow inward and in the third-pass with flow outward also depends on model orientation with differences up to 40 to 50 percent. However, the differences for the other passages and with ribs are less. In addition, the effects of buoyancy and Coriolis forces on heat transfer in the rotating passage are decreased with the model at 45° compared to the results at 0°.

Dutta and Han[18] used high performance ribs in a rotating two-pass square channel. These broken V-ribs have shown to be better performing in several rib configurations tested in a stationary channel. Three different channel orientations were used and these orientations are shown in Figure 4. These high performance ribs have their own secondary flows. Figure 4 shows the schematics of the secondary flows developed by rib orientation and rotation. The channel orientation with respect to the rotation axis influences the secondary vortices. The secondary flow developed by ribs can interact with the secondary flow of rotation and a new flow condition may be established. Results show that the effect of rotation on Model C is less than that on Model A. Results also show that the broken V-ribs are better than the 60° angled ribs. Dutta et al.[19] studied the effects of channel orientation on the triangular duct heat transfer. Results show that the model orientation can significantly influence the heat transfer pattern.

JET IMPINGEMENT COOLING

Among all heat transfer enhancement techniques, jet impingement has the most significant potential to increase the local heat transfer coefficient. Jet impingement heat transfer is most suitable for the leading edge of an airfoil, where the thermal load is highest and a thicker cross-section of this portion of the airfoil can suitably accommodate impingement cooling. There are several arrangements possible with cooling jets and different aspects need to be considered before optimizing an efficient heat transfer design. There are some studies focused on the effects of jet-hole size and distribution, cooling channel cross-section, and the target surface shape on the heat transfer coefficient distribution. However, most

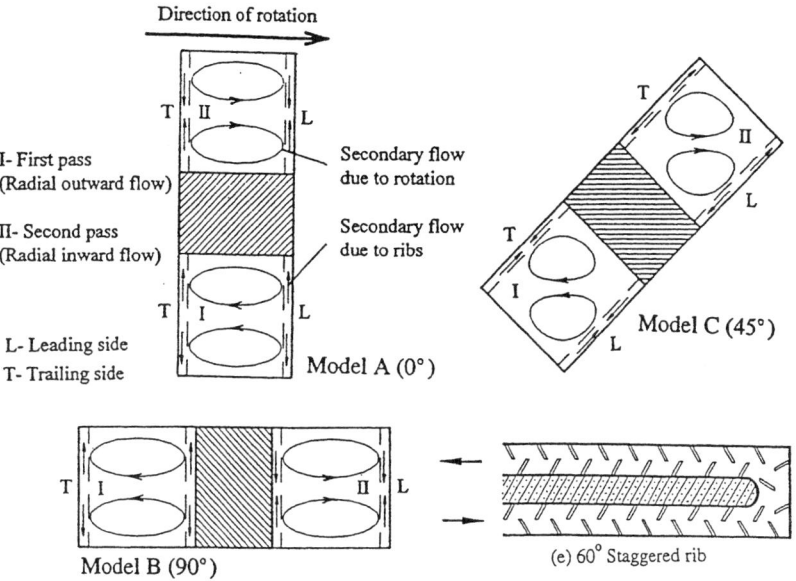

Fig. 4. Combined effects of rotating channel orientation and broken V-shaped ribs (Dutta and Han[18]).

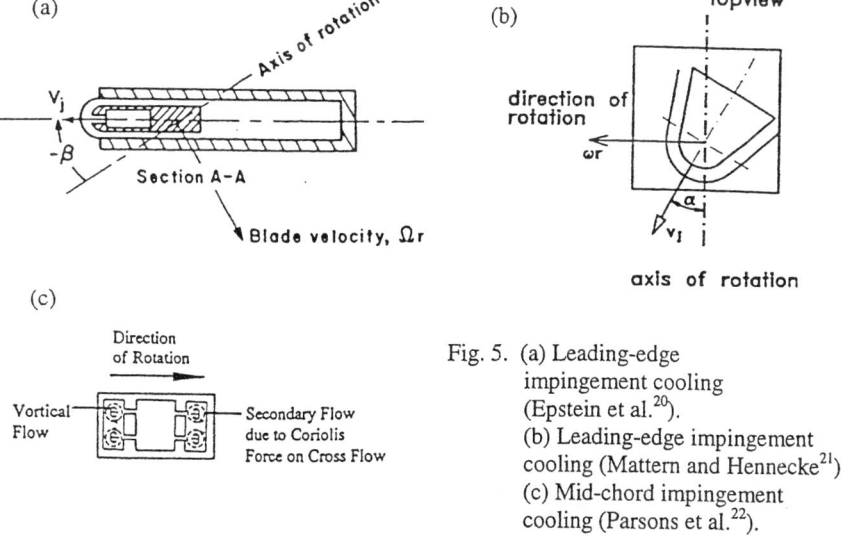

Fig. 5. (a) Leading-edge impingement cooling (Epstein et al.[20]).
(b) Leading-edge impingement cooling (Mattern and Hennecke[21]).
(c) Mid-chord impingement cooling (Parsons et al.[22]).

impingement cooling studies are for non-rotating blades, only a few studies focus on rotor blade impingement cooling.

Rotation Effects on Leading Edge Impingement Cooling: Epstein et al.[20] studied the effects of rotation on impingement cooling. Figure 5a shows schematics of their test facility and impingement arrangement. Note that the impingement direction is not aligned with the rotation direction. Results indicate that the rotation decreases the impingement heat transfer, but the effective heat transfer is better than a smooth rotating channel. The zero stagger of cooling jets in Figure 5a also shows lower Nusselt numbers compared to that with a stagger angle.

Mattern and Hennecke[21] studied the impingement effects in the leading edge with a mass transfer analogy. They used naphthalene sublimation technique to get details of the surface mass transfer. Their experiments did not include the rotational buoyancy effect. Like the previous study, the jet direction has an offset angle with respect to the rotation direction (Figure 5b). For zero offset angle, the Coriolis creates a swirl action on the spent flow. The jet is not affected directly by the Coriolis force; whereas, the Coriolis force deflects the jet in its 90° offset position. The peak mass transfer coefficients correspond to the stagnation locations. The peaks are dispersed in the $\alpha = 0°$. This can be attributed to swirl generated by the Coriolis effect. The peaks are preserved at $\alpha = 90°$. Results show that the effect of rotation is least on $\alpha = 45°$. However, all stagger angles show that rotation reduces the mass transfer compared to that with stationary channel mass transfer. A maximum of 40% reduction in the mass transfer was noted by rotation at $\alpha = 0°$

Rotation Effect on Mid-Chord Impingement Cooling: Parsons et al.[22] used a rotating channel to study the effect of rotation on impingement cooling (Figure 5c). However, their impingement configuration simulated the cooling requirements for the mid-chord region of an airfoil. A central chamber serves as the pressure chamber and impingement jets are released in either direction to impinge on two heated surfaces. Note that the impingement flow directions have different orientations with respect to the direction of rotation. In general, the trailing side shows a decrease in the heat transfer with rotation. Both side-walls of the trailing side show a significant reduction in the heat transfer coefficient. Parsons and Han[23] studied the effect of film coolant extraction from their rotating impingement channel. Results show that rotation decreases the impingement heat transfer coefficient by up to 25%.

PIN-FIN COOLING

Pins are mostly used in the narrow trailing edge of an airfoil where impingement and ribbed channels cannot be accommodated due to manufacturing constraint. Pin-fins commonly used in turbine cooling have pin height-to-diameter ratio between ½ and 4. Heat transfer in turbine pin-fin array combines the cylinder heat transfer and endwall heat transfer. Due to the turbulence enhancement caused by pins, heat transfer from endwalls is higher than smooth wall conditions; however, mounting pins may cover a considerable surface area, and that area needs to be compensated for by the pin surface area. Like cylinders in a cross-flow, pins shed wake at downstream flow. Beside this wake shedding, a horseshoe vortex originates just upstream of the base of the pin and wraps the pin around causing more flow disturbances. Boundary layer also separates if the pin in placed as a three-dimensional protrusion. These partial length pins or three-dimensional protrusions do not extend to the top surface and heat transfer enhancement occurs on the pin-mounted surface. Interactions of all these flow disturbances (wakes, horseshoe vortex, boundary layer separation) increase heat transfer from the pin-mounted surface. In addition to flow disturbances, pins conduct thermal energy away from the heat transfer surface. Long pins can increase the effective heat transfer area and perform better than short pins. There have been many studies that evaluated the effects of pin size, distribution, shape, and pin-fin-cooling with extraction on the heat transfer coefficient distribution (Armstrong and Winstanley[24]). However, all pin-fin-cooling studies so far are for non-rotating blades, none for rotor blade pin-fin-cooling.

Pin Array and Partial Length Pin Arrangement: There are two common array structures mostly used. One is the inline array and the other is the staggered array. Metzger et al.[25] used staggered arrays of circular pins with 1.5 to 5 pin diameter spacing in a rectangular channel. A closer spaced array (smaller x/D) shows a higher heat transfer coefficient. Results show that the staggered pin arrangement performs better in the inter-pin region. Their observations clearly indicate that addition of pin-fins significantly enhances the heat transfer coefficient. However, the addition of pins also increases the pressure drop in the flow channel. Chyu et al.[26] showed that the heat transfer coefficient on the pin surface for both arrays is consistently higher than that of the endwall. The pin surface heat transfer is observed to be 10 to 20 percent higher for the presented case. Arora and Abdel-Messeh[27] studied the effects of partial length pins in a rectangular channel. The surface containing pins are not affected by the pin tip clearance. Whereas the other surface, that does not have pins, shows a decrease in heat transfer coefficient with an increase in the pin tip clearance. The friction factor is lower for partial pins compared to full-length pins. In general, heat transfer coefficient decreases in partial length pins.

Effect of Flow Convergence and Turning: The flow channel in the trailing edge of an airfoil has a reducing cross-section, and therefore, the flow in the channel accelerates. The results are row averaged and the accelerating flow shows an increase in the heat transfer coefficient (Metzger et al.[28]). Chyu et al.[29] used mass transfer technique to study the effect of perpendicular flow entry in two pin-fin configurations. They show that the turning inlet configuration always results in lower average Sherwood numbers. The reduction is about 40-50% for the inline array and 20-30% for the staggered array.

Effect of Pin Shape and Array Orientation: Metzger et al.[30] studied the effects of pin shape and array orientations. They reported the effect of flow incident angle on oblong pins. All incident angles except 90 yield higher Nusselt numbers than circular pins. The $\gamma = 90°$ array yields significantly lower Nusselt number, especially toward the lower end of the Reynolds number range. The $\gamma = \pm 30°$ array has the highest Nusselt numbers, about 20 percent higher than the circular pin array on the average. Except for $\gamma = 90°$, the pressure drop for oblong pins are significantly higher than circular pins. This increase in the friction factor is associated with the flow turning caused by oblong pins. The pin shapes mostly studied are straight cylinders. However, the casting or other manufacturing processes cannot make perfect cylinders and these manufacturing imperfections may affect the heat transfer performance. Chyu[31] studied the effect of a fillet at the base of the cylindrical pin. Straight cylinders in staggered array formation have the highest heat transfer followed by fillet cylinders in the staggered formation. It is interesting to note that the fillet cylinder inline formation has better heat transfer than the straight cylinders in inline formation. Though staggered array gives higher heat transfer coefficient, performance of the inline straight cylinders is best among the group and the fillet cylinders in staggered formation are the worst. In a different experimental work, Goldstein et al.[32] studied the effect of stepped diameters on mass transfer coefficients. The diameter of the pin is axially varied. The base diameter is greater than the center diameter and no fillet radius is provided. The array configuration is staggered. Results show that the mass transfer increases or remains the same compared to a straight cylinder pin array when the radius is varied, but the pressure drop reduces significantly for the stepped diameter cylindrical pins. Chyu et al.[33] used cube and diamond shaped pins to enhance the heat transfer coefficient from a surface. The cube-shaped pins have the highest mass transfer coefficients among the shapes considered and round pins have the lowest mass transfer coefficients. Corresponding pressure loss coefficients are higher for the cube and diamond shaped pins relative to the circular pins.

Pin-Fin Cooling With Ejection: The trailing edge pin-fin channel normally has ejection holes through which the spent coolant exhausts to the main stream flow. Kumran et al.[34] investigated the effects of the length of coolant ejection holes on the heat transfer coefficient in pin-fins. The length of the ejection hole can significantly alter the discharge rate of coolant. More coolant ejection reduces the Nusselt number significantly from no ejection. This decrease in the heat transfer coefficient can be explained by the fact that coolant mass is extracted from the coolant channel before its cooling capacity is fully utilized. Results indicate that the correlation based on local Reynolds number can predict the heat transfer coefficient distribution for lower coolant ejection but not so good for higher coolant ejection rates.

COMPOUND AND NEW COOLING TECHNIQUES

Several internal heat transfer enhancement techniques are discussed in previous sections. Most common methods of heat transfer augmentation in gas-turbine airfoils are ribs, pins, and jet impingement. It is shown that these enhancement techniques increase heat transfer coefficients, but can combining these techniques increase the heat transfer coefficient more? Several researchers have combined these heat transfer enhancement techniques to improve the heat transfer coefficient. However, it is not always recommended to combine more than one heat transfer augmentation technique. Besides compounding more than one heat transfer enhancement technique, there are attempts to incorporate new concepts, e.g., jet swirlers, dimpled surfaces, heat pipes, in the turbo-machinery cooling. Several studies are available on new cooling techniques, application, and introductory concepts in that regard are also reviewed.

Impingement on Ribbed, Pinned, and Dimpled Walls: Akella and Han[35] included skew ribs in the target surfaces of their rotating impingement test facility. The target plates are 45° rib-turbulated and two streamwise rows of jets are used for impingement. Results show that the increase in Nusselt number by the addition of ribs is non-linear with respect to the log of jet Reynolds number. The enhancement is more at higher jet Reynolds numbers. It can be argued that the cross-flow developed by spent jets is stronger for higher jet Reynolds numbers and ribs are more effective in heat transfer coefficient enhancement with a stronger cross-flow. Like earlier impingement configurations, rotation reduces the heat transfer coefficient. Azad et al.[36-37] studied the impingement effect on dimpled and pinned surfaces. Dimples are circular depressions whereas pins are circular protrusions on the target surface. At lower Reynolds number the pinned surface performs better than the dimpled surface. At higher Reynolds numbers, the dimpled surface performs better than the pinned surface for a certain flow orientation.

Moon et al.[38] studied the convective heat transfer enhancement in rectangular cooling channels with dimples (concavity imprinted) on one wall. They found that the heat transfer can enhance 2.1 times with 1.6-2.0 times pressure drop penalty for Reynolds numbers between 12,000-60,000. It is expected that the pressure drop would be double if dimples are placed on two opposite walls.

Combined Effect of Ribbed Wall With Grooves or Pins: Zhang et al.[39] studied the heat transfer and friction in rectangular channels with rib-groove combination. The Stanton numbers for the ribbed-grooved walls are higher than that for the only ribbed walls at similar rib spacing values. Metzger et al.[40] indicated that the addition of ribs does not change the heat transfer coefficient from the pin-mounted surface. However, the heat transfer coefficient on the rib surface is significantly higher than the pin-mounted surface.

Combined Effect of Swirl Flow and Ribs: Kieda et al.[41] experimentally investigated the single-phase water flow and heat transfer in a rectangular cross-sectioned twisted channel. Several aspect ratios and twist pitches were used. Results indicate that in cooling application, this twisted channel performs similar to a ribbed pipe. Zhang et al.[42] used different types of inserts to study the combined rib and twisted tape inserts in square ducts. Four test configurations were used: twisted tape, twisted tape with interrupted ribs, hemi-circular wavy tape, and hemi-triangular wavy tape. Twisted tape with interrupted ribs provides a higher overall heat transfer performance over twisted tape without ribs and hemi-circular wavy tape. The performance of the hemi-triangular wavy tape is comparable with the twisted tape plus interrupted ribs. Hemi-circular wavy tapes show the lowest heat transfer performance in this group.

Combined Effect of Swirl and Impingement: A new jet impingement and swirl was investigated by Glezer et al.[43]. A preliminary test showed significant improvement in the heat transfer performance. Based on that study, a new airfoil has been designed with swirling impingement in the leading edge. This new airfoil is tested in the hot cascade test section. Results indicate that screw shaped swirl cooling can significantly improve the heat transfer coefficient over a smooth channel and this improvement is not significantly dependent on the temperature ratio and rotational forces. Moreover, it was concluded that optimization of the internal passage geometry in relation to location and size of the tangential slots is very important in achieving the best performance of the screw-shaped swirl in the leading edge cooling. Pamula et al.[44] studied the heat transfer enhancement by a combination of impingement and cross flow-induced swirl in a two-pass channel. Results show that the new impingement system, from the first pass to second pass, using cross flow injection holes produce significantly higher heat transfer on the second pass walls.

New Cooling Concepts: Heat pipes have very high effective thermal performance (Zuo et al.[45]). Therefore, they can transfer heat from high temperature to the low temperature regions. This concept may be used in the airfoil cooling. Heat is removed from the initial stage stator airfoils and the heat is delivered at a later stage to heat up the main flow. This way the heat extracted can be recycled to the main flow. Another cooling concept developed by Yamawaki et al.[46]. In this concept, the heat is conducted away from the hot airfoil to the fin assembly. This passive heat extraction reduces the required cooling air. Most heat pipe applications are designed for the stator airfoils, where it is easier to mount the connecting pipes or fins. Recently, Kerrebrock and Stickler[47] proposed a design to incorporate heat pipe in the rotor.

The concept of cooled cooling air systems, through a heat exchanger, for turbine thermal management was reported by Bruening and Chang[48]. Results show that the use of a cooled cooling air system can make a positive impact on overall engine performance for land-based turbines. Commonly a closed loop steam cooled nozzle with thermal barrier coatings (TBC) is used in order to reduce the hot gas temperature drop through the first stage nozzle (Corman and Paul[49]). A closed looped mist/steam cooling was reported by Guo et al.[50]. Results show that an average heat transfer enhancement of 100% can be achieved with 5% mist (fine water droplets) compared to the steam cooling.

CONCLUDING REMARKS

For a typical cooling arrangement in a modern rotor blade, impingement cooling is used in the leading edge and pin-fin-cooling is used in the narrow trailing edge, and mid-section is cooled by rib-turbulated convection. More studies are needed for rectangular ribbed passages with and without film holes under realistic coolant flow, thermal, and rotation conditions. Meanwhile, more studies are needed for rotating impingement cooling as well as rotating pin-fin-cooling in order to guide the efficient rotor blade internal cooling designs. Also, investigation on compound and new cooling concepts should be continued, in order to further improve cooling efficiency.

REFERENCES

1. Han, J.C., 1988, "Heat Transfer and Friction Characteristics in Rectangular Channels With Rib Turbulators," *ASME Journal of Heat Transfer*, Vol. **110**, No. 2, pp. 321-328.
2. Han, J.C. and Park, J.S., 1988, "Developing Heat Transfer in Rectangular Channels With Rib Turbulators," *International Journal of Heat and Mass Transfer*, Vol. **31**, No. 1, pp. 183-195.
3. Han, J.C., Ou, S., Park, J.S. and Lei, C.K., 1989, "Augmented Heat Transfer in Rectangular Channels of Narrow Aspect Ratios with Rib Turbulators," *International Journal of Heat Mass Transfer*, Vol. **32**, No. 9, pp. 1619-1630.
4. Han, J.C. and Dutta, S., 1995, "Internal Convection Heat Transfer and Cooling - An Experimental Approach," Von Karman Institute for Fluid Dynamics, Lecture Series 1995-05 on Heat Transfer and Cooling in Gas Turbines, May 8-12, Belgium.
5. Taslim, M.E. and Lengkong, A, 1998, "45° Round-Corner Rib Heat Transfer Coefficient Measurements in a Square Channel," International Gas Turbine and Aeroengine Congress and Exhibition, Stockholm, Sweden, June 2-5, ASME Paper No. 98-GT-176.
6. Park, J.S., Han, J.C., Huang, Y., Ou, S., and Boyle, R.J., 1992, "Heat Transfer Performance Comparisons of Five Rectangular Channels With Parallel Angled Ribs," *International Journal of Heat Mass Transfer*, Vol. **35**, No. 11, pp. 2891-2903.
7. Han, J.C. and Zhang, Y.M., 1992, "High Performance Heat Transfer Ducts with Parallel and V-Shaped Broken Ribs," *International Journal of Heat Mass Transfer*, Vol. **35**, No. 2, pp. 513-523.
8. Han, J.C., J. Joy Huang, and C. Pang Lee, 1993, "Augmented Heat Transfer in Square Channels with Wedge-Shaped and Delta-Shaped Turbulence Promoters," *Journal of Enhanced Heat Transfer*, Vol. **1**, No. 1, 1993, pp. 37-52.
9. Han, J.C. and Zhang, P., 1991, "Effect of Rib-Angle Orientation on Local Mass Transfer in A Three-Pass Rib-Roughened Channel," *ASME Journal of Turbomachinery*, Vol. **113**, pp. 123-130.
10. Shen, J.R., Wang, Z., Ireland, P.T., Jones, T.V., and Byerley, A.R., 1996, "Heat Transfer Enhancement Within a Turbine Blade Cooling Passage Using Ribs and Combinations of Ribs With Film Cooling Holes," *ASME Journal of Turbomachinery*, Vol. **188**, pp. 428-433.
11. Ekkad, S.V., Huang, Y., and Han, J.C., 1998, "Detailed Heat Transfer Distributions in Two-Pass Smooth and Turbulated Square Channels With Bleed Holes," International Journal of Heat and Mass Transfer, Vol. **41**, No. 13, pp. 3781-3791.

12. Wagner, J.H., Johnson, B.V., and Kopper, F.C., 1991, "Heat Transfer in Rotating Serpentine Passages With Smooth Walls," *ASME Journal of Turbomachinery*, Vol. **113**, pp. 321-330.
13. Dutta, S. and Han, J.C., 1997, "Rotational Effects on the Turbine Blade Coolant Passage Heat Transfer," *Annual Review of Heat Transfer*, Vol. **9**, pp. 269-314.
14. Han, J.C., Zhang, Y.M., and Kalkuehler, K., 1993, "Uneven Wall Temperature Effect on Local Heat Transfer in a Rotating Two-Pass Square Channel with Smooth Walls," *ASME Journal of Heat Transfer*, Vol. **114**, No. 4, pp. 850-858.
15. Johnson, V.V., Wagner, J.H., Steuber, G.D., and Yeh, F.C., 1992, "Heat Transfer in Rotating Serpentine Passages with Trips Skewed to the Flow," ASME Paper No. 92-GT-191, *ASME Journal of Turbomachinery*, Vol. **116**, No. 1, pp. 113-123.
16. Wagner, J.H., Johnson, B.V., Graziani, R.A., and Yeh, F.C., 1992, "Heat Transfer in Rotating Serpentine Passages With Trips Normal to the Flow," *ASME Journal of Turbomachinery*, Vol. **114**, pp. 847-857.
17. Johnson, B.V., Wagner, J.H., Steuber, G.D., and Yeh, F.C., 1994, "Heat Transfer in Rotating Serpentine Passages with Selected Model Orientations for Smooth or Skewed Trip Walls," ASME Paper No. 93-GT-305, *ASME Journal of Turbomachinery*, Vol. **116**, pp. 738-744.
18. Dutta, S. and Han, J.C., 1996, "Local Heat Transfer in Rotating Smooth and Ribbed Two-Pass Square Channels With Three Channel Orientations," *ASME Journal of Heat Transfer*, Vol. **118**, No. 3, pp. 578-584.
19. Dutta, S., Han, J.C., and Lee, C.P., 1996, "Local Heat Transfer in a Rotating Two-Pass Ribbed Triangular Duct With Two Model Orientations," *International Journal of Heat and Mass Transfer*, Vol. **39**, No. 4, pp. 707-715.
20. Epstein, A.H., Kerrebrock, J.L., Koo, J.J, and Preiser, U.Z., 1985, "Rotational Effects on Impingement Cooling," GTL Report No. 184.
21. Mattern, Ch. and Hennecke, D.K., 1996, "The Influence of Rotation on Impingement Cooling," presented at the International Gas Turbine and Aeroengine Congress and Exhibition, Birmingham, UK, June 10-13, ASME Paper No. 96-GT-161.
22. Parsons, J.A., Han, J.C., and Lee, C.P., 1998, "Rotation Effect on Jet Impingement Heat Transfer in Smooth Rectangular Channels With Four Heated Walls and Radially Outward Crossflow," *ASME Journal of Turbomachinery*, Vol. **120**, No. 1, pp.79-85.
23. Parsons, J.A. and Han, J.C., 1996, "Rotation Effect on Jet Impingement Heat Transfer in Smooth Rectangular Channels With Heated Target Walls and Film Coolant Extraction," presented at the International Mechanical Engineering Congress and Exhibition, Atlanta, Georgia, November 17-22, ASME Paper No. 96-WA/HT-9.
24. Armstrong, J. and Winstanley, D., 1988, "A Review of Staggered Array Pin Fin Heat Transfer for Turbine Cooling Applications," *ASME Journal of Turbomachinery*, Vol. **110**, pp. 94-103.
25. Metzger, D.E., Berry, R.A., Bronson, J.P., 1982, "Developing Heat Transfer in Rectangular Ducts With Staggered Arrays of Short Pin Fins," *ASME Journal of Heat Transfer*, Vol. **104**, pp. 700-706.
26. Chyu, M.K., Hsing, Y.C., Shih,T.I.P., and Natarajan, V., 1998, "Heat Transfer Contributions of Pins and Endwall in Pin-Fin Arrays: Effects of Thermal Boundary Condition Modeling," International Gas Turbine and Aeroengine Congress and Exhibition, June 2-5, Stockholm, Sweden, ASME Paper No. 98-GT-175.
27. Arora, S.C. and Abdel-Messeh, W., 1989, "Characteristics of Partial Length Circular Pin Fins as Heat Transfer Augmentors for Airfoil Internal Cooling Passages," Gas Turbine and Aeroengine Congress and Exposition, June 4-8, Toronto, Ontario, Canada, ASME Paper No. 89-GT-87.

28. Metzger, D.E., Shephard, W.B., and Haley, S.W., 1986, "Row Resolved Heat Transfer Variations in Pin-Fin Arrays Including Effects of Non-Uniform Arrays and Flow Convergence," International Gas Turbine Conference and Exhibit, Dusseldorf, West Germany, June 8-12, ASME Paper No. 86-GT-132.

29. Chyu, M.K., Natarajan, V., and Metzger, D.E., 1992, "Heat/Mass Transfer from Pin-Fin Arrays With Perpendicular Flow Entry," ASME HTD-Vol. **226**, Fundamentals and Applied Heat Transfer Research for Gas Turbine Engines.

30. Metzger, D.E., Fan, S.C., and Haley, S.W., 1984, "Effects of Pin Shape and Array Orientation on Heat Transfer and Pressure Loss in Pin Fin Arrays," *ASME Journal of Engineering for Gas Turbines and Power*, Vol. **106**, pp. 252-257.

31. Chyu, M.K., 1990, "Heat Transfer and Pressure Drop for Short Pin-Fin Arrays With Pin-Endwall Fillet," *ASME Journal of Heat Transfer*, Vol. **112**, pp. 926-932.

32. Goldstein, R.J., Jabbari, M.Y., and Chen, S.B., 1994, "Convective Mass Transfer and Pressure Loss Characteristics of Staggered Short Pin-Fin Arrays," *International Journal of Heat and Mass Transfer*, Vol. **37**, Suppl. 1, pp. 149-160.

33. Chyu, M.K., Hsing, Y.C., and Natarajan, V., 1998, "Convective Heat Transfer of Cubic Fin Arrays in a Narrow Channel," *ASME Journal of Turbomachinery*, Vol. **120**, pp. 362-367.

34. Kumaran, T.K., Han, J.C. and Lau, S.C., 1991, "Augmented Heat Transfer in a Pin Fin Channel with Short or Long Ejection Holes," *International Journal of Heat Mass Transfer*, Vol. **34**, No. 10, pp. 2617-2628.

35. Akella, K.V. and Han, J.C., 1999, "Jet Impingement Cooling in Rotating Two-Pass Rectangular Channels With Ribbed Target Walls," *AIAA Journal of Thermophysics and Heat Transfer*, Vol. **13**, No. 3, pp. 364-371.

36. Azad, GM S., Huang, Y., and Han, J.C., 2000, "Jet Impingement Heat Transfer on Dimpled Surfaces Using a Transient Liquid Crystal Technique," *AIAA Journal of Thermophysic and Heat Transfer*, Vol. **14**, No. 2, pp. 186-193.

37. Azad, GM S., Huang, Y., and Han, J.C., 2000 "Impingement Heat Transfer on Pinned Surfaces Using a Transient Liquid Crystal Technique," Proceedings of the 8th International Symposium on Transport Phenomena and Dynamics of Rotating Machinery, Honolulu, Hawaii, March 26-30, Vol. **2**, pp. 731-738.

38. Moon, H.K., O'Connell, T., and Glezer, B., 1999, "Channel Height Effect on Heat Transfer and Friction in a Dimpled Passage," presented at the International Gas Turbine and Aeroengine Congress and Exhibition, Indianapolis, Indiana, June 7-10, ASME Paper No. 99-GT-163.

39. Zhang, Y.M., Gu, W.Z., and Han, J.C., 1994, "Heat Transfer and Friction in Rectangular Channels With Ribbed or Ribbed-Grooved Walls," *ASME Journal of Heat Transfer*, Vol. **116**, No. 1, pp. 58-65.

40. Metzger, D.E. and Fan, C.S., 1992, "Heat Transfer in Pin-Fin Arrays With Jet Supply and Large Alternating Wall Roughness Ribs," HTD-Vol. **226**, Fundamental and Applied Heat Transfer Research for Gas Turbine Engines, pp. 23-30.

41. Kieda, S., Torii, T., and Fujie, K., 1984, "Heat Transfer Enhancement in a Twisted Tube Having a Rectangular Cross Section With or Without Internal Ribs," ASME Paper No. 84-HT-75.

42. Zhang, Y.M., Azad, GM S., Han, J.C., and Lee, C.P., 2000, "Heat Transfer and Friction Characteristics of Turbulent Flow in Square Ducts with Wavy, and Twisted Tape Inserts and Axial Interrupted Ribs," *Journal of Enhanced Heat Transfer*, Vol. **7**, pp. 35-49.

43. Glezer, B., Moon, H.K., Kerrebrock, J., Bons, J., and Guenette, G., 1998, "Heat Transfer in a Rotating Radial Channel With Swirling Internal Flow," presented at the International Gas Turbine & Aeroengine Congress and Exhibition, Stockholm, Sweden, June 2-5, ASME Paper No. 98-GT-214.

44. Pamula, G., Ekkad, S.V., and Acharya, S., 2000, "Influence of Cross-Flow Induced Swirl and Impingement on Heat Transfer in a Two-Pass Channel Connected by Two Rows of Holes," presented at the International Gas Turbine and Aeroengine Congress and Exhibition, Munich, Germany, May 8-11, ASME Paper No. 2000-GT-0235.
45. Zuo, Z.J., Faghri, A., and Langston, L., 1997, "A Parametric Study of Heat Pipe Turbine Vane Cooling," presented at the International Gas Turbine and Aeroengine Congress and Exhibition, Orlando, Florida, June 2-5, ASME Paper No. 97-GT-443.
46. Yamawaki, S., Yoshida, T., Taki, M., and Mimura, F., 1997, "Fundamental Heat Transfer Experiments of Heat Pipes for Turbine Cooling," presented at the International Gas Turbine and Aeroengine Congress and Exhibition, Orlando, Florida, June 2-5, ASME Paper No. 97-GT-438.
47. Kerrebrock, J.L and Stickler, D.B., 1998, "Vaporization Cooling for Gas Turbines, the Return-Flow Cascade," presented at the Gas Turbine and Aeroengine Congress and Exhibition, Stockholm, Sweden, June 2-5, ASME Paper No. 98-Gt-177.
48. Bruening, G.B. and Chang, W.C., 1999, "Cooled Cooling Air Systems for Turbine Thermal Management," presented at the International Gas Turbine and Aeroengine Congress and Exhibition, Indianapolis, Indiana, June 7-10, ASME Paper No. 99-GT-14.
49. Corman, J.C. and Paul, T.C., 1995, "Power Systems for the 21st Century "H" Gas Turbine Combined Cycles," GE Power Systems, Schenectady, New York, GER-3935, pp. 1-12.
50. Guo, T., Wang, T., and Gaddis, J.L., 1999, "Mist/Steam Cooling in a Heated Horizontal Tube, Part 2: Results and Modeling," Presented at the International Gas Turbine and Aeroengine Congress and Exhibition, Indianapolis, Indiana, June 7-10, ASME Paper No. 99-GT-145.

Heat Transfer Technology for Internal Passages of Air-Cooled Blades for Heavy-Duty Gas Turbines

BERNHARD WEIGAND*, KLAUS SEMMLER, JENS VON WOLFERSDORF**

* *Institute for Aerospace Thermodynamics, University of Stuttgart, 70569 Stuttgart, Germany*
** *Heat Transfer Group, ALSTOM POWER, 5401 Baden, Switzerland*

ABSTRACT: The present review paper, although far from being complete, aims to give an overview about the present state of the art in the field of heat transfer technology for internal cooling of gas turbine blades. After showing some typical modern cooled blades, the different methods to enhance heat transfer in the internal passages of air-cooled blades are discussed. The complicated flows occuring in bends are described in detail, because of their increasing importance for modern cooling designs. A short review about testing of cooling design elements is given, showing the interaction of the different cooling features as well. The special focus of the present review has been put on the cooling of blades for heavy-duty gas turbines, which show several differences compared to aero-engine blades.

INTRODUCTION

During the last thirty years, the inlet temperature for industrial gas turbines has steadily been increased in order to rise thermal efficiency of the turbine. On the other hand, the allowable material temperature for the gas turbine blades increased on a much slower rate. Therefore, the cooling requirement for gas turbines has become more demanding over time to compensate the difference in hot gas temperature rise and allowable component temperature. This caused a rapid development of methods for the external protection of the blade surface by film cooling and the application of thermal barrier coatings (TBC), usually as thin ceramic coatings, as well as the development of different enhancement methods of the internal heat transfer[1-5]. Because of the very high hot gas temperature in modern gas turbines the highly loaded vanes and blades of the first stage are cooled by a combination of film cooling and internal cooling. Sometimes additionally a thermal barrier coating is applied at the outside of the blade. Therefore, the external flow field is complicated by the influence of film cooling on the airfoil and the endwall surfaces and by the leakage of secondary air purging into the primary gas path flow. The turbine operates in a most agressive environment in terms of high pressure and temperature. It is subjected to strong thermo-mechanical loading which affects the durability of the turbine. The design of a modern industrial gas turbine requires expert knowledge of thermodynamics, aerodynamics, heat transfer, stress analysis, material science and manufacturing technology. The designer attempts to balance the often conflicting requirements of heat transfer against those of aerodynamics, stress, ease of manufacture and costs. For each discipline, it is important that physical phenomena are well understood and captured in the design tools and procedures. Focusing on the heat transfer perspective, this means that the external and internal flow and heat transfer phenomena have to be well understood and that correlations and tools are available for predicting these characteristics.

The aim of the present work is to review the current status of the heat transfer technology for internal passages of turbine blades. Special focus will be given to the heavy-duty gas turbines. Compared to the heat transfer requirements for aero – engine blades, the cooling of blades for an industrial gas turbine differs in some aspects. An important difference is that blades for heavy-duty gas turbines are subjected to higher Reynolds numbers in the internal cooling passages mainly due to their much larger dimensions. The larger size of the blade and also the internal cooling passages give much more freedom for designing the shape of cooling and heat transfer enhancement features inside the blade which can be cast in. On the other hand, the much bigger dimensions of the blade lead to larger casting tolerances, influencing the heat transfer characteristics and component temperatures. Most important, the durability of the heavy-duty gas turbine and the guaranteed long life of the blades of about 50000 operating hours require a very safe design of the blades. Because of the long operation time and the large dimensions of the blades, creep problems can be much more severe than for aero-engine blades. Of course, there are also a lot of similarities in the heat transfer behaviour between blades used for aero-engines and industrial engines. It can be seen that for modern industrial gas turbine blades the general view of the internal cooling scheme is quite similar compared to aero-engine blades.

It would be impossible to give a complete overview of the heat transfer technology in the present paper. Therefore, the present review tries to explain briefly the state of the art from the authors point of view. The references given in the paper are far from being complete. They have been selected to show examples and help the reader to get familiar with the subject.

PASSAGES OF AIR COOLED BLADES

In order to review the heat transfer technology for internal passages for heavy-duty gas turbine blades it is important to first look on the actual design of such blades. Fig. 1 shows typical examples for a first stage vane and a first stage rotating blade of a modern industrial gas turbine[6].

For vanes, impingement cooling is used very often together with pins and turbulators. For rotating blades mostly turbulators are preferred together with pins in order to cool the blade efficiently. Both cooling schemes can use additionally film cooling if the external hot gas temperature is too high for the application of a pure internal convection cooling system. Blades and vanes operating at lower hot gas temperature levels as second and third stages normally have much simpler cooling schemes. If the blade has been optimised aerodynamically, the internal cooling channel can have a funny looking shape with sharp corners. Fig. 2 shows an example for such a cooling arrangement. Special care has to be taken for such non rectangular shaped cooling channels to cool them efficiently and to guide cooling air into the sharp corners of the cooling channels.

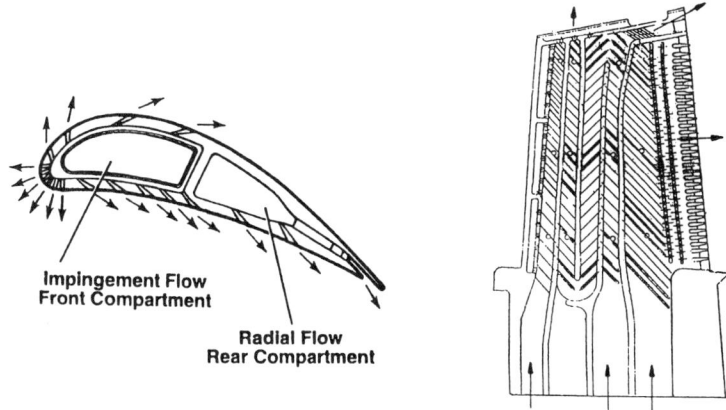

Fig. 1: Typical cooling features for stationary and rotating blades of a modern industrial gas turbine (Schulenberg et al.[6], reprinted with permission of VDI).

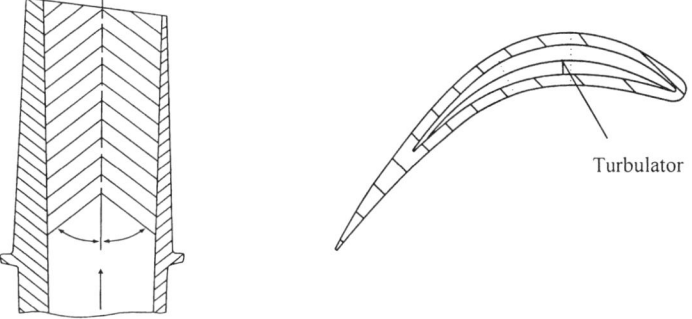

Fig. 2: Cooling scheme for a rotating blade at later stages (Hall et al.[7])

EXTERNAL HEAT TRANSFER

The external heat transfer of the blade provides the boundary conditions for the internal cooling problem. Therefore, the present review paper will also focus briefly on this subject. If one considers the external heat transfer at midspan of a blade which is subjected to pure convective heat transfer without film cooling, a lot of experimental data are available[1]. For such flow situations, boundary layer codes can provide quite accurate external heat transfer data, if they are adopted to the cases of interest. Good reviews about this subject can be found in Kays and Crawford[8] and in Sieger[9]. This is even true if we consider additionally roughness effects on the blade surface[10].

If the blade is cooled additionally by applying film cooling, the situation is much more complicated. The heat flux into the surface is now given by

$$q_F = h_F \left(T_{aW} - T_{Surf} \right) \tag{1}$$

where h_F is the heat transfer coefficient in the presence of film cooling, T_{aW} is the adiabatic wall temperature and T_{Surf} is the surface temperature of the component. The heat flux q_F should always be smaller than q_0, which is the heat flux without film cooling

$$q_0 = h_0 \left(T_{r\infty} - T_{rW}\right) \qquad (2)$$

where h_0 is the heat transfer coefficient in the absence of film cooling and T_{rW} is the recovery temperature at the wall. Otherwise, film cooling would have a negative impact on the overall cooling of the blade. Injecting cold air into the boundary layer will basically have two different effects: On the one side, the coolant injected into the main stream will reduce the driving temperature difference for the heat transfer. This effect can be expressed, in a non dimensional form, by the adiabatic film cooling effectiveness

$$\eta_{ad} = \left(T_{aW} - T_{r\infty}\right) / \left(T_C - T_{r\infty}\right) \qquad (3)$$

On the other side, the cooling jets entering and disturbing the boundary layer will increase the heat transfer coefficient, so that h_F is usually larger than h_0.

Because of the growing importance of film cooling for turbine blade design, the subject has been studied extensively over the past 35 years and a large part of the results is available in the open literature. Several review articles are published on this subject[11-13]. Most of the studies concentrate on flat plate configurations with film injection through slots, cylindrical or shaped holes. See for example[14-17]. Fewer publications are available concerning film cooling on airfoil type flows. Examples are given in[18-21] or more general on curved surfaces[22]. Because of the large number of parameters influencing the film cooling process (e.g. cooling hole geometry, blowing and momentum flux ratio, density ratio, temperature ratio, angles of the cooling holes, curvature effects, upstream boundary layer effects, main stream turbulence effects, effect of main stream pressure gradient and Mach number, effect of surface roughness and system rotation, ...) numerical methods and correlations have been developed to predict the adiabatic film cooling effectiveness and the increase in heat transfer coefficients. Several models can be found in literature[1-3,23-27] which have been used for this purpose for different types of applications.

INTERNAL HEAT TRANSFER

After the external boundary conditions are known, the internal heat transfer can be discussed. For the internal heat transfer it is important to understand the methods for enhancing the internal heat transfer coefficients first. As already mentioned above, the common methods for doing this are: turbulators (ribs), impingement, pins and combinations thereof. The designer has to choose a certain method or a combination of several methods based on the cooling requirements, internal Reynolds number, system rotation, space limitations and manufacturing aspects[1-5].

Impingement Cooling

This type of cooling is very often used for vanes. Impingement cooling is very effective, if the jet can impinge onto the surface without being deflected too much by any cross-flow. Therefore, this type of cooling arrangement is preferred very often for internally cooling the leading edge of a vane. A comprehensive review on this subject is given by Pagenkopf[28]. In this thesis, the author cited more than 60 relevant publications about impingement cooling. Because the geometry, the number of the impingement holes and the target surface (planar or curved) are very important for the development of the jet and the effectiveness of the cooling method, the literature in[28] has been systematically ordered by taking this into account. There are also a number of design correlations for impingement cooling available in literature[29-30]. A summary of some useful correlations for this subject can be found in[3]. Nowadays impingement cooling is used also in combination with pin plate target surfaces[31]. This arrangement is of particular interest for the cooling of highly loaded regions, e.g. leading edges and endwalls of vanes. By using cast pins on the target surface, an enhancement of heat transfer in the order of 10 to 15% could be observed by Höcker et al.[31]. Of course this increase is very much dependent on the pin material and the arrangement used (e.g. pin spacing to pin diameter x/d and impingement distance to pin diameter z/d, pin height and pin shape).

Pin-fin Heat Transfer

Pin-fins are commonly used to increase the internal heat transfer to a turbine blade or vane, especially in the trailing edge of the blade. They increase the internal wetted surface area and the flow turbulence in the passage. The pins can be placed in various configurations. Mostly they are arranged as in-line or staggered arrays. Generally staggered arrays of pin-fins provide a higher heat transfer rate and are preferred for the cooling of gas turbine blades[32]. The heat transfer coefficient in pin arrays increases with the number of rows up to the third to fifth row and then decreases slightly towards an asymptotic value [34]. A good review about staggered pin-fin arrays is given by Armstrong and Winstanley[35]. They present available data on heat transfer and pressure loss for such arrangements. Generally, it can be said that pins are most effective for lower Reynolds numbers. For Reynolds number smaller than 20 000 the enhancement in heat transfer coefficient can be of the order of 50-100 %. With increasing Reynolds number the increase in heat transfer coefficient drops and reaches enhancement levels of about 10-30% above the smooth passage value for a Reynolds number of about 100000. The effect of streamwise acceleration is important for pin-fin arrays, if they are used in the converging trailing edge section of the blade. Metzger et al.[34] investigated this effect and proposed a correlation for the effect of flow acceleration on heat transfer. The present authors have the feeling that more experimental and computational studies would be needed on the subject of pressure drop and heat transfer in pin-fin arrays, especially for accelerating flow with high local Mach numbers.

Turbulators

One of the most commonly used features to increase heat transfer in gas turbine blades are turbulators, as they are easy to manufacture and result in a good heat transfer augmentation. The turbulators trip the boundary layer periodically and cause repeating flow pattern within the passage which lead to a high turbulence level in the core flow. It has been found that the flow achieves a periodic "fully developed" state after about five ribs. A large body of work has been done in the past on the effect of different turbulators on the pressure drop and on the heat transfer in a square or a rectangular channel[36-43]. Good summaries are given in[1,3,44,45]. In the literature cited above also the influence of rib spacing, rib angle and rib height has been investigated for rectangular channels. Fig. 3 shows some experimental results for the local heat transfer in a rectangular duct with V-shaped ribs. On the left side the secondary flow field and the associated effect on the heat transfer is depicted for ribs pointing downstream in flow direction, whereas the figure on the right side shows the situation for V-shaped ribs pointing upstream. From the two figures it can be seen that the heat transfer can be locally adapted by selecting appropriate

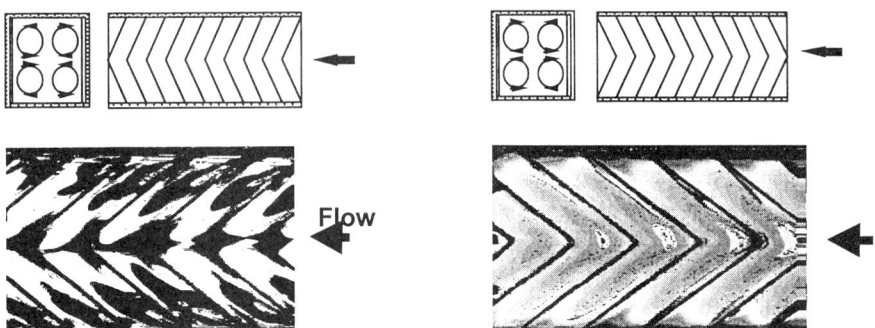

Fig. 3: Internal heat transfer and secondary flow field in a ribbed channel (Weigand [92]).

turbulators. For example for the arrangement shown on the left side of Fig. 3, the secondary motion takes fluid from the center region in the passage upwards and transports this fluid towards the sidewalls. Near the sidewalls a downwash of the fluid can be observed, resulting in high heat transfer at this location.

In the last years effort has been taken to numerically predict the pressure drop and the heat transfer in ribbed channels. Examples of 3D numerical predictions of such arrangements can be found in [46-48]. In general it can be noted that it is possible to predict the pressure drop within the channels quite well. On the other hand it is very difficult to accurately predict the heat transfer. Mostly only qualitative results can be obtained for the heat transfer characteristics. This is caused by the fact that the prediction of the heat transfer coefficient requires a very accurate resolution of the near wall region and appropriate turbulence models. Because the 3D numerical predictions could help in optimizing ribbed channels, more work is needed in this area.

A lot of the cooling channels in gas turbine blades can not be approximated by a rectangular shape. Good examples are the cooling channels near the leading and trailing edge of the blade or channels of aerodynamically optimized blades for later stages (see Fig. 2). All these ducts might be approximated by triangular channels. In contrast to the large number of papers available for the heat transfer augmentation in rectangular channels, only some few publications can be found in the open literature on the subject of heat transfer enhancement in triangular ducts. In triangular shaped channels without ribs it can be observed that laminar flow in the apex area might exist, even if the flow in the core area is fully turbulent[49-50]. This is caused by the high friction coefficients in the apex resulting in much slower flow in this area. Ribbed triangular channels have only been analyzed in[51-52]. Metzger et al.[51] investigated a triangular channel with two ribbed walls. The channel had three apex angles of 60°. The experimental investigation showed that the heat transfer was best for crossed ribs. Zhang et al.[52] investigated a triangular channel with apex angles 90°, 55°, 35°. It could be shown that a channel with two ribbed walls resulted in about 10% higher heat transfer than the channel with three ribbed walls. Today also 3D shaped turbulators can be manufactured for industrial gas turbine blades. With such turbulators the local heat transfer can be influenced drastically. Fig. 2 shows such an arrangement. The rib height is gradually decreased with decreasing channel height. Hall et al.[7] used the criteria that the local height of the rib divided by the local channel height is constant. By using this criteria, the local friction in the channel is equalized between the core region and the apex area. This promotes the secondary flow to exchange air between the core area and the apex area, leading to higher heat transfer coefficients in the edges. 3D shaped turbulators can also be used for effectively cooling the leading edge

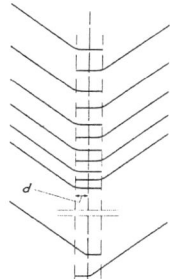 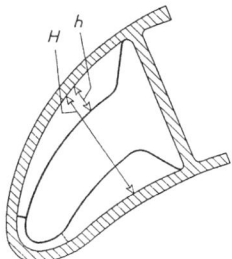

Fig. 4: 3D shaped turbulators for the leading edge of an industrial gas turbine blade (Beeck et al. [53]).

of a blade. Fig. 4 shows an example of such an arrangement. The authors have the feeling that a lot of more work is needed in this area in order to optimize further the rib arrangement for different kinds of cooling channels. As stated before CFD plays here a key role in order to find such optimum arrangements.

Bends

As it can be seen from Fig. 1 the radial aligned coolant channels are connected by 180° turns. Current trends are towards increasing number of channels, so that the flow and heat transfer characteristics around bends and turns are of high interest. Fig. 5 shows a flow visualization (oil painting) for a 180° bend. It can be seen that the flow characteristics in the bend region are quite complex. Secondary flow and separation zones lead to complex three-dimensional flow structures. Additionally the flow behavior in the bend is highly geometry dependent. Pressure drop characteristics are given for example in [54-55] for different bend geometry used for gas turbine applications. Because of the deflection of the flow in the bend area, secondary motion moves the fluid from the center of the channel in the direction to the outer wall and from there back to the center of curvature. The resulting flow pattern are shown in Fig. 5. Because the flow is unable to follow the sharp changes in direction, flow separation occurs within the bend region. Guide vanes or other flow guiding devices help to prevent big separation zones and can decrease the pressure drop in the bend up to 80%[56-57]. If turbulators are used in the channels linked to the bend, the secondary motion within the bend can be drastically changed[58-62]. This has, of course, also strong implications on the heat transfer. In general, the maximum heat transfer is obtained at the point where the flow impinges on the outer wall of the downstream corner region of the bend (see Fig. 5). The value of the heat transfer coefficient at this area can be several times larger than the one for a smooth channel[63-67]. Detailed flow measurements have been conducted by Schabacker[68] using a stereo PIV system. With this method it was possible to obtain very accurate and high resolved measurements of the velocity field in the bend. Because the flow and heat transfer in bends are very much dependent on the actual geometry, 3D CFD predictions can be very helpful in order to improve the bend geometry. A lot of work has been done on this subject within the last ten years. A good summary of the work is given by Launder and Iacovides[69]. Mostly only U-bends are considered. On the other hand, the data obtained by Schabacker[68] provide a good basis for improving 3D CFD predictions for

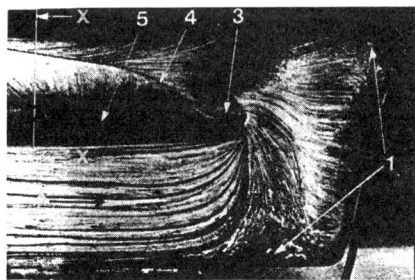

1 Regions of low velocity
2 Flow direction
3 Area of flow separation
4 Separation line
5 Flow attachment point

Fig. 5: Flow visualization (oil painting) for a flow in a bend.

180° bends. Results of a numerical study done with FLUENT have been shown in Schabacker[68]. For the bend connected to smooth channels the comparison between calculations and measurements is quite promising. On the other hand bigger differences between calculation and measured data could be observed for the case where the bend has been connected to ribbed channels.

Rotation Effects

Rotation induces additional forces on the flow field and alters the flow pattern and heat transfer distribution in the internal cooling channels. Coriolis forces introduce cross stream secondary flows, which cause different heat transfer pattern on the individual channel walls. Buoyancy forces are important at high rotational speeds and

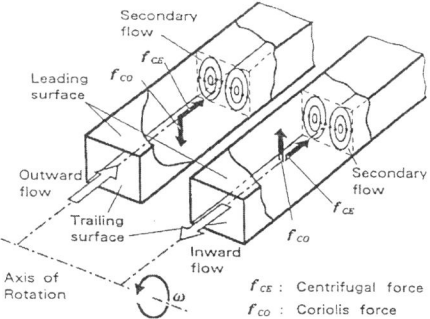

Fig. 6: Effect of Coriolis and centrifugal forces on the flow in rotating passages (Mochizuki et al.[70], reprinted with permission of ASME).

coolant temperature gradients and cause radial secondary flows. On the "high pressure side" heat transfer is generally increased above the stationary value and on the "low pressure side" heat transfer is generally decreased. For a serpentine cooling configuration the "high pressure side" of the passage is the trailing side for flow outward and the leading side for flow inward. The "low pressure side" is the leading side for flow outward and the trailing side for flow inward (Fig. 6). The dimensionless correlating parameters governing the flow field and therefore the heat transfer in the rotating frame are the Reynolds number, $Re=\rho UD/\mu$, the Rotation number $Ro=\omega D/U$, and the rotational

Rayleigh number, $Ra=\beta\Delta TH(ReRo)^2Pr/D$, where ω is the rotational speed; β the thermal expansion coefficient; ΔT the difference between mean wall temperature and mean fluid temperature; and H the mean radius of rotation. Wagner et al.[71] suggested to use the buoyancy parameter, which is for an ideal gas $Ra/(PrRe^2)$ as correlating parameter. They showed further that the heat transfer on surfaces, where it is increased with rotation and buoyancy, varied by as much as a factor of four. On the surfaces, where heat transfer is decreased with rotation, the variation is as much as down to one third of the non-rotating case. Because of this big impact numerous experimental studies have been performed on this subject. Good reviews are given in[45,72]. The parameter ranges covered by these experimental studies are given in[73] with 0<Ro<0.6 and 2500<Re<100000. These parameter ranges are typical for aircraft engines[74]. As already mentioned above, heavy-duty gas turbine blades operate more on the upper end with respect to Reynolds number, but in the lower to medium rotation number range and at smaller buoyancy parameters. Therefore it is felt, that the available information can be equally applied to the design of large rotor blades. Further, Zhang et al.[75] showed, that the Reynolds number has a minor effect on the Nusselt number ratio Nu_r/Nu_o, where Nu_r ist the Nusselt number with rotation and Nu_o is the stationary value. Nevertheless experimental studies at high Reynolds and rotation numbers, which are difficult to achieve simultaneously in an experiment under laboratory conditions, are not yet available. Other factors which strongly influence the heat transfer distribution in internal cooling channels are the flow entrance conditions, thermal boundary conditions, channel shape, aspect ratio and orientation. Of course also the interaction of the rotation induced secondary flows with the secondary flows due to the geometry of the turbulator configuration and the turn regions alter the heat transfer distribution significantly. The optimum rib arrangement determined under stationary conditions needs not to be the best choice in the rotating case. Therefore a large number of experimental investigations addressing these factors have been performed. The effect of wall heating conditions and model orientation were investigated for example in[76-78]. Only a few studies investigated the rotation effects in triangular[79] and narrow rectangular ducts[80] typical for leading and trailing edge channels. Generally most of the studies perform heat transfer measurements. Therefore the information on pressure drop characteristics is still limited[81,82]. Due to the large number of geometrical and flow parameters numerical investigations are appearing more and more to address this complex flow and heat transfer problem. Some examples are given in[46,83-85]. Large eddy simulations were applied for example in [86,87]. Although the accuracy in the heat transfer predictions is still limited, the continuous development of these methods is needed. For these developments detailed experimental information on the flow field in rotating channels are emerging using PIV[88] and Laser Doppler methods[73,90] complementary to the detailed heat transfer data.

Component Tests/Engine Tests

After addressing the different basic elements for the internal cooling system, it is important to understand how they interact in a gas turbine blade. For this normally component test are performed. Here large scale models of ribbed serpentine turbine cooling passages are used. Different methods can be used to visualize heat transfer in such cooling systems. There is some work available in the open literature which deal with this sort of component tests [47,90-93]. Abuaf and Kercher[47] investigated the heat transfer of a scaled model (ten times larger than the original) of a ribbed serpentine turbine cooling passage using the thermal liquid crystal technique. For industrial gas turbines typical scalings are of the order of 3 to 5 times the original size to achieve Reynolds number similarity at ambient laboratory conditions. Fig. 7a shows a typical view of a large scale model used for such heat transfer tests. The resulting heat transfer is indicated in Fig. 7b [93]. For this model, the surface under investigation was sprayed with narrow band liquid crystals. The model is cold at the beginning of the test and, as hot air enters, a video record of the liquid crystal color change reflects the local heat transfer coefficients. Nowadays, such component tests are heavily supported by 3D CFD using unstructured grids. Of course also component tests in the lab can not guarantee to match all the required conditions in the real engine. Therefore, finally engine tests are needed to continuously improve the heat transfer designs[4,92,95-97]. For heavy-duty gas turbines such tests are very expensive.

Fig. 7a: Large Scale Model   Fig. 7b: Heat Transfer Results

Fig. 7: Large scale model and test results for the heat transfer inside a modern gas turbine blade (Fig. 7b: Jennions et al.[93], reprinted with permission of ASME).

SPECIAL PROBLEMS IN COOLING HEAVY-DUTY GAS TURBINE BLADES

This chapter focuses on some special problems arising in cooling industrial gas turbine blades. These problems can be subdivided into three main categories:

1. **Larger blades:**
Heavy-duty gas turbine blades are normally much larger than blades of aero-engines. This is especially true for the blades of later stages in the turbine. Because of the use of industrial gas turbines as power supply, the blades have to be in service for a very long period (50 000 operating hours). This increases the importance of creep elongation and creep bending investigations of such blades during the design phase. Special care has to be taken that the internal cooling of such blades is optimized to prevent creep problems. This can be done by locally adapting the cooling effectiveness from hub to tip (e.g. by increasing the rib height), so that the highly loaded blade is sufficiently cooled in the area needed. Larger external dimensions of the components usually result in bigger wall thickness as well. This is mainly driven by mechanical stress requirements and manufacturing constraints. As the thermal stresses tend to increase with increasing wall thickness, special effort has to be taken during the design to limit the thermal stresses and achieve the required durability of the part.

The much longer inspection intervals and guaranteed life of industrial gas turbines compared to aero-engines, requires special care in designing for failure tolerance. The designs need to survive the exposure to dust and dirt carried along with the coolant for a long time without resulting in partially or totally blocked cooling systems. Further on, the long inspection intervals require cooling designs to be very tolerant against damages of the blading and cooling system due to foreign objects travelling through the hot gas path.

Nowadays, the trend is to produce even larger industrial gas turbines. This will lead also to special problems concerning the augmentation of the internal heat transfer coefficients because of growing internal Reynolds numbers. As it was stated before, the augmentation of heat transfer, compared with the value of a smooth duct, decreases with increasing Reynolds numbers for pin-fins and turbulators. Therefore, new methods have to be thought of in order to rise the internal heat transfer also to a high level for such blades. One approach is to use many small coolant passages located close to the blade surface to provide effective near wall cooling or to increase the number of coolant passages and therefore to reduce their individual cross sectional area.

2. **Fuel Flexibility:**

The need to have flexible industrial gas turbines, which can use different fuels is also a difficulty for the cooling of the blades. This is caused by different reasons. If for example crude oil needs to be burned in the engine, there might be limitations in using film cooling for the first stage because of potential blocking of these holes. Otherwise, if film holes are allowed for such applications, they might be much bigger in size, as for an engine using gas as fuel. If coal-derived fuels are used, sulfur, halides and low level of metallic elements, such as vanadium, will be carried by the coal-derived fuel and will present a significant corrosion attack to the blades[98]. Especially the resistance of coatings and TBC (mainly of the bond coat) against the corrosion attack due to hot gas composition and atmosphere conditions are very important.

3. **Materials and Casting**

For small first stage aero-engine blades it is quite common to use single crystal blades which offer much higher allowable metal temperatures and/or stress levels than a conventionally cast or directional solidified blade. For industrial gas turbines, however, the bigger dimensions of the blades offer here a real challenge to produce large single crystal blades. On the other hand the use of single crystal blades for the first stage allows the cooling engineer to save a substantial amount of cooling air, which will rise cycle efficiency of the turbine. For blades used for later stages, similar challenges exist to produce very large cores and cast blades.

CLOSURE

The preceding discussion on the heat transfer technology for internal passages of air-cooled blades for heavy-duty gas turbines, although far from being complete, should serve as an overview of the current state. The present paper tried to describe the different elements of a cooling design which are present in modern gas turbine blades. The internal cooling designs rely also today very much on experiments made for basic geometries, components or engine tests. On the other hand 3D CFD becomes more and more important for the optimization of cooling geometry and helps in selecting the final design. Therefore, it is the opinion of the authors that more work is needed on integrated studies using 3D CFD and experimental results together for optimizing actual cooling designs.

This will also require a massive development of 3D CFD in order to produce reliable heat transfer predictions.

ACKNOWLEGDGEMENT

The authors J. von Wolfersdorf and K. Semmler would like to thank ALSTOM POWER Ltd for permission to publish this paper. The authors also would like to thank the ASME for permission to reprint Figs. 6,7b and VDI, Düsseldorf, for the permission to reprint Fig. 1.

REFERENCES

1. Lakshminarayana, B. 1996. Fluid Dynamics and Heat Transfer of Turbomachinery, John Wiley & Sons, Inc., New York.
2. Metzger, D.E. 1985. Cooling Techniques for Gas Turbine Airfoils – A Survey, Proceedings of AGARD Conference, Bergen, Norway, 1 – 13.
3. Harasgama, S.P. 1995. Aerothermal Aspects of Gas Turbine Flows. VKI-LS 1995-05. Heat Transfer and Cooling in Gas Turbines.
4. Hennecke, D.K. 1984. Heat Transfer Problems in Aero-Engines. Heat and Mass Transfer in Rotating Machinery. Hemisphere. Washington. 353-379.
5. Yeh, F.C. & F.S. Stepka. 1984. Review and Status of Heat-Transfer Technology for Internal Passages of Air-Cooled Turbine Blades. NASA Technical Paper 2232: 1-33.
6. Schulenberg, T., Kopper, F. & J. Richardson. 1995. An Advanced Blade Design for V84.3 Gas Turbines, VDI Berichte, **1185**: 257-275.
7. Hall, K., Johnson, B., Weigand, B. & Pey-Shey Wu. 1999. Coolable Blade. United States Patent, Patent Number: 5919031.
8. Kays, W.M. & M.E. Crawford. 1993. Convective Heat and Mass Transfer. Mc Graw-Hill, Inc., New York.
9. Sieger, K. 1993. Vergleich der Leistungsfähigkeit erweiterter k-ε Turbulenzmodelle bei der Berechnung transitionaler Grenzschichten an Gasturbinenschaufeln, PhD Thesis, Karlsruhe.
10. Tarada, F. 1990. Prediction of rough-wall boundary layers using a low Reynolds number k-ε model. Int. J. Heat and Fluid Flow **11**: 331-345.
11. Goldstein, R.J. 1971. Film Cooling. In Advances in Heat Transfer. Irvine, T.F. and J.P. Hartnett Eds. Academic Press. New York 7: 321-379.
12. VKI Lecture Series. 1982. Film Cooling and Turbine Blade Heat Transfer. VKI-LS 82-02.
13. Leontiev, A.I. 1999. Heat and Mass Transfer Problems for Film Cooling. J. Heat Transfer. **121**: 509-527.
14. Sinha, A.K, Bogard, D.G. & M.E. Crawford. 1990. Film Cooling Effectiveness Downstream of a Single Row of Holes with Variable Density Ratio. 90-GT-43.
15. Wittig, S., Schulz, A. Gritsch, M. & Thole, K.A. 1996. Transonic Film-Cooling Investigations: Effect of Hole Shapes and Orientations. 96-GT-222.
16. Jabbari, M.Y. & R.J. Goldstein. 1978. Adiabatic Wall Temperature and Heat Transfer Downstream of Injection Through Two Rows of Holes. J. of Eng. for Power. **100**: 303-307.
17. Forth, C.J., Loftus, P.J. & T.V. Jones. 1980. The Effect of Density Ratio on the Film Cooling of a Flat Plate. AGARD CP 390. Bergen.
18. Nicolas, J. & A. Le Meur. 1974. Curv. Effects on a Turbine Blade Cooling Film. 74-GT-156.
19. Ito, S., Goldstein, R.J. & E.R.G. Eckert. 1978. Film Cooling of a Gas Turbine Blade. J. Eng. for Power. **100**: 476-481.
20. Takeishi, K., Aoki, S., Sato, T. & K. Tsukagoshi. 1992. Film Cooling on a Gas Turbine Rotor Blade. J. of Turbomachinery. **114**: 828 – 834.
21. Drost, U. 1998. An Experimental Investigation of Gas Turbine Airfoil Aero-Thermal Film Cooling Performance. PhD Thesis. Lausanne, Switzerland.
22. Lutum, E., von Wolfersdorf, J., Weigand, B.& K. Semmler. 2000. Film Cooling on aconvex surface with zero pressure gradient flow. Int. J. Heat Mass Transfer. 43: 2973-2987.
23. Crawford, M.E. 1986. Simulation codes for calculation of heat transfer to convectively cooled turbine blades, VKI-LS, Convective heat transfer and film cooling in turbomachinery.

24. Walters, D.K. & J.H. Leylek. 1997. A Detailed Analysis of Film-Cooling Physics. Part I: Streamwise Injection with Cylindrical Holes. 97-GT-269.
25. Le Grives, E. & J.J. Nicolas. 1977. Method Nouvelle de Calcul de L'Efficacite de Refroidissement des Aubes de Turbine par Film d'Air. AGARD CP 229.
26. Weigand, B., Bonhoff, B. & J. Ferguson. 1997. A comparative study between 2D boundary layer predictions and 3D Navier-Stokes calculations for a film cooled vane. In Proc. of the U.S. National Heat Transfer Conference, Batimore, HTD **350**: 213-221.
27. Garg, V.K. 1997. Comparison of Predicted and Experimental Heat Transfer on a Film-Cooled Rotating Blade using a Two-Equation Turbulence Model. 97-GT-220.
28. Pagenkopf, U. 1996. Untersuchung der lokalen konvektiven Transportvorgänge auf Prallflächen. PhD Thesis, Darmstadt, Germany.
29. Kercher, D.M., & W. Tabakoff. 1970. Heat transfer by a square array of round air jets impinging perpendicular to a flat surface including the effects of spent air. J. of Eng. of Power **73** :73-82.
30. Florschuetz, L.W., Metzger, D.E., Su, D.E. & Y. Isoda. 1982. Jet array impingement flow distributions and heat transfer characteristics. NASA Contractor Report-3630.
31. Hoecker, R., Johnson, B.V., Hausladen, J., Rothbrust, M. & B. Weigand. 1999. Impingement Cooling Experiments with Flat Plate and Pin Plate Target Surfaces. 99-GT-252.
32. Sparrow, E.M. & Molki, M. 1982. Effect of a missing cylinder on heat transfer and fluid in an array of cylinders in cross flow. Int. J. Heat and Mass Transfer. **25**: 449-456.
33. Van Fossen, G.J. 1982. Heat transfer coefficients for staggered arrays of short pin fins. J. of Eng. for Power. **104**: 268 – 274.
34. Metzger, D.E., Shepard, W.B. & S.W. Haley. 1986. Row resolved heat transfer variation in pin fin arrays including the effects of non-uniform arrays and flow convergence. 86-GT-132.
35. Armstrong, J. & D. Winstanley. 1988. A Review of Staggered Array Pin Fin Heat Transfer for Turbine Cooling Applications. J. of Turbomachinery. **110**: 94 – 103.
36. Han, J.C. 1984. Heat transfer and friction in channels with two opposite rib-roughened walls. J. Heat Transfer. **106**: 774-781.
37. Han, J.C. ,Park J.S. & C.K. Lei. 1985. Heat Transfer enhancement in channels with turbulence promoters. J. Engng. Gas Turbines Power. **107**: 629-635.
38. Han, J.C. & J.S. Park. 1988. Developing heat transfer in rectangular channels with rib turbulators. Int. J. Heat Mass Transfer. **31**: 183-195.
39. Han J.C., Ou, S., Park, J.S. & C.K. Lei. 1989. Augmented heat transfer in rect. channels of narrow aspect ratios with rib turbulators. Int. J. Heat Mass Transfer. **32**: 1619-1630.
40. Lau, S.C., Kukreja, R.T. & R.D. Mc Millin. 1991. Effects of V-shaped ribs on turbulent heat transfer and friction of fully developed flow in a square channel. Int. J. Heat Mass Transfer. **34**: 1605-1616.
41. Han, J.C., Zhang, Y.M. & C.P. Lee. 1992. Influence of surface heat flux ratio on heat transfer augmentation in square channels with parallel, crossed and V-shaped angled ribs. J. of Turbomachinery. **114**: 872-880.
42. Hong,Y.J. & S.S. Hsieh. 1993. Heat transfer and friction factor measurements in ducts with staggered and in-line ribs. J. of Heat Transfer.**115**: 58-65.
43. Rau, G. 1998. Einfluss der Rippenanordnung auf das Strömungsfeld und den Wärmeübergang in einem Kühlkanal mit quadratischem Querschnitt. PhD Thesis. Darmstadt. Germany.
44. Bergles, A.E. & R.L. Webb. 1985. A guide to the literature on convective heat transfer augmentation. Advances in Heat Transfer. Shenkman et. al. Eds. ASME HTD - **43**: 81-89.
45. Han, J.C. & S. Dutta. 1995. Internal Convection Heat Transfer and Cooling - An Experimental Approach, VKI-LS, Heat Transfer and Cooling in Gas Turbines.
46. Prakash, C. & R. Zerkle. 1993. Prediction of Turbulent Flow and Heat Transfer in a Ribbed Rectangular Duct with and without Rotation. 93-GT-206.
47. Abuaf, N. & D.M. Kercher. 1994. Heat Transfer and Turbulence in a Turbulated Blade Cooling Circuit. J. of Turbomachinery. **116**: 169-177.
48. B. Bonhoff, B., Boelcs, A., Johnson, B.V., Leusch, J., Parneix, S. & J. Schabacker (1998): Experimental and numerical study of developed flow and heat transfer in coolant channels with 45 deg rib arrangement, Turbulent Heat Mass Transfer, Manchester, UK.

49. Bandopadhayay, P.C. & J.B. Hinwood. 1973. On the coexistence of laminar and turbulent flow in a narrow triangular duct. J. Fluid Mechanics. **59**: 775-783.
50. Altemani, C.A.C. & E.M. Sparrow. 1980. Turbulent heat transfer and fluid flow in an unsymmetrically heated triangular duct. J. Heat Transfer. **102**: 590-597.
51. Metzger,D.E. & R.P. Vedula. 1987. Heat transfer in triangular channels with angled roughness on two walls. Experimental Heat Transfer. **1**: 31-44.
52. Zhang, Y.M., Gu,W.Z. & J.C. Han. 1994. Augmented heat transfer in triangular ducts with full and partial ribbed walls, J. of Thermph. and Heat Transfer. **8**: 574-579.
53. Beeck, A., Johnson, B. ,Weigand, B. & Pey-Shey Wu. Cooling system for the leading edge of a hollow blade for a gas turbine. EP0892149.
54. Metzger, D.E. ,Plevich, C.W. & C.S. Fan. 1984. Pressure loss through sharp 180-deg turns in smooth rectangular channels. J. of Engng. for Gas Turbines and Power. **106**: 677-681.
55. Abuaf, N. Gibbs, R. & R. Baum. 1986. Pressure drop and heat transfer coefficient distributions in serpentine passages with and without turbulence promoters, Heat Transfer Conference. **6**: 2837-2845.
56. Frey, K. 1934. Verminderung des Strömungsverlustes in Kanälen durch Leitflächen. Forschung im Ingenieurwesen. **5**: 105-117.
57. Plevich, C.W. 1985. Effects of Turning Vanes, Radial Ribs and Corner Fillets on Flow Patterns and Pressure Losses in Rectangular Duct 180-deg Turns. M.S. Thesis. Arizona State University.
58. Han, J.C. & P. Zhang. 1989. Pressure loss distribution in three-pass rectangular channels with rib turbulators. J. of Turbomachinery. **111**: 515-521.
59. Han, J.C., Chandra, P.R. & S.C. Lau. 1986. Local heat/mass transfer distribution around sharp 180 deg. turn in multipass rib-roughened channels. 86-GT-114.
60. Chandra, P.R., Han, J.C. & S.C. Lau. 1987. Effect of rib angle on local heat/mass transfer distribution in a two-pass rib-roughened channel. 87-GT-94.
61. Han, J.C., Chandra, P.R. & S.C. Lau. 1988. Local heat/mass transfer distrib. around sharp 180 deg. turns in two-pass smooth and rib-roughened channels. J. Heat Transfer. **110**: 91-98.
62. Han, J.C. & P. Zhang. 1991. Effects of rib-angle orientation on local mass transfer distribution in three-pass smooth and rib-roughened channels. J. of Turbomachinery. **113**: 123-130.
63. Metzger, D.E. & M.K. Sahm. 1986. Heat Transfer around sharp 180-deg. Turns in smooth rectangular channels. J. of Heat Transfer. **108**: 500-506.
64. Fan , C.S. & D.E. Metzger. 1987. Effects of channel aspect ratio on heat transfer in rectangular passage sharp 180-deg. turns. 87-GT-113.
65. Chyu, M.K. 1991. Regional heat transfer in two-pass and three-pass passages with 180-deg. sharp turn. J. of Heat Transfer. **113**: 321-330.
66. Wang, T.S. & M.K. Chyu . 1994. Heat convection in a 180-deg. turning duct with different turn configurations. J. of Thermophysics and Heat Transfer. **8**: 595-601.
67. Hirota, M., Fujita, H., Syuhada, A., Araki, S., Yoshida, T. & T. Tanaka (1999): Heat/mass transfer characteristics in two-pass smooth channels with sharp 180-deg. turn. Int. J. Heat Mass Transfer. **42**: 3757-2770.
68. Schabacker, J. 1998. PIV Investigation of the Flow Characteristics in Internal Coolant Passages of Gas Turbine Airfoils with two Ducts connected by a sharp 180-deg. Bend. PhD Thesis. Lausanne. Switzerland.
69. Launder, B.E. & H. Iacovides. 1995. CFD Applied to Internal Cooling. VDI Ber. **1185**: 1-34.
70. Mochizuki,S.; J. Takamura, S. Yamawaki, W.J. Wang . 1992. Heat transfer in serpentine flow passages with rotation. 92-GT-190.
71. Wagner, J.H., Johnson, B.V., Graziani,R.A. & F.C. Yeh. 1991. Heat Transfer in Rotating Serpentine Passages With Trips Normal to the Flow. 91-GT-265.
72. Taslim, M.E. 2000. Convective Cooling in Non-Rotating and Rotating Channels - Experimental Aspects, VKI-LS 2000-03, Aero-Thermal Performance of Internal Cooling Systems in Turbomachines.
73. Hsieh, S.S., Chen, P.J. & H.J. Chin. 1999. Turbulent Flow in a Rotating Two Pass Smooth Channel. J. of Fluids Engineering. **121**: 725-734

74. Hwang, G.J., Tzeng, S.C. & C.P. Mao. 1999. Heat Transfer of Compressed Air Flow in a Spanwise Rotating Four-Passage Serpentine Channel. J. of Heat Transfer. **121**: 583-591.
75. Zhang, N., Chiou,J., Fann, S. & W.J. Yang. 1993. Local Heat Transfer Distribution in a Rotating Serpentine Rib-Roughened Flow Passage. J. of Heat Transfer. **115**: 560-567.
76. Parsons, J.A., Han, J.C.. & Y. Zhang. 1995.Effect of model orientation and wall heating condition on local heat transfer in a rotating two-pass square channl with rib turbulators. Int. J. Heat Mass Transfer. **38**: 1151-1159.
77. Park, C.W., Yoon, C. & S.C. Lau. 2000. Heat (Mass) Transfer in a Diagonally Oriented Rotating Two-Pass Channel With Rib-Roughened Walls. J. of Heat Transfer. **122**: 208-211.
78. Dutta, S. & J.C. Han. 1996. Local Heat Transfer in Rotating Smooth and Ribbed Two-Pass Square Channels With Three Channel Orientations. J. of Heat Transfer. **118**: 578-584.
79. Dutta, S., Han, J.C. & C. P. Lee. 1995. Experimental Heat Transfer in a Rotating Triangular Duct: Effect of Model Orientation. J. of Heat Transfer. **117**: 1058-1061.
80. Willett, F.T. & A.E. Bergles. 2000. Heat Transfer in Rotating Narrow Rectangular Ducts With Heated Sides Oriented at 60° to the R-Z-Plane. 2000-GT-224
81. Prahbu, S.V. & R.P. Vedula. 1997. Pressure Drop Characteristics in a Rib Roughened Rotating Square Duct With a Sharp 180 Deg Bend. Exp. Heat Transfer, Fluid Mechanics and Thermodynamics 1997, M.Giot, F.Mayinger, G.P.Celata (Eds.), Edizioni ETS, 1483-1490.
82. Prahbu, S.V. & R.P. Vedula. 2000. Pressure Drop Characteristics in a Rotating Smooth Square Channel With a Sharp 180° Bend. Exp. Thermal and Fluid Science, 21, 198-205.
83. Iacovides, H., 1998. Computation of flow and heat transfer through rotating ribbed passages. int. J. of Heat and Fluid Flow, 19, 393-400.
84. Jang, Y.J., Chen, H.C. & J.C. Han. 2000. Flow and Heat Transfer in a Rotating Square Channel with 45° Angled Ribs by Reynolds Stress Turbulence Model. 2000-GT-229
85. Bonhoff, B., Tomm, U., Johnson, B.V. & I. Jennions. 1997. Heat Transfer Predictions for Rotating U-Shaped Coolant Channels With Skewed Ribs and With Smooth Walls. 97-GT-162
86. Murata, A. & S. Mochizuki. 1999. Effect of cross-sectional aspect ratio on turbulent heat transfer in an orth. rotating rect. smooth duct. Int. J. Heat Mass Transfer. **42**: 3803-3814.
87. Murata, A. & S. Mochizuki. 2000. Large eddy simulation with a dynamic subgrid-scale model of tubulent heat transfer in an orthogonally rotating rectangular duct with transverse rib turbulators. Int. J. Heat Mass Transfer. **43**: 1243-1259.
88. Bons, J.P. & J.L. Kerrebrock. 1999. Complem. Velocity and Heat Transfer Measurements in a Rotating Cooling Passage With Smooth Walls. J. of Turbomachinery. **121**: 651-662.
89. Liou,T.M. & C.C. Chen. 1999. Heat transfer in a rotating two-pass smooth passage with a 180° rectangular turn. Int. J. Heat Mass Transfer. **42**: 231-247.
90. Clifford, R.J. 1985. Rotating Heat Transfer Investigations on a Multipass Cooling Geometry. AGARD CP 390.
91. Farmer, R. & K. Fulton. 1995. Design 60% net efficiency in Frame 7/9H steam cooled CCGT. Gas Turbine World. Mai-June. 12-20.
92. Weigand, B. 1998. What should we measure? The industrial – engine heat transfer perspective. In Proc. of the XIV Bi-annual Symp. On Measuring Techniques in Transonic and Supersonic Flow in Cascades and Turbomachines. Invited Lecture. Limerick. Ireland.
93. Jennions, I.K., Sommer, T. Weigand, B. & M. Aigner. 1998. The GT24/26 Low Pressure Turbine, ASME 98-GT-29.
94. Johnson, B.V., Wagner, J.H., Steuber, G.D. & F.C. Yeh. 1994. Heat Transfer in Rotating Serpentine Passages With Trips and Skewed to the Flow. J. Turbomachinery. **116**: 113-123.
95. Scheurlen, M. 1999. Aero-thermische Auslegung und Erprobung für eine fortschrittliche Gasturbinengeneration. VGB KraftwerksTechnik. **3**: 41-45.
96. Aoki, S., Tsukuda, Y., Akita, E., Terazaki, M., McLaurin, L.D. & M. Kizer. 1994. Uprated 501F Gas Turbine, 501FA. 94-GT-474.
97. Sato, M., Kobayashi, Y., Matsuzaki, H., Aoki, S., Tsukuda, Y. & E. Akida. 1995. Final Report of the Key Technology Development Program for a Next Generation High Temperature Gas Turbine. 95-GT-407.
98. Bannister, R.L., Cheruvu, N.S., Little, D.A. & G. Mc Quiggan. 1994. Development requirements for an advanced gas turbine system. 94-GT-388.

Cooling Systems for Ultra-High Temperature Turbines

TOYOAKI YOSHIDA

Aircraft Propulsion Research Center, National Aerospace Laboratory
7-44-1 Jindaiji-higashi, Chofu, Tokyo 182-8522, Japan

ABSTRACT : This paper describes an introduction of research and development activities on steam cooling in gas turbines at elevated temperature of 1500 C and 1700 C level, partially including those on water cooling. Descriptions of a new cooling system that employs heat pipes are also made. From the view point of heat transfer, its promising applicability is shown with experimental data and engine performance numerical evaluation.

INTRODUCTION

In the middle of the nineties, turbine inlet temperature (TIT) reached 1500 C level in some advanced practical turbine engines not only of aircraft propulsions but also of big-scale industrial gas turbines. Recently, the Japanese research turbo-fan engine for a hypersonic transport propulsion, HYPR verified its applicability under the 1700 C level gas temperature condition by the demonstrator engine tests. This hot topic is to be presented in this symposium by Yamawaki, one of the members engaged in the development [1]. High-temperature turbines in all of these are air-cooled.

A recent notable trend of cooling technology is to employ steam or even liquid water other than air as coolant for hot parts. Various research and development projects on this technology have been conducted. It is well known that steam-cooled combustor and turbine sections are put into practical use in some representative big-scale industrial gas turbines. This kind of activities is mainly aimed for the considerable increase in total thermal efficiency of the engine system.

The author's group has been conducting extensive studies on a future cooling system for ultra-high temperature turbines. The system utilizes heat pipes for turbine vane cooling in its ultimate case together with a heat exchanger as a heat-releasing device. The author once introduced the basic concept of the system [2].

This paper describes an introduction of research and development activities on steam cooling in gas turbines at ultra-high temperature up to 1700 C level. Some of those include water cooling. As the main part of the contents, a series of the work on the said cooling system are summarized. Those include formation of the system, analytical and experimental studies of the individual component of the system, and total engine performance evaluation by numerical simulation for a target engine with the present cooling system.

STEAM- COOLED GAS TURBINE

The modern high temperature gas turbines have utilized sophisticated air cooling technologies and have elevated those working gas temperature to increase thermal efficiency. However, the increase of turbine inlet temperature (TIT) results in the increase of cooling air consumption and this reduces the specific output of gas turbine and decreases the thermal efficiency by cold coolant mixing with hot mainstream gas. Cooling air increase also makes it difficult to keep the air requirement for dry low nitrogen oxide (NOx) combustor design. Therefore, further TIT and thermal efficiency increase by the open circuit air-cooled (OCAC) gas turbine becomes difficult.

One of the solutions to this problem is to use steam which is generated by the Heat Recovery Steam Generator (HRSG) in the combined cycle system as a coolant for high temperature gas turbine. Since steam has higher cooling ability (higher specific heat, density and thermal conductivity at high-pressure) than that of air, it is suitable for the higher temperature gas turbine coolant. In the case of combined cycle, steam can be reintroduced to the lower pressure stage of steam turbine after cooling the gas turbine hot parts to generate more power. This system is called as closed-circuit steam cooling (CCSC) and is thought to be very effective for large-scale combined cycle power plant. Another benefit of CCSC against OCAC is that CCSC can realize relatively low combustor exit gas temperature (approximately 110C for 1500C class gas turbine) under the same turbine 1st stage blade inlet gas temperature. This is because that 1st stage nozzle has designed without film cooling ejection to hot gas path. This CCSC cycle concept is not new and was reported by Alderson et.al. [3] and Stambler [4]. Of course the development and the design of CCSC blades strongly depend on the development of materials such as single crystal base metal and high temperature thermal barrier coating (TBC). Since, CCSC blades are realized by internal steam cooling and replacing the film cooling by TBC.

1500 C Class Steam-Cooled Gas Turbine

Several research and development projects had been performed on the development of steam-cooled high temperature turbines.

A joint research project between Tohoku Electric Power Co. and three companies in Japan (Hitachi Limited, Mitsubishi Heavy Industries and Toshiba Corp.) had aimed the development of elemental technologies for the realization of a high efficiency (50% HHV) 1500 C class high temperature gas turbine in 1989-1995.

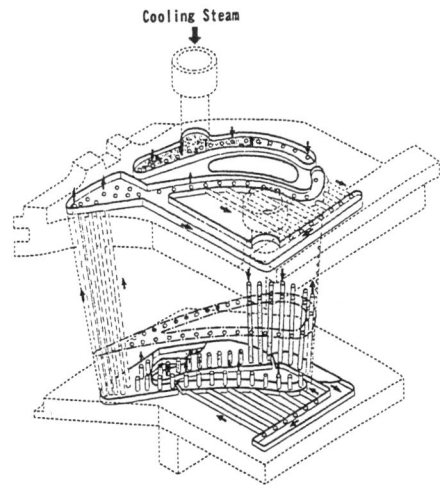

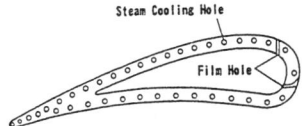

Fig.1 Steam-cooled vane for 1500 C class gas turbine

Results can be found in the open literatures for steam-cooled nozzle vane [5,6] and steam-cooled combustor and transition piece [7]. The research and development for base metal(single crystal) and thermal barrier coatings(TBC) were also included in the project [8].

Fig.1 shows a cross-section and schematic cooling construction of a turbine vane developed by Tohoku Electric Power Co. and Toshiba Corp.[6]. The main specifications of the combined cycle gas turbine are as follows; Output: 246MW, TIT: 1450C, Cycle efficiency: 56.8%(LHV). The scale model was chosen for the verification of a steam-cooled vane to a scale of 60MW, Chord length: 117mm, Span height: 92mm, 30 holes cooling passages with the diameter of 2mm.

TBC is coated over the whole surface and further, film cooling by air is adopted in the leading edge suction side region to reduce excess temperature gradient in the relevant area. Closed cycle steam cooling is introduced. The main specifications of flow conditions are as follows ; TIT: 1450 C, Total pressure of hot gas: 1.8 MPa, Temperature

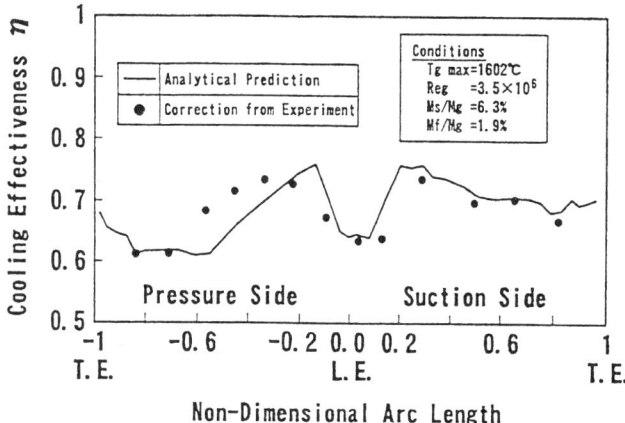

Fig.2 Cooling effectiveness at the design point of the vane

of steam:435C, Pressure of steam : 10.3 Mpa, Mass flow rate of steam:6.3%, Mass flow rate of film air : 1.9%. Fig.2 shows resulting cooling effectiveness.

Experimental data are corrected to those at the design point. The analytical prediction is also shown in the figure. Both agree well with each other. The cooling effectiveness is located at excellent value, considering the fact that it is defined by outer metal temperature.

The feasibility study was also conducted using intermediate pressure steam in this project [9].

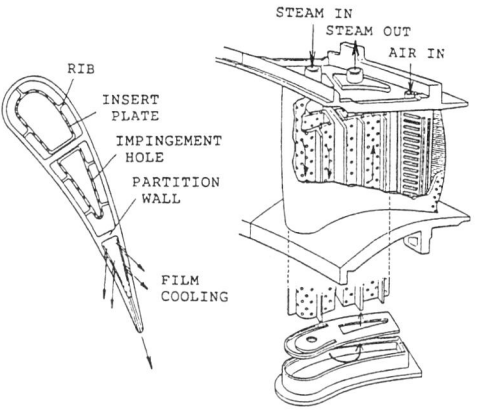

Fig.3 Steam-cooled vane by intermediate pressure steam

In the trailing edge region, air cooling is also applied with film cooling and blow-off. Cooling effectiveness was measured by two-dimensional model vanes in a hot wind tunnel. The obtained average effectiveness was about 5 percent higher than that for an air-cooled vane.

In the United States, U.S.Department of Energy has been supporting the Advanced Turbine System (ATS) technology development projects since 1993. In this project several high-energy efficient gas turbine systems have been developed. They are simple-cycle industrial gas turbines less than 20MW in capacity for distributed generation and cogeneration market and combined-cycle utility gas turbines greater than 20MW in size for central base-load power station. Steam cooling has been applied for the large scale utility gas turbine named H system by General Electric Company. H system may become one of the largest machines in the world, 7H and 9H type are announced to have more than 55% HHV with the power output of 400MW and 480MW, respectively.

1700 C Class Steam-cooled Gas Turbine

A research and development of 1700C class hydrogen combustion turbine had been carried out in Japan as a part of the World Energy Network Program (WE-NET) entrusted by New Energy and Industrial Technology Development Organization (NEDO) in 1993-1998. This is called the PHASE-I of WE-NET project. The target of the WE-NET project is to realize the very large (500MW) power station with very high thermal efficiency (>60% HHV) by the closed circuit cooling topping recuperated cycle. The results for PHASE-I are summarized by Mouri et. al [10]. In this phase of the project, Hitachi Ltd. has developed a water-cooled vane and a steam-cooled blade [11]. While Toshiba Corp., a steam-cooled vane and a blade with least film cooling [12]. They have tested those in the hydrogen/oxygen combustion test stand up to 1700C environment.

Hitachi Ltd. [11] analyzed the performances of a thermal plant with three different types of cooling systems for a 1700C class hydrogen-fueled combustion gas turbine, 1) Closed-circuit water-cooled nozzle vanes and steam-cooled rotor blades, 2) Closed-circuit steam-cooled nozzle and rotor blades, 3) Open-circuit steam-cooled nozzle and rotor blades. The results indicated that closed-circuit cooling systems were more effective than conventional open-circuit systems.

Fig.4 shows a schematic cooling construction of a designed water-cooled vane. It is a 50% scale model of the target plant (500MW). The main flow conditions are as follows; TIT: 1700C, inlet total pressure: 4.66Mpa, cooling water inlet temperature: 100C, pressure: 9.81Mpa. TBC with the total thickness of 0.3mm are covered over the entire surface. The diameter of 62 passage circular holes and those pitches are 2.0 and 4.0 mm, respectively. Heat conduction numerical analysis showed that the maximum wall temperature was 313 C, meeting the allowable limit of the copper alloy adopted as core material. The average cooling effectiveness was 0.91 at the cooling flow

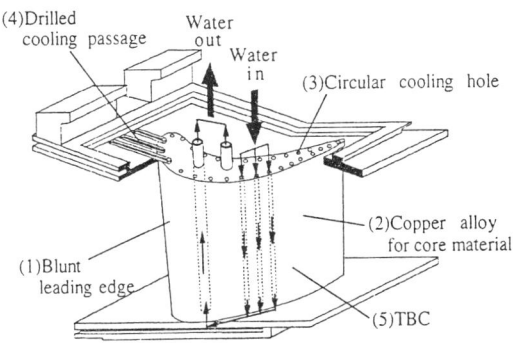

Fig.4 Water-cooled vane for 1700 C class gas turbine

ratio of 16.3%. Heat flux passing through the TBC was about 7.4MW/m².

Fig.5 shows a sketch of a cooling construction of a designed steam-cooled blade. It is minimally allocating film cooling holes at the leading edge region to keep closed-circuit cooling as much as possible. TBC is covered over the surface. The main flow conditions are as follows; inlet average relative total temperature: 1622C, inlet relative pressure: 3.69Mpa, cooling steam inlet temperature: 300C, inlet pressure: 5.88Mpa. Calculated blade metal temperatures ranged from 800C to 1000C (average cooling effectiveness was 0.585), meeting the allowable limit of a single crystal superalloy. While temperatures on the TBC outer surface ranged from 1130C to 1350C. The maximum temperature gradient and heat flux were 400C in 0.2mm and 3.6MW/m², respectively. These are about three times as much as current usage.

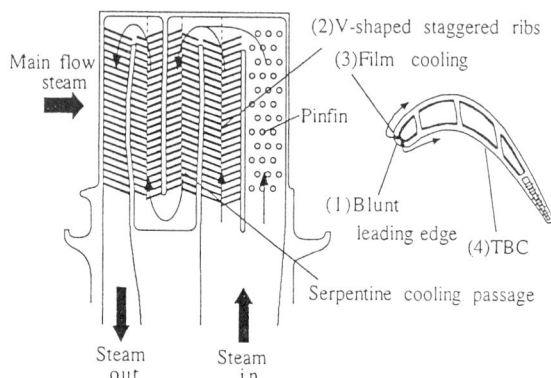

Fig.5 Steam-cooled blade for 1700C class gas turbine

Toshiba Corp.[12] has developed a pair of closed-circuit steam-cooled vane and blade with least film cooling (hybrid-cooling). Fig.6 shows schematic cooling constructions of the scale models (56.3%). The main flow conditions at those verification tests are as follows; inlet mean temperature: 1700C (vane), 1570C (blade), inlet pressure: 2.44MPa (v & b), cooling flow inlet temperature: 350C (v & b), inlet pressure: 2.81MPa (v), 3.63MPa (b), coolant mass flow ratio: 6.7% (v), 7.0% (b). Dimension; path height: 40mm (v & b), chord length: 79mm (v). 36.6mm (b). Both are TBC coated. Diffused shaped film cooling holes (v) and fan shaped holes (b) are provided at the leading edge region, respectively.

High-temperature heat transfer tests were carried out in the hydrogen-oxygen combustion cascade test facility. The obtained cooling effectiveness of the test vane and blade ranged from 0.6 to 0.8 in either model. Those were compared with numerical predictions and good agreements between those were obtained respectively.

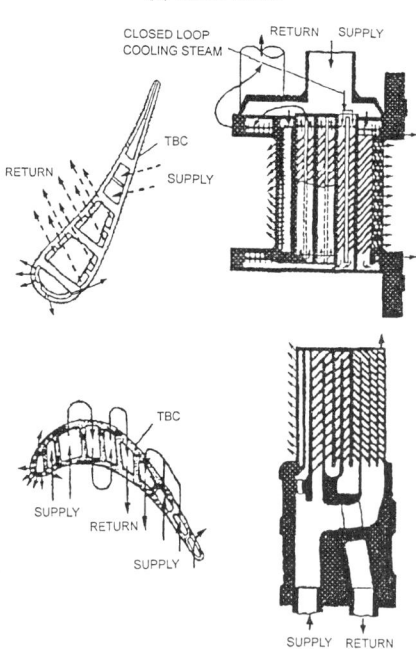

Fig.6 Steam-cooled vane and blade for 1700C level gas turbine

The research and development of the WE-NET project is now in the PHASE-II of the program. The project target has updated to the methane combustion carbon dioxide recovery system with 1700C gas turbine. This target setting is to consider the near term realization of this highly efficient system.

Introducing a high-temperature heat pipe system for turbine cooling is very challenging. It is well known that the heat transfer rate from hot to cold section in a heat pipe is far greater than that by heat conduction or convection. While, there are some barriers that should be overcome, such as sensitivity to gravitational force, tempera-ture response time until high-temperature working condition and safety against water.

Silverstein et al. carried out a feasibility study of heat pipe vane cooling as shown in Fig.7 [13]. This was aimed for its application to aero-engines with a bypass air system that was allocated to heat sink portion. They remarked benefits of the heat pipe use in view of SFC improvement or decrease of hot end temperature and thus extension of engine life.

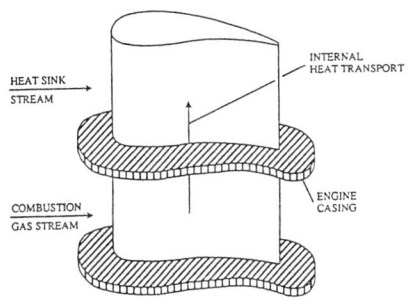

Fig.7 Concept of a heat pipe cooled turbine vane

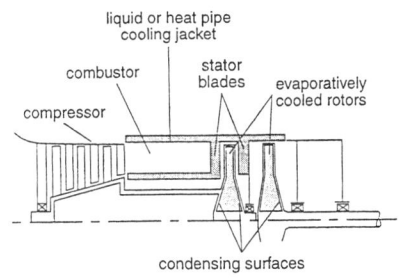

Fig.8 Concept of a vaporization-cooled gas turbine

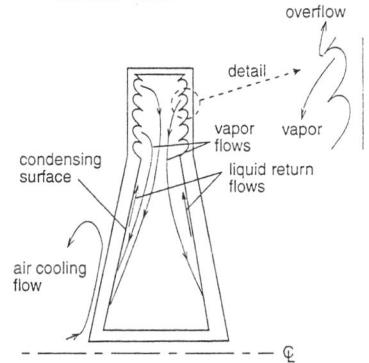

Fig.9 Concept of a vaporization cooling system for rotor

Kerrebrock et al. presented a concept of vaporization cooling for gas turbines [14]. The idea involves cooling of all the hot parts of engines, combustor, stator and rotor blades (Fig.8). In the rotor portion, evaporating cooling is applied in the blade, while condensation of coolant is done in the disk (Fig.9). This configuration is considered as a heat pipe. Making use of a phase change for cooling obviously provides a powerful mean for hot parts cooling of a gas turbine. However much remains to be done.

NEW COOLING SYSTEM

The author's group created a new concept for a combined cooling system [15]. It is shown schematically in Fig.10. It consists of a heat exchanger, coolant transportation lines and ultra-high temperature turbine vanes. Cooling air is pre-cooled and/or heat pipes are introduced to the lines, in either case, the heat exchanger is used as a heat-releasing

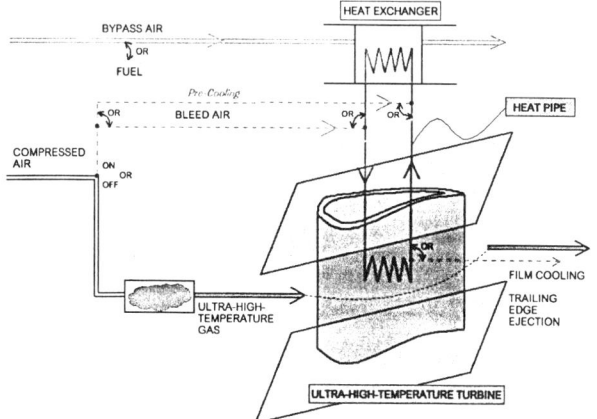

Fig.10 Concept of a new combined cooling system

device. This system is based on the idea that coolant mass flow rate can be reduced when the coolant is pre-cooled by introducing a heat exchanger. Ultimately, cooling air discharged from a compressor may become unnecessary if a heat pipe is utilized as a heat transportation device and works well enough for turbine cooling. Either bypass air or fuel can be assigned as a heat sink for the heat exchanger. This application will enable turbine vanes to endure gas flow temperature of 1700 C level or higher and thus it will contribute drastic improvement in specific output power and thermal efficiency of gas turbine engines.

In order to verify advantages of the above concept, following researches were done; a total system performance simulation of the HYPR target engine for the case with the present cooling system, and fundamental works on manufacturing techniques, mechanical strength evaluations and heat transfer characteristic tests of the system components, ultra-high temperature turbine element (vane) and a heat pipe system.

Ultra-high temperature vane

A new general concept for the turbine element was created as the main component of the combined cooling system [16]. Figure 11 shows a schematic cross-section of a symmetrical blade shape vane model. It consists of three layers, a thermal barrier coating (outer surface), substrate (advanced superalloy) and high-heat conductive

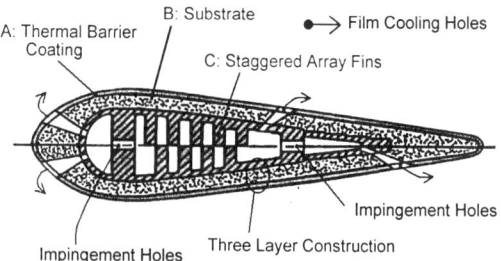

Fig.11 Schematic construction of an ultra-high temperature turbine vane

material (inner liner). The TBC layer at the outer surface restrains intense heat flux from ultra-high temperature gas stream. Regarding structural core materials, any kind of high-heat-resistant materials superior to existing superalloys can be used. Among those, oxide dispersion strengthened (ODS) mechanical alloys are considered promising for application in the near future. For powerful convective heat transfer on the inside of the core, some appropriate fine fins and/or impingement cooling constructions will be formed. This inner portion should be made of high-heat-conductive material such as copper so that intense heat flux may be dispersed into relatively lower temperature regions, resulting in an averaging of the temperature distribution. No film cooling is introduced in the present analytical and experimental study so far, although its application is possible and sketched in the figure.

Two types of practical high-temperature test models without film cooling holes were successfully manufactured (basic circular models and vane models shown in Fig.11). Based on heat conduction analysis of the circular models and those high-temperature wind tunnel tests, heat transfer characteristics of the individual three layers and pre-cooling condition of cooling air were quantitatively disclosed [16].

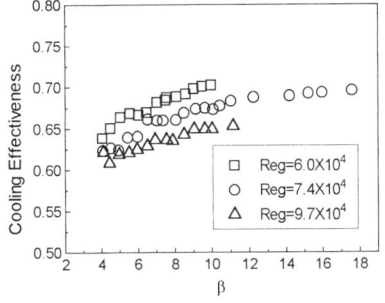
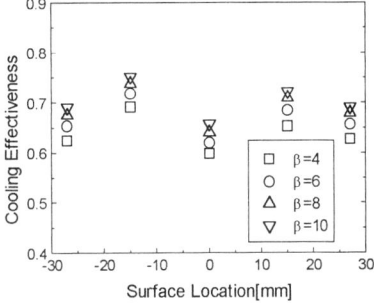

Fig.12 Cooling effectiveness vs. cooling air flow rate ratio (%)

Fig.13 Cooling effectiveness chordwise distributions

Figs. 12 and 13 show representative results of cooling effectiveness from high-temperature wind tunnel tests of the vane models [17]. The average cooling effectiveness reached an order of 0.67 with the design air flow ratio of 7.5% without film cooling (Fig.12). Its deviation over the surface remained less than 0.1 in the chordwise direction (Fig.13). Hence this type of vanes are considered applicable to ultra-high temperature turbines if cooling air is moderately pre-cooled.

High-temperature heat pipe

A concept of the present work is schematically shown in Fig.14. In parallel with the heat pipe cooling, conventional air cooling may also be applied as shown in the figure. A series of high-temperature heat pipes were prepared that would match the HYPR target engine temperature condition [18]. Fig.15 shows a cross-section of the experimental model D made of Ni-superalloy. This utilizes sodium as coolant. Wick construction is made of nickel sintered powder metal. At the bottom of the heat pipe shell, an artery is placed for the promotion of return flow of liquid phase coolant. Non-condensable gas (argon) is sealed to shorten startup time. These were manufactured by Thermacore. Heat transfer experiments were conducted in the high-temperature wind tunnel at NAL.

In the heat transfer tests, evaporator portion was heated with electric lamps (19kW max) in a high-temperature furnace in stead of hot gas flow, while at the condenser portion, natural convection or wind tunnel exit flow was applied for cooling. Heat transfer data were taken with a change of electric power input at the individual heat pipe angle, 0deg: horizontal, 90deg: vertical (top: evaporator, bottom: condenser) and others.

Figs.16 and 17 show representative results of heat pipe performance. The main dimensions of heat pipe B are as follows; evaporator:100mm, adiabatic section:100mm,

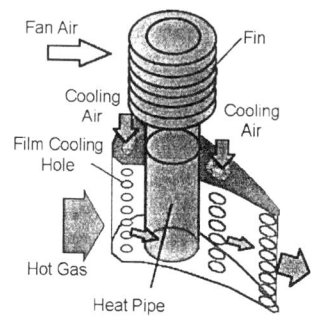

Fig.14 Concept of turbine vane cooling by heat pipe

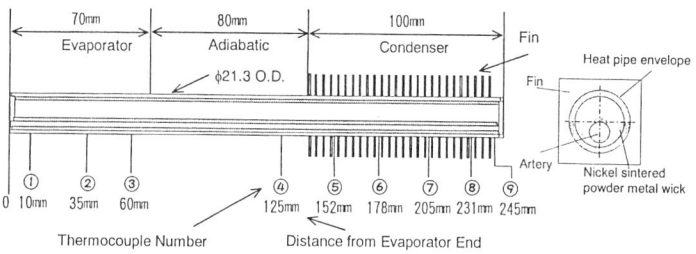

Fig.15 Schematic construction of the heat pipe model D

condenser:80mm with smaller fins. Startup time (Fig.16) and heat flux (Fig.17) indicate relatively weak dependence on a heat pipe set angle, namely the effect of gravitational force is small. This can be considered as a good feature for the practical application. In a representative heat pipe test model, the startup time was an order of 8 minutes and the highest heat flux could transport heat by about 40 W/cm^2 under the present test maximum heat load condition. According to the discussion on the limits of test facility heat load and heat pipe's various proper characteristics such as wick limit, sonic limit entrainment limit and boiling limit, the maximum heat transport was predicted at about 3kW in total and 64W/m^2 in heat flux.

Under the condition of the HYPR target engine, the advantage of heat pipe applications is becoming superior to the engine with advanced air cooling system only when coolant flow rate can be saved by 20%. Fig.18 shows a summary of evaluation of fuel consumption in the present study.

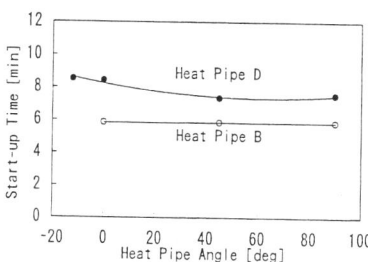

Fig.16 Startup time of heat pipe models

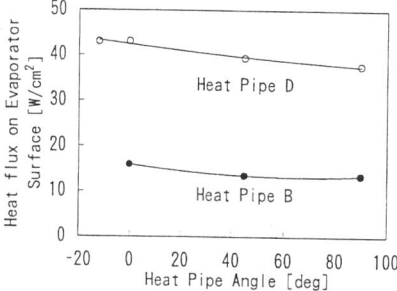

Fig.17 Heat flux of heat pipe models

Since heat flux over the vane of the target engine is estimated at 125 W/m², it can be said that coolant air flow is saved by 51% by the present heat pipe system. Consequently fuel consumption may be saved by approximately 600kg/h under the engine maximum operating condition.

Engine performance simulation

Numerical simulation was done on the total performance of the HYPR target engine for the case with the present cooling system [19] (see Fig.10). The effect of heat pipe application to turbine cooling system on engine fuel

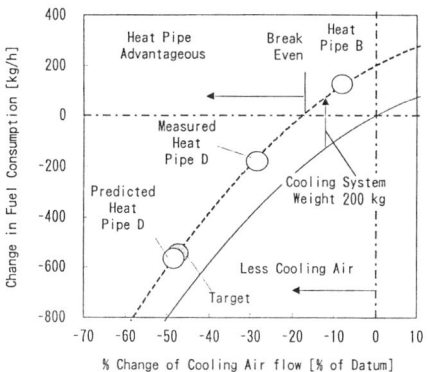

Fig.18 Evaluation of fuel consumption for heat pipe application to an aero-engine

consumption and specific output were mainly discussed. Figs.19 and 20 show representative performance maps of the engine at Mach 2.5 climb and Mach 3.0 cruise condition, respectively. In the each figure, lower net map stands for the case with the present cooling system, where heat pipe system and pre-cooling of air are assumed (Heat exchanger + Heat pipe), while upper one is for the case with advanced air-cooled turbines (Normal HYPR). When performances (SFC and specific thrust) are compared at the design point (with the same TIT and engine pressure ratio (EPR)), a specific thrust increases by about 15% but SFC remains almost unchanged at Mach 2.5 climb condition. While a drastic increase in specific thrust, say 20% and an improvement of SFC by about 2% can be achieved by the present system at Mach 3.0 cruise condition. In all the flight conditions, a considerable increase in specific thrust is obtained, while SFC depends on the conditions.

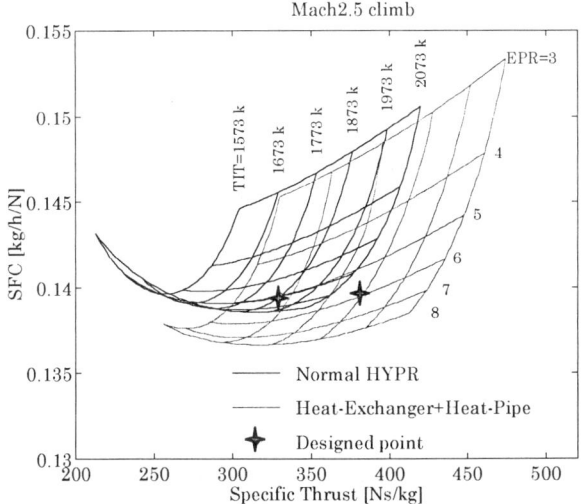

Fig.19 Turbo-fan engine performance with and without heat pipe cooling system at Mach 2.5 climb condition

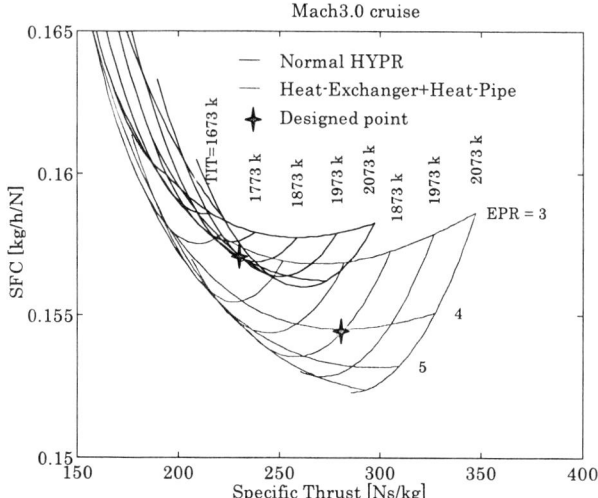

Fig.20 Turbo-fan engine performance with and without heat pipe cooling system at Mach 3 cruise condition

Once we have freedom for the design of a turbo-fan engine, apart from the HYPR target engine, optimization of performance improvement can further be made by the introduction of an appropriate heat pipe and pre-air-cooling system.

CONCLUDING REMARKS

In modern and future ultra-high temperature turbines, heat flux from hot gas flow to turbine vanes and blades is getting intense. In this sense, the role of a passive cooling by TBC is becoming more important than ever. Pre-cooling of cooling air and/or heat pipe application are very effective for the improvement in gas turbine performance. Although these applications need additional component such as a heat exchanger, it is very worth while to consider these cooling technology. For industrial gas turbines, introducing steam or water other than air as a turbine coolant is very promising because it is effectively utilized from a steam turbine part in the combined system.

The author expresses his sincere gratitude to Dr.Y.Fukuyama for his devoted collaboration in preparing this material.

REFERENCES

1. Yamawaki, S. 2000. Heat Transfer in Gas Turbine Systems, ICHMT Turbine 2000, Cesme, Turkey, August 14-18.
2. Yoshida, T. 1997. *In* Proceedings.of Heat Transfer in Turbomachinery, ICHMT 1992: 19-32. Begell House.
3. Alderson,E.D., Scheper,G.W. & Cohn,A. 1987. 87-JPGC-GT-1.
4. Stambler, I. 1989. Gas Turbine World, February: 28-30.
5. Matsuzaki,H., Shimomura,K., Fukuyama,Y., Araki,T., Ishii,J., Yamamoto,M., Shibuya,S. & Okuhara,I. 1992. ASME Paper 92-GT-240.

6. Nomoto,H., Koga,A., Itoh,S., Fukuyama,Y., Otomo,F., Shibuya,S., Sato,M., Kobayashi,Y. & and Matsuzaki,H. 1997. ASME Jr. of Engineering for Gas Turbine and Power. **113**: 624-632.
7. Sato,M., Kobayashi,Y., Matsuzaki,H., Aoki,S., Tsukuda,Y. & Akita,E. 1995. ASME Paper 95-GT-407.
8. Matsuzaki,H., Suto,T., Kanazawa,Y., Sato,M., Kobayashi,Y. & Shimomura,K. 1996. ASME Paper 96-GT-294.
9. Amagasa,S., Otomo,F., Fukuyama,Y. & others. 1991. JSME Annual Conf. No.920-17B: 408-410 [in Japanese].
10. Mouri,K., Arai,N., Taniguchi,H. & Maekawa,H. 1998. IJPGC'98, Vol.1: 433-437.
11. Kizuka,N., Sagae,K., Anzai,S., Marushima,S., Ikeguchi,T. & Kawaike,K. 1998. ASME Paper 98-GT-345.
12. Okamura,T., Koga, A., Itoh,S. and Kawagishi,H. 2000. ASME Paper 2000-GT-615.
13. Silverstein, C.C., Gottschlich, M.J. and Meininger, M. 1994. ASME Paper 94-GT-306.
14. Kerrebrock,J.L. & Sticker,D.B. 1998. ASME Paper 98-GT-177.
15. Yoshida, T. & others. 1999. *In* Proceedings of the 3rd International Symposium on Japan's National Project for a HYPR system, Tokyo, Japan, May 1999: 125-130.
16. Yoshida, T., Kumagai,T., Taki,M., Taguchi,H. & Matsuki,M. 1995. *In* Proceedings of the 12th ISABE, Melbourne, Australia, Sept. 1995: 1113-1120.
17. Sakida,T., Kumagai,T., Taki,M. & Yoshida,T. 1997. *In* Proceedings of the 12th Gas Turbine Autumn Meeting, Nara, Japan: 169-174 [in Japanese].
18. Yamawaki, S., Yoshida,T., Taki,M. & Mimura,F. 1998. ASME Jr.of Engineering for Gas Turbine and Power, **120**: 580-587.
19. Tagashira, T. & Yoshida, T. 1999. *In* Proceedings of the 14th ISABE, 99-7198, Florence, Italy, Sept. 1999.

Some Current Research in Rotating-Disc Systems

J MICHAEL OWEN AND MICHAEL WILSON

Department of Mechanical Engineering
University of Bath
Bath BA2 7AY, UK

ABSTRACT: Rotating-disc systems are used to model the flow and heat transfer that occurs inside the cooling-air systems of gas-turbine engines. In this paper, recent computational and experimental research in three systems is discussed: rotor-stator systems, rotating cavities with superposed flow and buoyancy-induced flow in a rotating cavity. Discussion of the first two systems concentrates respectively on pre-swirl systems and rotating cavities with a peripheral inflow and outflow of cooling air. Buoyancy-induced flow in a rotating cavity is one of the most difficult problems facing computationalists and experimentalists, and there are similarities between the circulation in the Earth's atmosphere and the flow inside gas-turbine rotors. For this case, results are presented for heat transfer in sealed annuli and in rotating cavities with an axial throughflow of cooling air.

NOMENCLATURE

a,b	inner, outer radius of cavity
C_W	nondimensional flow rate (= $\dot{m}/\mu b$)
$\tilde{g}$	acceleration
G	gap ratio (=s/b)
Gr	Grashof number (= $\tilde{g}\, l^3\, \beta\Delta T/\nu^2$)
k	thermal conductivity
l	characteristic length
$\dot{m}$	mass flow rate
Nu	Nusselt number (=$ql/k\Delta T$)
Pr	Prandtl number (=ν/α)
q	heat flux
r, ϕ, z	radial, tangential and axial coordinates
r_m	mean radius of annulus (=½ (a+b))

Ra   Rayleigh number (=PrGr)
Re_ϕ, Re_z rotational Reynolds number (=$\Omega b^2/\nu$), axial Reynolds number (=$W l/\nu$)
Ro   Rossby number (=$W/\Omega a$)
s   axial gap between discs
T   absolute temperature
V_r, V_ϕ, V_z time-averaged radial, circumferential, axial components of velocity
W   bulk-average axial velocity
x nondimensional radius (=r/b)
x_a radius ratio of cavity (=a/b)
α thermal diffusivity
β T_{ref}^{-1}, volume expansion coefficient
β_p pre-swirl ratio (= $V_{\phi,p}/\Omega r_p$)
ΔT temperature difference
ε turbulent energy dissipation rate
Γ ratio of speed of slower disc to that of faster one
λ_T turbulent flow parameter (= $C_w/Re_\phi^{0.8}$)
ρ density
μ, ν dynamic viscosity, kinematic viscosity (= μ/ρ)
Ω angular speed of cavity

Subscripts

b   blade-cooling air
d   disc-cooling air
l   inlet value
o   disc surface
p   pre-swirl air
ref   reference value

1 INTRODUCTION

Rotating-disc systems are used to represent the flow and heat transfer in the internal cooling-air systems of gas-turbine engines, and Fig. 1 shows a schematic diagram of a typical cooling and sealing system.

The essential features of these complex systems can be modelled using plane rotating and stationary discs, as shown in Fig. 2. It is convenient to classify these systems using the parameter Γ, which is the ratio of the angular speed of one disc to that of the other: $\Gamma = 0$ is the rotor-stator system; $\Gamma = +1$ is the rotating cavity; $\Gamma = -1$ is the contra-rotating disc system. Contra-rotating discs have been described by Owen[1], and Owen and Rogers[2,3] have described many of the characteristics of rotor-stator systems and rotating cavities. This paper sets out to review recent work on these systems.

In Section 2, the rotor-stator system is discussed. Some areas of current research interest include pre-swirl cooling systems and hot gas ingress from the external mainstream. Section 3 is concerned with the case of a rotating cavity with a peripheral inflow and outflow of cooling air. Buoyancy-induced flow in a rotating cavity, which is discussed in Section 4, has features in common with the flow in the Earth's atmosphere, not least the occurrence of

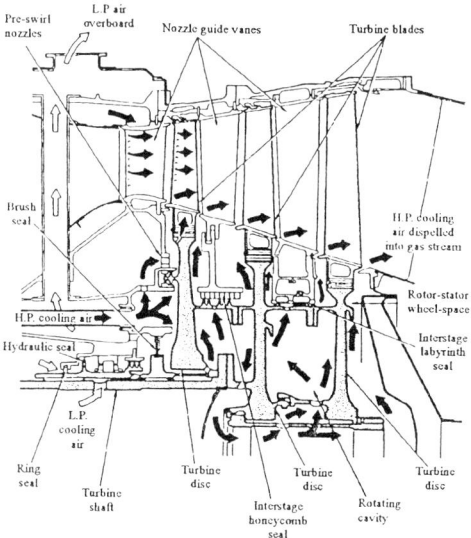

Fig. 1 Schematic diagram of a gas turbine cooling and sealing system
(from Rolls-Royce "The Jet Engine")

cyclonic and anti-cyclonic circulations, and it is one of the most difficult problems facing computationalists and experimentalists alike.

2 ROTOR-STATOR SYSTEMS

2.1 Superposed flows and ingress

The ingress of hot mainstream gas radially inward into the wheelspace of a gas turbine is a problem of major importance for the turbine designer. In extreme cases, hot-gas ingress can cause overheating, and failure, of the rim of the turbine disc. For the background and a review of ingress, the reader is referred to Owen and Rogers[2] and Johnson et al.[4] Recent contributions have been made by Chew et al.[5], Reichert and Lieser[6] and Bohn et al.[7]

A radial outflow of disc-cooling air can be used to reduce ingress. Wilson et al.[8] summarised work which showed that the flow and heat transfer in systems with a superposed radial outflow can be computed with reasonable accuracy using k-ε turbulence models, and some validation of commercial codes has been carried out (Scott et al.[9]). In general, heat transfer computations benefit from the use of low-Reynolds-number turbulence models. Iacovides et al.[10] tested more sophisticated low-Reynolds-number differential stress models and obtained some improvements to flow-field predictions.

Studies of systems with a superposed radial inflow are also of interest in the ingress context. Iacovides et al.[10] tested the Launder-Sharma[11] low Reynolds-number k-ε turbulence model against experimental data for $Re_\phi = 1.47 \times 10^6$ and $C_w = -7389$ (by convention, $C_w < 0$ when the cooling air is directed inward). The empirical Yap correction term for the ε equation, based on near-wall turbulence equilibrium assumptions and which is beneficial for

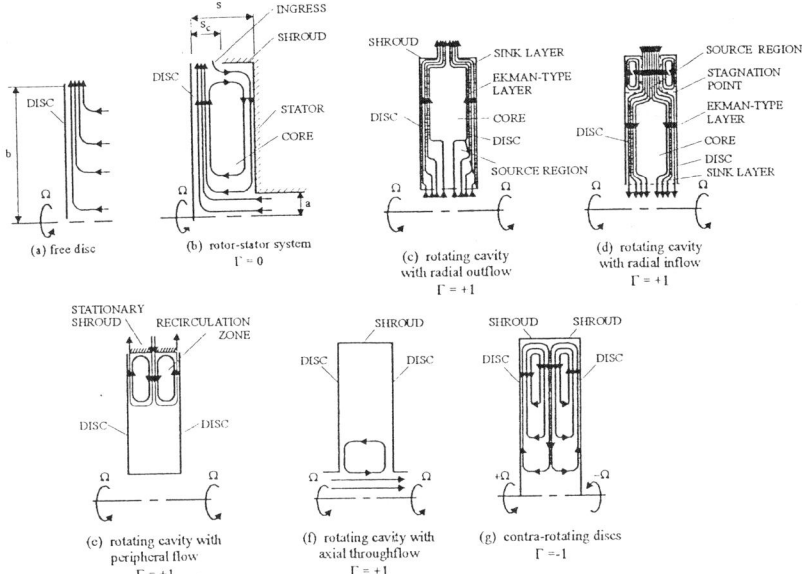

Fig. 2 Schematic diagram of rotating-disc systems

computations without throughflow ($C_w = 0$), behaved badly in the inflow situation, where turbulent flow from the outer part of the system is convected radially inward to less turbulent regions. Further experimental work on radial inflow has recently been completed in Marseille (Gassiat[12]) which may be suitable for the testing of improved computational models.

2.2   Pre-Swirl Systems

Fig. 3a and Fig. 3b illustrate alternative systems used to supply air for the internal cooling of turbine blades through holes near the periphery of the turbine disc. In the "direct transfer" system shown in Fig. 3a, cooling air is delivered, at a high radius, from angled ("pre-swirl") nozzles in a stationary casing. The swirl imparted to the air in the direction of rotation reduces the relative total temperature of the air entering the rotating holes on the disc.

The flow between the discrete stationary pre-swirl nozzles and the blade-cooling holes on the rotating disc is complex, three-dimensional and unsteady. Meierhofer and Franklin[13] measured air temperatures inside rotating blade-cooling holes, and Owen and Rogers[2] reviewed research into direct-transfer systems under adiabatic conditions. Wilson et al.[14] summarised some of these findings, and carried out computations and experiments for an idealised direct-transfer rotor-stator system. Axisymmetric steady flow computations (using the Launder-Sharma low-Reynolds-number turbulence model and with annular slots representing the discrete pre-swirl nozzles and blade-cooling holes) consistently under-predicted "blade-cooling air" temperatures measured inside the holes on the experimental rig. The computed mixing (between the pre-swirl cross-flow and a radial outflow of disc-cooling air) is illustrated in Fig. 4, for the pre-swirl chamber formed by inner shrouds and a

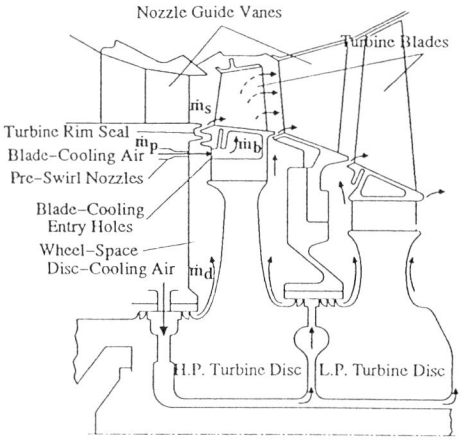 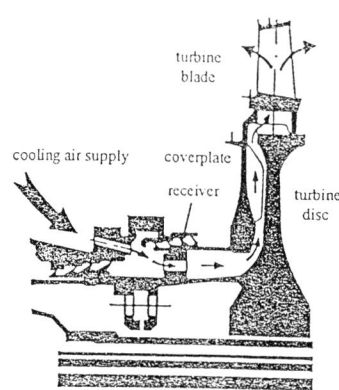

Fig. 3a A direct-transfer pre-swirl system (from Wilson et al.[14])

Fig. 3b A cover-plate pre-swirl system (from Popp et al.[19])

"rim seal". Table I shows the conditions considered by Wilson et al and comparisons between computed and measured values of ΔT, the change in relative total temperature between the stationary pre-swirl nozzles and the rotating blade-cooling holes. For the direct transfer system, any hot gas ingested via the rim seal can cause serious thermal contamination of the blade-cooling air.

Fig. 3b shows a "cover-plate" system in which air, delivered from pre-swirl nozzles at a low radius, flows radially outward between the turbine disc and a cover-plate attached to it. Ingested mainstream gas is prevented from mixing with the pre-swirl cooling air by a seal at the base of the cover-plate; convective and windage heating of the pre-swirl air will occur as it flows outward over the hot turbine disc.

The flow and heat transfer in cover-plate systems, such as that illustrated in Fig. 3b, has proved tractable to experiment, computation and idealised theoretical analysis. At the high pre-swirl flow-rates expected to occur in engines, Karabay et al.[15] found that free-vortex flow occurred in the rotating cavity between the cover-plate and the disc on an experimental rig. Karabay et al.[16] used this result to estimate the pressure drop and adiabatic temperature change between the inlet nozzles and the rotating holes. Measured tangential velocities and theoretical pressure distributions were predicted accurately by axisymmetric computations, using similar modelling to that described above for direct-transfer work.

Pilbrow et al.[17] described measured and computed heat transfer results for the same cover-plate system. Comparisons between computed and measured Nusselt numbers on the heated disc were reasonably good, however blade-cooling air temperatures were again under-predicted. Fig. 5 shows computed secondary flow streamlines for a typical computation, and Table II gives the results for ΔT obtained by Pilbrow et al. Much of the discrepancy between computations and measurements can be attributed to three-dimensional

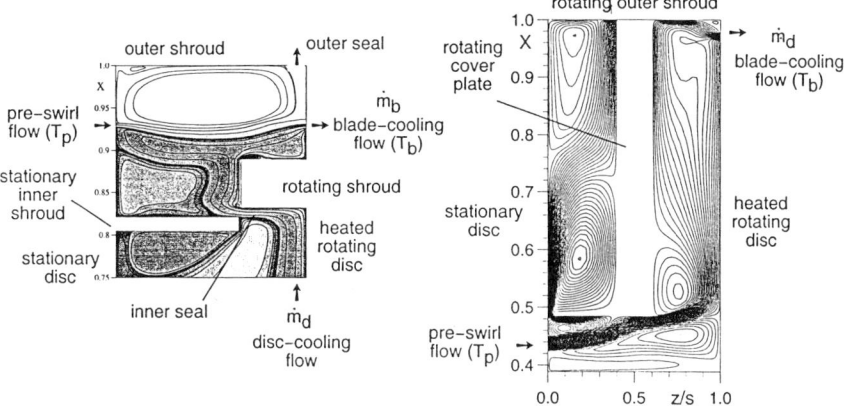

Fig. 4 Computed secondary flows in a direct-transfer system (from Wilson et al.[14]) See Table I case 3 for flow conditions

Fig. 5 Computed secondary flows for a cover-plate system (from Pilbrow et al.[17]) See Table II case 2 for flow conditions

heat transfer effects close to the discrete holes on the disc, which can be represented by three-dimensional steady models but not by axisymmetric models. The theoretical results for cover-plate systems have been developed further by Karabay et al.[18], including an expression for the temperature difference ΔT in terms of system geometry, the adiabatic work done on the pre-swirl air and the heat transfer from the surface of the disc. The latter term involves the average Nusselt number over the heated disc surface, which is shown to have a minimum value for a particular (optimal) value of inlet pre-swirl for a given system.

Unsteady computational studies of pre-swirl systems (and evaluation of simplified quasi-steady models) are a component of the ICAS-GT integrated research projects now being carried out by European industries and universities, under the Brite-EuRam Framework IV programme (due to be completed in December 2000). The work includes optimisation of pre-swirl nozzle designs, using detailed three-dimensional simulations such as those described by Popp et al.[19] for the nozzles in a cover-plate system, and with data provided from complementary experimental research. Cross-validation of commercial CFD codes and standard turbulence models for both flow and heat transfer is an important aspect of this work for the industrial partners.

$Re_\phi \div 10^6$	β_p	m_b/m_d	λ_T	$\Delta T = T_b - T_p$	
				(comp)	(exp)
1.27	0.99	4.4	0.03	6.5	8.7
1.27	1.98	9.0	0.03	-3.4	-1.7
1.23	0.99	2.2	0.06	2.9	6.0
1.23	1.98	4.5	0.06	-2.8	-1.4

Table I Blade-cooling air temperature rise in a direct-transfer system (Wilson et al.[14])

$Re_\phi \div 10^6$	β_p	λ_T	$\Delta T = T_b - T_p$	
			(comp)	(exp)
0.535	1.110	0.173	12.2	21.9
1.490	1.267	0.175	11.9	16.2
0.542	1.537	0.176	11.4	18.0
0.898	2.049	0.351	6.6	10.6
0.965	2.866	0.349	6.2	6.5
0.588	3.059	0.353	6.6	7.1

Table II Blade-cooling air temperature rise in a cover-plate system (Pilbrow et al.[17])

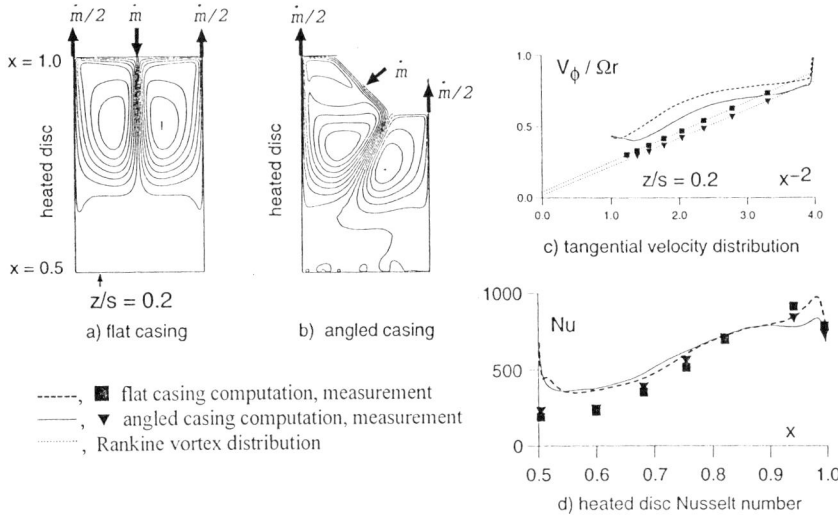

Fig. 6 Flow and heat transfer in a rotating cavity with a stationary outer casing (Jaafar et al.[26]) (a/b = 0.5, G = 0.3); $Re_\phi = 1.5 \times 10^6$, $\lambda_T = -0.034$

3 ROTATING CAVITY WITH A STATIONARY OUTER CASING

In some gas-turbine cooling systems, a peripheral flow of cooling air is introduced either through nozzles in a stationary casing, as in Fig. 6a,b, or through holes in a rotating disc as in Fig. 7a; the air leaves through clearances in the stationary casing as illustrated. In Fig. 6a, tangential shear at the casing gives rise to flow being pumped radially outward on the discs, resulting in a free shear flow inward between the discs, as shown, and combined free and forced (or Rankine) vortex flow outside the boundary layers (Gan et al.[20]). For a closed system (i.e. with no superposed cooling flow) Gan et al. found that the inward penetration of the recirculating secondary flows reduced with increasing Re_ϕ. Increasing the magnitude of the turbulent flow parameter, λ_T, increased the inward penetration (by convention, $\lambda_T < 0$ when the cooling air is directed inward).

Owen[1] reviewed research into the flow structure in a closed rotating cavity with a stationary flat casing (such as Fig. 6a for $\lambda_T = 0$). The closed system is relevant to the space between rotating components in computer disc drives, and instabilities in the free shear layer can give rise to unsteady three-dimensional flow (see Herrero et al.[21]) and Randriamampianina et al.[22] as examples of work in the USA and in Europe, respectively). The effect of the superposed peripheral flow on such instabilities has not yet been studied.

Mirzaee et al.[23] made heat transfer measurements for the flat casing system with one disc heated (Fig. 6a) and carried out axisymmetric computations, using the Launder-Sharma low-Reynolds-number k-ε turbulence model. Mirzaee et al.[24] described a similar study for the stepped-casing configuration (Fig. 7a) and identified different secondary flow recirculations for low or high values of $|\lambda_T|$.

Extensive experimental studies of flow and heat transfer for the angled casing geometry, shown in Fig. 6b and Fig. 7b, have also been carried out at Bath. Axisymmetric

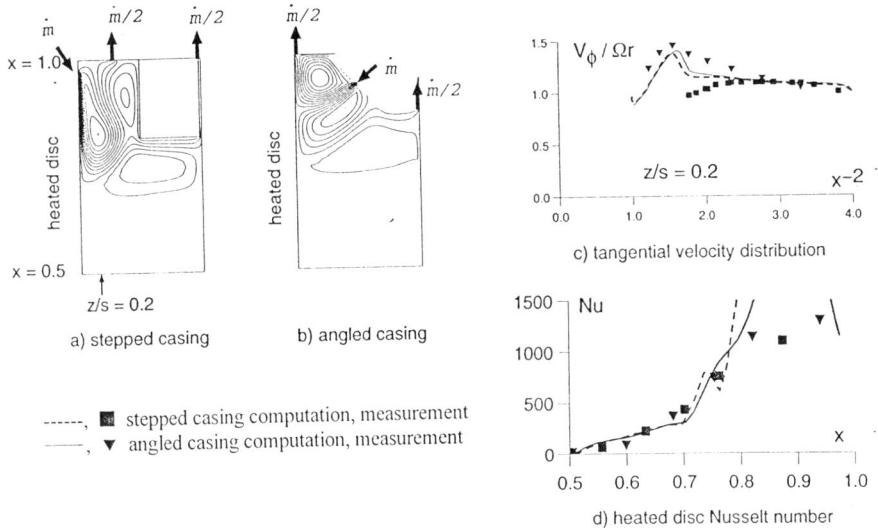

Fig. 7 Flow and heat transfer in a rotating cavity with a stationary outer casing (Jaafar et al.[25,26]) (a/b = 0.5, G = 0.3); $Re_\phi = 7.5 \times 10^5$, $\lambda_T = -0.475$

computations for this and the other two systems have been made by Jaafar et al.[25,26], using a general-purpose code (PHOENICS) and a high-Reynolds-number k-ε turbulence model with wall-functions. The "flat casing" and "stepped casing" computations confirmed the results of earlier studies. For all of the systems, it was found, both computationally and experimentally, that Nusselt numbers for the heated disc increased with both Re_ϕ and $|\lambda_T|$.

Jaafar et al.[26] discussed the fluid dynamics for the angled casing configuration, and Fig. 6 shows a comparison of both flow and heat transfer results for the flat and angled casings at one condition, $Re_\phi = 1.5 \times 10^6$ and $|\lambda_T| = 0.034$. For the angled casing, the inlet flow is entrained into the boundary layer on the angled surface (Fig. 6b). The inward shear flow formed where the boundary layers meet at the inner edge of the casing is closely related to that for the "flat casing" in Fig. 6a, as are the secondary flow recirculations. For both configurations, the measured radial distribution of tangential velocity (outside the disc boundary layers) follows a Rankine vortex structure, as described above and shown in Fig. 6c, but this is not reproduced by computations using k-ε turbulence models. This is discussed further by Mirzaee et al.[23] Fig. 6d shows that there is little difference between values of Nu for the two systems either for the measurements or the computations.

Fig. 7 shows a comparison of results for the stepped and angled casings at $|\lambda_T|$ = 0.475 (and at an inlet swirl ratio close to unity). For "high" values of $|\lambda_T|$ (> 0.1 approx.), Jaafar et al.[26] characterised the secondary flows as "inertially-dominated", and in Fig. 7b the powerful inlet flow controls the secondary flow recirculations in the outer part of the cavity. The trend of the measured tangential velocity distribution (at z/s = 0.2), Fig. 7c, is reasonably well predicted for the angled casing configuration. The stepped casing computations, however, show none of the sensitivity to the changes in geometry and inlet location suggested by the measurements.

Heat transfer rates are again very similar for the two configurations, Fig. 7d, and for these cases the computations agree well with the data in the inner part of the system. In the outer region, the computational results are affected by stagnation of the radial flow on the heated disc surface (separating regions of radially inward and outward flow); the computed location of this region may be inaccurate, due to the deficiencies and simplifications of the axisymmetric model.

The discrepancies between measured and computed velocity distributions illustrated above suggest that the stationary casing problem poses challenges for developers of improved computational and turbulence models for rotating-disc flows. Work is now being carried out in a collaboration between Bath and IRPHE, Marseille, to apply a differential Reynolds-stress turbulence model (Elena and Schiestel[27]) to the closed "flat casing" configuration, for both steady and unsteady flow. The importance of three-dimensional effects, whether caused by instability in the flow or the presence of discrete inlet nozzles for cases involving superposed flow, has not yet been properly addressed.

4 BUOYANCY-INDUCED FLOW IN A ROTATING CAVITY

In gas-turbine engines, cooling air often flows through the centre of a stack of rotating compressor discs on its way to the turbine section of the engine. The rotating cavity with an axial throughflow (see Fig. 2f) provides a simple model of the flow between a pair of the corotating compressor discs.

In the absence of a superposed axial throughflow, the temperature differences between the rotating surfaces and the air in the cavity create buoyancy forces, which give rise to free convection. The axial throughflow creates secondary flow, in the form of a toroidal vortex, which interacts with the buoyancy-induced flow. The subsequent flow structure is usually unsteady and three-dimensional, which makes the problem difficult to solve experimentally or computationally. Before studying the axial throughflow case, it is useful to consider the case of a sealed rotating annulus.

4.1 Sealed Rotating Annulus

In a stationary annulus, the gravitational acceleration, g, controls the flow; in a rapidly-rotating annulus (where $\Omega^2 r \gg g$), the centripetal acceleration, $\Omega^2 r$ is controlling.

Geophysicists are interested in the case where $\Omega^2 r$ is the same order-of-magnitude as g, which corresponds to the conditions in the Earth's atmosphere. Hide and his co-workers (see, for example, Hide[28,29], Hide and Mason[30], Read[31]) have used the model of an annulus, filled with water and rotating about a vertical axis, to simulate atmospheric flow. When the outer cylindrical surface is hot, the inner surface cold and the two discs adiabatic, then a number of interesting phenomena are observed. For a given value of ΔT (the temperature difference between the outer and inner cylinders), the structure changes from axisymmetric to "wavy" flow as the rotational speed is increased. When viewed from an axial direction, a sinuous (nonaxisymmetric) stream of fluid meanders circumferentially around the annulus (in a manner similar to that of the jet stream in the atmosphere) transporting heat from the outer to the inner cylinder. Irregular waves, which appear as the speed is increased further, are associated with "geostrophic turbulence" which, unlike turbulence in a stationary frame of reference, tends to be two-dimensional or stratified.

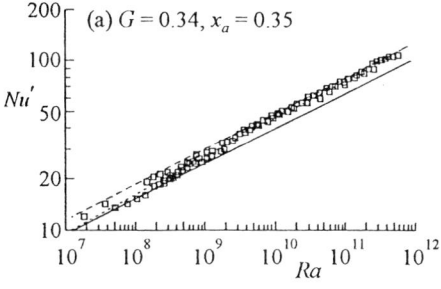
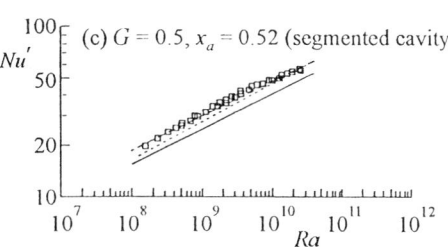
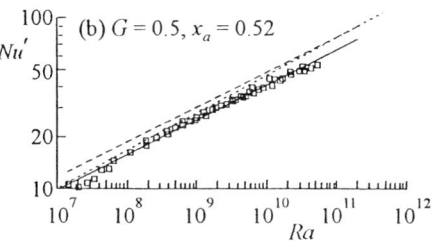

Fig. 8 Variation of measured values of Nu' with Ra in a rotating annulus, for three geometries

------ correlation, equation (4.1)
——— correlation, equation (4.2)
- - - - correlation, equation (4.3)
☐ experimental data of Bohn et al.[32]

Although the range of rotational Reynolds numbers and Grashof numbers in geophysical flows are much smaller than those found in gas turbines, some of the flow phenomena are believed to be related.

Bohn et al.[32] conducted a combined computational and experimental study for the case of a hot outer cylindrical surface and a colder inner one, and with the two discs insulated. Tests were carried out, for $10^7 < Ra < 10^{12}$, for three different geometries, and the correlations are given below and shown in Fig. 8.

Geometry A: axisymmetric annulus, $G = 0.34$, $x_a = 0.35$

$$Nu' = 0.246\, Ra^{0.228} \tag{4.1}$$

Geometry B: axisymmetric annulus, $G = 0.5$, $x_a = 0.52$

$$Nu' = 0.317\, Ra^{0.211} \tag{4.2}$$

Geometry C: annulus with eight 45° segments, $G = 0.5$, $x_a = 0.52$

$$Nu' = 0.365\, Ra^{0.213} \tag{4.3}$$

For the above correlations

$$Ra = 2\, \frac{1-x_a}{1+x_a}\, Pr\, \beta\Delta T\, \frac{\Omega^2 r_m^2 b^2 (1-x_a)^2}{\nu^2} \tag{4.4}$$

and Nu' is the ratio of the convective flux to conduction in a stationary fluid. It is interesting to note that for laminar free convection from a stationary surface, Nu' $\propto$ Ra$^{0.25}$, and the above results show no sign of transition to turbulent flow even at values of Ra in excess of Ra = 10^{11}. It should also be noted that, for similar conditions in a sealed rotating annulus, the Nusselt numbers for a radial flow of heat are significantly greater than those for an axial flow of heat.

Bohn et al.[32] also solved the unsteady, 3D, laminar, elliptic equations for the case of a 45° segment, corresponding to geometry C. The partitions and unheated surfaces were taken to be adiabatic, and two types of heating were used: *axial heat flow* with a hot and a cold disc; *radial heat flow* with a hot outer cylindrical surface and a cold inner one. For the axial heat flow, the computed flow structure reached a steady state in a few seconds of "simulated" time. For the radial heat flow, no steady state was found within the computational time available to the authors. Bohn and Gier[33,34] carried out unsteady 2D and 3D computations, for the 45° segment with a radial heat flow, using the Launder-Sharma low-Reynolds number k-ε turbulence model. They concluded that, compared with the laminar computations, turbulence led to an increase in the average heat transfer and to an increase in the ratio of the local maximum to minimum heat flux on the cylindrical surface.

Bohn et al.[35] conducted a combined computational and experimental study for the case of *axial heat flow* in an air-filled rotating annulus. Their measurements were correlated, for G = 0.5, x_a = 0.52, 2 x 10^8 < Ra < 5 x 10^{10}, by

$$\mathrm{Nu}' = 0.346\ \mathrm{Ra}^{0.124} \tag{4.5}$$

Their axisymmetric laminar computations were in good agreement with this correlation for Ra < 2 x 10^9. Divergence between computations and measurements at larger values of Ra was attributed to a significant radial flow of heat through the insulated cylindrical surfaces of the experimental rig.

4.2 Rotating Cavity with Axial Throughflow

Vortex breakdown can take place in a swirling, diverging jet of fluid. Under these conditions, which can occur in a rotating cavity with an axial throughflow, the structure of the flow changes dramatically, and formerly steady, axisymmetric flow can become unsteady and three-dimensional. This increases the exchange of fluid between the central jet and the rotating cavity, and it consequently has a strong effect on the heat transfer from the heated surfaces to the cooling air. The radial extent of the recirculation region created by the throughflow tends to decrease as the Rossby number, Ro, and the gap ratio, G, decreases. The reader interested in vortex breakdown is referred to Owen and Pincombe[36] and Farthing et al.[37]

Heat transfer tests were conducted for axial throughflow in a number of different rigs (see Farthing et al.[37,38]). Flow visualisation revealed that, when the discs were hot and the air cold, the resulting buoyancy-induced flow was nonaxisymmetric. Fig. 9 shows a sequence of photographs taken after (white) smoke had been injected into the air entering the cavity. ("Symmetrical heating" means both discs had the same radial distribution of temperature, and ΔT is the difference between the maximum temperature on the discs and the temperature of the air at inlet to the cavity.) Fig. 9a shows a "radial arm" of smoke; Fig. 9b shows the outline of a cyclonic and anti-cyclonic vortex at the end of the radial arm; Fig. 9c shows the recirculation spreading to most, but not all, of the cavity. The cyclonic (low

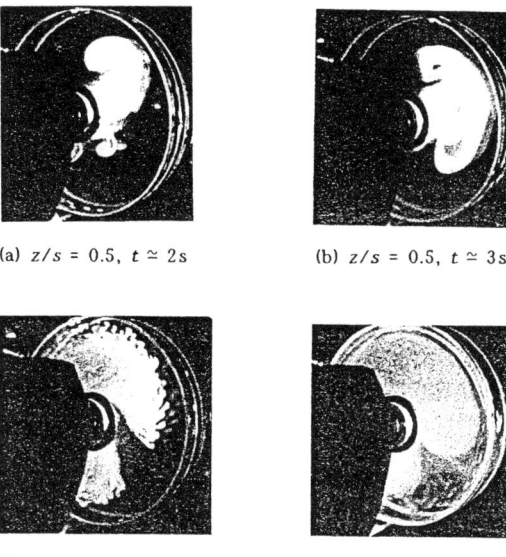

Fig. 9 Photographs of flow structure in a symmetrically heated rotating cavity with axial throughflow: $G = 0.267$, $Re_z = 2180$, $Re_\phi = 1.3 \times 10^4$, $Ro_z = 8.4$, $\Delta T = 55°C$, anticlockwise rotation. (Farthing et al.[37]) [t is the approximate lapsed time from when the smoke was injected into the air.]

pressure) and anti-cyclonic (high pressure) vortices create the Coriolis forces required for the flow to move radially outward in the radial arm. Velocity measurements showed that the core of fluid between the two discs rotated at a slower speed than the discs themselves, and the difference between the angular speeds of the core and disc, which could be up to 10%, increased as ΔT and G increased.

For the symmetrically-heated case, Farthing et al.[38] correlated their measured local Nusselt numbers on the rotating discs by

$$Nu = 0.0054 \, Re_z^{0.30} \, Gr^{0.25} \tag{4.6}$$

where

$$Nu = \frac{q(b-r)}{k(T_0 - T_1)} \tag{4.7}$$

and

$$Gr = \frac{\rho^2 \beta (T_0 - T_1) \Omega^2 r (b-r)^3}{\mu^2} \tag{4.8}$$

Fig. 10 shows the results for $G = 0.138$, $Re_z = 2 \times 10^4$ and $4 \times 10^5 < Re_\phi < 1.6 \times 10^6$. Also shown are the correlations for free convection for a stationary vertical plate, where for laminar flow

$$Nu = 0.36 \, Gr^{0.25} \tag{4.9}$$

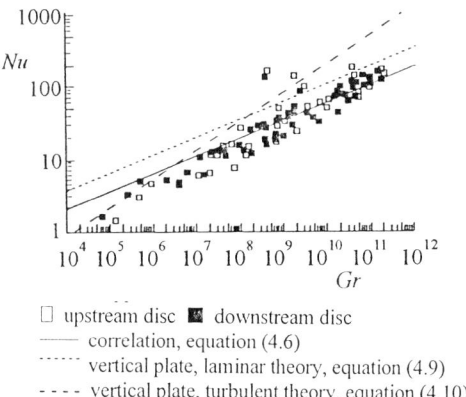

Fig. 10 Variation of Nu with Gr for a symmetrically heated rotating cavity with axial throughflow distribution: $G = 0.138$, $Re_z = 2 \times 10^4$ experimental data of Farthing et al[38].

and for turbulent flow

$$Nu = 0.022 \, Gr^{0.4} \tag{4.10}$$

Not surprisingly, the local Nusselt numbers, which were obtained from fluxmeters located at eight radial locations on each disc, show considerable scatter. However, even at Grashof numbers in excess of 10^{11}, there is no obvious sign of transition from laminar to turbulent flow.

Long[39] and Long and Tucker[40] presented heat transfer measurements for the rotating cavity with a circular inlet for the case where the discs and the peripheral shroud could be heated. They also made measurements of the air temperature inside the rotating cavity, from which they deduced the percentage of axial throughflow that was ingested into the cavity. The percentage depended on the Rossby number: for $Ro < 1$, around 50% of the throughflow entered the cavity; for $Ro > 10$, it was around 10%.

Long and Tucker[41] solved the unsteady, 3D, laminar, elliptic equations for one of the geometries studied by Farthing et al., with $Re_\phi = 1.3 \times 10^4$ and $Re_z = 2180$. The computed flow structure, which was three-dimensional and periodic, depended strongly on the radial distribution of the surface temperature of the discs. The computed flow structures were also similar to those observed experimentally by Farthing et al., with radial arms and cyclonic and anti-cyclonic vortices. When the shroud was unheated, the number of radial arms increased from one to two to three as the maximum temperature on the disc was moved radially outward. When the shroud was heated, even more radial arms appeared. The computed flow structure could also change with time: for example, a "one-arm structure" could change to or from a "two-arm structure".

Bohn et al.[42] carried out flow visualisation and heat transfer measurements in a rotating cavity with G = 0.2 and a/b = 0.3 for $2 \times 10^5 < Re_\phi < 8 \times 10^5$ and $2 \times 10^4 < Re_z < 7 \times 10^4$. The two discs were heated to produce a radially-increasing temperature distribution with a maximum temperature of 105° C for an air-inlet temperature of 25° C. Their flow visualisation revealed a structure similar to that observed by Farthing et al.[37], in which a pair of cyclonic and anti-cyclonic vortices was observed, and the core of fluid rotated at an angular speed around 88% to 90% of that of the discs.

In summary, buoyancy-induced flow tends to be unsteady and three-dimensional. The flow structure comprises a number of cyclonic and anti-cyclonic vortices, and the core of the fluid rotates at a speed slower than that of the discs. Even for values of Gr up to 10^{11} (a value associated with turbulent flow in a stationary system), $Nu \sim Gr^{1/4}$ (a correlation usually associated with laminar flow). It is probable that the large Coriolis accelerations suppress or modify the turbulence inside the rotating cavity, and it is unclear that "conventional" turbulent flow occurs at even at the very large Grashof numbers found inside gas-turbine engines. To complicate the above problem even further, an axial throughflow of cooling air can create vortex breakdown which interacts with the buoyancy-induced flow in the cavity.

5 CONCLUDING COMMENTS

This paper summarises some recent rotating-disc research having applications to the flow and heat transfer in gas-turbine engines: ingress of hot mainstream gases, pre-swirl cooling-air supply systems and cooling air flows in rotating cavities. It is hoped that the references cited here will adequately direct the interested reader to more details of existing published work. Much further research is currently being carried out for the above systems which should improve the detailed understanding of these complex three-dimensional phenomena.

References

1. Owen, J. M. (2000) Flow and heat transfer in rotating-disc systems: some recent developments, Proc. 3rd Int. Symp. on Turbulence, Heat and Mass Transfer, Nagoya, pp 33-58
2. Owen, J.M. and Rogers, R.H. 1989. Flow and heat transfer in rotating-disc systems, Vol.1: Rotor-stator systems. Research Studies Press, Taunton. (John Wiley, New York.)
3. Owen, J.M. and Rogers, R.H. 1995. Flow and heat transfer in rotating disc systems: Vol. 2, Rotating cavities. Research Studies Press, Taunton, UK (John Wiley, New York).
4. Johnson, B.V., Mack, G.J., Paolillo, R.E., and Daniels, W.A. (1994). Turbine rim seal gas path flow ingestion mechanisms. AIAA Paper No. 94-2703.
5. Chew, J.W., Green, T., and Turner, A.B. (1994). Rim sealing of rotor-stator wheelspaces in the presence of external flow. ASME Paper No. 94-GT-126.
6. Reichert, A.W. and Lieser, D. (1999). Efficiency of air-purged rotor-stator seals in combustion turbine engines. ASME Paper No. 99-GT-250.

7. Bohn, D., Rudzinski, B., Surken, N. and Gartner, W. (2000). Experimental and numerical investigation of the influence of rotor blades on hot gas ingestion into the upstream cavity of an axial turbine stage. ASME Paper No. 2000-GT-284.
8. Wilson, M., Chen, J. X. and Owen, J. M. (1996) Computation of flow and heat transfer in rotating-disc systems, Trans IMechE 3rd Int Conf on Computers in Reciprocating Engines and Gas Turbines, pp 41-49
9. Scott, R. M., Childs, P. R. N., Hills, N. J. and Millward, J. A. (2000) radial inflow into the downstream cavity of a compressor stator well, ASME Paper No. 2000-GT-0507
10. Iacovides, H., Nikas, K. S. and TeBraak, M. A. F. (1996) Turbulent flow computations in rotating cavities using low-Reynolds-number models, ASME Paper No. 96-GT-159
11. Launder, B.E. and Sharma, B.I., 1974. Application of the energy dissipation model of turbulence to the calculation of the flow near a spinning disc, Letters in Heat and Mass Transfer, **I**, 131-138.
12. Gassiat, M. R. (2000) Etude experimentale d'ecoulements centripetes avec prerotation d'un fluide confine entre un disque tournant et un carter fixe, PhD thesis, Universite de la Mediterranee Aix-Marseille II, France
13. Meierhofer, B. and Franklin, C. J. (1981) An investigation of a preswirled cooling airflow to a gas turbine disk by measuring the air temperature in the rotating channels, ASME Paper No. 81-GT-132
14. Wilson, M, Pilbrow, R. and Owen, J. M. (1997) Flow and heat transfer in a pre-swirl rotor-stator system, J. Turbomachinery, **119**, pp 364-373
15. Karabay, H., Chen, J. X, Pilbrow, R., Wilson, M. and Owen, J. M. (1999) Flow in a "cover-plate" pre-swirl rotating-disc system, J. Turbomachinery, **121**, pp 160-166
16. Karabay, H., Pilbrow, R., Wilson, M. and Owen, J. M. (1999) Performance of pre-swirl rotating-disc systems, ASME Paper No. 99-GT-197
17. Pilbrow, R., Karabay, H., Wilson, M. and Owen, J. M. (1999) Heat transfer in a "cover-plate" pre-swirl rotating-disc system, J. Turbomachinery, **121**, pp 249-256
18. Karabay, H., Wilson, M. and Owen, J. M. (2000) Predictions of effect of swirl on flow and heat transfer in a rotating cavity, Submitted to Int. J. Heat Fluid Flow
19. Popp, O., Zimmermann, H. and Kutz, J. (1998) CFD analysis of coverplate receiver flow, J. Turbomachinery, **120**, pp 43-49
20. Gan, X., Mirzaee, I., Owen, J. M., Rees, D. A. S. and Wilson, M. (1996) Flow in a rotating cavity with a peripheral inlet and outlet of cooling air, ASME Paper No. 96-GT-309
21. Herrero, J., Giralt, F. and Humphrey, J. A. C. (1999) Influence of the geometry on the structure of the flow between a pair of corotating disks, Phys. Fluids **11**, 86-96.
22. Randriamampianina, A., Schiestel, R. and Wilson, M. (1999) Spatio-temporal behaviour in an enclosed corotating disc pair , submitted to J. Fluid Mech.
23. Mirzaee, I., Gan, X., Wilson, M. and Owen, J. M. (1998) Heat transfer in a rotating cavity with a peripheral inflow and outflow of cooling air , J. Turbomachinery, **120**, pp 818-823
24. Mirzaee, I., Quinn, P., Wilson, M. and Owen, J. M. (1999) Heat transfer in a rotating cavity with a stationary stepped casing, J. Turbomachinery, **121**, pp 281-287
25. Jaafar, A. A., Motallebi, F., Wilson, M. and Owen, J. M. (2000) Flow and heat transfer in a rotating cavity with a stationary stepped casing, ASME Paper No. 2000-GT-281
26. Jaafar, A. A., Gan, X., Wilson, M. and Owen, J. M. (2000) Flow in a rotating cavity with a stationary angled casing, Proc. 3rd Int. Symp. on Turbulence, Heat and Mass Transfer, Nagoya, pp 653-660

27. Elena, L. and Schiestel, R. (1996) Turbulence modeling of rotating confined flows, Int. J. Heat Fluid Flow, **17**, 283-289
28. Hide, R. (1977) Experiments with rotating fluids. Quart. J. R. Met. Soc., **103**, 1-28.
29. Hide, R. (1988) Studies of geostrophic turbulence, chaos and other non-linear phenomena in rotating fluids: the role of combined laboratory and numerical experiments. Met. Mag., **117**, 33-34.
30. Hide, R. and Mason, P.J. (1975) Sloping convection in a rotating fluid. Adv. in Phys., **24**, 1, 47-100.
31. Read, P.L. (1988) The dynamics of fluids: the 'philosophy' of laboratory experiments and studies of the atmospheric general circulation. Met. Mag., **117**, 35-45.
32. Bohn, D., Deuker, E., Emunds, R. and Gorzelitz, V. (1995) Experimental and theoretical investigations of heat transfer in closed gas filled rotating annuli. J. Turbomachinery, **117**, 175-183.
33. Bohn, D. and Gier, J. (1997) The effect of turbulence on the heat transfer in closed gas-filled rotating annuli. ASME Paper No. 97-GT-242.
34. Bohn, D. and Gier, J. (1998) The effect of turbulence in closed gas-filled rotating annuli for different Rayleigh numbers. ASME Paper No. 98-GT-542. (To be published in TransASME.)
35. Bohn, D., Edmunds, R., Gorzelitz, V. and Kruger, U. (1996) Experimental and theoretical investigations of heat transfer in closed gas-filled rotating annuli II. J. Turbomachinery, **118**, 11-19.
36. Owen, J.M. and Pincombe, J.R. (1979) Vortex breakdown in a rotating cylindrical cavity. J. Fluid Mech., **90**, 109-127.
37. Farthing, P.R., Long, C.A., Owen, J.M. and Pincombe, J.R. (1992) Rotating cavity with axial throughflow of cooling air: flow structure. J. Turbomachinery, **114**, 237-246.
38. Farthing, P.R., Long, C.A., Owen, J.M. and Pincombe, J.R. (1992) Rotating cavity with axial throughflow of cooling air: heat transfer. J. Turbomachinery, **114**, 229-236.
39. Long, C.A. (1994) Disk heat transfer in a rotating cavity with an axial throughflow of cooling air. Int. J. Heat Fluid Flow, **15**, 307-316.
40. Long, C.A. and Tucker, P.G. (1994) Numerical computation of laminar flow in a heated rotating cavity with an axial throughflow of air. Int. J. Num. Meth. Heat Fluid Flow, **4**, 347-365.
41. Long, C.A. and Tucker, P.G. (1994) Shroud heat transfer measurements from a rotating cavity with an axial throughflow of air. J. Turbomachinery, **116**, 525-534.
42. Bohn, D.E., Deutsch, G.N., Simon, B. and Burkhardt, C. (2000) Flow visualisation in a rotating cavity with axial throughflow. ASME Paper No. 2000-GT-280.

Selection of a Turbine Cooling System Applying Multi-Disciplinary Design Considerations

BORIS GLEZER

Optimized Turbine Solutions, San Diego, California 92130, USA

ABSTRACT: The presented paper describes a multi-disciplinary cooling selection approach applied to major gas turbine engine hot section components, including turbine nozzles, blades, discs, combustors and support structures, which maintain blade tip clearances. The paper demonstrates benefits of close interaction between participating disciplines starting from early phases of the hot section development. The approach targets advancements in engine performance and cost by optimizing the design process, often requiring compromises within individual disciplines.

INTRODUCTION

The constantly increasing gap between rapidly rising operating gas temperatures and temperature capabilities of the materials available for life-limiting gas turbine components has been filled by the cooling of these components. Introduction of the turbine cooling air further downstream of combustor in an air based open cycle cooling system increases its negative impact on cycle performance. At the same time, reduction of the amount of the air available for the combustor makes a liner cooling task and emission control more difficult. This results in major challenge for designer of the cooling system: select a system which requires minimal amount of the cooling air and produces the smallest negative impact on engine durability, performance, weight (particularly for aero engines), emission (particularly for industrial engines), cost and fabrication complexity. Such a task can benefit significantly from a multi-disciplinary optimization effort, with each discipline effectively contributing to a successful design. The importance of this approach began to get appropriate attention from the international gas turbine community [1].

DISCUSSION

Traditional "step-by-step" engine hot section design process which has been following the sequential loop of thermodynamic cycle analysis - aerodynamic design - mechanical design - cooling design (and back to the cycle analysis) is being recognized as outdated due to boundaries historically established within each discipline. Significant improvement in the development process for advanced gas turbine engines can be achieved through close interaction between various disciplines participating in the development. This approach often requires compromises within each of the disciplines to accommodate major interdisciplinary constraints described in Figure 1. The engine development program starts with a specification of the application, performance, cost, emission limits and size/weight targets, and later progresses through thermodynamic cycle analysis and gas path geometry definition. Even at this early stage of development, the cooling issues play an equally important role with aerodynamic and structural considerations affecting blade tip to hub

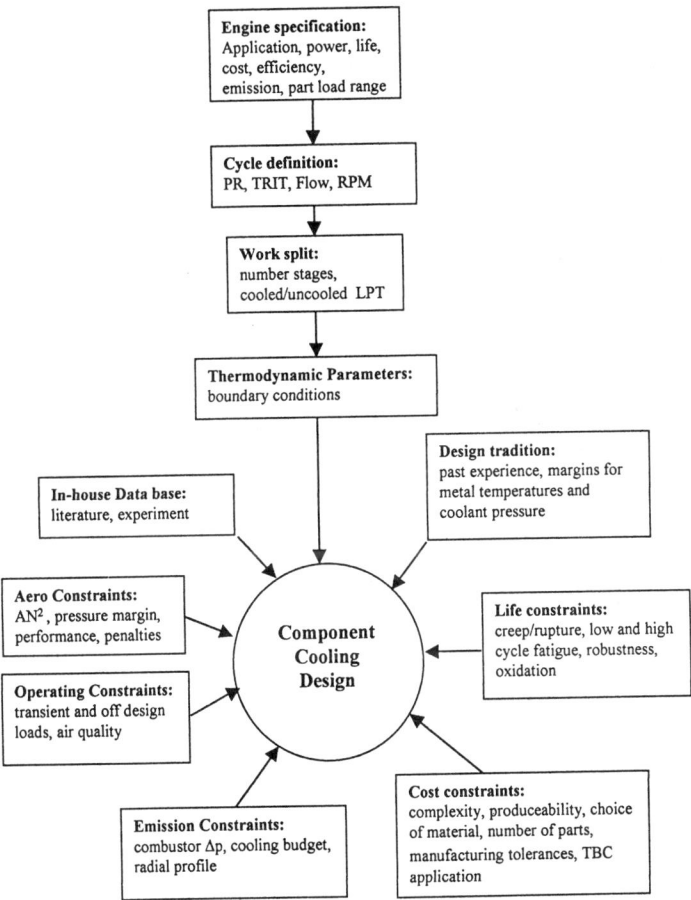

Fig. 1 **Multi-Disciplinary Factors Affecting Turbine Cooling Design Selection**

diameter ratio, work splits between stages, combustor liner surface to volume ratio, etc. For a development program to be successful, a true concurrent engineering process is essential during these early conceptual and preliminary design phases. Risk-sharing between disciplines which leads to justified safety margins and is based on probabilistic risk analysis, has to replace traditional more conservative approach which is based on over-conservative summation of margins defined within each discipline. This change in design culture is critical for the development of a high performance, cost effective engine.

High turbine inlet temperature is required to increase specific power of an engine and achieve corresponding reduction in size and weight of the engine. Optimized compressor pressure ratio has to increase with higher gas temperature. The increased pressure ratio is a major factor in improving engine efficiency. Unfortunately, higher gas temperature and increased temperature of the cooling air, resulting from the higher pressure ratio, require significantly larger turbine cooling flows producing a diminishing effect on engine performance improvement at these higher temperatures. For the industrial engines, which are not as sensitive to size and weight of the engine as are the aero engines, this factor quite often leads to application of moderately high turbine inlet temperatures. Cost of the engine is expected to rise resulting from application of advanced materials, more complicated cooling systems and associated manufacturing processes, thermal barrier and oxidation resistant coatings, more sophisticated and usually more complicated design of the hot section components. Figure 2 shows typical trend in increasing engine efficiency, specific cost and turbine cooling flow requirements corresponding to rising turbine inlet temperatures. An accelerated increase of the

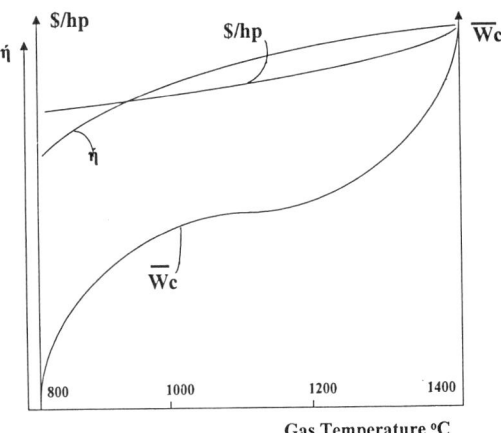

Figure 2. Effect of Turbine Inlet Temperature on Engine Cost, Efficiency and Turbine Cooling Budget

cooling air budget at gas temperatures of above 1200 -1300 °C is driven by the necessity to use extensive amount of the air for film cooling, particularly when compressor pressure ratio exceeds 30-35 and cooling air temperature increases to above 600 °C.

Preliminary cooling design of turbine components can start after the engine thermodynamic parameters and gas path geometry are conceptually defined. At this stage, a designer of cooling system should consider a number of major multi-disciplinary factors: company past design history, life limiting factors, manufacturing and material cost affecting factors, emission limit imposed constraints, field operating conditions, aero design imposed constraints and availability of in-house data base and facility to validate advanced cooling methods. A company with achieved close cross-disciplinary interaction is more likely to establish an integrated design methodology and experience-based optimization algorithms linking together different factors.

Figure 3 presents schematically a generic turbine cross-section with non-shrouded blades

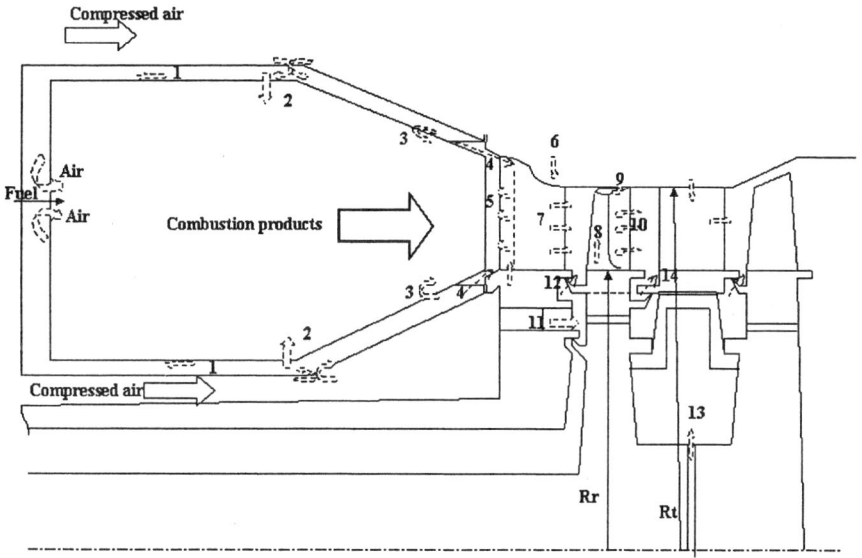

Figure 3. Typical Elements of Turbine and Combustor Cooling System:

1- back side cooling in-series with combustion air, 2- dilution air, 3- liner exit film, 4- nozzle endwall film, 5- nozzle shower head film, 6- tip clearance control modulated air, 7- nozzle trailing edge discharge, 8- blade leading edge air, 9- blade tip discharge, 10- blade trailing edge discharge, 11- preswirled air, 12- disc forward cavity buffer air/ blade platform film, 13- interstage buffer air, 14- disc aft cooling air/ stage 2 nozzle endwall film and shows major components of a typical cooling system. One of the most critical design decisions affecting not only selection of the cooling system, but also the entire turbine hot section design, is related to a turbine stage 1 work load and the stage corresponding pressure ratio. Higher stage 1 loading leads to a proportionally greater difference between stage 1 blade total and relative temperatures resulting in lower cooling flow requirements for the blade and possibility of using only the internal blade cooling for moderately high turbine inlet gas temperatures, without assistance of film cooling.

But even more importantly, this cross-disciplinary team decision can open opportunity to reduce significantly cooling of the downstream turbine components avoiding in some cases the stage 2 blade cooling at all. The higher stage 1 pressure ratio might also lead to a fewer turbine stages resulting in a larger gas path divergence angle. This leads to a larger gas path area for the blades, which should be optimized between longer blades and larger tip diameter holding acceptable AN^2 stress parameter limit (where A is the gas path plane area and N is a rotational speed of the rotor) . Longer blades are less sensitive to performance losses associated with tip leakages for the same tip clearance. This makes the option of a highly loaded stage 1 with longer blades/smaller tip diameter more attractive for a single stage high pressure turbine when the rotor speed can be maintained at a stress limiting maximum. However, longer blades with smaller tip and hub diameters have own, primarily mechanical stress limitations associated with smaller circumferential disc spaces available for the blade fir tree attachments and the disc posts separating them. The larger gas path divergence angle also should be taken in consideration regarding control of blade tip clearances associated with transient axial displacements in a rotor- stator system with axially tapered blades.

In cases where the turbine rotor speed is limited by the tip speed of the larger diameter of the downstream stages, an option with shorter stage 1 blades and larger blade tip diameter can be more beneficial for the turbine aerodynamic efficiency. This option typically requires a higher blade count and larger cooling flow for the stage blade-disc system, associated with increased disc pumping, which might negatively affect gains in the turbine thermal efficiency. The larger blade tip diameter also results in a larger and usually heavier and more costly engine package. Optimization between these options has to involve the cross-disciplinary team considering primarily aerodynamic, stress, heat transfer and cost factors. The defined as result of this optimization blade height and tip diameter determine height of the stage 1 nozzle exit and also affect to a great extend the height of the nozzle leading edge and also the shape of combustor liner exit transition. Establishment of correlations between influencing factors and development of physically and/or statistically proven algorithms are the necessary steps toward fully optimized turbine multi-disciplinary design system. A number of multi-variable design optimization tools using deterministic or stochastic approaches [2,3,4] have been developed and applied in certain areas of gas turbine engine design. However, development of the algorithms which can be unique for each turbine manufacturer due to the differences in design criteria for each company, continues to be a major obstacle for application of these optimizers.

A proper combustor design also can not be performed in isolation without members of the turbine cross-disciplinary team. Higher fuel to compressed air flow ratios, required for higher combustor exit temperatures, as well as a need to control nitride oxide emission by lowering the flame temperature, demand a larger amount of compressor discharge air to be introduced in the combustor primary zone. This limits the cooling air budget available for the combustor liner and for cooling of the turbine components, particularly when the liner cooling air circuit is in parallel (not in series) with primary air for the fuel injectors. Back side liner convective cooling methods are becoming more preferred to avoid relatively cold carbon oxide formation zones resulting from traditional film cooling method. Particularly interesting for this application are the techniques [5] based on a low pressure drop convective back side cooling allowing to use the spent liner cooling air in-series with the primary combustion air . With growing demands for amount of air in the primary combustion zone, to control nitrogen oxide emission, the amount of air available for dilution reduces. This leads to a flatter combustor exit radial temperature profile and results in higher gas temperatures near the endwalls.

Existing challenges associated with cooling of the nozzle endwalls and combustor liner exit transition walls require close interaction between combustor and turbine section designers. Maintaining a thin boundary layer at the combustor exit by constantly converging liner walls toward the nozzles, and then converging the nozzle endwalls by their contouring help to reduce secondary flow losses originated at the nozzle endwall. Recent research [6] shows that introduction of endwall film cooling upstream of the leading edge in combination with enwall contouring suppresses formation of the horseshoe vortex at the leading edge providing noticeable reduction in aerodynamic losses and preventing the cooler film layer from being diverted from the endwall. The studies also show that higher film blowing ratios utilizing maximum pressure head available from the combustor pressure losses, produces significant improvement in the endwall film coverage.

As a result of mentioned earlier flatter radial temperature profile, turbine component cooling design strategy has to change starting with the stage one nozzle, which typically consumes nearly 50% of the total turbine cooling budget used in high temperature engines. Tightening emission control requirements justify efforts in reduction of the nozzle cooling flows by using spent air from combustor liner cooling and also by applying advanced thermal barrier coatings in combination with nozzle internal convective cooling, instead of traditional full coverage film cooling. These options represent trade-off between product cost and engine efficiency to satisfy emission and life requirements. Reduced amount of the turbine nozzle cooling flow also assists in reduction of combustor exit temperature for a fixed rotor inlet temperature.

Similar considerations regarding emission, flatter radial temperature profiles and product cost have to be applied during selection of a blade cooling design. Recent advances in internal blade cooling, particularly for highly thermally loaded leading edges of the blades, including a technique based on a swirling flow [7], as well as increased confidence in durability of blade thermal barrier coatings, improve potential for a non-film-cooled blade leading edge option for the turbine inlet temperatures up to 1300 deg C. A four-quadrants chart shown in figure 4 illustrates a logical sequence for preliminary cross-disciplinary selection of the blade leading edge cooling . Two applications are considered in parallel: one for 15,000 hours of operation, representing typical aero-engine blade life, and another for 60,000 hours of operation, representing industrial engine blade life expectancy. Both engines assumed to have similar thermodynamic cycle, similar design and materials. Quadrant I, based on the creep-rupture data for a selected blade material, shows a correlation between initially assumed blade stresses _ and maximum allowable metal temperatures T_m for both specified applications. The assumed difference in blade life corresponds to about 27°C difference in blade metal temperatures (as calculated from Larson-Miller parameter for advanced blade alloys). Moving vertically down to the quadrant II to the intersection with specified blade inlet relative gas temperature line **Tgas** and then horizontally to the quadrant III until intersection with specified blade cooling air temperature line **Tc** and finally to quadrant IV, a designer arrives to a point where choice has to be made between few available cooling options to satisfy a required cooling effectiveness $\eta_c = (T_{gas} - T_m)/(T_{gas} - T_c)$. The quadrant IV represents a typical correlation between cooling effectiveness and blade leading edge cooling flow parameter $FP = (W_c \times c_p)/(A_{gas} \times h_{gas})$ for four different blade leading edge cooling techniques, including 1- trip-strip augmented passage, 2- impingement of the leading edge without cross flow, 3- swirl cooling technique (based on data from [7]) and 4- shower head film. Curves for the internally cooled techniques include an effect of the TBC coating.

As it is seen from the chart, the 15,000 hours application may use any of the techniques 2, 3 or 4; as the 60,000 hours application is limited only to techniques 3 and 4. Swirl cooling option (3) coupled with TBC can match cooling effectiveness of the shower head cooled blades (4) for moderately high turbine

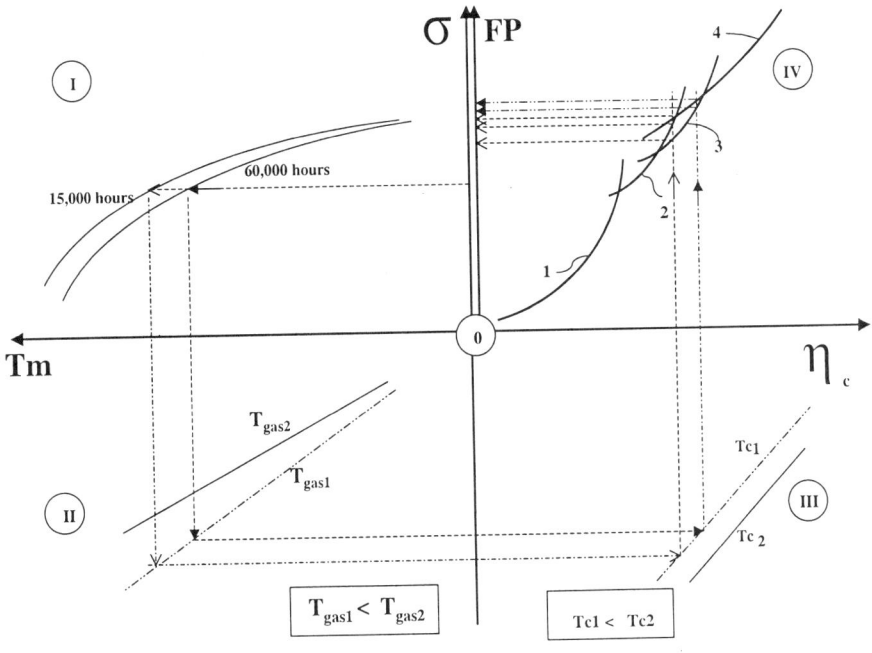

Figure 4. Conceptual Selection of a Blade Leading Edge Cooling

inlet temperatures . In addition to reduced cooling flow this option, may also provide improved cost and durability associated with elimination of the rows of shower head film cooling holes. Comparing with jet impingement (2) this technique provides almost uniformly high heat transfer coefficient along the whole circumference of the leading edge from the cooling side. This addresses a concern for off-design operation with the impingement cooling (2) when a nominal design stagnation point on the leading edge shifts away due to a change in the incidence angle.

Mentioned earlier flatter radial gas temperature profiles, more typical for the industrial engines, produce higher thermal loads in the blade root, platform and tip, creating significant challenges for a cooling designer. Interdisciplinary considerations might allow higher metal temperatures in the blade root section without changes in the creep-rupture life, for example, by tapering blade wall thickness. An increased tapering of whole airfoil toward the tip combined with more favorable ratio between root and tip cross-sectional areas might be required in certain cases to meet life targets. Figure 5 illustrates this design strategy. Figure 5a) shows a generic local inertial

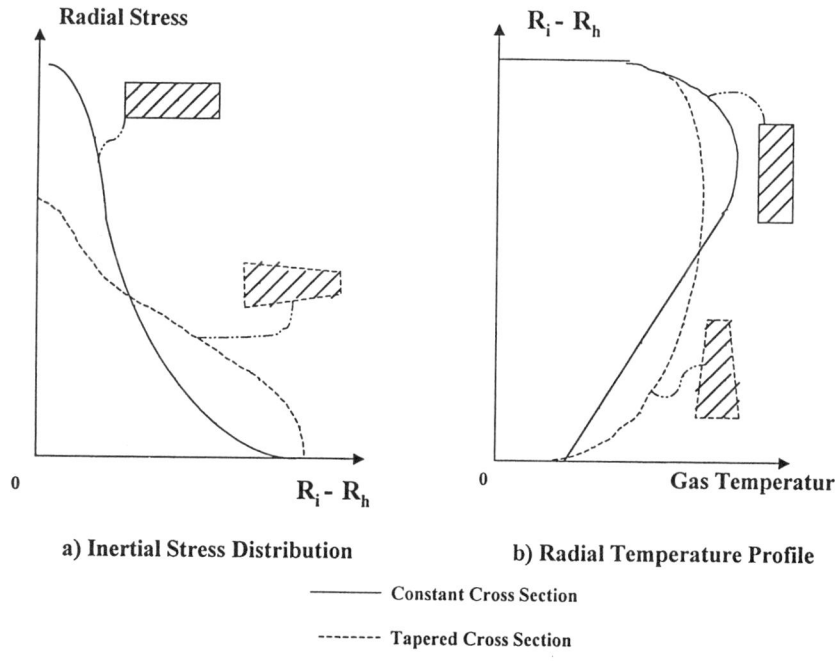

Figure 5. Local cross-section-average blade inertial stresses and expected gas temperature profiles for constant and tapered cross sections

stress distribution along the height of a non-shrouded blade with a constant radial (cylindrical) cross-sections (solid line) and with tapered radial blade sections (broken line). Figure 5b) presents corresponding gas temperature radial profiles acceptable from stress stand point for a blade with constant cross-section (solid line) and for a blade with cross-sections tapered toward the tip (broken line). The increased gas temperatures near the blade tips, which are usually oxidation life limiting (for non-shrouded blades) due to high local heat transfer coefficients and unfavorable heating to cooling surface ratio, can be counterbalanced by a redistribution of aerodynamic loading away from the tips.

Non-shrouded blade castings variation and occasional blade tip rubs through the wall of the tip resulted in design of a relatively tall radial tip fences around blade tip cavity (squealer tip). Cooling of this fence is recognized as a significant challenge. Better control of the tip clearances during transient operation and improved blade casting tolerances provide partial answer to this challenge. Turbine blade tip leakages represent one of the largest sources of turbine efficiency loss. Maintaining tight operating tip clearances without tip rubs is another major task for a cross-disciplinary design. Proper thermal management of the entire engine rotor-to-stator transient behavior, particularly for the non-shrouded blades in the diverging gas path, creates an extra ordinary challenge. Both thermal and inertial transient displacements as well as some aerodynamic loading effects should be considered. Passive tip clearance control, which is based on close matching between transient growth of rotor and stator, requires typically a bulky stationary structure, unacceptable for aero engines. Active and semi-active [8] tip clearance controlling techniques, based on modulation of the cooling air supply to the stationary structure during transient operation, present an alternative solution when lighter structures are required. The subject of design treatment for the blade tips combining effective cooling with minimized tip leakages and reduced aerodynamic losses continues to present a major challenge for the cross-disciplinary turbine design teams and gas turbine research community. Some recently introduced forms of blade tip geometry treatment [9] with so called "winglets" might be beneficial for both thermal loading of the tips and improvements in aerodynamic performance.

Although turbine airfoils are exposed to higher thermal loads, turbine disc should be treated as the most critical component when failure is considered. Because disc alloys have temperature capabilities significantly lower than airfoil materials, their cooling and hot gas ingress prevention present another challenge for a multi-disciplinary design team. It has been recognized that the disc hot gas ingress mechanism, compensating disc pumping outflow, is mainly associated with a circumferential steady pressure variation along the inner endwall propagating downstream of the nozzle trailing edges and also unsteady pressure variation propagating forward from the blade leading edges. Some techniques intended to reduce these pressure variations, have been developed recently to confront these effects [10]. However, their implementation requires close interaction between different disciplines. Certain aerodynamic performance compromises between blade tip to hub radius ratio affecting disc size for a specified hot gas flow and number of blades to be inserted in the disc rim, are required even during early conceptual design phase. Compromises between cost and efficiency using established life target are also required when disc material and disc cavity sealing design configurations are considered. Required axial overlapping seal, preventing hot gas ingress in to the disc rim plenum, results in increased rotor span affecting entire engine rotor dynamics. All these factors have to be addressed in a true multi-disciplinary design optimization process.

CONCLUSIONS

1. The paper emphasized an importance of close multi-disciplinary interaction in selection of optimal turbine cooling system design which is expected to provide significant effect on a new generation of advanced gas turbine engines.
2. Critical technical and market defined constraints are specified for major disciplines contributing to development of the cooling system.
3. Practical cross-disciplinary design factors are discussed to illustrate their impact on selection of the cooling system for turbine airfoils, combustor liners, turbine discs and tip clearance maintaining structures.
4. Development of the interdisciplinary correlations and proven by practice algorithms are specified as a next necessary step for the process optimization.

NOMENCLATURE

A	annular area of gas path
A_g	airfoil outer surface
FP	flow parameter for cooling air
h_{gas}	hot side heat transfer coefficient
N	rotor speed
PR-	compressor pressure ratio
R_i	blade current radius
R_t	blade tip radius
R_h	blade hub radius
T_{gas}	gas temperature
T_c	cooling air temperature
T_m	metal temperature
W_c	cooling flow
η	thermal efficiency
η_c	cooling effectiveness
σ	rupture stress limit

REFERENCES

1. Multi-Disciplinary Turbine Design Optimization. *Panel Session , ASME Turbo- Expo'99 Indianapolis, USA, Chair/ Vice Chair*: Glezer, B., Solar Turbines,/ Scrivener, C., Rolls-Royce; *Panelists*: Dulikravich, G., *Penn State U.*; Junod, L., *Allison*; Khalatov, A., *Cardiff U.*; MacArthur, C., *Wright-Patterson AFB*; Shelton, M., *General Electric*; Staubach, J., *Pratt & Whitney* .
2. Goel, S., Cofer, J., Singh, H., 1996. Turbine Airfoil Design Optimization. *ASME Paper 96-GT-158.*

3. Egorov, I., Kretinin, G., Leshchenko, I., Kostiuk, S., 1998. The Technology of Multipurpose Optimization of Gas Turbine Engines and Their Components. *ASME Paper 98-GT-512.*

4. Tappeta, R., Nagendra, S., Renaud, J., 1998. A Multidisciplinary Design Optimization Approach for High Temperature Aircraft Engine Components. 35th Joint Jet Propulsion Conference, *AIAA Paper # 98-1819.*

5. Moon, H-K., O'Connell, T., Glezer, B., 1999. Channel Height Effect on Heat Transfer and Friction in a Dimpled Passage. *ASME Paper 99-GT-163*

6. Oke, R., Burd, S., Simon, T., Vahlberg, R., 2000. Measurements in a Turbine Cascade Over a Contoured Endwall: Discrete Hole Injection of Bleed Flow. *ASME Paper 2000-GT-214*

7. Glezer, B., Moon, H-K., 1996. A Novel Technique for the Internal Blade Cooling. *ASME Paper 96-GT-181*

8. Glezer, B., Bagheri, H., 1998. Turbine Blade Clearance Control System. *US Patent #5779436*

9. Harvey, N., Ramsden, K., 2000. A Study of a Novel Turbine Rotor Partial Shroud. *ASME Paper 2000-GT-668*

10. Glezer, B., Fox, M., 1998, Turbine Ingress Prevention Method and Apparatus. *US Patent #5759012.*

Analysis of Particle Laden Flow and Heat Transfer in Cascade and Rocket Nozzle

H. H. CHO*, W. S. KIM*, M. S. YU* and J. C. BAE**

*Department of Mechanical Engineering
Yonsei University, Seoul 120-749, KOREA*

**Agency for Defense Development
Yusung, Taejon 305-600, KOREA*

ABSTRACT: This paper presents results for the calculation of particle trajectories in a cascade and a rocket nozzle using a Lagrangian method. When the floating particles collide to the components, the component surface is damaged severely. The surface erosion rate is strongly dependent on a particle size, a particle impact angle and a surface material. For a compressor cascade, the particle impact rate increases proportionally with the flow inlet angle and the erosion rate on the pressure side surface of blade are related to the surface or coating materials. For a solid rocket nozzle, the particle free zone in the nozzle divergent section increases quickly with increasing particle size and the maximum heat transfer density occurs at the starting region of nozzle convergent section. The Al_2O_3 droplet breaks up around the nozzle throat due to the high velocity difference between the droplet and gas stream, resulting in the big change of particle free zone.

INTRODUCTION

Analysis of Particle laden flow is considered in some machine components of gas turbines and rocket nozzles for predicting damage of components and performance of engines. For aircraft gas turbine engines, it is inevitable that particles floating in the atmosphere enter the engine due to the difficulty of inlet filtering. Particularly, relatively large particles enter the engine in a volcanic zone, an industrial zone and deserts and these particles bring about serious problems such as drops of engine efficiency, blockage effects in cooling holes and erosion of blades.[1-3] These damages are not serious initially, but eventually they can be decisive factors for airplane accidents and additional expenses in mending or repair. The numerical analysis is helpful to trace particle trajectories in the device and predict where the damage will occur and how severe it will be, because it is difficult to conduct experiments.

For the solid rockets, propulsion material contains aluminum powder to increase thrust. During the combustion process, this powder forms oxidized aluminum droplets and they float in the combustion gas. The melted Al_2O_3 droplets make serious damages to the nozzle wall and the jet vane installed in the nozzle.[4] Therefore, two phase flow analysis and prediction of erosion rates are essential to control thrust vector and to predict the change of the rocket motor efficiency. The particle motion in particle laden flow is determined by the gas and particle

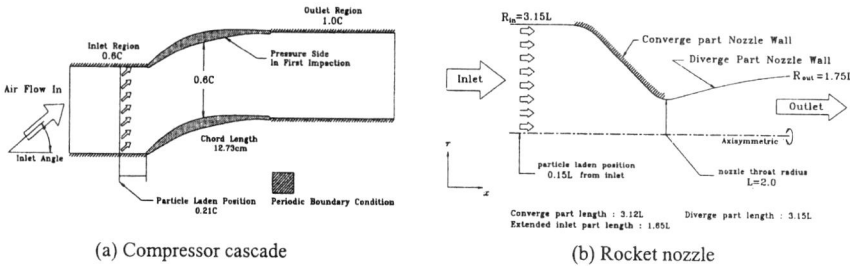

(a) Compressor cascade   (b) Rocket nozzle

Fig. 1 Computational domain.

interaction and particle and boundary impaction. The erosion rate is strongly depended on the particle velocities, size, density and impact angles. The oxidized aluminum particles have as many as about 20% fraction in the combustion gas of solid rocket motors.

In the present study, two phase flow analysis in a compressor cascade and a solid rocket nozzle are conducted to predict the local erosion rates and heat transfer on the surface for various flow conditions and particle sizes.

NUMERICAL ANALYSIS METHOD

If the mass ratio of particles to combustion gas is less than one, the laden particles are known to be negligible effects on the gas flow. Therefore, the gas flow field in the nozzle is analyzed first, and then particles are floated in the calculated flow field to trace their trajectories by Lagrangian method.[5] The governing equation of particle motion in gas flow is given by[6]

$$\frac{d\vec{V}_P}{d\theta} = C_D \frac{\text{Re}_P}{24\,Stk}(\vec{V}_P - \vec{V}) \quad (1)$$

where

$$C_D = \frac{24}{\text{Re}} \qquad \text{Re} < 0.1$$

$$C_D = \frac{24}{\text{Re}}(1 + 0.0916\,\text{Re}) \qquad 0.1 \le \text{Re} < 5.0$$

$$C_D = \frac{24}{\text{Re}}(1 + 0.158\,\text{Re}^{\frac{2}{3}}) \qquad 5.0 \le \text{Re}$$

$$Stk = \frac{\rho_P d_P^2 V C_S}{18\mu L}$$

The flow field analysis is accomplished by the SIMPLER algorithm with a low Reynolds number k-ε turbulence model and the periodic boundary condition is applied for the compressor cascade.[7] For the rocket nozzle, the compressible flow calculation is applied due to changing from a subsonic to a supersonic flow in the computational domain. Figure 1 shows the computational domains of the cascade and the rocket nozzle.

RESULTS FOR COMPRESSOR CASCADE

Particle Trajectory

Figure 2 shows streamlines and particle trajectories for several particle sizes, when the flow inlet angle is the same value with the designed blade inlet angle of 40°. An air inlet velocity and an initial particle velocity are fixed at 85 m/s and so the Stokes number represents the particle size with the fixed particle density and air viscosity. In the case of low Stokes number, Stk = 0.188, trajectories of floating particles follow to the streamlines as shown in Fig. 2(b). As the Stokes number increases, the particles have relatively large inertia forces and momentums and so the trajectories swerve from the streamlines. Therefore, the particles impact on the pressure side surface of blade including the leading edge. It is considered that the large particle is less influenced by the drag force but is controlled mainly by its inertia force. If the Stokes number is large enough, there is much more erosion on the pressure side of blade by the large number of impacted particles and the large inertia force.

Figure 3 presents particle trajectories for the different flow inlet angles for the particle size of 10 μm. In the case of flow inlet angle of 35°, particles collide slightly less (about 10%) on the pressure side surface than those in the case of 40° as shown in Fig. 3(a), (b). Figures 3(c) and 3(d) show particle trajectories for the flow inlet angles of 46° and 50°, respectively. If the flow inlet angle is larger than the designed blade inlet angle, there is a small separation bubble on the suction side at the leading edge region and a large separation flow at the trailing edge for the flow inlet angle of 50°. The flow pattern affects largely the particle trajectories resulting in strong impact of particles on the pressure side surface as shown in Fig. 3(d).

Therefore, it is concluded that the more particles collide on the pressure side surface for the larger Stokes number and flow inlet angle.

Particle Impact Efficiency

To calculate the particle impact possibility on the blade, the particle impact efficiency is defined as

$$\eta = \frac{number\ of\ particles\ colliding\ on\ the\ blade}{total\ number\ of\ particles\ floating\ in\ the\ flow} \quad (2)$$

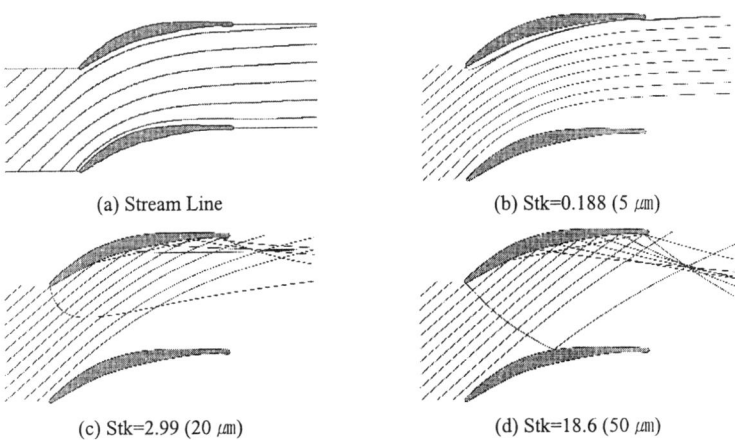

(a) Stream Line (b) Stk=0.188 (5 μm)

(c) Stk=2.99 (20 μm) (d) Stk=18.6 (50 μm)

Fig. 2 Particle trajectories for compressor cascade at flow inlet angle of 40°.

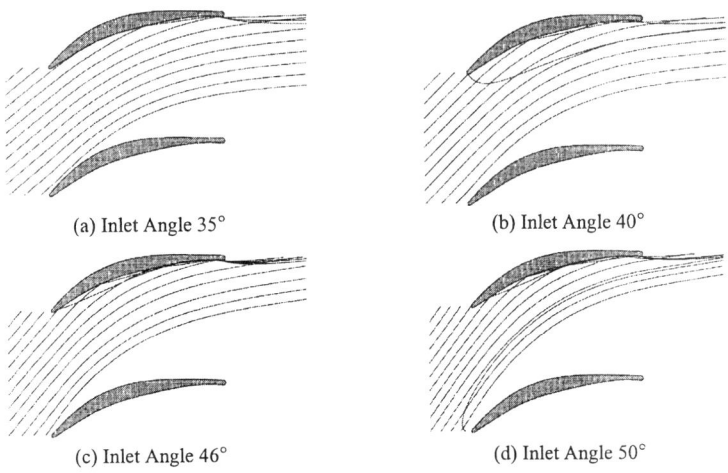

(a) Inlet Angle 35°   (b) Inlet Angle 40°

(c) Inlet Angle 46°   (d) Inlet Angle 50°

Fig. 3 Particle trajectories for various inlet angle at Stk=0.753 (d_P=10 μm).

In Eq. (2), the denominator presents the total number of particles entering the calculational domain and the numerator presents the number of particles colliding on the pressure side surface of blade. Figure 4 presents the particle impact efficiency for several flow inlet angles. As the particle size (Stokes number) increases, the particle impact efficiency increases rapidly, especially for the large flow inlet angle. In the ideal case, the particle impact efficiency has a form of step function and there is a critical value of a particle size determining whether the particle impact occur. However, actually the particle impact efficiency cannot follow the step function as shown in Fig. 4. Therefore, the 50% cut-point is defined and that is represented by the Stokes number at η =0.5. When the flow inlet angle is larger than the designed blade inlet angle, the particle impact efficiency increases rapidly at small Stokes numbers. The 50% cut-points of $Stk^{0.5}$ are 1.9, 1.38 and 1.2 for flow inlet angles of 40°, 46° and 50°, respectively. It means that most floating particles impact to the pressure side surface of blade for $Stk^{0.5} \geq 2$ if the flow inlet angle is larger than the designed blade inlet angle. In the case of 35° flow inlet angle, the 50% cut-point is 3.1 and the particle impact efficiency rises slowly after that point. These results will guide the particle impact possibility on the blade for various inlet angle and particle sizes.

Prediction of Erosion Rate

Figure 5 shows the erosion rates of two different materials, such as soft metal and ceramic for the blade surface. It is note that the erosion rate presented in the figure is not a real value but a relative value for comparison. The results show that the erosion rates are strongly dependent on the surface material, particle impact angles and particle sizes. Figure 5(a) presents erosion rates along the pressure side surface of blade for two different materials. The erosion rates are very high at the leading edge due to impact angle and large momentum change of the impact particles. The ceramic material has approximately 4 times higher erosion rates at the leading edge because

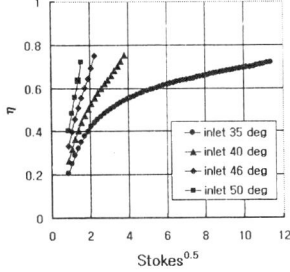

Fig. 4 Particle impact efficiency.

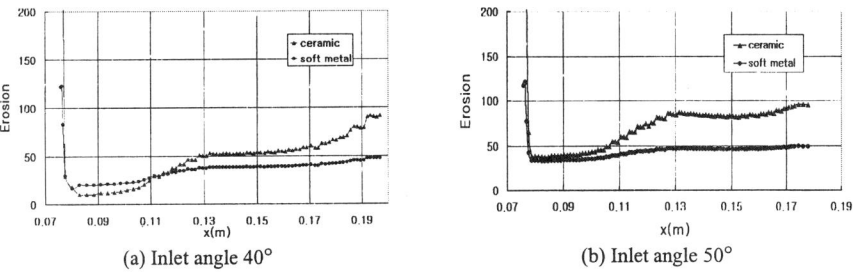

(a) Inlet angle 40° (b) Inlet angle 50°

Fig. 5 Erosion rates on the pressure side surface for two different materials at Stk=18.6.

brittle materials generally have a high erosion rate with large impact angles (about 90°). The erosion rates of ceramic are lower around the front region of blade and increase slightly around middle region. However, the erosion rates of soft metal change little along the whole blade except the leading edge region. Therefore, these results can guide to choose the blade surface materials or coating region. For the flow inlet angle of 50°, the erosion rates are changed due to the large number of impact particles and the increased impact angles.

RESULTS FOR ROCKET NOZZLE

Particle Trajectory

Figure 6 presents the particle trajectories in the rocket nozzle for four different particle sizes. For the case of Stk = 0.00131 (the smallest tested particle size), the particles tend to follow streamlines very well because of their small inertia. For the case of Stk = 0.0188, as shown in Fig. 6(b), the particle free zone in the nozzle divergent region becomes wider than that of Stk = 0.00131 because the particle starts to be separated from streamlines by its inertia force. Figure 6(c) presents the particle trajectories for the case of Stk = 0.115 with colliding to the nozzle wall in the convergent section. Although many floating particles in the nozzle don't impact on the nozzle wall, the particle free zone becomes wider in the divergent region due to increasing

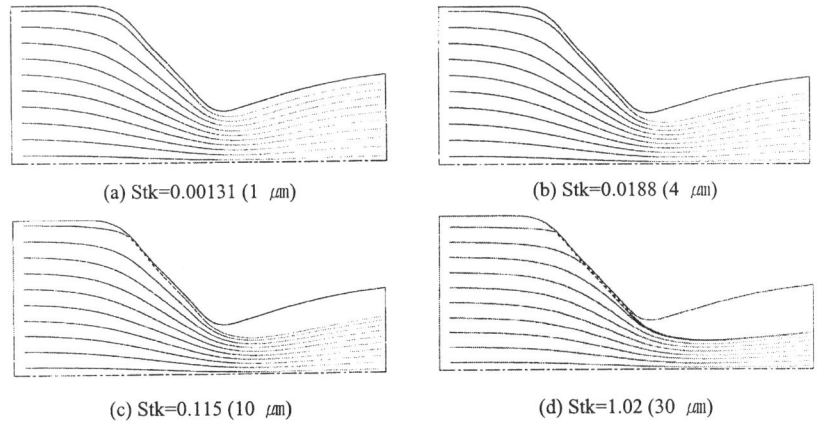

(a) Stk=0.00131 (1 μm)   (b) Stk=0.0188 (4 μm)

(c) Stk=0.115 (10 μm)   (d) Stk=1.02 (30 μm)

Fig. 6 Particle trajectories in the rocket nozzle.

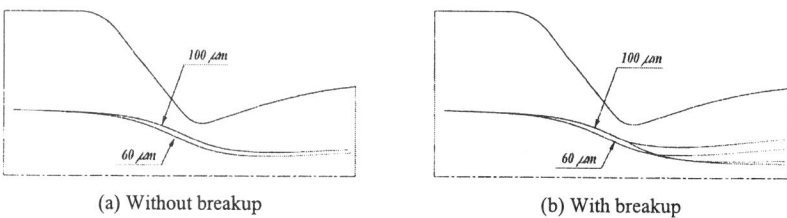

(a) Without breakup (b) With breakup

Fig. 7 Droplet break-up trajectories for Stk=4.08 (d_P=60 μm) and Stk=11.3 (d_P=100 μm).

particle inertia. As the particle size increases more, the number of colliding particle increases rapidly and the particle free zone is wider as shown in Fig. 6(d). The reason is that the particle inertia force becomes a major factor determining the particle motion in the gas flow. The large particles concentrate around the nozzle centerline rather than stay around the nozzle wall. It agrees well with the results of Hwang's analysis for the two phase flow.[8]

Particle Breakup

Oxidized aluminum (Al_2O_3) particles in the hot gas exist as a liquid phase and the droplets break up in the high velocity nozzle flow with a large velocity lag between particles and flow. The criterion of droplet break-up is that the Weber number is larger than 16;

$$We = \frac{\rho_P[(u-u_p)^2 + (v-v_p)^2]d_P}{\sigma} \geq 16 \qquad (3)$$

where σ is a surface tension of Al_2O_3 droplet. The particle starts to break up at the particle diameter of 50 μm in this study.

Figure 7 presents particle trajectories in the nozzle whether the particles break up or not. The particle break-up occurs near the nozzle throat because the flow velocity increases rapidly resulting in a large velocity lag with the particle as shown in Fig. 7(b). The good expectation of particle free zone is important to predict the erosion rates of jet vane installed in the nozzle wall of divergent section. The particle break-up changes largely the particle free zone in the divergent section.

Heat Transfer by Particle Deposition

The deposition of high temperature particle transfers a large heat flux to the nozzle wall resulting in thermal erosion of the wall. The present study analyzes the heat transfer rate using the results from T2-VIZA code[9] which predicts a wall temperature profile.

It is known that the particle temperature in the convergent section is little different from the gas temperature.[10] Therefore, it is assumed that the particle temperature in the convergent section is the same as the gas temperature of 3550K.

Figure 8 presents the heat transfer rates from the collided particle to the nozzle wall for four different particle sizes. In a nozzle entrance region, the wall temperature is relatively high due to the thermal equilibrium between the combustion gas and the nozzle surface, so the heat transfer rates by the particle deposition are low.

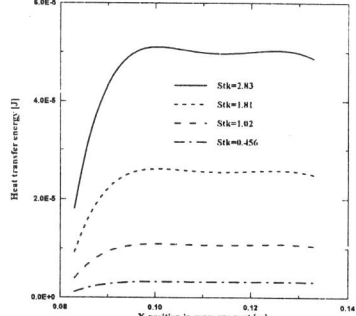

Fig. 8 Heat transfer to the converge part.

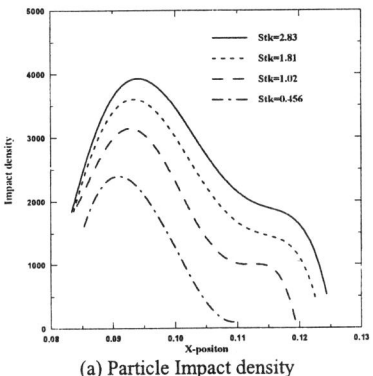
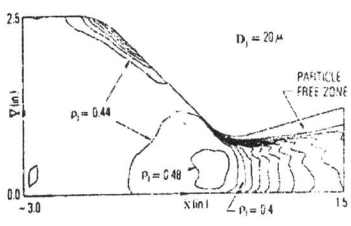

(a) Particle Impact density   (b) Particle distribution density by Chang[11]

Fig. 9 Particle impact probability on the nozzle convergent section.

The high heat transfer by particle deposition occurs on the large portion of nozzle convergent section up to the nozzle throat, so the thermal load on this region is high.

Particle Impact Probability

The particle impact density, which is defined by relative probability of particle deposition on the convergent nozzle surface, is calculated from the particle trajectory analysis. Figure 9(a) presents the particle impact density on the convergent section for four different particle sizes. The impact density has a maximum value at $x = 0.095$ that is the region just after starting the convergent part of nozzle. The impact density decreases rapidly until the nozzle throat because the floating particles tend to be accelerated and concentrated into the nozzle center near the nozzle throat region. As the particle size increases, the particle impact density increases over the entire convergent region due to the increased particle inertia force.

Figure 9(b) presents the particle distribution density in a 45°-15° nozzle by Chang.[11] The particle distribution density is very high near the nozzle wall at the beginning of convergent section and this result is very similar to the present study (Fig. 9(a)).

Heat Transfer Probability

The heat transfer density on the nozzle convergent wall is defined by the heat transfer rate of one particle multiplied by the particle impact density. The heat transfer density on the wall has a maximum value near $x = 0.095$ with the highest particle impact density as shown in Fig. 10. The pattern of heat transfer density is similar to the particle impact density as shown in Fig. 9(a). The region with high heat transfer density is considered to be damaged with high thermal erosion.

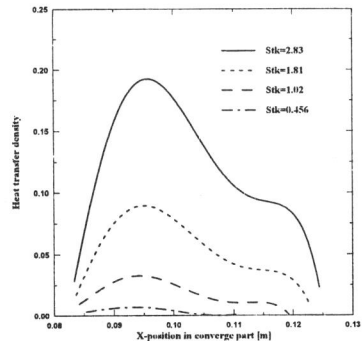

Fig. 10 Heat transfer density on the nozzle convergent wall.

CONCLUSIONS

For the compressor cascade, the particle impact rates relate closely with the flow inlet angle and the particle size. As the flow inlet angle increases from the designed blade angle, the particle impact rates increase proportionally. The particle erosion rate is influenced mainly by the blade material, particle impact angle and particle size. The erosion rate of a ceramic material is much higher than that of a metal at the leading edge region, so it is not a desirable way to coat the blade leading edge with the ceramic material.

For the solid rocket nozzle, the particle free zone in the nozzle divergent section increases rapidly with increasing particle size. The Al_2O_3 droplet with high shear force breaks up around the nozzle throat resulting in the change of particle free zone. The heat transfer density has a maximum value at the beginning region of nozzle convergent section, so it is needed to use the proper heat-resistant materials on this region.

ACKNOWLEDGEMENT

This research was supported by Agency for Defense Development under the grant No. TECD-401-991141 and by Ministry of Science & Technology through their National Research Laboratory program.

REFERENCES

1. Hamed, A., Effect of Particle Characteristic on Trajectories and Blade Impact Patterns, J. of Fluid Engineering, Vol. 110, pp. 33-36, 1988.
2. Dunn, M. G., Baran, A. J. and Miatech, J., Operation of Gas Turbine Engines in Volcanic Ash Clouds, Trans. ASME, J. of Engineering for Gas Turbines and Power, Vol. 118, pp. 724-731, 1996.
3. Metwally, M., Tabakoff, W. and Hamed, A., Blade Erosion in Automotive Gas Turbine Engine, Trans. ASME, J. of Engineering for Gas Turbines and Power, Vol. 117, pp. 213-219, 1995.
4. Bae, J. C., Jet Vane Thrust Vector Control System, Seminar Material, KAIST, Daeduk, Korea, 1998.
5. S Michael James, 1987, Numerical Prediction of Fluid and Particle Motions in Flow Past Tubes, University of California Berkeley, Ph. D thesis.
6. Friedlander, S. K., 1977, Smoke, Dust and Haze, Wiley-Interscience, New York.
7. Karki, K. C., Patankar, S. V., 1988, Calculation Procedure for Viscous Incompressible Flows in Complex Geometries, Numerical Heat Transfer, Vol. 14, pp. 295~307.
8. Hwang, C. J., Numerical Study of Gas-Particle Flow in a Solid Rocket Nozzle, AIAA Journal, Vol. 26, No. 6, pp. 682-689, 1988.
9. Bae, J. C., Boundary Layer and Heat Transfer Parameters Calculation - Two Dimensional Heat Conduction Calculation, ADD Report, 1998.
10. Debendra K Das, Moore, G. R. and Boyer, C. T., A Numerical Study of Two-Phase Flow and Heat Transfer in a Nozzle, AIChE Symposium Series, Heat Transfer, Vol. 83, No. 257, pp. 243-248, 1987.
11. Chang, I.S., 1980, One and Two-Phase Nozzle Flows, AIAA Journal, Vol. 18, pp. 1455~1461.

Convective Heat Transfer on An Inlet Guide Vane

M.-L. HOLMER[1], L.-E. ERIKSSON[2] and B. SUNDEN[3]

[1]*Thermo and Fluid Dynamics, Chalmers University of Technology, 41296 Göteborg, Sweden,* [2]*Volvo Aero Corporation, 46181 Trollhättan, Sweden,* [3]*Division of Heat Transfer, Lund Institute of Technology, Box 118, 22100 Lund, Sweden*

ABSTRACT: The flow and temperature fields around an inlet guide vane are determined numerically by a CFD method. Outer surface temperatures, heat transfer coefficient distributions, and static pressure distributions are presented. Three different thermal boundary conditions on the vane are analysed. The computed results are compared with experimental data. The governing equations are solved by a finite-volume method with the low Reynolds number version of the k-ω turbulence model by Wilcox implemented. It is found that the calculated results agree best with measurements if a conjugate heat transfer approach is applied and thus this wall condition is recommended for future investigations of film cooling of guide vanes and turbine blades.

INTRODUCTION

Published numerical calculations of convective heat transfer on the outer surfaces of guide vanes and turbine blades are carried out for either a uniform wall heat flux or a uniform wall temperature boundary condition. However, this may lead to significant errors in the calculated heat loads because the internal cooling of the vane or blade is not considered. The present work aims to investigate the importance of the thermal wall boundary condition and numerical calculations are performed for a) a uniform wall temperature on the outer vane surface, b) a non-uniform wall temperature distribution based on experiments, and c) a conjugate heat transfer wall condition. The conjugate condition means that the heat transfer coefficient and bulk temperature distributions on the inner cooling ducts of the vane are prescribed. The heat conduction across the vane material is also taken into account. The vane outer surface temperature is then found as part of the numerical solution. The work also aims to establish a numerical calculation procedure which can be used in engineering applications with sufficient accuracy.

PROBLEM STATEMENT

The geometry considered is an inlet guide vane in the first turbine stage of a gas turbine. In this particular case, the turbine guide vane consists of 50 blades, i.e., 7.2 degrees of blade partition. The inner radius of the vane, the hub, is at a radius of about 610 mm and the outer radius of the vane, the tip, is at a radius of about 695 mm. The thickness to chord ratio for the vane geometry is, $t/c \approx 0.17$ and the pitch to chord ratio is, $s/c \approx 0.73$. The geometry is identical to that in an experimental investigation by AAP (ABB Alstom Power) in Finspong, Sweden, see Rådeklint et al.[1].

The governing equations of the compressible flow around the vane are solved numerically subject to the three different boundary conditions for the wall surface mentioned in the preceding section. The surface temperature, pressure and convective heat transfer coefficient distributions are of particular interest.

GOVERNING FLOW EQUATIONS

The following equations must be solved for a compressible flow in general.
The continuity equation describes conservation of mass:

$$\frac{\partial \rho}{\partial t} + \frac{\partial}{\partial x_j}[\rho u_j] = 0 \tag{1}$$

The momentum equation:

$$\frac{\partial}{\partial t}(\rho u_i) + \frac{\partial}{\partial x_j}[\rho u_i u_j + p\delta_{ij} - \tau_{ji}] = 0 \tag{2}$$

The energy equation:

$$\frac{\partial}{\partial t}(\rho e_0) + \frac{\partial}{\partial x_j}[\rho u_j e_0 + u_j p + q_j - u_i \tau_{ij}] = 0 \tag{3}$$

The total internal energy, e_0, is defined by:

$$e_0 = c_v T + \frac{u_i u_i}{2} \tag{4}$$

The stress tensor, τ_{ij}, for compressible flow reads

$$\tau_{ij} = \mu\left(\frac{\partial u_i}{\partial x_j} + \frac{\partial u_j}{\partial x_i}\right) + \zeta \frac{\partial u_k}{\partial x_k}\delta_{ij} \tag{5}$$

where δ_{ij} is the Kronecker's delta and ζ is the second viscosity, which is calculated as $\zeta = -2/3\mu$.

In the energy equation the heat flux vector, q_j, is given by Fourier's law:

$$q_j = -k\frac{\partial T}{\partial x_j} \tag{6}$$

The conjugate heat transfer boundary condition:

Figure 1 shows a sketch of the heat transfer from the hot gas side to the interior of the vane. The following variables are assumed known: T_{1b} the bulk temperature of the air inside the vane, T_{2cell} the air temperature at the centre of the first cell outside the vane surface. t_{wall} is the wall thickness of the vane, h_1 the heat transfer coefficient on the inside of the vane, k_{wall} the wall thermal conductivity, k_{air} the thermal conductivity of air, D the height of the first cell outside the vane surface.

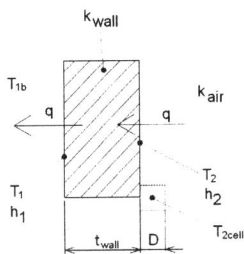

Figure 1. Heat flux through the vane wall.

An energy balance for the model in Fig. 1 gives:

$$q = \frac{T_{2cell} - T_{1b}}{\frac{D}{2k_{air}} + \frac{1}{h_1} + \frac{t_{wall}}{k_{wall}}} \quad (7)$$

NUMERICAL SOLUTION PROCEDURE

The governing equations are solved by a 3D finite-volume method developed by Eriksson[2] at Volvo Aero Corporation. It is an explicit time marching 3D method for compressible flow based on Navier-Stokes equations. For discretization of the equations the cell-centered finite-volume method is used. A TVD (Total Variation Diminishing) scheme is implemented to achieve oscillation-free solutions. The software is called G3DFLOW. The low Reynolds number version of the k-ω turbulence model by Wilcox[3] is implemented to enable calculations of turbulent flow cases. A realizability constraint is applied to reduce the generation of unphysical turbulent kinetic energy, particularly close to the leading edge.

The solution domain is a multi-block-structured grid organized in five subdivisions or blocks. Each block is defined as a structured grid. The block interfaces are matched. The mesh consisted of almost 250 000 nodes. Figure 2 shows the generated grid of the flow passage at a section half the distance from the hub to the shroud.

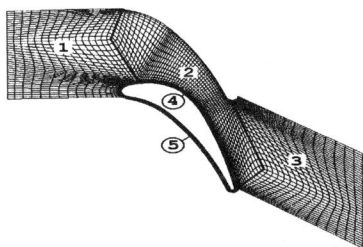

Figure 2. Generated grid of the flow passage, including sub-grid areas.

The areas close to the vane, subgrids 4 and 5, consist of 62 cells each in the flow direction and 28 in the normal direction. The grid spacing is non-uniform in the normal direction. The y+ value for the cell closest to the vane surface is about 0.8 (+/- 0.4).

As a computational stability criterion, the Courant-Friedrich-Lewy criterion, the so-called CFL criterion, has been used. The CFL number was set to 0.7 to ensure stable converged solutions.

As a convergence criterion for flow calculations, the mass flow difference between the inlet and the outlet surfaces is used. The convergence for the heat transfer coefficients was controlled by considering different solutions until no significant difference in the latest results and the previous ones (e.g., 5000 iterations ahead) could be found.

Boundary conditions:

At the inlet boundary, the absorbing boundary conditions for subsonic inflow are used. At the outlet boundary, the mixed boundary conditions for subsonic outflow are used. The static pressure at the outlet surface is specified in four subsections varying in the radial directions.

Either of the three boundary conditions, as described in section Introduction, is applied at the vane surfaces. For the hub and shroud surfaces uniform wall temperature was set.

No-slip boundary conditions were used at all wall surfaces. Periodic boundary conditions were assumed and therefore only one vane had to be modelled.

Heat transfer coefficient distribution on the inner surface of the vane:

The heat transfer coefficient distribution at the inside of the vane, thermal conductivity for the vane material and the wall thickness distribution are required as input values. Figure 3 shows a section of the vane. The heat transfer coefficient distribution on the inside of the vane was supplied by AAP (ABB Alstom Power), see Fig. 4 and Petrunin[4].

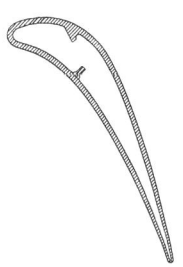

Figure 3. Section of the vane.

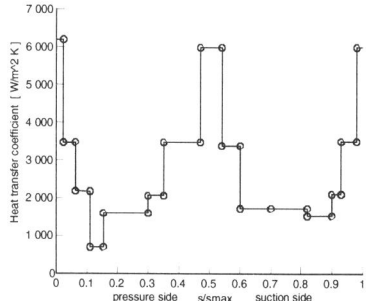

Figure 4. Heat transfer coefficient distribution applied to the inner surface of the vane.

Physical data at the inlet boundary:

The total temperature, the total pressure and the speed at the inlet boundary are set to $T_0=1026$ K, $P_0=1.026\times10^6$ Pa and $U_{ax}=80$ m/s. The turbulence intensity is set to 0.05 (dimensionless). All these data are based on measurements by Rådeklint et al.[1]. For the turbulence model, the turbulent kinetic energy and the specific frequency are set to k=24 m^2/s^2 and ω=5100 s^{-1}, respectively. These values are in accordance with the turbulence intensity and the estimated length scale at the inlet boundary.

RESULTS AND DISCUSSION

Figure 5 shows the Mach number and the static pressure distributions through the vane passage. The flow at the inlet boundary starts at a low subsonic Mach number, Ma ≈ 0.1, increases through the whole passage and reaches its maximum far downstream on the suction side with the Mach number at the sonic level, i.e., Ma ≈ 1.0.

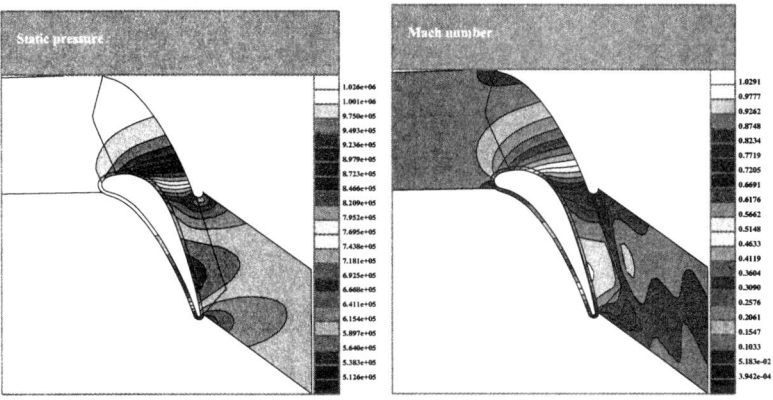

Figure 5. Mach number (left) and static pressure (right) distributions through the vane passage.

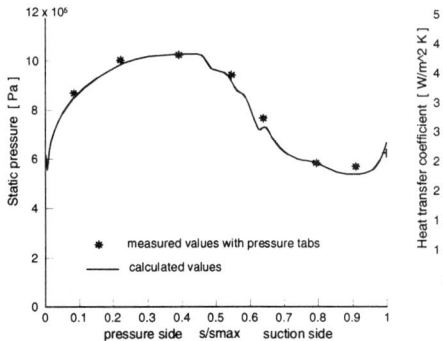

Figure 6. Calculated pressure distribution compared to measurements.

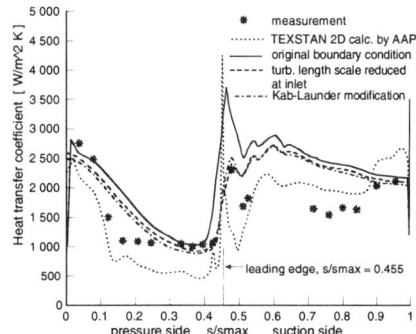

Figure 7. Heat transfer coefficients with different modifications of the turbulent kinetic energy production.

Pressure distribution along the outer wall surface of the inlet guide vane:

The calculated vane surface pressure distribution agrees well with the experimental results particularly on the pressure side, see Fig. 6. Both experimental and calculation results are presented at a section cut at 25 per cent of the height from hub to shroud. On the suction side the acceleration in speed is smooth almost all the way downstream to the trailing edge. However, from the leading edge, s/smax=0.455, to s/smax=0.65 there are some discrepancies between measurements and calculations. The differences appear as sudden jumps in the calculated pressure distribution. The reason for these sudden jumps are small deviations in the description of the vane surface geometry.

Heat transfer for a uniform wall temperature:

Calculated heat transfer coefficients for a uniform wall temperature over the vane surface are presented in Fig. 7. Both experimental and calculated results are presented at a section cut at 50 per cent height from hub to shroud. On the pressure side the calculated and the experimental results agree well, although the transition back to turbulence occurs earlier for the calculated results than for the experimental results. For the suction side the calculated values are much higher than the experimental ones, which has been found in the literature by others as well, see, e.g., Garg and Rigby[5], Garg and Ameri[6], and Dahlander[7]. The sudden jumps due to imperfections in the blade surface description can also be found here for 0.455 < s/smax < 0.65.

In the area close to the leading edge, large differences in the heat transfer coefficients are found between calculations and measurements.

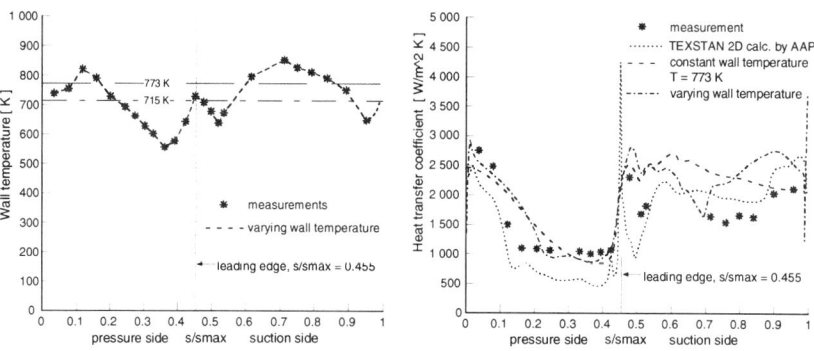

Figure 8. Wall temperature distribution around the vane.

Figure 9. Calculated heat transfer coeffcents with varying wall temperature.

Reducing the turbulent length scale close to the leading edge:

To improve the calculation results, at the leading edge, reduction of the turbulent kinetic energy production can be introduced. Kato and Launder[8] suggested a model for this reduction but in the present work also a simpler method, namely reduction of the turbulent length scale, was applied. In principle both methods provided similar results, i.e., bringing the leading edge heat transfer coefficient much closer to the experimental value, see Fig. 7.

Heat transfer coefficients for a non-uniform wall temperature distribution:
The wall temperature along the vane varies considerably. The measurements by Rådeklint et al.[2] indicate that the biggest wall temperature difference in the main flow direction for the midsection of the vane is almost 300 K, see Fig. 8.

Applying a non-uniform wall temperature distribution, based on measurements (see Fig. 8), gives reasonable results for the pressure side, see Fig. 9. For the suction side the trends from measurements are better captured than with a uniform wall temperature.

The TEXSTAN calculations presented are run with non-uniform wall temperatures. Compared to TEXSTAN 2D calculations, the G3DFLOW results are much closer to the measurements on the pressure side. An exception is the location of the transition point. In TEXSTAN this has to be specified. For the suction side it is obvious that a non-uniform wall temperature distribution should be used.

Calculation of vane outer surface temperatures by a conjugate heat transfer approach:
Figures 10 and 11 show the influence of the non-constant value of the inside heat transfer coefficent and the wall thickness.

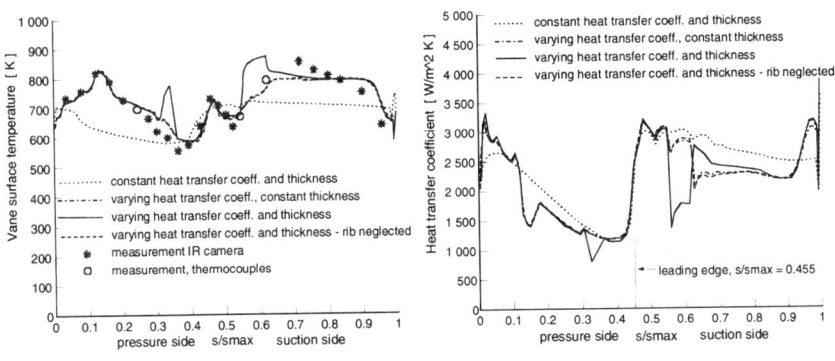

Figure 10. Calculated vane outer wall temperature distribution with conjugate heat transfer boundary condition.

Figure 11. Calculated heat transfer coefficients with conjugate heat transfer boundary condition.

From Figs. 10 and 11 it is obvious that applying a varying heat transfer coefficient distribution on the inside of the vane gives better results compared to measurements than applying a uniform distribution.

Applying a varying wall thickness distribution in the calculations has no significant influence on the results except in the areas close to the stiffeners (or ribs). The areas influenced are, at the pressure side for $s/s_{max} \approx 0.30$-0.35 and at the suction side mainly for $s/s_{max} \approx 0.55$-0.62. On the suction side the stiffener has some influence on the results almost all the way to the trailing edge.

CONCLUSIONS

The results presented show that the calculated pressure distribution along the inlet guide vane surface agrees well with the experimental results.

The common method to use a constant wall temperature as boundary condition is not very accurate. The variation in wall temperature distribution around the vane cannot be neglected. The non-uniform wall temperature distribution used as a wall boundary condition for the blade is in preference to a uniform wall temperature boundary condition.

It is shown that the calculated results agree best with measurement if the conjugate heat transfer approach is applied and thus this wall condition is recommended for related future investigations of film cooling of guide vanes and turbine blades.

For the suction side some kind of relaminarization is found in all the calculated results except when a uniform wall temperature is applied. The indication of relaminarization can also be found in the experimental results. This emphasizes the importance of not using a uniform wall temperature as a wall boundary condition in the calculations.

ACKNOWLEDGEMENT

This work was carried out as part of a joint project on Turbine Cooling Performance (TCP) within the Swedish Gas Turbine Centre (GTC). This centre is sponsored by the Swedish Energy Authority (STEM), Volvo Aero Corporation and ABB Alstom Power.

The authors would like to thank the GTC for financial support.

REFERENCES

1. Rådeklint, U.R., Hjalmarsson, C., Rubensdörffer F. and Annerfeldt, M. 1999. Experimental investigation of the external heat transfer on a nozzle guide vane. *In* IMechE Conf. Trans. 1999. C557/119. Vol. **B**: 937-949. Institution of Mechanical Engineers, UK.
2. Eriksson, L.E. 1995. Development and validation of highly modular flow solver versions in G2DFLOW and G3DFLOW series for compressible viscous reacting flow. Volvo Aero Corporation Technical Report 9970-1162.
3. Wilcox, D.C. 1998. Turbulence Modeling for CFD. 2nd edn. DCW Industries Inc.
4. Petrunin, D. 1997. GT10C cooling development. Q3D cooling code development. Technical Report. ABB Uniturbo, Moscow, Russia.
5. Garg, V.K. and Ameri, A. 1997. Comparison of two-equation turbulence models for prediction of heat transfer on film-cooled turbine blades. Num. Heat Transfer. Part A. Vol. **31**: 347-371.
6. Garg, V.K. and Rigby, D.L. 1998. Heat transfer on a film-cooled blade - effective of hole physics. ASME 98-GT-404.
7. Dahlander, P. 1998. An engineering tool for film cooling simulations. Thesis for the degree licentiate of engineering 98/10. Department of Thermo and Fluid Dynamics, Chalmers University of Technology, Göteborg, Sweden.
8. Kato, M. and Launder, B.E. 1993. The modelling of turbulent flow around stationary and vibrating square cylinders. *In* Ninth Symposium on Turbulent Shear Flow. Vol. **1**: 10-4-1 – 10-4-6. Kyoto.

Effect of Reynolds Number, Turbulence Level and Periodic Wake Flow on Heat Transfer on Low Pressure Turbine Blades.

DMITRY SUSLOV, ACHMED SCHULZ, SIGMAR WITTIG

Lehrstuhl und Institut für Thermische Strömungsmaschinen
Universität Karlsruhe(TH), D-72168 Karlsruhe, Germany
e-mail: *Dmitri.Souslov@its.uni-karlsruhe.de*

ABSTRACT

The development of effective cooling methods is of major importance for the design of new gas turbines blades. The conception of optimal cooling schemes requires a detailed knowledge of the heat transfer processes on the blade´s surfaces. The thermal load of turbine blades is predominantly determined by convective heat transfer which is described by the local heat transfer coefficient.

Heat transfer is closely related to the boundary layer development along the blade surface and hence depends on various flow conditions and geometrical parameters. Particularly Reynolds number, pressures gradient and turbulence level have great impact on the boundary layer development and the according heat transfer. Therefore, in the present study, the influence of Reynolds number, turbulence intensity, and periodic unsteady inflow on the local heat transfer of a typical low pressure turbine airfoil is experimentally examined in a plane cascade.

NOMENCLATURE

c Chord
h Heat transfer coefficient
κ Acceleration parameter
L Integral length scale
n Distance normal to surface
St Stanton number
s Surface distance from leading edge
Tu Free stream turbulence level

T_C Cooling water temperature
T_S Blade's surface temperature
T_∞ Gas flow temperature
U Flow velocity

γ Intermittency
λ_S Thermal conductivity

INTRODUCTION

Increasing turbine inlet temperatures necessitate detailed investigations of heat transfer phenomena in convectively cooled low pressure turbines. Convective heat transfer is described by the local heat transfer coefficient and the associated boundary layer along the blade's surface. The boundary layer development depends on various flow parameters, predominantly the Reynolds number, pressure gradient, and turbulence level, which vary both locally and with time. A tremendous impact on heat transfer is attributed to periodical fluctuations caused by moving blade wakes. The wakes are transported by the flow downstream through consecutive rows of airfoils. Accurate computation of this complex unsteady flow is still difficult, although considerable progress was achieved in numerical modelling through recent years. Especially the treatment of near wall regions requires further efforts. Since detailed experimental data on local heat transfer coefficients of low pressure turbine airfoils is not enough available in the open literature, the present study aims on the generation of a database for developing new models and correlations describing the time averaged heat transfer under steady and fluctuating flow conditions.

EXPERIMENTAL TEST FACILITY AND MEASUREMENT SYSTEM

Experimental studies were conducted in a linear cascade under subsonic flow conditions in an open loop low pressure test facility of the Institut für Thermische Strömungsmachinen (ITS). Figure 1 shows a schematic of the test facility. The radial compressor delivers a maximum air flow of 3.0 kg/s at a pressure of 1.5 bar. Before entering the test section, the air is heated up to 393 K by an electrical heater.

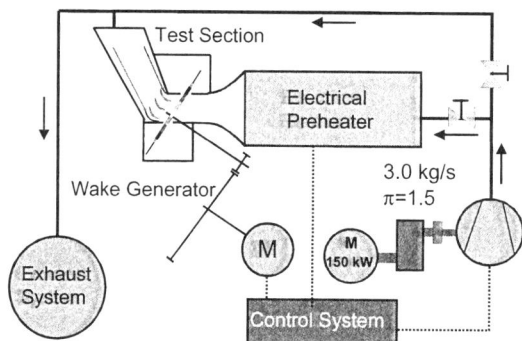

Fig. 1: Schematic of the test facility

Cascade inflow Mach numbers up to 0.4 and Reynolds numbers up to 700000 can be realized which is in accordance to the parameters of real low pressure turbines. The test section (see Figure 2) is installed between electrical heater and exhaust system. For simulating turbine like unsteady inflow conditions, a wake generator is implemented into the cascade entrance. As wake generator, a rotating disk with radially mounted cylindrical bars is utilized. The bars of the rotating disk are passing the cascade inflow through slots in the side walls of the inflow duct parallel to the entrance plane of cascade. The axis of disk is located outside of test section. The maximum rotational speed is 7000 rpm. In varying the number of bars along the circumference of the disk and the rotational speed, a wide range of Strouhal numbers can be realized. The diameter of the cylindrical bars is chosen to cause the same aerodynamic resistance as an airfoil. The superposition of elevated background turbulence and wake induced turbulence can be investigated by inserting turbulence grids into the cascade inflow duct. Level and integral length scale of background turbulence in the cascade inflow duct are summarized in Table 1.

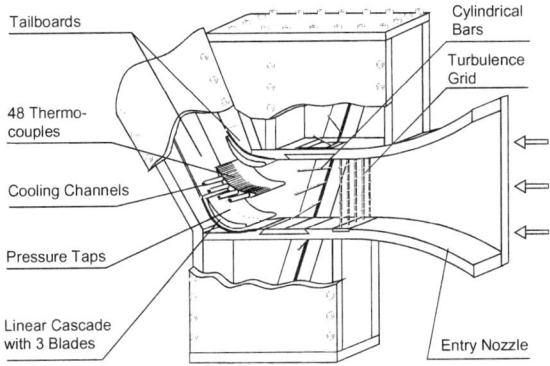

Fig. 2: Schematic of test rig

Table 1: Turbulence parameters in the cascade inflow duct

turbulence grid	Tu, %	L, mm
without turbulence grid	2	7
grid 1	2.7	2.5
grid 2	5	4
grid 3	8	5

The flow and turbulence parameters downstream of the wake generator were measured with a hot wire system in the cascade entrance plane.

The linear cascade consists of three blades. The pressure and velocity distributions of the blades show a typical aftloaded behaviour with an extended region of accelerated flow on the suction surface. Figure 3 shows the acceleration parameter k of the blade´s boundary

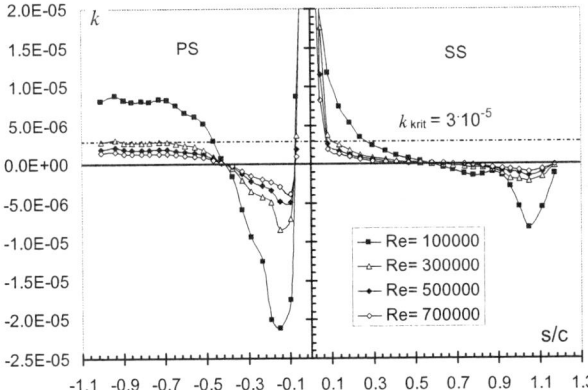

Fig. 3: Acceleration parameter distribution on blades surface

layer for different Reynolds numbers.

$$k = \frac{v}{U_\infty^2} \cdot \frac{dU_\infty}{dx}$$

The abscissa is dimensionless surface distance starting from the leading edge. Negative values correspond to the pressure surface, positive values to the suction surface.

To establish periodic flow conditions with three blades only, the cascade is bounded with flexible side walls and tailboards. The inner blade of the cascade is instrumented for heat transfer measurements. The method for determining time averaged heat transfer coefficients is based on surface temperature measurements and a heat flux calculation using finite elements method.

To introduce heat flux, the blade is convectively cooled by water from a temperatur-controlled circuit through 11 cylindrical cooling holes. Heat transfer coefficient in the cooling channels are determined by a correlation for turbulent pipe flow. Therefore, coolant mass flux and mean temperature of the coolant is measured for each individual cooling channel. The surface temperature, necessary as outer boundary condition for the finite element code, is measured by 48 thermocouples of 0.25mm diameter embedded into small grooves in the surface. After installation of the thermocouples the blade's surface is given a smooth finish. The method allows an aerodynamic smooth surface at minimal distortion of the temperature field due to the small dimensions of the thermocouples. The calculated temperature field in the blade cross section at midspan is then used to determine local heat transfer coefficients by a simple heat balance at the blade surface.

$$h = -\frac{\lambda_S (dT/dn)}{T_\infty - T_{Surface}}$$

dT/dn- calculated temperature gradient normal to surface

Figure 4 shows the numerical grid and a calculated temperature field in the blade cross section. The measuring method was first established by Schulz (1985) and proofed its excellent suitability in steady (Schiele 1995) and unsteady (Dullenkopf, Mayle 1990) investigations.

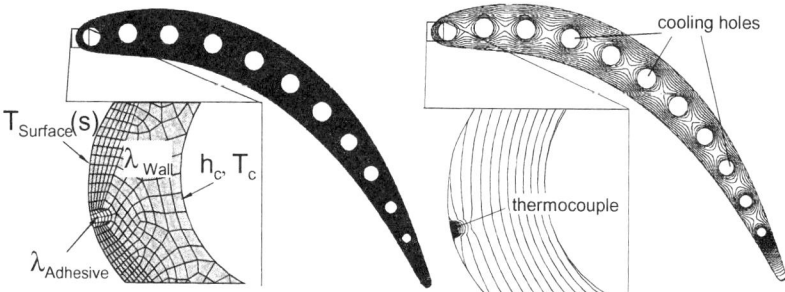

Fig 4: FE- calculation grid and temperature field in the test blade's cross section

RESULTS

Steady flow results
In the present study the effects of Reynolds number and turbulence level on the boundary layer development are investigated. In Figure 5, the local Stanton number and acceleration parameter on the blade's surface is plotted for different Reynolds numbers.

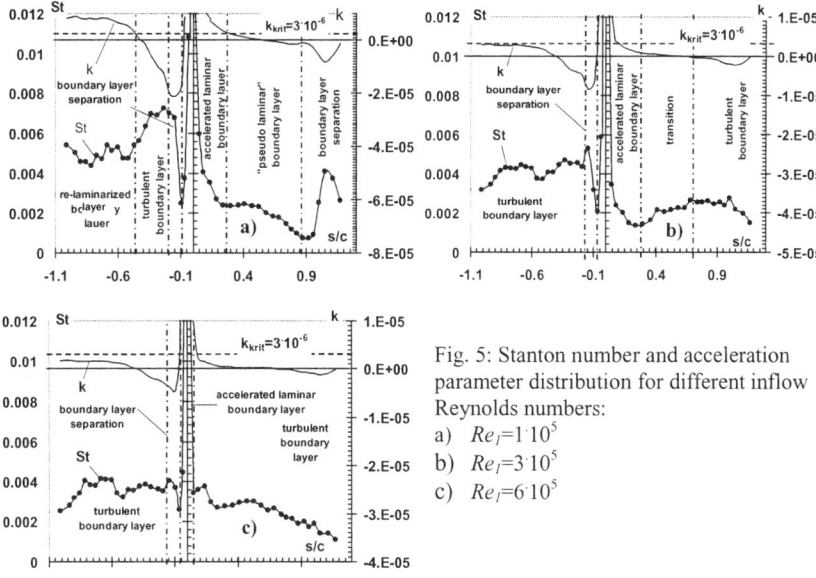

Fig. 5: Stanton number and acceleration parameter distribution for different inflow Reynolds numbers:
a) $Re_l = 1 \cdot 10^5$
b) $Re_l = 3 \cdot 10^5$
c) $Re_l = 6 \cdot 10^5$

Leading Edge Region. At the stagnation point on the leading edge a typical peak in heat transfer can be observed. Downstream of the leading edge the heat transfer coefficients strongly decrease on pressure and suction surface in accordance with the drop of the acceleration parameter. In this area the behaviour of the Stanton number can be described by the wellknown correlation of Hartree (1937) for accelerated flow with a power development of velocity: $U = a^m$ (see Fig. 7).

$$St(x) = 0.56 \frac{(\beta + 0.2)^{0.1}}{(2-\beta)^{0.5}} \cdot Re_x^{-0.5} Pr^{-0.667 + 0.067 \cdot \beta - 0.026 \cdot \beta^2} \quad \text{with} \quad \beta = \frac{2m}{m+1}.$$

m is determined by best fit of the local velocity distribution.

Pressure Surface. In the front region of the pressure surface in an area with high negative acceleration a local separation bubble occurs, which makes the downstream boundary layer turbulent and drastically increases heat transfer (see Fig. 5). For Reynolds numbers $Re_l < 3 \cdot 10^5$ the acceleration parameter exeeds the kritical value $k = 3 \cdot 10^{-6}$ for re-laminarization at the position $s/c = -0.47$. Downstream of this location a re-laminarized accelerated boundary layer exists. To demonstrate the re-laminarization of the boundary layer on the pressure surface, Figure 6 shows the heat transfer coefficient and acceleration parameter at $s/c = -0.92$ as a function of inflow Reynolds number. It can be clearly seen

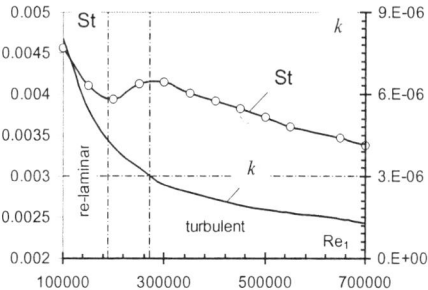

Fig 6: Dependency of Stanton number on inflow Reynolds number at $s/c = -0.92$ (pressure side)

that with increasing accelaration parameter, i.e. decreasing inflow Reynolds number, the Stanton number begins to drop, when the acceleration parameter exceeds the critical value.

Suction Surface. As indicated earlier, the front area of the suction surface shows an extended region of accelerated flow. The boundary layer's development in this area may be divided into follow sections according to the inflow Reynolds number Re_1.

a) $Re_1 < 2 \cdot 10^5$ (see Fig. 5a). The boundary layer stays laminar almost on the entire suction surface. At the location $s/c = 0.24$ the boundary layer changes to a "pseudo laminar" state (Dyban 1985). Near the trailing edge, at high deceleration, the boundary layer separates thus leading to a sudden increase of heat transfer. Increasing main flow turbulence level, leads to more filled velocity profiles with increased stability of the "pseudo laminar" boundary layer. Therefore, Stanton number at the reattachment point is reduced (Fig. 7).

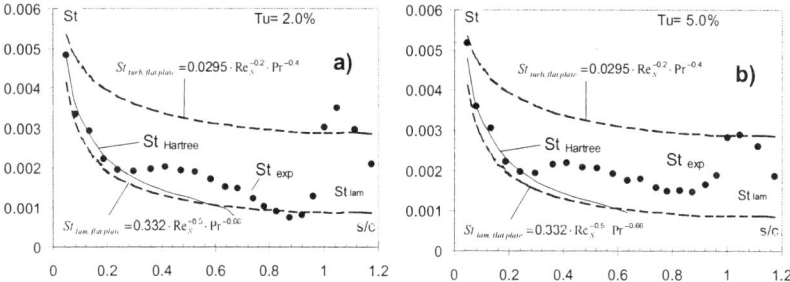

Fig. 7: Effect of turbulence level on heat transfer on the suction surface for $Re_1 = 1.5 \cdot 10^5$

b) $2 \cdot 10^5 < Re_1 < 3.5 \cdot 10^5$ (see Fig. 5b). The heat transfer on the suction surface is predominantly determined by the position and the development of the laminar-turbulent transition. Increasing Reynolds number and turbulence level lead to an upstream shift of the onset of transition and to a reduction in transition length. Under the present experimental conditions, the transition is in the bypass mode (Mayle 1991).

The major parameter of this kind of transition is the *intermittency* γ, i.e. the fraction of time that the boundary layer will be turbulent (Hodson 1992), $\gamma = 0$ corresponds to laminar and $\gamma = 1$ to fully turbulent boundary layer state. Figure 8 shows *intermittency* distribution for varying inflow Reynolds numbers at a turbulence level of $Tu = 2\%$. For Reynolds numbers $Re_1 < 3.0 \cdot 10^5$ the boundary layer separates near the trailing edge. The stability increase of the boundary layer due to transition is not sufficient to prevent separation. For Reynolds numbers $Re_1 > 3.0 \cdot 10^5$ the laminar-turbulent transition prevents flow from separation in an area with negative acceleration near the trailing edge.

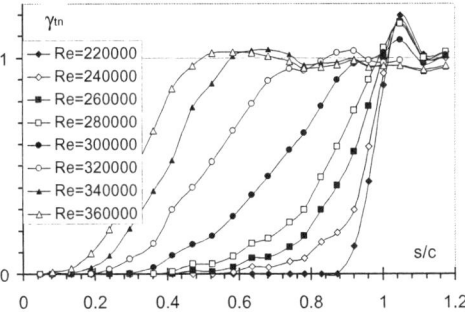

Fig. 8: Intermittency distribution on the suction surface for different inflow Reynolds numbers at a turbulence level of $Tu = 2\%$.

c) $3.5 \cdot 10^5 < Re_1$ (see Fig. 5c). The boundary layer is turbulent on almost the entire suction surface. Due to the large curvature of the airfoil the turbulent boundary layer is almost unaffected by the main flow turbulence level.

Unsteady flow results
Wakes are characterised by their high turbulence intensity, which in general exceedingly surpasses the background turbulence. Since the affecting time of the passing wake is

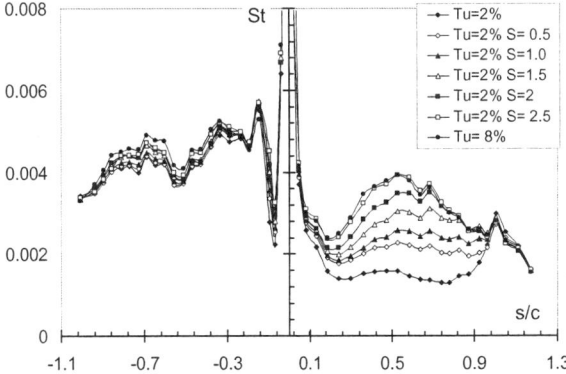

Fig. 9: Influence of Strouhal number on the heat transfer ($Re_1 = 250000$)

higher than the reaction time of the boundary layer, position and development of the laminar-turbulent transition is highly influenced by the wake flow. In this part of the study the effect of the Strouhal number on the local heat transfer on a low pressure turbine blade is investigated at a background turbulence of $Tu= 2\%$. Figure 9 shows distributions of heat transfer coefficients for different Strouhal numbers. On the suction surface, the strong impact of unsteady wake flow on the boundary layer state, and hence heat transfer, becomes obvious. At low Strouhal numbers the wake induced transition is completed by a natural transition. For reasons of comparison a heat transfer distribution at 8% background turbulence is shown as well. On the pressure side the effect of wake flow on the heat transfer is weak.

CONCLUSIONS

At low Reynolds numbers, increasing turbulence leads to the formation of a „pseudo laminar" boundary layer on surface areas with positive acceleration parameters. For acceleration parameter $k > 3 \cdot 10^{-6}$ was found the re-laminarization effect on the pressure surface near the trailing edge. Increasing Reynolds number and turbulence level leads to an upstream shift of the onset of transition and to a reduction in transition length on the suction surface. The laminar-turbulent transition increases the stability of the boundary layer and hence prevents flow separation in an area with negative acceleration near the trailing edge. Unsteady wake flow results in a similarly effect as high turbulence level and leads to an upstream shift of the onset of transition and to a reduction in transition length on the suction surface.

REFERENCES

1. Bradshaw, P. 1969. The analogy between streamline curvature and buoancy in turbulent shear flow. *In* Journal of Fluid Machanik, Vol. 36, pp. 177-191
2. Dullenkopf, K. Schulz A., Wittig, S. 1991.The Effect of Incident Wake Conditions on the Mean Heat Transfer an Airfoil. *In* ASME Journal of Turbomachinery, Vol. 113, pp. 412-418.
3. Dyban, E., Epick, E. 1985. Thermal/Mass transport and Hydrodynamics of Turbulent Flows, Kiev, (in Russian)
4. Hartree, D.R. 1937.On an equation occuring in Falkner and Skan's approximate treatment of the equationsof the boundary layer. *In* Proc. Camb. Phyl. Soc., Vol. 33, N 11, p. 223-231
5. Hodson, H.P., Addison, J.S., Shepherdson, C.A. 1992. Models for unsteady wake-induced transition in axial turbomachines. *In* Journal Physique, Vol. 2, pp. 545-574
6. Mayle, R.E., Dullenkopf K. 1990. A Theory for Wake-Induced Transition. *In* ASME Journal of Turbomachinery, Vol. 112, pp. 188-195.
7. Mayle, R.E. 1991.The Role of Laminar-Turbulent Transition in Gas Turbine Engines. *In* ASME Journal of Turbomachinery, Vol. 113, pp. 509- 532.
8. Schiele, R., Sieger, K., Schulz, A., Wittig S. 1995. Heat Transfer Investigations on a Highly Loaded, Aerothermally Designed Turbine Cascade. 12th International Symposium on Air Breathing Engines, pp 1091-1101, Melbourne, Australia.

SURFACE TEMPERATURE MAPPING OF GAS TURBINE BLADING BY MEANS OF HIGH RESOLUTION PYROMETRY

Stefan L.F. FRANK

Siemens Power Generation, Huttenstrasse 12, 10553 Berlin, Germany

Email: stefan.frank@blnh.siemens.de

Introduction

For more than a decade Siemens has been using optical pyrometry[1,2,3] for the testing of its gas turbine prototypes in the Berlin test bed[3]. The main objective is the experimental evaluation of both standard and prototype blade designs under real base load conditions. Furthermore, pyrometry is a valuable tool for quality assurance, since the temperature distribution of each individual blade is carefully determined. This paper describes the application of a newly developed high resolution pyrometer[4] on the latest prototype, the V84.3A 60Hz 180MW gas turbine[5]. A pragmatic approach to the different sources of error, such as flame radiation, limited resolution or exceeded incidence angle will be discussed in more detail.

System Setup

A schematic of the setup is shown in figure 1. Two traversing units are mounted on heavy duty gas turbine in order to position cylindrical probes (figure 2) into the hot gas path. High pressure water is used for probe cooling and nitrogen for purging of the optical components. The cooling power is in the magnitude of 50kW for fully inserted probes at base load.

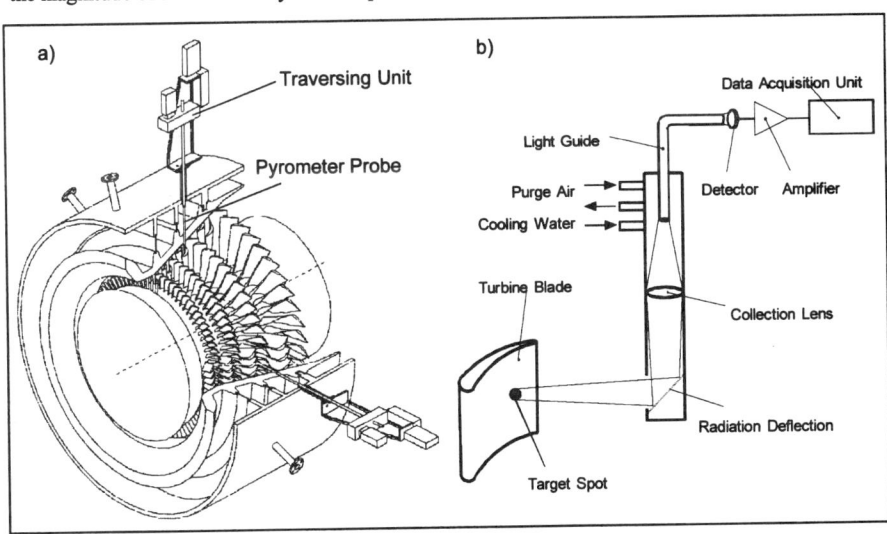

Figure 1: a) Pyrometer system mounted on a heavy duty gas turbine b) Measurement principle

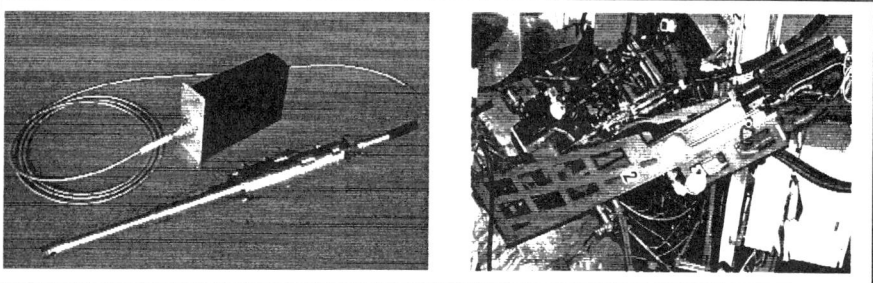

Figure 2: High resolution pyrometer probe with electronic receiver[4] and probe traverse unit with pyrometer probe mounted on a V84.3A gas turbine

In order to measure surface temperatures the heat radiation emitted from the desired target is led via a mirror (radiation deflection), a collection lens and a light guide, onto the detector, where the optical signal is converted into an electrical current. After amplification the signal is led to the data acquisition unit. Figure 3 shows a schematic of the system used: a host PC controls both the traversing unit for radial and rotational movement of the cylindrical probe as well as the "satellite" PC for data acquisition, positioned adjacent to the gas turbine. The temperature data is transferred via ethernet to the host PC in the remotely located control room.

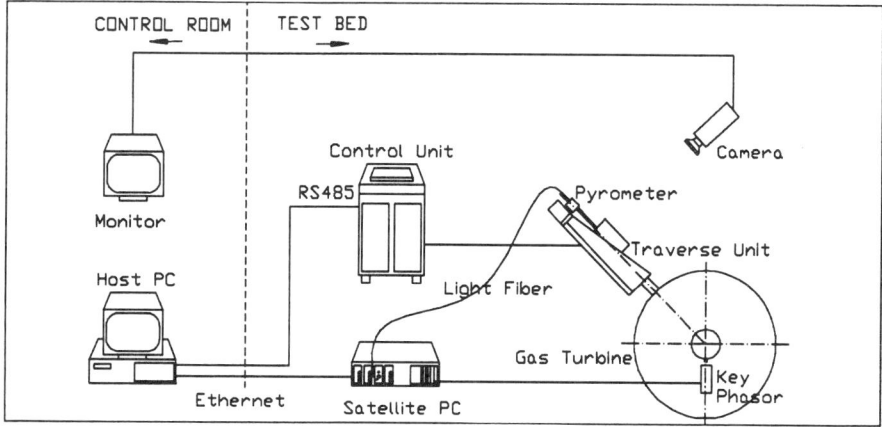

Figure 3: Schematic of the pyrometer system

The system is fully automated and up to four satellite PC and pyrometer probes can be operated at the same time. Alternatively, probes which are equipped with thermocouples may be used in order to determine the radial gas temperature profile. The single wavelength spectral band pyrometer ($\lambda = 0.85$ μm) has a minimum field of view of 1mm and a temporal resolution of 1 μs.

With regard to the uncertainty of measurement, traversable probes have one important advantage over non-intrusive methods: the spatial resolution is equally small for each data point of the blade regardless of traverse depth, whereas the effective field of view for non-intrusive scanning pyrometers[6,7] increases with insertion depth. Furthermore, the intrusive type pyrometer allows the viewing angle to be changed, hence, different zones of the blade may be scanned at an acceptable incidence angle and with a minimum field of view. Thus, in the test bed, surface temperature mapping is performed in three different ways:

- discrete blade measurement, i.e. the probe is traversed at a certain radius and viewing angle, before blade data is acquired, subsequently the radius and viewing angle may be changed
- continuous blade measurement, i.e. the probe is traversed slowly from the tip to the hub, e.g. with 2mm/revolution, and data is acquired continuously; subsequently the pyrometer is withdrawn and the process is repeated for another viewing angle
- continuous vane measurement, i.e. the probe is traversed quickly from the tip to the hub in order to not disturb the temperature distribution, e.g. with 1.2m/s, and data is acquired continuously during the feed action only; afterwards the probe is moved to its start position until another measurement with a new viewing angle begins.

The main disadvantage of the intrusive mode, however, is the potential risk for the gas turbine and the pyrometer itself: the latter is highly loaded by thermo-mechanical stress, whereas the former is jeopardized by vibration problems, due to the blockage of the flow channel between two vanes when the probe is fully inserted (see figure 4a). In order to overcome this problem continuous blade measurements are preferred. Hence, the necessary time for measurement and thus for vibration excitement of the turbine is minimized. In a period of not more than one or two seconds, the whole blade surface is mapped at maximum resolution, thus providing some five thousand different data points per blade and viewing angle.

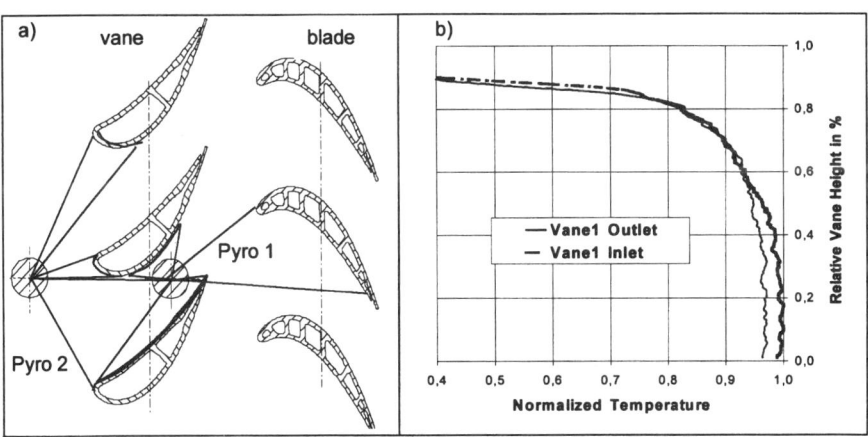

Figure 4: a) Schematic of the pyrometer probe position relative to the first stage blading
b) Radial temperature profiles at two different positions at base load using natural gas
Pyro1 = Vane1 outlet, Pyro2 = Vane1 inlet

Experimental Results

Optical pyrometry has been used for various measurements on the first stage vanes and blades under design conditions, i.e. at base load. Figure 4a shows a schematic of the pyrometer positions relative to the vanes and blades respectively. Please note that the position of pyrometer 2 has been turned into the drawing plane. Actually, the two positions are circumferentially separated by about 120°. Since stationary vane measurements are accomplished, it is essential to determine the actual hot gas temperature in the measuring positions. This is performed by thermocouple probes that are traversed slowly into the flow channel. In figure 4b the normalized temperature is plotted against the relative vane height for the two positions, exhibiting a very homogeneous temperature distribution for natural gas at base load.

The V84.3A gas turbine is equipped with an annular ring combustor that is suitable for different fuels such as natural gas and fuel oil. Flame radiation of the combustor is directly emitted onto the blading; it is one of the most challenging items for radiative measurements in the first stage. The maximum influence of the flame radiation in the near infrared may be estimated by Planck's law with Wien's simplification[8] as a function of temperature and wavelength:

$$L = \frac{c_{1L}}{\lambda^5 \cdot \exp(c_2/(\lambda \cdot T))} \quad (1)$$

where:
c_{1L}: 1st rad. constant $1.1910439 \ast 10^{-16}$ W·m^2
c_2: 2nd radiation constant 0.0143876 m·K
λ: wavelength
T: object temperature

$T_{combustor}$	1423 K	1473 K	1523 K
$\varepsilon=0.85$ & $\rho=0.15$	37 K	57 K	79 K
$\varepsilon=0.90$ & $\rho=0.10$	26 K	41 K	58 K
$\varepsilon=0.95$ & $\rho=0.05$	14 K	22 K	33 K

Table1: Maximum error in the pyrometer reading due to combustor radiation for an object temperature of 1273K

The total radiation measured by the pyrometer consists of the blade radiation according to its temperature and emissivity ε and the ambient radiation (i.e. especially combustor radiation) according to the ambient temperature and the reflectance ρ of the blade:

$$L_{brightness} = \varepsilon_{blade} \cdot L_{blackbody}(T_{blade}) + \rho_{blade} \cdot L_{blackbody}(T_{ambient}) \quad (2)$$

With eq. (1) and (2) the brightness temperature as indicated by the pyrometer can be calculated:

$$T_{brightness} = \frac{c_2}{\lambda \cdot \ln(c_{1L}/(L_{brightness} \cdot \lambda^5))} \quad (3)$$

Table1 shows the maximum error between the true object temperature (1273K) and the pyrometer temperature for different emissivities and reflectances for a wavelength of 0.85 µm. For $\varepsilon=0.9$ & $\rho=0.1$ and a combustor radiation that corresponds to a black body temperature of 1473K the measured temperature is 1041°C, i.e. the temperature reading is +41 K in the worst case. Since the combustor is a spectral emitter, the blades are partly blocked against radiation by the vanes and most of the measured surfaces are not perpendicular to the combustor outlet, the influence is actually considerably lower than indicated here. Nevertheless, flame radiation is one of the major source of error for measurements in the first stage.

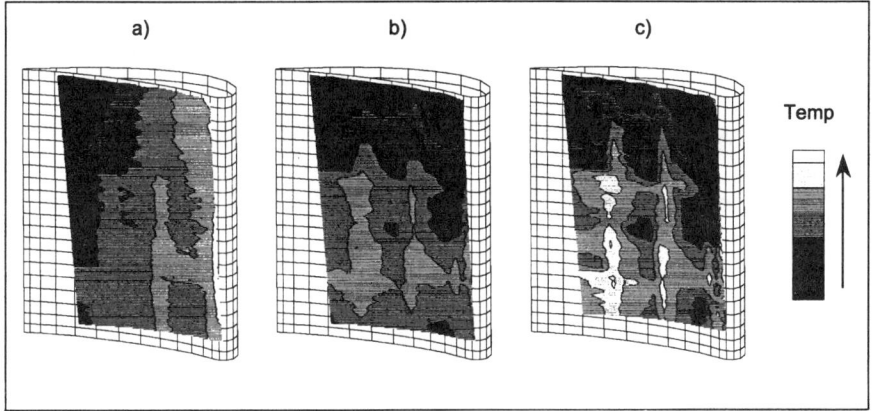

Figure 5: Thermal images of one particular first stage prototype vane at base load. a) Liquid fuel, conventional pyrometer, b) natural gas, conventional pyrometer, c) natural gas, high resolution pyrometer

Figure 5 shows the thermal images of a prototype vane under different test and operating conditions, but always at base load. Figure 5a shows the result for the conventional pyrometer for liquid fuel operation. The minimum field of view is 3 to 4 mm. In comparison to this Figure 5b exhibits already more temperature details, using the same pyrometer as before but natural gas operation. Hence, the difference in the resolution results from flame radiation only. Figure 5c shows the best image using a high resolution pyrometer with a minimum field of view of less than 1.5mm, i.e. the area of the measuring spot is at least four times smaller as is for the conventional probe. All details of the fairly complex mechanical structure of the prototype vane are now clearly visible and even individual cooling holes may be identified. Thus, the new probe design provides much more detailed information that gives a more realistic image of the true temperature conditions. Statistic evaluation of a set of forty measurements of one particular blade at a constant viewing angle yields a standard deviation of 0.2%. This clearly indicates that there is no soot or particle interference. Hence, no picking of minimum values[6, 7] or averaging has to be applied.

For vane measurements, only the improved spatial resolution is advantageous, whereas for rotating blades spatial and temporal resolution of the probes is of equal importance. Extensive measurements of the first stage blades have been performed with the new pyrometer system under varying test conditions. During evaluation, the 3D-software allows for visualization of each individual blade.

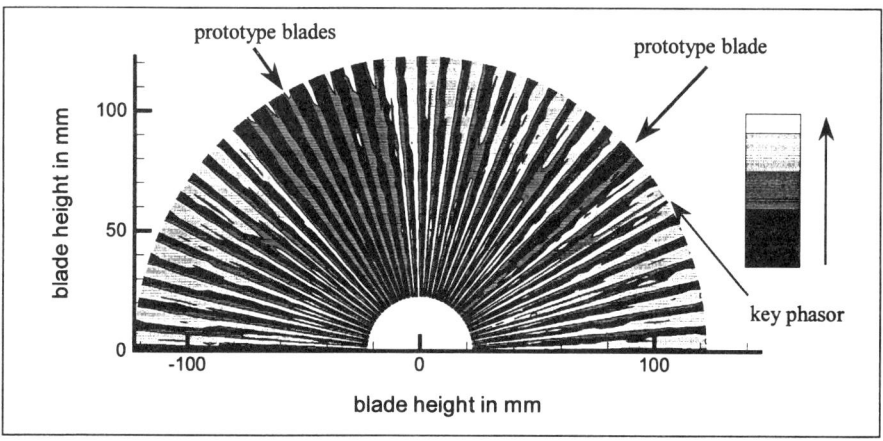

Figure 6: Temperature distribution of a set of 32 standard and 8 prototype blades

Figure 6 gives an overview of the test results for the upper 40 blades (relative to the key phasor) at one specific viewing angle (test time is 1.5 s). The eight metal surface prototype blades can easily be distinguished from the 32 TBC-coated standard blades. Please note that figure 6 shows a highly reduced data set that over-emphasizes both minimum and maximum temperature values, in order to allow for easy detection of cold or hot spots.

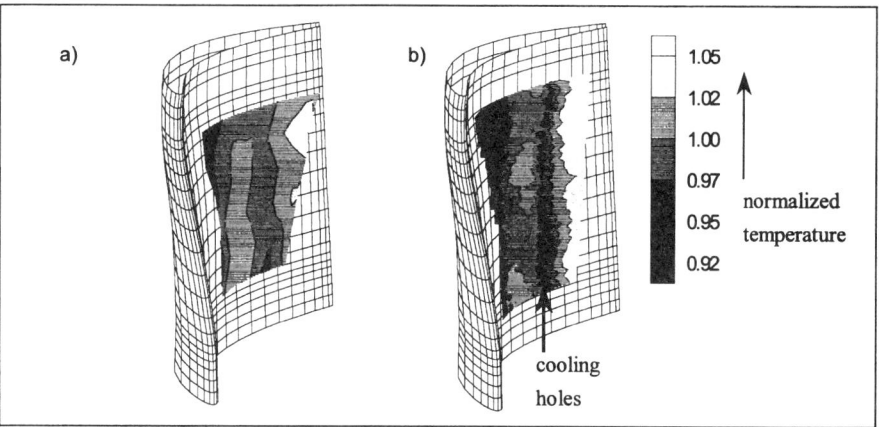

Figure 7: Temperature distribution (almost axial viewing angle) of the same first stage prototype blade under different test conditions, but constant turbine inlet temperature
  a) fuel oil diffusion mode, commercial pyrometer (min. field of view 3mm, sample frequency 100kHz), discrete measurements, i.e. six radii with 20 data points each
  b) natural gas premix mode, high resolution pyrometer (min. field of view 1mm, sample frequency 500kHz), continuous measurement, i.e. 50 radii with 100 data points each

Statistical evaluation of discrete blade measurements at one certain radius allows for the assessment of random errors and provides the following results: the standard deviation of temperature at a certain point on a particular blade for a given set of revolutions amounts to about 0.6% for fuel oil, whereas it is only 0.2% for natural gas. Moreover, the average blade temperature is slightly higher for fuel oil than it is for natural gas at the same gas turbine inlet temperature. This can be explained by the broad band characteristic of the silicon photodiode that is sensitive in a wavelength range from 0.4 to 1.1µm, i.e. it is sensitive to visible light as emitted by oil flames. In order to reduce the flame radiation influence, suitable high pass filters that block the visible portion are applied. Consequently the standard deviation of the liquid fuel measurements do not exceed those for natural gas. Figure 7 displays the temperature distribution of one particular prototype blade for different operating and measuring conditions. The left-hand distribution has been determined by means of a commercial pyrometer using fuel oil premix mode and discrete blade measurements, while on the right the distribution is shown to be determined by high resolution pyrometry in conjunction with continuous blade measurements using natural gas. Obviously, Figure 7b provides a much more detailed thermal image than figure 7a. This is due to the considerable radiation emitted by oil flames, and only partly as a matter of pyrometer resolution. However, in order to detect very small structures such as film cooling holes, high resolution pyrometry has to be applied.

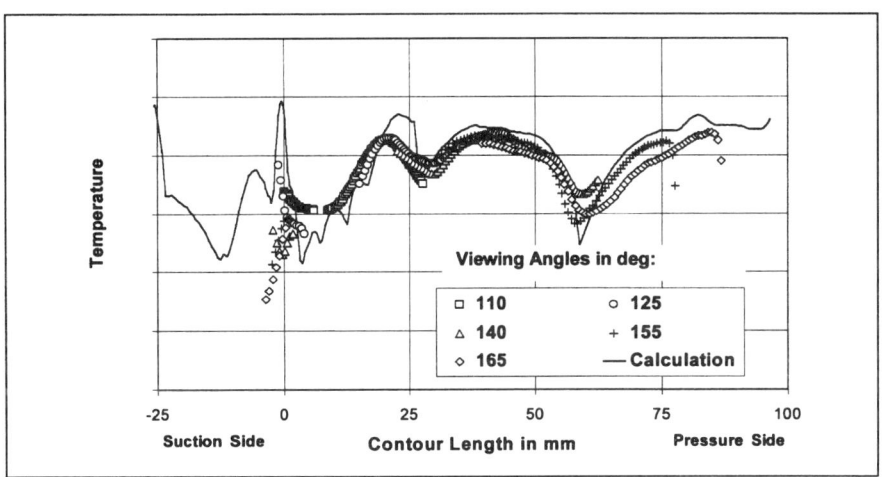

Figure 8: Comparison between pyrometer measurements and numerical calculations of temperature at mid-span

A comparison between measurement values and those from numerical calculations[2] is shown in figure 8. Qualitatively the results are in close agreement. The slightly higher values of the calculation result from a higher assumption of the firing temperature than there has actually been during the measurement. There is a certain redundancy in the experimental information because different viewing angles have been measured. However, in certain positions, the values deviate from each other which is mainly due to exceeded incidence angle and lower resolution[9]. Thus the measurements have to be interpreted thoroughly.

Measurements on thermal barrier coated parts (TBC) in the near infrared are generally regarded as highly unreliable due to low emissivity, semi-transparency and high reflectivity[10, 11]. Long infrared wavelength, however, seems to be very promising, since the TBC behaves almost as a black body radiator. Therefore, simultaneous long[12] and short infrared wavelength pyrometer measurements have been performed on both coated and uncoated blades under base load conditions using both fuels. The different data sets are currently being evaluated and seem to be reasonable and consistent. First results will be presented soon.

Conclusions

Taking into account the latest test results it can be concluded that
- Pyrometer measurements in the presence of natural gas flame radiation yield reasonable results, whereas for fuel oil a suitable flame radiation correction is essential
- For natural gas no averaging is necessary since the standard deviation is very low, i.e. any of a set of several revolutions is representative for the actual temperature distribution
- Using high resolution pyrometry even smallest structures such as individual film cooling holes can be easily detected in the thermal image of the blade

Thus, using new pyrometer probes in conjunction with a continuous data acquisition mode allows reliable, high resolution blade surface temperature measurements, where, at the same time, errors and potential risks for the gas turbine are minimized.

The author would like to thank Mr. A. Gamlin and Mr. M. Tram for their assistance during the measurements, Mr. A. Klix for the calibration of the pyrometers and Mr. M Luedtke for the preparation of the 3D blade graphics. Special thanks to Mr. H. Bals for the comparison of calculated and measured values.

References

1. Bals, H.; Schulenberg, T.; H.: Blade Temperature Measurements of Model V84.2 100MW/60Hz Gas Turbine, ASME Paper 87-GT-135.
2. Haendler, M.; Raake, D.; Scheurlen, M.: Aero-Thermal Design and Testing of Advanced Turbines Blades. ASME Paper 97-GT, Orlando, 1997.
3. Seume, J.: 25 years of experimentally verified gas turbine design. Power-Gen Asia, Singapore, 1997.
4. Eggert, T.: Turbine pyrometer with high spatial and temporal resolution. Ph.D. Thesis, Technical University Berlin (in German). Wissenschaft und Technik Verlag, Berlin, February. 2000.
5. Boehm, W. et al: Testing the model V84.3A gas turbine – experimental techniques and results. ASME Paper96-TA-14
6. Suarez, E.; Prziremebel, H.R.: Pyrometry for Turbine Blade Development. AIAA 88-3036, 1988.
7. Sellers, R.R. et al.: The Use of Optical Pyrometers in axial Flow Turbines. AIAA 89-2692, 1989.
8. De Witt, D.P.; Nutter, G.D.: Theory and Practice of Radiation Thermometry. John Wiley & Sons, New York, 1988.
9. De Lucia, M. et al.: Temperature Measurements in a Heavy Duty Gas Turbine Using Radiation Thermometry Technique: Error Evaluation. ASME 99-GT-311, Indianapolis, 1999
10. Alaruri, S. et al.: Effective Spectral Emissivity Measurements of Superalloys and YSZ Thermal Barrier Coating at High Temperatures Using a 1.6 micrometer Single Walvelength Pyrometer. AGARD '97 Brussels, 1997.
11. Ruud, J.A. et al.: Surface Temperature Measurement of Thermal Barrier Coatings Using Infrared Pyrometry. ASME 96-GT-281, 1996.
12. Latvakoski, H. et al.: Measurement of advanced ceramic coated superalloys with a long wavelength pyrometer. AIAA 2000-2212, 2000.

STUDIES ON FREE STREAM TURBULENCE AS RELATED TO GAS TURBINE HEAT TRANSFER A REVIEW OF AUTHORS' PAST WORK AND FUTURE IMPLICATIONS

SAVAS YAVUZKURT* and GANESH R. IYER**

*Department of Mechanical and Nuclear Engineering
The Pennsylvania State University, University Park, PA 16802 USA

**Caterpillar Inc.
100 N.E. Adams Street, Peoria, IL 61629

ABSTRACT: A review of the past work done on free stream turbulence (FST) as applied to gas turbine heat transfer and its implications for future studies are presented. It is a comprehensive approach to the results of many individual studies in order to derive the general conclusions that could be inferred from all rather than discussing the results of each individual study. Three experimental and four modeling studies are reviewed. The first study was on prediction of heat transfer for film cooled gas turbine blades. An injection model was devised and used along with a 2-D low Reynolds number k-ε model of turbulence for the calculations. Reasonable predictions of heat transfer coefficients were obtained for turbulence intensity levels up to 7%. Following this modeling study a series of experimental studies were undertaken. The objective of these studies was to gain a fundamental understanding of mechanisms through which FST augments the surface heat transfer. Experiments were carried out in the boundary layer and in the free stream downstream of a gas turbine combustor simulator, which produced initial FST levels of 25.7% and large length scales (About 5-10 cm for a boundary layer 4- 5 cm thick). This result showed that one possible mechanism through which FST caused an increase in heat transfer is by increasing the number of ejection events. In a number of modeling studies several well-known k-ε models were compared for their predictive capability of heat transfer and skin friction coefficients under moderate and high FST. Two data sets, one with moderate levels of FST (about 7%) and one with high levels of FST (about 25%) were used for this purpose. Although the models did fine in their predictions of cases with no FST (baseline cases) they failed one by one as FST levels were increased. Under high FST (25.7% initial intensity) predictions of Stanton number were between 35-100% in error compared to the measured values. Later a new additional production term indicating the interaction between the turbulent kinetic energy (TKE) and mean velocity gradients was introduced into the TKE equation. The predicted results of skin friction coefficient and Stanton number were excellent both in moderate and high FST cases. In fact these model also gave good predictions of TKE profiles whereas earlier unmodified models did not predict the correct TKE profiles even under moderate turbulence intensities. Although this new production term seems to achieve the purpose, it is the authors' belief that it is the diffusion term of the TKE equation, which needs to be modified in order to fit the physical events in high FST boundary layer flows. The results of these studies are currently being used to come up with new diffusion model for the TKE equation.

INTRODUCTION

Free stream turbulence is the turbulence in the approach stream. It is experienced in many applications. For example, nozzle guide vanes and the rotor blades in a gas turbine are exposed to the high levels of free stream turbulence. FST has an important influence on the surface heat transfer. Under high levels of turbulence (10-20%), there is an appreciable increase in the heat transfer rate regardless of the character of the boundary layer. Some representative results in this area can be found in the studies by Brown and Burton[1], Bradshaw and Simonich[2], and Blair[3]. A detailed review of this literature is given by Moffat and Maciejewski[4]. They conclude that the FST levels up to 10% cause a proportional increase in heat transfer for constant velocity and accelerating turbulent boundary layers. It is indicated that large effects on the average value may result if the turbulence affects the location of the transition and if the heat transfer data are compared at constant x-Reynolds numbers. One of the important observations from the literature in this area is that under the same levels of turbulence, different researchers found different enhancement of heat transfer rates. This leads to the speculation that not only the velocity scale but also the length scale of the turbulence is important. In fact, Moffat and Maciejewski[4] relate to this fact and suggest that the effect of the length scale should be investigated. Most of the studies that were discussed used, one way or another, grid generated turbulence in their experiments where the length and velocity scales are usually small. More recently, the flow fields of jets and wall jets have been used in order to simulate high FST encountered in turbomachinery. Moffat and Maciejewski[5] used a circular wall jet in order to obtain FST intensities up to 48%. They measured Stanton numbers, which are as much as 350% above the standard, zero free stream turbulence correlations. Ames and Moffat[6] have investigated the effects of FST (created by 2.5 inches diameter jet injection into a main flow in the plenum chamber followed by a wind tunnel test section) on the heat transfer to a flat plate boundary layer. This study used autocorrelations to measure the length scales. In another recent study, MacMullin et al.[7] investigated the effects of FST from a circular wall jet on a flat plate boundary layer heat and momentum transfer with turbulence intensities 7-18%. They also observed increased Stanton numbers and skin friction coefficient with increasing turbulence intensities. They used autocorrelations for the determination of length scales. The influence of length scale on the Stanton numbers was not conclusive.

Yavuzkurt and Batchelder[8] used the same model combustor rig used by Ames and Moffat[6] to measure all the normal Reynolds stresses and their length scales in three directions both in the free stream and the boundary layer. A FST number was defined using only the length and velocity scales of the FST. Excellent correlation between Stanton numbers and FST numbers were obtained.

A number of experimental studies concerning the effect of high FST (velocity scale and length scale) on turbulent boundary layers have been carried out in the last two decades. To name a few, the works of Hancock and Bradshaw[9], Blair et. Al.[10,11], Ames and Moffat[6], Thole and Bogard[12] have shown the effects of high FST on enhancement of skin friction coefficient and Stanton number in a turbulent boundary layer. It is only in the recent past that attention has also been focused on simulating these effects (Kwon and Ames[13]). Recently, Volino[14] developed a model which incorporated a free stream induced viscosity besides the eddy viscosity and molecular viscosity. This viscosity is modeled with the aid of empirical data and scales

STUDIES ON FREE STREAM TURBULENCE

Prediction of Heat Transfer with Film Cooling under High FST

Model development for the prediction of film cooling under high FST was attempted by Tafti and Yavuzkurt[15,16,17]. This necessitated the development of a two-dimensional injection model to distribute the coolant into the boundary layer flow, and use of a low Reynolds number k-ε model. In order to perform calculations for the leading edge film cooling geometries, stagnation region flow over a cylinder was modeled within the framework of the k-ε model and extended to flow over turbine blades. A method was proposed for the generation of initial profiles of velocity, temperature, TKE and dissipation rate of TKE. An additional source term for the production of TKE in free stream form of the TKE equation was introduced based on strong streamwise velocity gradients. Predictions of heat transfer over cylinders and airfoil were very dependent on the predicted levels of free stream turbulence. The accuracy of the predictions of heat transfer was good for FST levels up to 7%.

Experimental Studies

Following this modeling study, a series of experimental studies were undertaken at The Center for Turbulence Research at Stanford University and Wright Labs at Wright Patterson Air Force Base by the first author[8,18,19]. The objective of these studies was to gain a fundamental understanding of mechanisms through which FST augments surface heat transfer. It was theorized that FST must stir the near wall layer of the turbulent boundary layer, thereby increasing heat transfer. It was hoped that this understanding would lead to improvement of the turbulence models used to predict heat transfer coefficients under high FST. Experiments were carried out in the boundary layer and in the free stream[8,18] downstream of a gas turbine combustor simulator, which produce initial FST levels of 25.7% and large length scales (about 5-10 cm for a boundary layer 4-5 cm thick). Using two triple hot wire probes, all three components of mean velocity and 6 Reynolds stresses were obtained along with length scales of normal Reynolds stresses in vertical and transverse directions using space correlation and in the axial direction via auto correlations. Figure 1. shows space correlations in the vertical direction of turbulent velocity fluctuations in axial, vertical and transverse direction to the wall within the boundary layer and the free stream. As can be seen form this figure the length scales are quite large – in the order of 10 cm. In these studies a correlation for Stanton number as a function of a newly defined FST number "α" was obtained. "α" represents the normalized total TKE of the flow. It is obtained from the multiplication of length scales of each fluctuating velocity component in three directions with the mean square value of that component, summed over all three components and normalized with the boundary layer thickness and the free stream velocity. This correlation is shown in figure 2. This quantity did not contain any viscosity and correlated well with Stanton number data. This result also lead to the conclusion that in a high FST (> 10%) boundary layer, heat transfer is mainly determined by the FST and not by the production of turbulence near the wall due to mean shear. Also from quadrant plots of axial and vertical components of turbulent fluctuating velocities, the number of ejection events near the wall were determined. This data and its analysis showed that the number of ejection events increased with increased FST turbulence intensity levels. This is shown in figure 3. Increased ejection events should lead to an increase in heat transfer. This result showed that one possible mechanism through which FST caused an increase in heat transfer is by increasing the number of ejection events. This result was later confirmed by a study carried out in a wall jet[19]. During this study instantaneous heat transfer rates at the wall were measured by a hot film probe positioned flush with the surface at several axial locations in the wall jet flow. It was seen that an increase in the turbulence intensity lead to an increase in the high peaks observed in the instantaneous wall heat transfer. This result is shown in Figure 4. As a result of these studies it was theorized that the only way FST can affect the processes near the wall is through diffusion of this high FST through the boundary layer towards the wall. This diffusion is mainly caused by normal Reynolds stress components and particularly the normal Reynolds stress in the direction perpendicular (vertical) to the wall. This idea was later used in the improvement of k-ε models of turbulence for the prediction of heat transfer and skin friction under high FST.

k-ε Models of Turbulence under High FST

In a number of modeling studies[20,21,22] several well known k-ε models were compared for their predictive capability of heat transfer and skin friction coefficients under moderate and high FST. Two data sets, one with moderate levels of FST and one with high levels of FST were used for this purpose. Although the models did fine in their

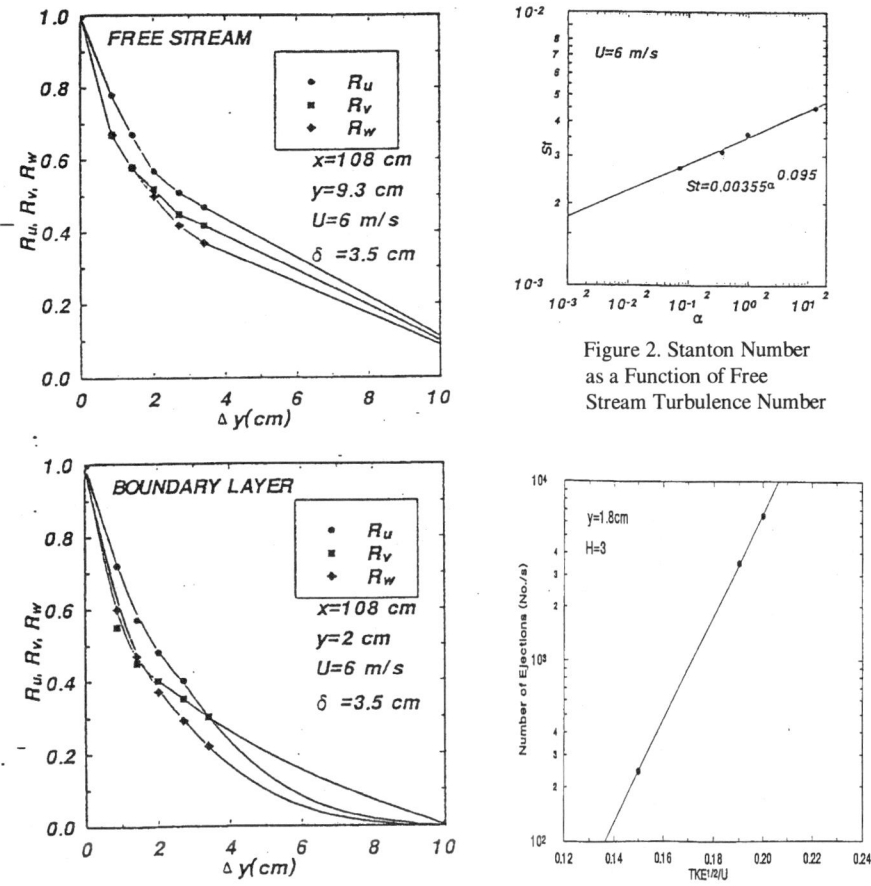

Figure 1. Space Correlation Coefficients for Turbulent Velocity Fluctuations in The vertical direction (y) to the wall

Figure 2. Stanton Number as a Function of Free Stream Turbulence Number

Figure 3. Number of Ejections per Unit Time as a Function of Turbulence Intensity

predictions of cases with no FST (baseline cases), they failed one-by-one as FST levels were increased. Under high FST (25.7% initial intensity) predictions of Stanton number were between 35-100% in error compared to the measured values. These results are shown in figure 5. The models utterly failed to predict TKE levels. Later a new additional production term indicating the interaction between the TKE and mean velocity gradients

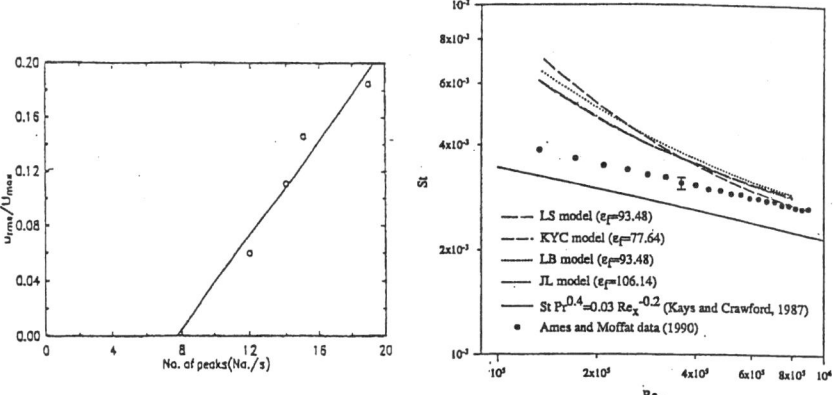

Figure 4. Increase in the Number of Peaks Per Unit Time in the Instantaneous Heat Transfer as a Function of Turbulence Intensity

Figure 5. Stanton Number Predictions by well-known k-ε Models under High FST

was introduced into the TKE equation $\left(P = -\overline{u'v'}\partial U/\partial y + Cfk\,\partial U/\partial y\right)$. In this equation $\overline{u'v'}$ is the Reynolds shear stress, U is the free stream velocity, y is the vertical direction to the wall, k is TKE, C is a constant which is function of free stream length scales and f is a damping function near the wall. The predicted results of skin friction coefficient and Stanton number were excellent both in moderate and high FST cases. Figure 6 shows predicted Stanton numbers by the new model (FSKYC) under high FST. In fact these models also gave good predictions of TKE profiles whereas earlier unmodified models did not predict the correct TKE profiles even under moderate turbulence intensities. Although this new production term seems to achieve the purpose, it is the authors' belief that it is the diffusion term of the TKE equation, which needs to be modified in order to fit the physical events in high FST boundary layer flows. Diffusion modeling is usually not very rigorous in many modeling applications since under low or no FST, diffusion terms are not that important compared to production and dissipation terms. However, in high FST flows it is believed that the diffusion term is a more important term in the TKE equation. In fact this is the mechanism, which causes high FST to affect the events near the wall. The results of these studies are currently being used to develop a new diffusion model for the TKE equation.

A New Model for Diffusion of TKE

Although it is important to come up with a new diffusion model by understanding the fundamentals, it is also important to develop a model that could be used easily for industrial design purposes. For this study, one such model proposed is to write the model of diffusion of TKE in 2-D as

$$\frac{\partial}{\partial y}\left[\frac{\mu_t}{\sigma_k}\frac{\partial k}{\partial y}\right] + \frac{\partial}{\partial y}\left(\rho\, C f\, kU\right)$$

The first term is the usual gradient diffusion term and the second term is the new term added which represents distribution of k by large eddies which move with the mean speed of the flow. The f in front of kU is a function similar to $f\mu$ used to damp turbulent viscosity near a wall. C is a function of the length scales of free-stream turbulence. Work on this new model is continuing.

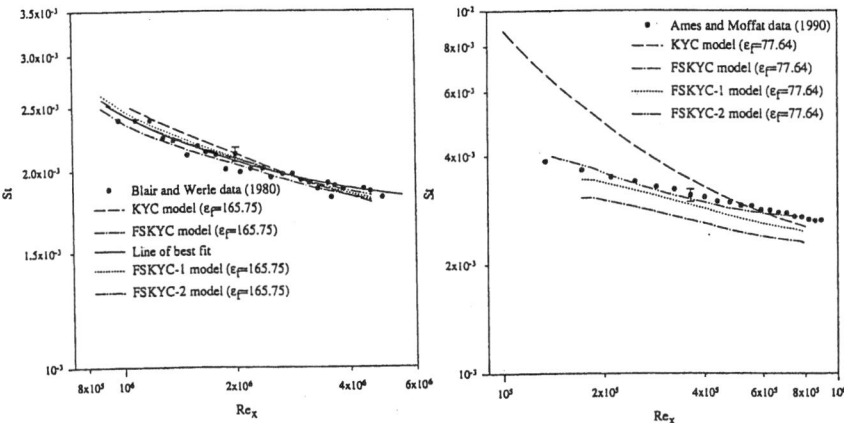

Figure 6. Stanton Number Predictions by the New Model under Moderate (6.5 %) and High (25.7%) FST

CONCLUSIONS

A review of the studies carried out by the authors in the area of free steam turbulence is presented. Step by step approach to understanding the fundamentals of how FST affects surface heat transfer and use of this knowledge in modeling is demonstrated. One important conclusion of the studies is that under high FST well-known k-ε models do not and cannot perform well. Diffusion of high FST from free stream towards the wall has to be included in TKE equation. Moving from this point, a new preliminary model of diffusion of TKE is presented.

REFERENCES

1. Brown, A., & Burton, R. C., 1978, "The Effects of Free Stream Turbulence intensity and velocity distribution on heat transfer to curved surfaces," J. of Engr. For Power. 100, 159 168.
2. Bradshaw, P., Simonich, P., 1978, "Effect of free stream turbulence on heat transfer through a turbulent boundary layer," ASME J. of Heat Transfer. 100-4.
3. Blair, M. F. 1983, "Influence of Free Stream Turbulence on Turbulent Boundary Layer Heat Transfer," - Parts I and II, ASME J. of Heat Transfer, Feb.
4. Moffat, R. J., & Maciejewski, P. K. 1984, "Effects of Very High Turbulence on Convective Heat Transfer, " NASA Conference Publication 2339, NASA LeRC, 381-388.
5. Moffat, R. J., Maciejewski, P.K. 1985, "Heat Transfer with Very High Free Stream Turbulence, " Proc. of HOST Conference, NASA conference publication 2405, NASA LeRC, 203-215.

6. Ames, F. E., and Moffat, R. J., 1990 "Heat Transfer With High Intensity, Large Scale Turbulence: The Flat Plate Turbulent Boundary Layer and the Cylindrical Stagnation Point," Report No. HMT-44, Thermosciences Division, Department of Mechanical Engineering, Stanford University, California.
7. MacMullin, R., Elrod, W., & Rivir, R. 1989, "Free Stream Turbulence from a Circular Wall Jet on a Flat Plate Heat Transfer and Boundary Layer Flow," Trans. ASME J. of Turbomachinery, iii, 78-86.
8. Yavuzkurt, S., and Batchelder, K., 1993, " A Correlation for Heat Transfer under High Free Stream Turbulence Conditions," *Ninth Symposium on "Turbulent Shear Flows"*, Kyoto, Japan, pp. P104-1-P104-4.
9. Hancock, P. E., and Bradshaw, P., 1983, "The Effect of Free-Stream Turbulence on Turbulent Boundary Layers,", *ASME Journal of Fluids Engineering*, Vol. 105, pp. 284-289.
10. Blair, M. F., and Werle M. J., 1981, "Combined Influence of Free-Stream Turbulence and Favorable Pressure Gradients on Boundary Layer Transition and Heat Transfer," UTRC Report R81-914388-17.
11. Blair, M. F., and Edwards D. E., 1982, "The Effects of Free-Stream Turbulence on the Turbulence Structure and Heat Transfer in Zero Pressure Gradient Boundary Layers," UTRC Report R82-915634-2.
12. Thole, K. A., and Bogard, D. G., 1996 "High Freestream Turbulence Effects on Turbulent Boundary Layers," *ASME Journal of Fluids Engineering*, Vol. 118, pp. 276-284.
13. Kwon, O., and Ames, F. E., 1995, "Advanced k-epsilon modeling of Heat Transfer," NASA Contractor Report 4679, Allison Engine Company, Indianapolis.
14. Volino, R. J., 1997, "A New Model for Free-Stream Turbulence Effects on Boundary Layers, " *ASME Paper 97-GT-122*, pp. 1-9.
15. Tafti, D. K. and Yavuzkurt, S., Prediction of Heat Transfer Characteristics of Discrete Hole Film Cooling One-Row of Injection into a Turbulent Boundary Layer, paper presented at the 1988 ASME WAM, *ASME publication Heat Transfer in Gas Turbine Engines and Three Dimensional Flows*, Chicago, Illinois, pp. 45-52, Nov. 27-Dec. 2, 1988.
16. Tafti, D. K. and Yavuzkurt, S., Prediction of Heat Transfer Characteristics for Discrete Hole Film Cooling for Turbine Blade Application, paper presented at the ASME Intl. Gas Turbine and Aerospace Congress and Exposition, Toronto, Canada, June 4-8, 1989, *ASME paper No. 89-GT-139*.
17. Tafti, D. K., Prediction of Heat Transfer Characteristics for Discrete Hole Film Cooling on Flat Plate and Turbine Blades. *Ph.D. thesis,* Dept. of Mech. Engr., Penn State Univ., Aug. 1989.
18. Yavuzkurt, S., Seo S., Effect of High Free Stream Turbulence on Ejection Events in a Boundary Layer, *Proc. of 10th Symp. on Turbulent Shear Flows*, University Park, PA, Aug. 14-16, 1995, pp. P2-31, P2-30.
19. Yavuzkurt, S., Effects of Free Stream Turbulence on the Instantaneous Heat Transfer in a Wall Jet Flow, *ASME Paper No. 95-GT-43*, presented at the ASME Intl. Gas Turbine and Aerospace Congress and Exposition, Houston, TX, June 5-8, 1995.
20. Iyer, G. R., Yavuzkurt, S., Comparison of k-ε Models in Predicting Heat Transfer and Skin Friction Under High Free Stream Turbulence, *ASME Paper No. 96-GT-537*, presented at the ASME Intl. Gas Turbine and Aerospace Congress and Exposition, Birmingham, UK, June 10-13, 1996.
21. Iyer, G. R., Yavuzkurt, S., A New Low-Reynolds Number k-ε model for Simulation of High FST Flows, ASME paper published in the proceedings of Fluids Engineering Division Summer Meeting in Washington, DC. (*ASME Paper No. FEDSM98-4851*), June 1998.
22. Iyer, G. R., Yavuzkurt, S., Comparison of Low-Reynolds Number k-ε models in Simulation of Momentum and Heat Transport Under High Free Stream Turbulence, *International Journal of Heat and Mass Transfer*, Vol. 42, No. 4, pp. 723-737, 1998.

A Conjugate Heat Transfer Procedure for Gas Turbine Blades

GIULIO CROCE

DIEM - Dipartimento di Energetica e Macchine, Università di Udine, 33100 Italy

Abstract: A conjugate heat transfer procedure, allowing for the use of different solvers on the solid and fluid domain(s), is presented. Information exchange between solid and fluid solution is limited to boundary condition values, and this exchange is carried out at any pseudo-time step. Global convergence rate of the procedure is, thus, of the same order of magnitude of stand-alone computations.

INTRODUCTION

The temperature distribution in the blade metal is the result of the combined effects of convective internal heat transfer, external convection, and conduction through the metal itself. As a consequence, although stand-alone external flow heat transfer results yield important information for the designer, only a fully coupled conjugate heat transfer (CHT) computation allow the correct evaluation of metal temperature.

Such a fully coupled approach involves the solution of different physical problems: transonic flow in the blade passage, low Mach number flow in the cooling passages, conduction in the metal. It could be possible to use the same solver for all these problems, maybe using preconditioning schemes to handle the different Mach number regimes, and using the (modified) energy equation of the NS solver to get the conduction solution.

However, since structural codes for conduction analysis are widely available in industry, the most rational approach for conjugate heat transfer requires the development of an interface between existing codes, rather then embedding a solid analysis capability in the flow solver. Furthermore, this can also allow the use of different codes optimised for the high and low Mach number domains.

It is quite important to keep the interface procedure as independent as possible from the details of the single solvers, and to avoid any change in the solvers. This will not probably lead to the most efficient possible CHT procedure, but will allow greater flexibility. Such flexibility is of significant interest, especially for an industrial environment, where often different codes are used for different problems, and the user may not have access to the source of some commercial code.

FLOW AND SOLID SOLUTIONS

An implicit Navier Stokes solver[1], whose accuracy in gas turbine heat transfer application has previously been assessed[2], was used for external flow computation. The Navier-Stokes equations are written in a general curvilinear coordinate system ξ, η:

$$U_t + (F - F_v)_\xi + (G - G_v)_\eta = 0 \qquad (1)$$

The conservative variable vector and the inviscid flux vector are defined as in the following:

$$U = J^{-1}[\rho, \rho u, \rho v, e] \qquad (2)$$

$$F = J^{-1}\left[\rho\bar{u}, \rho\bar{u}u + \xi_x p, \rho\bar{u}v + \xi_y p, (e+p)\bar{u}\right] \qquad (3)$$

$$G = J^{-1}\left[\rho\bar{v}, \rho\bar{v}u + \eta_x p, \rho\bar{v}v + \eta_y p, (e+p)\bar{v}\right] \qquad (4)$$

where ρ is the density, u and v the velocity components, p the pressure, t the time, J the Jacobian of the coordinate transformation. Viscous flux vectors Fv and Gv have the standard form described in the paper[1].

An ADI factorisation, within an hybrid finite volume-finite difference frame, is adopted, while turbulence can be taken into account through a standard Baldwin-Lomax model or an algebraic RNG based one[1], and structured grids are used[3].

The conduction equation within the solid has been discretized with a finite volume formulation on an unstructured triangular grid[4]. Integral formulation on each triangular cell yields:

$$\int_\Omega \frac{\partial T}{\partial t} = \int_{\partial\Omega} \frac{k}{\rho c} \nabla T \cdot n \qquad (5)$$

where T is the cell temperature, k the thermal conductivity and c the thermal capacity of the solid and the discretized Euler implicit formulation is given by:

$$\Delta T^n = \Delta t \sum_{i=1,3} A_i q_I^{''} = \Delta t \sum_{i=1,3} A_i \frac{\rho c}{k} \left[\nabla(T^n + \Delta T^n)\right]_i \cdot n_i \qquad (6)$$

where $\Delta T^n = T^{n+1} - T^n$ and A_i is the area of cell face i. Neumann boundary conditions are easily implemented, due to the integral finite volume formulation. For 2D computation, an heat transfer rate coefficient is assumed for blade/coolant interface, allowing for a non-uniform temperature distribution along the internal surfaces; for 3D computation, internal flow could be solved with a suitable code, such as an artificial compressibility algorithm. However, it should be remarked that the coupling procedure is fully independent from the details of the chosen solution algorithms.

CHT PROCEDURE

The coupling between fluid and solid is obtained via an exchange of boundary conditions. Fluid solutions for both internal and external flow are obtained with a guessed temperature distribution along the wall; the heat flux computed by the two solvers is then used as a boundary condition for the solid conduction problem, which gives back a new temperature distribution along the walls. Consistent, conservative interpolation procedures can be used to handle mismatching grids. The procedure can, thus, be summarised as in the following:

1. solve the Navier Stokes equations (1) $f_f(H_0, u_i, p) = 0$ in the fluid domain Ω_f with Dirichlet boundary conditions at the interface Γ:

$$\begin{cases} f_f(H_0, u_i, p) = 0 & \text{on } \Omega_f \\ T = T^k & \text{on } \Gamma \end{cases} \quad (7)$$

2. Evaluate surface heat transfer on Γ from the flow solution.
3. Solve the Poisson conduction equations (6) $f_s(T) = 0$ in the solid domain Ω_s with Neumann boundary conditions at the solid/fluid interface:

$$\begin{cases} f_s(T) = 0 & \text{on } \Omega_s \\ k_s \dfrac{\partial T}{\partial n} = -q_f & \text{on } \Gamma \end{cases} \quad (8)$$

4. Evaluate surface temperature T_Γ on solid/fluid interface Γ from the conduction solution:
5. Update Dirichlet boundary condition on Γ for the Navier Stokes solution, introducing a relaxation parameter ω:

$$T^{k+1} = (1-\omega)T^k + \omega T_\Gamma \quad \text{on } \Gamma \quad (9)$$

6. Go to step 1.

In the above expression k_s, k_f and represent the fluid and solid thermal conductivities. Such an approach have been widely used in some simplified way in industrial environment: usually a few complete cycles (7-9) are carried out, using converged or nearly solution for each domain problem (7),(8). Here we use a stronger coupling, calling the interface procedure at each pseudo-time step. This approach, thus, corresponds to the Shur Complement algorithm for domain decomposition of partial differential equations described by Funaro et al.[5], where convergence properties of the methods are demonstrated for elliptic problems. In the same paper[5], some guidelines are given on the optimal choice of the relaxation parameter ω.

Using different equation on the two sides of the interface, as in the CHT case, the boundary condition choice is imposed by stability consideration. According to a simplified one dimensional finite differences analysis[6], in fact, the use of a heat flux b.c. for the solid and a temperature b.c. for the fluid provides stability for the global problem, while a Neumann condition for the fluid and Dirichlet for the solid lead to instabilities[6]. Similar algorithms have been used to couple fluid and solid energy equations in incompressible flow (where flow field is decoupled from thermal one), CHT[7,8].

As stated above, the whole procedure is not dependent on the details of the flow solver code: the same interfacial procedure has been successfully applied by the author to a finite-element codes for anti-icing problem analysis[9,10], including the cooling effect of runback water film.

Advancing the solution at the same time as the flow solution allows the solution even if the solid is completely surrounded by flow domain (i.e. Neumann condition on all solid boundaries), as we are actually solving on a single domain spanning the fluid and the solid, and the correct flow b.c. ensures the convergence to the steady-state solution. Furthermore, the strong coupling implies that the number of time steps required for the CHT computation is of the same order of magnitude of the time required for a stand alone computation. However, as we use an implicit algorithm for the fluid and conduction solver the global convergence rate is slightly reduced, in comparison to the stand-alone computation, as the coupling is not linearized[9,10].

A useful feature of the outlined procedure is the ability to deal with non-matching grids. This can be obtained using interpolation techniques consistent with a conservative FVM formulation in order to transfer the temperature and heat fluxes from one side of the interface to the other.

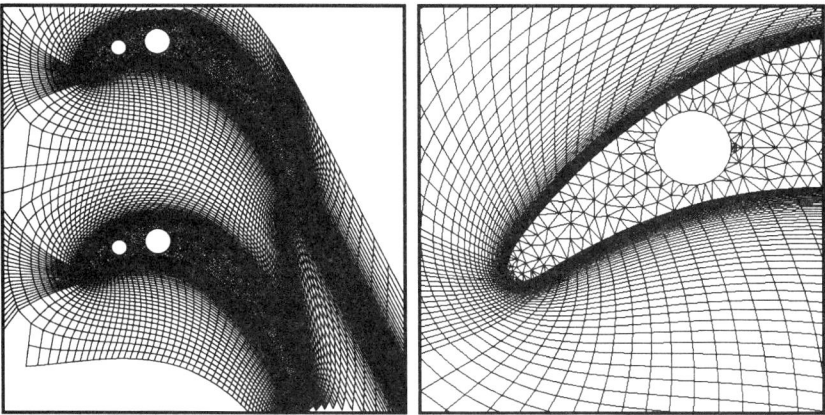

Fig.1 Computational grids

FORTRAN routines have been written to handle interpolations and boundary data exchange, while the driver of the whole procedure is a simple shell script calling the two solvers (for the conduction and flow problem, respectively) and the CHT program. In this way we avoid any change in the single solver, keeping the maximum flexibility in their choice. Obviously, since we have to restart the fluid solver at each time step, we could have significant overhead due to the initial readings (restart and grid files). If access to the source is available (this can be a problem with some commercial codes), different and more efficient approaches are possible, ranging from the use of FIFO (first in/first out) input files to the use of C language sockets routines.

RESULTS AND DISCUSSION

The procedure was tested on the gas turbine blade shown in fig.1, whose external geometry has been used as the AGARD Test case E-CA6[11]. Grid nodes don't match exactly at the interface due to the automatic refinement from the unstructured grid generator[5], and a suitable (and conservative) interpolation and integration routine is used to handle the mismatch. A picture of the two different grids, for the fluid and solid domain, is given in Figure 1, with details of the leading edge region.

In order to test the stability of the coupling process, we considered a simple cooling system with radial ducts inside the blade, as described, as an example, in reference[12]. Three different configurations, with different duct positions and sizes, have been tested. The solid grids for each of the geometries are shown in Figure 2.

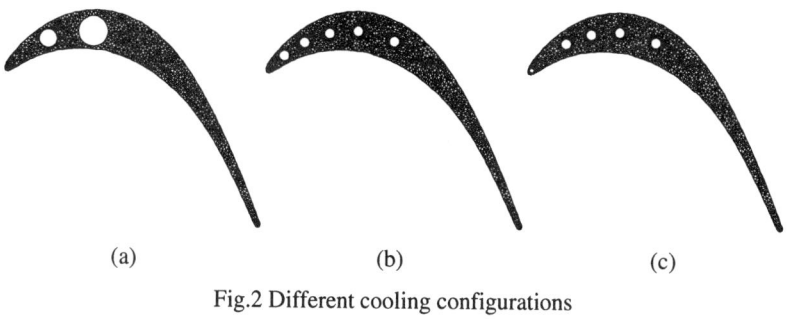

(a)   (b)   (c)

Fig.2 Different cooling configurations

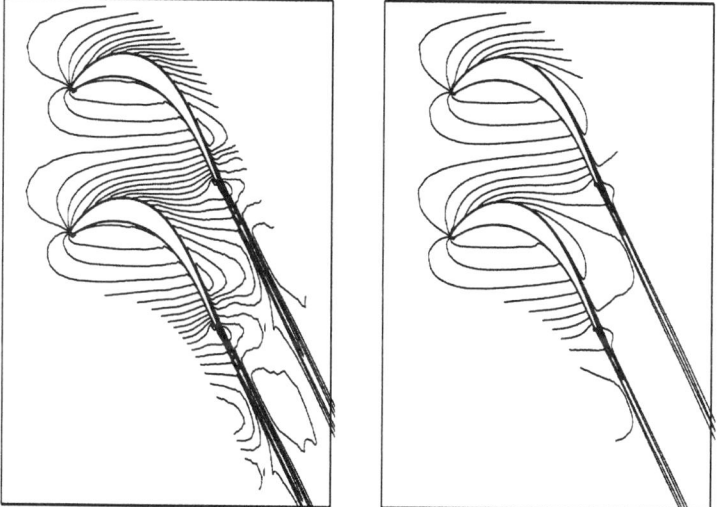

Fig.3 External flow results, Mach number contours. Left: Ma_{2is}=0.95; right: Ma_{2is}=0.59

Computations have been carried out for two different external flow isentropic exit Mach number, Ma_{2is}=0.59 and $Ma2_{is}$=0.95. Mach number contours for blade geometry (b) (Fig.2) and both values of $Ma2_{is}$ are presented in Fig.3. The flow grid is probably not enough refined to sharply capture the shock at the highest Mach, as is shown in Fig.4.

Cooling configurations have little or no effect on the pressure distribution, at least at low Mach number, as can be observed from the closely superimposed wall pressure coefficient presented in Fig.5. Different cooling system, however, have obviously a strong impact on wall temperature distribution, as shown in Fig.6.

On the internal side we assumed the same average velocities and the same correlations for the heat transfer coefficient. From Fig.6, thus, it appears that configurations (b) and (c) allows better cooling around the leading edge, with a smaller mass flow rate with respect to geometry (a), due to the reduced duct area.

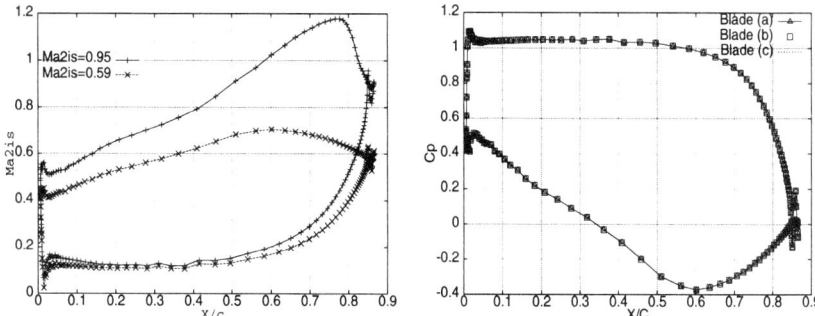

Fig.4 - Wall isoentropic Mach number   Fig.5 - Wall pressure coefficient, $Ma_{2is}=0.59$

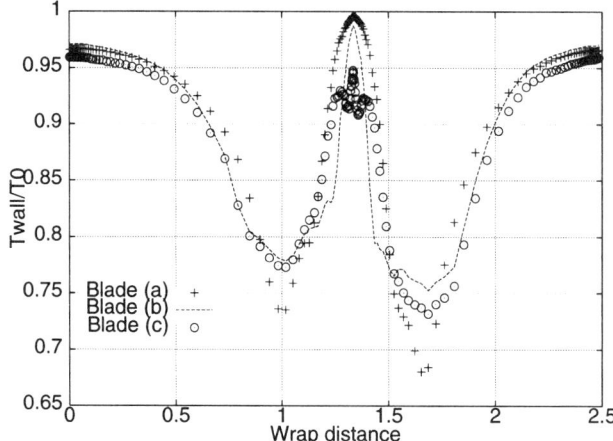

Fig.6 Wall temperature vs wrap distance. Suction side to the left, pressure side to the right

In the following Figures 7,8 sample temperature fields for different geometries and flow conditions are presented.

In Fig.9, finally, we show a convergence history for the flow problem. Computation was started from adiabatic solution, $Ma2_{is}=0.95$. The number of pseudo times step required for final convergence is of the same order of magnitude of the number required for a stand-alone computation. On a Linux PC at 700 Mhz each pseudotime step requires about one second of CPU time. This means that the present closely coupled CHT interaction is more efficient than the standard approach involving several fully converged or nearly converged solutions of eq.(**8,9**).

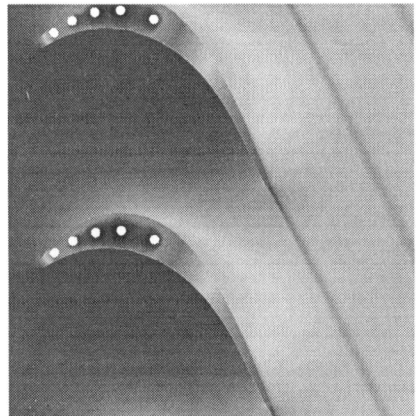 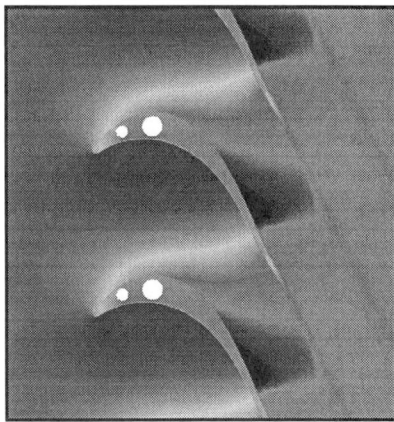

Fig.7 Temperature field. Left, Ma_{2is}=0.59, case(b); right: Ma_{2is}=0.95, case (a)

 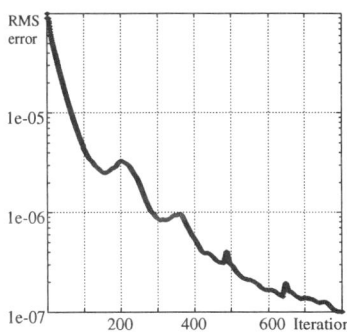

Fig.8 Temperature field, Ma_{2is}=0.95, case (a)   Fig.9 Convergence history, Ma2is=0.95, case (a)

CONCLUSIONS

The above outlined conjugate heat transfer procedure allows an easy coupling of different existing codes. The procedure, in fact, does not rely on the details of the numerical conduction/convection solutions; furthermore, interaction with conduction/convection solution is achieved only through their input and output files. This allow the use of any kind of solvers (FEM, FVM, FDM, in-house codes, commercial ones ...) simply by changing the input/output format. Finally, interpolation techniques allow the use of non-matching grids, obtaining the same flexibility also in the choice of grid generation. In the present paper, example of the utility of such flexibility is given by the use of both structured and unstructured grids.

Results show that the convergence rate of the procedure is of the same order of magnitude of stand-alone computation, and, thus, can be considered fully satisfactory. Finally, even in a simple test case the CHT computation has shown its utility in order to clearly identify the effect of different cooling system design.

REFERENCES

1. Croce, G.. 1995. Viscous 3D Cascade Flow Analysis Using an RNG Algebraic Turbulence Model. ASME Paper 95-CTP-78, Presented at ASME Cogen Turbo-Power Conference. Vienna, Austria, Aug. 23-25
2. Croce, G. 1996, Heat Transfer Analysis in a 2D Turbine Cascade. *In* Advanced Computational Methods in Heat Transfer IV. L.C. Wrobel, G. Comini, C.A. Brebbia, A.J. Nowak, Eds. WIT Press. Southampton, UK.
3. Marini, M., 1991. An Elliptic Technique to Generate Orthogonal Grids Along Boundaries, *FED, Fluid Machinery Forum*, Book G00607.
4. Shewchuk, J.R. 1996. Triangle: Engineering a 2D Quality Mesh Generator and Delaunay Triangulator. *In* Applied Computational Geometry: Towards Geometric Engineering, M. C. Lin, D. Manocha eds., Springer Verlag.
5. Funaro, D., A. Quarteroni, P. Zanolli. 1988. An iterative procedure with interface relaxation for domain decomposition methods. SIAM J. Numer. Analysis. **25**(6): 1213-1236.
6. Giles, M.B. 1997. Stability Analysis of Numerical Interface Conditions in Fluid-Structure Thermal Analysis. Int. J. for Num. Meth. in Fluids. **25**: 421-436.
7. Chen, Y., M. Fiebig, N.K. Mitra. 1998. Conjugate Heat Transfer of a finned oval tube. Part A: Flow Patterns. Numer. Heat Transfer, Part A. **33**(4): 371-385.
8. Cheng, C.H., J.H. Yu. 1999. Conjugate Heat Transfer in an Inclined Slab with an Array of Horizontal Channels. Numer. Heat Transfer, Part A. **35**(7): 779-796.
9. Croce, G., W.G. Habashi, G. Guevremont, F. Tezok, 1998. 3D Thermal Analysis of an Anti Icing Device using FENSAP-ICE. AIAA Paper 98-0193. 36° Aerospace Meeting and Exhibition, Reno.
10. Croce, G., W.G. Habashi, 2000. Thermal Analysis of Wing and Nacelle Anti-Icing Devices. *In* Computional Analysis of Convection Heat Transfer. B. Sunden, G. Comini, Eds. WIT Press. Southampton, UK. In press.
11. Fottner L., Ed. 1990. Test Cases for Computation of Internal Flows in Aero Engine Components. AGARD AR 275: 17-20
12. Carcasci C., Facchini B. & Ferrara G. 1995. A Rotor Blade Cooling Design Method for Heavy Duty Gas Turbine Applications. ASME Paper 95-CTP-90. Presented at ASME Cogen Turbo-Power Conference. Vienna, Austria, Aug. 23-25

Heat/Mass Transfer Characteristics on Turbine Shroud with Blade Tip Clearance

H. H. CHO, D. H. RHEE and J. H. CHOI

Department of Mechanical Engineering
Yonsei University, Seoul 120-749, KOREA

ABSTRACT : The present study is conducted to investigate th local heat/mass transfer characteristics on the shroud with blad tip clearances. The relative motion between blade and shroud i neglected, and four-bladed linear cascade is used in this study. naphthalene sublimation method is employed to determine th detailed local heat/mass transfer coefficients on the surface o shroud. The tip clearance is changed from 0.66% to 2.85% of th blade chord length. The flow enters the gap between the blade ti and shroud at the pressure side due to the high pressur difference. Therefore, the heat/mass transfer characteristics o the shroud are changed significantly from those with endwall. A first, high heat/mass transfer occurs along the profile of blade a the pressure side due to the entrance effect and acceleration of th gap flow. Then, the heat/mass transfer coefficients on the shrou increase along the suction side of the blade because tip leakag vortices are generated and interact with the main flow. Th results show that the heat/mass transfer characteristics ar changed largely with the gap distance between the tip of turbin blade and the shroud.

INTRODUCTION

The clearance between the tip of rotating component (compressor or turbine blade) and the shroud is unavoidable in gas turbine engines. The leakage flow across the gap is caused by the pressure difference between the pressure side and suction sides of blade. The leakage flow can have significant effects both on the stage aerodynamic performance and also on the structural durability of the blade[1]. In the case of turbine blades, hot spots may occur in the blade tip and shroud regions. Therefore, the cooling schemes, such as film cooling, are applied to improve the durability of the components. Thus, the understanding of heat transfer characteristics at the blade tip and shroud is imperative for effective cooling of components required to improve durability. However, very limited experimental data have been reported in either stationary or rotating cascade environments.[2]

Metzger et al.[1] studied heat transfer on the blade tip and shroud, and provided the numerical model for early design estimates of tip and shroud heat transfer. Bunker et al.[2] and Ameri and Bunker[3] investigated the heat transfer characteristics on the flat and round-edge blade tip. They found that low transfer region is formed at leading edge region, and separation vortex at pressure side and leakage flow affect the heat transfer on blade tip region significantly. They also reported that the heat transfer with round tip geometry is higher than that with flat tip due to higher allowed tip leakage flow. Azad et al.[4,5] studied the flow/heat transfer characteristics on the blade tip region for flat tip and recessed tip geometry. They reported that overall heat transfer increases as tip clearance increases, and heat transfer for the recessed tip is lower than that for the flat tip for the same tip clearance. Dunn and Haldeman[6] used heat flux gages to investigate time-resolved heat flux on a blade tip region for the rotating turbines with the variations of spacing between stator and rotor.

The present study is conducted to investigate the local heat/mass transfer characteristics on the turbine shroud with blade tip clearance. The stationary linear turbine cascade is used to investigate heat/mass transfer characteristics on the shroud with different tip clearances between the blade tip and the shroud. Many researchers have reported that the relative motion has little influence on the overall heat transfer characteristics.[7] The relative motion between the blade and shroud is neglected in this study. A naphthalene sublimation method is employed to determine the detailed local heat/mass transfer coefficients on the surface of shroud. This technique eliminates the conduction error inherent in heat transfer experiments. The surface boundary condition is analogous to an isothermal surface in a corresponding heat transfer problem.

EXPERIMENTAL APPARATUS AND PROCEDURE

Experimental Apparatus

The wind tunnel used for this experiment is an open circuit and blowing type. The test section is 300 mm wide by 195 mm high and the area ratio of the contraction section to this is 6:1. A trip wire of diameter 3.0 mm is attached at the beginning of the test section to ensure fully developed turbulent boundary layer in the test section. Nominal velocity of inlet flow is maintained at 8 m/s during the experimental tests, and exit flow velocity at approximately 20 m/s. The corresponding Reynolds number based on chord length and exit flow velocity is 2.1×10^5. Average turbulence intensity of mainstream is measured as 0.7% without grid. Thin guide plate is installed at the flow passage between blade 1 and 2, and adjustable trailing edge tailboards are placed to obtain identical flow patterns in passages as shown in Fig. 1(a).

The gap distance is changed from 0 mm to 4.56 mm to investigate the effect of tip clearance. The blade is fixed to the aluminum plate, which is positioned below the bottom wall of test section. Desired tip clearance is obtained by traversing the plate in vertical direction as shown in Fig. 1(b). Naphthalene coated test plate is installed at the top of the test section to measure the heat/mass transfer coefficients on the shroud, and the geometry of shroud is flat.

The pressure taps are placed at the mid-span and the height of 95% of the span of blade 2 to measure the static pressure distributions at the mid-span and near the tip region.

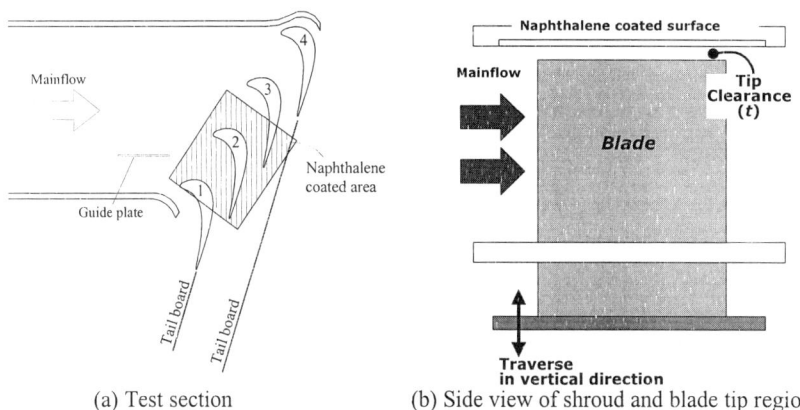

(a) Test section (b) Side view of shroud and blade tip region
Fig. 1 Schematic view of experimental apparatus.

Table 1 Cascade geometry data and operating conditions

Number of blade	4
Chord length of blade (C)	160 mm
Axial chord length (C_x)	112 mm
Pitch to chord ratio (p/C)	0.7
Aspect ratio (l/C)	1.22
Blade inlet angle	35°
Blade exit angle	-72.5°
Inlet flow velocity(U_1)	8 m/s (Mach number, M_1=0.024)
Turbulence intensity of incoming flow	~0.7%
Exit flow velocity (U_2)	20 m/s (Mach number, M_2=0.059)
Reynolds number based on U_2 and C (Re_C)	2.1×10^5
Tip clearance (t)	0 ~ 4.56 mm (0~2.85% of chord length)

To check the identity of flow patterns in each passage, 7 pressure taps are placed at the mid-span of blade 3. Cascade geometry and operating conditions are listed in Table 1.

Data Acquistion

In order to obtain local mass transfer coefficients, the profile of the naphthalene surface coated on the test plate is scanned by an automated surface measuring system before and after exposure to air flow. Sublimation depth during the run is calculated from the difference of the surface profiles. The measuring system consists of a depth gauge, a linear signal conditioner (LUCAS ATA-101), a digital multimeter (Keithley model 2001), two stepping-motor driven positioners, a motor controller, and a personal computer with GPIB (IEEE-488) board. The depth gauge is a Linear Variable Differential Transformer (LVDT) made by Schaevitz Engineering (LBB-375TA-020), which has a resolution of 0.025 μm. Error of the LVDT measurements on a flat plate is within 1% of averaged sublimation depth of 40 μm during the run. The automated system typically obtains more than two thousand data points in an hour.

Heat/Mass Transfer Coefficient

The local mass transfer coefficient is defined as:

$$h_m = \frac{\dot{m}}{\rho_{v,w} - \rho_{v,\infty}} = \frac{\rho_s (dy/d\tau)}{\rho_{v,w}} \quad (1)$$

since inlet air contains no naphthalene, $\rho_{v,\infty}=0$ in the present study. Therefore, the mass transfer coefficient is calculated from the local sublimation depth of naphthalene (dy), run time (dτ), density of solid naphthalene (ρ_s), and naphthalene vapor density ($\rho_{v,w}$). The naphthalene vapor pressure is obtained from a correlation of Ambrose et al.[8]. Then the naphthalene vapor density, $\rho_{v,w}$, is calculated from the perfect gas law.

The Sherwood number can be expressed as:

$$\text{Sh} = h_m C / D_{naph} \quad (2)$$

D_{naph} is based on the discussion of naphthalene properties given by Goldstein and Cho[9].

The mass transfer coefficients can be converted to the heat transfer coefficients using the heat and mass transfer analogy[10].

$$\text{Nu/Sh} = (\text{Pr}/Sc)^{0.4} \quad (3)$$

Uncertainty of the Sherwood numbers using Kline and McClintock's[11] method for single sample experiments, considering the measured temperature, depth, position and correlation equations, is within 7.1% in the entire operating range of the measurement, based on a 95% confidence interval. This uncertainty is attributed mainly to the uncertainty of properties of naphthalene, such as the naphthalene saturated vapor pressure (3.8%), and diffusion coefficient of naphthalene vapor in the air (5.1%). However, uncertainty due to the sublimation depth measurement is only 0.7%. The other uncertainties are 0.2%, 1.1% and 4.9% for T_w, ρ_s and h_m, respectively.

RESULTS AND DISCUSSION

Static Pressure Measurement

Figure 2 presents static pressure coefficients at the mid-span of blade 2 and 3. The static pressure coefficient is defined as :

$$Cp_s = (p_{01} - p_s)/(0.5 \rho_1 U_1^2) \quad (4)$$

where p_{01} is the inlet total pressure. Figure 2 verifies that the flow patterns in the passages are almost identical.

The distributions of static pressure coefficients at 95% of blade span for each tip clearance are shown in Fig. 3. Higher Cp_s means lower static pressure, and vice versa. The leakage flow is generated due to the pressure difference between the pressure side and suction side as shown in Fig. 3. Static pressure coefficients at the pressure side show almost the same values for all the tip clearances, however a little higher values are observed with the larger tip gap due to the flow entrance effect. At the suction side, the peak value moves downstream as the tip clearance increases, and this may be due to the redistribution of leakage flow and increase.

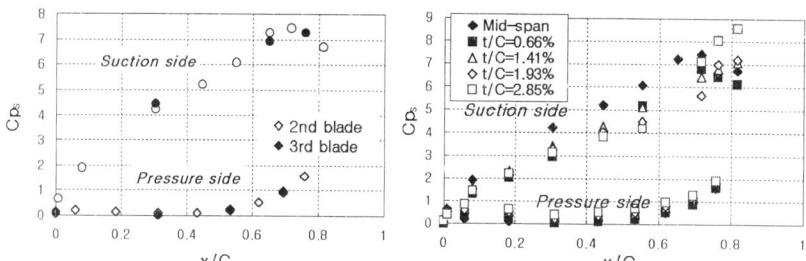

Fig. 2 Static pressure coefficient at mid-span on blade 2 and 3.

Fig. 3 Static pressure coefficient at 95% of blade span on blade 2.

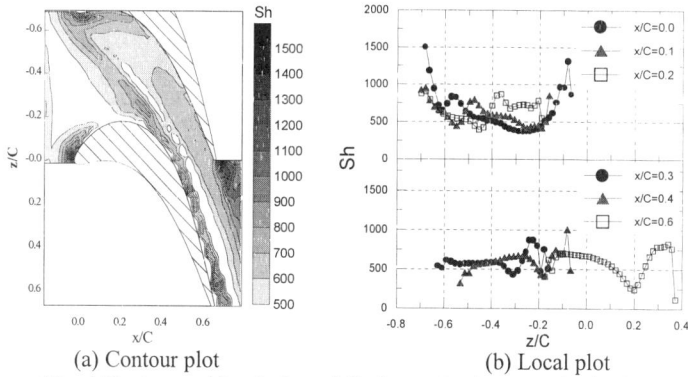

(a) Contour plot (b) Local plot
Fig. 4 Contour and local plots of Sh for no tip clearance (endwall).

Heat/Mass Transfer Measurement

Figure 4 presents the contour plot and local distributions of Sh on the endwall of blade. The heat/mass transfer characteristics with no gap distance have been reviewed by many researchers.[12-14] As one can expect, the heat/mass transfer enhancement due to horseshoe vortex is observed near the leading edge of blade, and its maximum value is about 1500. As the flow goes, a passage vortex is generated, and the region with high heat/mass transfer is formed along the suction side of blade. Additional peak values caused by the generation of wake at the trailing edge are observed.

Some spots and wavy pattern are shown along the suction side of blade in the contour plot (Fig. 4(a)). This is due to the problem in conversion of local data into the contour plot, and not real situation.

With the gap distances between the blade tip and shroud, the flow enters the gap at the pressure side due to the pressure difference. Therefore, heat/mass transfer characteristics are changed significantly with the tip clearance, and the heat/mass transfer coefficients are much higher than those on the endwall with no tip clearance.

Figure 5 shows the contour plots of Sh for various tip clearances. Heat/mass transfer is enhanced near the leading edge due to the horseshoe vortex, and Sh at this region increases as the tip clearance decreases. For extremely small tip clearance (Fig. 5(a)), horse-shoe vortex legs are observed.

For the extremely small tip gap, $t/C=0.66\%$ (Fig. 5(a)), heat transfer patterns near the

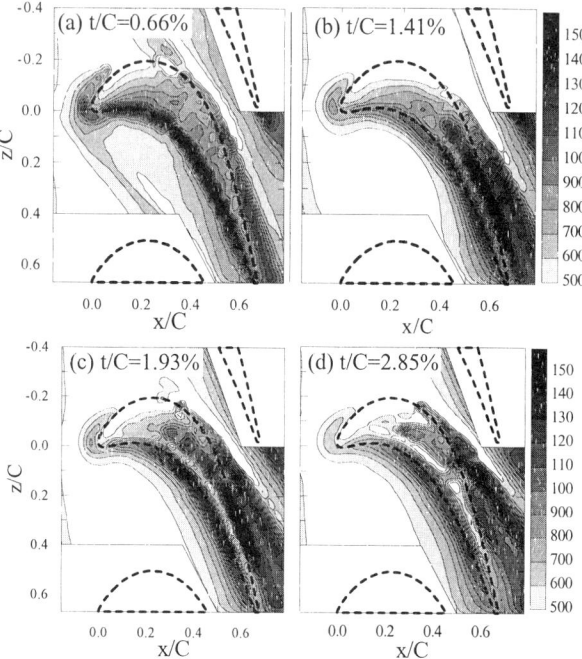

Fig. 5 Contour plots of Sh for various tip clearances.

pressure side are similar to that for no tip gap (Fig. 4) due to flow resistance against the leakage flow. In addition, high heat/mass transfer coefficients are observed along the profile of pressure side due to the separation vortex at the blade tip. Heat/mass transfer coefficients in the region between the blade tip and shroud are low, and the effect of leakage flow on heat/mass transfer near the suction surface is weak because the amount of leakage flow is very small. Whereas, the passage vortex originated from the leading edge region of adjacent blade affects the heat/mass transfer near the suction side, as shown in Fig. 5(a).

For the moderate tip clearances (Figs. 5(b), (c) and (d)), high heat/mass transfer region is observed along the pressure side due to the entrance effect and the acceleration of the tip gap flow. For the relatively small tip clearance (Fig. 5(b)), the enhanced region due to flow entrance is smaller than that for the large tip clearance. However, the peak value at this region is higher for small tip clearance. This pattern may be related to the local velocity of the entering flow. Complex heat transfer patterns are observed in the region between the blade tip and shroud due to the flow separation and reattachment for the moderate tip clearances.

The entering flow exits through the suction side of blade, and interacts with mainstream. Therefore, large-scale tip leakage vortex is generated on the suction side and heat/mass transfer enhanced region is observed along the tip leakage vortex path. For the small tip clearance (Fig. 5(b)), the flow starts to exit the tip gap at $z/C=-0.1$ and $x/C=0.4$. However, for large tip clearance, the position at which the flow starts to exit moves upstream; at $z/C=-0.15$ and $x/C=0.3$ for $t/C=2.85\%$. The reason is that the amount and

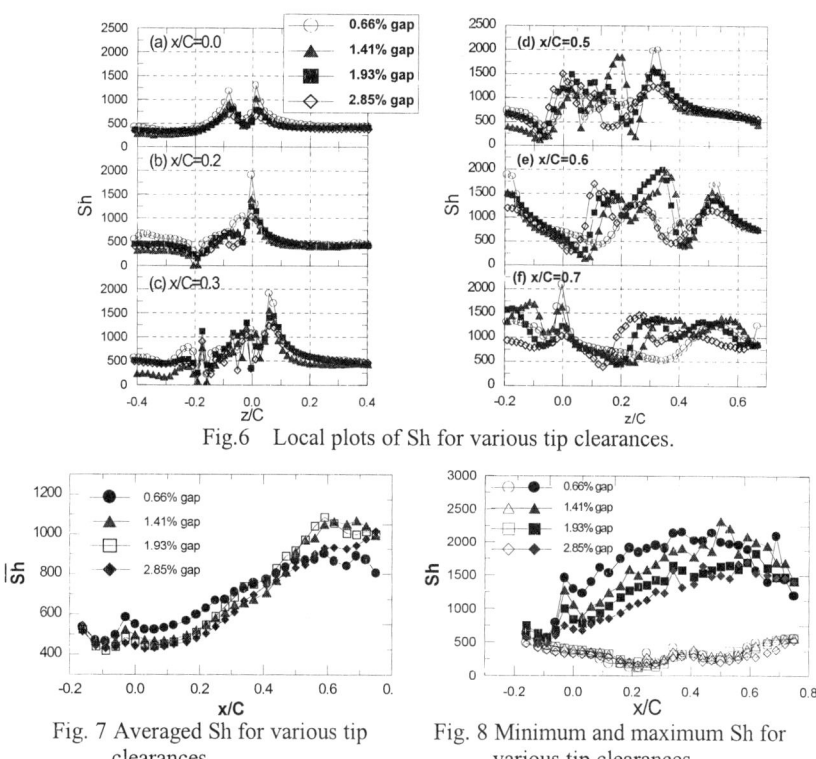

Fig.6 Local plots of Sh for various tip clearances.

Fig. 7 Averaged Sh for various tip clearances.

Fig. 8 Minimum and maximum Sh for various tip clearances.

momentum of the leakage flow increase as tip clearance increases, therefore the leakage flow can easily follow the inlet flow angle. Also, the trajectory of leakage flow moves in the direction of negative z/C, and the region affected by entering/leakage flow increases as the tip gap increases.

Figure 6 presents the local distributions of Sh for various tip clearances. As mentioned before, the effect of horseshoe vortex is decreased as the tip clearance increases (Fig. 6(a)). For t/C=0.66%, the effect of leakage flow is hard to observe. As the tip clearance increases, Sh near the pressure side is decreased and the peak value due to the leakage vortex moves in the direction of negative z/C.

Figure 7 shows the averaged Sh for various tip clearances. The averaged values are obtained by numerical integration in one blade-to-blade pitch. As one can expect, the averaged Sh shows a minimum value at about x/C=0.1, and increases gradually. Then the averaged Sh shows a maximum value at about x/C=0.6 because the effect of leakage vortex increases. The averaged Sh shows higher value for t/C=0.66% at the upstream region (0.0<x/C<0.4) due to the effect of horseshoe vortex. However, at the downstream region, the averaged Sh shows lower value than those for other cases due to the weak effect of leakage flow. The tip clearance has little influence on the averaged value for moderate tip clearances, however a little higher value is obtained with relatively smaller tip clearances.

Shroud region is subjected to periodic thermal load due to the rotation of turbine blade.

The peak heat transfer occurs as the blade tip and its overtaking leakage flow passes the shroud[1], while the low heat transfer occurs at the mid-way region between adjacent two blades. The minimum and maximum values of Sh on the shroud are shown in Fig. 8. The maximum values are about 4~5 times higher than the minimum values for each tip clearance and the difference increases as x/C increases due to the effect of leakage vortex.

CONCLUSIONS

Heat transfer characteristics on the shroud surface for various tip clearances are investigated. The results are summarized as follows:
(1) High heat/mass transfer region is observed along the pressure side due to the entrance effect and the acceleration of the tip gap flow. Complex heat transfer patterns are observed in the region between the blade tip and shroud due to the flow separation and reattachment. A large-scale tip leakage vortex is generated as the leakage flow exits through the gap and the heat/mass transfer enhanced region is observed along the tip leakage vortex path.
(2) As the tip gap increases, the trajectory of leakage flow moves in the direction of negative z/C and the affected region is increased due to the increase of leakage flow. However, the tip clearance has little influence on the averaged value.
(3) Maximum values of Sh on the shroud are about 4~5 times higher than minimum values for each tip clearance. Therefore, the shroud may be subjected to severe thermal stresses due to the wide variation of heat transfer coefficients.

ACKNOWLEDGEMENT

This research was supported by Korea Research Foundation under the grant No. KRF-99-042-E0005 and by Ministry of Science & Technology through their National Research Laboratory program.

REFERENCES

1. Metzger, D. E., Dunn, M. G. and Hah, C., 1991, Turbine Tip and Shroud Heat Transfer, *Trans. ASME, Journal of Turbomachinery*, Vol. 113, pp. 502-507.
2. Bunker, R. S., Bailey, J. C. and Ameri, A. A., 1999, Heat Transfer and Flow on the First Stage Blade Tip of a Power Generation Gas Turbine Part 1:Experimental Results, *ASME Paper* No. 99-GT-169.
3. Ameri, A. A. and Bunker, R. S., 1999, Heat Transfer and Flow on the First Stage Blade Tip of a Power Generation Gas Turbine Part 1:Simulation Results, *ASME Paper* No. 99-GT-283.
4. Azad, Gm S., Han, Je-Chin, Teng, S. and Boyle, R. J. 2000, Heat Transfer and Pressure Distributions on a Gas Turbine Blade Tip, *ASME Paper* No. 2000-GT-194.
5. Azad, Gm S., Han, Je-Chin and Boyle, R. J. 2000, Heat Transfer and Flow on the Squealer Tip of a Gas Turbine Blade, *ASME Paper* No. 2000-GT-195.
6. Dunn, M. G. and Haldeman, C. W. 2000, Time-aeraged Heat Flux for a Recessed Tip, Lip, and Platform of a Transonic Turbine Blade, *ASME Paper* No. 2000-GT-197.
7. Mayle, R. E. and Metzger, D. E., 1982, Heat Transfer at the Tip of an Unshrouded Turbine Blade, *Proceeding of 7th International Heat Transfer Conference*, Hemisphere Pub., pp. 87-92.
8. Ambrose, D., Lawrenson, I. J. and Sparke, C. H. S., 1975, The Vapor Pressure of Naphthalene, *J. Chem. Thermo.*, Vol. 7, pp. 1173-1176.
9. Goldstein, R. J. and Cho, H. H., 1995, A Review of Mass Transfer Measurement Using Naphthalene Sublimation, *Experimental Thermal and Fluid Science*, Vol. 10, pp. 416-434.
10. Eckert, E. R. G., 1976, Analogies to Heat Transfer Processes, in *Measurements in Heat Transfer*, ed. Eckert, E. R. G. and Goldstein, R. J., pp. 397-423, Hemisphere Pub., New York
11. Kline, S. J. and McClinetock, F., 1953, Describing uncertainty in single sample experiments, *Mech. Engineering*, Vol. 75, pp. 3-8.
12. Jabbari, M. Y., Goldstein, R. J., Marston, K. C. and Eckert, E. R. G. 1992, Three dimensional flow at the junction between a turbine blade and end-wall, Wärme- und Stoffübertragung 27, pp. 51-59.
13. Langston, L. S., Nice, M. L. and Hooper, M. R. 1977, Three Dimensional Flow Within a Turbine Cascade Passage, *Trans. ASME, Journal of Engineering for Power*, Vol. 99, pp. 21-28.
14. Goldstein, R. J. and Spores, R. A. 1988, Turbulent Transport on the Endwall in the Region Between Adjacent Turbine Blades, Trans. ASME, Journal of Heat Transfer, Vol. 110, pp. 862-869.

Heat Transfer and Flow Characteristics on a Gas Turbine Shroud

Masakazu OBATA[1], Masaya KUMADA[2] and Nobuaki IJICHI[3]

[1] Division of Mechanical Engineering, Kanazawa Institute of Technology
Ohogigaoka, Nonoichi, Ishikawa 921-8501, Japan
[2] Department of Mechanical Engineering, Gifu University
Yanagido, Gifu 501-1112, Japan
[3] Industrial Machine and Plant Development Centre, Ishikawajima-Harima
Heavy Industries Co., Ltd.
Toyosu, Koto-Ku, Tokyo 135-8732, Japan

ABSTRACT: The work described in this paper is an experimental investigation of the heat transfer from the main flow to a turbine shroud surface, which may be applicable to ceramic gas turbines. Three kinds of turbine shrouds are considered with a flat surface, a taper surface and a spiral groove surface opposite to the blades in an axial flow turbine of actual turbo-charger. Heat transfer measurements were performed for the experimental conditions of a uniform heat flux or a uniform wall temperature. The effects of the inlet flow angle, rotational speed, and tip clearance on the heat transfer coefficient were clarified under on- and off-design flow conditions. The mean heat transfer coefficient was correlated to the blade Reynolds number and tip clearance, and compared with an experimental correlation and measurements of a flat surface. A comparison was also made for the measurement of static pressure distributions.

INTRODUCTION

The thermal efficiency and specific power output of modern gas turbines have been improved by an increase in the turbine inlet temperature and through improvement of the aerodynamic performance. However, the higher inlet temperature produces a larger increase of the cooling air flow-rate to the metallic components, and subsequently the increase of overall thermal efficiency is not previously expected to be larger. One attempt to decrease the cooling air flow-rate and to increase the turbine inlet temperature is to develop ceramic gas turbines, in which fine ceramics adapt to various turbine components within acceptable limits[1,2]. In the development of high temperature ceramic gas turbines, a thorough knowledge

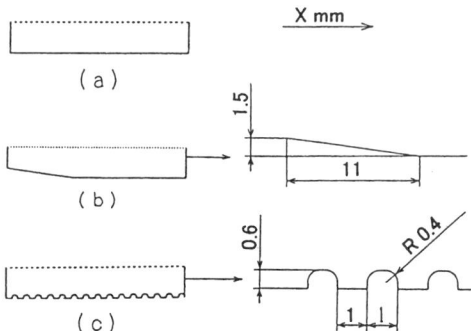

Fig. 1 Configuration of shroud surfaces opposite to the blade tip.

of turbine heat transfer and aerodynamics is necessary to establish a new technology to absorb the thermal expansion between ceramic and metallic parts, and to protect high thermal heat loads. Ceramics have a superior thermal resistance compared with super-alloys in the temperature up to approximately 1,400°C, but they have a tender property in fracture toughness, thermal shock, and scatter of strength. Therefore, the gas turbine with ceramic turbine blades and a ceramic abradable shroud is especially required to avoid the collision between the blade tip and the shroud surface, and to keep the tip clearance to a minimum.

Several efforts have been made to measure the heat transfer coefficient on a shroud surface opposite to turbine blades aiming at the development of a ceramic abradable shroud for axial turbines[3,4,5], to clarify the complex flow phenomenon between the blade tip leakage flow and the blade passage flow intending the improvement of aerodynamic performance of turbines[6,7,8]. In these studies, the configuration of the shroud was considered to be a flat surface with a constant tip clearance applicable for metallic turbines. However, this seems to be unsuitable for a shroud surface of ceramic turbines, for which an inconstant surface is more suitable to reduce the contact between the blade tip and the shroud ring surface.

In the present study, three types of shroud surfaces in an axial turbine were considered and these are a flat surface (a), a taper surface (b) and a spiral groove surface (c) as shown in Fig.1. On the flat surface, the heat transfer and flow characteristics have been presented in the previous paper[4] by authors. The result of tip leakage flow measurements for the flat shroud surface was reflected in the selection of the taper surface and the spiral groove surface. The taper surface has a larger tip clearance in the axial region where the equivalent surface of the shroud between the leading-edge region of blades and the middle region shows a relatively little tip leakage flow. The spiral groove surface has spiral grooves with a slant 20-degrees hollow to the circumferential direction to keep the tip leakage flow as little as possible and to reduce a contact area with blade tips as well.

The shroud test facility provides a well-known aerodynamic condition at the inlet to the rotor in an axial flow turbine. Heat transfer measurements were performed for the experimental conditions of a uniform heat flux or a uniform wall temperature. The effects of the inlet flow angle, rotational speed and tip clearance on the heat transfer coefficient were clarified under on- and off-design flow conditions. The mean heat transfer coefficient was correlated to the blade Reynolds number and tip clearance, and compared with an experimental correlation and measurements of the flat surface. A comparison was also made for the measurement of static pressure distributions.

EXPERIMENTAL APPARATUS AND PROCEDURE

An axial-flow turbine of the IHI VTR-161 type turbocharger was provided for all measurements. Figure 2 shows a cross-sectional view of the test turbine which has a rotor of an outer diameter 169 mm, 53 blades and a hub to tip ratio of 0.67. The blade chord length is 13.14 mm at the tip section. The principal experimental conditions and the rotor blade geometry are shown in Table 1, and the fundamental velocity diagram is indicated in Fig. 3. The profile is consistent with the tip section of the test turbine and the blade itself has a three-dimensional shape.

The air to drive the rotor was supplied by a blower through a scroll chamber and inlet nozzle guide vanes. The mass flow rate was linearly related to the rotational speed and the inlet flow angle of the rotor blade was controlled by cutting off the impeller of the compressor and/or by partial sealing with a thin aluminum tape at the inlet of the compressor. The rotational speed N was measured by means of a pulse counter.

Figure 4 (a) shows the measuring section with the coordinate system for the taper shroud. The measurement of the local heat transfer coefficient on the surface was performed by applying a well-established thin-film technique, which gives a uniform heat flux condition. Stainless steel foils of 30 μm thick were glued on the taper and flat surfaces of an acrylic block shroud and were directly heated by a stabilized DC source. The local wall temperature was measured with nine Cu-Co thermocouples of 0.07 mm in diameter soldered to the back of the foil. The temperature difference ΔT to calculate the heat transfer coefficient h was obtained from the difference between the temperatures of the foil when it is heated and also when it is not heated, for a heat drop caused by the air passing through the rotor was taken into consideration.

The measuring section with the coordinate system for the spiral groove shroud is also shown in Fig. 4 (b). As a thin-film technique could not easily apply for this configuration, the heat transfer measurement was conducted by a constant wall-temperature technique that can measure the mean heat transfer coefficient. Figure 5 shows the overview of the test shroud, in which the heating element is made of copper block and heated by electric heaters installed inside. The heat loss from the circumferential and back-end surfaces was reduced carefully by means of controlling the temperature of dummy heaters in the copper block and back insulation. The wall temperature on the shroud surface was measured with four Cu-Co

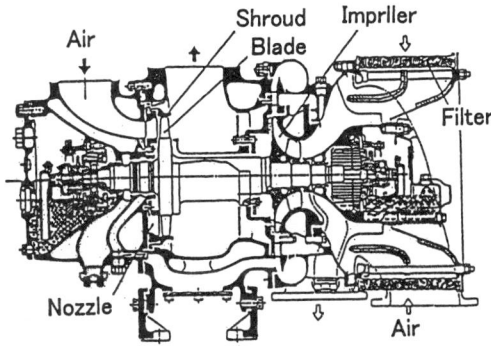

Fig. 2 Cross-sectional view of test turbine.

thermocouples. The mean heat transfer coefficient was calculated by using the wall temperature difference in the heated and unheated test conditions in consideration of the heat drop of the air passing through the rotor.

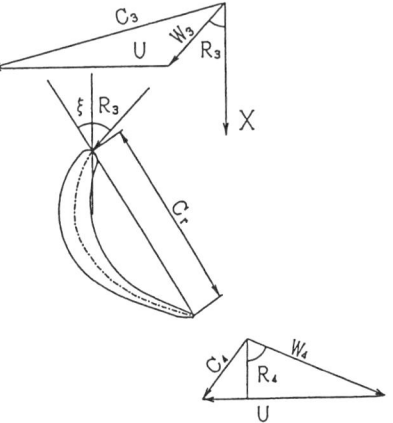

Table 1 Principal experimental conditions and rotor blade geometry.

$Re = C_m C_r / \nu$	$1.39 \times 10^4 \sim 7.03 \times 10^4$
N [rpm]	2,000~14,000
U [m/s]	17.7~123.9
C_r [mm]	13.14
C_3 [m/s]	30.8~112.2
C_4 [m/s]	7.4~80.1
ξ [deg]	36

Fig. 3 Velocity diagram.

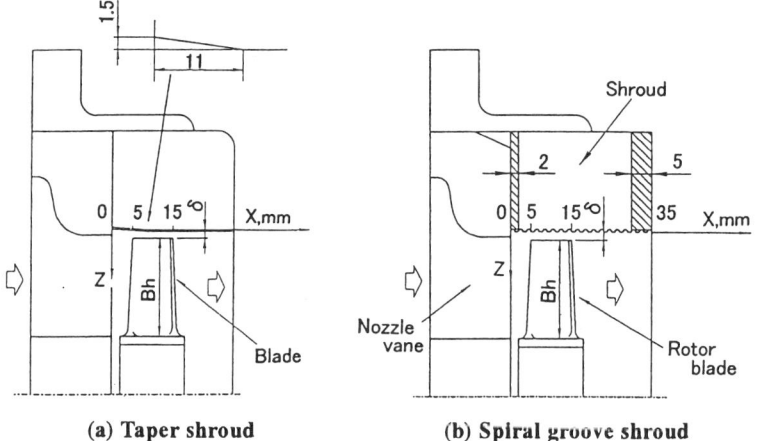

(a) Taper shroud (b) Spiral groove shroud

Fig. 4 Measuring section and coordinate system.

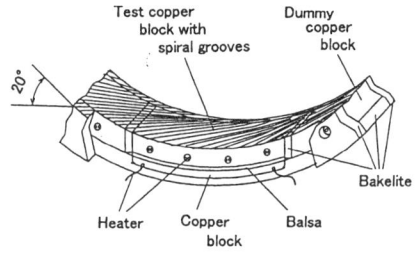

Fig.5 Overview of spiral groove test shroud.

EXPERIMENTAL RESULTS AND DISCUSSION

Local Heat Transfer Coefficient

Local heat transfer coefficients h measured on the taper shroud surface are shown in Figs. 6 (a), (b) and (c). These results are the typical distributions measured by changing the blade Reynolds number Re for the three kinds of non-dimensional tip clearance δ_n at the inlet flow angle R_3 of on-design condition. In comparison with these results, a result for the flat shroud[5] is also shown in Fig. 6 (d). The shroud surface opposite to the rotor blade tip is the region equivalent from 5 mm of X to 15 mm. It is found that the results of the taper shroud indicate a similar distribution in which the changes of the rotational speed and tip clearance have little effect, although the values of local heat transfer become higher as the blade Reynolds number takes a larger value. This tendency is similar to the flat shroud and was also similar to the condition of the inlet flow angle. On the taper shroud, generally, the values of the local heat transfer coefficient in the tapered region decrease compared with those of the flat shroud so that the value of the first peak is evidently defined and the location of the peak moves to the forward region. It is considered that the behavior of a secondary flow caused by tip leakage flow changes substantially.

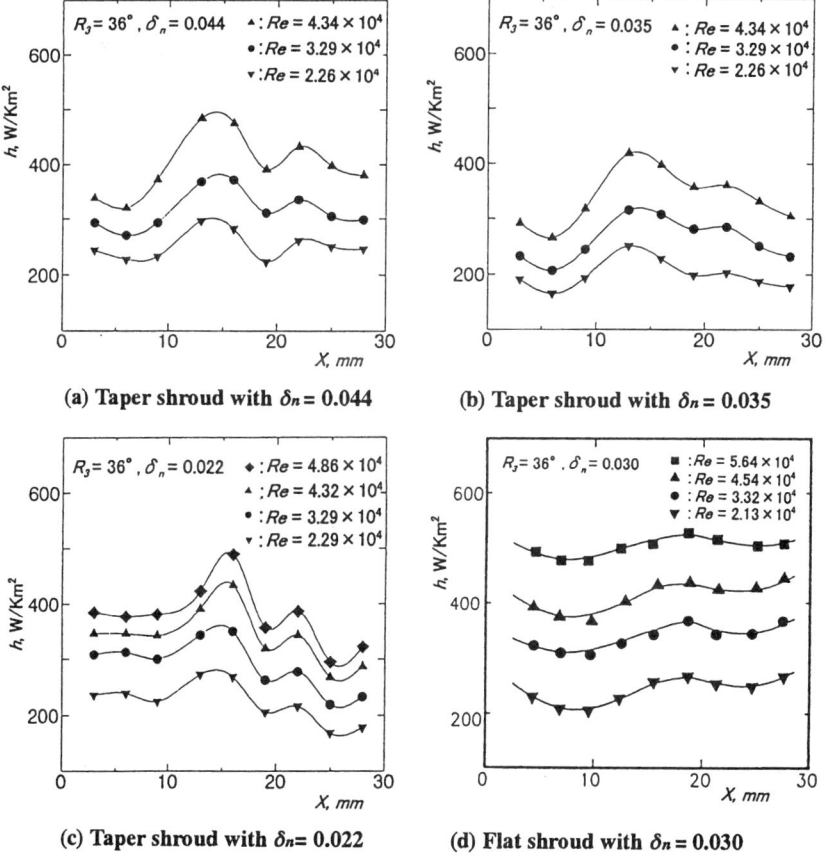

(a) Taper shroud with $\delta_n = 0.044$

(b) Taper shroud with $\delta_n = 0.035$

(c) Taper shroud with $\delta_n = 0.022$

(d) Flat shroud with $\delta_n = 0.030$

Fig. 6 Distribution of local heat transfer coefficient under on-design condition.

Mean Heat Transfer Coefficient

Mean heat transfer coefficients Nu_m measured on three kinds of shroud surfaces for a different tip clearance δ_n in the operating condition of the on-design inlet flow angle are shown against the change of blade Reynolds number in Figs. 7 (a) and (b). Comparable results in the off-design condition are also shown in Figs. 8 (a) and (b) and each result is equivalent to the turbine operating condition of deceleration and acceleration at the same tip clearance shown in Fig. 7 (a). Here, the characteristic length for the definitions of Reynolds number Re and Nusselt number Nu is the chord length at blade tip section. The mean values for the taper shroud and the flat shroud are also calculated from the numerical integration of their local heat transfer values. In these figures, an empirical correlation to a flat surface reported by Karimova et al.[3] is plotted for comparison; the expression is

$$Nu_m = 0.052 Re^{0.8}(1 - 2\delta_n^{0.8}) \qquad (1)$$

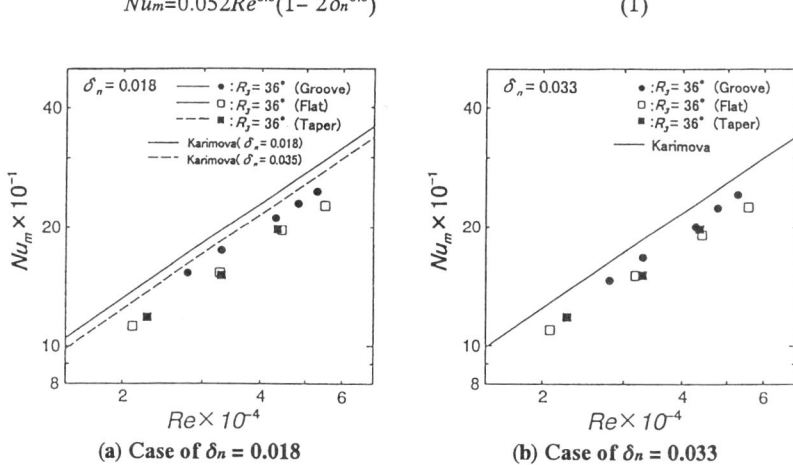

Fig. 7 Mean Nusselt number against blade Reynolds number at on-design inlet flow angle condition.

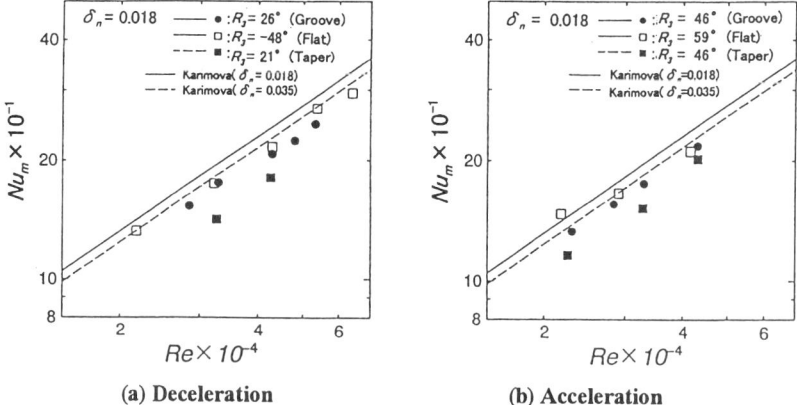

Fig. 8 Mean Nusselt number against blade Reynolds number in off-design operating conditions.

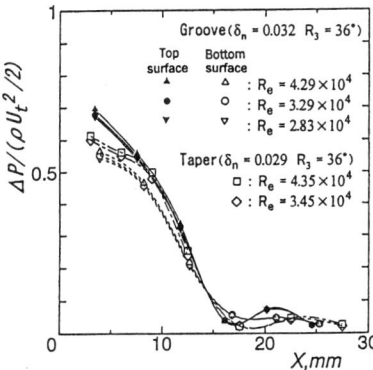

Fig. 9 Comparison of static pressure distributions.

Although a little scatter may be seen in the measured values in Figs. 7 and 8, it is found that they can be fundamentally correlated to the 0.8 power of blade Reynolds number, as shown in the correlation of Karimova et al.[3] which was obtained through an experimental apparatus and a procedure different from those of the present study. This interesting characteristic does not depend on the operating conditions, nor, it seems, on the configuration of the blade surface, at least in the present study.

In general, the measured values in the present experimental conditions seem to be quantitatively smaller than those calculated by the correlation of Karimova et al.[3]. The effect of the tip clearance in the on-design condition was not remarkable and rather the quantitative values were slightly affected by the configuration of the shroud surface as shown in Fig. 7. In the off-design condition, the taper shroud was recognized to suffer a relatively large effect, while the groove shroud and the flat shroud gave a relatively little effect as shown in Fig. 8.

Static Pressure Distribution

Figure 9 shows a comparative static pressure distribution on the taper shroud and groove shroud surfaces at the on-design inlet flow angle, in which the static pressure difference ΔP between wall surface and atmosphere is normalized by the dynamic pressure based on the tip speed. It is found that the wall static pressures do not depend on the blade Reynolds number Re and they decrease rapidly in the region from 5 mm of X to 15 mm opposite to the blade tips with nearly the same distribution. To take the taper shroud for example, the region where the static pressure decreases rapidly coincides with the region where the local heat transfer coefficient increases rapidly. The reason is considered to be that there exists a strongly accelerated flow in this region. Therefore, the flow and heat transfer on the groove shroud may also be characterized by nearly the same flow situation as the taper shroud.

CONCLUSIONS

(1) The local heat transfer coefficient on the taper shroud surface increases as blade Reynolds number Re increases, the axial distribution is similar, and the value decreases in the tapered region compared with that on the flat shroud.

(2) The average heat transfer coefficients on the taper shroud surface and on the spiral shroud surface can be correlated to the 0.8 power of Re similar to a flat plate.

(3) The values increase with the increase of blade Reynolds number Re, decrease with the increase of tip clearance, and are affected a little by the inlet flow angle of blade.

(4) The static pressure distribution on the shroud wall surfaces shows a rapid decrease from the nozzle outlet to the downstream and Re has little effect on the pressure distribution.

NOMENCLATURE

B_h : mean blade height, mm
C : absolute velocity at blade tip, m/s
C_m : mean absolute velocity $=(C_3+C_4)/2$, m/s
C_r : chord length at blade tip, mm
H : heat transfer coefficient, W/(m²K)
N : rotational speed, rpm
Nu : Nusselt number $= hC_r/\lambda$
R : inlet flow angle, deg
Re : Reynolds number $= C_m C_r/\nu$
ΔT : temperature difference between wall temperature while heated and unheated, K
U : blade tip speed, m/s

W : relative velocity at blade tip, m/s
X : distance of axial direction, mm
δ : blade tip clearance, mm
δ_n : relative blade tip clearance $= \delta/B_h$
λ : thermal conductivity, W/(m k)
ν : kinetic viscosity, m²/s
ξ : stagger angle, deg

Subscripts

3 : rotor inlet
4 : rotor outlet
m : average or mean
t : tip

REFERENCES

1. Hara, Y. et al., 1991. Development of a 20 MW Class Ceramic Gas Turbine for Power Generation, In Proceedings of 1991 Yokohama Int. Gas Tubine Congress I : 135-142.
2. Machida, T. et al., 1991. Development of Ceramic Rotor for Industrial Gas Turbine, In Proceedings of 1991 Yokohama Int. Gas Tubine Congress I : 165-170.
3. Karimova, A.G., Lokai, V.I. & Tkachenko, N.S., 1973. Investigation of Heat Release from a gas to the Elements of Turbine body, In Izvestya VUZ aiatsinnaya Tekhnika 16: 114-119.
4. Kumada, M., Iwata, I., Obata, M. & Watanabe, O., 1994. Tip Clearance Effect on Heat Transfer and Leakage Flows on the Shroud-Surface in an Axial Flow Turbine, In ASME Journal of Turbomachinery 116: 39-45.
5. Guenette, G.R., Epstein, A.H., Norton, R.J.G. & Yozhang, C., 1985. Time Resolved Measurements of a Turbine Rotor Stationary Tip Casing Pressure and Heat Transfer Field. AIAA 85-1220.
6. Inoue, M. & Kuroumaru, M., 1989. Structure of Tip Clearance Flow in an Isolated Axial Compressor Rotor, In ASME Journal of Turbomachinery 111: 250-256.
7. Lakshminarayana, B., Pouagare, M. & Davino, R., 1982. Three-Dimensional Flow Field in the TipRegion of a Compressor Rotor Passage, In ASME Journal of Engineering for Power 104: 760-771.
8. Pouagare, M. & Delaney, R.A., 1986. Study of Three-Dimensional Viscous Flows in an Axial Compressor Cascade Including Tip Leakage Effect Using a SIMPLE-Based Algorithm, In ASME Journal of Turbomachinery 108: 51-58.

A Novel Digital Image Processing System for the Transient Liquid Crystal Technique applied for Heat Transfer and Film Cooling Measurements

GREGORY VOGEL, ALBIN BOELCS

Laboratoire de Thermique Appliquée et de Turbomachines (LTT)
Swiss Federal Institute of Technology
1015 Lausanne, Switzerland

ABSTRACT: This paper is dedicated to the transient liquid crystal technique measurements for multiple view access by using a novel digital recording and image processing system.

The transient liquid crystal technique is widely used for heat transfer investigations in turbomachinery. It has been applied in our laboratory in several test facilities such as a linear cascade for external film cooling measurements or on a ribbed squared duct for internal cooling measurements.

The data analysis as well as the measurement equipment is described, with a special focus on the newly developed computerized image processing system suitable to capture the liquid crystal signal.

INTRODUCTION

Heat transfer is an important topic in the development of higher efficiency gas turbines. The principal problem is that the gas temperatures in the first turbine stages (nozzle guide vanes and blades) can largely surpass the metal melting temperatures, which then implies an extensive cooling of these components. Usually a combination of internal convection cooling and external film cooling is employed. The cooling design of these parts have to be of high performance, and due to the complexity of the phenomena, which is quite difficult to correctly model, high quality experimental data plays an important role.

The experimental set-up available at the LTT is able to provide data of internal and external cooling by the use of transient liquid crystal techniques.

This paper is focused on the external film-cooling situation, and on the simultaneous determination of the heat transfer coefficient and film cooling effectiveness within one set of experiments.

Basic Principle

The transient technique measurement consists in monitoring the surface temperature evolution in time triggered by a heat pulse. This heat pulse can be generated by electric heater foils on surfaces or with heater grids in the flow, or by rapidly exposing a preconditioned model to a different temperature level. Transient experiments have the advantage of avoiding conduction problems and to be usually of short duration, hence allowing to perform more measurements in a limited amount of time compared to steady state techniques.

The transient surface temperature can be accurately measured by thermocouples or thin film gauges (on discrete positions) but with a known drawback of a poor spatial resolution. In order to obtain a higher spatial resolution, field methods need to be employed such as the thermo-chromic liquid crystal technique that consist in monitoring the surface temperature evolution by acquiring the color signal of a liquid crystal coating.

By using a single layer of narrow-band thermo-chromic liquid crystals, the evolution of the local surface temperature ($T_{surface}(x,y,time) = T_{LC}$) is obtained from a hue capturing technique of the color play. A data analysis based on a transient one-dimensional heat conduction into a semi-infinite model allows then to determine the heat transfer coefficient and film cooling effectiveness as functions of the following main variables: *time*, T_{LC}, $T_{initial}$, $T_{recovery\ gas}$, $T_{coolant}$. This method has been previously used on flat plates, cylinders and on blade models and has already been described by Drost[1] and Reiss[2].

With the goal of further increasing the quality of the measurements, the hardware as well as the image processing system, used for this transient technique, has recently been completely updated. The use of multi digital camera signals combined with some powerful image processing software allows now to obtain even larger detailed liquid crystal surface measurements in a more user-friendly and versatile way compared to the old system.

DIGITAL IMAGE PROCESSING SYSTEM

Hardware

During the transient experiment, the color variation of the liquid crystal signal coating is captured with several miniature color CCD cameras (TELI Tokyo Electronic Industry Co) through access windows present around the test section. Placement of the small camera heads (diameter 12mm, length 40mm) in reduced area is simplified, as they are separate from the signal processing electronic module boxes. On cascade test facilities for example, four camera views are often used in order to capture the entire blade surface; and for turbine airfoil, in case of film cooling rows, additional zoom-in views on the near hole region of the coolant injection stations are also required.

However, multiple camera use can become problematic as the amount of data collected during the measurements increases rapidly. In the past, specific very expensive devices were used for two camera views only and the liquid crystal signal had to be filtered in real time during the experiment as the computers were limited by the amount of data to be treated. In some cases, images were even lost due to non-sufficient computer capacity and if the signal was of too poor quality or the filtering parameters not well adjusted, the experiment had to be repeated.

Because of all these problems, the real time signal processing has been avoided and replaced by a new system where each camera view is independently recorded on a digital support during the transient experiment. The use of the DV (Digital Video) format storage has been chosen for the advantage of ensuring precise color image signal restitution at a constant image frequency and without any noise generation, which was not the case for an analogue VHS system. Moreover DV tapes automatically contain the time and recording date information that simplifies search of a specific sequence and storing of the different data measurements.

It is only after the transient experiments that the digital video sequences of each camera are transferred one after the other to a computer where the image processing and the data reduction will be performed.

Signal Format

The liquid crystal signal format from the camera view to the computer is resumed in figure 1.

The CCD camera chip is composed of 752 x 582 pixels with a mosaic structure of complementary color filters (Cyan, Magenta, Yellow plus Green). True color is obtained on each pixel location by combining interpolated color plane values of the neighboring pixels, hence color value resolution depends essentially on the proximity of the different color filtered pixels (Theuwissen[3]). In case of a "3 x CCD" camera, (a CCD chip for each component) resolution is higher but the size of such camera is yet too large to be installed on small optical access of test facilities.

The liquid crystal color play, acquired by the CCD chip is converted into an analogue signal. The electronic camera device restitutes this information out to a so-called "S-video" signal. For standard PAL video format, the S-video signal has a resolution of 720 x 576 pixels at a frame rate of 25 images per second.

This S-video signal is then transmitted to a digital video recorder device where it is converted into a digital 24-bits RGB mode, compressed and recorded on a tape in the new standard DV format.

After the transient experiment, DV format sequences are transferred on a computer equipped with a DV in/out IEEE 1394 standard data bus card. DV transfer is done in real time by re-playing from the tapes the recorded sequences. DV in/out software (given by the data bus card manufacturer) often allows to chose between different video formats and spatial resolutions. The standard one is an AVI movie file, coded in a PAL DV format i.e. 25 images per second with a resolution of 720 x 576. In order to be compatible with the image-processing program, the movie file is finally converted into a set of image files in a (24-bits RGB color coded) TIFF format with the full or reduced spatial resolution.

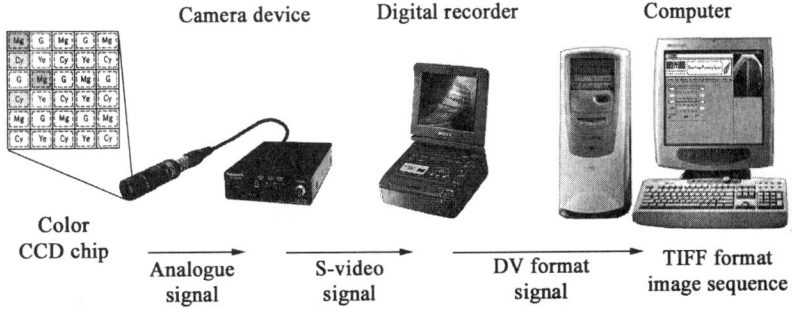

Figure 1: Liquid Crystal Signal Format.

DATA REDUCTION

An image treatment of the digital video sequences then allows to automatically convert the color signal into a temperature level and follows up its evolution in time.

In the past, data reduction had to be executed by powerful workstations. But thanks to the recent evolution of computer performance, the liquid crystal signal data reduction is now performed on a personal computer with in-house programs written in the commercial LabView® language and using the IMAQ™ image processing functions (National Instrument[4]). Powerful coordinate transformations and image filtering functions (Jähne[5]) have been programmed and are applied on the digital video sequences for noise signal filtering of undesired zones or light in-homogeneities. A hue color value threshold of the calibrated liquid crystal signal gives very precise temperature levels of the recorded surface. The local heat transfer coefficient on the test surface is determined by measuring the time it takes to reach a specific temperature value along the transient experiment.

For film cooling measurements, using a multi-regression process on different coolant temperature experiments allows to determine the film cooling effectiveness.

Figure 2 shows a typical transient liquid crystal TIFF image sequence that has been recorded during a transient experiment where, in this case, the heat pulse has been generated by the rapid insertion of a preconditioned cold blade model into a hot mainflow. This sequence represents one of the views that were acquired in parallel during the experiment. It shows the leading edge of a nozzle guide vane model in a baseline (uncooled) configuration. First image of the sequence shows a dummy blade part with a grid reference, which is replaced by the test model at the start of the transient experiment. Surface of the model is coated with 30°C narrow-band reacting encapsulated thermochromic liquid crystals (Hallcrest™).

The color signal first appears on the leading edge around the flow stagnation line where heat transfer coefficient is very high. The front color level then progressively appears on the rest of the surface, as heat transfer coefficient is lower. The complete sequence of the liquid crystal signal evolution is then noise filtered (if required), hue threshold and data reduced to finally obtain precise local values of the heat transfer parameters.

Image Color Format

As post-image processing is performed on computer, image signal can be coded in different color modes. The standard RGB mode, often used in video systems, is based on a combination of red, green and blue color planes. A more physical color-coding is the HSI mode based on hue, saturation and intensity planes (Kunt[6]). Both color modes are coded in 24-bits i.e. 8 bits per plane corresponding to 255 discrete levels for each component.

Noise Reduction and Hue Filtering

In order to eliminate light reflections and in-homogeneities, a solution consists in performing a difference operation of all the images of the sequence with a reference image that does not contain any liquid crystal signal (taken before the start of the experiment, or just after the blade insertion). The best result is obtained by performing the difference on the saturation and intensity planes only in HSI mode so that the hue information of the liquid crystals is not altered. Figure 3 shows such a substraction operation for one image of the sequence.

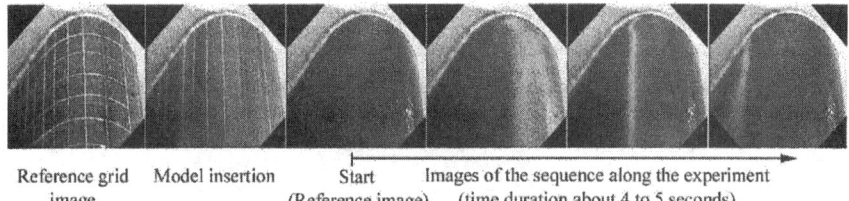

Reference grid image | Model insertion | Start (Reference image) | Images of the sequence along the experiment (time duration about 4 to 5 seconds)

Figure 2: Typical transient liquid crystal TIFF image sequence.

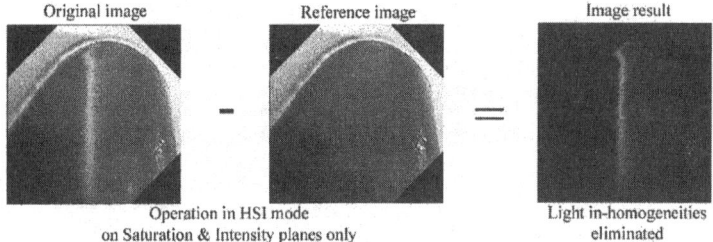

Figure 3: Substraction operation for one image of the sequence.

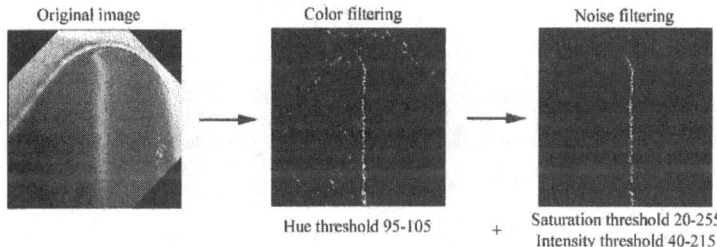

Figure 4: Hue, Saturation and Intensity filtering for one image of the sequence.

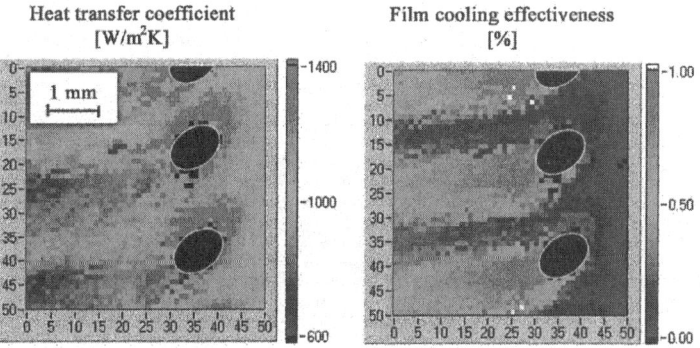

Figure 5: Typical results of heat transfer coefficient and film-cooling effectiveness.

By using a steady state calibration, a specific hue value of the liquid crystal signal can be related to a specific temperature level. Typically the green color signal with a hue level of around 100 (on a scale of 0 to 255) and a margin of ± 5 is taken into account. For narrow-band liquid crystals this gives accuracy on the surface temperature of ± 0.15°C.

Hue filtering can be combined with a saturation and intensity threshold in order to avoid too low or too high levels that are often sources of noise. Figure 4 shows both the simple hue filtered image result and the combined complementary saturation and intensity threshold case on which noise signal has been eliminated.

Hue filtering can either be applied on the original sequence images or on the previously obtained light in-homogeneities filtered sequence. In this case a combined saturation and intensity threshold is not necessary, as noise signal has normally already been eliminated.

With all the color play of the liquid crystal signal being acquired by this new digital system, hue threshold can also be applied on several hue values with adjustable margins so that in most cases better quality results are obtained compared to the old real-time hardware based image processing procedure.

Event Detection

Once the sequence contains only events of the chosen hue filtered color (hence of a specific temperature level) the time lap between the start of the transient experiment and a local event appearance needs to be extracted. As color band filtered sequence images are of good quality, time event detection can be performed for each pixel position, and not on a cell of pixels as before, leading to a higher spatial resolution. Typically, a zoom-in camera view around the film cooling holes can easily reach a spatial resolution of 20 pixels/mm.

In the past, due to poor and non noise-filtered data, the hue filtered signal needed to appear over a consecutive number of images in order to be considered as an event. The new software based image processing procedure allows to take into account every event of individual pixels on each image hence providing a better acquisition of fast evolution signals which for example appear only on one image of the sequence.

Coordinate Transformation

The camera views inevitably contain a perspective and/or deformed image representation, and it is therefore practical to perform a coordinate transformation so that each pixel position of the image is referenced to a surface coordinate of the model. Each view will have its own coordinate transformation but on the same final coordinate system so that they can be assembled all together for a complete surface representation.

In case of the preconditioned blade model rapid insertion, a simple square-type of 5mm spacing reference grid on a dummy blade is acquired by each camera view before the transient experiment. These grid lines are then used to define a set of splines interpolation functions allowing to transform the views into a surface reference coordinate system usually corresponding to the unwrapped model surface. Coordinate transformation can either be performed on all images of the original sequence, or only on the final data reduced time event detection, it gives the same result. In both cases, data obtained from the video sequences have to be compatible with other variables such as the initial surface and the recovery temperature distributions already defined in the unwrapped coordinate system, so that heat transfer parameters can correctly be determined.

DETERMINATION OF HEAT TRANSFER COEFFICIENT AND FILM COOLING EFFECTIVENESS

Once the local time event of a specific temperature level *(time, T_{LC})* has been sorted out from the liquid crystal video sequences, heat transfer data can be determined. The data analysis used to obtain the heat transfer coefficient and the film-cooling effectiveness has been described in Drost[1] and consists in performing a multiple regression analysis on different measurements where the coolant gas temperature level has been varied. Figure 5 shows typical results of heat transfer coefficient and film-cooling effectiveness behind a row of 0,5mm diameter shaped holes located on the suction side of a nozzle guide vane. Six experiments with varying the coolant gas temperature were used for the regression analysis. Measurements were performed with the new digital image processing system giving a high spatial resolution of 10 points (pixels) /mm (results in the figures represent only one portion of 50 x 50 points of the total area).

Results of the different views, often with various spatial resolutions, can be assembled all together on a same coordinate system by the use of standard data representation program such as Tecplot®. Good measurement quality verification consists then in checking the overlapped zones of the different views that should normally give the same results.

ADVANTAGES AND INCONVENIENCES OF THE NOVEL DIGITAL IMAGE PROCESSING SYSTEM

All the aforementioned image processing operations of the liquid crystal video sequences were performed on a computer by the use of the so called "DIPS" (for Digital Image Processing System) in-house program specially developed with a focus on flexibility and easy use for different users on various test facilities and experiments where liquid crystal technique is concerned. This LabView® written program is moreover compatible with any computer platform supporting this language.

The new digital hardware setup allows to easily increase the amount of views that need to be recorded during one experiment. The old system was limited to record at most 2 views in parallel, hence requiring to perform a large number of experiments whenever more than 2 views were required. Performing only one series of tests is a considerable gain of time and it ensures identical test conditions for all the views.

As image processing is performed on a computer after the measurements, powerful filtering functions allow to delete noise of light in-homogeneities so that hue threshold detection is done for each pixel and not for a cell of pixels as before, hence with a higher spatial resolution. Another consequence of the very well-filtered noise signal is that the event detection does not need to be checked on several consecutive images hence resulting in a better acquisition of fast evolutions of liquid crystal signals that for example only appear on one image of the sequence. Moreover, as all the color information of the liquid crystal sequence is available (not only one or two hue levels as before) it is possible to carefully adjust the filtering parameters on different hue calibrated levels so that there is more chance to extract a good signal from the image sequence.

An inconvenience of the new digital image processing system is that post processing operation is slightly longer compared to real-time treatment, but it is ridiculous compared to the gain of time generated by drastically reducing the number of test required.

CONCLUSIONS

A novel digital image acquisition and processing system for the transient liquid crystal technique was presented.

The new digital video equipment allows to easily perform experiments with multiple cameras. Powerful post-processing filtering and data reduction are performed on a computer with the so-called "DIPS" (Digital Image Processing System) new developed program.

Principal advantage is that due to the very low noise filtered signal obtained, data analysis can be performed on each pixel hence resulting in an even higher spatial resolution and a better acquisition of fast hue evolution signals.

This powerful and flexible tool for various transient liquid crystal measurement techniques is being successfully used in our laboratory on several heat transfer test facilities and is these days, for example, applied for the investigation of platform cooling geometries equipped with heater foils.

ACKNOWLEDGEMENTS

The work specific to this paper has been subsidized by a project collaboration with ALSTOM Power and a participation of the CTI (Commision pour la Technologie et l'Innovation in Switzerland).

ABBREVIATIONS

AVI	Audio Video Interleave *movie format on computer*
CCD	Charge-Coupled Devices
DV	Digital Video *format*
HSI	Hue Saturation Intensity *color-coding*
IEEE	Institute of Electrical and Electronics Engineers
PAL	Phase Alternating Line *video format*
RGB	Red Green Blue *color-coding*
S-video	Super-video *format*
TIFF	Tagged Image File Format *on computer*
VHS	Video Home System

REFERENCES

1. Drost, U. *et al.* 1997. Utilization of the Transient Liquid Crystal Technique for Film Cooling Effectiveness and Heat Transfer Investigations on a Flat Plate and a Turbine Airfoil. IGTA, Orlando, Florida.
2. Reiss, H. *et al.* 1998. The Transient Liquid Crystal Technique Employed for Sub- and Transonic Heat Transfer and Film Cooling Measurements in a Linear Cascade. 14th bi-annual symposium on Measurement Techniques in Transonic and Supersonic Flow in Cascades and Turbomachines.
3. Theuwissen, A.J.P. 1995. Solid-State Imaging with Charge-Coupled Devices. Kluwer Academic Publishers. Dordrecht, NL.
4. National Instruments. 1997. BridgeView and LabView®, IMAQ™ Vision for G Reference Manual, National Instruments Corporation.
5. Jähne, B. 1993. Digital Image Processing. Springer-Verlag, London.
6. Kunt, M. 1993. Traitement numérique des images. Presse Polytechniques et Universitaires Romandes, Lausanne.

Contribution of Heat Transfer to Turbine Blades and Vanes for High Temperature Industrial Gas Turbines Part 1 : Film Cooling

KEN-ICHIRO TAKEISHI and SUNAO AOKI

*Takasago Research and Development Center, Mitsubishi Heavy Industries Ltd.,
2-1-1, Shinhama, Arai-Cho, Takasago City, 676-8686, Japan
Tel. +81-794-45-9705, Fax. +81-794-45-6089, E-mail: takeishi@wl.trdc.mhi.co.jp*

ABSTRACT: This paper deals with the contribution of heat transfer to increase the turbine inlet temperature of industrial gas turbines in order to attain efficient and environmentally benign engines. High efficiency film cooling, in the form of shaped film cooling and full coverage film cooling, is one of the most important cooling technologies. Corresponding heat transfer tests to optimize the film cooling effectiveness are shown and discussed in this first part of the contribution.

INTRODUCTION

There is a strong demand for efficient electric power generation to meet environmental regulations and energy saving requirements. Large LNG (Liquid Natural Gas) burning gas-steam combined cycle power plants fulfill these requirements. One of the most powerful means of achieving higher efficiency in industrial gas turbine engines is to raise the turbine inlet temperature (TIT). For this reason, high-temperature industrial gas turbines have been actively developed over the last thirty years as Figure 1 indicates[1].

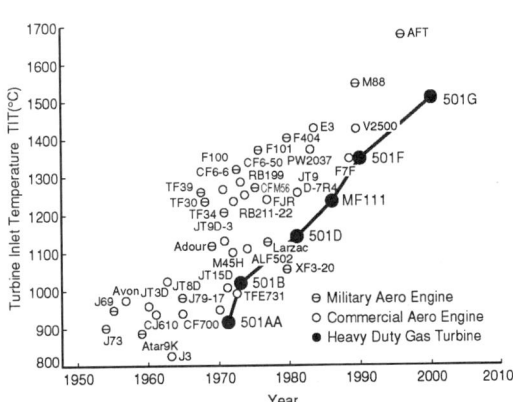

Figure 1 Evolution of turbine inlet temperature of heavy-duty gas turbines

The thermal efficiency of a combined cycle power plant with a 1300°C class gas turbine is about 54% (LHV (Low Heat Value) level), but for a 1500°C class gas turbine it exceeds 58%.

The key improvements for turbine blades and vanes for high-temperature industrial gas turbines are made in the areas of advanced cooling technology, the development and use of advanced thermal resistant material and the application of anti–oxidation / thermal barrier coatings. However, the effect of enhanced heat transfer to raise the TIT is the most significant factor. Figure 2 shows the improvement of cooling technologies applied to the first blades and vanes for Mitsubishi's 1350°C class F-type and 1500°C class G-type gas turbine[2][3]. In the development of the turbine blades and vanes of the G-type gas turbine, advanced cooling technologies such as full coverage film-cooling, shaped film-cooling, and angled turbulence promoters for serpentine flow passages etc. have been introduced, while the base concept of the F-type gas turbine was kept. Shaped film-cooling in particular played an important role in improving the film-cooling effectiveness, as it leads to an enhanced thermal efficiency by reducing the coolant flow rate.

To decrease the heat load from the hot gas, shaped film cooling holes with FC/FC (Full Coverage Film Cooling) are distributed in the gas path surface of the first row vane in addition to the internal impingement cooling. In such high temperature gas turbine vanes and blades, high efficiency film cooling schemes should be adopted to decrease the heat flux and to minimize the cooling air flow rate. This can be achieved by blowing the

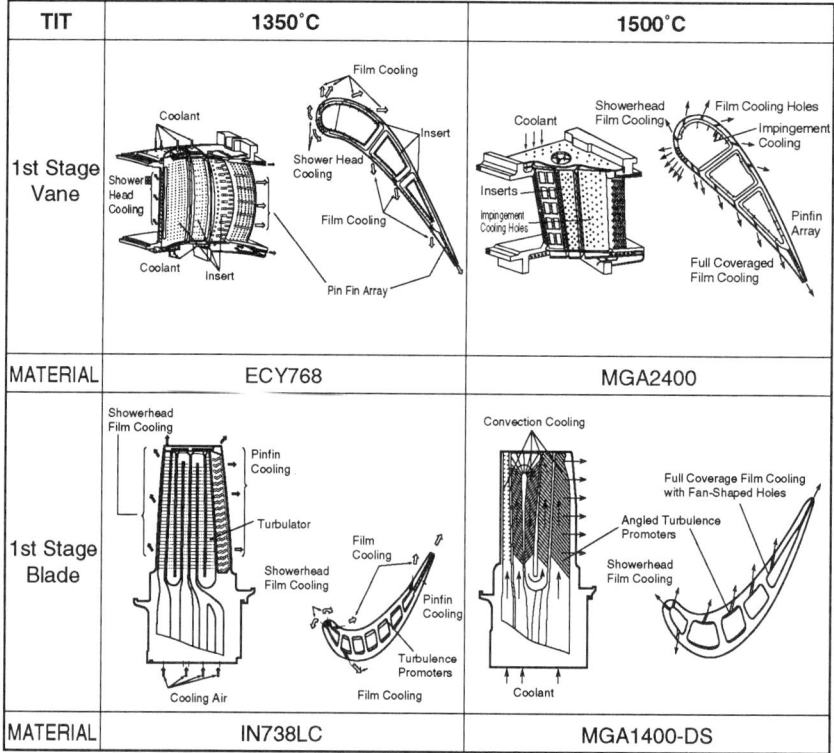

Figure 2 Cooling scheme of the first stage vane

coolant through shaped film cooling holes as shown in Figure 2.

The present experimental works have been performed to study the film cooling effectiveness of shaped film cooling holes on turbine airfoils. Typical test results are presented and discussed in this paper. Another important factor, the influence of the rotational movement on the heat transfer in an internal flow passage, is outlined in the Part 2 of this contribution.

NOMENCLATURE

d	=	Film cooling hole diameter
M	=	Mass flux ratio (=$\rho_a U_a / \rho_g U_g$)
s	=	Maximum surface length
T	=	Temperature
U	=	Velocity
x	=	Distance
η_f	=	Film cooling effectiveness ($=(T_\infty - T_{aw}) / (T_\infty - T_{f0})$)
ρ	=	Density

Subscripts:
- aw : Adiabatic wall
- ex : Vane exit
- f : Film
- f_o : Film cooling hole exit
- ∞ : Main stream
- a : Air
- g : Gas

FILM COOLING FOR TURBINE VANE AND BLADE

Numerous experimental and numerical studies have been conducted to attain higher film cooling effectiveness. Goldstein[4], Goldstein and Eckert[5], Makki and Jakubowski[6], Watanabe et al.[7], and Gritsch et al.[8] investigated so-called fan-diffused film cooling geometries, and Papell et al.[9] studied the film cooling improvement by using curved film cooling holes. In recent years, numerical investigations to solve the complicated flow phenomena of cooling jets in diffused holes have been published (Kohli and Thole[10]). However, there are only few data available for the optimization of the film cooling geometry of diffused holes on turbine airfoils.

Low Speed Wind Tunnel Cascade Test
The experiments to investigate the influence of the film cooling hole geometry on the film cooling effectiveness were conducted in a low-speed wind tunnel cascade, where the main stream was kept at room temperature, while the temperature of the film cooling air was about 50K higher. Figure 3 shows a schematic diagram of the test apparatus. The characteristics of the low-speed wind tunnel are listed in Table 1, and the cascade geometry is shown in Figure 4 and Table 2.
The profile of the model vane and blade as well as the location and the geometry of the film cooling holes are indicated in Figure 5 and Table 3. The model vane and blade, a typical 1st row vane and blade of modern industrial gas turbines, are made of "Bakelite", a low thermal conductivity material.
The film cooling effectiveness η_f defined in the nomenclature is obtained by measuring the adiabatic wall temperature using thermocouples embedded in the wall at the downstream side of the film cooling hole.

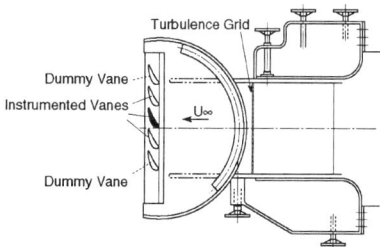

Table 1 Low-speed cascade test facility	
Dimension of the Test Section	300W × 600H mm
Maximum Mainstream Exit Velocity	50 m/s
Main Stream Temperature	Room Temp.
Main Stream Pressure	0.101MPa
Main Stream Turbulence Intensity	0.5~8%
Film Air Temperature	Room Temp. +50°C

Figure 3 Schematic geometry of the low-speed cascade

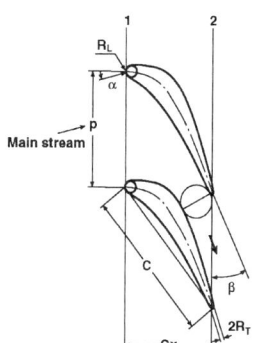

Figure 4 Cascade geometry

Table 2 Cascade parameter

			Vane	Blade
Chord	C	(mm)	186.89	233.01
Axial chord	C_x	(mm)	90.98	192.35
Pitch	P	(mm)	159.27	177.50
Span	S	(mm)	300	300
Aspect ratio		(–)	1.61	1.29
L.E. Radius	R_L	(mm)	14.42	11.0
T.E. Radius	R_T	(mm)	1.79	4.06
Inlet angle	α	(deg.)	0	18.0
Outlet angle	β	(deg.)	41.0	23.7

Table 3 Film cooling hole geometry

Location		S_1	P_1	P_2	S_2
Non-dimensional Distance from L.E.	x/s	0.475	0.257	0.575	0.240
Diameter	d (mm)	1.4	1.4	1.4	2.5
Blowing angle	θ_1 (deg)	35	35	35	30
Depth of fan-shaped hole	t (mm)	2.3	1.6	2.0	3.9
Angle of fan shaped holes	θ_2 (deg)	22	12	24	20
Diffusion angle of fan-shaped holes	ϕ (deg)	9.2	12	8.5	9
Pitch	p (mm)	4.26	4.26	4.26	6.25

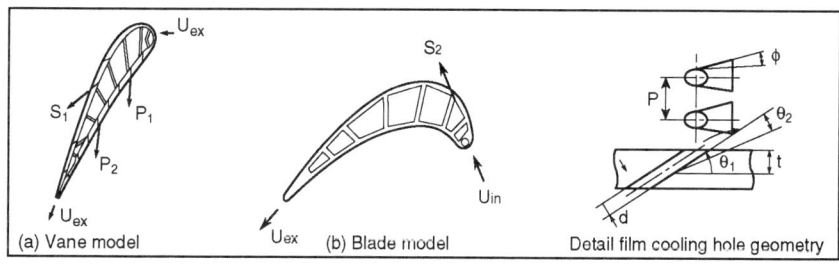

Figure 5 Location of film cooling holes on vane and blade surface

Test Results and Discussion on Film Cooling

Figure 6 shows the measured non-dimensional velocity distribution calculated from the measured static pressure distribution around the model vane. The location of the film cooling holes is also plotted in the same figure. The film-cooling hole S_1 is located on the suction surface, while P_1 and P_2 are positioned within the flow acceleration zone of the pressure surface.

The film cooling effectiveness of S_1 measured on the suction surface is shown in Figure 7. It is understood from this figure that the film cooling effectiveness for a shaped film cooling hole is about 20~30% higher than for a conventional cylindrical one. Figure 7 also indicates that there is little mixing between the main stream and the film cooling jet, as a difference in film cooling effectiveness can be seen for measurement points between two cooling holes compared to those just downstream of a cooling hole. The spatial adiabatic wall temperature distribution downstream of the film cooling hole S_1 was also obtained using an infrared camera. The radiant emission recorded by the IR camera was calibrated against the wall temperature simultaneously measured by the embedded thermocouples. A typical measurement result is shown in Figure 8. It is observed that fan-shaped film cooling persists much longer than the conventional circular film cooling. It is also noted that the cooling air from the fan-shaped holes tends to cover a larger area in the span-wise direction than what is observed for circular holes. The film cooling effectiveness measured on the pressure surface obtained for cooling air blowing through one row of film cooling holes P_1 and P_2 is shown in Figures 9 and 10, respectively.

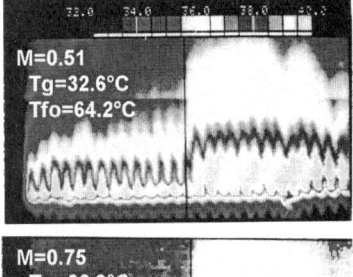
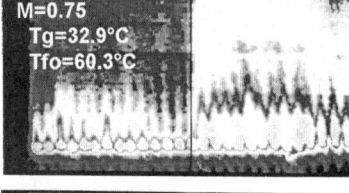
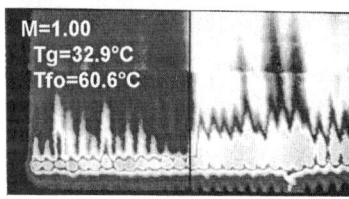

Figure 8 Distribution of the adiabatic wall temperature measured by the infrared camera

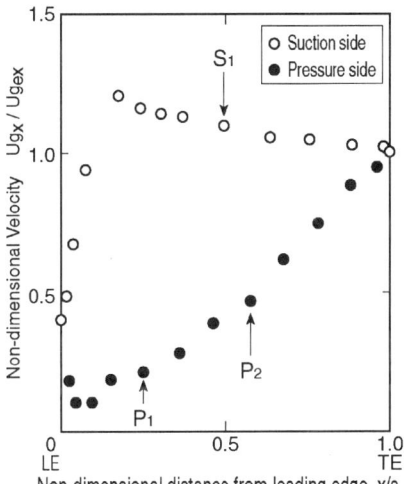

Figure 6 Velocity distribution around 1st model vane

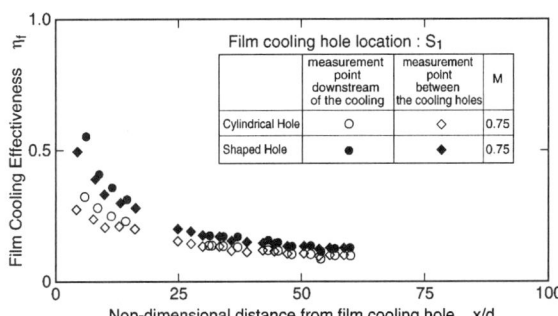

Figure 7 Film cooling effectiveness on suction surface of model vane (S_1 film cooling hole)

On the pressure surface, the film cooling effectiveness decreases more rapidly with increasing distance from the cooling hole than on the suction surface. Another characteristic is that the difference in film cooling effectiveness for measurement locations downstream of the film cooling hole and between the holes, which was observed on the suction surface, does not exist on the front portion of the pressure surface. This is interpreted as the result of a strong mixing of the coolant and the main stream in this region (P_1). In this zone, fan-shaped film cooling holes have a lower cooling effectiveness than the cylindrical ones. This is believed to be caused by a higher degree of mixing between the coolant and the main stream flow. The coolant layer in the case of shaped film cooling is thin because it is relatively wide-spread, which favors stronger mixing with the main flow. Further downstream (location P_2) the film cooling effectiveness of the fan-shaped holes is again higher than that of the cylindrical ones, and the difference in film cooling effectiveness between measurement points downstream of the film cooling hole and between the holes is observed again (Figure 10).

Figure 11 shows the non-dimensional velocity distribution around the model blade, in identical fashion as for the vane. The location of the film-cooling hole S_2 is also plotted in the same figure. The film-cooling hole S_2 is located in the acceleration zone of the suction surface.

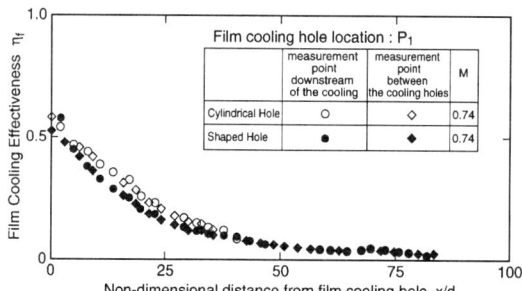

Figure 9 Film cooling effectiveness on pressure surface of model vane (P_1 film cooling hole)

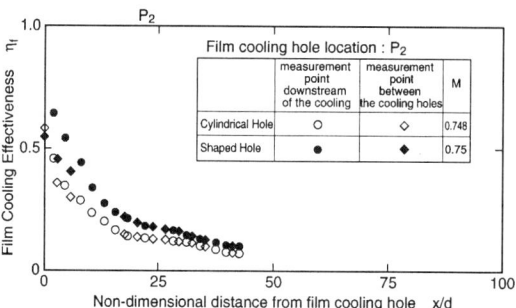

Figure 10 Film cooling effectiveness on pressure surface of model vane (P_2 film cooling hole)

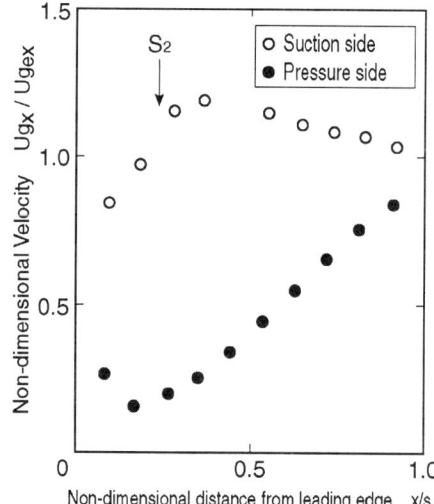

Figure 11 Velocity distribution around 1st model blade

The film cooling effectiveness for a shaped film-cooling hole S_2 is about twice as higher than that for a conventional cylindrical one as shown in Figure 12. Due to the acceleration of the main stream and the convex surface curvature at S_2 the diffused film cooling air adheres tightly on the blade surface, which leads to high film cooling effectiveness.

Figures 13(a) and 13(b) show the film cooling effectiveness as a function of the coolant mass flux ratio M. The results show that, for circular film cooling holes, near to the hole exit (x/d=5) the film jet increasingly penetrates into the main stream with increasing flux ratio M, which reduces η_f. For shaped film cooling holes, on the other hand, a nearly constant film cooling effectiveness is maintained in spite of an increasing flux ratio M. Far down from the film cooling hole exit (x/d=25), the film cooling effectiveness has maximum at M=0.5 for circular holes, while it reaches a maximum at M=0.75 for shaped film cooling holes. From previous works it is known that a maximum film cooling effectiveness can be expected at about M=0.5 for circular holes and at M=1.0 for two-dimensional slots. The film cooling characteristics of the shaped film cooling holes are thus similar to those of a two-dimensional slot.

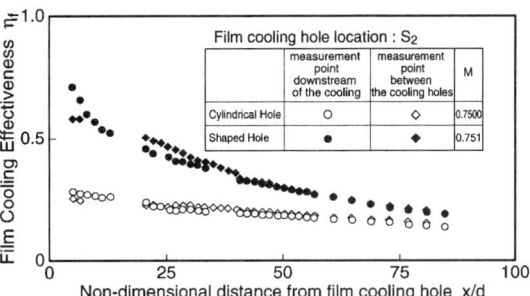

Figure 12 Film Cooling Effectiveness on suction surface of model blade (S_2 film cooling hole)

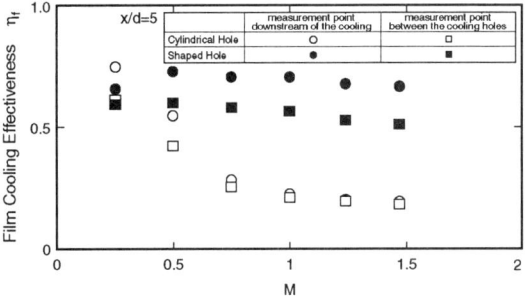

Figure 13(a) Film cooling effectiveness as a function of blowing rate M (x/d=5)

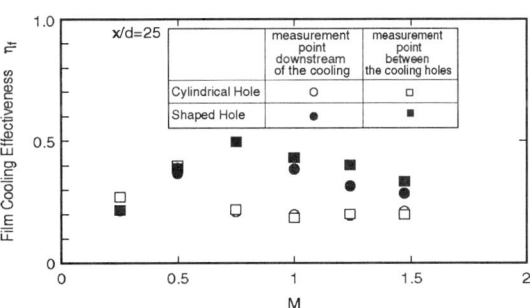

Figure 13(b) Film cooling effectiveness as a function of blowing rate M (x/d=25)

As shown in Figure 9, for a single film coolant hole, the film cooling effectiveness decreases rapidly on the pressure surface. But if FC/FC is adopted on this surface, a high value for η_f can be maintained similarly to what was shown in Figure 12. In Figure 14, a comparison is made between the film cooling effectiveness with a single circular film cooling hole and that with FC/FC keeping the blowing rate of each cooling hole at approximately M=0.3.

The test results show that the film cooling hole geometry strongly influences the film cooling effectiveness and that higher film cooling effectiveness is attained for properly applied expanded (i.e., fan-shaped) hole geometry on the airfoil.

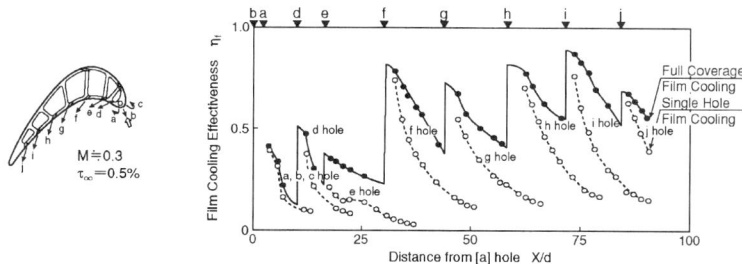

Figure 14 Film cooling effectiveness on suction surface of 1st model blade

CONCLUSIONS

The following conclusions were obtained through the measurements of film cooling effectiveness on turbine airfoils.

1) Fan-shaped film cooling has an advantage in increasing the film cooling effectiveness, especially on the suction side and the rear part of the pressure side.
2) The mixing between the main stream and the film cooling air is small for the suction side and the rear part of the pressure side. This is the main reason for an increased film cooling effectiveness.
3) Full coverage film cooling has an advantage in increasing the film cooling effectiveness, even on the pressure side.

ACKNOWLEDGMENTS

The authors wish to express their gratitude to Mitsubishi Heavy Industries, Ltd. for permission to publish this paper and also would like to thank Prof. Dr. H. Kleine of Tohoku University for valuable comments on the first draft.

REFERENCES

1. Amagasa, S. *et al.* 1994. Journal of Turbo-machinery, **116**: 597-604.
2. Takeishi, K. 1994. Heat Transfer in Turbomachinery, Edited by R.J. Goldstein, Begell House, Inc. 3-12.
3. Nishida, M. *et al.* 1998. 22nd CIMAC International Congress on Combustion Engines, Session 14.
4. Goldstein, R.J. 1971. Advances in Heat Transfer, **7**: 321-371. Academic Press, New York.
5. Goldstein, R.J. and Eckert, E.R.G. 1974. Int. J. Heat Transfer, **17**: 595-607.
6. Makki, Y.H. and Jakubowski, G.S. 1986. AIAA Paper, AIAA-86-1326.
7. Watanabe, K. *et al.* 1999. IGTC '99 Kobe TS-48.
8. Gritsch, M., Schulz, A., and Witting, S. 1998. ASME Paper, 98-GT-28.
9. Papell, S.S. *et al.,* 1982. NASA TP-2062.
10. Kohli, A. and Thole, K.A. 1997. HTD-Vol.350, National Heat Transfer Conference, **12**: 223-232.

Experimental Investigation of Film Cooling Flow Induced by Shaped Holes on a Turbine Blade

Sylvain BARTHET and François BARIO

Laboratoire de Mécanique des Fluides et d'Acoustique – UMR CNRS 5509
Ecole Centrale de Lyon, BP 163
69131 Ecully Cedex, France

ABSTRACT : The present study is the second half of a piece of work carried out in collaboration with SNECMA. It investigates shaped hole film cooling, numerically and experimentally. The aim of this paper is the experimental analysis of shaped hole film cooling on a large scale turbine blade (1.4 m chord).

The test section is a large scale turbine inlet guide vane cascade. The test airfoil is equipped with a row of nine 50° sloped shaped holes. They are located on the suction side at 20% of the curvilinear length of the blade from the stagnation point. The inlet film cooling hole diameter is 12 mm.

The jet flow is heated to 55°C above the crossflow temperature. Velocity and temperature field measurements have been done to obtain mean and fluctuating values. The results are compared to those obtained by Béral[1] on the same experimental apparatus and in the same test conditions, for a row of cylindrical holes.

INTRODUCTION

The high inlet temperatures of modern gas turbines necessitate the use of sophisticated cooling schemes to protect the exposed components. Common techniques include a combination of elaborate internal convective schemes and film cooling. In the film cooling technique, coolant air is injected through the turbine blade surface producing a protective layer between the surface and the hot mainstream gas. The coolant air is distributed through rows of closely spaced discrete holes. Current designs of turbine airfoils incorporate film cooling holes near the leading edge as well as on both suction and pressure sides. The distribution of holes over the components creates a cooling skin to protect the airfoil from the hot mainstream gas.

The injection of discrete film cooling into a two dimensional boundary layer induces a complicated three-dimensional flow field which depends on many geometrical and fluid dynamic parameters, see Andreopoulos[2]. As a direct consequence, the heat transfer is also dictated by the same parameters.

Few studies have investigated the influence of the shape of the hole. In most cases, the holes are cylindrical with a diffuser part at the exit. Thus, Goldstein and Eckert[3] examined holes with lateral expansion of 10° at the hole's exit portion, while Makki and Jakubowski[4] tested holes with a trapezoidal cross-sectional area at the exit. Thole et al.[5] tested three different hole shapes. One of these was forward-lateral expanded, similar to the one we discuss in this paper.

For the same injected mass flow, the increased cross-sectional area at the hole exit, compared to a standard cylindrical hole, leads to a reduction of the mean velocity and so of the jet momentum flux. Therefore, the penetration of the jet into the main flow is reduced, resulting in an increase of the cooling effectiveness. Haven et al.[6] show two additional factors to explain the better effectiveness of shaped holes. Their study is based on vortex dynamics. They specially bring to light the presence of anti-kidney vortices in the upper zone of the jet.

In our experiment, temperature fields show that film cooling with shaped holes is more effective than with cylindrical ones. Stagnation pressure measurements reveal the presence and location of kidney shaped vortices when PIV measurements bring into light the anti-kidney vortices. LDA measurements allow investigations close to the wall and in the hole.

EXPERIMENTAL APPARATUS
PROCEDURES AND TEST CONDITIONS

The wind tunnel has a 3.4 m x 3.4 m x 2.4 m settling chamber. Honey combs and two fine grids are used to reduce the turbulence. The outlet section of the convergent is 0.85 m wide and 1.80 m high. The test section is 0.80 m wide and 1.80 m high ; leakage at the top and the bottom leading edges and on the side walls is thus created.

A turbine inlet guide vane cascade is used (figure 1). The blade chord length C is 1.4 m long. The first and third blades are the limiting surfaces of the test section. At their leading edges, a leakage flow is used to simulate the stagnation region. At the trailing edges, two straight walls are used to obtain equivalent cascade conditions. The second blade is the test airfoil.

The blade (SNECMA ESCA CD01 airfoil) is hydraulically smooth, the stagger angle is $\gamma=43°$, the relative gap $C/g=1.6$ and the relative span $b/C=0.57$. The outlet angle is $\alpha=63°30'$. Note that these values are not optimum. The purpose of the arrangement is to obtain continuous acceleration on the pressure side and acceleration on the suction side followed by slight deceleration. We have used a Katsanis[7] blade to blade inviscid flow calculation to check the flow. The settings of the flaps and streamline shaped top and bottom walls were changed in order to converge measurement results (pressure distributions on the walls and inlet velocity profile) on the calculated results. 2D flow conditions are easily obtained by setting the bleed ratio into the bleed slots ; then a new boundary layer begins on each side wall. Because of the slow evolution of this boundary layer in the negative pressure gradient and the low value of its thickness compared to the blade span, the flow is 2D (check with flow visualisations and measurements). The blockage effect of the side wall boundary layer is 0.5% at the exit of the blade channel.

Since the mass flow injected through the orifices is low, the adjustment of the cascade is not modified when cooling jets are present.

The measurements have been made with the natural residual turbulence of the wind tunnel. The turbulence level is $Tu=0.6\%$ at mid-height between the leading edges at point B of figure 1. The Reynolds number $Re = U_\infty \cdot C/\nu$ (where U_∞ is the outlet velocity of the cascade, ν is the kinematic viscosity) was kept constant and equals $Re_\infty = 1.31 \cdot 10^6$ for the entire study. The holes are drilled in the wall (figure 2). The diameter of the cylindrical part of the hole is D=12 mm (D/C~1/100). The location of the orifice is the same as that used in Béral[1]. The distance between the centres of two adjacent orifices is 6D. The angle of the axis of the cylindrical pipe of the orifice with the tangent to the wall is 50° in the streamwise direction. In the plane normal to the streamwise direction, this angle is zero (no yaw). The row has 9 holes. The mass flow injected in each hole is such that the product of the density ρ_{jet} and the jet mean velocity in the cylindrical pipe V_{jet} is equal to the product of the density ρ_{out} and the velocity V_{out} of the external flow at the location of injection $\rho_{jet} V_{jet} = \rho_{out} V_{out}$.

The blade is not heated (athermane wall). The jet flow is heated at $\theta=55\pm0.1°C$ above the main flow temperature in the channel. The inversion of the temperature compared to the real case, realised for economical reasons, is possible because the buoyancy Richardson number is low (Ri=g $\Delta\theta$ $\Delta\rho$ D / ρ U_{out}^2 = 4 10^{-3}).

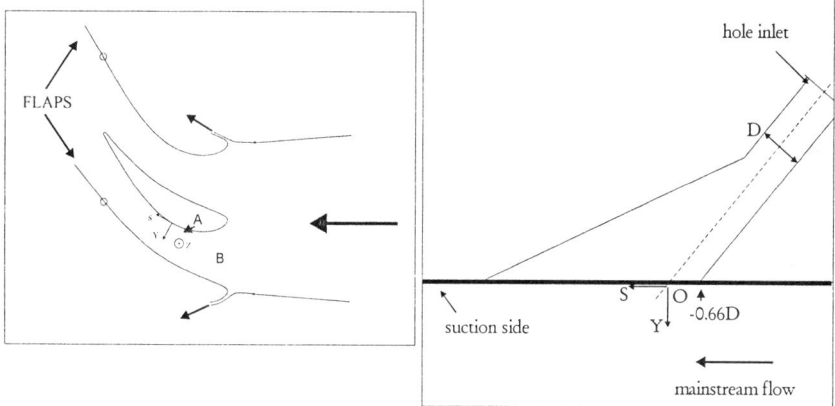

Figure 1 : Test section (not to scale)   Figure 2 : The injection hole (sketch)

The velocity field has been obtained with a PIV technique (Dantec Measurement Technology). Near the wall, PIV measurements are difficult because of the noise generated by the reflecting laser light. In the hole they are impossible. Therefore an LDA technique is used in these zones (Aerometrics DSA 2D system). The seeding of the main flow is homogenous because spectacle smoke is introduced far upstream of the wind tunnel at the fan inlet. The jet is seeded with micro droplets of olive oil. The stagnation pressure measurements are obtained with a Pitot tube, whose head diameter is 1mm (0.08 D). The temperature field is obtained with a 2.5μm diameter and 1mm long cold wire.

The accuracy of measurements is hereafter presented : for PIV measurements, 1000 images are used for mean and fluctuating velocity measurements. The accuracy of the location of each point of measurement (including calibration, data reduction process) is ±0.2mm. For LDA, 1000 samples have been taken for each measurement. The measurement volume is 1mm long for a 65μm diameter. The bias error and the error induced by the necessary tilting of the optical system are fully presented in Bario and Béral[8]. The accuracy of the location of the measurement point is ±0.2mm (including all the measurement process) for LDA, temperature and pressure measurements. Temperatures are given with an accuracy of ±0.2°C. The accuracy of the stagnation pressure measurement is poor in the region where the flow angle quickly changes (first part of the orifice) because the probe is only aligned with an averaged flow and not exactly with the local mean velocity, so qualitative use of results only is possible in that region.

RESULTS

S' denotes the curvilinear coordinate (S) normalised by the inlet hole diameter. The origin of the curvilinear coordinate (S) is located at the intersection point between the cylinder revolution axis (inlet hole) and the wall (figure 2).

Stagnation Pressure measurements

The stagnation pressure Pt was measured to show the near field aerodynamic characteristics. The cross-sectional investigations concern the injection region, (-3.58< S'< 13). The stagnation pressure coefficient presented $Cpt = (Pt_{out} - Pt)/0.5 \cdot \rho U_{out}^2$ is representative of the stagnation losses (where the subscript "out" defines values taken at the inlet of the test section). The blockage effect appears at S'=-1.08 (figure 3a) where the upstream boundary layer thickness close to Z=0 increases. This phenomenon is due to the adverse pressure gradient induced in the vicinity of the jet which acts like a supple body (Baker[9]). It leads to the three-dimensional separation of the boundary layer. At S'=-0.66 (figure 3b) (the leading edge of the injection) the upstream separated boundary layer is divided into two symmetrical parts with Z=0. This is induced by the jet which dragged up the central part of the boundary layer (close to Z=0). Figure 4a shows the rise of the counter-rotating vortices (sidewall vortices) for the station S'=1.08. In this section the hole is 18mm wide at the exit. These vortices are induced by the annular boundary layer rolling up and occur close to the lateral walls of the hole. The distance between the centres of these two vortices is 10mm. Up to S'=5.50, the jet lifts off. In this cross section which is located downstream of the injection, it seems that the two counter-rotating vortices have merged (figure 4b). Moreover, they have lifted off the wall. Their rotation ($\pm\Omega_X$) is such that one vortex, by mutual induction, lifts the other off the surface, Haven et al. (1996). We can also observe, figure 4b, the presence of high losses below the kidney structure, close to the wall, due to the boundary layer separation. Werle[10] observed this phenomenon which is characterised by two small counter-rotating vortices. These structures have been revealed in the numerical part of our work (Barthet et Kulisa[11]).

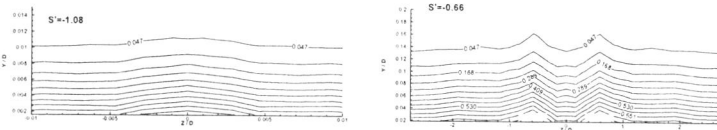

Figure 3 : Stagnation pressure coefficient : a : S'=-1.08 (left) and b : S'=-0.66 (right)

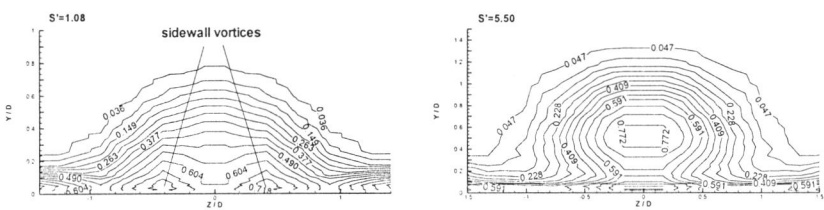

Figure 4 : Stagnation pressure coefficient : a : S'=1.08 (left) and b: S'=5.50 (right)

Temperature Measurements

The temperature coefficient considered here is defined as : $\theta = (T - T_{out})/(T_{jet} - T_{out})$ where T is the measured temperature, $T_{jet}=T_{out}+55°C$. Figure 5 presents the mean temperature distribution in the median jet plan. Far downstream, at 15 diameters, the maximum temperature in the jet is up to $\theta = 0.30$. It is greater than the temperature obtained by Béral[1] : at 13 diameters this value is less than $\theta=0.28$ for the cylindrical case. The temperature fields for both shaped and cylindrical hole cases for S'=13D (figure 6) show the better effectiveness of shaped holes : the normalized temperature θ is higher in the core of the jet and at the wall for the shaped hole jet. The area covered by the jet is also greater.

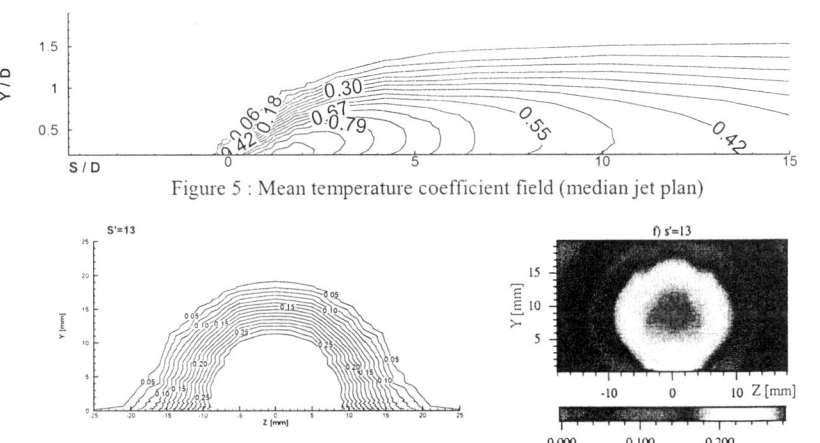

Figure 5 : Mean temperature coefficient field (median jet plan)

Figure 6 : Mean temperature coefficient cross field for expanded (left) and cylindrical cases (right)

PIV Measurements

The better effectiveness of the film cooling induced by shaped holes has been shown above. It can be explained by the reduced momentum of the cooling flow. Another reason may be considered. It concerns the rise of new swirling structures. They have been observed and called the anti-kidney vortices by Haven et al.[6]. These unsteady structures arise in the upper edge of the jet.

We know (Andreopoulos[2]) that a succession of vortices ($+\Omega_Z$) are located at the leading edge of a jet in a crossflow. They are induced by the vorticity of the annular boundary layer at the leading side of the jet. This zone undergoes the strong shear of the mainstream flow and a shedding (unsteady) vortex rings system is created. The vortex rings are stretched due to the high acceleration of the mainstream flow which is diverted by the jet.

The presence of the four structures shown in figure 7 may be tied up with this phenomenon. The low momentum exit flow (compared to a cylindrical injection) involves the warp of the leading edge of the jet. The warp of the ring vortex (shed vortex) leads to the birth of the anti-kidney pair and the second pair just below. The first one induces an opposite motion on the jet lift off due to the kidney vortices. The film cooling is closer to the wall and therefore the thermal effectiveness is increased. The anti-kidney structures may be explained by the sketch presented in figure 8.

These structures are shown by our PIV measurements. Figure 7 presents the mean vorticity field in the cross section S'=0. A pair of anti-kidney vortices and the associated lower structures (Ω'_X) are shown. The vortex pair (Ω'_X) located just below has the same vorticity levels ($\Omega_X \cong \Omega'_X$).

For the station S'=5.5 two small zones of opposite vorticity sign are seen close to the wall at the lateral sides of the jet (figure 9). They are the two arms of the horseshoe vortex diverted by the jet. The formation mechanism of the horseshoe vortex, described by Baker[9], is due to the upstream boundary layer rolling up. The same cross-section has been visualised (figure 10). No seeding was introduced in the main flow. This implies that some jet particles have been absorbed by the horseshoe vortex. Backward velocities have been detected with LDA measurements at the leading edge of the hole exit close to the wall (figure 11). The feeding of the horseshoe vortex by injected flow has also been observed in our numerical work for a cylindrical geometrical configuration

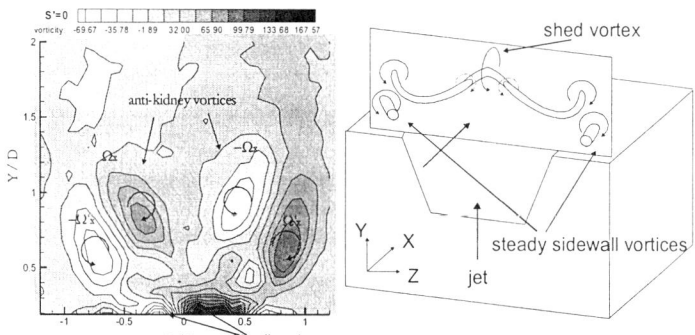

Figure 7 : Mean vorticity field, S'=0   Figure 8 : Concavity of the leading edge

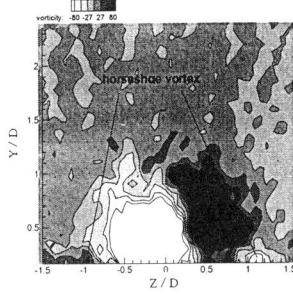

Figure 9 : Mean vorticity field, S'=5.5   Figure 10 : Horseshoe vortex visualization

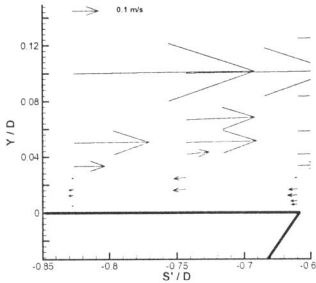

Figure 11 : Backward velocity zone – leading edge of the hole

In-Hole Investigations

Inside the hole, considerable separation at the leeward side of the injection hole is shown by numerical calculation, figure 12. The vorticity ($\pm\Omega_S$) is also brought to light by LDA measurements (figure 13). A three-dimensional separation occurs.

The separation of the jet flow within the hole is due to the large expansion angle and to the break in the slope of the leeward side.

CONCLUSION

A selection of measurement data has been analysed for shaped hole film cooling. The experimental set-up is an inlet guide vane turbine blade cascade and the coolant flow is injected near the leading edge on the suction side. The ratio of the diameter of the hole (inlet part) to the blade laminar boundary layer is 6/1. The results of the temperature measurements are compared with those obtained on the same blade but with a cylindrical hole. The effectiveness of cooling is better with the shaped holes. This is firstly due to the diffusion of the flow in the second part of the orifice. The momentum of the coolant fluid being lower, the jet flow is closer to the wall and cooling is increased. Secondly, anti-kidney vortices have been found in the upper zone of the jet. They help to reduce the lift-off motion of the jet due to the rotation of the kidney vortices. The birth of these vortices is certainly related to the in-hole inner leading edge boundary layer. In the shaped hole tested, strong 3D flow separation occurs. The important diffusion of the walls of the hole associated with a sudden change of the slope of the leeward wall are probably the causes of this separation. It can be supposed that controlling this separation with a new design of the shape of the hole will give better cooling (Thole et al.[5]) with a jet core closer to the wall. All the trends found in the experimental analysis were predicted by numerical calculation (except the horseshoe vortex), see Barthet and Kulisa[11].

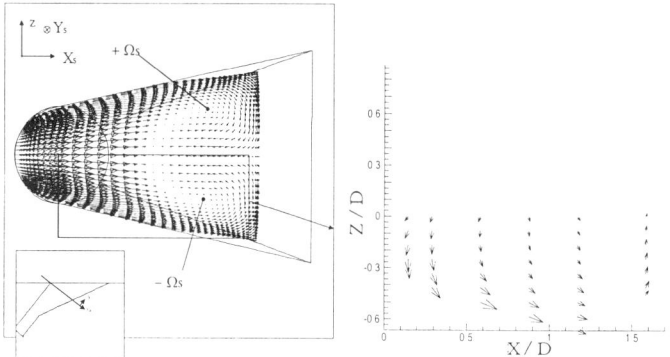

Figure 12 : Separation zone – Numerical result　　Figure 13 : Separation zone – LDA result

REFERENCES

1. Béral, C. 1996. Etude expérimentale des écoulements se développant sur un aubage de turbine en absence et en présence d'injections pariétales de refroidissement. Thesis. Ecole Centrale de Lyon.
2. Andreopoulos, J. 1985. On the structure of jets in a crossflow. JFM. **157** : 163-197.
3. Goldstein, R.J.& E.R.G. Eckert. 1974. Effects of hole geometry and density on three-dimensional film cooling. J. Heat Mass Transfer. **17** : 595-607.
4. Makki, Y.H. & G.S. Jakubowski. 1986. An experimental study of film cooling from diffused trapezoïdal shaped holes. AIAA paper n° 86-1326.
5. Thole, K., M. Gritsch, A. Shulz & S. Wittig. 1996. Flowfield measurements for film cooling holes with expanded exits. ASME paper. 96-GT-174.
6. Haven, B.A., D.K. Yamagata & M. Kurosaka. 1997. Anti-kidney pair of vortices in shaped holes and their influence on film cooling effectiveness. ASME paper. 97-GT-45.
7. Katsanis, T. 1968. NASA TND 4525.
8. Bario, F. & C. Béral. 1996. Cooling jets and boundary layer interaction on the suction side of a turbine inlet guide vane. Proceed. of the 3rd Int. Symp. On Aerothermodynamics of Internal Flows : 615-621.
9. Baker, C.J. 1979. The laminar horseshoe vortex. JFM. **95** : 347-367.
10. Werle, H. 1977. Ecoulement le long d'une paroi plane comportant des jets normaux ou inclinés de section circulaire. Papport technique ONERA. N° 65/710ENA.
11. Barthet, S. & P. Kulisa. 2000. Numerical investigation of film cooling flow induced by cylindrical and shaped holes. International symposium on heat transfer in gas turbine systems. Cesme, Turkey.

The authors thank the MENRT (Ministère de l'Education Nationale de la Recherche et de la Technologie) for its financial support of the PhD thesis and the SNECMA for its financial support of the experimental work.

Film cooling from rows of holes - effect of cooling hole shape and row arrangement on adiabatic effectiveness

J. DITTMAR*, I. S. JUNG**, A. SCHULZ*, S. WITTIG*, J. S. LEE**

*Institut für Thermische Strömungsmaschinen
Universität Karlsruhe (TH), 76128 Karlsruhe, Germany
e-mail: jan.dittmar@its.uni-karlsruhe.de
**Turbo and Power Machinery Research Center
Seoul National University, Seoul, Korea

ABSTRACT: In the present study the film cooling performance in terms of the adiabatic film cooling effectiveness on the scaled suction side model of an actual guide vane was investigated. An infrared thermography measurement system was used to determine highly resolved distribution of models surface temperature. Two different film cooling hole configurations were investigated: a single row of fanshaped holes and a double row of cylindrical holes in staggered arrangement. The influence of blowing rate and mainstream turbulence level on effectiveness was investigated in a wide range for both of the injection configurations.

NOMENCLATURE

Streamwise velocity	u_∞	*Subscripts*
Turbulence level	$Tu = u'_\infty / u_\infty$	c: coolant
Kinematic viscosity	$v = \mu/\rho$	∞: mainstream
Blowing ratio	$M = \rho_c u_c / \rho_\infty u_\infty$	aw: adiabatic wall
Density ratio	$DR = \rho_c / \rho_\infty$	r: state of recovery
Film cooling effectiveness	$\eta = (T_{aw} - T_{\infty,r})/(T_c - T_{\infty,r})$	
Film cooling hole diameter	d	
Streamwise distance	x	
Reynolds-number	$Re(x) = u_\infty(x) L / v_\infty$	
Acceleration parameter	$k = (\partial u_\infty / \partial x) v_\infty / u_\infty^2$	

INTRODUCTION

In modern gas turbine sophisticated cooling schemes including film cooling are widely used to protect the vanes and blades of the first turbine stages from failure and to achieve high life-cycles. In film cooling applications, injection from discrete holes is commonly used to generate a coolant film on the blade's surface. In an attempt to improve the cooling process, recent attention has been given to contouring the injection hole geometry. Modern manufacturing technologies like precise electric-discharge machining or laser drilling enable to form the injection hole in more complex shapes.

Many of the earlier studies of film cooling considered injection from a single row of discrete holes. Due to the three dimensional character of the flow field downstream of the coolant injection the cooling effectiveness decreases compared to injection from a continuous slot. In order to improve the lateral distribution of the injected coolant and to approach a two dimensional film cooling situation, more studies were focussed on the injection from a double row of cylindrical holes[1-3]. In general the results show that for the same injected mass flow rate per unit span the double row arrangement provides better cooling effectiveness compared to injection from a single row. Staggered rows show better performance than rows with inline arrangement. The gain in effectiveness is attributed to the lower penetration of the coolant jets because of lower momentum flux ratio resulting from the increased injection area and better lateral spreading of the cooling air. Increasing the distance between the two rows give a significant decrease to both local and lateral averaged effectiveness, especially close behind the downstream row. A compound angle orientation of the holes, especially of the 2nd row, increases cooling effectiveness[4].

Recent studies on film cooling holes with a diffuser shaped expansion at the exit portion of the hole ("fanshaped" holes) have shown a promising improvement of the film cooling performance. Various research groups [5-10] investigated film cooling effectiveness with injection from different hole shapes, including holes with a lateral or forward expanded exit part. They all found higher effectiveness values for the shaped holes compared to cylindrical holes. The lateral expanded holes show much better lateral spreading of the injected coolant and hence more uniform distribution of effectiveness. Due to the reduced jet exit momentum, shaped holes show less penetration of the coolant jet into the mainstream and reduced velocity gradients in the mixing region[11].

The present study aims to compare a single row cooling configuration including shaped holes with a double row of cylindrical holes in a realistic flow field, typical for a turbine giude vane.

EXPERIMENTAL APPARATUS
Wind Tunnel and Test Section

The experiments presented in this paper were conducted on a model of a suction side of an actual turbine guide vane assembled in an open loop atmospheric wind tunnel, see Figure 1. Its contoured shape was designed in order to simulate almost realistic distribution of Reynolds-number and acceleration parameter k on the model's surface. The Reynolds-number based on the chord length was about 5.5×10^5 and about 7400 based on the injection hole diameter at the location of injection.

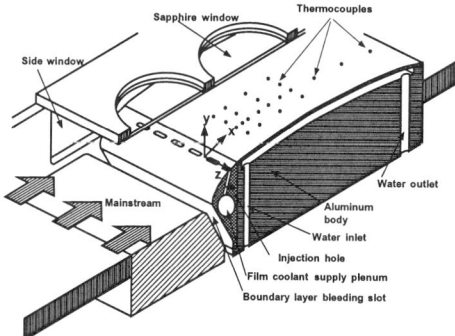

Figure 1: Suction side model used for the film cooling experiments

The model basically consists of three different parts: A base block, an interchangeable injection module and a contoured test plate for surface temperature measurements. The test plate was made of a high temperature plastic (TECAPEEK©) with a low thermal conductivity of about 0.35 W/m·K. In the top wall three sapphire windows are inserted to enable optical access to the test model.

In this study the cooling performance of a single row of 8 fanshaped holes was compared to the performance of a double row of standard cylindrical holes in staggered arrangement including 16 holes in total. Details of the hole geometries are shown in Figure 2. The holes are placed at 10% of the chord length from the stagnation point of the blade model. Both hole types have a diameter of 4 mm at the inlet part of the hole and are inclined 45° in streamwise direction.

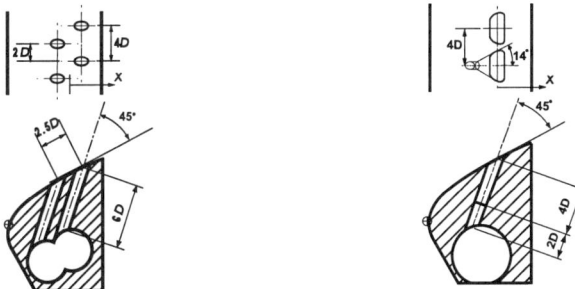

Figure 2: Investigated film cooling hole configurations: Double row of cylindrical holes (left) and single row of fanshaped holes (right)

Measurement Technique and Data Processing

Surface temperature measurements were performed by means of an infrared thermography system (AGEMA© Thermovision 900 SW). The IR-system consists of a optical scanner which directs the incoming infrared radiation line by line on the detector working in a wavelength bandwidth of 2-5.4 µm. The scanner is mounted above the test section of the wind tunnel and reads the surface temperature information through the sapphire windows. The output signal of the IR-detector is digitised by a processing computer in a frame of 136×272 Pixels corresponding to a spatial resolution of 0.6×0.6 mm with the optical setup used in the experiments. The surface of the contoured test plate is covered with a black paint with a well-known emissivity of 0.95.

In order to get quantitatively accurate wall temperature data, the radiation data from the IR-system was re-calibrated using 12 thermocouples embedded flush with the surface. A calibration routine was developed taking reflected ambient radiation as well as transmission losses into account. The total amount of infrared radiation detected by the IR-system I_{tot} can be summarized as follows:

$$I_{tot} = \tau \cdot \varepsilon \cdot I_w + \tau \cdot (1-\varepsilon) \cdot I_{sur} \qquad (1)$$

where I_w is the emitted infrared energy from the test plate and I_{sur} is the consolidated radiation from all the surrounding that is reflected from the test plate (reflectivity=(1-ε)). The parameter τ is the overall transmission factor (surface to detector) and ε is the emissivity of the test plate (covered with black paint). It is assumed that no infrared radiation is transmitted through the test plate and any infrared radiation from the hot mainflow itself is neglected. To correlate surface temperature and the emitted infrared radiation I_w a semi-empirical relation is used:

$$I_w = \frac{R}{e^{B/T_w} - F} \quad . \tag{2}$$

The factors R, B and F are calibration factors provided with the AGEMA© IR-system, taking the transmission behaviour of different lenses and various measuring ranges of the detector into account. Combining Eqns. (1) and (2) the surface temperature from the test plate can be calculated if the radiation of the surrounding I_{sur} and the overall transmission factor τ are known:

$$T_w = \frac{B}{\ln\left[\dfrac{R \cdot \tau \cdot \varepsilon}{I_{tot} - \tau \cdot (1-\varepsilon) \cdot I_{sur}} + F\right]} \quad . \tag{3}$$

For accurate measurements a more sophisticated procedure is needed especially in situations were the parameters τ and I_{sur} are difficult to determine because of their spatial variation. Therefore, additional temperature data obtained with embedded thermocouples is used to re-calibrate the IR-data. For each thermocouple location, values for τ and I_{sur} were derived by minimizing the difference of thermocouple and IR temperature value. The averaged values for τ and I_{sur} that give the best result in n thermocouple locations are determined by using a least error squares method:

$$\Psi^2 = \sum_{i=1}^{n}\left[T_{TC,i} - T_{w,i}\left(\begin{array}{c}\tau\\I_{sur}\end{array}\right)\right]^2 = \min. \quad \Rightarrow \quad \frac{\partial \Psi^2}{\partial(\tau, I_{sur})} = 0 \quad . \tag{4}$$

To calculate the best fitting values for τ and I_{sur}, the system of n non-linear equations is solved by using a robust numerical algorithm based on the Levenberg-Marquardt procedure. A detailed description of this mathematical procedure is given in Press et al.[12].

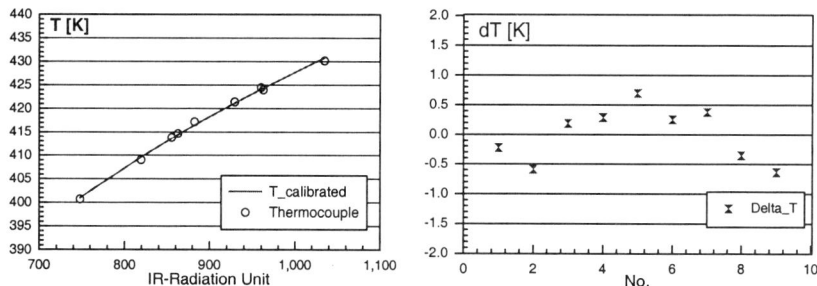

Figure 3: *Typical result of the IR-calibration procedure (left: T_w calculated with Eq. (3) and compared to thermocouple data; right: difference between IR and thermocouple temperature)*

An example result of the IR-calibration procedure for a typical case with coolant injection is shown in Figure 3. The wall temperatures of the test plate measured with the help of the IR system after calibration show very good agreement with those determined by the thermocouples. The temperature difference is less than 1% for the vast majority of measured test cases.

In order to get adiabatic wall temperature data and to account for the remaining heat losses through the test plate, a three dimensional remnant heat flux calculation is performed. A finite element analysis has been conducted using the measured wall temperatures as boundary condition and taking the radiative heat exchange with the surrounding into account. The results show that the influence of remnant heat flux is generally small for the lateral averaged effectiveness. At the centerline and right between the injection holes the effectiveness data can be influenced by about 10 – 20% close behind the injection mainly due to lateral heat exchange inside the test plate. The overall uncertainties in adiabatic effectiveness have been calculated to a maximum of 12% at low effectiveness values.

RESULTS AND DISCUSSION

Using the measurement technique and post processing procedure described above, adiabatic film cooling effectiveness data have been determined for both of the injection hole geometries. During the experiments the blowing ratio M was varied in the range of 0.2 to 1.5 in case of the double row with cylindrical holes and in the range of 0.25 to 2.83 for the single row with fanshaped holes. The mainflow temperature was set to 450 K and the ratio between coolant and mainflow density was kept constant at 1.3 during the experiments.

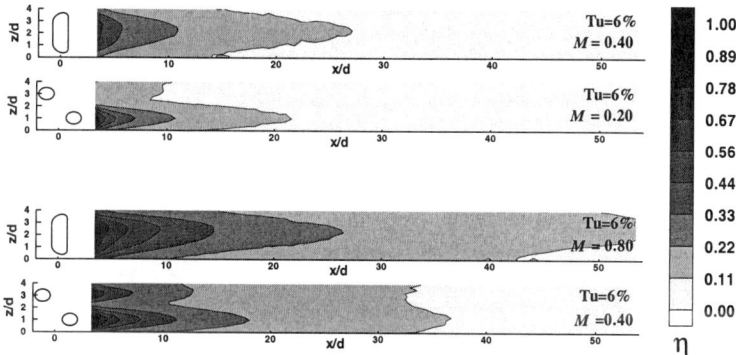

Figure 4: Adiabatic film cooling effectiveness for low blowing ratios.

Figure 4 and Figure 5 show the distribution of adiabatic effectiveness at a mainstream turbulence intensity of 6% downstream of the point of injection up to x/d=54 for low and higher blowing rates, respectively. In case of the shaped holes, an increasing blowing rate causes increasing effectiveness for almost the total range of blowing rates studied. At very high blowing rates (2.5 to 3) the effectiveness decreases slightly. On the one hand more cooling air is injected but on the other hand the lateral spreading became worse due to coolant flow separation inside the diffuser part of the shaped holes. The double row of cylindrical holes shows a good performance in the low and medium blowing rate range (up to M=0.75). Due to the staggered arrangement of the two rows, the coolant is well spread in lateral direction and a quite uniform distribution of effectiveness is achieved. At higher blowing rates (M>1) the cooling jets start to separate from the wall which is more pronounced for the first row due to a blockage effect of the second row. A wake region right behind the injection is established and hot mainstream gas is transported to the wall caused by complex vortex generation, mainly in the shear layer between jet and mainstream. In this blowing rate band the adiabatic effectiveness is decreasing drastically in the near holes region as far as $x/d=15$.

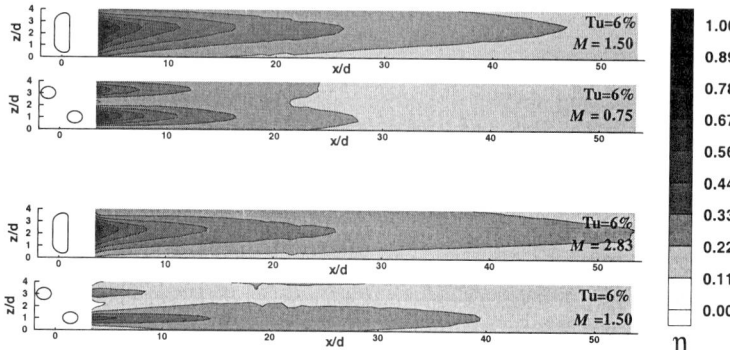

Figure 5: Adiabatic film cooling effectiveness for high blowing ratios.

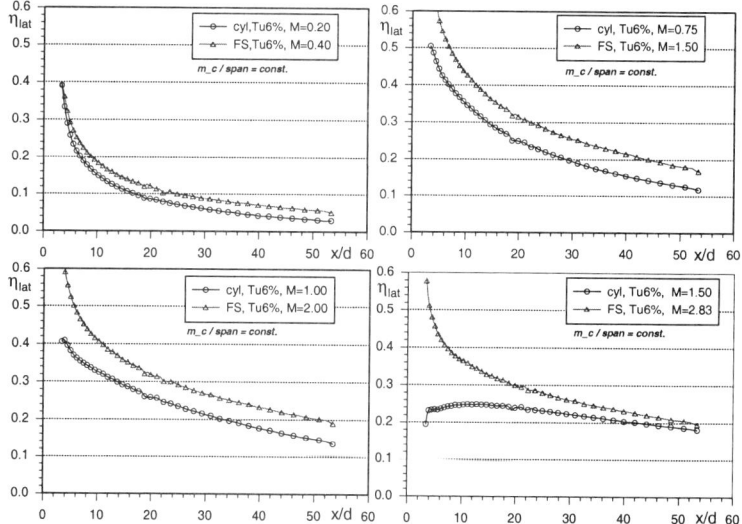

Figure 6: Lateral averaged adiabatic effectiveness for fanshaped (FS) and cylindrical (cyl) holes. Each plot shows data for equivalent coolant mass flow per spanwise distance injected.

The plots are arranged in such a way as they show couples with the same amount of injected coolant mass flow per spanwise distance. Comparing the two different configurations indicates the superior cooling performance of the shaped holes. In the low blowing rate range the cylindrical holes show likewise cooling effectiveness but at higher blowing ratios the shaped holes do have clear advantages due to the coolant jet separation of the cylindrical hole injection. The superior performance of the shaped holes at moderate and high blowing rates especially up to $x/d=20$ is emphasized looking at the lateral averaged effectiveness data, see Figure 6. The improvement in adiabatic film cooling effectiveness when using shaped holes is caused mainly by the reduction of the jet exit momentum. The reduced penetration of the jet into the mainstream and the improved lateral spreading of the jet results in increased cooling effectiveness.

The adiabatic effectiveness measurements have been made at three different mainstream turbulence intensities (4%, 6% and 10%), adjusted by inserting different grids made of rectangular bars upstream of the test model.

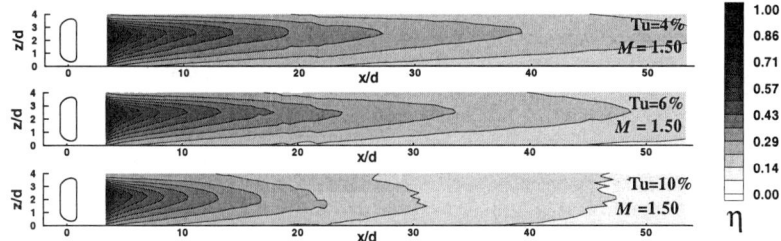

Figure 7: Influence of mainstream turbulence level on local adiabatic film cooling effectiveness for fanshaped holes ($M=1.5$).

Results for the shaped at a blowing rate of $M=1.5$ are shown in Figure 7. As supposed the effectiveness in decreased at enhanced mainstream turbulence level for both injection configurations. The higher mainstream turbulence level enforces the mixing between coolant and mainstream and a faster decay in effectiveness in streamwise direction, at least at moderate blowing rates was observed.

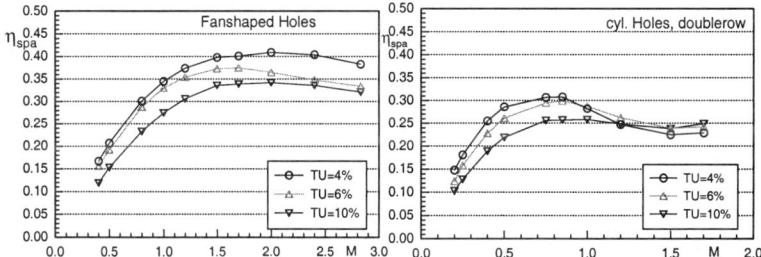

Figure 8: Space averaged adiabatic effectiveness for both hole types with varying mainstream turbulence level.

Figure 8 show space averaged data of adiabatic effectiveness with varying mainstream turbulence intensity. The space averaged value was calculated by using the lateral averaged effectiveness data in a streamwise range up to $x/d=30$. Enhanced mainstream turbulence intensity causes a drop in effectiveness for both cooling configurations at least up to moderate blowing ratios. In case of the cylindrical holes at high blowing ratios the influence of turbulence level seems to be much weaker or even reverse. In case of jet separation the enhanced mixing due to higher turbulence transports colder mixed air back to the wall which reduces the wall temperature compared to high momentum injection into a mainstream with low turbulence.

CONCLUSIONS

Adiabatic film cooling effectiveness measurements have been conducted using a scaled model of a typical guide vane suction side. The key findings of the experiments can be summarized as follows:

A double row of cylindrical holes provides nearly similar adiabatic film cooling effectiveness values compared to a single row of fanshaped holes only at small blowing ratios. At moderate and high blowing ratios the fanshaped holes show clearly superior cooling effectiveness. Enhanced mainstream turbulence intensity reduces film cooling effectiveness due to higher diffusion and leads in general to a faster decay of effectiveness in streamwise direction. For cylindrical holes at high blowing ratios the turbulence intensity has only little effect on adiabatic effectiveness because of jet separation on the one hand and enhanced coolant diffusion on the other hand.

REFERENCES

1. Jabbari, M.Y. & Goldstein, R.J. 1978. Adiabatic wall temperature and heat transfer downstream of injection through two rows of holes. *In* Journal of Engineering for Power, Vol. 100, April 1978, pp. 303-307
2. Jubran, B. & Brown, A. 1985. Film cooling from two rows of holes inclined in the streamwise and spanwise directions. *In* Journal of Engineering for Gas Turbine and Power, Vol. 107, January 1985, p.p. 84-91
3. Jubran, B.A. & Maiteh, B.Y. 1999. Film cooling and heat transfer from a combination of two rows of simple and / or compound angle holes in inline and / or staggered configurations. *In* Heat and Mass Transfer, 34 (1999), pp. 495-502
4. Ligrani, P.M., Wigle, J.M., Ciriello, S. & Jackson, S.M. 1994. Film-cooling from holes with compound angle orientations: Part1- Results downstream of two staggered rows of holes with 3d spanwise spacing. *In* Journal of Heat and Mass Transfer, Vol. 116, May 1994, pp. 341-352
5. Goldstein, R.J., Eckert, E.R.G. & Burggraf, F. 1974. Effects of hole geometry and density on three-dimensional film cooling. *In* International Journal of Heat and Mass Transfer, 1974, Vol. 17, pp. 595-607
6. Gritsch, M., Schulz, A. & Wittig, S. 1997. Adiabatic wall effectiveness measurements of film cooling holes with expanded exits. ASME Paper 97-GT-164
7. Gritsch, M., Schulz, A. & Wittig, S. 1998. Heat transfer coefficient measurements of film cooling holes with expanded exits. ASME Paper 98-GT-28
8. Makki, Y. & Jakubowski, G. 1986. An experimental study of film cooling from diffused trapezoidal shaped holes. AIAA-Paper 86-1326
9. Yu, Y., Yen, C-H., Shih, T. I-P., Chyu, M.K. & Gogineni, S. 1999. Film cooling effectiveness and heat transfer coefficient distribution around diffusion shaped holes. ASME Paper 99-GT-34
10. Reiss, H. & Bölcs, A. 1999. Experimental study of showerhead cooling on a cylinder comparing several configurations using cylindrical and shaped holes. ASME Paper 99-GT-123
11. Thole, K., Gritsch, M., Schulz, A. & Wittig, S. 1996. Flow field measurements for film cooling holes with expanded exits. ASME-Paper 96-GT-174
12. Press, W.H., Flannery, B.P., Teukolsky, S.A. & Vetterling, W.T. 1988. Numerical Recipes. *In* Cambridge University Press

Effects of Bulk Flow Pulsations on Film Cooling with Shaped Holes

HONG-WOOK LEE and JOON SIK LEE

Institute of Advanced Machinery & Design
School of Mechanical & Aerospace Engineering
Seoul National University, Seoul 151-742, Korea
Fax: +82-2-883-0179, E-mail: jslee@gong.snu.ac.kr

ABSTRACT: Experimental results are presented which describe the effects of bulk flow pulsations on injectant behavior and film cooling performance around shaped holes with compound angle orientations. The shaped holes have a 15° forward expansion with an inclination angle of 35°. The orientation angles considered are 0°, 30°, and 60°. The pulsation frequency is fixed at 32 Hz, but changes in the time-averaged blowing ratios of 0.5, 1.0, and 2.0 produce three different coolant Strouhal numbers. Flow visualization shows that bulk flow pulsations cause significant periodic variations of the injectant flow rate, and produce severe crossflow ingestion at a high coolant Strouhal number. Although shaped holes show improved film coverage compared to round holes under steady condition, shaped holes are so sensitive to flow pulsations that a drastic reduction in the film cooling performance is observed at a high coolant Strouhal number.

INTRODUCTION

Discrete hole film cooling is an effective cooling method applied to gas turbine blades. Most of previous researches have concentrated on film cooling with round holes. Nowadays people are getting more interested in film cooling with shaped holes because it is supposed to provide more uniform and improved film coverage compared to that with simple round holes.

The importance of film cooling pertaining to shaped holes has been recognized only recently. Among earliest studies, Goldstein et al.[1] used a 35°-inclined axial injection hole with an initially round cross-section widened to each side by 10°. They reported a significant increase in the film cooling effectiveness in the near hole region as well as an improved lateral spread of the coolant. More recently, Schmidt et al.[2] and Sen et al.[3] examined the performance of forward-expanded holes with compound angle injection. They showed that the film-cooling performance with expanded holes is improved in the case of large momentum ratios and density ratios. In the measure of flowfield for expanded holes, Thole et al.[4] found that by expanding the exits of the cooling holes, the penetration of the injectant and the intense shear regions are significantly reduced relative to a round hole. McGrath and Leylek[5] conducted numerical simulations for the 15° forward expanded hole with a 60° orientation angle. They reported that the forward expanded hole provides improved film cooling performance, while simultaneously producing undesirable crossflow ingestion in the film hole.

Flow unsteadiness is another very important factor which might cause dramatic changes in film cooling performance. Film cooling flows are subject to bulk flow pulsations which result in important variations of the static pressure near turbine airfoil surfaces as blade rows move relative to each other. As a result, coolant flow rates pulsate at film hole exits, and injectant trajectory and coverage vary with time downstream of the film holes. Abhari and Epstein[6] indicated that flow pulsations cause the time-averaged heat transfer rate to increase by 12% on the suction surface and to decrease by 5% on the pressure surface, compared to non-pulsation values. Recent investigations examined the effects of bulk flow pulsations on film cooling from two rows of holes (Sohn and Lee[7]), from different length injection holes (Seo et al.[8]), and from holes with compound angle orientations (Jung[9]). Of these studies, Jung[9] reported that the adiabatic film cooling effectiveness decreases substantially at a high coolant Strouhal number regardless of the orientation angle. However, most of previous studies on flow pulsations are conducted with round holes and there has been no open literature that investigates the effects of flow pulsations on film cooling with shaped holes.

Present study investigates the effect of bulk flow pulsations on flow and film cooling characteristics around shaped holes. The study consists of injectant flow visualization and measurements of adiabatic effectiveness distributions. The shaped holes have a 15° forward expansion with an inclination angle of 35°, while the orientation angle takes on values of 0°, 30°, or 60°. The blowing ratios considered are 0.5, 1.0, and 2.0. All studies are conducted both under a steady condition and with bulk flow pulsations. The pulsation frequency is fixed at 32 Hz.

EXPERIMENTAL FACILITIES AND TECHNIQUES

A schematic of the wind tunnel and the injectant supply system is shown in Fig. 1. The wind tunnel is an open-circuit and subsonic one, with a 6.25 to 1 contraction ratio nozzle. At a free-stream velocity of 10 m/s, flow at the test section inlet shows excellent spatial uniformity with spanwise velocity variations less than 0.3%, and a turbulence level less than 0.3 %. A boundary layer trip wire of 2.4 mm diameter is located on the test plate just downstream of the nozzle exit. The air, used as the injectant, first flows through an orifice followed by two heat exchangers that control the air temperature. The air is then ducted to a plenum chamber and discharged through the injection holes.

The geometry of shaped film cooling holes and the coordinate system are shown in Fig. 2. The origin of the coordinate system is located at the trailing edge (TE) of the central hole. The film hole plates are prepared for each orientation angle of 0°, 30°, and 60°.

Static pressure pulsations are produced in the test section using an array of six shutter blades located at the exit of the wind tunnel test section and driven by a system of gears and a DC motor. In producing these bulk flow pulsations of 32 Hz, the coolant Strouhal number, $St_c = 2\pi f L/U_c$, where f is the pulsation frequency, L is the hole length

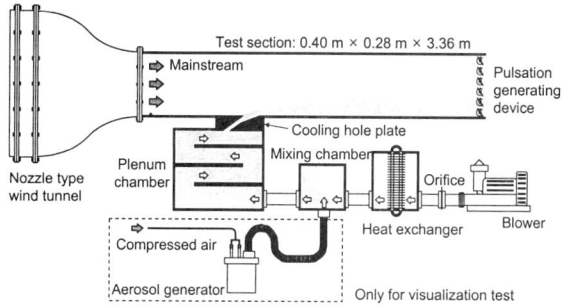

Fig. 1. Schematic view of the film cooling test facility

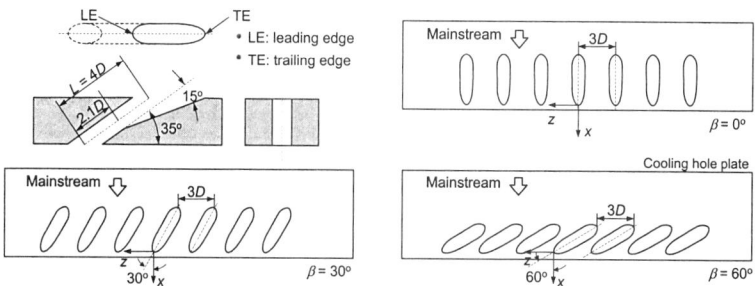

Fig. 2. Shaped hole geometry and orientation angle

and U_c is the time-averaged injectant velocity, varies from 0.6 to 2.4 for the hole with diameter $D = 15$ mm. Typical values for operating turbines range from 0.2 to 6.0.

Experiments are conducted at a fixed free-stream mean velocity of 10 m/s. The Reynolds number, $Re_X = U_\infty X / \nu$, is 488,000, where U_∞ is the time-averaged free-stream velocity, X is the distance between the trip wire and the hole center and ν is the kinematic viscosity. The boundary layer thickness (based on $0.99 U_\infty$) at the hole center is 24 mm (= $0.8D$ when $D = 15$ mm). The shape factor of the boundary layer is 1.4, which is the typical value of a fully developed turbulent boundary layer.

Flow Visualization

Figure 3(a) shows the experimental setup for the visualization test. Enlarged film cooling holes with a diameter of 30 mm are used for the purpose of flow visualization. To track the injectant and to determine its distribution at the hole exit plane, the injectant air is contaminated with aerosol. Oil aerosols produced by an aerosol generator[10] are mixed with the injectant in a mixing chamber. The cross-section of the injectant trajectory at the hole exit plane is illuminated by the laser sheet. The injectant motions are captured by a high speed camera, which is aligned perpendicular to the test surface 350 mm away.

The capturing speed is fixed to 60 frames per second with a exposure time of 1/125 s. 546 frames are recorded in the digital memory for a single capturing process. Each output image has a fairly fine resolution of 512 × 480 pixels with 256 levels of gray. The captured images are downloaded as standard TIFF files via the SCSI-2 port directly to a personal computer. Image-processing steps for averaging and image correction are then

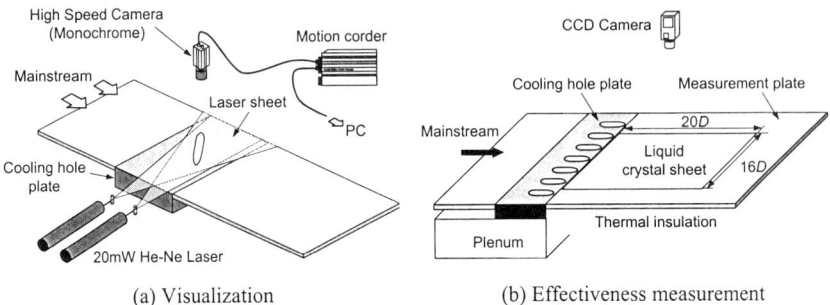

(a) Visualization (b) Effectiveness measurement

Fig. 3. Experimental set-up for visualization and effectiveness measurement

performed.

Adiabatic Film Cooling Effectiveness

For the film cooling effectiveness measurements, a row of seven holes with 15 mm in diameter is used. The hole spacing between the hole centers is $3D$. As described in Fig. 3(b), the test section consists of a film hole plate and a measurement plate. To measure the temperature distributions on the entire surface downstream of the film holes, a thermochromic liquid crystal (TLC) sheet is used. The TLC sheet covers the test plate from $x/D = 0.4$ to 20.4, and from $z/D = -8.0$ to 8.0. A CCD camera is used to capture TLC color images. Among many techniques, the steady-state, hue capturing method is adopted in this study. The free-stream and the injectant temperatures are measured by thermocouples that are calibrated in a constant temperature bath with a precision platinum resistance thermometer.

The uncertainty analysis is evaluated on 20 to 1 odds (95 percent confidence level). All the uncertainty values are evaluated from the method of single-sample experiments proposed by Kline and McClintock[11]. The uncertainty of the adiabatic film cooling effectiveness η is 6.8 percent at a typical η value of 0.2.

RESULTS AND DISCUSSIONS

Bulk Flow Pulsation Characteristics

The periodic characteristics of free-stream velocity and static pressure difference between plenum and free-stream at the pulsation frequency of 32 Hz are shown in Fig. 4. Figure 4(a) shows that the free-stream pulsation is produced with a sinusoidal waveform and the peak-to-peak amplitude of about 12 %.

Typical pulsation characteristics of the phase-averaged free-stream static pressure are illustrated in Fig. 4(b). The data are presented for time-averaged blowing ratios of $M = 0.5, 1.0$, and 2.0 at a pulsation frequency of 32 Hz. Here, $\Delta \tilde{P}$ indicates the difference in the phase-averaged static pressure between the plenum and free-stream, and $\Delta \overline{P}$ is the difference in time-averaged static pressures. The pressure data show quantitative alterations to wave forms which occur as M is changed. Such variations evidence important coupling between the pulsations and the film coolant in the holes and within the plenum. The amplitude of the pressure difference increases as the blowing ratio decreases. Another important fact is that when the blowing ratios are 0.5 and 1.0, a

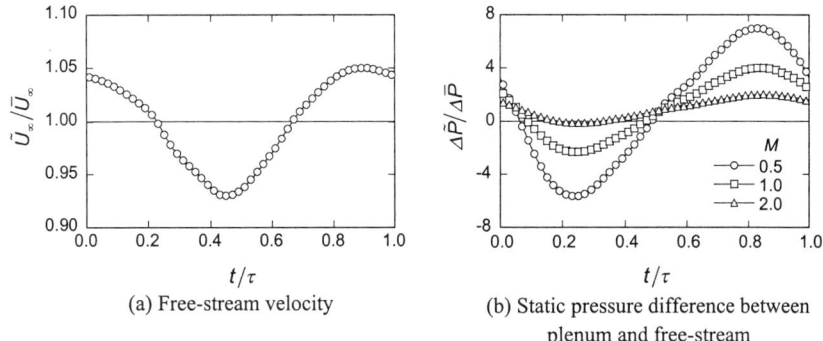

(a) Free-stream velocity (b) Static pressure difference between plenum and free-stream

Fig. 4. Phase-averaged free-stream velocity and static pressure difference

duration of negative values of $\Delta \widetilde{P}$ is observed in a pulsation period. At $M = 0.5$, negative values are present over almost half a pulsation period. This implies that the free-stream static pressure could be higher than the plenum pressure, which might cause flow ingestion into the holes.

Flow Visualization

Periodic variations of the injectant behavior at $M = 0.5$ are shown in Fig. 5. Each period is divided into five phases. The phase at which the injectant flow rate is considered minimum corresponds to $t = 4/5\,\tau$, where τ is the pulsation period. The phase-averaged images are presented at each phase. As expected in Fig. 4, the most significant change in the injectant behavior occurs at $M = 0.5$, which corresponds to the highest coolant Strouhal number of 4.8. The dark image in the hole indicates the region where the injectant is blocked by the crossflow.

Fig. 5(a) shows the injectant coming out discontinuously from the simple angle hole ($\beta = 0°$). As time goes on in a pulsation period, the portion at the hole exit where the injectant comes out (bright region) moves from the leading edge side to the trailing edge side. Such a drastic change of the injectant flow behavior is caused by the very large periodic variation of static pressure difference between the mainstream and the plenum.

At $t = 4/5\,\tau$, the injectant is shown only near the trailing edge and most part of the hole exit area is filled with the crossflow. Under the steady condition with no pulsations, the injectant near the leading edge retains its high momentum protecting the low momentum injectant near the trailing edge from the crossflow penetration. With pulsations, however, the occurrence of the negative static pressure difference as shown in Fig. 4(b), causes the consequent crossflow ingestion into the film hole. Although the mechanisms are different, the crossflow ingestion into film holes occurs more severely under the unsteady condition, even at simple angle orientation. As described by McGrath and Leylek[5], and Lee et al.[12], crossflow ingestion occurs even with no pulsations when the orientation angle is large enough such as 60°. In this case, the crossflow blocks the injectant near the trailing edge at the hole exit. This flow behavior results from the geometric peculiarity of the hole. The injectant coming out from the trailing edge side of the film hole slows down near the exit because the trailing edge is made of a diffusion section (expanded trailing edge), while the injectant from the leading edge side retains its momentum. Mainstream thus flows across the hole exit plane near the trailing edge where

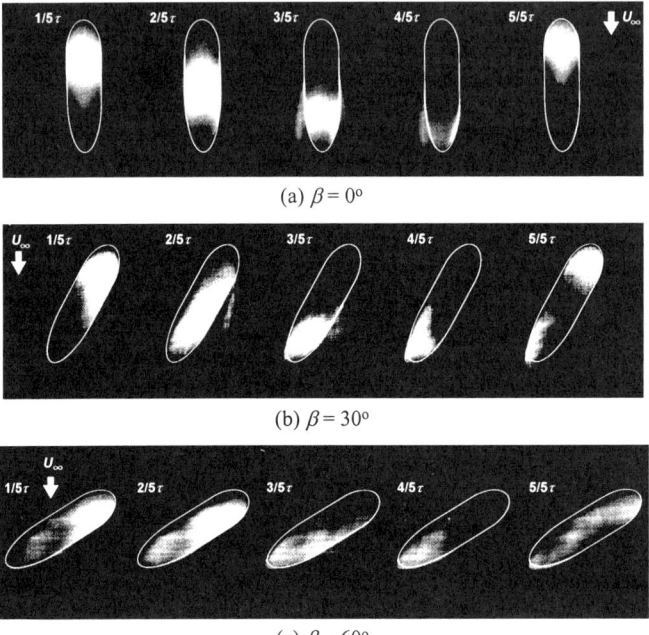

Fig. 5. Periodic variations of injectant behavior at $M = 0.5$

the injectant momentum is relatively low. Figure 5(b) and 5(c) show the phase-averaged images for compound angle injections with $\beta = 30°$ and $60°$, respectively. The general trend is similar to that with simple angle injection. The above visualization indicates that bulk flow pulsations cause significant periodic variations of the injectant flow rate and produces crossflow ingestion at a high coolant Strouhal number in addition to the hole geometric effect.

Adiabatic Film Cooling Effectiveness

Figure 6 shows the spanwise-averaged effectiveness distributions at $M = 0.5$ in comparison with the round hole data[9]. Although the coolant Strouhal numbers are not exactly the same, but those data are sufficient to make a qualitative comparison.

As expected from flow visualizations, the effectiveness values drop drastically due to the pulsations regardless of orientation angles. Shaped holes with simple angle injection do not provide substantial improvement in the film cooling performance compared to round holes. However, shaped holes exhibit a notable improvement with compound angle injection especially in the near hole region.

Figure 7 shows the space-averaged effectiveness over the full range tested. This quantity is defined as the integral average of the local effectiveness from $x/D = 0.6$ to 15.6 and from $z/D = -1.5$ to 4.5. The amount of reduction in the averaged effectiveness value due to pulsations is greatest at $M = 0.5$. On the other hand, the averaged effectiveness does not affected by pulsations at $M = 2.0$ regardless of hole shapes. The amounts of reduction in the space-averaged effectiveness due to pulsations are 62%, 63%,

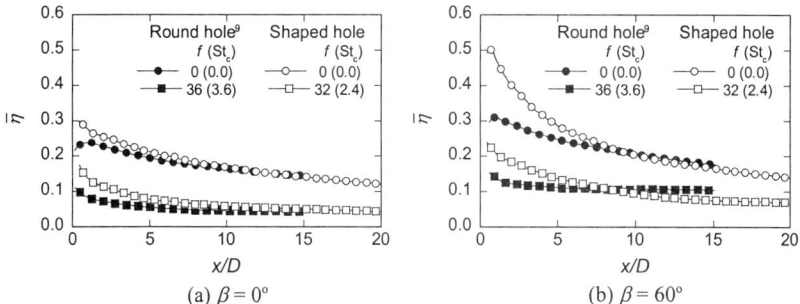

Fig. 6. Spanwise-averaged adiabatic effectiveness distributions compared to round hole data at $M = 0.5$

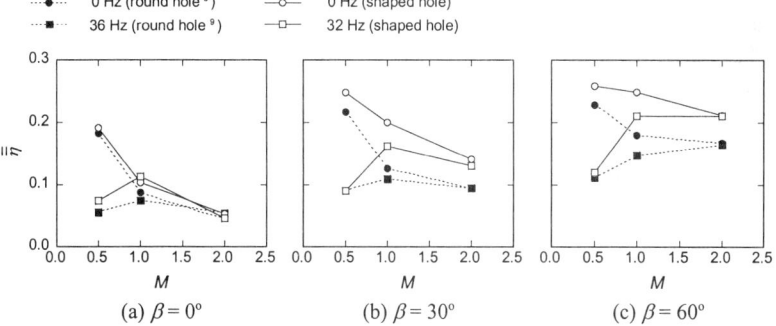

Fig. 7. Space-averaged adiabatic effectiveness compared to round hole data

and 53% for $\beta = 0°$, 30°, and 60°, respectively, at $M = 0.5$.

An important fact to note is that with pulsations the blowing ratio where the maximum space-averaged effectiveness occurs depends on the orientation angle. The maximum occurs at $M = 1.0$ for $\beta = 0°$ and 30°. For $\beta = 60°$, however, the space-averaged effectiveness increases with the blowing ratio to a maximum value at $M = 2.0$. Previous studies conducted under steady conditions, including a part of this study, showed that best film cooling performance could be obtained at the lower blowing ratio such as 0.5. Unsteady environment as in the real engine operation, however, leads to a completely different conclusion.

Compared to round holes, shaped holes with simple angle injection provide little benefit to film coverage for all blowing ratios. Although the range over which the high effectiveness levels are maintained can be expanded in the case of compound angle injection, the reduction in the effectiveness at a high coolant Strouhal number is inevitable.

It is well known that round holes provide higher effectiveness at lower blowing ratios. However, when bulk flow pulsations are imposed, round hole film cooling is significantly affected by the pulsation at lower blowing ratios. In the real engine operation where several kinds of flow unsteadiness are generated, round holes might yield poor performance over the range of blowing ratios. On the contrary, shaped holes with compound angle injection maintains high effectiveness at high blowing ratios regardless of flow pulsations. The best protection by film cooling in a real turbine can thus be achieved when shaped holes with compound angle injection are incorporated at high blowing ratios.

CONCLUSIONS

The effects of flow pulsations on film cooling with shaped holes are investigated. Flow visualization and adiabatic film cooling effectiveness measurement are conducted with forward-expanded holes under both steady and unsteady with bulk flow pulsation conditions. Some important observations are noted and summarized below.

1. Flow visualization reveals the periodic variation of the injectant flow rate with pulsations. The crossflow ingestion into film holes is observed at a high coolant Strouhal number, even with simple angle injection.

2. Shaped holes with compound angle orientation exhibit a notable improvement in the film cooling effectiveness compared to round holes when flow pulsations are imposed. The adiabatic film cooling effectiveness values are dramatically reduced due to pulsations at lower blowing ratios for shaped holes as well as for simple round holes.

3. At moderate and high blowing ratios, shaped holes provide a relatively good film cooling performance regardless of flow pulsations. The shaped holes are thus advantageous in delivering good film cooling performance at a high blowing ratio despite the flow unsteadiness.

REFERENCES

1. Goldstein, R.J., E.R.G. Eckert & F. Burggraf. 1974. Effects of Hole Geometry and Density on Three-Dimensional Film Cooling. Int. J. Heat Mass Transfer. **17:** 595-607.
2. Schmidt, D.L., B. Sen & D.G. Bogard. 1996. Film Cooling With Compound Angle Holes: Adiabatic Effectiveness. ASME Journal of Turbomachinery. **118:** 807-813.
3. Sen, B., D.L. Schmidt & D.G. Bogard. 1996. Film Cooling With Compound Angle Holes: Heat Transfer. ASME Journal of Turbomachinery. **118:** 800-806.
4. Thole, K., M. Gritsch, A. Schulz & S. Wittig. 1998. Flowfield Measurements for Film-Cooling Holes With Expanded Exits. ASME Journal of Turbomachinery. **120:** 327-336.
5. McGrath, E.L. & J.H. Leylek. 1998. Physics of Hot Crossflow Ingestion in Film Cooling. ASME Paper No. 98-GT-191.
6. Abhari, R.S. & A.H. Epstein. 1994. An Experimental Study of Film Cooling in a Rotating Transonic Turbine. ASME Journal of Turbomachinery. **116:** 63-70.
7. Sohn, D.K. & J.S. Lee. 1997. The Effect of Bulk Flow Pulsations on Film Cooling From Two Rows of Holes. ASME Paper No. 97-GT-129.
8. Seo, H.J., J.S. Lee & P.M. Ligrani. 1998. The Effect of Injection Hole Length on Film Cooling With Bulk Flow Pulsations. Int. J. Heat Mass Transfer. **41:** 3515-3528.
9. Jung, I.S. 1998. Effects of Bulk Flow Pulsations on Film Cooling with Compound Angle Injection Holes. Ph.D. Thesis. Seoul National University. Seoul. Korea.
10. Echols, W.H. & J.A. Young. 1963. Studies of Portable Air-Operated Aerosol Generators. NRL Report No. 5929.
11. Kline, S.J. & F.A. McClintock. 1953. Describing Uncertainties in Single Sample Experiments. Mechanical Engineering. **75:** 3-8.
12. Lee, H.-W., J.J. Park & J.S. Lee. 2000. Flow Visualization and Film Cooling Effectiveness Measurements around Shaped Holes with Compound Angle Orientation. presented at 12th International Symposium on Transport Phenomena. Istanbul. Turkey.

Film Cooling:
Case Of Double Rows Of Staggered Jets

E. DORIGNAC[+] & J.J. VULLIERME[+]
P. NOIRAULT[*], E. FOUCAULT[*] & J.L. BOUSGARBIÈS[*]

[+]*Laboratoire d'Etudes Thermiques (UMR 6608)*
ENSMA, Teleport 2, BP 40109, 86961 FUTUROSCOPE Chasseneuil Cedex. FRANCE.

[*]*Laboratoire d'Etudes Aérodynamiques (UMR 6609)*
Boulevard 3, Teleport 2, BP 179, 86961 FUTUROSCOPE Cedex. FRANCE.

ABSTRACT: An experimental investigation of film cooling of a wall in a case of double rows of staggered hot jets (65°C) in an ambient air flow. The wall is heated at a temperature value between the one of the jets and the one of the main flow. Experiments have been carried out for different injection rates, the main flow velocity is maintained at 32 m/s. Association of the measures of temperature profiles by cold wire and the measures of wall temperature by infrared thermography allows us to describe the behaviour of the flows and to propose the best injection which assures a good cooling of the plate.

INTRODUCTION

This work completes the studies done before in our laboratories about film cooling behind injection through one row of jets (Deniboire[1], Dorignac[2]). This experimental study (Noirault[3]) is an attempt to understand better the behaviour of the flow obtained after double rows of staggered jets. The injected flow meets the main flow at the room temperature and has to protect the wall from the main flow. This study conducted in collaboration with Snecma Moteurs is a part of the programme of improvement of the film cooling efficacy in order to protect walls put through severe conditions. Mean temperature profiles are measured using the cold wire technique for two different values of the injection ratio. The results are compared with those obtained from the examination of the wall temperature map measured with an infrared camera. The latter technique is then used to extend the study to a larger range of injection ratios.

DESCRIPTION OF EXPERIMENTS

The experiments are carried out in an open circuit wind tunnel with the ambient air flow. The flow configuration consists in the interaction of 11 air injections through the tunnel floor with a subsonic turbulent boundary layer (boundary layer thickness

$\delta = 15$ mm at $x = 0$). The considered hole configuration (figure 1) is of two rows of circular staggered holes. The holes have a diameter D equal to 5 mm, they are laterally spaced by 3D on the row and angled at 45 degree to the mainstream. The two rows are spaced by 3D.

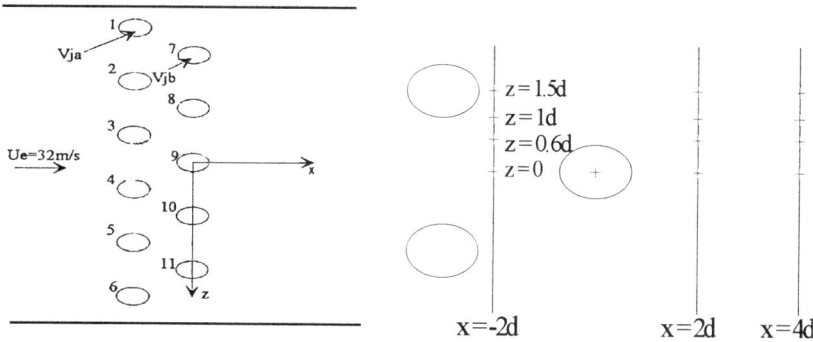

Figure 1 : (x, z) coordinate system and hole the pattern

Figure 2 : (x, z) positions of the temperature profiles

    The velocity and the temperature of the main flow are uniform and maintained at 32 m/s and between 14 and 19 °C respectively. In the injection flow the temperature is around 65 °C, the range of the injection rate (First row : $R_{ja} = V_{ja}/U_e$, Second row : $R_{jb} = V_{jb}/U_e$) is between 0.16 and 1.56. Flows temperature is controlled with type K thermocouples, flows velocity by the Pitot tube for the main flow velocity and by flowmetres for the injection velocity.

    During the test, the plate is maintained at a uniform heat flux density. In order to obtain this condition, we use the technique of printed circuit to have a surface comparable to a fluxmetre. We impose an electric dissipated power around 2000 W/m² which gives the mean surface temperature of approximately 45°C for the chosen main flow velocity and the chosen range of injection rates. At thermal balance, the electric power is dissipated by convection in the fluid, by radiation and by conduction in the plate. The plate is isolated on the back face; we have checked experimentally that the losses through the back side are weak (less than 100 W/m²).

    The wall surface temperature map is established using an infrared camera (AGEMA 880 SWB). This technique supposes a precise calibration that requires correct evaluation of the wall emissivity. In order to increase the value of the wall emissivity, we paint the wall surface black. The measure of the emissivity value is done at the laboratory and is equal to $\varepsilon = 0.94 \pm 0.01$. The temperature profiles are obtained with cold wire probes, and its characteristics are the 4 mm length and the 2.5 µm diameter. They are first connected to the DISA temperature bridge (type 55M20) and then to a PC calculator. Each measure point is issued from an average of 400 instantaneous values.

RESULTS

Temperature Profiles

The thermal field detailed study is realised for two different values of couple of the injection ratio ($R_{ja} = R_{jb}$) respectively equal to 0.63 and 1.56. The mean temperature profiles are plotted with a non - dimensional variable defined as :

$$T^* = \frac{T - T_e}{T_j - T_e},$$

where T is local mean temperature measured by cold wire, T_e, ambient temperature and T_j injection temperature.

Figures 3 and 4 present these variation profiles carried out along the normal at the wall at $x = -2d$ and $x = +4d$ positions respectively (see figure 2 to visualise the positions along x and z). Their analysis allows us to reveal the leading part of each row.

At the upstream position, $x = -2d$ (just after the first row of jets), the temperature profile corresponds to the one found behind a jet of a unique row (Deniboire[1]). For $0 < z < 0.4d$, the mean temperature decreases regularly from the wall temperature value to the one of the main flow whatever the injection rate. These profiles look like boundary layer profile. From $z = 0.6d$, we start to see the influence of the first row for the largest injection rate. The dimensionless temperature variation shows a maximum value at $y \approx 0.5d$. We see the same phenomenon for the injection rate $R_{ja} = R_{jb} = 0.63$ from the $z = 0.8d$ position. And at this position, for the injection rates $R_{ja} = R_{jb} = 1.56$, the profile slope is negative close to the wall. This, before the maximum value, we have here a minimum value. This minimum existence was explained (Dorignac[2]) by a coming up of the cold main flow which, caught by the contrarotative vortex, slips under the injection flow. We can observe as well that the presence of this minimum collapses at the $z = 1.5d$ position, which means that the coming up movement of the cold air does not reach the considered abscissa.

At the downstream position, $x = +4d$ (after the second row of jets), the temperature profiles show a maximum value that corresponds to the presence of the central jet of the second row for $0 \leq z \leq 1$ whatever the injection rate. We can notice that for the largest injection rate, the y position of the maximum is higher. Indeed, the injection rate being important, the jet rises more from the wall. Beyond $z/d = 1$, there is not a well defined maximum, the profiles have just levelled off. At this abscissa, we can consider that there is a mixing layer. The figure 5 is a detail of the zone close to the wall, the first point at $y/d = 0$ is issued from the infrared measurements, other points are issued from the cold wire measurements. On the figure 5, we can observe that the heat exchange very close to the wall (that means for $y/d < 0.1$ mm) is very high. We have verified by a calculation from the slope that we find the same heat flux density value we imposed at the plate by electric heating. After this temperature fall, the temperature gradient becomes rapidly invariant and close to 0.6 times $(T_j - T_e)$. Nevertheless, it is not easy to conclude that this mixing layer is an efficient protection of the wall. Indeed, the wall is heated and so the dimensionless wall temperature is around 0.5 without injection. That means that the existence of injection leads to an increase of around 20% of the wall temperature, at least four diameters downstream the injection area. The question is now to know whether the heating level of the plate has some influence on this value of 0.6 times $(T_j - T_e)$. If this value is invariant with the heating of the plate, it would be interesting to use it in the modelling of the convective transfer, and to compare this modelling with the one using adiabatic wall temperature (Jabbari[4]).

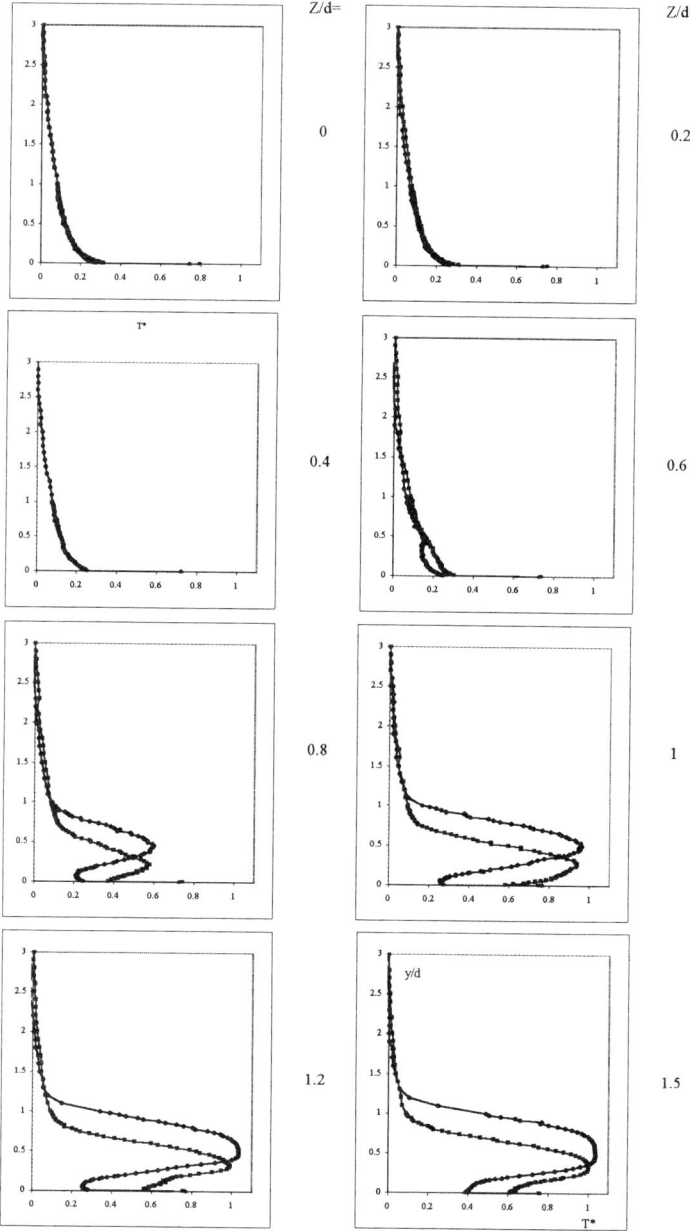

Figure 3 : Dimensionless mean value temperature profiles in several locations (z/d) in the plane x = -2d. Injection rates : ◆ $R_{ja} = R_{jb} = 0.63$; ■ $R_{ja} = R_{jb} = 1.56$

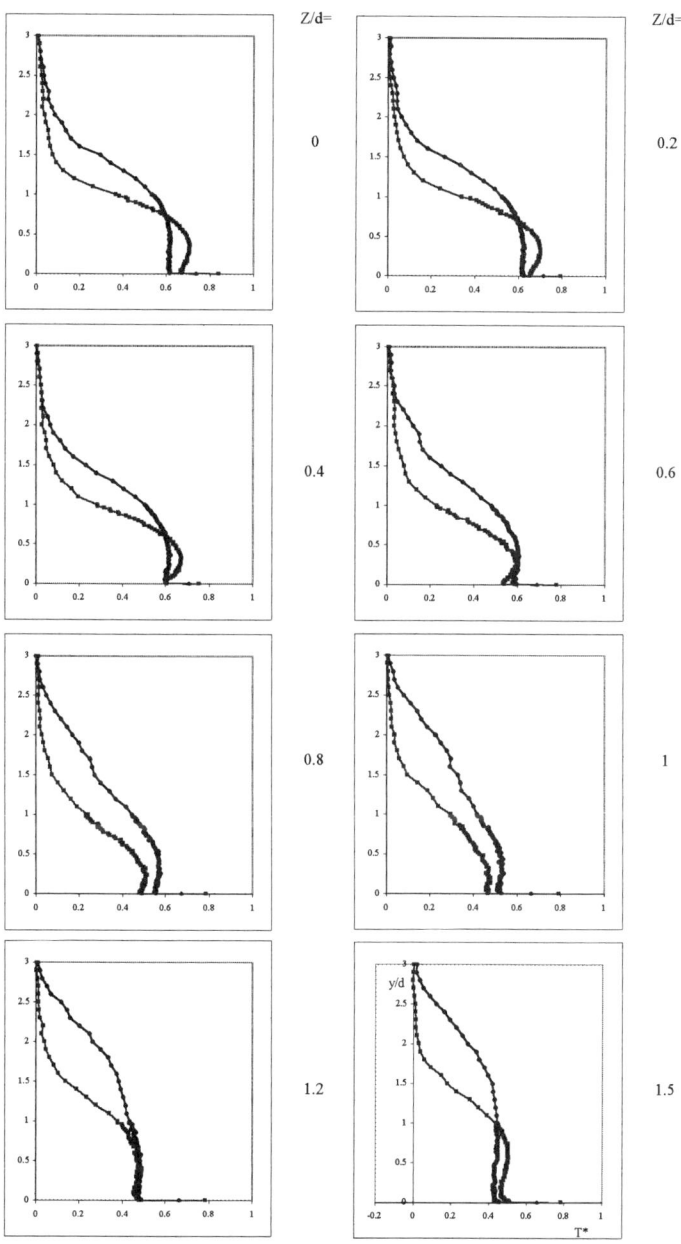

Figure 4 : Dimensionless mean value temperature profiles in several locations (z/d) in the plane x = +4d. Injection rates : ♦ $R_{ja} = R_{jb} = 0.63$; ■ $R_{ja} = R_{jb} = 1.56$

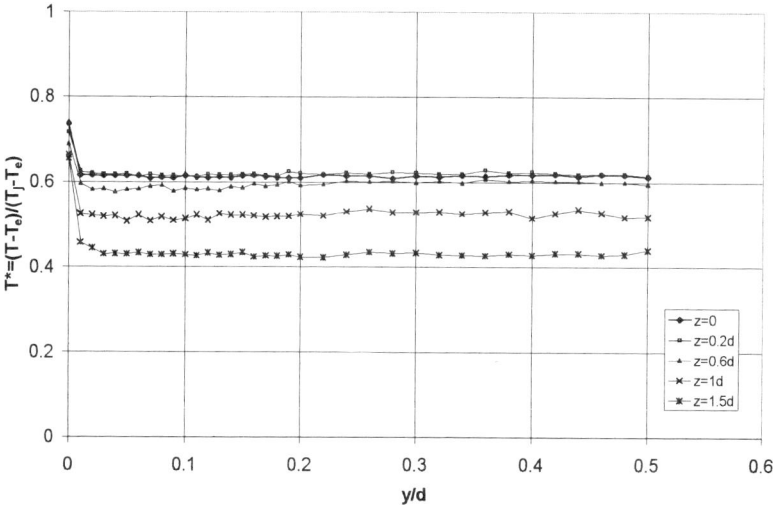

Figure 5 : Detail of dimensionless mean temperature profiles in several locations z/d in x/d = 4d plane. Injection rate : $R_{ja} = R_{jb} = 1.56$

Wall Temperature Variations

The examination of the different profiles close to the wall shows that from the two considered test conditions, it is the flow generated with the couple $R_{ja} = R_{jb} = 0.63$ which ensures the best overlap of the plane wall. Indeed, for this couple of values, the temperature close to the wall is almost always greater than that registered in the case where $R_{ja} = R_{jb} = 1.56$. The wall temperature maps (Figures 6 and 7) obtained with the infrared camera confirm this result. The latter technique is then used to register the wall temperature. The injection velocity of one of the rows is kept constant while the injection velocity of the second row is varied in the range of 5 m/s to 50 m/s. Figures 6 and 7 present examples of injection cases where R_{ja} are different than R_{jb}. The analysis of the obtained temperature maps allows us to determine the injection conditions ensuring more efficiently the plane surface overlap. In order to analyse the IR map, we choose to present the results with another dimensionless variable defined as :

$$T^{**} = \frac{T - T_e}{T_{upstream} - T_e},$$

where $T_{upstream}$ is the wall temperature value 10 diameters before the injection, the location not disturbed by the injection. Indeed, as the heating of the plate is not strictly the same from one experiment to another, the initial wall temperature without injection is not the same either. Thus, the dimensionless variable proposed above allows us to compare the different experimental cases. Figure 8 presents the T^{**} variation along the wall surface at lateral position z = 0 for different injection rate couples. These values are extracted from infrared maps. The injection zone is not represented on these curves because of the existence of the injection tubes which disturb the wall temperature variation. Indeed, the conductive heat transfer in this area is three dimensional and not negligible. Beyond this zone, the observation of these curves confirms that when the injection velocity ratio R_{ja} and R_{jb} are equal, there is a similar protection along the whole length of the plate. This is true whatever the value of R_{ja} and R_{jb} since it is greater than 0.5. If these two velocities are not equal, it is then better to have $R_{ja} < R_{jb}$ to ensure the protection of the plate. Indeed, the curves where $R_{ja} < R_{jb}$ are always above the one $R_{ja} > R_{jb}$.

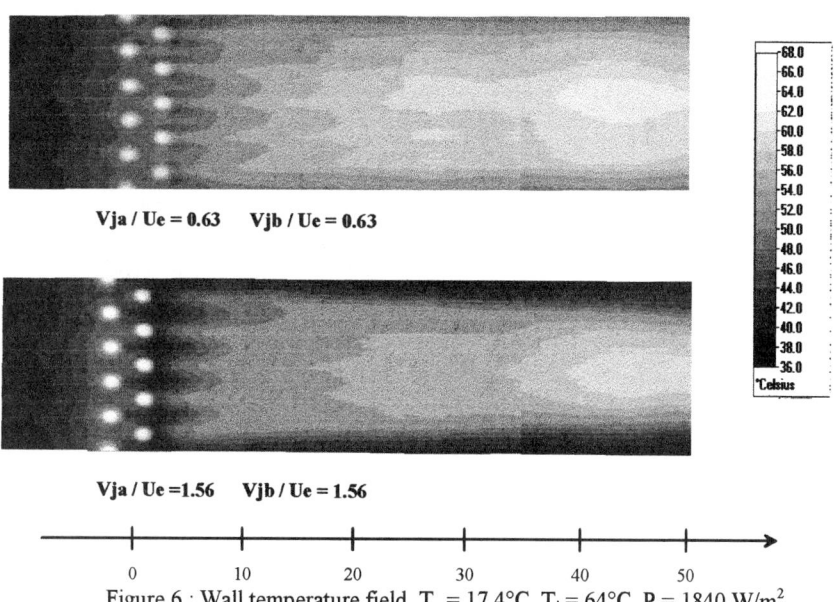

Figure 6 : Wall temperature field. $T_e = 17,4°C$, $T_j = 64°C$, $P = 1840$ W/m^2

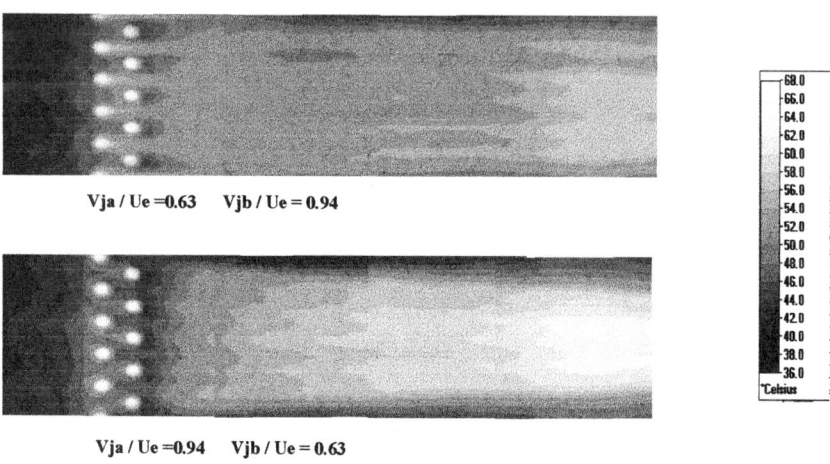

Figure 7 : Wall temperature field. $T_e = 15,5°C$, $T_j = 65°C$, $P = 2330$ W/m^2

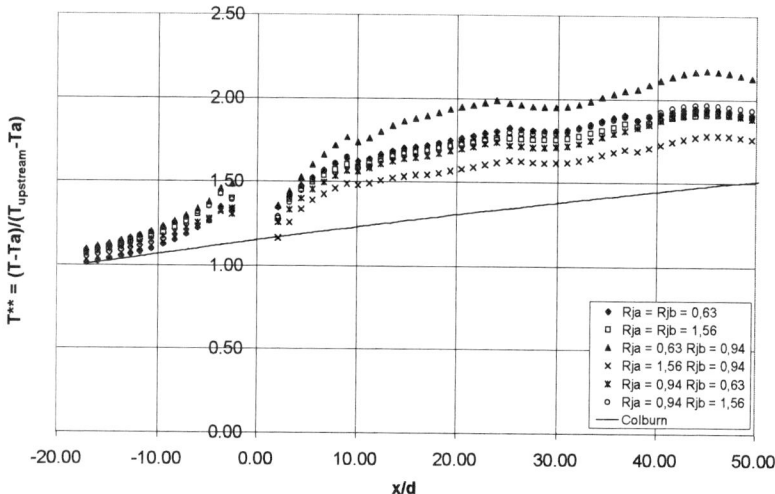

Figure 8 : Dimensionless wall temperature for different injection rates. $T_j = 65°C$

CONCLUSION

Finally, the association of the measures of temperature profiles by cold wire and the measures of wall temperature by infrared thermography seems to put in obviousness that the wall heat transfer has a weak influence on the behaviour of the flows beyond a certain distance downstream the injection. Nevertheless, complementary experiences are needed to confirm the hypothesis of the presence of a mixing layer. The temperature of this mixing layer could be used to model wall heat transfers in these complex thermal cases.

REFERENCES

1. Deniboire, P. 1993, Contribution à l'étude aérothermique de jets turbulents débouchant au sein d'une couche limite transversale, Thèse de doctorat, Université de Poitiers

2. Dorignac, E. 1990, Contribution à l'étude de la convection forcée sur une plaque plane en présence de jets pariétaux dans un écoulement subsonique. Thèse de doctorat, Université de Poitiers

3. Noirault, P. 1999. Interaction entre une couche limite plane et une double rangée de jets en quinconce : application au refroidissement de parois. Thèse de doctorat, Université de Poitiers.

4. Jabbari, M.Y., Goldstein, R.J. 1978, Adiabatic wall temperature and heat transfer downstream of injection through two rows of holes, J. of Engineering for Power, **100**: 303-307.

Acknowledgement : This work conducted in collaboration with Snecma Moteurs was supported by la Direction des Recherches, Etudes et Techniques.

Characteristics of Various Film Cooling Jets Injected in a Conduit

HIRONORI TAKAHASHI*, CHAYUT NUNTADUSIT*, HIDEO KIMOTO*,
HIDESHI ISHIDA*, TAKEOMI UKAI** and KENICHIRO TAKEISHI***

*Mechanical Science Division, Graduate School of Engineering Science, Osaka University, Osaka, 560-8531, Japan
**Daikin Industry Ltd., Osaka, 530-8323, Japan
***Takasago R & D Center, Mitsubishi Heavy Industry Ltd, Takasago, 676-8686, Japan

ABSTRACT: In the present study, film cooling characteristics by the jets through various easy-to-make straight holes and slots have been investigated. In this experiment, seven kinds of injection geometries were used. They were circular, rectangular, elliptic and oval holes and slots, respectively.

INTRODUCTION

Various techniques on film cooling have been proposed and a lot of researches on film cooling have been reported. In the reports the effect of blowing ratio, density difference between the film jet and the mainstream, geometry of injection hole, boundary layer characteristics on film cooling effectiveness and so on have been investigated experimentally and numerically.[1-6]

In the present study on film cooling, characteristics by a row of cooling jets through various easy-to-make straight holes were experimentally examined in a rectangular conduit under the conditions of several blowing (mass-flux) ratios. Their geometries were a circle, three kinds of rectangles, an ellipse, and two kinds of ovals. In addition, the mixing process between film jet and mainstream for each hole geometry was investigated by measuring the temperature field in the conduit. In the present study, it should be noted that in order to make the experimental apparatus simple the film cooling jet was heated and the temperature of the cooling jet was higher than the mainstream temperature,.

EXPERIMENTAL APPARATUS AND METHOD

The schematic of rectangular conduit is shown in Fig.1(a). The test section of the conduit was made of acrylic plate and the sizes were 60mm high and 100mm wide. The mainstream was passed through a diffuser and honeycomb cells before arriving at the test section.

Injection holes and slots were opened on a plate as shown in Fig.1 (b). The pitch between the centers of each hole was a constant of 24mm. The details of each injection holes and slots are

Table.1 Geometry of injection holes

	Cross section		D_z [mm]	S [mm]	Area [mm²]
A	Rectangle		8.5	6.68	56.8
B	Circle		8.5	8.50	56.8
C	Rectangle		12.0	4.73	56.8
D	Ellipse		12.0	6.00	56.8
E	Oval		12.0	6.00	64.3
F	Rectangle		18.0	3.15	56.8
G	Oval		18.0	6.00	100.3

shown in Table.1. In the table the cross sections of A, B, C, D and F have the same area, and their equivalent diameters are corresponding to a circle of 8.5mm diameter. The film cooling jet was provided by a blower. It was previously heated by a pre-heater, and then was passed though a laminar flow meter. The jet was injected into the mainstream in the conduit at an angle of 30 degrees. The distribution of temperature in the conduit was obtained by a comb-shaped thermal probe, which had nine Cu-Co thermocouples set 6 mm pitch in a row. The thermal probe was supported and fixed to a three-dimensional traverse equipment, controlled by a computer. The data sampling of temperature was conducted 15 times at each measuring point. The temperature deviation in the data was within ± 0.3 degree centigrade and the averaged value has been saved. Therefore the uncertainty of the cooling effectiveness in the present paper is estimated approximately one percent.

In the present experiment, the mainstream temperature was around room temperature of $20\,°C$. The velocity of mainstream at the center of conduit was fixed at 8.26m/s and its turbulent intensity was about 1.01%. The temperature of film jet was $50\,°C$ and as mentioned above it was about $30\,°C$ higher than the mainstream temperature. In the experiment, the geometry of film jet and the blowing (mass flux) ratio M were changed.

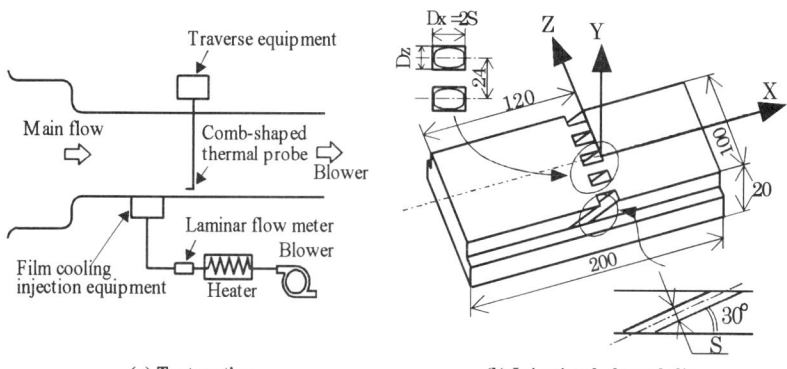

(a) Test section   (b) Injection hole and slit

Fig.1 Schematic of experimental apparatus

RESULTS AND DISCUSSION

From the local temperature distribution measured for the present film heating experiment, the film cooling effectiveness in the conduit is defined as follows;

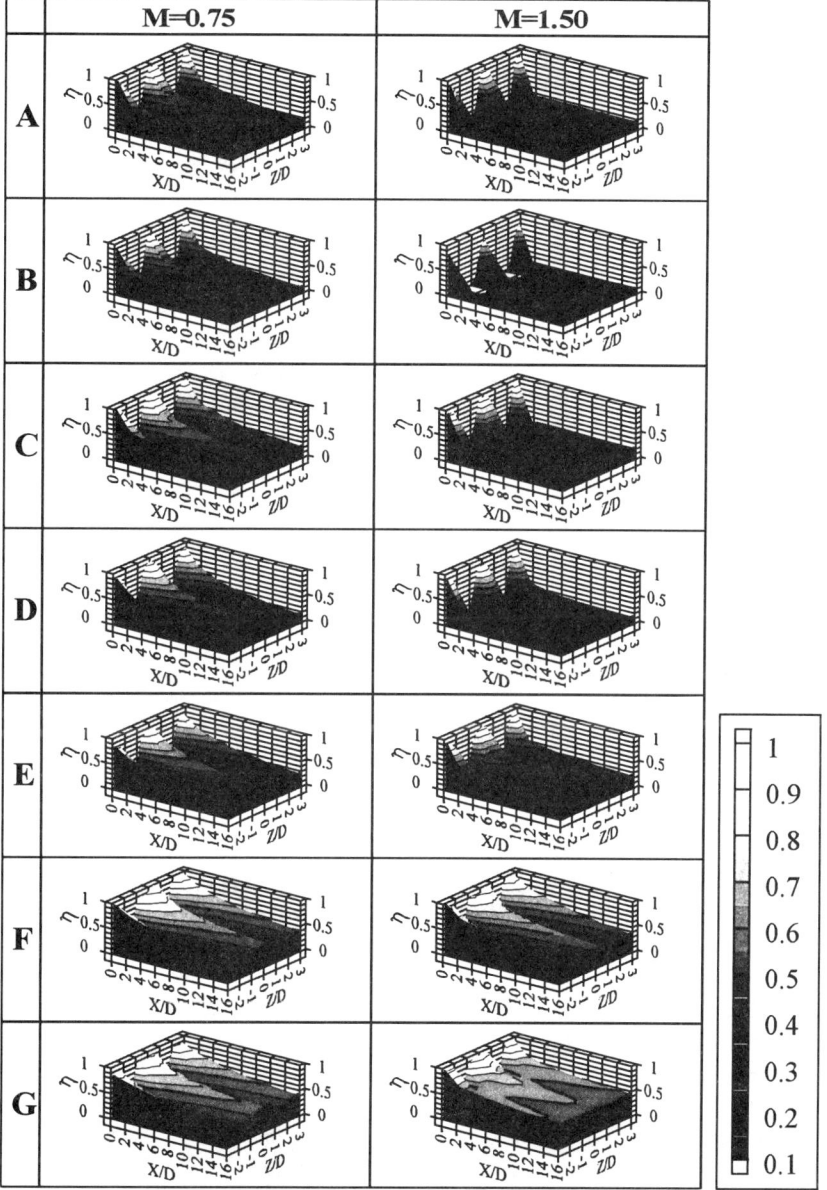

Fig.2 Distribution of film cooling effectiveness on Z-X plane (Y=2mm)

$$\eta = (T_f - T_\infty)/(T_j - T_\infty),$$

where T_∞ is the mainstream temperature, T_j is the temperature of injected film jet and T_f is the local temperature in the conduit.

Distribution of Film Cooling Effectiveness on the Cooling Surface

The typical profiles of film cooling effectiveness near the cooling surface (Y=2mm) are shown as three-dimensional graphs in Fig.2. In general for all of the injection holes, the effectiveness is relatively high along each downstream centerline of the injection holes (Z/D=0), and it is relatively low along the lines downstream between the holes (Z/D=+1.4, -1.4). From the figure it is confirmed that as the film jet flows downstream it mixes with mainstream and spreads toward Z-direction. This tendency becomes stronger as the blowing ratio M increases.

Comparing the rectangular hole of F with the oval hole of G in Fig.2, which have the same hole width D_Z, film jet from the rectangular holes in the case of F spreads in Z-direction more slowly than that from the oval holes in the case of G. In the case of rectangular hole F, the cooling effectiveness is higher on each downstream centerline of injection holes than that along the downstream between the holes. In the case of oval hole G, downstream effectiveness is almost even in Z-direction.

Meanwhile, comparing the case of rectangular hole A with the case of circular hole B for M=1.5 in Fig.2, in the case of B, there is an area where the cooling effectiveness becomes nearly zero along the lines downstream between injection holes, but in the case of A such a tendency doesn't appear.

These results stated above are noteworthy. In the cases of rectangular hole A and circular hole B the injection holes have the same width (D_Z=8.5mm), and in the cases of rectangular hole F and oval hole G the injection holes have also the same width (D_Z=18.0mm). So from the point of view of the injection hole geometry it is expected that the tendency of these two results are similar. However, the relation between the cases of A and B are different from the result in the cases of F and G, with regards to the spreading of jets. The reason is caused by the difference of the width of holes D_Z and the pitch between the hole edges.

Comparing the rectangular hole of C with the oval hole of E, both cases have similar result of spreading the film jets toward Z-direction. This result is also noteworthy. The injection holes of C and E have the same width (D_Z =12.0mm), which is larger than the holes of A and B and smaller than the holes of F and G. From these facts, the cooling effectiveness in the case of rectangular hole C becomes intermediate value between the cases of A and F, and the cooling effectiveness in the case of oval hole E becomes intermediate value between the cases of B and G.

The cooling effectiveness in the case of elliptic hole D is lower than that in the cases of rectangular hole C and oval hole E. This point will be minutely discussed in the following section.

Distribution of Film Cooling Effectiveness on Y-Z plane

Distributions of film cooling effectiveness on Y-Z plane, which is normal to main flow direction, are shown in Fig.3. The film cooling effectiveness is high just downstream of every injection holes. As the jet flows downstream it spreads toward Y- and Z-directions and mixes with mainstream.

In the cases of rectangular hole A and circular hole B, the width of injection holes is comparatively small and the distributions of film cooling effectiveness are like a kidney-shape as shown in Fig.3. The tendency is common for all range of M in the cases of A and B.

In the cases of rectangular hole C, elliptic hole D and oval hole E the same tendency is appear for $M \geq 1.25$. On the other hand, in the cases of rectangular hole F and oval hole G, the width of injection holes is comparatively large and no kidney-shaped area appears in Fig.3. In these cases the film jet is leaned against the wall by the main flow and as a result it becomes very thin layer. Another reason is the fact that the pitch between the hole edges is small enough for the main flow to wrap up the film jet.

Comparing the flow characteristics of the jets in the case of A with those in the case of B, both jets are engulfed by the surrounding main flow and leave away from the wall surface. However the circular jets of B are engulfed by the mainstream more strongly than the rectangular jets of A. As a result, the jets in the case of B mix more gradually with mainstream and spread in spanwise direction more slowly than the jets in the case of A. Meanwhile comparing the flow characteristics of the jets in the case of F with in the case of G, both jets stick closely to the wall surface. It is obvious in Fig.3 that the oval jets of G are leaned against wall surface and widened toward Z-direction more easily than the rectangular jets of F and this tendency becomes clear enough when the blowing ratio M becomes large.

As discussed in the previous section, the geometry of injection holes in the cases of A and F is rectangular and has corners of right angle, but the injection holes in the cases of B and G have no sharp corners. From this point of view, the results in the cases of A and B are completely opposite in comparison with the cases F and G. The reason for this phenomenon is as follows. In the case of

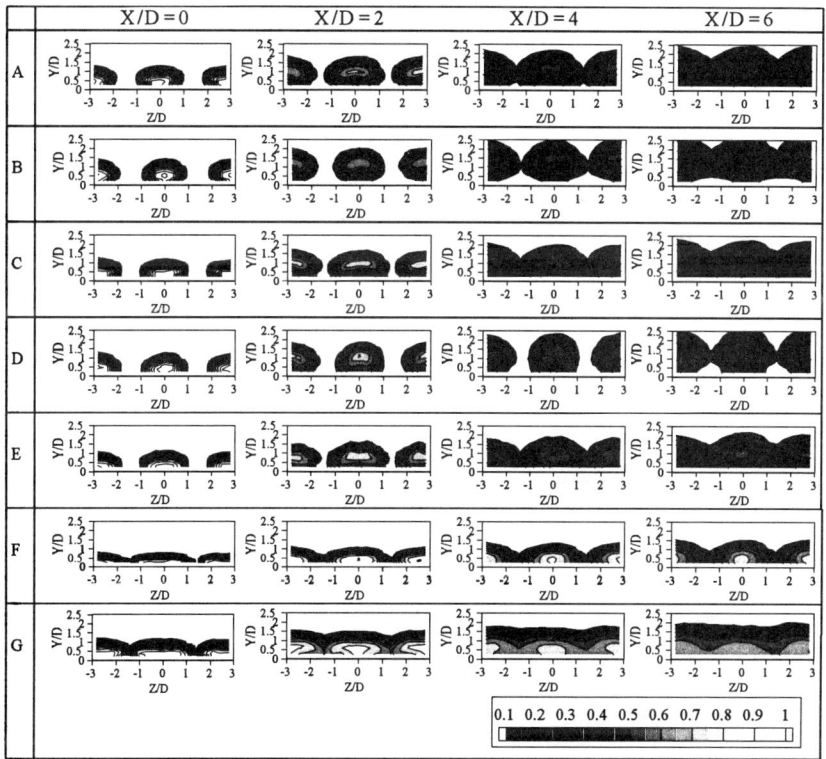

Fig.3 Distribution of film cooling effectiveness on Y-Z plane (M=1.5)

circular hole B the main stream completely covers the film jets. On the other hand in the case of oval hole G, D_Z is so large that the film jet sticks to the wall more easily, and the distance between the injection hole edges is so small that the main flow can hardly go into the sides of the film jets. In addition, in the case of G the cross sectional area of the injection hole is larger than in the cases of B, C, D and F, and the difference of D_X of the holes in the cases of F and G is larger than those of the holes in the cases of A and B.

Comparing the circular holes of B with the elliptic holes of D, the film jets in the case of D become widespread more slowly than those in the case of B, although D_Z of the injection holes in the case of D is larger than in the case of B. It is clear from Fig.3 that the thermally distributed area is long in Y-direction, although the injection holes of D have larger D_Z than in case of B. This is because the injection holes of D have elliptic shape and have their long axis in spanwise direction. So the elliptic jets of D spread in spanwise direction at first, but the thickness of the jets is thin at their sides. Therefore it is easily contracted in Z-direction by the mainstream as they flow downstream. The elliptic film jet of D is vulnerable to mainstream and its diffusion in Z-direction is prohibited by the mainstream.

Distribution of Film Cooling Effectiveness on X-Y plane

Typical distributions of film cooling effectiveness on X-Y cross-section (at Z=0) for M=1.5 are shown in Fig.4 by the contour lines of equal effectiveness.

Comparing the cases of F and G in Fig.4, in the upstream region ($0 \leq X/D \leq 8$) each contour line in the case of rectangular hole F extends downstream further than in the case of oval hole G, but in the region of X/D>8 the contour line of G goes further downstream than in the case of F. The reason for this phenomenon is as follows. As explained in the former section, the oval film jets of G mix with mainstream faster because they have no corners of right angle at their sides and are easily engulfed by the mainstream. But after the film jet and mainstream mix at X/D>8, the mixed and stable film layer is formed over the wall surface and the high cooling effectiveness is kept. On the other hand, the rectangular jets of F are leaned against the wall surface and continuously exposed to mainstream, without the mixing of the film jets and the mainstream.

The similar argument holds good with regard to the cases of C and E. In the upstream region ($0 \leq X/D \leq 8$), each contour line in the case of rectangular hole C extends downstream further than in the case of oval hole E, but the contour line of E goes further downstream in the region of X/D>8.

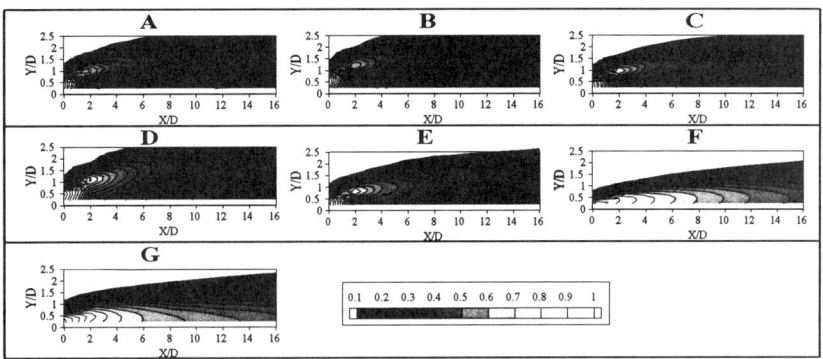

Fig.4 Distribution of film cooling effectiveness on X-Y(Z=0) plane (M=1.5)

As discussed in the previous section, it is also recognized that the circular film jets of B leave away from the wall surface faster than the rectangular jets of A, because they are engulfed by the mainstream stronger.

Effect of Blowing Ratio M

Relationships between the averaged film cooling effectiveness on the wall surface $\bar{\eta}_Z$ and the blowing ratio M are shown in Fig.5. Here, $\bar{\eta}_Z$ is the averaged value of $\eta_Z(\eta$ at Y=2mm) in Z-direction.

Considering the cases of A, C and F, which have the same cross sectional area and are all rectangular shape, for the cases of A and C, the cooling effectiveness is higher at downstream region (X/D>8) than at upstream region ($0 \leq X/D \leq 8$) when M is relatively large. The high region of the averaged cooling effectiveness starts at $M \cong 1.25$ for the case of A, but it starts at about $M \cong 1.5$ for the case of C. This phenomenon, as discussed in 3.3, indicates that the effect of mainstream on the effectiveness near the wall becomes weak as the width of the holes D_Z becomes wider. In general, $\bar{\eta}_Z$ becomes larger as the blowing ratio M increases. When M excesses certain value, however, $\bar{\eta}_Z$ begins to decrease. From this point of view, in upstream region ($0 \leq X/D \leq 4$) in Fig.5, $\bar{\eta}_Z$ decreases or remains constant for all geometries of holes.

In the case of F the peak is between M=1.00 and M=1.25 and in the case of C it is around M=0.75. In the case of A it seems to have a peak at M of smaller than 0.5. The rectangular hole becomes similar to a slit as the hole width D_Z becomes larger and consequently the effectiveness becomes higher. This result fits very well with the reports that the film cooling effectiveness has a peak around M=0.5 for the circular holes and has a peak at about M=1.0 for two-dimensional film cooling from a slit.[1]

Comparing the rectangular hole with the circular hole as in cases of A and B, which have the same area and the same width, in the case of circular hole A the cooling effectiveness has lower value than in the case of rectangular hole B in all range of M, because the jets from the circular holes are easily engulfed by the mainstream than the jets from the rectangular holes. Consequently

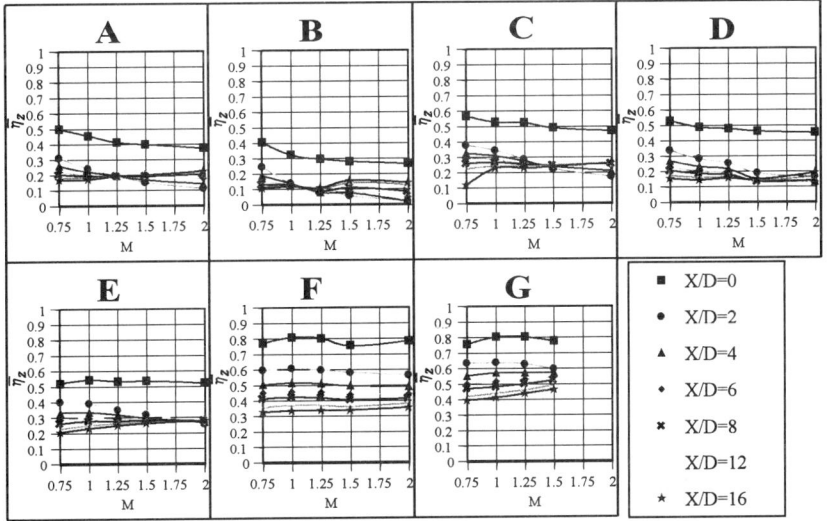

Fig.5 Averaged film cooling effectiveness

the jets from the holes of B do not lean on the wall easily.

In the case of oval hole E, there is not any phenomena that the effectiveness at upstream region ($0 \leq X/D \leq 8$) indicates lower value than that at downstream region ($X/D>8$). As a whole, cooling jet injection in the case of oval hole E has a little higher effectiveness than that in the case of rectangular hole C, which has the same width as the oval hole E.

Changing the blowing ratio M the film cooling effectivenesses on the wall surface (η_{CL}) were measured along the downstream centerline of the hole (Y=2mm, Z=0). In the cases of F and G, whose hole width is large, η_{CL} is higher than $\overline{\eta_Z}$, because the mainstream do not penetrate into the downstream centerline of the holes. In the case of rectangular hole F, the dependence of η_{CL} upon M is very similar to that of $\overline{\eta_Z}$. In the case of oval hole G the dependence of η_{CL} upon M has little fluctuations and remains stable.

When M is small, in the cases of A, B, C, D and E, η_{CL} is greater than $\overline{\eta_Z}$, but when M is large enough, η_{CL} is nearly equal to $\overline{\eta_Z}$. This is because the mixing effect of the film jets and the mainstream becomes stronger as M increases.

CONCLUSION

In this experiment, using the seven kinds of geometries of injection holes, which have the same width, the film cooling effectiveness was measured on condition that the main flow velocity is constant. The main results obtained in the experiment are as follows:

(1) Film cooling jet through a circular hole does not spread over the downstream wall. So the cooling effectiveness by the film cooling jet through circular holes is lower than by the film cooling jet through rectangular holes, whose width is the same as the circular holes.
(2) Among the rectangular holes and slots, which have the same cross sectional area, the highest cooling effectiveness is obtained for the film cooling jet through the widest rectangular slot.
(3) The mass flux ratio M, which gives the highest cooling effectiveness, increases to 1.0 as the injection hole geometry becomes similar to the so-called "slit".
(4) The film cooling jets through the oval slots are leaned against the wall by the mainstream and they spread over the downstream wall earlier than those through the rectangular slots. So the oval slot is the most effective hole for the film cooling jet and it is an easy geometry to open through the real blade wall.

REFERENCES

1. Goldstein, R.J. 1971. Film Cooling. *In* Advances in Heat Transfer. 7: Academic Press, New York, 321-371.
2. Goldstein, R.J. and Eckert, E.R.G. 1974. Effect of hole geometry and density on three-dimensional film cooling. Int. J. Heat Transfer. 17: 595-607.
3. Tillman, E.S. and Jen, H.F. 1984. Trans.ASME, J.of Eng. For Gas Turbine and Power. 106:
4. Makki, Y.H. and Jakubowski, G.S. 1986. An experimental study of film cooling from diffused trapezoidal shaped holes. AIAA Paper, AIAA-86-1326.
5. Ou, S. and Han, J.C. 1991. ASME Paper, 91-GT-254.
6. Watanabe, K., et al., 1999. An experimental study on the film cooling effectiveness with expanded hole geometry. *In* Proc. IGTC'99 Kobe.

FILM COOLING PERFORMANCE ON CURVED WALLS WITH COMPOUND ANGLE HOLE CONFIGURATION

Ping-Hei Chen[+], Min-Sheng Hung[*], and Pei-Pei Ding[*]
Department of Mechanical Engineering
National Taiwan University, Taipei 10617, Taiwan
Fax: 886-2-23644871; E-mail: phchen@ccms.ntu.edu.tw

ABSTRACT: In order to explore the effect of compound angle holes on film cooling over a convex wall and a concave wall, the present study adopts the transient liquid crystal thermography for conducting the film cooling measurement on simple hole and expanded-hole configurations. Two compound angles of 0 and 45 deg are tested at an elevated mainstream turbulence condition (Tu) of 3.8 %. The test pieces have the different radius of curvature ($2r/D$) of 92.5 on convex and 86.5 on concave, and the same pitch to diameter ratio (P/D) of 3 on both convex and concave walls. All measurements were conducted under the mainstream Reynolds number (Re_d) of 1700 on convex and 2300 on concave with the density ratio between coolant and mainstream (ρ_c/ρ_m) of 0.98. In current study, the effect of blowing ratio (M) on film cooling performance is investigated by varying the range of blowing ratio from 0.5 and 2.0. The present measured results show that the forward-expanded hole injection provides better surface protection than the simple hole injection. As far as the injection angle is concerned, compound angle injection provides higher film effectiveness than simple angle injection. However, the forward-expanded hole injection ($\beta = 0^0$) has the best performance on both convex and concave surfaces.

KEYWORDS: curved walls, liquid crystal thermography, compound angle

INTRODUCTION

Film cooling is a technique of cooling gas turbine blades to protect them from high temperature gases. The technique can be done by injection of a film of cooling air onto the blade surface through discrete holes. These holes are typically inclined at approximately 30^0 to 40^0 with respect to the surface. The flow through turbine passages will experience strong curvature effect, which is not observable in flow over flat surfaces.

Several studies have concentrated on film cooling effectiveness by using different methods. Schwarz et al. (1991)[1] and Goldstein et al. (1997)[2] measured the film cooling effectiveness on both convex and concave surfaces by a foreign gas injection technique, too. They studied the effects of blowing ratio, density ratio, injection angle, and different relation strength of curvature on both convex and concave surfaces. The flow field was also visualized by using an ammonia vapor injection and a carbon dioxide-water vapor injection. For simple

---

[+]Professor
[*]Graduate Student

injection hole configuration on both concave and convex surfaces, Goldstein et al.[2] found that at very low blowing ratios, the injection flows were ejected gently into the boundary layer of mainstream, and the film cooling effectiveness would increase with blowing ratio independently of injection angle. In curved flows, Schwarz et al.[1] found that cross-stream pressure gradient tended to move film cooling jets onto a convex surface and away from a concave surface.

Drost et al. (1999)[3] used the transient liquid crystal technique and a multiple regression procedure to measure film cooling effectiveness and heat transfer on a turbine airfoil. They also showed the detailed η and heat transfer ratios for near hole region. They reported that a double row of injection holes can improve film coverage and delayed jet lift-off at high blowing ratio on the suction surface. Lutum et al. (2000)[4] also used thermochromic liquid crystal to measure film cooling effectiveness and heat transfer on a convex surface. They considered the constant Mach number over the test section to investigate the film cooling performance under the injection flow through three cylindrical holes and two shaped holes.

About the effect of compound angle hole, most of the studies were concerning film cooling performance on flat surfaces. Sen et al. (1996)[5] presented the effect of compound angle injection on heat transfer coefficients by using thermocouples and IR camera measurements. Ekked et al. (1997a)[6] and Ekked et al. (1997b)[7] studied local heat transfer coefficients and film cooling effectiveness over a flat surface with one row of injection holes inclined streamwisely at 35^0 and three spanwise injection angles (β) of 0^0, 45^0 and 90^0 by using a transient liquid crystal image method. McGovern et al. (2000)[8] studied the detailed field results (in the supply plenum, film hole and cross flow regions) and surface results (adiabatic effectiveness and heat transfer coefficient) with compound angle injection by a computational fluid dynamics (CFD) model. The consistent trend among all the studies of inclined, compound angle injection is found to increase film cooling effectiveness and heat transfer coefficient because of the increased interaction between injection flows and mainstream, and lateral spreading and uniformity.

Drost et al. (1999)[3] presented the film cooling performance with a compound angle configuration on the pressure side on a turbine airfoil. They showed that very little heat load reduction at low blowing ratios with a compound angle injection due to strong heat transfer enhancement. Lutum et al. (2000)[4] considered the effect of compound angle at $\beta = 60^0$ on a convex surface. They found that the injection with compound angle had the better film cooling effectiveness, but slightly higher heat transfer compared to simple configuration injection.

In this study, spanwise averaged heat transfer coefficient and film cooling effectiveness distributions are demonstrated over a convex surface and a concave surface by employing transient liquid crystal thermography (Vedula and Metzger, (1991)[9]). For the present experimental measurement, the streamwise injection angle (γ) is 35^0 and the spanwise injection angle are 0^0 and 45^0. Simple angle injection configuration is called in the present study for tests at $\beta = 0^0$.

THEORY

The present experiment used the thermochromic liquid crystal for the surface temperature measurement. The local heat transfer coefficient over the liquid crystal coated surface without cooling film can be obtained by one-dimensional semi-infinite model. A detailed analysis can be found from Chen et al. (2000)[10] and give:

$$k\frac{\partial^2 T}{\partial z^2} = \rho_s C_{p_s} \frac{\partial T}{\partial t} \tag{1}$$

Parameter Authors	P/D	L/D	ρ_c/ρ_m	2r/D	γ	β	δ_l/D	M
Concave surface								
Goldstein et al. (1997)	2.92	10	0.95	86.5	45°	0°	1.19	2.0
The present study	3	3.5	0.98	86.5	35°	0°	0.16	2.0
Convex surface								
Schwarz et al.(1991)	3	10	0.95	94	35°	0°	0.9	1.01
The present study	3	3.5	0.98	92.5	35°	0°	0.12	1.0

Table 1 Test conditions for the measured results shown in Figure 1.

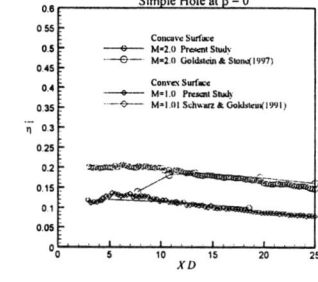

Fig. 1 Comparison with prior measured results of spanwise averaged film cooling effectivenes.

$$\text{at } z = 0, \quad -k\frac{\partial T}{\partial z} = h(T_w - T_f); \quad \text{as } z \to \infty, \quad T = T_i \quad (2)$$

$$\text{at } t = 0, \quad T = T_i \quad (3)$$

where k, Cp_s, and ρ_s are respectively the thermal conductivity, the specific heat, and the density of test piece. The analytic solution of Eq. (1) on the surface ($z = 0$) is:

$$T_w - T_i = \left[1 - exp\left(\frac{h^2\alpha t}{k^2}\right) erfc\left(\frac{h\sqrt{\alpha t}}{k}\right)\right] \times [\eta T_c + (1-\eta)T_m - T_i] \quad (4)$$

Besides the heat transfer coefficient and film cooling effectiveness, heat flux ratio is another important parameter used to quantify the film cooling performance. The present study adopts the expression of heat flux ratio as defined by Ekkad et al. (1997b)[7] given by

$$\frac{q}{q_o} = \frac{h(T_w - T_{aw})}{h_o(T_w - T_m)} = \frac{h}{h_o}(1 - \frac{\eta}{\phi}) \quad (5)$$

where ϕ is the overall cooling effectiveness. In current study, $\phi = 0.6$ is used.

EXPERIMENTAL APPARATUS AND PROCEDURES

The experimental investigation was conducted in a wind tunnel with rectangular cross-section of 10 cm × 5 cm and a bend of 135°. The wind tunnel consisted of a mix section, a uniform development section, and a curved test section, and the detailed experimental process can be found from Chen et al. (2000)[10]. The mainstream velocity (u_m) was kept at a constant value of 9.1 m/s. The injection flow was injected through a single row of injection holes with hole spacing of 3D along the spanwise direction. The ratio of hole length to hole diameter (L/D) and the streamwise injection angle (γ) were 3.5 and 35° respectively.

The operating conditions are listed in Table 1. The mainstream velocity measured by hot-film anemometer was 9.1 m/s at 25 cm upstream of injection holes. The measured turbulence intensity was 3.8 %. Measurements were conducted at two different blowing ratios of 0.5 and 2.0.

The estimated uncertainty in the effective data was ± 5 percent estimated by the root mean square method by Moffat (1988)[11]. For the experiments done in this study, the total uncertainties of h and η were respectively around 7.2 % and 10.4 % of their nominal values (h = 29 W/m^2K and η = 0.32). Furthermore, the estimated uncertainties for the Reynolds number,

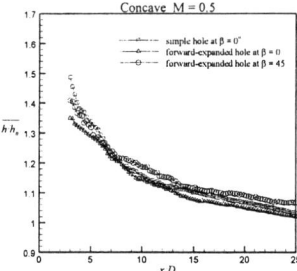 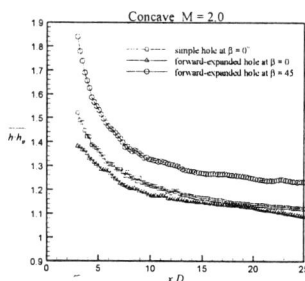

Fig. 2 Effect of compound angle on the spanwise averaged heat transfer coefficient on concave surface at (a) $M = 0.5$ (b) $M = 2.0$.

blowing ratio, mainstream turbulence intensity, and boundary layer displacement thickness were respectively 3.4 %, 2.8 %, 4.3 %, and 5.6 %.

RESULTS AND DISCUSSION

In Fig. 1, the spanwise averaged film cooling effectiveness results on both two surfaces are compared with the experimental results of Goldstein et al. (1997)[2] and Schwarz et al. (1991)[1]. The comparison on $\bar{\eta}$ of the concave surface and convex surface in the figure showed similar trends except the near hole region.

CONCAVE SURFACE RESULTS

Figure 2 illustrates the effect of injection hole configuration and blowing ratio on the spanwise averaged heat transfer ratio ($\overline{h/h_0}$) for $2.9 < x/D < 25$. For all tested cases, forward-expanded hole configuration provides lower $\overline{h/h_0}$ values than simple hole configuration. The lower magnitude of $\overline{h/h_0}$ for forward-expanded hole configuration shows that the diffused hole can reduce the normal momentum of coolant jets effectively. This evidence that the forward-expanded hole configuration can reduce the penetration of coolant flow into the mainstream boundary and perform better surface protection just after injection. For forward-expanded hole configuration, Fig. 2 also presents that compound angle injection produces higher $\overline{h/h_0}$ values than simple angle injection at $M = 2.0$. Compound angle injection increases lateral spreading of the jets and causes of higher heat transfer coefficients on the concave surface, particularly near the injection hole region ($x/D < 4$).

The results of spanwise averaged film cooling effectiveness ($\bar{\eta}$) at various blowing ratios and exit hole configurations are shown in Fig. 3. Similar trend in both $\bar{\eta}$ and $\overline{h/h_0}$ can be found in both Figs. 2 and 3 that the higher magnitude of $\bar{\eta}$ within a measured region for forward-expanded hole configuration indicates the better spanwise coverage of cooling film. However, minimal penetration into mainstream and better surface coverage are yet needed for high film cooling performance. It is obvious that the diffused exit of forward-expanded hole can attain these insistances simultaneously for all test cases with blowing ratios of 0.5 and 2.0.

At $x/D < 15$ for all configurations, a steep descent in $\bar{\eta}$ for blowing ratio of $M = 0.5$ can be observed. At $M = 2.0$, increasing blowing ratio provides higher momentum and mass flux ratio that yield higher effectiveness at downstream when injection flows return to the concave

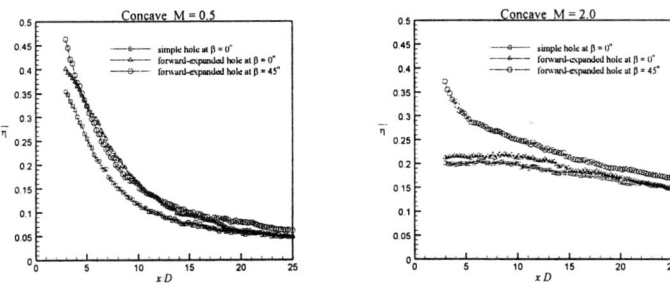

Fig. 3 Effect of compound angle on the spanwise averaged film cooling effectiveness on concave surface at (a) $M = 0.5$ (b) $M = 2.0$.

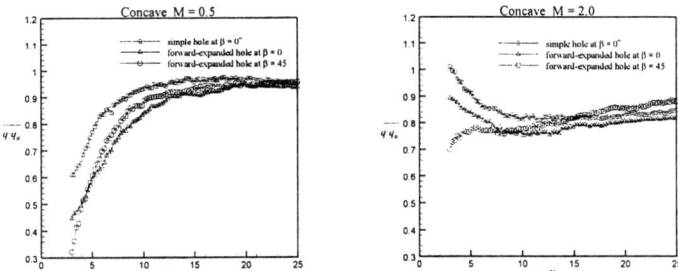

Fig. 4 Effect of compound angle on the overall heat flux ratio on concave surface at (a) $M = 0.5$ (b) $M = 2.0$

surface. For compound angle injection, an obvious increase in $\bar{\eta}$ for all blowing ratios at $x/D < 7$, and a steep descent farther downstream can be observed. It is that the lateral interaction of the injection flows with the mainstream for compound angle injection could be attributed to the higher and more uniform effectiveness near the injection hole region.

The film cooling performance of test surface is best indicated by the result of spanwise averaged heat flux ratio shown in Fig. 4. The magnitude of $\overline{q/q_0}$ lower than 1.0 at all tested blowing ratios for all test pieces indicate the better film cooling performance compared to the surface without film injection. For concave surface, it is effective to reduce the heat load on the surface with cooling film as shown in Fig. 4. In particular the obvious reduction in the heat flux ratio indicated that the film coverage effect of injection flows performs better in forward expanded hole configuration than the simple hole configuration for all blowing ratios. At compound angle injection, it yields a higher heat flux reduction near the injection hole ($x/D < 5$) at $M = 0.5$ and 2.0, and decays to a lower reduction as compared to injection at 0 deg.

At $M = 0.5$, the injection flows reduce the heat flux ratio effectively at $x/D < 5$ because the low normal momentum of ejected injection flows tend to lead the flows to stay close to the surface. Nevertheless, the low mass flux of injection flows is unable to provide film protection at further downstream, and therefore increase the magnitude of $\overline{q/q_0}$ at $x/D > 10$.

The results obtain with forward-expanded hole configuration without compound angle injection indicate generally a reduction heat flux ratio for all blowing ratio tests. This is attributed to the reduced momentum of the injection fluid at the hole exit due to the hole exten-

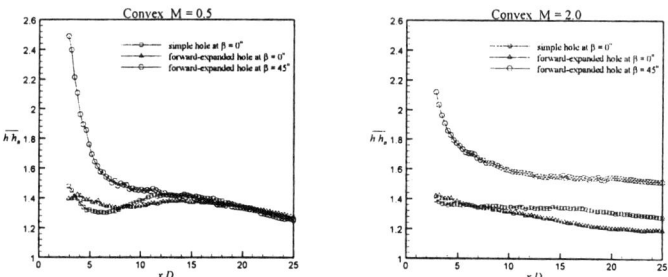

Fig. 5 Effect of compound angle on the spanwise averaged heat transfer coefficient on convex surface at (a) $M = 0.5$ (b) $M = 2.0$

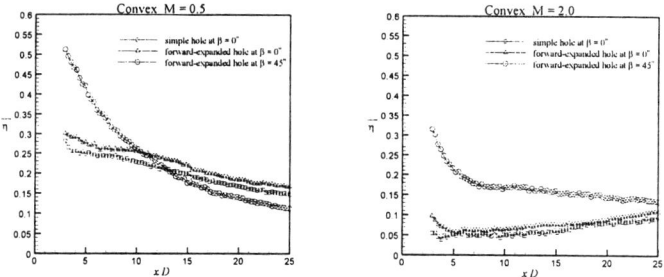

Fig. 6 Effect of compound angle on the spanwise averaged film cooling effectiveness on convex surface at (a) $M = 0.5$ (b) $M = 2.0$.

sion in streamwise direction, and therefore minimal penetration into mainstream and better surface coverage provide high film cooling performance and low heat transfer coefficient.

CONVEX SURFACE RESULTS

Figure 5 compares lateral averaged heat transfer coefficient ratio obtained with simple, forward-expanded and forward-expanded with compound configurations at different injection blowing ratios $M = 0.5$ and 2.0. For forward-expanded hole configuration, it is obvious that h/h_0 for compound angle injection increases just downstream of injection hole as compared to simple hole injection. However, the increased heat transfer decreased with downstream distance. This is caused by injection flows staying close to the convex wall after leaving the injection hole at a lower momentum ratio.

The heat transfer ratios obtained with compound angle injection increases with increasing blowing ratio. The heat transfer ratios obtained with compound angle are higher than those values obtained with both simple hole and forward-expanded hole at $\beta = 0°$. It is obvious that there is a quite strong increase in heat transfer due to injection flow close to the injection location. The compound angle injection increases lateral interaction between injection flows and mainstream. This increased lateral spreading of the jets for compound angle injection could be the cause of higher heat transfer coefficient on the convex surface.

Figure 6 compares lateral averaged film cooling effectiveness results obtained with simple,

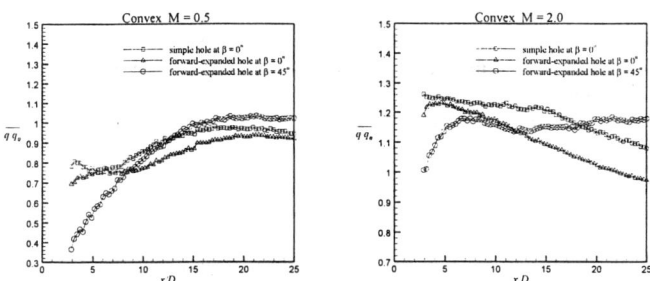

Fig. 7 Effect of compound angle on the overall heat flux ratio on convex surface at (a) $M = 0.5$ (b) $M = 2.0$.

forward-expanded at $\beta = 0°$ and forward-expanded with compound angle injection configurations at different blowing ratios $M = 0.5$ and 2.0. At blowing ratio of 0.5, the spanwise averaged effectiveness is the best among all tested blowing ratios. At downstream regions, mixing with the mainstream dilutes the injection flows, thereby lower $\bar{\eta}$ obtained. Also, the lower injection flow mass flux of injection flow for $M = 0.5$ causes the steep descent in $\bar{\eta}$ at further downstream.

For compound angle injection, the $\bar{\eta}$ is greatly higher at $x/D < 5$ than both two configuration injections. The lateral spreading of injection flow provides more coverage and increases effectively in $\bar{\eta}$. When blowing ratio $M = 2.0$, $\bar{\eta}$ is also higher than the other two configuration injections because of the increase of mass and more coverage at downstream.

The film cooling performance of test surface is best indicated by the result of spanwise averaged heat flux ratio shown in Fig. 7. At $M = 0.5$, the injection flows reduce the heat flux ratio effectively over most measured regions for all test cases because the low normal momentum of ejected injection flows tend to lead the flows to stay close to the surface. But the low mass flux of injection flows is unable to provide film protection at further downstream, and therefore increase the magnitude of $\overline{q/q_0}$ at $x/D > 15$. When M is further increased to 2.0, the increased normal momentum will promote lift-off and therefore perform poor protection just after injection.

CONCLUSIONS

The present study demonstrates the influence of compound angle injection on convex and concave surface film cooling performance using transient liquid crystal thermography. On the convex surface, experimental results show that high blowing ratio at $M = 2.0$ has no film cooling protection on the surface at all cooling performance. At high blowing ratio, the injection flows will lift away from the surface and will also reduce the film cooling effectiveness on the convex surface. However, the increasing blowing ratio on the concave surface is that the larger mass of injection flow ejected into the mainstream causes adjacent injection flows to merge far downstream of injection holes and increases the film cooling effectiveness. In the present study, the compound angle configuration injection is not able to offer better protection on both two surfaces than forward-expanded hole with simple angle injection for $x/D > 12$.

ACKNOWLEDGMENT

The authors deeply appreciate the financial support by NSC under the grant number 86-2212-E-002-080. The work in this study could not be achieved without their support.

NOMENCLATURE

- D injection hole diameter on the inlet plane, $[m]$
- I momentum flux ratio, $= \rho_c u_c^2 / \rho_m u_m^2$
- M blowing ratio, $= \rho_c u_c / \rho_m u_m$
- P pitch of injection holes, $[m]$
- r radius of curvature of convex surface, $[m]$
- Re_d mainstream Reynolds number based on the inlet dia-meter of injection hole, $= \rho_m u_m D / \mu$
- t time, $[s]$
- T temperature, $[K]$
- X axial distance, $[m]$
- Y spanwise coordinate, $[m]$
- z coordinate axis perpendicular to the test surface, $[m]$
- α thermal diffusivity of test surface, $[m^2/s]$
- δ_1 displacement thickness, $[m]$
- γ injection hole angle with respect to the test surface as projected into the streamwise/normal plane (inclination angle), $[deg.]$

REFERENCES

1. Schwarz, S. G., Goldstein, R. J. & Eckert, E. R. G., 1991. The Influence of Curvature on Film Cooling Performance. ASME *J. of Turbomachinery*, Vol. 113, pp. 472-478.
2. Goldstein, R. J. & Stone, L. D., 1997. Row-of-Holes Film Cooling of Curved Walls at Low Injection Angles. ASME *J. of Turbomachinery*, Vol. 119, pp. 574-579.
3. Drost, U., BÔlcs A., 1999. Investigation of Detailed Film Cooling Effectiveness and Heat Transfer Distributions on a Gas Turbine Airfoil. ASME *J. Turbomachinery*, Vol. 121, pp. 233-242.
4. Lutum E., Wolfersdorf J. von, Weigand B. & Semmler K., 2000. Film Cooling on a Convex Surface with Zero Pressure Gradient Flow. *Int. J. Heat and Mass Transfer*, Vol. 43, pp. 2973-2987.
5. Sen, B., Schmidt, D. L. & Bogard, D. G., 1996. Film Cooling with Compound Angle Holes: Heat Transfer. ASME *J. of Turbomachinery*, Vol. 118, pp. 800-806.
6. Ekkad, S. V., Zapata, D. & Han, J. C., 1997a. Heat Transfer Coefficients over a Flat Surface with Air and CO_2 Injection Through Compound Angle Holes Using a Transient Liquid Crystal Image Method. ASME *J. Turbomachinery*, Vol. 119, pp. 580-586.
7. Ekkad, S. V., Zapata, D. & Han, J. C., 1997b. Film Effectiveness over a Flat Surface with Air and CO_2 Injection Through Compound Angle Holes Using a Transient Liquid Crystal Image Method. ASME *J. Turbomachinery*, Vol. 119, pp. 587-593.
8. McGovern, K. T. & Leylek, J. H., 2000. A Detailed Analysis of Film Cooling Physics: Part II-Compound-Angle Injection with Cylindrical Holes. ASME *J. Turbomachinery*, Vol. 122, pp. 113-121.
9. Vedula, R. J. & Metzger, D. E., 1991. A Method for the Simultaneous Determination of Local Effectiveness and Heat Transfer Distributions in Three-Temperature Convection Situations. ASME Paper, No. 91-GT-345.
10. Chen, P. H., Hung, M. S. & Ding, P. P., 2000. A Transient Method Using Liquid Crystal for Film Cooling over a Convex Surface. *Proceedings of the 8th Internation Symposium on Transport Phenomena and Dynamics of Rotating Machinery*, Vol. II, pp. 666-673.
11. Moffat, R. J., 1988. Describing the Uncertainties in Experimental Results. *Experimental Thermal and Fluid Science*, Vol. 1, pp. 3-17.

The Variation of Heat Transfer Coefficient, Adiabatic Effectiveness and Aerodynamic Loss with Film Cooling Hole Shape

J.E. SARGISON[1], S.M. GUO[1], M.L.G. OLDFIELD[1], A.J. RAWLINSON[2]

1. Department of Engineering Science, University of Oxford, Parks Rd, Oxford, OX1 3PJ, UK
2. Rolls Royce plc, Derby, DE24 8BJ, UK

ABSTRACT

The heat transfer coefficient and adiabatic effectiveness of cylindrical, fan shaped holes and a slot are presented for the region zero to 50 diameters downstream of the holes. Narrow-band liquid crystals were used on a heated flat plate with heated air coolant. These parameters have been measured in a steady state, low speed facility at engine representative Reynolds number based on hole diameter and pressure difference ratio (ideal momentum flux ratio).
The aerodynamic loss due to each of the film cooling geometries has been measured using a traverse of the boundary layer far downstream of the film cooling holes.
Compared to the cylindrical holes, the fan shaped hole case showed an improvement in the uniformity of cooling downstream of the holes and in the level of laterally averaged film cooling effectiveness. The fan effectiveness approached the slot level and both the fan and cylindrical hole cases show lower heat transfer coefficients than the slot and non film cooled cases based on the laterally averaged results. The drawback to the fan shaped hole was that the aerodynamic loss was significantly higher than both the slot and cylindrical hole values due to inefficient diffusion in the hole exit expansion.

NOMENCLATURE

B Blowing ratio, $B = \dfrac{\rho_c v_c}{\rho_m v_m}$

d Hole diameter
h Heat transfer coefficient
H Height over which the velocity profile is measured (fig. 2).
I_{actual} Momentum flux ratio (MFR):

$$I_{actual} = \dfrac{\rho_c v_c^2}{\rho_m v_m^2}$$

I_{ideal} Ideal MFR:

$$I_{ideal} = \dfrac{\rho_c v_c^2}{\rho_m v_m^2} = \dfrac{p_{0c} - p_c}{p_{0m} - p_m}.$$

q Heat transfer rate
t Time

v Flow velocity
T Temperature
x Distance downstream from edge of slot or hole
z Lateral distance from hole centre

Greek symbols

ρ Density
η Film cooling effectiveness
θ Dimensionless temperature: $\theta = \dfrac{(T_c - T_m)}{(T_w - T_m)}$

Subscripts

aw Adiabatic wall
c Coolant
m Gas, mainstream
mw Property between wall and mainstream
w Surface or wall
0 Total or stagnation property

1. INTRODUCTION

The use of film cooling to allow turbine entry temperatures well above the melting temperature of turbine components is well established and considerable research has been applied to optimise film cooling design to maximise the thermal protection of the surface[1]. However, the cost of film cooling is reduced aerodynamic efficiency due to the interaction of coolant and mainstream flows. Thus the optimisation of designs must include consideration of increased aerodynamic loss with any improvements in heat transfer and adiabatic effectiveness.

An attractive film cooling configuration is a straight or shaped slot[2], which has the advantage of providing laterally uniform cooling across the downstream surface. The two dimensional flow of coolant from the slot is free of the typical mixing of mainstream and coolant flows by the production of vortices, and hence the coolant effectiveness can be higher and more lasting downstream of the cooling row. The structural impracticability of the slot in turbine blade and vane film cooling has led to the development of alternative arrangements of cylindrical holes in staggered rows, cylindrical holes at lateral angles to the mainstream, and shaped holes.

Heat Transfer Coefficient and Adiabatic Effectiveness

The main quantities used to measure the thermal performance of film cooling configurations are the convective heat transfer coefficient h and the film cooling effectiveness η. h is defined:
$$q = h(T_w - T_{aw}) \tag{1}$$

Note that, in the experiments to be described, hot "coolant" was used, so this definition is convenient. In the engine, T_w is less than T_{aw} and so q, as defined here, is negative, although h remains positive.

With film cooling, where there are two flows present, T_{aw} is intermediate between the coolant T_{0c} and mainstream T_{0m} total temperatures and depends upon the geometry and degree of mixing between these gases upstream of the point of interest on the surface. To eliminate the temperature dependence a dimensionless adiabatic film cooling effectiveness is defined
$$\eta = \frac{T_{aw} - T_m}{T_c - T_m} \tag{2}$$

where T_{aw} corresponds to the temperature of an isothermal wall which is adiabatic at the local point of interest. A dimensionless temperature, θ, can also be defined, which reduces the three temperature variables in the experiment to a single parameter.
$$\theta = \frac{T_c - T_m}{T_w - T_m} \tag{3}$$

Aerodynamic Loss

Aerodynamic loss[3,4] is defined as 1-ε, where the aerodynamic efficiency, ε, is defined:
$$\varepsilon = \{Actual\ Kinetic\ Energy\}_{mixingplane} / \{Theoretical\ Kinetic\ Energy\}_{available} \tag{4}$$

The mixing plane is defined as the hypothetical plane at which the coolant and mainstream flows are fully mixed and the velocity, temperature and pressure are uniform. The mixing plane is mathematically equivalent to the measurement plane and the velocity and static pressure at this plane are found by applying the laws of conservation of mass, momentum and enthalpy between the boundary layer traverse measurement plane and the mixing plane. The loss experiments described here were conducted at uniform temperature and so the enthalpy equation is not required.

2. ANALYTICAL METHOD
Heat Transfer Coefficient and Adiabatic Effectiveness

If equations 2 and 3 are substituted into equation 1, and the resulting equation rearranged, then:

$$h_{mw} = \frac{q}{(T_w - T_m)} = h - h\eta\theta \qquad (5)$$

For constant h and η, this expression indicates that the variable h_{mw} is linear in θ. From a line relating h_{mw} to θ, h and η can be found, as h is the slope of the line and the intercept on the $\theta(x)$ axis is $1/\eta$.

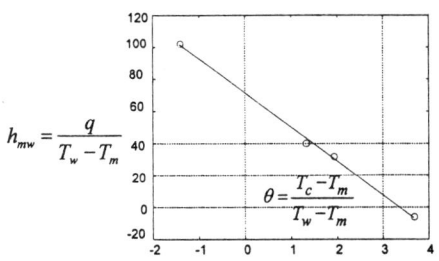

Figure 1: Typical set of data points and fitted straight line used in *Matlab* data manipulation

The linearity of h_{mw} with θ was tested as part of the analysis sequence, written using the *Matlab* package, by plotting h_{mw} against θ along with the straight line fitted to the data, for a number of positions on the plate as shown in Figure 1.

The computational method outlined can be applied to calculate η and h at any point on the flat plate downstream of the cooling holes. Referring to Figure 1, the adiabatic wall temperature at a point could be obtained directly by setting h_{mw} to zero. Similarly, the heat transfer coefficient could be measured directly, by setting the coolant temperature to be equal to the mainstream temperature. The reason that this method was not used and the superposition method developed was twofold. Firstly, the power required to obtain a change in liquid crystal temperature was in some cases above the recommended limit for the aluminium foil and above the capability of the direct current source used. In addition to this, the accuracy of the experiments was improved by using four experiments and hence four points to determine h, and η, rather than two.

At any distance downstream of the holes, the two parameters, h and η, will vary in the lateral direction across the cooling holes, depending on the geometry of the hole (with the most variation for the cylindrical holes, and the least for the slot). For design purposes, it is the lateral average of these parameters that is of the most interest, so that the laterally averaged heat transfer at a distance downstream of the holes can be determined.

Returning to the linear expression in θ, the correct method to calculate the lateral averages is now considered. Taking the average of both sides in equation 6:

$$\overline{\left[\frac{q}{T_w - T_m}\right]} = \overline{h} - \overline{h\eta}\theta \qquad (6)$$

Thus the two properties of the film cooling heat transfer that are required for the average heat transfer prediction are $\overline{h}$ and $\overline{h\eta}$. The correct way to calculate the laterally averaged effectiveness from the contour data, if it is to be used to determine the average heat transfer is:

$$\overline{\eta} = \frac{\overline{h\eta}}{\overline{h}} \qquad (7)$$

Aerodynamic Loss

The pressure and velocity profiles at the measurement plane are extended to a distance equal to the boundary layer thickness, δ, from the far wall of the low speed wind tunnel.

The loss is dependent on this distance H over which the velocity profile is measured. The purpose of the current experiments is to compare the aerodynamic loss due to the different film cooling hole geometries, so the actual value defined is not critical.

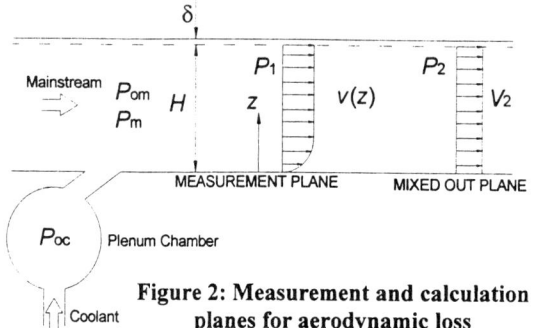

Figure 2: Measurement and calculation planes for aerodynamic loss

Applying conservation of mass, assuming constant density: $\int_0^H v\,dz = hV_2$ (8)

Applying conservation of momentum: $P_1 + \rho \int_0^H v^2 dz = P_2 + \rho h V_2^2$ (9)

From these two relationships, the actual kinetic energy of the flow at the mixing plane can be calculated: $KE_{actual} = \frac{1}{2}mV_2^2 = \frac{1}{2}\rho V_2^3$ (10)

The theoretical kinetic energy, based on the initial total pressure and the calculated static pressure at the mixing plane is: $KE_{theoretical} = \dot{m}_m(P_{om} - P_2) + \dot{m}_c(P_{oc} - P_2)$ (11)

Such that the loss is finally written;

$$Loss = 1 - \frac{\frac{1}{2}\rho h V_2^3}{\dot{m}_m(P_{om} - P_2) + \dot{m}_c(P_{oc} - P_2)} \quad (12)$$

3. EXPERIMENTAL APPARATUS

The experiments were performed in a low speed, induced flow wind tunnel. The mainstream (primary flow) air speed and temperature were measured upstream of coolant injection and these parameters were independent of the coolant flow.

The cooling air was supplied from the laboratory compressed air source and the mass flow rate of the air supplied to the plenum chamber was measured through a British Standard BS 1042 orifice plate (17.96mm dia.). The air was heated using a variable power inline heater and fed into a plenum chamber to settle and mix the air before it exited through film cooling holes mounted in the side of the wind tunnel. A set of interchangeable plates with the different film cooling hole geometries investigated were machined in plates of *Rohacell* type 51IG, a closed pore structural foam. This material was selected because of its very low thermal conductivity of 0.028 – 0.034 W/mK at 20°C, compared with air at 0.025 W/mK. The angle of inclination of the holes was 35 ° and the exit area per unit width was based on a row of five 20mm diameter cylindrical holes at a pitch of 60mm. The fan shaped holes had a total expansion angle of 25° and the slot height was 5mm.

A uniform heat flux flat plate, located downstream of the film cooling holes, was produced from a large sheet of Rohacell, covered with an electrically heated thin film of aluminised Mylar, bonded to the surface with high strength double-sided tape. The resistance of the heating element was two ohms per square, but it varied slightly with temperature, so the voltage and current were measured in order to calculate the power supplied to the plate. The total resistance of the plate was 5.4 Ω and the maximum heater power used was 200W.

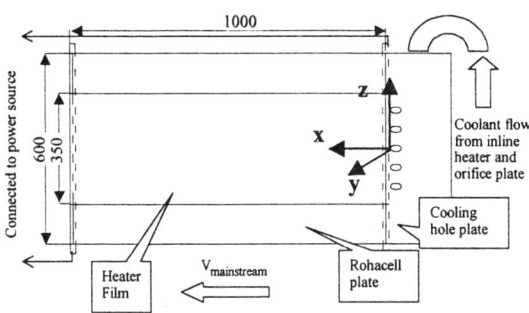

Figure 3: Tunnel wall showing layout of cooling holes and heated flat plate

The heating surface was painted black and narrow band thermochromic liquid crystals were used to measure temperature contours on the surface. A map of heat flux and coolant contours was constructed by changing the coolant temperature or heat flux to the plate and allowing the system to settle.

An automatic *Matlab* programme was developed to calculate the heat transfer coefficient and adiabatic effectiveness at each point on the plate. The temperature contours were extracted using the hue[10] signal in the digital image, and the linear superposition method outlined above was applied to sets of data to calculate the film cooling parameters h and η.

The measurement of aerodynamic loss was conducted using the equipment outlined above, except that there was no heat applied to either the heater film or the coolant. The boundary layer velocity and total pressure profiles were measured with a horizontally traversing pitot probe at the downstream edge of the plate, at positions downstream of a hole centre and halfway between two holes.

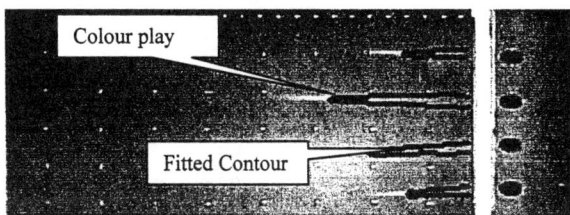

Figure 4: Example of fitted temperature contour on liquid crystal colour play

All measurements presented were conducted at engine representative Reynolds number based on hole diameter of 36000, ideal momentum flux ratio of 1.1, actual momentum flux ratio of 0.5 and density ratio of 1. The difference between the ideal momentum ratio and the actual momentum ratio is due to the discharge coefficient being less than unity.

4. RESULTS AND DISCUSSION

Lateral Variation in Effectiveness and Heat Transfer Coefficient

Figures 5a-c show the lateral variation of adiabatic effectiveness over one hole pitch, averaged over the two centre holes. It should be noted that the fan expanded to a width of 3d at hole exit and had a pitch of 3.5d, while the cylindrical holes had a pitch of 3d. The hole exit area per unit width was identical for all configurations. The result demonstrates that the fan shaped hole provides a significantly more uniform distribution in effectiveness from very close to the film cooling hole. The reduced mixing with the mainstream flow compared with the cylindrical hole results in a higher peak value of effectiveness for both the slot and the fan shaped hole, both of which have similar peak values although the slot is, as expected, more uniform.

As the results were measured on the tunnel side wall, rather than over the floor of the wind tunnel, there appears to be some buoyancy effects which cause the results to be slightly asymmetrical. Increasing z indicates increasing vertical height in the tunnel.

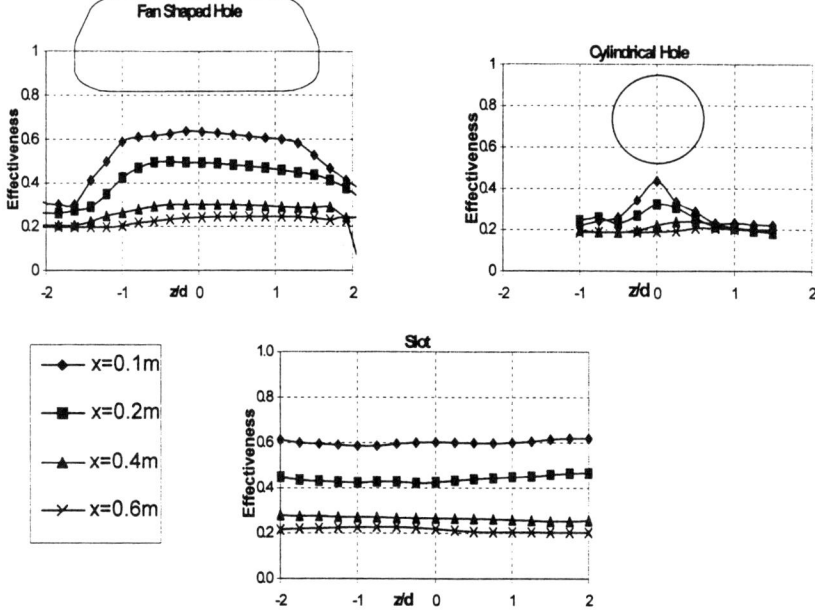

Figure 5 a-c: Lateral variation in adiabatic effectiveness

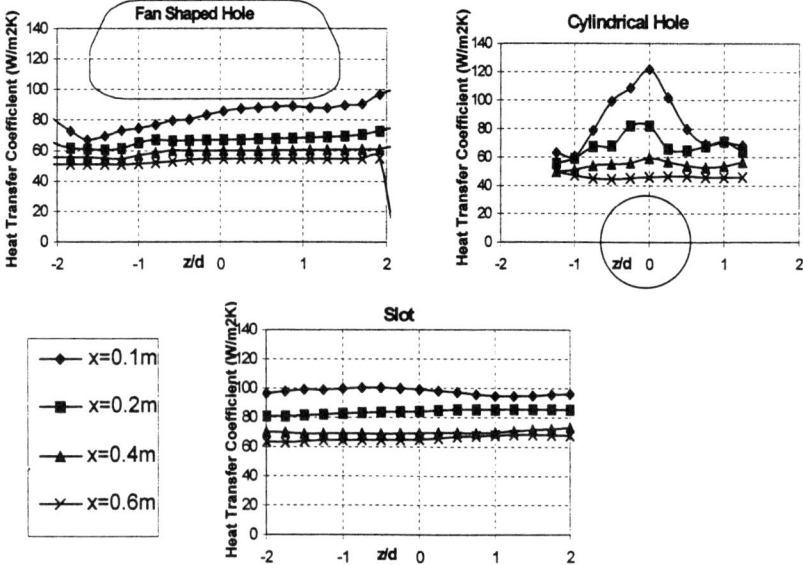

Figures 6 a-c: Lateral Variation in heat transfer coefficient

The heat transfer coefficient results in figures 6 a-c demonstrate that while the peak heat transfer coefficient downstream of the cylindrical hole centre is augmented by discrete hole film cooling, the regions between film cooling holes have lower heat transfer coefficient than the slot case. This difference is due to the general slowing and thickening of the boundary layer in the fan and cylindrical hole case, compared with the slot which causes the boundary layer to restart and act more like a uniform boundary layer that would be formed in the absence of film cooling. This flow regime explains why the laterally averaged heat transfer coefficient is lower for the fan and cylindrical hole configurations, while the laterally averaged heat transfer coefficient is similar for the slot and non film cooled configurations.

Laterally Averaged Heat Transfer Coefficient and Adiabatic Effectiveness

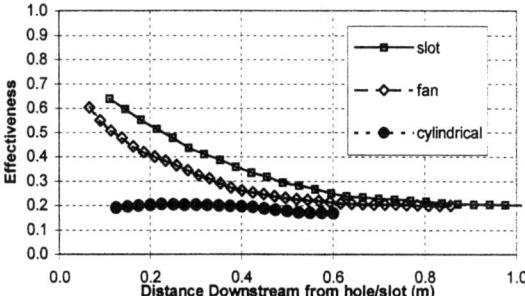

Figure 7: Laterally Averaged Adiabatic Effectiveness

The laterally averaged film cooling effectiveness demonstrates that shaping the cylindrical hole to form a fan shaped hole significantly improves the adiabatic effectiveness to approach the level of slot film cooling effectiveness. Note that the downstream distance is presented in metres, to avoid confusion between slot height and hole diameter. The slot height is 5mm and the hole diameter is 20mm.

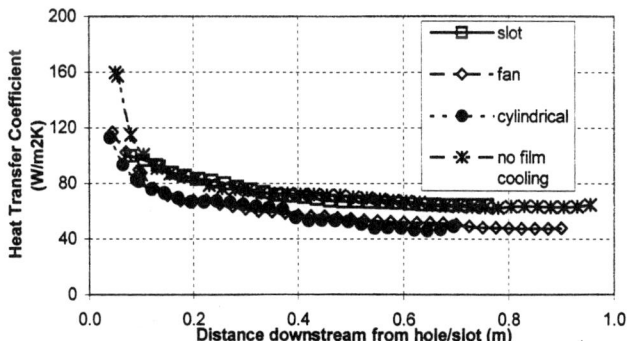

Figure 8: Laterally Averaged Heat Transfer Coefficient (W/m^{2}K)

Aerodynamic Loss

The aerodynamic loss measured for each film cooling configuration is presented in figure 9 and compared with the aerodynamic loss without film cooling. The fan shaped hole exhibits the highest film cooling loss, with the smallest loss due to the slot being only 0.2 % higher than the loss for the uncooled case. The plot clearly shows that despite the advantage in thermal performance, there is a loss penalty associated with the use of fan shaped film cooling holes.

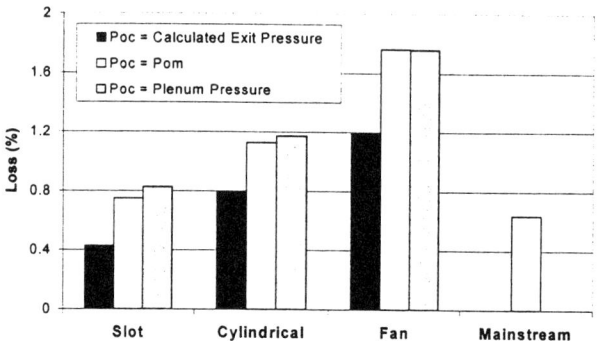

Figure 9: Comparison of Aerodynamic Loss Results at $I_{ideal} = 1.1$

5. CONCLUSIONS

The heat transfer coefficient and adiabatic effectiveness of the film cooling process have been measured and are considered in conjunction with the aerodynamic loss due to flow through the hole and mixing with mainstream air.

An automated data processing system has been developed to analyse a series of digital images and return results over the entire surface analysed of adiabatic effectiveness and heat transfer coefficient.

The thermal film cooling performance of fan shaped holes is a significant improvement on cylindrical shaped holes, and it approaches slot performance due to spreading of the coolant over the surface to form a closed film a short distance downstream of the holes.

The penalty associated with fan shaped holes is large aerodynamic loss due to inefficient diffusion in the expanding section of the hole.

ACKNOWLEDGEMENTS

This work is supported by Rolls-Royce plc, DERA, DTI CARAD and MOD ARP26c.

REFERENCES

1. Goldstein, R.J., 1971, Film Cooling, *In* Advances in Heat Transfer, (7), pp 321-379.
2. Farmer, JP, Searger, DJ and Liburdy JA 1997 The Effect of Shaping Inclined Slots on Film Cooling Effectiveness and Heat Transfer Coefficient, ASME 97-GT-399
3. Day, CRB, Oldfield MLG, Lock GD 1999, The Influence of Film Cooling on the Efficiency of an Annular Nozzle Guide Vane Cascade, *In* Journal of Turbomachinery, Transactions of the ASME, Jan 1999, Vol 121 No.1 pp 145-151
4. Jones, T.V., 1991, Definition of Heat Transfer Coefficient in the Turbine Situation, *In* Symposium in Turbomachinery: Latest Developments in a Changing Scene, IMechE Paper C423/046.
5. Day, C.R.B., Oldfield, M.L.G. and Lock, G.D., 2000, Aerodynamic Performance of an Annular Cascade of Film Cooled Nozzle Guide Vanes Under Engine Representative Conditions, *In* Experiments in Fluids, Vol. 29, August 2000.
6. Gonzalez, R.C. and Woods, R.E., 1993 Digital Image Processing, Addison-Wesley.

Numerical Investigation of Film Cooling Flow Induced by Cylindrical and Shaped Holes

Sylvain BARTHET and Pascale KULISA

Laboratoire de Mécanique des Fluides et d'Acoustique – UMR CNRS 5509
Ecole Centrale de Lyon
69131 Ecully Cedex, France

ABSTRACT : The present study is the second half of a two part work carried out in collaboration with SNECMA which tends to investigate a shaped hole film cooling experimentally and numerically. The aim of this paper is the numerical study of 3D phenomena induced by cylindrical and shaped hole film cooling on a flat wall.

The two calculations show up classical structures such as horseshoe or kidney vortices and their differences according to the shape configuration. A detailed study demonstrates their influence on the jet behaviour. Comparing both cases reveals the impact of shaping on the velocity field and vortex motions.

The calculations were performed by resolving the 3D Navier-Stokes equations associated with a k-ε turbulence model. The solver is the CANARI code developed by ONERA.

INTRODUCTION

The high inlet temperatures of modern gas turbines necessitate the use of sophisticated cooling schemes to protect the exposed components. Common techniques include rather elaborate internal convective schemes, film cooling and a combination of both. In the film cooling technique, relatively cool air is injected through the turbine blade surface producing a protective layer between the surface and the hot mainstream gas. The cool air is distributed through rows of closely spaced discrete holes. Current designs of turbine airfoils incorporate film cooling holes near the leading edge as well as on both the suction and pressure sides. The distribution of holes over the components creates a cool skin to protect the airfoil from the hot mainstream gas.

The injection of discrete film cooling into a two dimensional boundary layer induces a complicated three-dimensional flowfield which depends on many geometrical and fluid dynamic parameters, Andreopoulos[1]. As a direct consequence, the heat transfer is also dictated by the same parameters.

Few studies have investigated the influence of contoured film cooling holes. In most cases, the holes are cylindrical with a diffuser-shaped expansion at the exit segment of the hole. Thus, Goldstein and Eckert[2] examined holes with a lateral expansion of 10° at the hole exit portion, while Makki and Jakubowski[3] tested holes with a trapezoidal cross-sectional outlet area. Thole et al.[4] tested three different hole geometries. One of these was a forward-lateral expanded hole, similar to the one we discuss in this paper.

The increased cross-sectional area at the hole exit, compared to a standard cylindrical hole, leads to a reduction of the mean velocity and, thus, of the momentum flux of the jet exiting the hole. Therefore the penetration of the jet into the main flow is reduced, resulting in increased cooling efficiency. But Haven and al.[5] show two additional factors to explain the efficiency of shaped holes. Their study is based on vortex dynamics.

Our approach is to bring to light the main three-dimensional structures to explain the aerodynamic jet behaviour. Two simulations are presented concerning the development of a jet on a flat plate. The first one concerns a jet induced by a cylindrical hole with a 50° forward slope, the second a jet induced by a shaped hole. The results of this calculation will be compared to those of the cylindrical hole simulation to bring to light the effects of shaping on jet behaviour. In particular we will observe an important separation zone which occurs within the hole.

FLOW CHARACTERISTICS – NUMERICAL CALCULATION

The solver used for the simulation of steady three-dimensional viscous compressible flows is the CANARI code developed by ONERA which features a cell-centered approach for multi-domain structured meshes. We only focus on the aerodynamic aspect.

Governing Equations

The physical model is the compressible Reynolds-averaged Navier-Stokes set of equations associated with a turbulent model, Vuillot et al.[6], Cambier and Escande[7], Gleize et al.[8]. We write the equations in a Galilean frame and in a conservative form for cartesian coordinates.

A low Reynolds number k-ε model is implemented. We use the Jones and Launder[9] turbulence model because the damping functions depend on a turbulent Reynolds number but not on a distance to the wall. Thus we do not need to calculate the distance to the wall which could be a problem close to the exit area of the hole.

Numerical Scheme

A four steps Runge-Kutta scheme associated with a centered space finite volume discretization is used to solve averaged Navier-Stokes and k-ε equations. Second and fourth dissipative terms are added to ensure numerical stability. An implicit stage allows convergence acceleration.

The convergence state is reached for a three orders of magnitude loss of the residual terms. The mass flow conservation is verified. The calculations were performed at CFL=2.

COMPUTATIONAL CONFIGURATIONS

Two geometrical types of injection were computed. The cylindrical case concerns the flowfield calculation for a film cooling jet induced by a cylindrical hole, figure 1. The hole diameter, D, equals 3.2mm. The slope angle is 50° with the flat plate, in the forward direction. The expanded case concerns the flowfield calculation of a film cooling jet induced by a shaped hole, figure 2. The hole inlet diameter D and the slope angle are identical in both cases. Expanded directions are lateral and forward.

Two domains are considered for each simulation. On the flat plate the H-type grid numbers 340,032 points. The second one, inside the injection hole tube, is also an H-type grid with 39,546 and 85,176 points for the sloped and expanded case respectively. The mesh concerning the flat plate is refined all around the hole which is a zone characterised by important velocity gradients. Few nodes have been added in this region to ensure the grid independance solution. Moreover, the nodes distribution close to the wall has been refined to reproduce the boundary layer : $y^+=10^{-2}$ on the flat plate at X=-1.22.

The origin of the coordinate system (X,Y,Z) is the intersection point of the cylinder revolution axis and the flat plane, figure 1. The system values are normalized by hole diameter D.

Three main parameters will influence the hole near the flowfield. The first one is the blowing ratio, M, defined as : $M = \dfrac{\rho_{jet} V_{jet}}{\rho_\infty V_\infty}$, with M= 0.95 in both cases. The second parameter is the upstream boundary layer thickness, δ. At X=-1.22, ratio δ/D equals 1.83 and 1.65 in the cylindrical and the expanded case respectively. The last parameter is the Reynolds number based on the inlet diameter and the mainstream velocity above the hole : Re_d= 16115 (cylindrical case) and 16324 (expanded case). The turbulence intensity is fixed at 8% in the inlet section of the flat plane domain, and 6% at the inlet injection tube.

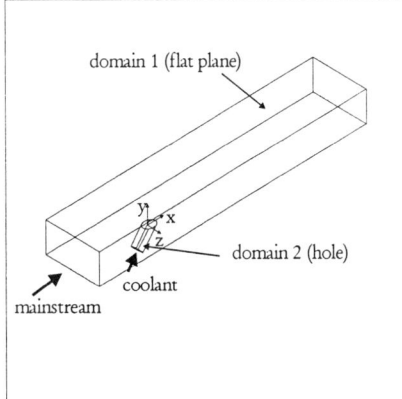
figure 1 : 3D view of the two domains

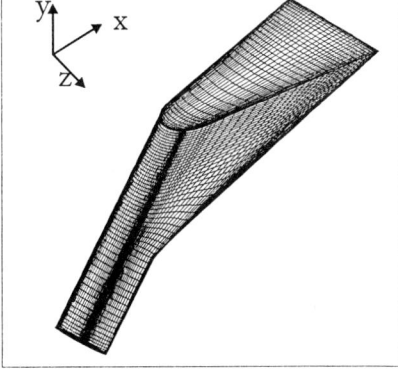
figure 2 : shaped hole grid

CYLINDRICAL CASE

This chapter presents the computation results of the cylindrical case, figure 1. We show the main vortex structures like the horseshoe vortex and the counter-rotating vortices known as the kidney vortices.

The horseshoe vortex is shown in figure 3. The generative mechanism was revealed by Baker[10]. The upstream flat wall boundary layer undergoes the adverse pressure gradient induced by the jet. A three-dimensional separation occurs. The separated boundary layer rolls up and is concentrated at point T_1, figure 4. A vortex system (two swirls) is generated downstream of the separation line and swept around the hole exit, close to the flat plane. It assumes the characteristic shape which has led to its name : the horseshoe vortex, figure 4.

The two swirls tend to get closer far downstream of the injection. Figure 3 shows they approach the plane Z=0, near the wall. This behaviour was observed by Charbonnier[11] but it depends on the width of the separation region located in the wake of the jet. Thus Bousgarbies et al.[12] observed that the two swirls are absorbed in the wake of the jet and integrated in the mini-tornado (T_5) structures.

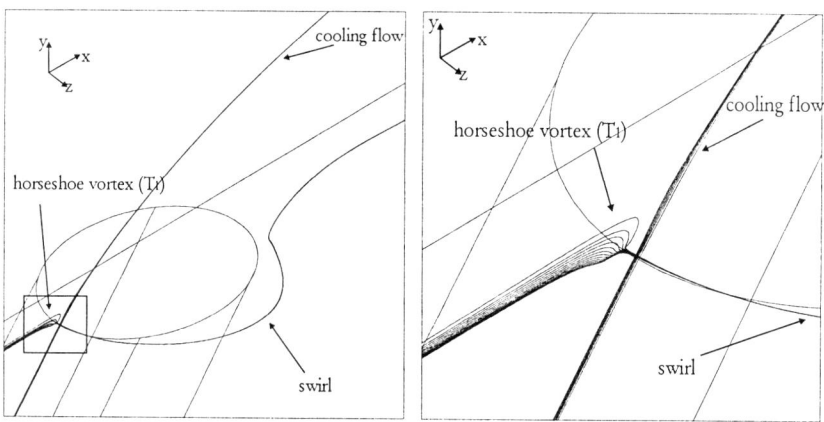

figure 3 : The horseshoe vortex   figure 4 : Focus on the roll up

The counter-rotating vortices, shown in figure 5, are ascribed to the rolling up of the hole boundary layer. Their rotation direction ($\pm\Omega x$) is such that one vortex, by mutual induction, lifts the other off the surface. Additionally, the rotational motion of the vortices pulls the crossflow around the jet and down toward the wall surface. Both coolant lift-off and the pulling in of the hot gas decrease the effectiveness of the film layer. They are dissipated by viscous diffusion far downstream.

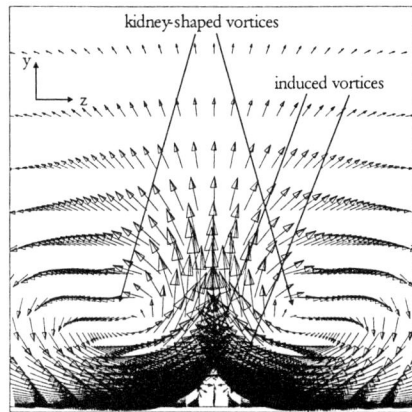 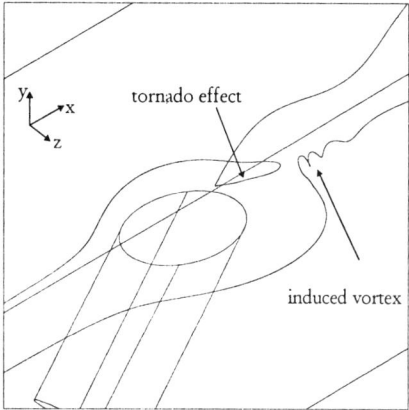

figure 5 : Vector field – X=1.5   figure 6 : Streamlines view

Small counter-rotating vortices are shown, figures 5 and 6, just below the kidney-shaped ones. We named them "induced vortices" because they result from the pulling in of the boundary layer by the kidney-vortices. They appear close to the wall, just downstream of the wake of the jet and only include mainstream fluid. They could have a harmful effect on thermal protection.

Tornado effects are located in the wake of the jet where the mainstream boundary layer is attracted in an upstream motion as far as the downstream edge of the hole. The shear brought about by the crossflow induces an ascending rotational motion like a tornado, figure 6.

SHAPED HOLE CASE

We now present the computation results for the shaped injection. We observe vortex structures such as kidney-shaped, tornado, induced vortices which have been shown before. Their formation mechanism is the same as for the previous configuration. The real nearfield flow difference is located at the leading edge of the outlet injection. Shaping increases the exit section area so that the velocity level decreases. The outlet momentum cooling flow level is thus less than for the cylindrical case. This leads to the formation of the bound vortex observed by Andreopoulos[1] and the disappearance of the horseshoe vortex. We can see the leading edge flowfield in figure 7.

The leading edge vortex is created in the same way as the horseshoe vortex. Thus, if we consider the leading edge hole, the boundary layer undergoes an adverse pressure gradient produced by the mainstream flow. It rolls up in the leading edge vortex at point T_2 point. Particles are concentrated at this point until they are ejected and dragged along the wall, within the hole. The peculiarity of this case is the ingestion of the mainstream fluid into the windward side of the cooling hole, figure 8. The mainstream fluid enters the cooling hole to be absorbed by the leading edge vortex. This phenomenon was observed by Thole et al. (1996) in a similar geometrical configuration. The ingestion reduces the effectiveness of the film cooling jet.

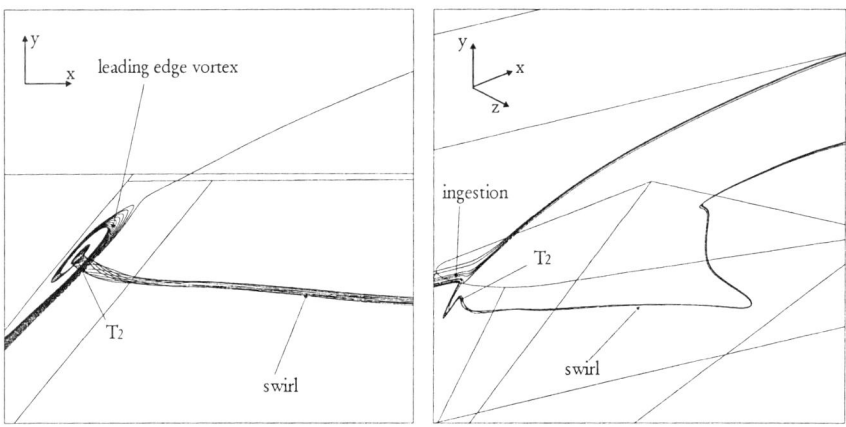

figure 7 : The bound vortex   figure 8 : The ingestion phenomenon

The large forward expansion angle produces a separation from the leeward side of the wall, figure 9. It leads to three-dimensional separation that induces a blockage effect and an increased velocity zone on the windward side. This part of the jet flow is thus ejected far into the mainstream, so the effect of shaping is not optimal.

We can observe in figure 10 (k normalized) that high turbulence levels occur at the exit of the expanded hole due to the separation, which induces a swirling motion and the coolant fluid is mixed. Here the thermal exchanges are important. Moreover, we have seen above that some mainstream flow is ingested and integrated by the leading edge vortex. But this vortex system is dragged downstream to the separation region. This leads to a mixing between some mainstream flow (swirl) and jet flow, within the hole, on the leeward side. We can see the differences in jet development in figures 11 and 12.

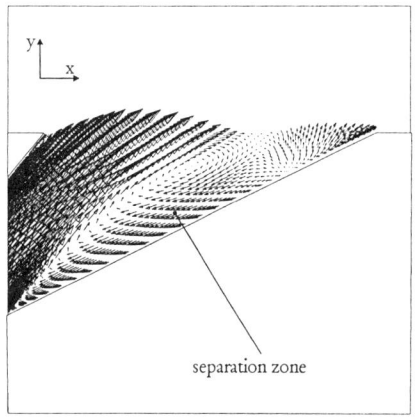

figure 9 : Separation zone   figure 10 : Turbulent energy in the hole

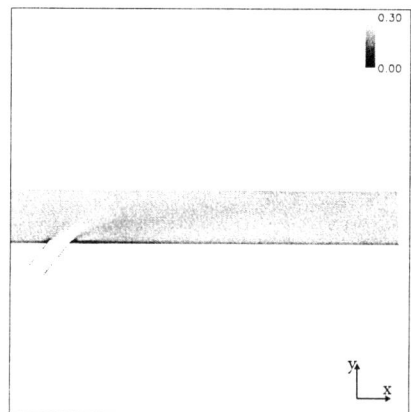 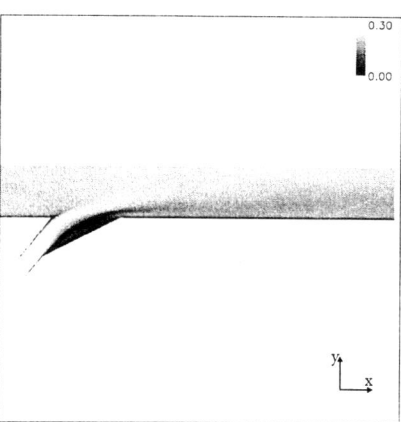

figure 11 : Isomach – cylindrical case   figure 12 : Isomach – expanded case

CONCLUSIONS

Two geometrical film cooling hole configurations have been studied in this paper. The simulations have been computed with a three-dimensional Navier-Stokes solver associated with a k-ε turbulence model. The code is CANARI, developed by ONERA.

The first computation concerns the flowfield jet induced by a cylindrical sloped hole, while the second one concerns a jet induced by a shaped sloped hole.

Classical swirly structures have been well reproduced by the computations. Their formation mechanisms have been analysed and the aerodynamic differences between the two cases have been revealed.

Moreover we observed the occurrence of a large separated region on the leeward side of the hole and specially the aerodynamic consequences in the nearfield of the injection. Their effect on cooling effectiveness has been evaluated. An expanded angle reduction should lead to a decrease of the separated zone.

REFERENCES

1. Andreopoulos, J. 1985. On the structure of jets in a crossflow. J. Fluid Mech. **157** : 163-197.
2. Goldstein, R.J.& E.R.G. Eckert. 1974. Effects of hole geometry and density on three-dimensional film cooling. J. Heat Mass Transfer. **17** : 595-607.
3. Makki, Y.H. & G.S. Jakubowski. 1986. An experimental study of film cooling from diffused trapezoïdal shaped holes. AIAA paper n° 86-1326.
4. Thole, K., M. Gritsch, A. Shulz & S. Wittig. 1996. Flowfield measurements for film cooling holes with expanded exits. ASME paper. 96-GT-174.
5. Haven, B.A., D.K. Yamagata & M. Kurosaka. 1997. Anti-kidney pair of vortices in shaped holes and their influence on film cooling effectiveness. ASME paper. 97-GT-45.
6. Vuillot, A.M., V. Couailler & N. Liamis. 1993. 3D turbomachinery Euler and Navier-Stokes calculation with multidomain cell-centered approach. AIAA paper n° 93-2576.

7. Cambier, L. & B. Escande. 1989. Calcul de l'écoulement tridimensionnel turbulent dans un aubage rectiligne de turbine. AGARD/PEP 527.
8. Gleize, V., R. Schiestel & V. Couailler. 1996. Multiple scale modeling of turbulent nonequilibrium boundary layer flows. Phys. Fluids. **8** : 2716-2732.
9. Jones, W.P. & B.E. Launder. 1973. The calculation of low-Reynolds number phenomena with a two equation model of turbulence. Int. Journal of Heat and Mass Transfer, **16** : 1119-1130.
10. Baker, C.J. 1979. The laminar horseshoe vortex. J. Fluid Mech. **95** : 347-367.
11. Charbonnier, J.M. 1992. Etude théorique et expérimentale d'un jet à travers un écoulement transversal. Thesis. Université de Poitiers.
12. Bousgarbies, J.L., L.E. Brizzi & E. Foucault. 1991. Vorticité au voisinage d'un jet circulaire débouchant perpendiculairement dans une couche limite ; cas des petits nombres de Reynolds. AUM. $10^{\text{ème}}$ Congrès Français de Mécanique. Paris.

The numerical calculations were performed on the NEC SX5 computer of the IDRIS center.

The authors thank the MENRT (Ministère de l'Education Nationale de la Recherche et de la Technologie) for their financial support of the PhD thesis and SNECMA for its financial support of the experimental work.

NUMERICAL INVESTIGATION OF HEAT TRANSFER ON FILM-COOLED TURBINE BLADES

P. Ginibre, M. Lefebvre and N. Liamis
Turbine Aero and Cooling Department
Snecma Moteurs
77550 Moissy-Cramayel, France

ABSTRACT: The accurate heat transfer prediction of film-cooled blades is a key issue for the aerothermal turbine design. For this purpose, advanced numerical methods have been developed at Snecma Moteurs. The goal of this paper is the assessment of a three-dimensional Navier-Stokes solver, based on the ONERA CANARI-COMET code, devoted to the steady aerothermal computations of film-cooled blades. The code uses a multidomain approach to discretize the blade to blade channel with overlapping structured meshes for the injection holes. The turbulence closure is done by means of either Michel mixing length model or Spalart-Allmaras one transport equation model. Computations of thin 3D slices of three film-cooled nozzle guide vane blades with multiple injections are performed. Aerothermal predictions are compared to experiments carried out by the von Karman Institute. The behavior of the turbulence models is discussed, and velocity and temperature injection profiles are investigated.

INTRODUCTION

For a long time in turbomachinery, the common trend to increase the temperature at the combustion chamber exit, in order to increase the engine thermodynamical efficiency and the thrust to weight ratio, has led to cool the High Pressure turbine blades to ensure sufficient blade life. The cooling is often achieved through film cooling, in which cool gas is ejected through thin holes or slots on the blades. The need for engine manufacturers to predict the performance of film cooling becomes critical as the turbine inlet temperature increases.
In the last decade, tremendous progress has been achieved in computation of film cooling configurations. With the development of supercomputers, fully 3D computations have become more and more frequent. Leylek and Zerkle[1] showed, using a film-cooled flat plate comprising the plenum, the injection pipe and the main domain, that the whole geometry is important in establishing the jet feature, in particular its profile at exit plane. Garg and Gaugler[2] modeled the injection with profiles at the wall and showed that the injection profile has a great influence in terms of heat transfer coefficient. With this model, Garg[3] performed a computation of a complete film-cooled turbine blade. In an intermediate case, Fougères and Heider[4] performed computations of thin 3D slices of film-cooled blades, using overlapping meshes to discretize the holes but without meshing the internal channel.
In the present paper, a 3D Navier-Stokes method (similar to that used by Fougères

and Heider) containing different turbulence models will be used to compute cooled 3D slices of nozzle guide vane blades. Results will be compared to experimental data. The injection profiles and the effect of turbulence model will be discussed.

NUMERICAL METHOD

Equations and turbulence closure

The equations to be solved are the Reynolds Averaged Navier-Stokes (RANS) equations with constant specific heat ratio. The molecular viscosity depends on temperature through the Sutherland formula.

Two different turbulence models based on the Boussinesq hypothesis are available in the code to compute the eddy viscosity. Neither model has an equation for the turbulent kinetic energy, therefore the turbulence rate cannot be evaluated.

The first one, noted M model afterwards, is the algebraic mixing length model described by Michel[5]. The values of both the boundary layer thickness and the turbulent to molecular viscosity ratio are limited to ensure that no excessive amount of turbulence is reached with the model.

The second one, noted SA model afterwards, is the one transport equation model proposed by Spalart and Allmaras[6] to take into account the transport of turbulence.

Transition is handled through an intermittency factor, according to Abu-Ghannam and Shaw[7]. Transition start and end are specified manually. The intermittency factor is used in the laminar and transitional zone to respectively cancel and reduce the eddy viscosity in both models and the source terms in the turbulence transport equation.

Discretization

The method is based on the 3D Navier-Stokes CANARI-COMET solver, developed at ONERA[8], which uses a node centered finite volume structured multi-domain technique. Time integration is performed through a four step Runge-Kutta scheme and space integration through a centered second order scheme with second and fourth order artificial dissipation. An implicit residual smoothing phase is applied in order to increase the time step and the robustness of the scheme. A local time step technique is used to accelerate convergence to steady state.

The turbulence transport equation is resolved separately using a first order Roe scheme for spatial integration and a four step Runge-Kutta scheme for time integration. The implicit residual smoothing phase and the local time step technique are also applied.

Boundary conditions

Inflow and outflow boundary conditions are specified following the characteristics theory. In this way, the total temperature and the flow angles are defined at inflow. The total pressure is also specified upstream of the blade whereas the massflow rate is specified at the injection entrance. Static pressure is prescribed downstream of the blade. The variables at the walls are computed using zero pressure gradient with a specified wall temperature, and boundaries between different mesh domains are handled through trilinear interpolations.

In the SA model, the turbulent variable is set to zero at walls and to a very small value at inlet planes. When explicitly stated, the turbulent to molecular viscosity

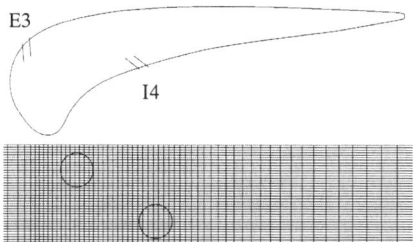

FIG. 1 – Test case 1: Geometry and mesh in the injection zone

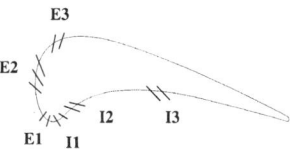

FIG. 2 – Test case 2: Injection location

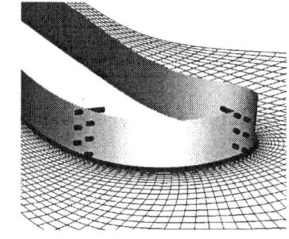

FIG. 3 – Test case 3: Configuration

	CASE 1	CASE 2	CASE 3
T_w (K)	296.6	298.25	295.35
T_∞ (K)	423.7	420.9	403.6
T_c (K)	330.	260 to 273	253.3

TAB. 1 – Temperature test operating conditions

ratio μ_t/μ is specified at the injection entrance. It is extrapolated from interior points at outflows.

Meshes

Taking advantage of the multi-domain approach of the code, the main flow domain is discretized with 3 domains: an H upstream of the blade, an O around the blade to easily handle the refinement of the boundary layer and the leading edge and trailing edge description, and an H downstream of the blade. The injection holes are discretized with overlapping cylindrical O meshes with a central axis. Therefore, the mesh generation is quite simple, since the only parameters to take care of are the location of the cooling injections on the blade, their angles and the hole diameter.

To ensure that interpolations between overlapping meshes are as conservative as possible, the O mesh of the main flow is refined near the injections and a minimum of 50 points of the blade wall mesh are located inside the hole injection area.

TEST CASES

Configurations

Three test cases have been investigated.

Experimental data were obtained in the Isentropic Light Piston Compression Tube CT2 (linear) or CT3 (annular) at the von Karman Institute (vKI) in the framework of Snecma private research contracts. Tests were conducted using short duration measurement techniques. Uncooled blades instrumented with pressure tapping were used for aerodynamic tests while ceramic blades intrumented with thin film gauges were used for thermal tests. The experimental results obtained are isentropic Mach numbers and span averaged wall heat fluxes. The temperature operating conditions are shown in Table 1.

The first test case is a linear HP nozzle guide vane with 2 double-rows of injections, one, called I4, on the pressure side and the other, called E3, on the suction side

(Fig. 1). The pitch of the rows is equal to 3 hole diameters. The injections are inclined in the streamwise direction with an angle of 50° to the blade profile.

The second test case is a linear HP nozzle guide vane with a total of 13 rows of injections on the pressure and suction sides of the blade with different pitches (Fig. 2). The injection angles in the streamwise direction cover a range from 90° near the leading edge to 35° on the pressure side and to 45° on the suction side.

The third test case is an annular HP nozzle guide vane with 2 double-rows of injections on the suction side of the blade (Fig. 3). The pitch of the rows is equal to 3 hole diameters. The injections are inclined streamwise with an angle of 70° to the blade profile for the upstream staggered row and 60° for the downstream row.

Periodicity

The span periodicity of the test configurations is used in order to minimize the span of the blade to be calculated and, thus, the mesh size and the CPU time.

For the first test case, the pitch of the double-row is representative of the span period. Therefore, only a period of the blade is computed. For the second test case, an assumption of periodicity is made on a span representative of 2 periods of the biggest pitch. For the third test case, 3 periods are computed.

RESULTS AND DISCUSSION

General features

The flow downstream of the injections displays the main features of jets in crossflow.

In the first test case on the suction side, the injections are located in an accelerated flow with a low blowing ratio ($M=0.4$), the jets remain close to the wall (Fig. 4 top). On the pressure side, as injections have a higher blowing ratio ($M=0.8$), the jets separate from the wall (Fig. 4 bottom). Periodicity of the flow and symmetry in respect to the central plane of each injection hole are well established (Fig. 5 for the suction side injections).

In the second test case with multiple injections on the blade, the jets display showerhead flow in the leading edge area (Fig. 6), whereas at other locations, the jets have a behavior similar to the first test case.

In the third test case, the blowing ratio is rather high, namely 1.7, and the jets separate from the wall (Fig. 7).

First test case

The computed isentropic Mach numbers with both turbulence models agree well

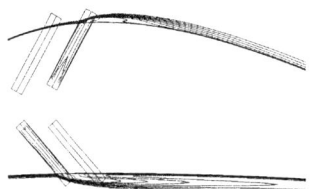

FIG. 4 – Test case 1: Total temperature

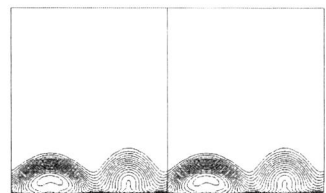

FIG. 5 – Test case 1: Periodicity of the flow

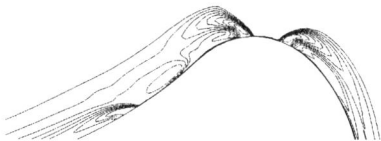

FIG. 6 – Test case 2: Total temperature

FIG. 7 – Test case 3: Total temperature

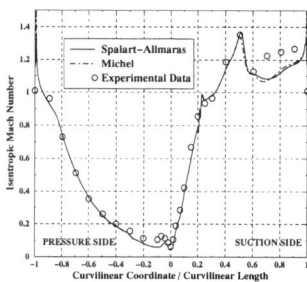

FIG. 8 – Test case 1: Isentropic Mach number

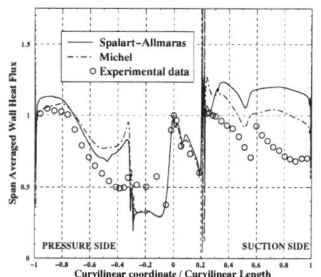

FIG. 9 – Test case 1: Wall heat flux

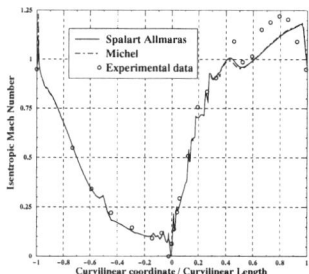

FIG. 10 – Test case 2: Isentropic Mach number

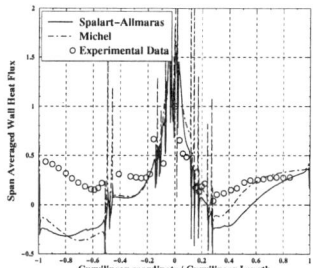

FIG. 11 – Test case 2: Wall heat flux

with experimental data except downstream of the shock on the suction side (Fig. 8). The computed wall heat fluxes reproduce the tendencies of experimental ones (Fig. 9). On the pressure side, as the jets lift off, hot gas from the main stream is pushed near the wall. Therefore, heat transfer is enhanced but this trend is overestimated. Near the leading edge, a transition followed by a relaminarization may occur but this transition phenomenon is not modeled in the computations. On the suction side, the cooling effect is underpredicted and the two models have different span averaged effects on the wall heat flux.

Second test case

The computed isentropic Mach numbers on the blade show reasonably good agreement with experiments (Fig. 10). The several injections are easily identified through jumps at the injection areas.

The computed span averaged wall heat fluxes display the main tendency of the

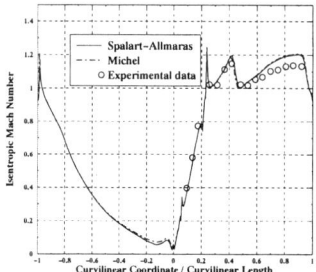
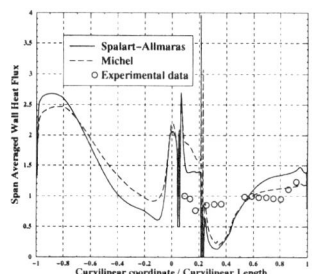

FIG. 12 – Test case 3: Isentropic Mach number   FIG. 13 – Test case 3: Wall heat flux

experimental span average heat fluxes (Fig. 11). The heat flux is underestimated on the pressure side and on the suction side, whereas it is overestimated in the leading edge area. The efficiency of the cooling is overestimated which hints that the cold jets in respect to the blade temperature remain too close to the wall.

Third test case

The computed isentropic Mach numbers agree very well with experiments (Fig. 12).
However, the heat fluxes display discrepancies between computations and experiments (Fig. 13): the first row cooling is underestimated whereas the second row cooling is overestimated. As only the total injection massflow rate is measured, the various massflow rates of each injection hole are not exactly known. The prescribed massflow rate may thus be wrongly distributed between the different holes.

Injection profiles

Leylek and Zerkle showed that the velocity profile shape at injection is dominated by two competing mecanisms: the jetting effect, due to the passage of the flow from the plenum to the injection pipe in the case of short length to diameter ratio, and the blockage effect, in the case of low blowing ratio. They produce upwards and downwards shifted velocity maxima respectively. In order to minimize the size of the domain to be computed, Garg used a 2D polynomial profile at the blade wall for velocity and temperature to model the shifted profiles.

In our computations, the jetting effect does not exist since the plenum is not discretized. The velocity and temperature profiles in the exit plane, defined in Fig. 14,

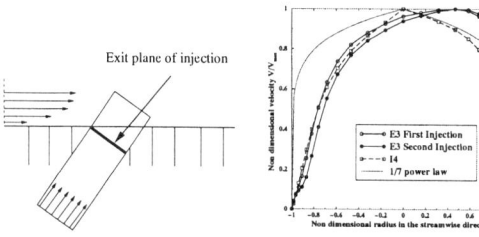

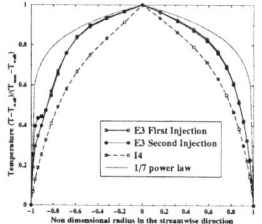

FIG. 14 – Exit plane   FIG. 15 – Velocity profiles   FIG. 16 – Temperature profiles

have been extracted from the first test case SA computation. The blowing ratios are 0.4 for the E3 injections and around 0.8 for the I4 injections. As stated by Leylek and Zerkle, the lower the blowing ratio is, the more the exit velocity profile is shifted downstream (Fig. 15). The temperature profiles at exit plane (Fig. 16) are less shifted downstream. They display a separation in the E3 second injection pipe at R=-0.9.

Looking at these phenomena, the specification of injection profiles at blade wall is a difficult problem to take into account the interaction between the main flow and the pipe flow, specially the contra-rotating vortices inside the pipe.

Turbulence model effect

Analysis of turbulence models focuses on the first test case E3 injection. Another SA computation using a higher eddy viscosity level at injection entrance ($\mu_t/\mu \sim 300$) differs little from the previous SA computation in terms of heat transfer.

The temperature maps at the vicinity of the blade show that the cooling effect is more important close to the injection with the SA model while it is better averaged downstream with the M model (Fig. 17). The M model produces turbulence in the wake of the jet by using the local flow conditions (in particular local vorticity) and, therefore, mixes up the jet close to injection, whereas the SA μ_t needs some distance to reach its maximum (Fig. 18). This is confirmed by the profiles perpendicular to the wall of eddy viscosity, velocity and temperature, one hole diameter downstream the second injection where the turbulence level achieved with the M model is higher and the velocity and temperature profile are smoother (Fig 19).

The lateral expansion of the jet is lower with the M model, as shown in the plane

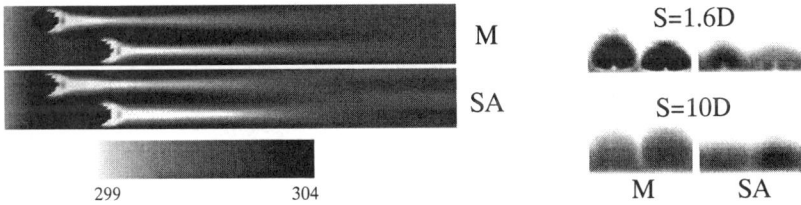

FIG. 17 – Test case 1: Static temperature near wall on the suction side

FIG. 18 – Eddy viscosity downstream suction side injections

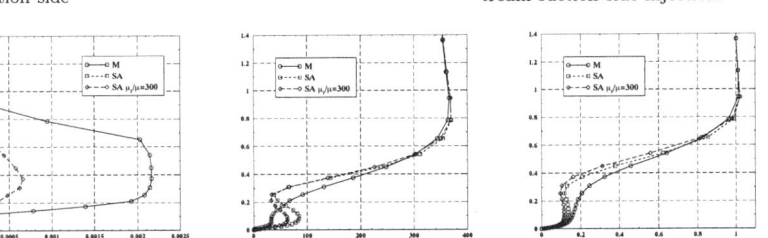

FIG. 19 – Eddy viscosity, Velocity and temperature $(T - T_w)/(T_\infty - T_w)$ versus y/ϕ, 1 diameter downstream suction side injection

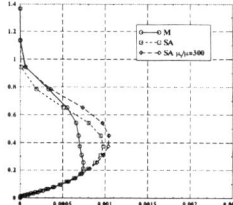 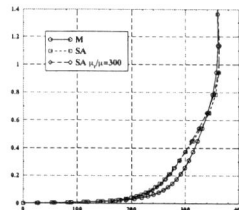 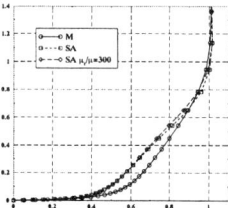

FIG. 20 – Eddy viscosity, Velocity and temperature $(T - T_w)/(T_\infty - T_w)$ versus y/ϕ 7 diameters downstream suction side injection in the plane between injections

centered between the injections, where the turbulence level is lower and the temperature and velocity are greater with the M model (fig. 20).

The turbulence models lead to very different features of the injection jets, but their effect on the span averaged wall heat flux is small. Finer validations are required to assess definitively the two turbulence models for the computation of film-cooled configurations.

CONCLUSION

In order to improve heat transfer prediction of film-cooled blades, advanced 3D Navier-Stokes methods have been developed at Snecma Moteurs. Their validation has been carried out with several computations performed on thin slices of film-cooled nozzle guide vane blades. The results prove the capability of the code to predict the aerothermal field around film-cooled turbine blades. The computed aerodynamic results agree well with the experimental data. The computed wall heat fluxes show the main tendencies of the experimental ones.

The use of injection pipe in the computational domain is necessary to handle a realistic injection profile in the main flowfield.

The two turbulence models do not behave the same way. The M model mixes up the jets near the injection area whereas the SA model mixes up the jets further downstream. Nevertheless, they lead to similar span averaged heat fluxes. However, due to lack of generality and robustness, the Michel model is less appropriate than the Spalart-Allmaras model to compute film cooled blades for industrial use.

ACKNOWLEGMENT

The authors gratefully acknowlege Snecma for the authorization to publish the present CFD and test results.

REFERENCES

1. Leylek J. H. & Zerkle R. D., ASME Journal of Turbomachinery, Vol. 116, 358-368, 1994.
2. Garg V. K. & Gaugler R. E., ASME 95-GT-2,1995.
3. Garg V. K., ASME 99-GT-44,1999.
4. Fougères J. M. & Heider R., ASME 94-GT-14,1994.
5. Michel R., Quemard C., & Durand R., ONERA NT No. 154, 1969.
6. Spalart P. R. & Allmaras S. R., La Recherche Aérospatiale, Vol. 1, 5-21, 1994.
7. Abu-Ghannam B. J. & Shaw R., J. Mech. Eng. Sc., Vol. 22, No 5, 1980.
8. Liamis N. & Couaillier V., ECCOMAS 94, 1994.

Comparison between two models of cooling surfaces using blowing

L. MATHELIN, F. BATAILLE, A. LALLEMAND

Institut National des Sciences Appliquées de Lyon
Centre de Thermique de Lyon, UMR 5008
Bât. 404, 20 av. A. Einstein, 69621 Villeurbanne cedex, FRANCE

ABSTRACT: To protect surfaces against high temperatures, the blowing through a porous material is studied. The geometry is that of a circular cylinder in cross-flow and the effectiveness of the blowing for the thermal protection is numerically investigated. Two models are developed for the blowing simulation and comparisons are made with experimental data obtained in a heated wind-tunnel. It is shown that the blowing strongly affects the dynamical and thermal profiles over the surface, thickening the boundary layers and decreasing the external transfer coefficients. It results in a lower viscous drag and thermal stress. The wall temperature dramatically decreases with blowing and the heat flux is also affected.

INTRODUCTION

Today's engines require still higher performances and efficiency in terms of fuel consumption reduction. To achieve this, the maximum operating temperature of the thermodynamic cycle must be as high as possible. Due to the strong thermal stress involved, most of the parts have to be protected to preserve their mechanical characteristics. Different ways are usually used to reduce walls temperature of the turbine blades or the combustion chambers, such as film cooling[1-3], impingement[4] or ablation of a removable coating. Nevertheless, these processes present disadvantages as the requirement of an important coolant flow rate or difficulties to obtain homogeneous cooling. An alternative possibility is the blowing through a porous surface. The part to be protected, or at least its external coating, is made of a porous material and a coolant flows through the wall. It is the limit case of the discrete injection through tiny pores. A first model has been developed and validated for blowing on a flat porous plate[5]. To reproduce others geometrical configurations, which can be present in turbines, we here investigate the blowing through a porous circular cylinder in cross-flow. The blowing impact is numerically studied through two different models in terms of the thermal protection effectiveness.

A sketch of the studied configuration is presented in Figure 1. The blowing considered is normal to the surface and applied all around the cylinder surface. The main flow is

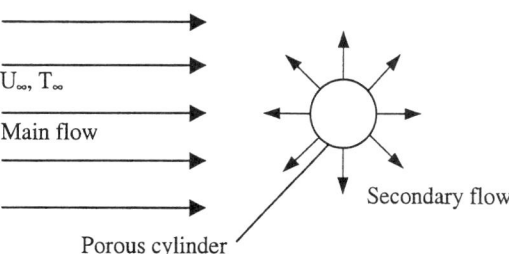

Figure 1 - Sketch of the configuration studied.

turbulent and the Reynolds number, *Re*, based on the outer cylinder diameter and upstream fluid properties is equal to 3900. The main flow temperature can vary from ambient to 200 °C, while the injected fluid remains at ambient temperature. The numerical results are compared to experimental works that we led in our subsonic heated wind tunnel[6]. Profile measurements are made using hot and cold wires probes. The velocity profiles are obtained using a single platinum-rhodium, 10 μm in diameter, hot wire probe. It has been preferred to X-wire probes which are more intrusive and too wide for accurate boundary layer measurements. The temperature profiles are acquired with a single Wollaston, 5 μm in diameter, cold wire probe. These probes are calibrated both in velocity and temperature. Surface temperatures are obtained with 0.1 mm, K-type thermocouples welded directly onto the surface. They have been proved not to distort the temperature field[7].

In a first time, the numerical approach and the models of the blowing through a porous wall will be presented. Next, the influence of the fluid injection on the dynamic and thermal profiles is exposed and the effectiveness of the blowing in the reduction of the thermal load is finally investigated.

NUMERICAL ASPECTS

A 2-D finite volume RANS solver code (Fluent) is used. Many different types of grid were tested and a number of their characteristics were varied including the cells density and the computational domain extent. In particular, the cells number around the cylinder was varied from 80 to 3000. The grid finally retained for computations is shown in Figure 2. It consists in a 50 000-triangular cells grid with a computational domain 5-diameter (*D*) long upstream of the cylinder axis, 15 *D* downstream and 10 *D* wide. Different numerical simulations have proved the retained grid is a good compromise between results accuracy and CPU time requirement. In particular, no important changes have been noted on global properties such as the Strouhal number or the velocity and temperature profiles when the grid was modified. Computations are carried-out with a Reynolds Stress Model (RSM) turbulence treatment, along with a low Reynolds number approach and an unsteady scheme. The time step was chosen to get around 1000 time steps per vortex shedding period, *i.e.* $\delta t^* = \delta t \, U_\infty / D = 3.10^{-3}$. All results presented in this paper are time-averaged over a shedding period.

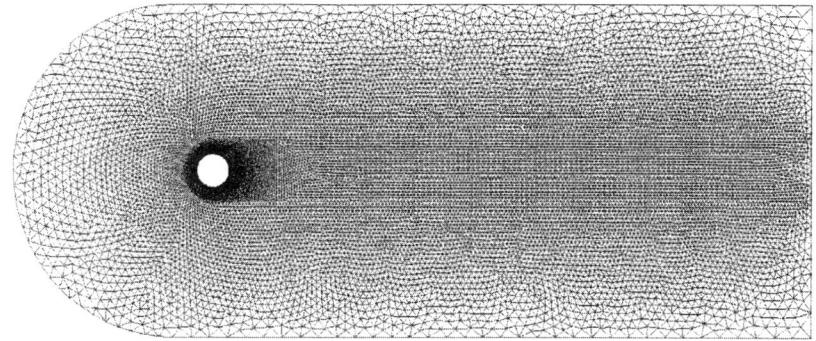

Figure 2 - Numerical grid used for computations.

MODELS OF BLOWING

In contrast with other models in the literature where the fluid governing equations must be modified near the surface according to the injection rate[8-14], we developed a model where the physical phenomena due to blowing are directly taken into account. They involve both a friction stress on the surface and a fluid injection in the boundary layer. The blowing is studied in terms of the blowing ratio, F, defined as $F = (\rho_{inj} U_{inj}) / (\rho_\infty U_\infty)$. The first model, hereafter denoted the holes model, consists in considering the porous wall as a succession of adiabatic wall segments and holes whose proportion is set according to the actual porosity of the material. The injected cold fluid is supplied through the holes. The size of the wall segments and the pores was numerically varied in a limited range and no influence on the results was noted. The number of cells within a pore was also varied from 1 to 3 and resulted again in no change. The second model of blowing, the sources model, consists in applying a mass, momentum and heat source at the first cell center above an impermeable adiabatic wall to account for the effect of the blowing. The cell size is small enough to consider the coolant to be provided at the wall surface location. A sketch of these two models is given in appendix.

RESULTS

Dynamical and temperature profiles

When blowing is applied, the velocity and the temperature profiles are strongly affected. In figure 3, the velocity magnitude profile along the vertical direction is plotted for three different injection rates at an angle of 65° (defined starting from the front stagnation point) and a Reynolds number of 3900. The numerical results are compared to experimental data but, due to experimental difficulties, the profiles are truncated below a certain distance to the wall. The boundary layer is thickened when blowing occurs and the velocity gradient above the surface decreases, leading to a lower viscous drag and the promotion of the boundary layer separation. A very good agreement between the numerical and the experimental results can be noted, allowing to validate our model. A similar trend can be observed in figure 4 where the velocity profile is plotted for an angle of 105°, beyond the separation. The profiles are S-shaped exhibiting a typical separated shear layer behavior. With blowing, the shape is changed, decreasing the gradients and increasing the shear layer thickness. Beyond the separation point, a dead fluid zone is

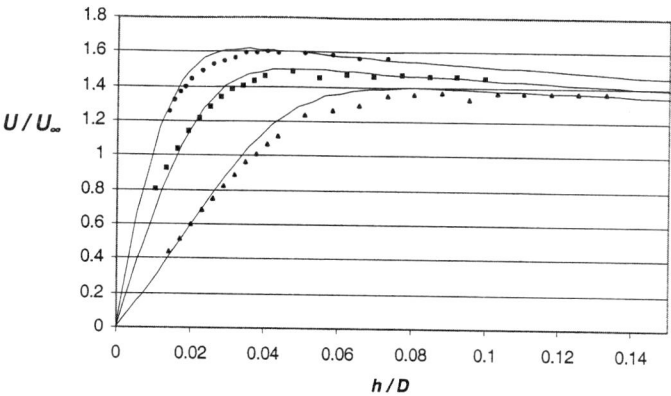

Figure 3 - Velocity magnitude profile along the vertical direction. $Re = 3900$, angle is 65°. Experimental data are symbols, numerical results are the corresponding solid lines. ●, no blowing, ■, $F = 2\%$, ▲, $F = 5\%$.

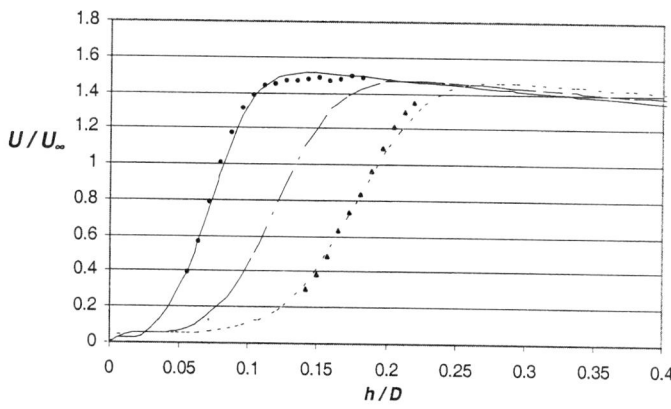

Figure 4 - Velocity magnitude profile along the vertical direction. $Re = 3900$, angle is 105°. Numerical results are the solid lines (from solid to dashed, $F = 0, 2$ and 5%). Experimental data are given for $F = 0\%$ (●) and $F = 5\%$ (▲).

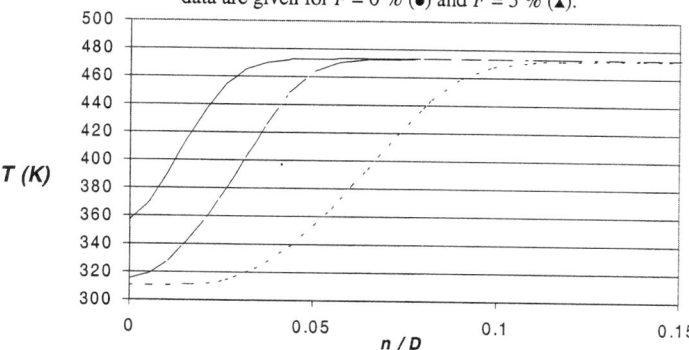

Figure 5 - Temperature profile along the normal direction. $Re = 3900$, angle is 65°. Solid line is for $F = 2\%$, semi-dashed line for $F = 5\%$ and dashed line for $F = 10\%$.

present below the separated shear layer, corresponding to the low velocity part in Figure 4. The numerical results are compared with experimental data for $F = 0$ and 5 %. The temperature profile is shown in Figure 5 for an injection rate of 2, 5 and 10 % at an angle of 65°. No experimental data are available. When blowing is applied, an important increase in the thermal boundary layer thickness occurs, reducing the transfer coefficients and leading to an efficient wall protection against the hot main flow. This phenomenon is observed for all locations around the cylinder even beyond the boundary layer separation point. It should be noted that the profiles were here presented using the holes model only as the curves perfectly collapse for both models. They give absolutely similar results, in terms of boundary layer thickness and seem to be very well adapted to account for the blowing impact and to determine the important parameters of the study.

Blowing effectiveness

The thermal effectiveness, defined as $\eta = (T_\infty - T) / (T_\infty - T_{inj})$ with T, T_∞ and T_{inj} the surface, main flow and coolant temperature respectively, is plotted in Figure 6 which exhibits its evolution as a function of the injection rate for an angle of 65° and a 200°C main flow. Both the holes and the sources model results are compared with experimental data. The wall temperature decrease is very important and only weak injection rates are necessary to obtain a significant thermal protection. For a 1 % injection, the thermal effectiveness is close to 50 % and reaches 90 % for 4 % of blowing. The numerical results are in good agreement with the experimental data and it allows to validate the blowing models and to predict the different temperatures. Nevertheless, the holes model is less accurate for weak injection rates: when blowing is applied, the boundary conditions change from a continuous isothermal wall to a non-isothermal discontinuous surface and leads to spurious results. When the blowing gets stronger, the temperature field above the surface is more uniform and this model becomes reliable.

Figure 7 shows the thermal effectiveness for both models as a function of the angle for 2 % of injection and a main flow temperature of 200°C. We can notice that the two models give similar results even if the effectiveness is slightly higher using the holes model, in particular for low angles. For the following, we will use the sources model only because it does not present discontinuities in the boundary conditions and is supposed to be more accurate for the determination of the effectiveness.

In the upstream part, the boundary layer is "squashed" onto the front part of the cylinder, leading to strong gradients, and thus high transfer coefficients. As the boundary layer grows along the cylinder surface, the heat transfer coefficient and the wall temperature decrease. Beyond the separation, a cold dead fluid zone is present below the separated shear layer, leading to a minimum in the wall temperature. At the back of the cylinder, recirculations draw back some hot fluid from the main flow and make the wall temperature to increase. These results are in full agreement with the first experimental data of Johnson & Hartnett[15].

When the injection rate is varied from 1 % to 5 %, the blowing effectiveness increases for all angles as plotted in figure 8 with the sources model. The evolution of the effectiveness along the surface remains the same whatever the blowing ratio while tending to a constant value for strong injection (5 %). The wall temperature difference between the front stagnation point and the most protected portion steadily decreases with the blowing.

Finally, the periphery-averaged convective heat flux is plotted for different injection rates in Figure 9. For the no-blowing case, the cylinder is supposed to be adiabatic and the incident heat flux is equal to zero. When blowing occurs, the internal heat exchange between the solid part of the wall and the coolant is accounted. The heat flux first increases before decreasing with the injection rate down to zero. For weak injection rates, the blowing lowers the external convective heat transfer coefficient together with the wall temperature. The temperature difference between the wall and the main flow thus grows up and induces an increase in the heat flux although the heat transfer coefficient decreases. Beyond a certain threshold (around 2 %), the heat transfer coefficient decrease cancels the increase in the temperature difference and the resulting incident heat flux begins to decrease. For a 10 % injection, the convective heat flux approaches zero and shows an excellent protection of the cylinder, both in terms of wall temperature and incident heat flux.

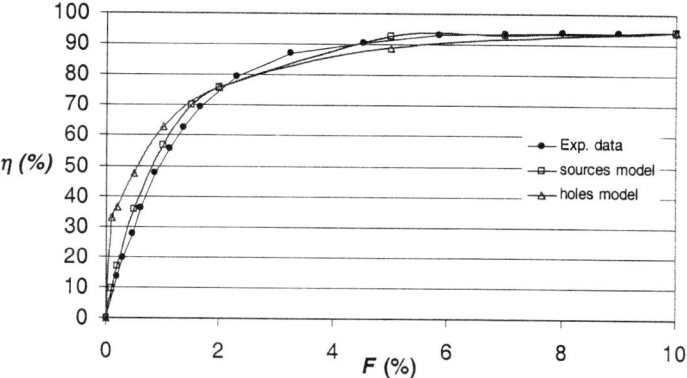

Figure 6 - Thermal effectiveness as a function of the blowing ratio. $Re = 3900$, $T_\infty = 473$ K, $\theta = 65°$.

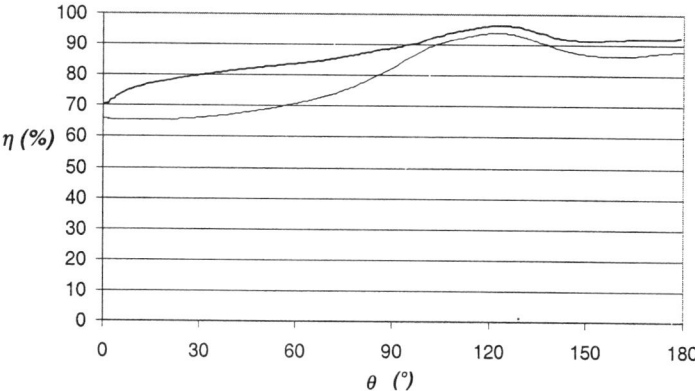

Figure 7 - Thermal effectiveness as a function of the angle. $Re = 3900$. $T_\infty = 473$ K, $F = 2$ %. Light line is the sources model, thick line is the holes model.

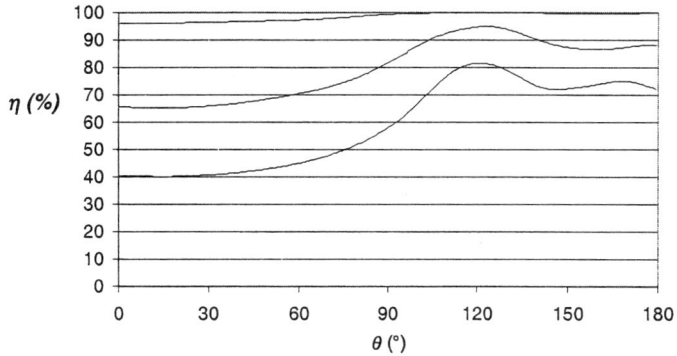

Figure 8 – Effectiveness evolution along the cylinder with the sources model. $Re = 3900$, $T_\infty = 473$ K, $\theta = 90°$. From bottom to top, $F = 1, 2$ and 5 %.

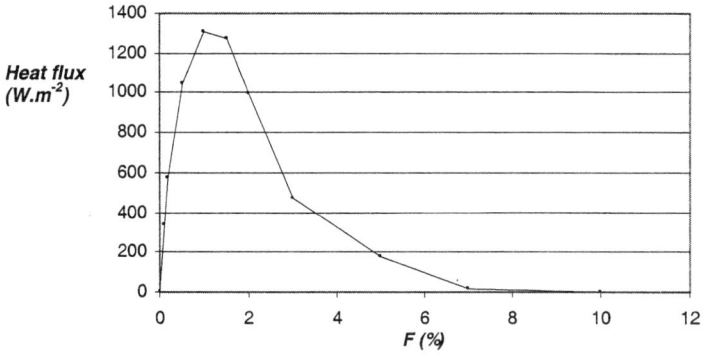

Figure 9 – Incident heat flux evolution with the blowing rate. $Re = 3900$. $T_\infty = 473$ K. Solid line is the sources model, dashed line is the holes model.

CONCLUSIONS

Two models of cooling parts using blowing through a porous surface were developed and compared to experimental data. The velocity and temperature profiles are found to be strongly affected by the blowing. The boundary layers thicken and the transfer coefficients decreases, leading to lower viscous stress and thermal load. The thermal effectiveness has been studied as a function of the cylinder angle and for different injection rates. An important decrease of the wall temperature and an excellent effectiveness of the cooling has been shown even for weak injection rates. Blowing is thus demonstrated to be a very attractive way of cooling surfaces subjected heavy thermal stress and can accurately and reliably predict the key parameters involve as the incident heat flux, the wall temperature and the blowing rate required for a given protection.

REFERENCES

1. Ekkad, S.V, Han, J.C. & Du, H. 1998. Detailled film cooling measurements on a cylindrical leading edge model: effect of free-stream turbulence and coolant density. J. Turbomach. **120**: 799-807.
2. Lee, S.W., Kim, Y.B. & Lee, J.S. 1997. Flow characteristics and aerodynamics losses of film-cooling jets with compounds angle orientations. J. Turbomach. **119**: 310-319.
3. Goldstein, R.J. & Stone, L.D. 1997. Row-of-holes film cooling of curved walls at low injection angles. J. Turbomach. **119**: 574-579.
4. Facchini, B., Ferrara, G. & Innocenti, L. 2000. Blade cooling improvement for heavy duty gas turbine: the air coolant temperature reduction and the introduction of stream and mixed steam/air cooling. Int. J. Therm. Sci. **39**(1): 74-84.
5. Bellettre, J., Bataille, F. & Lallemand, A. 1999. A new approach for the study of the turbulent boundary layer with blowing. Int. J. Heat Mass Transfer. **42**: 2905-2920.
6. Rodet, J.C., Campolina-França, G.A., Pagnier, P., Morel, P. & Lallemand, A. 1997. Etude en soufflerie thermique du refroidissement de parois poreuses par effusion de gaz. Rev. Gén. Therm. **37**: 123-136.
7. Bellettre, J., Bataille, F., Rodet, J.C. & Lallemand, A. 2000. Thermal behavior of porous plates subjected to air blowing. AIAA J. Thermophysics, in press.
8. Stevenson, T.N. 1968. Inner region of transpired turbulent boundary layers. AIAA J. **6**(3): 553-554.
9. Simpson, R.L. 1970. Characteristics of turbulent boundary layers at low Reynolds numbers with and without transpiration. J. Fluid Mech. **42**: 769-802.
10. Kays, W.M. 1972. Heat transfer to the transpired turbulent boundary layer. Int. J. Heat Mass Transfer. **15**: 1023-1044.
11. Landis, R.B. & Mills, F. 1972. The calculation of turbulent boundary layers with foreign gas injection. Int. J. Heat Mass Transfer. **15**: 1905-1932.
12. So, R.M.C. & Yoo, G.J. 1987. Low Reynolds number modeling of turbulent flows with and without wall transpiration. AIAA J. **25**: 1556-1564.
13. Silva-Freire, A.P. 1988. An asymptotic solution for transpired incompressible turbulent boundary layers. Int. J. Heat Mass Transfer. **31**: 1011-1021.
14. Campolina-França, G.A. 1996. Contribution à l'étude des écoulements parietaux avec effusion. Application au refroidissement de parois. PhD Thesis, CETHIL, INSA de Lyon, FRANCE.
15. Johnson, B.V. & Hartnett, J.P. 1963. Heat transfer from a cylinder in cross-flow with transpiration cooling. J. Heat Transfer.: 173-179.

APPENDIX

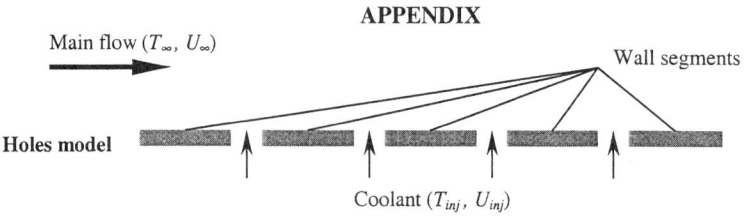

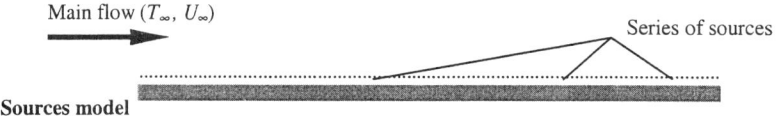

Finite Element Analysis of Flowfield in the Single Hole Film Cooling Technique

F. BAZDIDI - TEHRANI   A. A. MAHMOODI

Department of Mechanical Engineering,
Iran University of Science and Technology,
Narmak, Tehran 16844, Iran.

ABSTRACT : Film cooling is currently used in gas turbine hot sections, such as the combustor wall and the turbine blades, to prevent those sections from failing at elevated temperatures. In the single hole film cooling method, coolant air is injected from a hole into the mainstream and thus the flow is naturally three dimensional. In this paper, the Navier-Stokes and the energy equations are solved on a flat plate by the Finite Element Method (FEM) using brick elements. Algebraic equations are obtained by use of the Petrov-Galerkin method. The pressure term is removed from the momentum equations, by employing the Penalty method. The governing equations are transient and the flow is incompressible and turbulent. The model of turbulence in the near wall region is the wall function method, and in the fully turbulent region is the k-ε model. The system of the algebraic equations are solved by the Frontal method. The coolant injection angle and the blowing rate are among the parameters which are studied. In order to examine the present computer code, the results are compared with the Blasius (exact) solution and also with the empirical 1/7th power-law and good agreement is shown. Also, the optimum cooling performance is shown to be at 35 degree angle of coolant injection and the optimum blowing rate is 0.5. The film cooling effectiveness data, at the optimum conditions, is directly compared with the experimental results of Goldstein et al.[16] and good agreement is demonstrated.

INTRODUCTION

One of the important applications of film cooling is to protect the gas turbine hot sections, namely the combustor wall and turbine blades, against the hot gas mainstream. Considering the current maximum allowable alloy temperature in these sections, any further increase in the temperature would result in rapid wear and tear. Discrete hole film cooling is a relatively simple technique. The coolant air on its passage through the hole (or a row of holes) cools the wall convectively and then emerges as a film, protecting the surface in the region and downstream of injection.

Goldstein et al.[1] studied the variations of the adiabatic film cooling effectiveness, η [$= (T_\infty - T_{ad})/(T_\infty - T_2)$], with the blowing rate, M [$= \rho_2 w_2 / \rho_\infty w_\infty$], in the single hole film cooling technique. They reported that an increase in M, up to a certain limit, resulted in an increase in η, beyond which a reduction in η occurred. η is a dimensionless parameter, employed to evaluate the performance of the film cooling. T_{ad} is similar to the gas temperature adjacent to an adiabatic wall in studies of adiabatic film cooling, where it is taken as the limiting value of the wall temperature in the absence of heat transfer[2]. Subscripts ∞ and 2 represent conditions at mainstream and secondary (coolant) flow, respectively. Also, ρ and w stand for density and velocity, respectively.

Schmidt et al.[3,4] concluded that the use of the compound angle of coolant injection could increase η, and, if M was high enough, the difference between the compound angle and plane angle of injection would be noticeable. For the plane angle, as shown in Fig.1, the injection takes place toward the axial (downstream) direction, z, whereas for the case of compound angle the injection occurs in between the axial (z) and the lateral (y) directions.

On the basis of the experimental investigations[5], an increase in the mainstream turbulence intensity would decrease η. The same trend, based on the Finite Element analysis[6], has been reported for the intensity range 1-10%, unlike other numerical investigations where little attention has been paid to it.

Walters and Leylek[7] carried out a numerical analysis on film cooling, extending the solution domain such that the coolant supply plenum, the injection hole and the cross flow of injection were taken into account. This meant an extremely large nodal grid, which required a special parallel processing system. They reported that, according to Garg and Gaugler[8], almost 60% of the variation in the convective film heat transfer coefficient, downstream of injection, depended on the coolant velocity and temperature profiles at the hole exit. Also, Taeibi and Ebrahimi[9] used Direct Numerical Simulation, in which the Navier-Stokes equations were solved directly without the use of a turbulence model. This method also required an extremely large number of nodes for the laminar flow region (for Reynolds numbers less than 1000).

It should be noted that all the above researchers based their numerical simulation on Finite-Difference or Finite-Volume techniques, whereas the present work was based on the FEM. The advantage of the latter is that it requires fewer elements and hence much less CPU time is needed to solve for the flowfield.

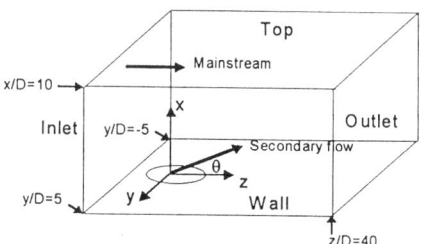

Fig.1 Overall computational domain.

GOVERNING EQUATIONS AND BOUNDARY CONDITIONS

The Navier-Stokes and the energy equations, in a reduced form, under transient, incompressible and viscous flow conditions are as follows:

$$\rho \frac{DV}{Dt} = -\nabla p + \mu \nabla^2 V \qquad (1)$$

$$\rho C_p \frac{DT}{Dt} = k \nabla^2 T \qquad (2)$$

Where, p represents pressure and ρ, C_p, μ and k stand for density, specific heat, viscosity and thermal conductivity. For the incompressible flow, the momentum equation (1) together with the continuity equation (i.e., $\nabla \cdot V = 0$) were solved first so as to compute the velocity vector, V. Then the temperature, T, was obtained from equation (2). Since the flow was considered as turbulent, using the $k - \varepsilon$ turbulence model, two more equations must be solved along with the above equations:

$$\rho \frac{\partial k}{\partial t} + \rho(V.\nabla)k = (\mu + \frac{\mu_t}{\sigma_k})\nabla^2 k + P - \rho\varepsilon \qquad (3)$$

$$\rho \frac{\partial \varepsilon}{\partial t} + \rho(V.\nabla)\varepsilon = (\mu + \frac{\mu_t}{\sigma_\varepsilon})\nabla^2 \varepsilon + (c_{\varepsilon 1}P - c_{\varepsilon 2}\rho\varepsilon)\frac{\varepsilon}{k} \qquad (4)$$

Where k, ε and P represent turbulent kinetic energy, dissipation energy and production term, respectively. The eddy viscosity, μ_t and the empirical constants in equations (3) and (4) were expressed as follows[10]:

$$\mu_t = c_\mu \rho \frac{k^2}{\varepsilon} \qquad (5)$$

$$\sigma_k = 1.0, \sigma_\varepsilon = 1.3, c_{\varepsilon 1} = 1.44, c_{\varepsilon 2} = 1.0, c_\mu = 0.09$$

Equations (3)-(5) were sufficient for determining the three unknowns (k, ε and μ_t).

In the region close to the wall, the wall function method was employed. Since in this region large gradients exist, the number of nodes near the wall should increase so as to enhance the degree of accuracy. But, of course, this would require a much larger CPU time and memory (RAM). One way to overcome this was to apply the wall function method or the two-layer model[10].

The overall computational domain is illustrated in Fig. 1. A uniform flow condition was considered at the inlet boundary, upstream of the round hole, and at the hole inlet. Since the outlet was far enough from the entering jet, all the gradients were assumed to be zero. Also since the top boundary was far enough from the wall (i.e., x=10D; D is the hole diameter) a symmetry condition could be applied there (i.e., zero gradient for all flow variables) without significantly affecting the flowfield created by the jet-in-crossflow interaction. A no-slip condition was used for the velocity at the wall, whilst zero pressure and temperature gradients were considered there. Finally, all the gradients were assumed zero for the front and back boundaries, in the y-direction.

Due to the presence of the convection term on the left side of equations (1) and (2), it was possible to have a series of fluctuations and hence unsteadiness in the flowfield, when the Peclet number, Pe $(= \|U\|.h/2\alpha)$, became greater than 2. For the momentum equation (1), the Reynolds number, Re $(= \|U\|.h/2\nu)$, would replace Pe under these circumstances. But, for the energy equation (2), Pe would be used. $\|U\|$ represents the magnitude of velocity resultant vector in the element ($\|U\| = \sqrt{u^2 + v^2 + w^2}$); h stands for element characteristic length; α and ν represent thermal diffusivity and kinematic viscosity. Now, in order to omit the fluctuations present, one could make Pe small by taking h as a small value. But this would not be appropriate, as this would increase the number of elements and hence a more costly computer execution time. Another approach could be to use the Upwind Method[11], whereby the weighting functions were not equal to the shape functions and these could be chosen in such a manner that the problem would tend toward steadiness. A suitable method for choosing the weighting function was the Petrov-Galerkin Method, in which the weighting functions were chosen as follows:

$$W = N + \frac{a.h}{\|U\|}\left(u\frac{\partial N}{\partial x} + v\frac{\partial N}{\partial y} + w\frac{\partial N}{\partial z}\right) \qquad (6)$$

where, W and N are weighting functions and shape functions vectors, respectively. a is the upwind parameter, chosen with respect to the Pe (here, Re). Its optimum value was estimated from the following equation:

$$a_{optimum} = \coth Re - \frac{1}{Re} \qquad (7)$$

Because of the presence of the pressure term in the momentum equation and its lack of presence in the continuity equation, some unsteadiness would occur. In a method, known as Mixed Formulation[12], pressure would be estimated along with velocity. In fact, for each node, there would be four variables (u, v, w and p, except k, ε and T). Another method not having the problem of unsteadiness, is known as the Penalty Method[12], in which the continuity equation is taken as a constraint for the governing equations and the pressure term is omitted from the momentum equations. In this method, the pressure is calculated as follows:

$$p = \lambda \nabla . V \qquad (8)$$

Where, λ is the penalty coefficient obtained from the following equation:

$$\lambda = Re.10^{9\sim12} \qquad (9)$$

The last step in the present finite element numerical method, employing brick elements having one node on each corner, was the solution of a system of algebraic equations, carried out by the Frontal Method.[13] One of the advantages of this algorithm was that it required a minimum amount of RAM due to its use of a smaller size of the coefficient matrix, in comparison with other existing methods.

COMPUTER CODE VERIFICATION

In order to test the validity of the computer code, both the laminar and turbulent boundary layer equations at Re=6700 and Re=2×10^6 (based on wall length), in the absence of injection, were solved. The FEM results for laminar flow were compared with the Blasius (exact) solution and the results for turbulent flow were compared with the empirical 1/7th power-law[14]. Fig. 2 shows good agreements for both the laminar and the turbulent flow comparisons, where the normalized vertical distance from the wall (x/L, where L is a characteristic length) versus the normalized velocity profiles (w/w_∞) are presented.

The code could then be used for the case of film cooling once the boundary conditions, according to Fig. 1, were applied. As an example, for a case of considering 1024 eight-node brick elements with a number of less than 200 iterations within an approximate time period of 2 hours, the difference between successive iterations met the prescribed criterion of less than 0.001. It should be noted that the results obtained were all grid - independent.

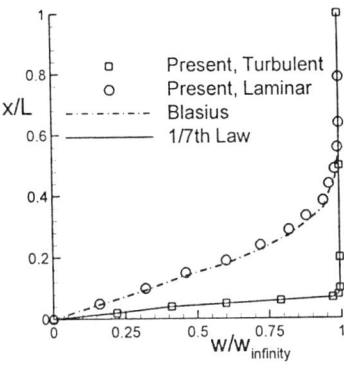

Fig.2 Comparisons of the present numerical results with those of Blasius (exact) solution and with the empirical 1/7 th power-law.

RESULTS and DISCUSSION

The axial variations of adiabatic film cooling effectiveness, η, at a fixed blowing rate value of 0.5 and along the central plane of flow (y/D=0), for a wide range of coolant injection angles, are illustrated by Fig. 3. The axial distance was normalized by the jet hole diameter, D. It can be seen that although η decreased considerably downstream of injection, but at $\theta = 35°$ some better results were demonstrated. The experimental results also verified the same angle of injection[15,16]. Fig. 4 shows that, under the same conditions as Fig. 3, by moving away from the hole in the lateral direction (y/D=0.75), η reduced significantly from a peak value of nearly 0.95 to 0.3, at z/D=0. But, $\theta = 35°$ remained as the optimum angle of injection.

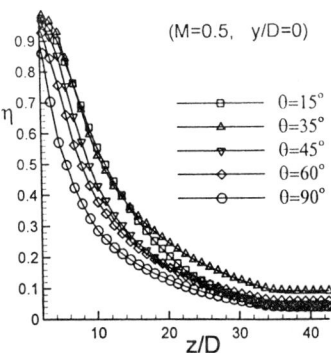

Fig.3 Axial variations of cooling effectiveness, at a fixed M, for a range of injection angles.

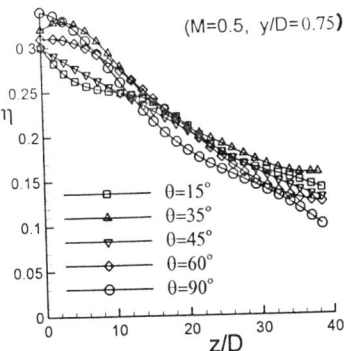

Fig.4 Axial variations of cooling effectiveness, at a fixed M, for a range of injection angles.

The significant decrease in η as a function of the lateral and axial directions, y/D and z/D, mentioned in Figures 3 and 4, was further emphasized by Fig. 5, where M and θ were fixed at values 0.5 and 35°, respectively. Fig 5 could be used to estimate roughly the spacing between adjacent holes in the single row film cooling, being around 2D to 3D.

The axial variations of η, at a fixed θ = 35° and y/D=0, for a wide range of blowing rates (0.2-1.0), are presented by Fig.6. It can be noticed that the best results for η occurred at M=0.5, which was hence considered as the optimum value for coolant injection into the mainstream. This value agreed with the experimental results reported in the past[1,16]. For higher M (toward 1.0), jet penetration occurs more in the vertical direction, x, failing to protect the wall surface against hot gas mainstream.

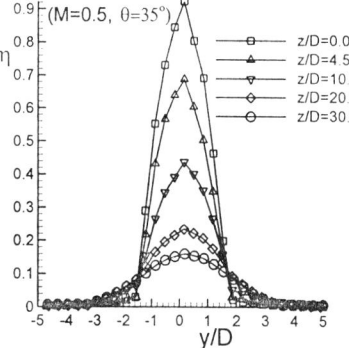

Fig.5 Lateral variations of cooling effectiveness, at a fixed θ and M, for a range of z/D.

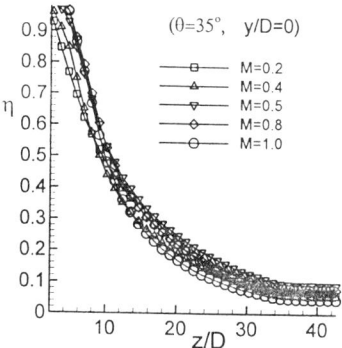

Fig.6 Axial variations of cooling effectiveness, at a fixed θ and y/D, for a wide range of blowing rates.

Fig.7 illustrates the comparison of present data on η with the experimental data of Goldstein et al.[16], at the optimum M and θ, for various lateral positions (y/D). The Reynolds number in both cases, defined on the basis of hole diameter, was equal to 58000. It can be seen that the agreement between the data was acceptable downstream from injection.

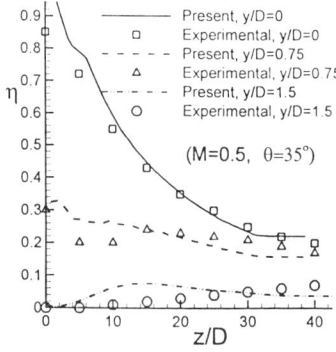

Fig.7 Comparison of present cooling effectiveness data with experimental data of Goldstein et al[16].

This was particularly the case for the data on the central plane, or on the injection hole axis (y/D=0). As y/D increased, the difference between data became larger. This could be due to the increase in the distance between adjacent nodes and hence an increase in the size of elements, in the lateral direction. This, however, could not be very significant as the average difference between the data increased from 7% at y/D=0 to approximately 30% at y/D=1.5.

Figures 8 and 9 show the variations of velocity and temperature with distance from the wall (all normalized), at a fixed M, θ and y/D, for various z/D. From Fig. 8, it can be noticed that velocity tended to approach that of the mainstream as x/D was increased. Beyond x/D=4, the velocity has reached its maximum limiting value. This figure also clearly demonstrates the velocity boundary layer development as z/D was increased. From Fig.9 it can be seen that by getting closer to the hole (i.e., as z/D was reduced), the adiabatic wall temperature (at x/D=0) was decreased. This indicated the presence of coolant injection from the hole. Also, at various z/D, the temperature in the vicinity of the wall (x/D<1) remained unchanged, showing the zero heat transfer situation at the wall (i.e., an adiabatic wall).

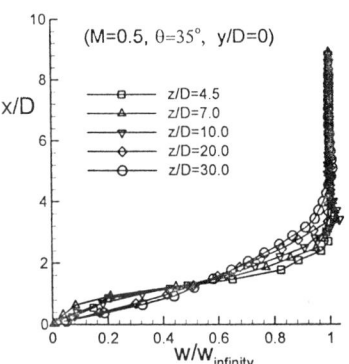

Fig.8 Velocity variations with normalized distance from the wall, at various z/D.

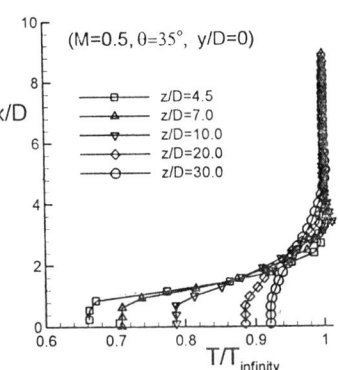

Fig.9 Temperature variations with normalized distance from the wall, at various z/D.

CONCLUSIONS

1. In the present work, the optimum film cooling performance was shown to be at a 35° angle of coolant injection and at a blowing rate of 0.5. This was in full agreement with the available experimental data in the literature.

2. The present cooling effectiveness data was compared with the experimental data of Goldstein et al.[16] and reasonable agreement was shown. Therefore, it is concluded that the present numerical technique, FEM, was capable of predicting the flowfield near and downstream of an injection hole.

REFERENCES

1. Goldstein, R.J., V.L. Eriksen & J.W. Ramsey. 1978. Flow and Temperature Fields Following Injection of a Jet Normal to a Cross Stream. In 6th International Heat Transfer Conference. Vol. 5: 255-260.
2. Bazdidi-Tehrani, F. & G.E. Andrews. 1994. Full-Coverage Discrete Hole Film Cooling: Investigation of the Effect of Variable Density Ratio. In ASME Journal of Eng. for Gas Turbines and Power. Vol. 116 : 587-596.
3. Schmidt, D.L., B. Sen & D.G. Bogard. 1996. Film Cooling with Compound Angle Holes: Adiabatic Effectiveness. In ASME Journal of Turbomachinery. Vol. 118: 807-813.
4. Schmidt, D.L., B. Sen & D.G. Bogard. 1996. Film Cooling with Compound Angle Holes: Heat Transfer. In ASME Journal of Turbomachinery. Vol. 118: 800-806.
5. Kadotani, K. & R. J. Goldstein. 1979. Effect of Mainstream Variables on Jets Issuing from a Row of Inclined Round Holes. In ASME Journal of Engineering for Power. Vol. 101: 298-304.
6. Bazdidi-Tehrani, F. & A. A. Mahmoodi. 2000. Investigation of the Effect of Turbulence Intensity and Injection Angle on the Flow and Temperature Fields in the Single Hole Film cooling Technique. To be printed in the International Journal of Engineering Sciences. Iran University of Science & Technology. Iran.
7. Walters, D. K. & J. H. Leylek. 1997. A Systematic Computational Methodology Applied to Three Dimensional Film-Cooling Flowfield. In ASME Journal of Turbomachinery. Vol. 119 : 777-785.
8. Garg, V.K. & R.E. Gaugler. 1997. Effect of Velocity and Temperature Distribution at the Hole Exit on Film cooling of Turbine Blades. In ASME Journal of Turbomachinery. Vol. 119: 343-351.
9. Taeibi-Rahni, M. & H.R. Ebrahimi-Kebria. 1999. Turbine Blade Film Cooling Numerical Simulations. In CFD99 Proceedings. Vol. 5 : 35-40. Canada.
10. Jansson, S., L. Davidson & E. Olsson. 1992. Film Cooling of Gas Turbine Combustion Chamber Walls: A Numerical Study Using a Two Layer Algebraic Stress Model. Chalmers University of Technology. Department of Thermo & Fluid Dynamics. Sweden.
11. Zienkiewicz, O.C. & R.L. Taylor. 4th Edition. 1989. The Finite Element Method. Vol. 2. McGraw-Hill. New York.
12. Hughes, Thomas J. R., 1987. The Finite Element Method. Prentice-Hall International. London.
13. Cheung, Y. K. & M.F. Yeo. 1980. A Practical Introduction to Finite Element Analysis. Pitman Publishing inc.
14. Fox, R. W. & McDONALD, A. T. 4th Edition. 1994. Introduction to Fluid Mechanics. John Wiley & Sons. New York.
15. Rohsenow, W. M., P. Hartnett & E. N. Ganic. 2nd Edition. 1985. Handbook of Heat Transfer Applications. McGraw-Hill. New York.
16. Goldstein, R.J., E. R. G. Eckert & J.W. Ramsey. 1968. Film Cooling with Injection Through Holes: Adiabatic Wall Temperatures Downstream of a Circular Hole. In Journal of Engineering for Power. Transactions ASME. pp. 384-395.

Effects of Entrance Crossflow Directions to Film Cooling Holes

CHRISTIAN SAUMWEBER, ACHMED SCHULZ, SIGMAR WITTIG,
MICHAEL GRITSCH[A]
Institut für Thermische Strömungsmaschinen
Universität Karlsruhe, 76128 Karlsruhe, Germany

ABSTRACT: Two-dimensional distributions of local adiabatic film cooling effectiveness as well as discharge coefficients have been measured to investigate the effect of different entrance crossflow orientations and magnitudes on film-cooling performance. Operating conditions have been varied in terms of hot gas Mach number (up to 0.6), coolant crossflow Mach number (up to 0.6), coolant crossflow orientation (perpendicular or parallel with respect to the mainflow), and blowing ratio (0.5 - 1.5). The temperature ratio of coolant and hot gas was kept constant at 0.56 for the effectiveness tests, leading to an enginelike density ratio of 1.8. Infrared thermography was applied to perform local measurements of the surface temperatures with high resolution. The results indicate that the impact of hot gas crossflow Mach number is not very pronounced within the range of Mach numbers investigated. In contrast to this finding, the effect of internal coolant crossflow is very pronounced and strongly depends on coolant crossflow orientation and the ejected mass flow rate.

NOMENCLATURE

C_D	discharge coefficient	$T_{rec,m}$	hot gas recovery temperature
D	hole diameter	T_{tc}	total coolant temperature
h_0	heat transfer coefficient, no ejection	T_w	wall temperature
L	film-cooling hole length	u	streamwise velocity component
M	blowing ratio	v_{jet}	mean velocity inside cooling hole
Ma_c	Mach number of coolant crossflow	x	streamwise coordinate
Ma_m	Mach number of external hot gas flow	y	lateral or vertical coordinate
$\dot{q}_{rad}$	radiative heat flux	z	lateral coordinate
$\dot{q}_w$	wall heat flux	η	adiabatic film cooling effectiveness
T_{aw}	adiabatic wall temperature		

INTRODUCTION

Besides simple cylindrical holes, film cooling holes with expanded exits have received an increased attention during recent years since they offer certain advantages which definitely improve film cooling efficiency. The enlarged cross-sectional area at the hole

A: ALSTOM Power Ltd., 5405 Baden-Dättwil, Switzerland

exit leads to a decreased mean velocity and thus a decreased momentum flux of the exiting coolant jet, resulting in a reduced penetration of the coolant into the hot gas flow. Therefore the interaction and the mixing of coolant and hot gas is less intense and the cooling efficiency increases. Furthermore, expanding the holes in lateral direction provides an improved lateral spreading of the jet, leading to a better coverage of the surface with coolant and higher laterally averaged film cooling effectiveness[1-4].

Most film cooling studies in the past have been performed using a stagnant plenum feeding the film cooling holes. A plenum, however, is not necessarily a correct means of representing the internal coolant supply passages of an airfoil. Especially in film cooling applications of turbine blades, crossflow velocities with Mach numbers up to $Ma_c=0.7$ may occur. Besides the coolant crossflow's magnitude its orientation has to be taken into account as well. For film cooling of turbine blades the coolant supply crossflow is typically oriented in radial direction of the blades and thereby perpendicular to the main flow, whereas for turbine vanes coolant in the supply passages and main flow might be oriented parallel to each other. Even for extremely low coolant crossflow velocities, the coolant supply direction might affect the surface adiabatic effectiveness[5]. Discharge coefficient measurements showed, that the throughflow characteristic of the hole is drastically altered when the coolant approach is perpendicular to the main flow[6,7]. Film cooling effectiveness tests in the close vicinity of a single hole revealed significant impacts of a perpendicular coolant supply direction as typically used in turbine blades[8]. However, the effect of internal coolant supply crossflow on film cooling performance has not yet been studied in detail sufficiently, particularly at elevated coolant crossflow Mach numbers and for rows of film cooling holes.

The objective of the present study therefore is to investigate to what extent coolant crossflow affects the performance of cylindrical and fanshaped holes. Results of a comprehensive experimental film-cooling investigation conducted at the University of Karlsruhe are presented. Adiabatic wall temperatures have been measured for a wide range of enginelike operating conditions. Highly resolved two-dimensional mappings of adiabatic film cooling effectiveness up to 22 hole diameters downstream of a row of holes are used to discuss the different effects of coolant crossflows magnitude and orientation.

EXPERIMENTAL APPARATUS AND MEASUREMENT TECHNIQUE

The present investigation was carried out in a continuous flow wind tunnel at the Institut für Thermische Strömungsmaschinen, University of Karlsruhe, Germany. The film-cooling test rig consists of a primary loop, representing the external flow, and a secondary loop, representing the internal flow of an airfoil. Primary and secondary loop were oriented parallel or perpendicular to each other, providing flow conditions typically found in turbine vanes or blades, respectively (Fig. 1). A more detailed description of the experimental facility is given by Wittig et al.[9].

Two different hole geometries have been considered. They comprise a cylindrical hole as reference case, and a fanshaped hole which is expanded in lateral direction (Fig.1). For both hole geometry the hole axis was inclined 30deg with respect to the hot gas flow and the length-to-diameter-ratio L/D was 6. The interior surfaces were aerodynamically smooth. Discharge coefficient measurements have been carried out using single scaled up film cooling holes with a hole diameter of 10mm. A detailed description of the discharge coefficient measurement technique and the applied data analysis is presented in Gritsch et al.[10]. For the effectiveness tests hole inserts with three scaled up holes in a row and a hole diameter of 5mm have been used. The hole pitch to diameter ratio was kept constant at 4, the temperature ratio was fixed at 0.56, leading to an enginelike density ratio of 1.8. Operating conditions have been varied in terms of hot gas Mach number (0.3, 0.6),

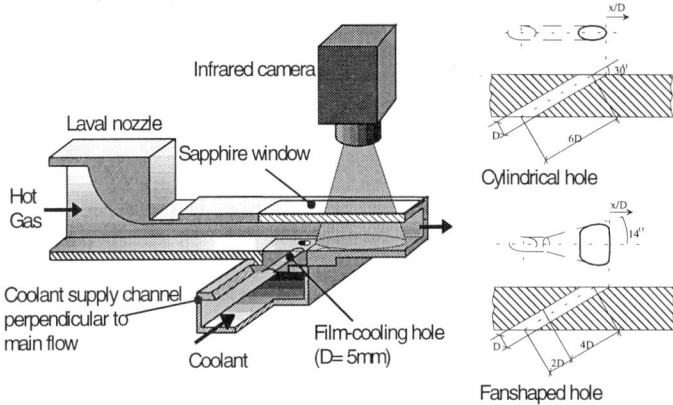

Fig. 1: Film cooling test section and hole geometry tested

coolant crossflow Mach number (0.0, 0.3, 0.6), coolant crossflow orientation (perpendicular or parallel with respect to the main flow), and blowing ratio (0.5–1.5). For the fanshaped hole the calculation of the blowing ratio is based on the cross-sectional area at the inlet. Thus, same blowing ratio is equivalent with same mass flow rate ejected through the film-cooling holes, which makes it more convenient to compare the different geometry and to evaluate the effect of expanding the hole exit.

To measure adiabatic film cooling effectiveness it is required to realize adiabatic thermal boundary conditions. The test plate used for these experiments was made from the high temperature thermoplastic material TECAPEEK, which has a very low thermal conductivity of 0.3 W/(mK) and a maximum operating temperature of about 570 K. Surface temperatures have been measured using an infrared camera system (AGEMA 900) that consists of a scanner, which is cooled by a Stirling motor, and a PC with the associated software to control the camera and to store the acquired thermographic data. The infrared camera provided a two-dimensional mapping of the temperatures that was digitized into an array of 272 x 136 pixels. Accounting for the optical setup used in the experiments this resulted in a spatial resolution of 0.8mm x 0.8mm per pixel. The surface of the test plate exposed to the hot gas was covered by a black paint with a well-known emissivity of 0.95. Seventeen 0.5mm NiCr-Ni-thermocouples have been embedded flush with the surface in order to recalibrate the infrared camera readings. Another fifteen 0.5mm NiCr-Ni thermocouples were mounted flush to the back of the test plate in order to control the remnant heat flux through the material. To account for radiative as well as conductive heat fluxes the measured surface temperatures from the calibrated IR camera images were corrected in the following manner[11]:

$$T_{aw} = T_w + \frac{\dot{q}_w - \dot{q}_{rad}}{h_0} \quad (1)$$

The radiative heat flux can be approximated using a simplified enclosed body approach. The conductive heat flux through the test plate is required to perform the correction of equation (1). Therefore the measured surface temperatures were interpolated onto a finite element (FE) model of the test plate. The model extended half a pitch to either side of the symmetry line of the test section to the midspan line between the center hole and each of the adjacent holes. A steady state thermal analysis was performed using the commercial FE code ABAQUS in order to calculate the three dimensional heat flux distribution. The correction according to equation (1) was performed for all nodes of the FE modell belonging to the surface exposed to the hot gas in order to finally acquire two-dimensional distributions of the corrected adiabatic film cooling effectiveness, which is given by equation (2):

$$\eta = \frac{T_{aw} - T_{rec,m}}{T_{tc} - T_{rec,m}} \quad (2)$$

RESULTS

In Fig. 2 a comparison of cylindrical and fanshaped holes in terms of discharge coefficients is shown. The orientation of the coolant crossflow was either parallel or perpendicular with respect to the hot gas flow. The coolant crossflow Mach number ranged from 0 to 0.6. No external hot gas flow was present for these tests. As expected, there is no effect of the coolant channel orientation if no coolant crossflow is present and the two lines for the same operating conditions basically collapse on the same curve. For the parallel orientation of the coolant channel and a given pressure ratio, an internal Mach number exists for which maximum discharge coefficients occur[12]. In case of a perpendicular coolant crossflow orientation, no such optimum internal Mach number exists since an increase of internal Mach number always results in decreased discharge coefficients. This fact can be explained by the existence of a separation region at the entry of the film-cooling hole with size and location strongly depending on internal crossflow conditions[13]. In case of parallel coolant crossflow orientation with plenum conditions, the separation region occurs at the downstream (with respect to the external crossflow) edge of the cooling hole. If the internal Mach number is increased, a velocity component in the direction of the hole axis is present and the size of the aforementioned separation region will decrease. A second separation zone will form at the upstream edge of the hole. For high internal Mach numbers the separation region will be found at the upstream edge only, its size increasing with internal Mach number. Between those two extremes, zero and high internal crossflow Mach number, there is a medium Mach number for which the overall size of the separation regions is at a minimum, resulting in minimum losses and maximum discharge coefficients. In case of a perpendicular orientation of coolant crossflow and hole axis, no velocity component of the internal crossflow in the direction of the hole axis exists and the coolant has to turn 90deg to enter the hole. A separation

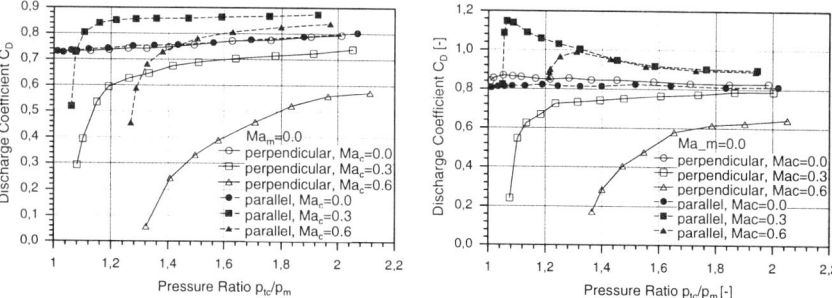

Fig. 2. Effect of coolant crossflow orientation on discharge coefficients; left: cylindrical hole; right: fanshaped hole

zone at the upstream edge of the hole is therefore always present and increases in size with the internal Mach number, resulting in increased pressure losses at the hole entry and therefore decreased discharge coefficients. A comparison of the two hole geometry shows a much larger dependence on coolant crossflow orientation for the fanshaped hole.

The results of the effectiveness tests are presented as surface contour plots of the corrected adiabatic film cooling effectiveness. For all figures the origin of the coordinate system is located at the holes trailing edge on centerline. Only the center hole is considered. In Fig. 3 effectiveness contours for three different blowing ratios and two hot gas Mach numbers downstream of the cylindrical holes are shown. The cooling holes have been fed by a stagnant plenum, which can be considered to be a reference case that has been widely reported in the literature. The well-known features of a cylindrical hole with an inclination of 30deg can be clearly recognized. When the blowing ratio is increased, the jet exhibits an increased tendency to lift-off and the overall effectiveness is decreased. Obviously the change in hot gas Mach number does not affect the effectiveness distributions significantly. For low and medium blowing ratios the area of intense cooling becomes slightly longer in streamwise direction when the hot gas crossflow Mach number is raised.

A stronger impact on the effectiveness patterns could be observed when a coolant crossflow at the hole entry was established, which was oriented parallel to the hot gas flow, Fig. 4. For the cylindrical hole the area of intense cooling is increased and the maximum is shifted in streamwise direction further downstream. In case of the fanshaped hole a benefit due to the presence of parallel coolant crossflow could be detected as well since, like for the cylindrical hole, the area of intense cooling is increased. This tendency can be explained by the results of some LDV measurements for the cylindrical hole under different parallel coolant crossflow conditions. Fig. 5. shows the axial velocity component inside the hole. It is obvious, that an increased parallel coolant crossflow Mach number shifts the velocity maximum towards the lower edge of the hole due to the change of size and location of the separation zones at the hole entry. Therefore the

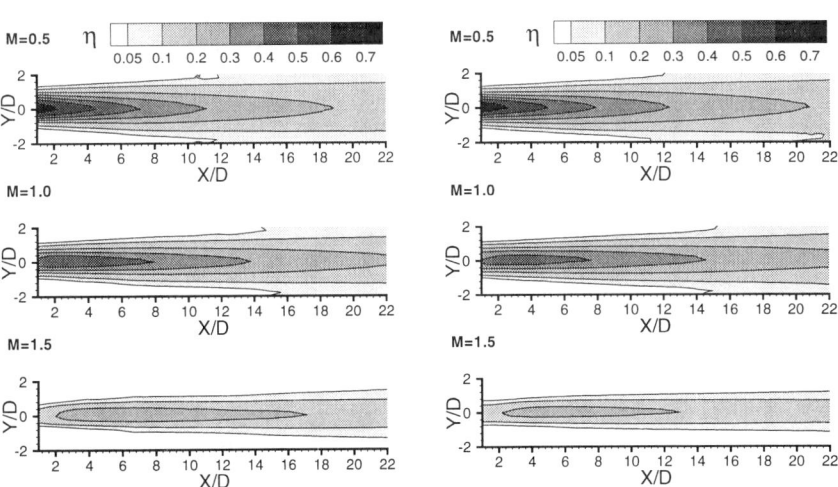

Fig. 3. Effect of hot gas Mach number on effectiveness without internal coolant crossflow ($Ma_c=0.0$); left: $Ma_m=0.3$; right: $Ma_m=0.6$

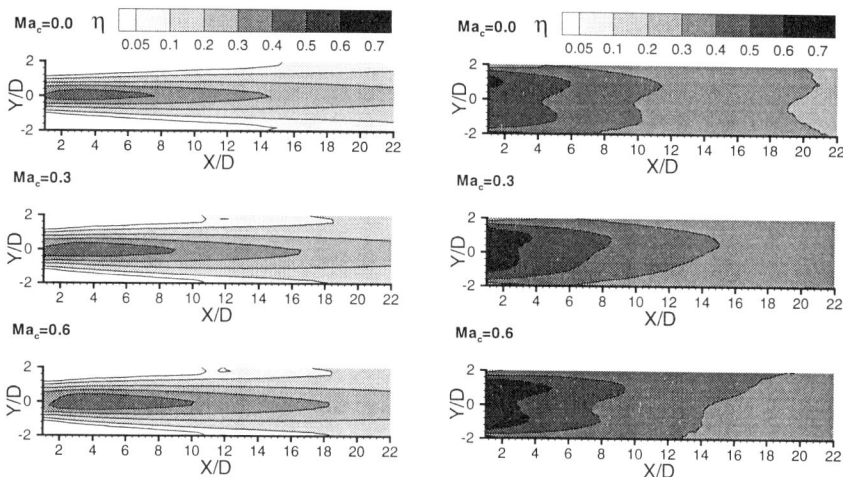

Fig. 4. Impact of parallel coolant crossflow at $Ma_m=0.6$ and $M=1.0$; left: cylindrical holes; right: fanshaped holes

coolant jet will be shifted in streamwise direction and exit the hole under a rather shallow angle. This leads to decreased mixing with the hot gas and a larger intense cooled area as just described above.

The effect of a perpendicular coolant crossflow is summarized in Fig. 6. For the cylindrical hole a pronounced reduction of the effectiveness levels compared to plenum conditions was observed with a coolant crossflow Mach number of 0.3. The perpendicular coolant crossflow causes a single swirling motion inside the cooling hole which forces the jet to exit the hole slighly skewed, resulting in a non-symmetric coolant distribution[14]. For the same blowing ratio however, an even larger coolant crossflow Mach number resulted in a benefit as compared to plenum conditions. The jet is kept closer to the wall and the lateral spreading is improved as well.

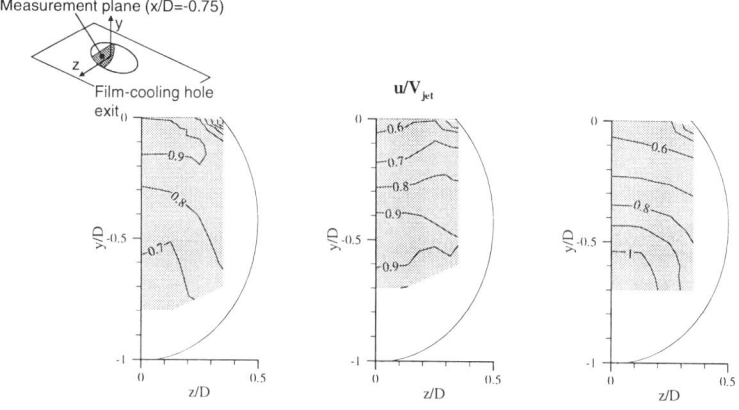

Fig. 5. Impact of parallel coolant crossflow at $Ma_m=0.6$ and $M=1.0$ on the distribution of the axial velocity component. Left to right: $Ma_c=0.0$, $Ma_c=0.3$, $Ma_c=0.5$

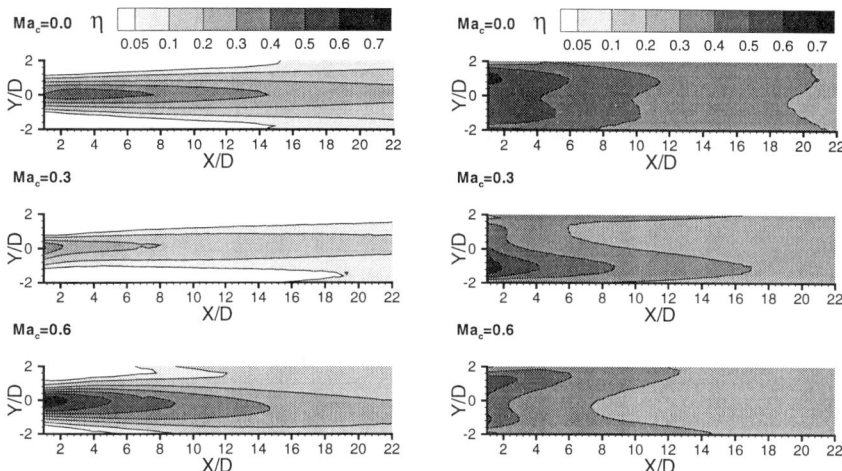

Fig. 6. Impact of perpendicular coolant crossflow at M=1.0; left: cylindrical holes; right: fanshaped holes

For the fanshaped hole however the effect of a perpendicular coolant crossflow is always detrimental. The flow entering the expanded part is severely disturbed and the diffuser will not work properly. The single vortex caused by the perpendicular coolant crossflow will break down in the expanded portion[15]. This leads to flow separation inside the diffuser part and a reduced lateral spreading of the coolant. The coolant distribution becomes non-symmetric across the span with increased temperature gradients in lateral direction. Additionally, the location of maximum effectiveness depends on coolant crossflow Mach number applied.

CONCLUSIONS

An experimental study was performed to investigate the effect of different coolant crossflow conditions at the entry to a film cooling hole. Operating conditions have been varied with respect to hot gas Mach number, coolant crossflow Mach number, coolant crossflow orientation and blowing ratio. Two different hole geometry have been considered, a cylindrical hole and a fanshaped hole. The following conclusions can be drawn:
- The effect of internal coolant crossflow strongly depends on coolant crossflow orientation, hole geometry and blowing ratio. Fanshaped holes are in general more sensitive to coolant crossflow effects, especially when the coolant crossflow is oriented perpendicular to the hole axis.
- For the cylindrical holes, the lateral spreading of the coolant is in general improved by the presence of a coolant crossflow – either parallel or perpendicular oriented with respect to the hot gas flow. With parallel coolant crossflow applied, the effectiveness peak is shifted downstream and the intense cooled area is stretched in streamwise direction. Perpendicular coolant crossflow might cause a delay of jet detachment and keep the coolant closer to the wall.
- For the fanshaped holes, film cooling effectiveness is increased in case of medium blowing ratios when a parallel coolant crossflow is present. In contrast to the

cylindrical holes, perpendicular coolant crossflow does have a detrimental effect on film cooling effectiveness for fanshaped holes. The coolant entering the expanded section is highly disturbed causing a poor performance of the diffuser. The effectiveness contours downstream the hole are skewed, leading to steep temperature gradients in lateral direction and reduced overall effectiveness.

ACKNOWLEDGEMENTS

This study was partly funded by the European Union through grant by the BRITE EURAM programs "Turbine Aero-Thermal External Flows" under Contract No. BRPR-CT97-0519 and "Investigation of the Aerodynamics and Cooling of Advanced Engine Turbine Components" under Contract No. AER2-CT92-0044.

REFERENCES

1. Goldstein, R.J., Eckert, E., Burggraf, F. 1974, Effects of Hole Geometry and Density on Three-Dimensional Film Cooling, Int. J. Heat Mass Transfer, Vol. 17, pp. 595-607.
2. Makki, Y., Jakubowski, G. 1986, An Experimental Study of Film Cooling from Diffused Trapezoidal Shaped Holes, AIAA Paper 86-1326.
3. Schmidt, D., Sen, B., Bogard, D. 1996, Film Cooling with Compound Angle Holes: Adiabatic Effectiveness, ASME Journal of Turbomachinery, Vol. 118, pp. 807-813.
4. Gritsch, M., Schulz, A., Wittig, S. 1998, Adiabatic Wall Effectiveness Measurements of Film Cooling Holes With Expanded Exits, ASME Journal of Turbomachinery, Vol. 120, pp. 549-556.
5. Burd, S.W., Simon, T.W. 1997, The Influence of Coolant Supply Geometry on Film Coolant Exit Flow and Surface Adiabatic Effectiveness, ASME Paper 97-GT-25.
6. Hay, N., Lampard, D., Benmansour, S. 1983, Effect of Crossflows on the Discharge Coefficient of Film Cooling Holes, ASME Journal of Engineering for Power, Vol. 105, pp. 243-248.
7. Gritsch, M., Saumweber, C., Schulz, A., Wittig, S., Sharp, E. 2000, Effect of Internal Coolant Crossflow Orientation on the Discharge Coefficient of Shaped Film Cooling Holes, ASME Journal of Turbomachinery, Vol. 122, pp. 146-152.
8. Gritsch, M., Schulz, A., Wittig, S. 1999, The Effect of Internal Coolant Crossflow on the Effectiveness of Shaped Film-Cooling Holes, Proceedings of the 33rd National Heat Transfer Conference, Albuquerque, New Mexico, August 15-17, Paper NHTC99-8.
9. Wittig, S., Schulz, A., Gritsch, M., Thole, K. 1996, Transonic Film-Cooling Investigations: Effects of Hole Shapes and Orientations, ASME Paper 96-GT-222.
10. Gritsch, M., Schulz, A., Wittig, S. 1998, Method for Correlating Discharge Coefficients of Film-Cooling Holes, AIAA-Journal, Vol. 36, June 1998, pp. 976-980.
11. Baldauf, S., Schulz, A., Wittig, S. 1999, High Resolution Measurements of Local Effectiveness by Discrete Hole Film Cooling, ASME Paper 99-GT-45.
12. Gritsch, M., Schulz, A., Wittig, S. 1998, Discharge Coefficient Measurements of Film Cooling Holes With Expanded Exits, ASME Journal of Turbomachinery, Vol. 120, pp. 560-567.
13. Thole, K. A., Gritsch, M., Schulz A., Wittig S. 1997, Effect of a Crossflow at the Entrance to a Film-Cooling Hole, ASME Journal of Fluids Engineering, Vol. 119, pp. 533-541.
14. Kohli, A., Thole, K.A. 1997, A CFD Investigation on the Effects of Entrance Crossflow Directions to Film-Cooling Holes, Proceedings of the 32nd National Heat Transfer Conference, Baltimore, MD, Aug. 10-12.
15. Kohli, A., Thole, K.A. 1998, Entrance Effects on Diffused Film Cooling Holes, ASME Paper 98-GT-402.

Mach Number Effect On Jet Impingement Heat Transfer

P. BREVET, E. DORIGNAC, J.J. VULLIERME

Laboratoire d'Etudes Thermiques (UMR CNRS 6608)
ENSMA, Teleport 2, BP 40109, 86961 FUTUROSCOPE Cedex, FRANCE

ABSTRACT: An experimental investigation of heat transfer from a single round free jet, impinging normally on a flat plate is described. Flow at the exit plane of the jet is fully developed and the total temperature of the jet is equal to the ambient temperature. Infrared measurements lead to the characterization of the local and averaged heat transfer coefficients and Nusselt numbers over the impingement plate. The adiabatic wall temperature is introduced as the reference temperature for heat transfer coefficient calculation. Various nozzle diameters from 3 mm to 15 mm are used to make the injection Mach number M vary whereas the Reynolds number Re is kept constant. Thus the Mach number influence on jet impingement heat transfer can be directly evaluated. Experiments have been carried out for 4 nozzle diameters, for 3 different nozzle-to-target distances, with Reynolds number ranging from 7200 to 71500 and Mach number from 0.02 to 0.69. A correlation is obtained from the data for the average Nusselt number.

INTRODUCTION

Impinging jets of air technique is a commonly used method to cool advanced aircraft turbines as it produces relatively large forced-convection heat transfer coefficients. Impingement is a complex function of many parameters: Reynolds number nondimensional nozzle to plate spacing (H/D), radial position from the stagnation point (r/D), nozzle geometry, flow confinement, jet outlet conditions (velocity profile, turbulence...), incidence, entrainment. That's why numerous experimental studies have been conducted to evaluate the flow and heat transfer characteristics associated with jet impingement on surfaces. Goldstein et al.[1] have notably contributed to a large extent to analyse the influence of those parameters and to evaluate the reference temperature which has to be used for heat transfer coefficient calculation. Several state-of-the-art literature reviews on impingement heat transfer were published between 1973 and 1993 and underline the main results of impingement studies (Livingood & Hrycak[2] 1973, Martin[3] 1977, Downs & James[4] 1987, Jambunathan et al.[5] 1992, Viskanta[6] 1993). Although all the studies on impingement heat transfer take into account the Reynolds number influence, they usually don't take into account the Mach number influence. In fact most studies bring in large injection diameters compared to those effectively used in turbojet engine turbine blades in order to obtain good accuracy in the measurements. The Reynolds number similarity is then introduced to link the experiments to the real injection conditions and dimensions of the turbojet engine. But keeping the Reynolds number constant and increasing the jet diameter lead to an important reduction of the injection Mach number in the experimental apparatus. As the Mach number can reach nowadays values as high as 0.4–0.5 in modern turbojet turbines, we have undertaken an experimental study in collaboration with *Snecma Moteurs* to estimate the Mach number effect on the impingement heat transfer coefficient.

EXPERIMENTAL TECHNIQUE AND APPARATUS

Our study is concerned with a circular free air jet impinging on a flat plate. The configuration is similar to the one used by Baughn & Shimizu[7]. The work of these authors is nowadays considered widely as a reference (especially for code validation) in so far as it deals with a rather simple geometry for which boundary conditions are well-known (single round free jet, fully-developped velocity profile, impingement on a flat plate). Baughn & Shimizu[7] made some heat transfer measurements and Cooper et al.[8] made some velocity and turbulence measurements for the same configuration which gives a complete set of data for one injection condition (Re=23750, M=0.04). The method we introduced here to investigate the Mach number influence is based on the use of four injection tubes with different interior diameters (D): 3 mm, 5 mm, 10 mm and 15 mm. It enables the evaluation of four Nusselt number distributions over the impingement plate with four identical injection Reynolds numbers but with four different injection Mach numbers. Comparisons including the Mach number influence are then directly possible. Each injection tube is tested over the same Reynolds number range: from Re=7.5 10^3 to Re=7.1 10^4 (except for the smallest tube for which the maximum Reynolds number is 5 10^4) which induces Mach numbers from 0.02 to 0.69 depending on the jet diameter. Three different jet-to-plate spacings are tested: H/D=2–5 and 10.

The test system we used is shown on Fig. 1. The injected flow is supplied by the building compressor. It is heated through the medium of an electric heater before going through a venturi. The jet flow is heated so that the jet total temperature is equal to the ambient temperature. The venturi gives the mass flow rate with the aid of a differential pressure converter (JUMO 4309-010/53). The venturi was designed to work in the incompressible range for all the considered tests and its discharge coefficient was measured over a wide range of mass flow rates. Flowmeters are used in conjunction with the venturi for very low flow rates for which the venturi is not perfectly adapted. Several Type-K-thermocouples measure the flow temperature along the air duct. A thermocouple for which the recovery factor was calibrated provides the jet center temperature at the jet exit. It is removed just before each test run not to interfere with the flow when temperature measurements are made on the test plate.

Injection tubes are pipes long enough (30D) to obtain a fully developed flow at the jet exit. This was confirmed by numerical simulations from both a commercial code and a research code. The turbulence intensity at the jet exit has been measured for each injection condition with the hot wire technique. Its variation range is 1% to 4%. Each injection tube has its own impingement test plate.

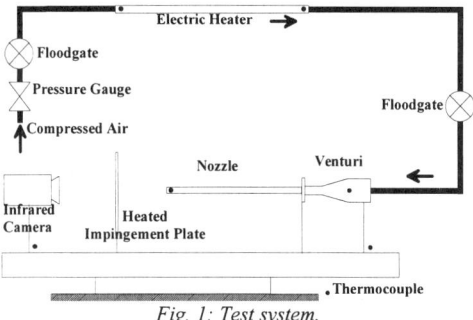

Fig. 1: Test system.

The test plates are circular and their diameter is proportional to the jet diameter (20D). They are made of epoxy resin (e=1.6 mm thick) which thermal conductivity was measured with the laboratory facilities (k_w=0.320±2.5% W/mK). An electric copper circuit (17.5 μm thick) is printed on each plate. It is located on the impingement side of the plate. It heats the plate by Joule effect. Its local resistivity has been calculated from the circuit design and experimentally checked. A correction is introduced to take the variation of resistivity with temperature into account during the test runs. The circuit is designed to provide a non-uniform axisymmetric heat flux density so that the plate surface temperature is nearly uniform (around 60°C) when it is cooled by the impinging air jet (Fig. 2). Thus radial heat conduction can be neglected. This was confirmed by previous numerical studies using the nodal method to model the conduction in the test plate. Furthermore the difference of temperature between the air and the plate surface can be kept as high as possible all over the plate (including the impingement region) without damaging the plate. It guarantees a very good accuracy on the entire surface in the local heat transfer coefficient calculation. The surface temperature distribution on the plate side opposite the impingement (back side) is measured using an infrared camera (AGEMA Thermovision 880 SWB) cooled by liquid nitrogen. As the temperature of the impingement wall is nearly uniform, a monodimensional conduction calculation is adequate to give the temperature field of the impingement side (front side). This allows a complete mapping of the heat transfer coefficient over the entire impingement plate. Both sides of the plate are painted in black to give the surfaces the high and uniform emissivity required by infrared camera measurements. This emissivity was experimentally calibrated (ε=0.96±0.02).

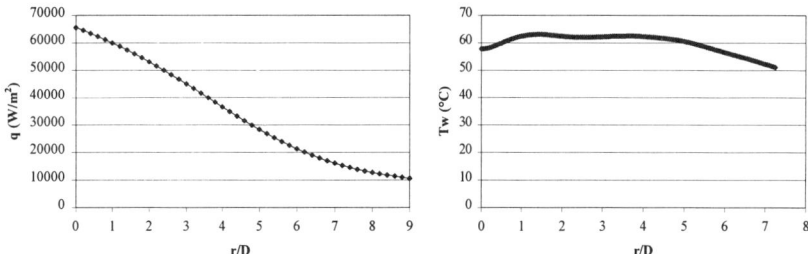

Fig. 2: Example of heat flux density q and wall temperature distribution T_w over the impinged plate.

Each test run is carried out in two parts. The first experiment provides the wall adiabatic temperature T_{aw} distribution by measuring the temperature distribution on the test plate when the wall heat flux is zero. As the tests plate thermal conductivity is low, the adiabatic assumption can be validated. The adiabatic wall temperature will be used as the reference temperature in the heat transfer coefficient calculation. It is indeed the only reference temperature which leads to heat transfer coefficients h which are independent of the boundary condition on the wall (h becomes independent of the heat power injected in the test plate).

The second experiment yields the wall temperature field T_{mes} on the plate back side when the plate is heated with the heat flux density distribution q(r). The wall temperature on the plate front side is then referred to as T_w and the ambient temperature as T_{amb}. The impingement side heat transfer coefficient h and corresponding Nusselt number Nu based on the jet diameter D are calculated taking into account corrections for radiation losses on the plate front side and radiation and convection losses on the plate back side. For these last corrections, a back side overall coefficient h_r is measured when the impingement jet is off. The following three equations gives respectively the expression of the front side temperature **(1)** and the heat transfer coefficient **(2)** and the Nusselt number **(3)**:

$$T_w = T_{mes} + \frac{e}{k_w} h_r (T_{mes} - T_{amb}) \quad (1)$$

$$h = \frac{q - \varepsilon\sigma(T_w^4 - T_{amb}^4) - h_r(T_{mes} - T_{amb})}{T_w - T_{aw}} \quad (2)$$

$$Nu = \frac{hD}{k_{air}} \quad (3)$$

RESULTS

Recovery and entrainment effects appear especially when the jet velocity is high In this case, the jet static temperature T_j becomes drastically lower than the ambient temperature because of the flow acceleration in the venturi. Although the jet total temperature is kept equal to the ambient temperature at the nozzle exit, it increases through mixing in the shear layer region as the "cold" jet develops and as "hot" fluid from the surroundings is entrained into the jet. Furthermore the jet velocity profile have also changed between the jet exit and the impinged plate. That's why the jet temperature profile in the impingement region is completely changed compared to the jet temperature profile at the jet exit. Fig. 3 shows recovery temperature profiles (measured into the jet by a thermocouple) at different distance y from the jet exit and illustrates that phenomenon.

When the jet impinges the plate, part of the total temperature is recovered by the test plate. When the impinged plate is not heated, the temperature recovered is the adiabatic wall temperature. This temperature can be written in a non-dimensionalized form referred to as the recovery factor r_c which is a function of the jet static temperature T_j and the jet total temperature T_j^0 (equation **(4)**). We can notice that recovery factor profiles are not easy to evaluate (Fig. 4). r_c can even be larger than unity when entrainment effects lead to a large increase in the total temperature between the jet exit and the impinged plate, which can happen when the impingement distance is high enough.

$$r_c = \frac{T_{aw} - T_j}{U_j^2/2c_p} = \frac{T_{aw} - T_j}{T_j^0 - T_j} \quad (4)$$

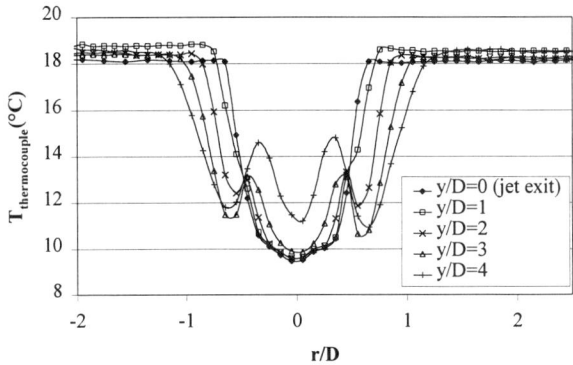

Fig. 3: Recovery temperature profile at different distances y from the jet exit (Re=71000 – M=0.6 – H/D=5 – D=5 mm).

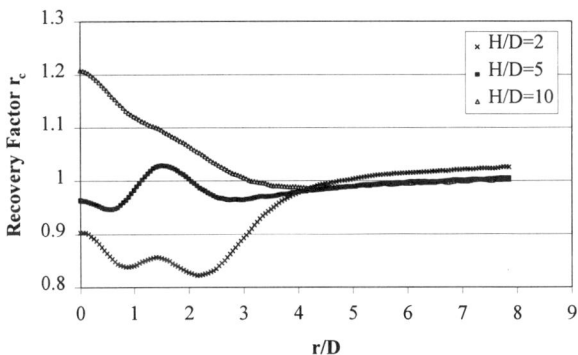

Fig. 4: Example of local recovery factor distribution (Re=71000 – M=0.6 – D=5 mm).

The local Nusselt number distribution have been obtained for each injection configuration (Fig. 5). The radial distribution of Nu has been averaged to give the mean Nusselt number:

$$\overline{Nu}(R) = \frac{2}{R^2} \int_0^R Nu(r) r \, dr \quad (5)$$

Fig. 6 shows the average Nusselt number (obtained from our experimental data for a radial position of 5D) as a function of the Reynolds number Re and the Mach number M. This particular radial position has been chosen for the visualization of the results because the main influence of the impingement on the Nusselt number is contained within that range.

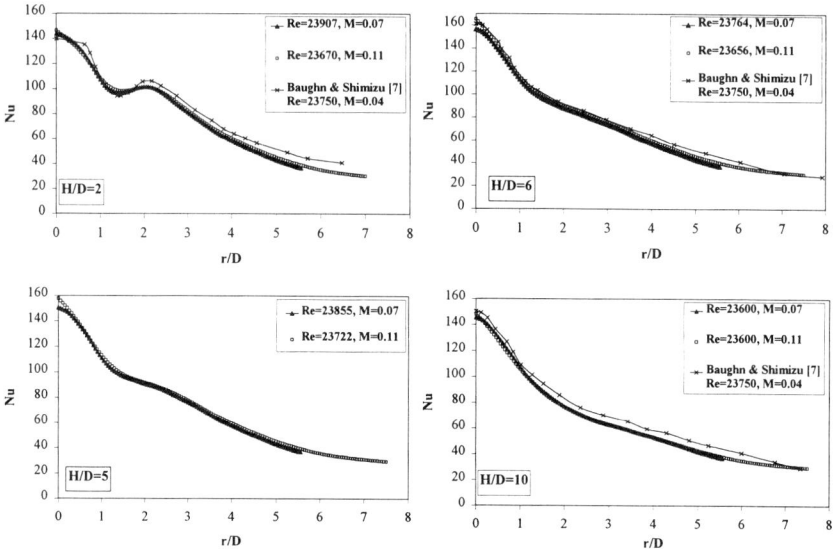

Fig 5: Examples of local Nusselt number distribution along the impingement surface.

A correlation linking the mean Nusselt number (calculated with R/D ranging from 2 to 8) to the Reynolds Number Re, the Mach number M and H/D (non dimensionalized impinging distance) has been computed from our experimental data (equations **(6)** to **(9)**) and fit the data within ±15%. This correlation regroups all the measurements and is plotted on Fig. 6 for the injection diameters D=3mm and 5mm.

$$\overline{Nu(R/D)} = 1.63 \left(\frac{R}{D}\right)^{-1.85} Re^a (1-M)^b \left(\frac{H}{D}\right)^c \quad (6)$$

$$a = 0.0014 \left(\frac{R}{D}\right)^3 - 0.0247 \left(\frac{R}{D}\right)^2 + 0.161 \frac{R}{D} + 0.311 \quad (7)$$

$$b = 0.0025 \left(\frac{R}{D}\right)^3 - 0.0428 \left(\frac{R}{D}\right)^2 + 0.245 \frac{R}{D} - 0.898 \quad (8)$$

$$c = 0.00047 \left(\frac{R}{D}\right)^4 - 0.01153 \left(\frac{R}{D}\right)^3 + 0.1028 \left(\frac{R}{D}\right)^2 - 0.3777 \frac{R}{D} + 0.3794 \quad (9)$$

Test range: 2<H/D<10; 7200<Re<71500; M<0.7

The results (Fig. 6) are consistent with those of Baughn & Shimizu[7] and Hollworth & Gero[9]. Local Nusselt number of Goldstein et al.[10] are slightly lower than the ones measured in the present study. The discrepancies can certainly be attributed to the differences in the experimental techniques and in the injection conditions as Goldstein et al. used an ASME nozzle whereas a long pipe with a fully turbulent flow is used here.

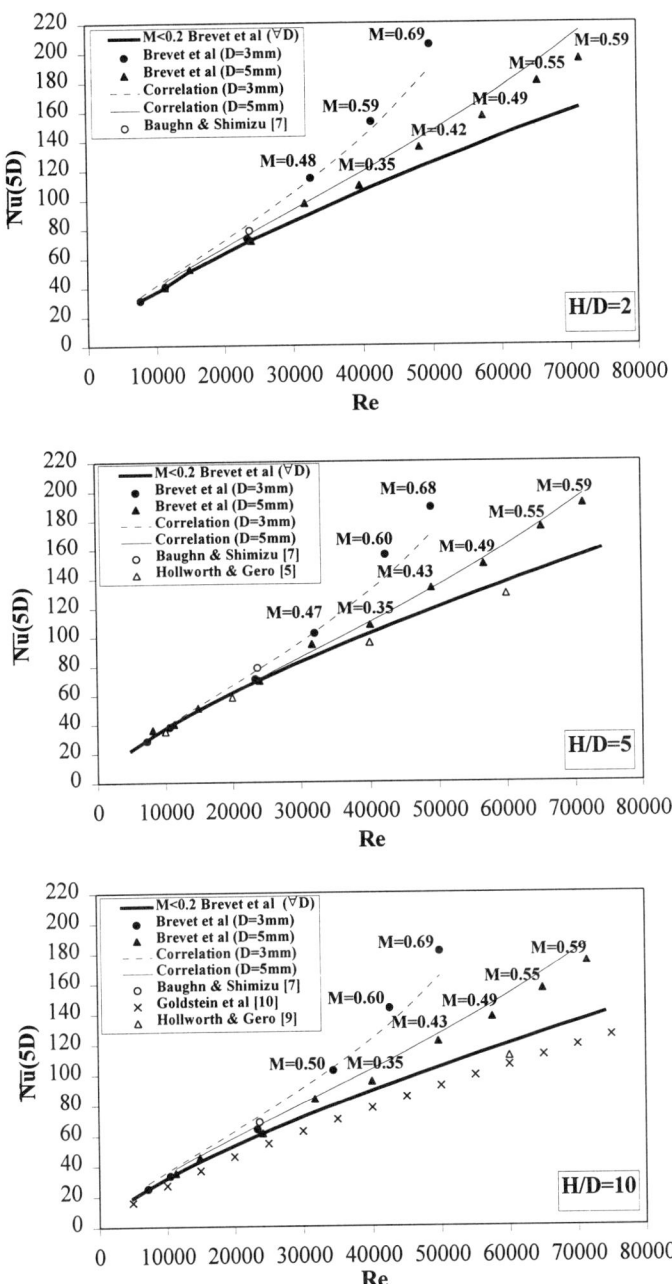

Fig 6: Variation of mean Nusselt number (R=5D) with Reynolds and Mach numbers.

At a given impingement distance H/D it appears that the mean Nusselt numbers fit the same function of Reynolds number Re independently of the jet diameter and independently of the Mach number M as long as M doesn't exceed a value of 0.2 (Fig. 6). The dependence of Nu(R=5D) can then be approximated by a power-law dependence based on $Re^{0.74}$, which is very close to the results of Goldstein et al[10] and Huang & El Genk[11] (who give $Nu \propto Re^{0.76}$). When M exceeds a value of 0.2, a divergence from the initial curve (the boldface curves on Fig. 6) has been observed: the mean Nusselt number is still mainly ruled by the Reynolds number but a high Mach number (>0.2) increases sensibly the mean Nusselt number. Our correlation fits perfectly the experimental results when the Mach number is below 0.2 and has not been plotted in this case on Fig. 6. It is slightly shifted from the data for higher Mach number and one shall be aware that this correlation is not perfectly satisfactory in so far as the polynomial fitting leads to coefficients which are alternatively positive and negative which means that they balance one another. But the correlation gives at least the general way of variation of the Nusselt number with the different parameters within the test range.

Finally one can see that the Mach number can have an influence on jet impingement heat transfer. In certain cases the Reynolds number is not enough to describe the mean heat exchange on the impinged surface. The Mach number rules the local values of the impingement Nusselt number as well. In our experiments the Mach number influence can be neglected as long as M is below a value of 0.2. But high Mach number tends to improve heat exchanges on the surface. Compressible effects should then be taken into account with high velocity impinging jets.

REFERENCES

1. Goldstein, R. J. & Seol, W. S. 1991. Heat Transfer to a Row of Impinging Circular Air Jets Including the Effect of Entrainment, *In* Int. J. Heat Mass Transfer, Vol. **34**: 8: 2133-2147.
2. Livingood, J. N. B. & Hrycak, P. 1973. Impingement Heat Transfer from Turbulent Air Jets to Flat Plates - a Literature Survey, *In* NASA TM X-2778.
3. Martin, H. 1977. Heat and Mass Transfer between Impinging Gas Jets and Solid Surfaces, *In* Adv. in Heat Transfer, Vol. **13**: 1-60.
4. Downs, S. J. & James, E. H. 1987. Jet Impingement Heat Transfer - A Literature Survey, *In* ASME paper n° 87-HT-35, The National Heat Transfer Conference, Pittsburgh, Pennsylvania.
5. Jambunathan, K., Lai, E., Moss, M. A. & Button, B. L. 1992. A Review of Heat Transfer Data for Single Circular Jet Impingement, *In* Int. J. Heat and Fluid Flow, Vol. **13**: 2: 106-115.
6. Viskanta, R. 1993. Heat Transfer to Impinging Isothermal Gas and Flame Jets, *In* Experimental Thermal and Fluid Science, Vol. **6**: 111-134.
7. Baughn, J.W. & Shimizu, S. 1989. Heat Transfer Measurements from a Surface with Uniform Heat Flux Impinging Jet, *In* Trans. ASME, J. Heat Transfer, Vol. **111**: 1096-1098.
8. Cooper, D., Jackson, D. C., Launder, B. E. & Liao, G. X. 1993. Impinging Jet Studies for Turbulence Model Assessment - I. Flow-Field Experiments, *In* Int. J. Heat Mass Transfer, Vol. **36**: 10: 2675-2684.
9. Hollworth B.R. & Gero, L.R. 1985. Entrainment Effects on Impingement Heat Transfer: Part II - Local Heat Transfer Measurements, *In* Trans. ASME, J. Heat Transfer, Vol. **107**: 910-915.
10. Goldstein, R.J., Behbahani, A.I. & Kieger Heppelmann, K. 1986. Streamwise Distribution of the Recovery Factor and the Local Heat Transfer Coefficient to an Impinging Circular Air Jet, *In* Int. J. Heat Mass Transfer, Vol. **29**: 8: 1227-1235.
11. Huang, L. & El-Genk, M.S. 1994. Heat transfer of an impinging jet on a flat surface, *In* Int. J. Heat Mass Transfer, Vol. **37**: 13: 1915-1923.

Acknowledgement: This work is supported by Snecma Moteurs.

Numerical Investigation of Combined Impingement and Convection Heat Transfer

A. ABDON and B. SUNDÉN

Division of Heat Transfer, Lund Institute of Technology, Box 118, 22100 Lund, Sweden

ABSTRACT: This paper concerns development of a prediction method for combined turbulent impingement and convection heat transfer. Firstly, a prediction of a single round unconfined impinging air jet without crossflow is investigated to assess the performance of linear and non-linear two-equation turbulence models. The results show that realizable and/or non-linear two-equation models may successfully be used for impinging jet heat transfer predictions but there are significant differences between the formulations. Among the models tested, a non-linear k-omega showed superior performance. Secondly, the influence of crossflow is considered and a plane confined air jet is investigated. For this case the models based on a frequency-equation (omega) show quite different and more reasonable predictions compared to the dissipation-equation (epsilon) based models.

INTRODUCTION

In the design of cooling systems of gas turbines impingement and forced convection cooling is preferable due to improved efficiency and reduced emission levels. Today this technology is applied for combustor walls and flame-holders and to some extent also for guide vane and blade cooling. However, influences of various design parameters like crossflow and surface enlargement are not well understood. In addition, reliable design methods are lacking for complex designs and only very few experimental data are available. As experiments on real applications are difficult and very costly, there is a request for reliable and cost-effective prediction methods. Such methods could be based on numerical solution of the Reynolds-averaged governing equations and some model for the turbulence field and such an approach is used in this work.

Several earlier investigations[1-3] have revealed that prediction of impinging jet heat transfer is a challenging task. Turbulence modelling is critical and the widely used linear two equation models suffer from a too high turbulence generation (and thus heat transfer) in stagnating flows. This problem may be avoided by using more advanced formulations like full Reynolds stress models or by application of a realizability constraint on two-equation models. In more complex flows a robust computational algorithm is very important which makes the latter approach attractive. Also, the poor reproduction of anisotropic turbulence (present in curved streamlines) of two-equation turbulence models may be improved by using non-linear (NL) or explicit algebraic stress (EASM) formulations. Thus only two-equation models will be considered here.

PROBLEM FORMULATION AND NUMERICAL METHOD

The test cases investigated here are a single unconfined impinging round air jet (cases 1-3: Re_D=23000 and 70000, H/D=2 and 6) and a confined plane air jet with crossflow (case 4: Re_W=Re_H=16700, u_j/u_c=5, H/W=5). Figure 1 shows the computational domains; the unconfined jet calculations (test cases 1-3) extend to H+2D and H+8D in the axial and radial directions, respectively, and for the crossflow geometry (test case 4) the domain is chosen as −20<x/W<70 (slot jet entering at x=0). As the jets were fully developed flows in the reference experimental setups[4-6] for test cases 1-3 this is also used as inlet condition in the calculations (profiles computed separately). For test case 4, the jet and crossflow inlet conditions present in the experiments[7] are not known in detail. But the experimental setup used is likely to yield a turbulent slot jet with near uniform velocity distribution and a crossflow halfway reaching a fully developed turbulent state. In the computations fully developed profiles are applied to the jet as well as the crossflow. A constant (uniform) value of the temperature is used at all the inlets. The free boundaries of the unconfined jets are subject to a constant pressure condition, and a zero gradient condition for outgoing flow is used for the transported variables. Incoming ambient flow is designated constant variable values to achieve low or zero turbulence and the temperature is set equal to the jet inlet temperature. A zero gradient condition is used at the outlet in test case 4. Constant fluid properties are applied and the flow is regarded as incompressible in all the calculations. For test cases 1-3 a constant heat flux is specified at the bottom wall and all other walls are thermally insulated. Test case 4 has a constant temperature bottom wall while the other walls insulated. Both these thermal boundary conditions are in line with those stated in the experiments[4-7].

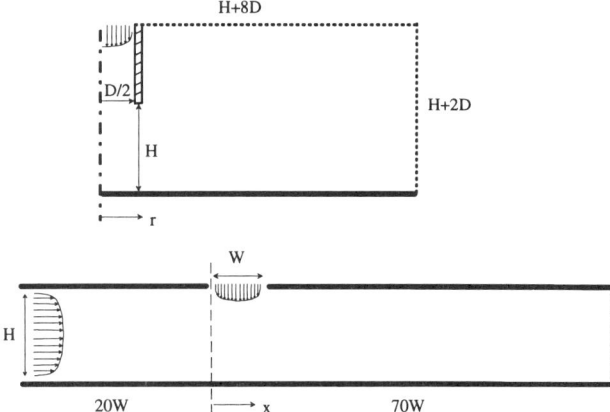

Figure 1. Computational domains of test cases 1-3 (top) and 4 (below).

The numerical method is based on the finite volume method[8] and uses a co-located non-uniform grid (code is CFX4.2). Cylindrical coordinates are used for the round jet (r=0 corresponds to a symmetry line). A hybrid differencing scheme is applied for the convective terms in the momentum and energy equations to reduce numerical instabilities. An investigation of grid dependence was carried out by doubling the number of nodes in each direction for the first (used grid: 180×120) and last (used grid: 275×120) test case. The result was a change less than 5% in local heat transfer rates which was considered acceptable (see Fig. 1 for case 1). The grids of the other cases have about the same density, and the first grid points are always at a dimensionless distance (y^+) less than 0.5 from the heated walls. In all computations the residuals are reduced about five orders of magnitude from an initial guess.

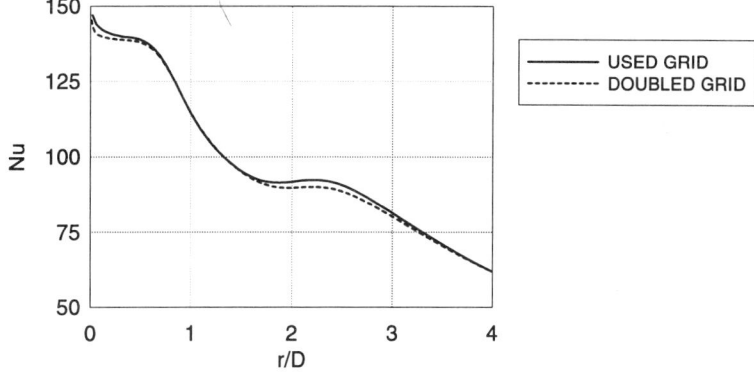

Figure 1. Grid dependence of Nu for test case 1 (Re_D=23000, H/D=2).

The considered turbulence models are all low Reynolds number models; Abe et al[9] ("k-eps"), Lien et al[10] ("NL k-eps"), Wilcox[11] ("k-omega"), Larsson[12] ("NL k-omega"), Rokni[13] ("EASM k-eps"). A realizability constraint[14] is applied on the linear models ("k-eps" and "k-omega") to prevent severe heat transfer overprediction. This constraint puts a limit on the timescale used in the eddy-viscosity and in the epsilon and omega equations:

$$T = \min\left(T_0, \frac{2\alpha}{3f_\mu C_\mu \sqrt{C_\alpha S_{ij}S_{ij}}} \right) \qquad (1)$$

where $\alpha = 0.6$, $T_0 = k/\varepsilon$ or $T_0 = 1/\omega$, $C_\alpha = 2$ in 2D and $C_\alpha = 8/3$ in 3D.

RESULTS AND DISCUSSION

In Figs. 2-6 the results for all test cases are displayed and compared to experimental data[4-7]. For the single unconfined jet very good results can be produced by a simple linear model at Re_D=23000 having the spread in experimental data in mind (local variation of 30% in case 2). However, the deviation from experiments is rather large for the case Re_D=70000. The model which has best overall performance for test cases 1-3 is the "NL k-omega", although it does not capture the minimum at r/D=1 of the Nu-profile for test case 3. This is on other hand clearly demonstrated by the "NL k-eps" and "EASM k-eps" models. Unfortunately, these models have a tendency to exaggerate the following secondary maximum of the profile (see Figs. 2 & 4). The "NL k-eps" model even produces a strong such a maximum for test case 2 which is not supported by the measurements.

All the computations using the linear turbulence models are subject to the realizability constraint in eq. (1). If this constraint is not present, the linear models will generate too much turbulence and consequently overpredict the heat transfer. This effect is shown in Fig. 5 for the linear k-ε model. The value of α in eq. (1) is the same as recommended by Behnia et al[1].

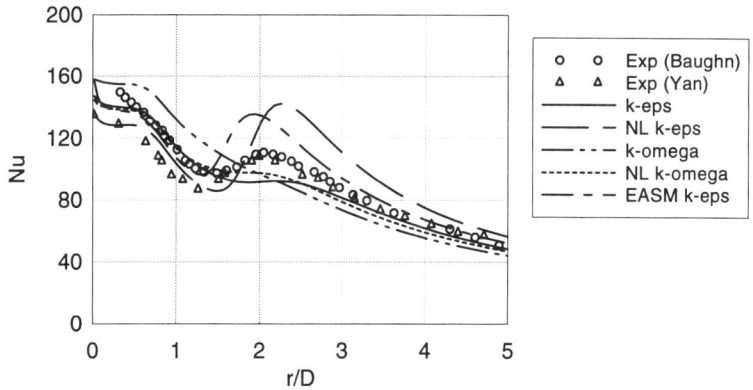

Figure 2. Test case 1: single round jet, Re_D=23000 and H/D=2.

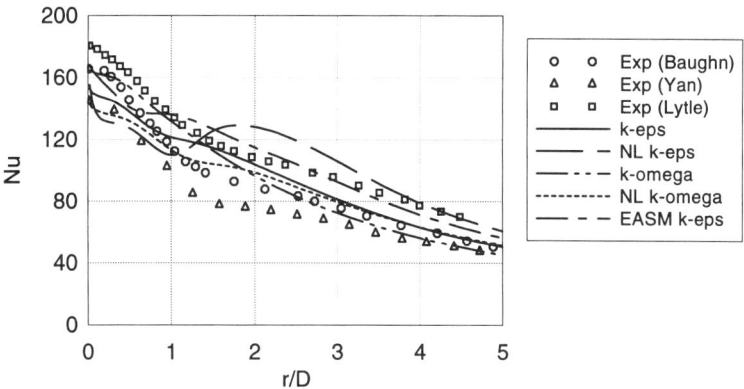

Figure 3. Test case 2: single round jet, Re_D=23000 and H/D=6.

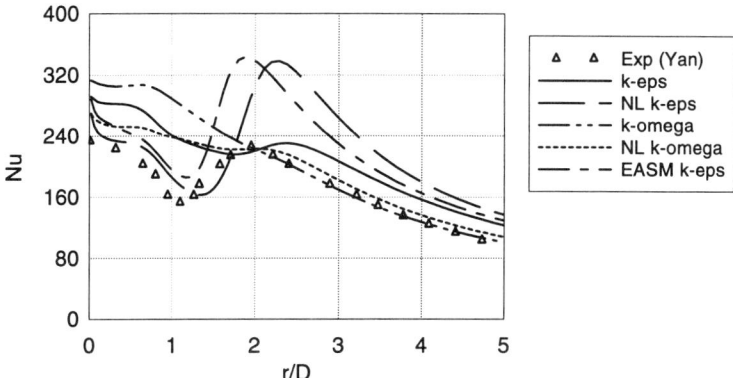

Figure 4. Test case 3: single round jet, $Re_D=70000$ and $H/D=2$.

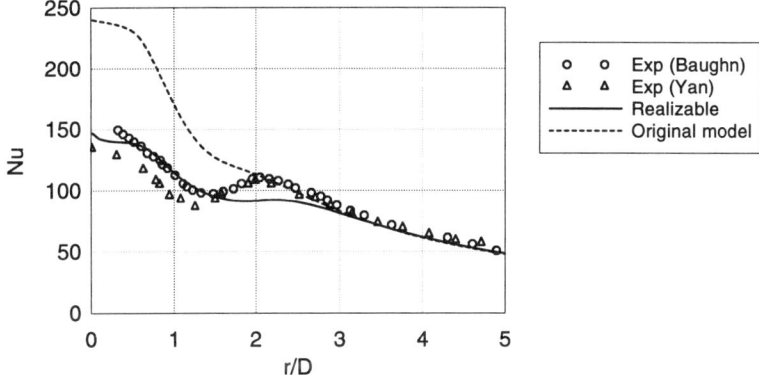

Figure 5. Test case 1: influence of realizability constraint on "k-eps" model.

Test case 4 shows severe differences between calculations and measurements, which may be due to overestimation of the heat transfer in the experiments[7] (used method is mass transfer and Chilton-Colburn analogy). In addition, the inlet boundary conditions of the computations may not be adequate as explained above. The results for this case reveal significantly different Nu-profiles for the turbulence models used. As seen in Fig. 6, the epsilon based models show a strange drop in heat transfer beginning around $x/W=4$. This seems to be caused by the equation determining the length scale as the omega models have no such drop. The explanation for this behaviour is not yet known but all models show that when the crossflow is squeezed under the jet the turbulent viscosity is heavily reduced in the boundary layer. As the flow continues downstream, the omega-based models recover from this much faster than the epsilon based ones. The same kind of heat transfer drop has been detected for lower values of u_j/u_c in the investigation by Chong[7], where the effect is attributed to boundary layer laminarization.

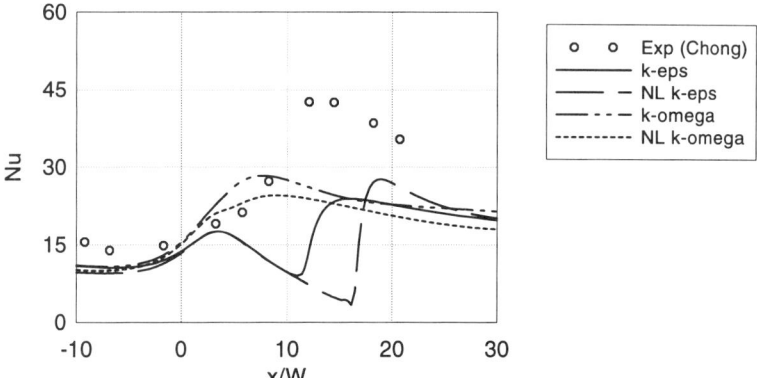

Figure 6. Test case 4: plane jet with crossflow, $Re_W=Re_H=16700$, $u_j/u_c=5$, $H/W=5$.

No converged solution of the "EASM k-eps" model could be found for case 4, but the unconverged results tend to be in line with the other epsilon based solutions (heat transfer drop beginning at x/W=4). These findings suggest further investigations of the performance of these models before they can be applied to more complex impinging flows.

CONCLUSIONS

The present investigation shows that realizable and/or non-linear two-equation models may be used for impinging jet heat transfer predictions with reasonable success, but there are significant differences between different formulations. Among the models tested, a non-linear k-omega showed superior performance.

NOMENCLATURE

2D	two dimensional flow	u_j	slot jet mean velocity
3D	three dimensional flow	u_c	crossflow mean velocity
C_α	realizability constant	u^*	friction velocity,$=(\tau_w/\rho)^{0.5}$
C_μ	turbulence model constant	W	slot jet width
D	round jet diameter	x	bottom wall coordinate
f_μ	viscosity damping function	y^+	distance to wall, $=u^*y/\nu$
H	nozzle-to-wall distance	α	realizability parameter
k	turbulent kinetic energy	ε	turbulent dissipation
Nu	Nusselt number	ρ	density
r	radial coordinate	τ_w	wall shear stress
Re	Reynolds number	ν	laminar viscosity
S_{ij}	mean-strain-rate,$=(\partial U_i/\partial x_j+\partial U_j/\partial x_i)/2$	ω	turbulent frequency
T	time scale		

ACKNOWLEDGEMENT

The present work was financed by the Swedish Gas Turbine Center, GTC.

REFERENCES

1. Behnia. M, Parneix, S. & P.A. Durbin. 1998. Prediction of heat transfer in an axisymmetric turbulent jet impinging on a flat plate. *In* Int. J. Heat Mass Transfer, Vol. **41**: 1845-1855.
2. Ashforth-Frost, S. & K. Jambunathan. 1996. Numerical prediction of semi-confined jet impingement and comparison with experimental data. *In* Int. J. Numerical Meth. Fluids, Vol. **23**: 295-306.
3. Craft, T.J., Graham, L.J.W. & B.E. Launder. 1993. Impinging jet studies for turbulence model assessment-II. An examination of the performance of four turbulence models. *In* Int. J. Heat Mass Transfer, Vol. **36**: 2685-2697.
4. Baughn, J., Hechanova, A. & X. Yan. 1991. An experimental study of entrainment effects on the heat transfer from a flat surface to a heated circular impinging jet. *In* J. Heat Transfer, Vol. **113**:1023-1025.
5. Yan, X. 1993. A preheated-wall transient method using liquid crystals for measurement of heat transfer on external surfaces and in ducts. Ph.D. Thesis. University of California, Davis.
6. Lytle, D. & B. Webb 1994. Air jet impingement heat transfer at low nozzle-plate spacings. *In* Int. J. Heat Mass Transfer, Vol. **37**: 1687-1697.
7. Chong, Y.K. 1979. Two-dimensional slot jet impingement in a confined cross-flow stream. M.Sc. Thesis. Cranfield Institute of Technology, U.K.
8. Versteeg, H.K. & W. Malalasekera. 1995. An introduction to computational fluid dynamics the finite volume method. Longman Scientific & Technical, Essex, England.
9. Abe, K., Kondoh, T. & Y. Nagano. 1994. A new turbulence model for predicting fluid flow and heat transfer in separating and reattaching flows-I. Flow field calculations. *In* Int. J. Heat Mass Transfer, Vol. **37**: 139-151.
10. Lien, F.S., Chen, W.L. & M.A. Leschziner. 1996. Low-Reynolds-Number eddy-viscosity modelling based on non-linear stress-strain/vorticity relations. *In* Proc. 3rd Symp. Engineering Turbulence Modelling and Measurements, Crete, Greece.
11. Wilcox, D.C. 1993. Turbulence modelling for CFD. DCW Industries Inc., La Cañada, USA.
12. Larsson, J. 1997. Two-equation turbulence models for turbine blade heat transfer simulations. ISABE Paper 97-7163. *In* Proc. 13th ISABE Conf., Chattanoga, Vol. **2**: 1214-1222.
13. Rokni, M. 1998. Numerical investigation of turbulent fluid flow and heat transfer in complex ducts. Ph.D. Thesis. Div. Heat Transfer, Lund Institute of Technology, Sweden.
14. Durbin, P.A. 1996. On the k-ε stagnation point anomaly. *In* Int. J. Heat Fluid Flow, Vol. **17**: 89-90.

Mass/Heat Transfer in Dimpled Two-Pass Coolant Passages with Rotation

FUGUO ZHOU AND SUMANTA ACHARYA
Mechanical Engineering Department
Louisiana State University
Baton Rouge, LA 70803

ABSTRACT: Mass/heat transfer measurements are made in dimpled (hemispherical depressions) inlet and outlet coolant flow passages using the naphthalene sublimation method. The leading and trailing surfaces are dimpled, while the side walls are kept smooth. Measurements are made at a Reynolds number of 21,000 and for Rotation numbers of 0 and 0.2. The measurements indicate that dimples enhance surface mass/heat transfer. This enhancement is stronger in the inlet passage than in the outlet passage. Peak mass/heat transfer occurs immediately downstream of the dimples, while the minimum mass/heat transfer occurs in the dimple region itself. Higher mass/heat transfer is also observed along the lateral edges of the dimple. The location of the Sherwood number peaks suggest the existence of streamwise vortical structures generated from the leading and lateral edges of the dimples.

NOMENCLATURE

d	dimple imprint diameter, mm
D	duct hydraulic diameter, mm
H	duct height, m
Nu	Nusselt number
Nu_o	Nusselt number for fully developed circular pipe flow
Pr	Prandtl number
Re	test-section Reynolds number based on hydraulic diameter($V_{av} D/\nu$)
Re_H	test-section Reynolds number based on channel height($V_{av} H/\nu$)
Ro	rotation number (O D/V_{av})
SD	diameter of the circle whose imprint forms the dimple, m
Sh	Sherwood number ($h_m DSc/\nu$)
Sh_o	Sherwood number for fully developed flow V_{av} average streamwise component of velocity in the coolant channel
X, Y, Z	coordinates in the channel (X follows the flow direction), m
O	Rotational speed, rad/s

INTRODUCTION

In the development of advanced turbine systems, there has been considerable effort directed toward increasing the turbine inlet temperature. This requires the development of more effective internal and external blade cooling strategies. In internal cooling, compressed air is circulated through ribbed serpentine passages, and discharged through the trailing edge of the blade. In recent years, efforts directed toward improving internal cooling have led to concepts that include the use of inclined turbulators[1], vortex generators[2,3], swirl-induced cooling or screw cooling[4,5], and the use of dimpled surfaces[6,7,8,9,10]. The present study deals with dimpled coolant turbine-blade passages and investigates the effect of rotation on the heat transfer from dimpled surfaces in both the radially-outward flow passage and the radially-inward flow passage.

Dimpled surfaces have been shown to enhance surface heat transfer by a variety of investigators[11,12,13], and under certain conditions, they have also been shown to reduce the drag coefficient[14]. This combination of enhanced heat transfer with minimum pressure drop penalty makes dimpled surfaces attractive from a turbine blade cooling perspective. For these reasons, several investigators have explored the use of dimpled internal coolant turbine blade passages. Schukin et al.[6] studied the effects of channel geometry (converging and diverging channels) on the heat transfer downstream of a single hemispherical cavity. Chyu et al.[7] reported overall heat transfer rates that are 2.5 times greater for the dimpled surface compared to a smooth surface. Lin et al.[8] presented corresponding flow and heat transfer predictions to help explain the observed heat transfer behavior. Moon et al.[9] investigated the effect of channel height on the heat transfer and friction in a dimpled passage. Heat transfer enhancements of the order of 2.1 over smooth surfaces were reported with pressure drop penalties in the range of 1.6-2.0 over smooth surfaces. Most recently, Mahmood et al.[10] have made detailed flow and heat transfer measurements on a dimpled plate, and have identified specific vortex structures responsible for augmenting heat transfer from the downstream rims of each dimple. Heat transfer enhancements ranging from 1.8-2.4 over smooth plates were noted.

The literature cited above for dimpled surfaces has been done for stationary channels. In turbine blades, data under rotating conditions are of interest, The only such data available is the recent work of Acharya and Zhou[15] where measurements in a dimpled rotating square passage was presented for a radially-outward-flow passage. This work extends the study in Ref. 15, and presents measurements in both the radial-inflow dimpled passage and the radial-outflow dimpled passage. Results are presented for one Reynolds numbers (Re=21,000) and two values of Rotation number (Ro=0 and Ro=0.2).

EXPERIMENTS

The experiments are performed in a test apparatus designed for the study of mass transfer (sublimation of naphthalene) in a two-pass internally square duct. The test section has a radially-outward flow passage (inlet channel) and a radially-inward flow passage (outlet channel). The present work specifically focuses attention on the leading and trailing walls of the duct. Therefore, only the leading and trailing surfaces in the inlet and outlet are dimpled, and the side surfaces are kept smooth. Mass transfer measurements permit the acquisition of detailed local distributions of the Sherwood number which can then be converted to Nusselt numbers using the heat-mass transfer analogy.

Apparatus

Figure 1 shows a schematic diagram of the test section. Compressed air is delivered through rotating seals into a hollowed rotating shaft, and is then redirected through a rigid tube connecting the hollow shaft to the test section. The air enters the test section through a conditioning plenum, and the naphthalene laden exhaust air is discharged through a flexible tube to a fume hood. The aluminum alloy test section consists of a 69.85mm tapered settling chamber; a frame that supports eight removable, hollow test section plates; and a removable 180 degree bend. These major components are secured in a flange-like manner, using O-rings between all parts to prevent air leakage. When assembled, the test section forms 25.4mm x 25.4mm x 304.8mm long inlet and outlet sections 38.1mm apart that are connected by the 180 degree, 25.4mm x 25.4mm square cross-section bend. Each test section plate is a reinforced recessed frame to accommodate casting of a naphthalene layer for mass transfer measurements. The two-pass test section is housed inside a small pressure-vessel, and prior to rotation, pressure inside the pressure vessel is equalized with the pressure in the two-pass coolant channel. The test-section assembly is connected to the hollow rotating shaft driven by a hydraulic motor. For load balancing, a dummy weight is placed opposite to the test-section assembly. The entire rotating arm is contained within a large pressure vessel for safety purposes.

Detailed surface profiles of the cast surfaces are required for local mass transfer results. These profiles are obtained by moving the walls under a fixed, linear variable differential transducer (LVDT) type profilometer. A custom written program run on a personal computer is used to control the motion of the traversing table through micro-step drive motors with a 0.00127mm step size.

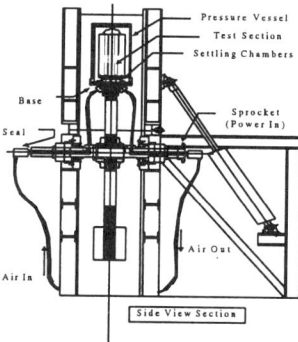

Figure 1: Schematic of the two-pass rotating coolant channel

A thermocouple is mounted into the pipe upstream of the test section to measure the free stream temperature, which is required for calculating the mass flow rate. To measure the napthalene surface temperature, two thermocouples are embedded into the naphthalene surface along one wall. The outputs of the thermocouples are saved in an Omega temperature data logger (OM-SP-1700). The recorded data is read out by a computer, and the average value is taken as the wall temperature of the naphthalene surfaces.

Casting of the Smooth and Dimpled Test Plates

Fresh, 99% pure naphthalene crystals are melted, and the molten naphthalene is quickly poured into the hollow cavity of the plate frame to fill completely the region between the frame walls and the stainless steel casting plate.

The pattern on the casting plates can be used to generate a specific pattern on the cast naphthalene surface. For a smooth surface the casting plate is polished to a mirror-smooth finish, and checked for flatness. For a dimpled surface, the casting plate surface has machined protrusions that are the inverse of the dimples desired. After eight hours of casting time, each wall is then separated from the casting plate (with the protrusions leaving the desired dimpled impressions on the naphthalene surface) and placed on the mounting plate for scanning. After scanning, the plates are stored in an air-tight container, saturated with naphthalene vapor, to hinder natural sublimation until the test section is assembled. After the experiment is over, the test section is disassembled and the walls are placed in the storage container until they are scanned again.

A schematic of the dimpled plates used is shown in Figure 2. Table 1 lists the critical dimple parameters (dimensionless) and shows the comparison with other published studies on dimpled surfaces. As may be seen from this table, the geometrical parameters used are in the range of parameters used by other investigators.

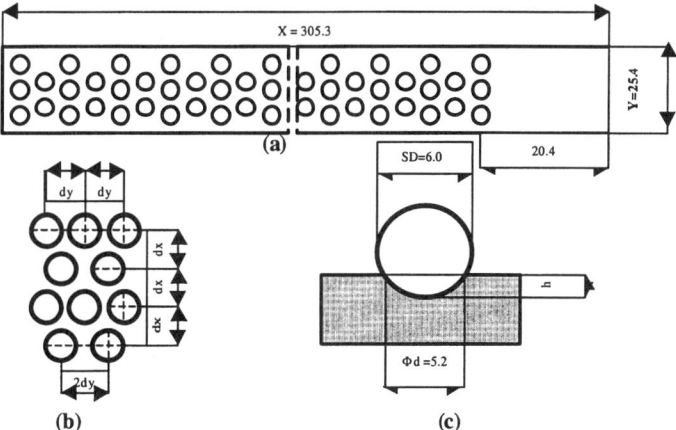

Figure 2: (a) Schematic of the dimpled test surface (b) Geometrical dimensions of the dimples in a plane parallel to the test surface (c) Geometrical dimension of the dimples in a plane normal to the test surface

Table 1: Comparison of parameters used in the present study with those reported in the

	d	H/d	SD/d	h/d	dx/d	dy/d	Re_D	Re_H
Present Study	5.2	4.88	1.15	0.29	1.17	0.81	7000, 21000	7000,21000
Mahmood et.al.[10]	50.8	0.50	1.45	0.20	0.81	0.81		1250~61500
Moon et al.[9]	17.1	0.35 ~ 1.46	1.49	0.19	1.06	0.63	29202 ~ 55545	
Lin et. al.[8]	8.3	2.30, 4.60	2.30	0.58	2.30	2.30		23000,46000

Data Reduction and Analysis

Mass flow rate in the meter run is calculated from measurements of temperature, pressure, and differential pressure using standard equations for concentric bore orifice meters [16, 17]. The naphthalene sublimation depth is calculated from the two surface profiles for each wall. Each profile is normalized with respect to a reference plane computed from three points scanned on the aluminum surface of the walls. The difference between the normalized profiles gives the local sublimation depth which can be converted into a local Sherwood number as described by Hibbs et al.[3]. Comparison of heat transfer and mass transfer results is done through the use of the heat-mass transfer analogy[17]. To facilitate comparison between mass and heat transfer results, a Sherwood number ratio, N_0, is defined as: $N_o = Sh / Sh_o = Nu / Nu_o$ where Nu_o and Sh_0 are the fully-developed Nusselt number and Sherwood number respectively. In the present study, Sh_0 is taken to be the measured fully developed average Sherwood number in the stationary smooth duct at Re=21,000. An average Sherwood number is computed as a simple arithmetic average of the data points scanned over the region over which averaging is performed.

Uncertainties for all computed values are estimated using the method of Kline and McClintock[19]. Volume flow rate and duct Reynolds number (Re) uncertainties are estimated to be less than 10 percent for Re > 6000. Experimental tests of accuracy and repeatability for the entire acquisition system indicate a sublimation depth uncertainty of 0.0038 mm. Sublimation depths are maintained at about 0.152 mm by varying the duration of the experiment for different Re. The resulting experimental duration for the present study was 120 minutes. Overall uncertainty in Sherwood number calculation is about 8 percent and varies slightly with Reynolds number (<1 percent).

To validate the experimental procedure, a verification test was carried out at Re=7,000 and Ro=0.2. The results for both the trailing and leading walls compare well with the results reported by Kandis and Lau[20]. This comparison is reported by the authors in Ref. 15, and provides the requisite validation of the experimental procedure used in this study.

RESULTS AND DISCUSSION

Results are presented for Reynolds number Re= 21,000 and two Rotation numbers Ro=0 and Ro=0.2 in the form of either local Sh/Sh_o distributions, or streamwise-averaged and spanwise-averaged Sh/Sh_o distributions. For the dimpled walls, contours of Sh/Sh_o are presented in the periodically fully-developed region. No pressure-drop measurements were made since, with rotation, they entail the use of slip rings, and the present facility is not currently equipped with slip-ring instrumentation.

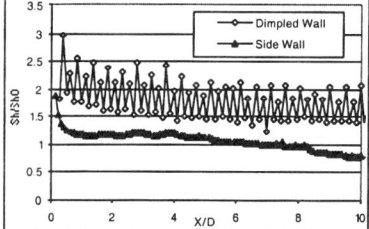
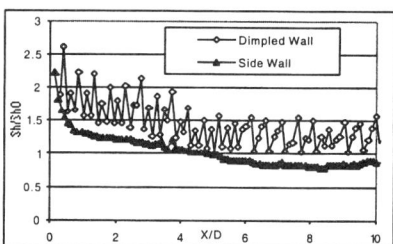

Fig.3: Spanwise-Averaged Sh/Sh_0 - Dimpled Inlet (left) and Outlet (right) Passages (Re=21,000, Ro=0)

Figure 3 shows the spanwise-averaged Sh/Sh_o distributions in the stationary dimpled coolant passages. Only one of the dimpled and one of the smooth surfaces are shown. The peak Sh/Sh_o are located just downstream of the rows with dimples, and the minimum Sh/Sh_o corresponds to the rows with dimples. Note that the successive peak values in the Sh/Sh_o distribution alternate in magnitude with higher values along the 3-dimple row, and lower values along the 2-dimple row. From Fig. 3 it can be seen that the dimples enhance the mass/heat transfer in the fully developed region by a factor of 1.5-2 in the radially-outward flow leg, and by a factor of 1.1-1.5 in the radially-inward flow leg. Moon et al.[9] have reported enhancements of the order of 2.1 in stationary dimpled passages.

Figure 4 presents the streamwise-averaged Sh/Sh_o distributions. The streamwise-averaging is performed by summing Sherwood number distributions along spanwise-rows that are at the same relative locations relative to the three-dimple row and dividing by the number of rows over which the summation is performed. The spanwise distributions of the averaged Sherwood numbers are shown at four streamwise locations: (1) along the row containing three dimples (the spanwise locations of the three dimples are Y/D=-0.33, 0, and 0.33), (2) downstream of the three-dimple row at a location midway between the three-dimple and the two-dimple row, (3) along the row containing two dimples (the spanwise locations of the two dimples are Y/D=-0.17, and 0.17), and (4) downstream of the two-dimple row at a location midway between the two-dimple and the

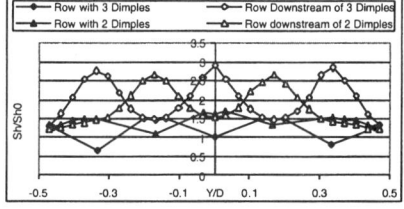 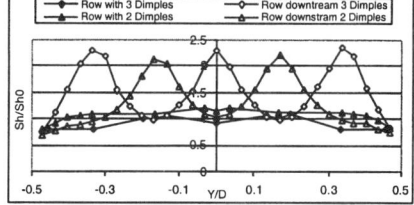

Fig.4 Streamwise-Averaged Sh/Sh_0. Dimpled Inlet (left) and Outlet (right) Passages (Re=21,000, Ro=0)

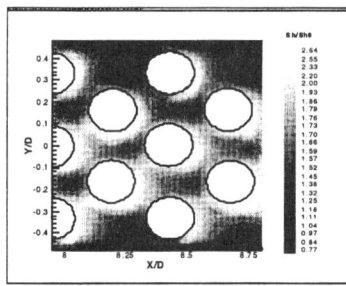 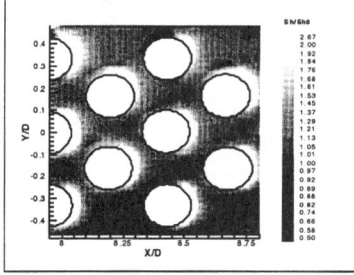

a) Dimpled Wall in the Inlet   (b) Dimpled Wall in the Outlet
Fig. 5: Sh/Sh_0 Contours in the Fully Developed Region on Dimpled Walls (Re=21,000 Ro=0)

three-dimple row. It is clear that the row downstream of the three dimples has the highest mass transfer, with the peak in the spanwise direction occurring downstream of the dimple. Along this row, the minimum in the spanwise direction corresponds to the location midway between the dimples. The row containing the dimples has lower heat transfer, with the minimum along the row coinciding with the dimple center. The maximum along this row occurs along the spanwise edges of the dimples. These observations are consistent with the observations of Moon et al.[9] and Mahmood et al.[10] who indicated that the dimple region itself is characterized by low heat transfer and that the maximum heat transfer rates occur immediately downstream of the dimples and along the spanwise edges of the dimples. The distributions in the outlet duct (with radially-inward flow) are similar to that observed in the inlet duct, except that the Sherwood numbers are somewhat lower.

The above-noted trends can be seen more clearly in Figure 5. The locations of the dimples are indicated in the contours, and regions containing both the two-dimple row and the three-dimple row are presented. The peak distributions exhibit a half-moon shape immediately downstream of the dimples. The contours of Sherwood number verify that the strongest mass/heat transfer occurs just downstream of each dimple, and along the lateral edges of the dimples.

The results for the dimpled walls with rotation are shown from Figure 6 through Figure 8. Comparing Figure 6 with the data from a smooth rotating channel (not shown in this paper), it is observed that the dimples enhance the mass/heat transfer on both leading and trailing walls. For example, in the inlet the Sh/Sh_0 in the fully-developed region increases from 2.5 to about 3.5 for the trailing wall, and increases from 1.5 to 2.5 for the leading wall. Figure 6 also shows that the enhancement in Sh/Sh_0 is larger in the inlet than that in the outlet. Surprisingly we note a greater divergence between the stabilized and destabilized surfaces in the outlet duct compared to that in the inlet duct.

Figure 7 presents the streamwise-averaged Sh/Sh_o distributions for the dimpled walls in the inlet passage. The distributions are similar to the stationary case shown in Figure 4, but the peak values of Sh/Sh_o has increased from 2.5 to more than 5. Figure 8 gives the corresponding Sh/Sh_o contours in the fully developed regime. The similarities between Figure 4 and Figure 7, and between Figure 5 and Figure 8 suggest that the Coriolis force does not alter the flow structures generated by the dimples.

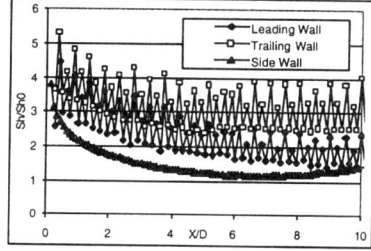 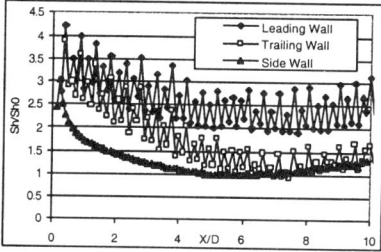

Fig.6: Spanwise-Averaged Sh/Sh_0 - Dimpled Inlet (left) and Outlet (right) Passages (Re=21,000, Ro=0.2)

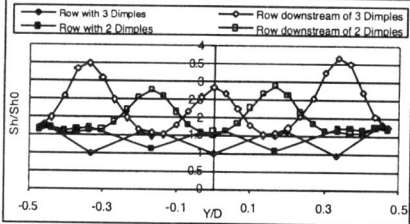 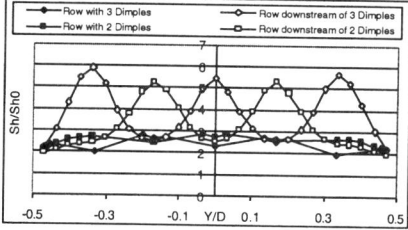

Fig.7: Streamwise-Averaged Sh/Sh_0 - Leading (left) and Trailing (right) Walls in the Dimpled Inlet Passage (Re=21,000 Ro=0.2)

CONCLUDING REMARKS

Measurements of surface mass/heat transfer rate are reported in a dimpled coolant flow passage in both the inlet and outlet. The experiments have been done at one Re=21,000 and two Rotation numbers of 0 and 0.2. The following major conclusions are observed:

1. The maximum mass/heat transfer rates are obtained downstream of the dimples. The minimum mass/heat transfer rates occur along the row containing the dimples
2. The dimpled walls lead to enhancements over smooth surfaces both in the stationary case and in the rotating case. These enhancement factors are less than 2.
3. The present results show that enhancements achieved with dimples are larger in the inlet passage compared to the outlet passage.
4. The Sherwood number distributions suggest the existence of three local peaks, with the strongest peak immediately downstream the dimples. These peaks appear to be related to the development of streamwise vortical structures generated from the dimples. The Sherwood number contours also support these observations.

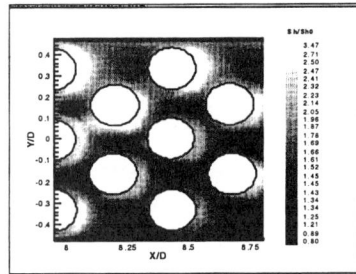 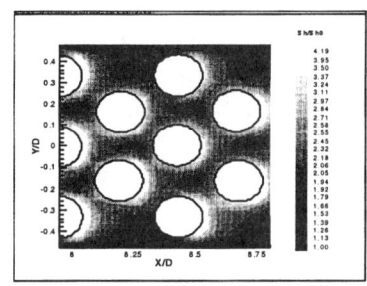

(a) Leading Wall in the Inlet   (b) Trailing Wall in the inlet
Fig.8 Sh/Sh_0 Contours in the Fully Developed Region on Dimpled Passges (Re=21,000 Ro=0.2)

REFERENCES

1. Han, J.C., Zhang, Y.M., and Lee, C.P., 1992, *Journal of Turbomachinery*, Vol. 114, pp. 872-880.
2. Myrum, T., Acharya, S., Sinha, S., and Qiu, X., 1996, *Journal of Heat Transfer*, Vol. 118, pp. 294-300.
3. Hibbs, R., Acharya, S., Chen, Y., Nikitopoulos, D., and Myrum, T., 1998, *Journal of Turbomachinery*, Vol. 120, No. 4, pp. 724-734.
4. Glezer, B., Moon, H. K., Kerrebrock, J., Bons, J., and Guenette, G., 1998, ASME Paper 98-GT-214.
5. Hedlund, C. R., Ligrani, P. M., Moon, H. K., and Glezer, B., 1998, ASME Paper 98-GT-466.
6. Schukin, A. V., Koslov, A. P., and Agachev, R. S., 1995, ASME Paper No. 95-GT-59, *ASME 40th Intl. Gas Tutrbine and Aero Congress*, Houston.
7. Chyu, M. K., Yu, Y., Ding, H., Downs, J. P., and Soechting, F., 1997, ASME Paper No. 97-GT-437, *ASME 42nd Intl. Gas Turbine and Aero Congress*, Orlando.
8. Lin, Y. l., Shih, T. I. P., and Chyu, M. K., 1999, ASME 99-GT-263, *ASME Turbo Expo*, 1999, Indianapolis
9. Moon, H. K., O'Connell, T., and Glezer, B., 1999, ASME 99-GT-163, *ASME Turbo Expo*, 1999, Indianapolis
10. Mahmood, G. I., Hill, M. L., Nelson, D. L., Ligrani, P. M., Moon, H. K., and Glezer, B., 2000 *ASME Turbo Expo*, 2000, Munich
11. Afanaseyv, V. N., Chudnovsky, Y. P., Leontiev, A. I., and Roganov, P. S., 1993, *Experimental Thermal and Fluid Science*, Vol. 7, pp. 1-8
12. Kesarav, V. S., and Kozlov, A. P., 1994, *Heat Transfer Research*, Vol. 25, No. 2, pp. 156-160
13. Gortyshov, Y. F., Popov, I. A., Amirkhanov, R. D., and Guiltsky, K. E., 1998, *Proc. Of the 11th Intl. Heat Transfer Congress*, Vol. 6, pp. 83-88
14. Bearman, P. W., and Harvey, J. K., 1993, *AIAA Journal*, Vol. 31, No. 10, pp. 1753-1756
15. Acharya, S., and Zhou, F., 2000, *ASME-IMECE Meeting*, Orlando, Nov. 2000
16. Stearns, R. F., Johnson, R. R., Jackson, R. M., and Larson, 1951, C. A., *Flow Measurement with Orifice Meters*, D. van Nostrand Co. Inc., Toronto
17. Miller, R. W., *Flow Measurement Engineering Handbook*, 2nd Ed., McGraw-Hill Pub. New York, 1989.
18. Sogin, H. H., and Providence, R. I., 1958, *Transactions of the ASME*, Vol. 80, 61-69.
19. Kline, S. J., and McClintock, F. A., 1953, *Mech. Engineering*, Vol. 75, No. 1, pp. 3-8.
20. Kandis, M., and Lau, S. C., 1994, *Fifth Intl. Symp. on Transport Phenomena and Dynamics of Rotating Machinery*, May 8-11, Maui, Hawaii.
21. Wagner, J. H., Johnson, B. V., and Hajek, T. J., 1991, *Journal of Turbomachinery*, Vol. 113, pp. 42-51.

Detailed Heat/Mass Transfer Distributions in a Rotating Two Pass Coolant Channel With Engine-Near Cross Section and Smooth Walls

L. RATHJEN[a], D. K. HENNECKE[a], S. BOCK[b] and R. KLEINSTÜCK[b]

[a] Gas Turbines and Flight Propulsion, Darmstadt University of Technology, 64287 Darmstadt, Germany, email: rathjen@gfa.tu-darmstadt.de
[b] MTU Aero Engines, 80995 München, Germany, email: stephan.bock@muc.mtu.de

ABSTRACT: This paper shows results obtained by experimental and numerical investigations concerning flow structure and heat/mass transfer in a rotating two-pass coolant channel with engine-near geometry. The smooth two passes are connected by a 180° U-bend in which a 90° turning vane is mounted. The influence of rotation number, Reynolds number and geometry is investigated. The results show a detailed picture of the flow field and distributions of Sherwood number ratios determined experimentally by the use of the naphthalene sublimation technique as well as Nusselt number ratios obtained from the numerical work. Especially the heat/mass transfer distributions in the bend and in the region after the bend show strong gradients, where several separation zones exist and the flow is forced to follow the turbine airfoil shape. Comparisons of numerical and experimental results show only partly good agreement.

NOMENCLATURE

D	mass diffusion coefficient for naphthalene vapor in air, m^2/s	Sh_{av}	spanwise averaged Sherwood number
d_h	hydraulic diameter of first pass, m	Sh_0	Sh of fully developed tube flow
h	heat transfer coefficient, $W/(m^2 K)$	T_b	bulk temperature, K
h_m	mass transfer coefficient, m/s	T_w	wall temperature, K
$\dot{m}$	naphthalene mass flux, $kg/(m^2 s)$	U	averaged air velocity, m/s
$\vec{N}$	normal vector for cross sections	$\vec{v}$	velocity vector
Nu	Nusselt number	x	streamwise coordinate, m
Nu_0	Nu of fully developed tube flow	x_u	circumferential coordinate
$\dot{q}$	heat flux, $W/(m^2)$	$\vec{x}$	location on the wall, $\vec{x} = (x, x_u)$
Re	Reynolds number, $Re = U d_h/\nu$	$\rho_{v,b}$	bulk naphth. vapor density, kg/m^3
Ro	Rotation number, $Ro = \omega d_h/U$	$\rho_{v,w}$	naphth. vapor density at wall, kg/m^3
Sc	Schmidt number	λ	thermal conductivity, $W/(mK)$
Sh	Sherwood number	ν	kinematic viscosity, m^2/s
		ω	angular velocity, rad/s

INTRODUCTION

A main objective in turbine blade cooling design is to achieve optimum heat transfer while minimizing the coolant flow rate. The influence of 180° bends, rib roughened walls, pressure loss and rotation is of particular importance for heat transfer research in internal turbine blade coolant channels.
In the literature there is a large number of investigations of idealised coolant channels, but only few publications can be found concerning measurements or numerical calculations on engine-near geometries. Therefore, this paper will show detailed results obtained from local mass transfer measurements and numerical calculations of the flow and Nusselt numbers in an engine-near coolant channel model. The influence of Reynolds number, Rotational number, engine-near geometry and turning vane on the flow field and heat/mass transfer is investigated.

TEST MODEL GEOMETRY

The test section consists of a first pass with trapezoidal cross section extending radially outward, a 180° bend with a 90° turning vane and a second pass with a larger trapezoidal cross section extending radially inward. Figure 1 shows a view of the model geometry and the annotation for reference to the different surfaces in the presentation of the results. The hydraulic diameter d_h is the value of the first passage which is used for the calculation of the Reynolds numbers. In the bend region the flow is not only directed through the 180° U-bend but also perpendicular to that following the turbine airfoil shape. The divider wall has a thickness of 0.3 d_h. For the experimental investigations the model was mounted to inlet and outlet sections at the rotating test rig which were simulated in the numerical calculation, too.

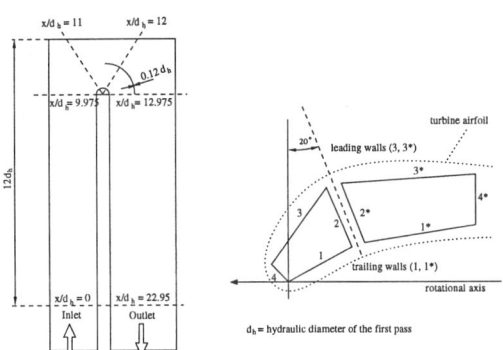

Figure 1: Test model geometry

ANALOGY OF HEAT AND MASS TRANSFER

The similarity between the gouverning differential equations for heat and mass transfer for incompressible flow can be used to establish an analogy as

$$\frac{Sh}{Sh_0} = \frac{Nu}{Nu_0} \quad \text{with} \quad Sh(\vec{x}) = \frac{h_m(\vec{x})d_h}{D} \quad , \quad Nu(\vec{x}) = \frac{h(\vec{x})d_h}{\lambda} \quad (1)$$

and $\quad Nu_0 = 0.023 Re^{0.8} Pr^{0.4} \quad , \quad Sh_0 = 0.023 Re^{0.8} Sc^{0.4}.$ (2)

The mass and heat transfer coefficients at each location $\vec{x}$ at the wall are

$$h_m(\vec{x}) = \frac{\dot{m}(\vec{x})}{\rho_{v,w}(\vec{x}) - \rho_{v,b}(x/d_h)} \quad \text{and} \quad h(\vec{x}) = \frac{\dot{q}(\vec{x})}{T_w(\vec{x}) - T_b(x/d_h)}.$$ (3)

The naphthalene vapor density at the wall, $\rho_{v,w}(\vec{x})$, is calculated with the ideal gas law and an empirical correlation of Presser[1]. The vapor density in the air stream, $\rho_{v,b}(x/d_h)$, is calculated with the cumulative mass of naphthalene in the fluid. Accordingly, at each cross section along the streamwise direction x/d_h the reference bulk temperature $T_b(x/d_h)$ can be calculated as a mass flux average

$$T_b(x/d_h) = \frac{\int_A T \rho \vec{v} \cdot \vec{N} dA}{\int_A \rho \vec{v} \cdot \vec{N} dA}.$$ (4)

EXPERIMENTAL AND NUMERICAL APPROACH

Generally the naphthalene sublimation technique is described by Goldstein et al.[2]. A detailed description of the experimental approach for the investigations shown in this paper is given by Rathjen et al.[3]. The maximum uncertainty of the Sherwood number is estimated to 10%. The uncertainty of the Reynolds number is found to be 2 %.

The mass transfer experiments were carried out for Reynolds numbers of 10000, 25000 and 50000. The Rotational number ranges from 0 to 0.2 for the first two Reynolds numbers and up to 0.1 for a Reynolds number of 50000.

In order to understand the mass transfer results being investigated experimentally and to compare these with numerical results, CFD calculations concerning fluid flow and heat transfer have been carried out using the commercial CFD code CFX-TASCflow. The code represents a Reynolds averaged Navier-Stokes solver with a $k - \epsilon$ turbulence model. The near wall portions of the flow were modeled using a low-Reynolds approach. Therefore, a relatively fine grid with approximately 700000 nodes was installed in order to properly resolve the near wall region. Two computations have been done, for $Ro = 0$ and $Ro = 0.2$, both at $Re = 25000$.

RESULTS

The results of the mass transfer experiments will be shown as Sherwood number ratios normalized by Sh_0. For the CFD calculations, Nusselt number ratios will be displayed accordingly. In addition to the streamwise coordinate x/d_h, a circumferential coordinate x_u is plotted in the contour plots. This coordinate can be determined by the absolute coordinate on a specific wall normalized by the absolute wall width. This dimensionless coordinate is summed up from wall 1 to wall 4 and from wall 1* to wall 4*, respectively.

Rathjen et al.[3] investigated the distribution of local values in detail for a Reynolds number of 50000 and Rotational numbers of 0 and 0.05. It could be stated, that the same characteristic regions concerning Sherwood/Nusselt number distributions can be found for the three different Reynolds numbers investigated. Thus, in this section representative results are shown from tests with $Re = 25000$ combined with $Ro = 0$ and $Ro = 0.2$.

Local Mass Transfer Distribution

Figure 2 shows the normalized Sherwood numbers on all investigated walls. In Figure 2(a) the values are plotted for the stationary case. From that picture it can be stated

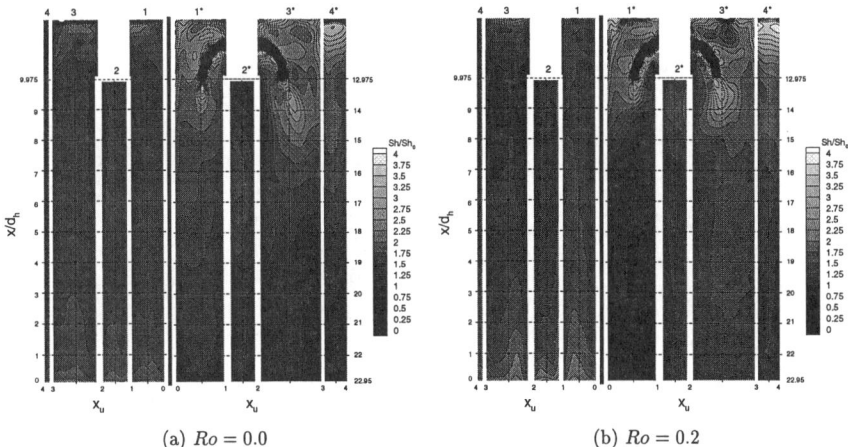

(a) $Ro = 0.0$ (b) $Ro = 0.2$

Figure 2: Local Sh/Sh_0 distribution for $Re = 25000$, $Ro = 0.0$ and $Ro = 0.2$

that in the two straight passages the decreasing Sherwood number ratios show lower values in the corners than on the wall center lines. In these areas the effect of the adjacent boundary layers, as described by Goldstein et al.[4], leads to a stronger decrease of the mass transfer coefficients. Above $x/d_h = 9.975$ the influence of the bend induced secondary flow motion is visible on the walls 1 and 3. Entering the second part of the bend, the air mass flow is divided into two parts and one can recognize strong gradients in the Sherwood number ratio distribution in this region affected by a separation bubble located on the turning vane (see Figure 5). The absolute maximum of Sherwood numbers can be found at wall 4* where the flow impinges. After the bend, at about $x/d_h = 13$ the two air mass flows start mixing again with different velocities and different directions. Thus, a strong effect at the walls 1* and 3* is visible, due to the enhancement of the wall shear stresses. However, the distribution of the measured values remains partitioned in two areas, especially at wall 1*. The effect of the separation bubble caused by the 180° bend can be recognized at wall 2*. The intended goal of the turning vane, namely to shrink or avoid such a separation zone, was not achieved. Generally the values in the second passage are smaller than in the first one, due to the lower Reynolds number of about 2/3 of the first passage value. In comparison to the stationary case, Figure 2(b) shows the distribution for a combination of $Re = 25000$ and $Ro = 0.2$. The rotation induced secondary flow pushes the fluid towards the trailing wall in the first pass and towards the leading wall in the second pass, respectively. The values in the bend region seem to be unaffected by the rotation. The partitioning of the distribution in the second pass is more pronounced than for the non-rotating case. It can be recognized that the separation bubble effect at wall 2* is displaced towards wall 3* and extends less in downstream direction. However, the influence of the bend and the turning vane at the model exit is still visible. The effect of secondary flow motion on heat/mass transfer values at leading and trailing walls increases with higher rotational numbers. Figure 3 shows the differences between leading and trailing wall values as percentage of the Sherwood number averaged over all four walls, Sh_{av}. The differences of each Rotational number were averaged over the three investigated Reynolds numbers. From the inlet of the test section up to the bend the deviation increases with the developing secondary flow. For $Ro = 0.1$ the difference is up to 10% in front of the bend. For the case of rotation the lines of leading and trailing wall values cross in the bend region. After the 180°-turn the influence of the geometry induced effects is visible: At about $x/d_h = 14$ the lines meets at the same value independent of Ro. Further downstream the rotational effect starts dominating again and leads to differences of about 30 % at $x/d_h = 23$. In this picture the effect of the engine-near geometry is visible, too. For $Ro = 0$ in the after bend region the trailing wall values are higher than the leading wall values caused by the swirled flow in the second pass (see Figure 4).

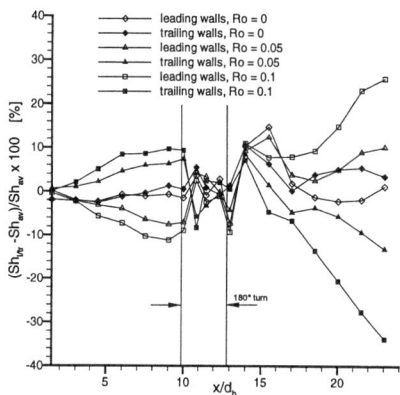

Figure 3: Averaged deviation of leading and trailing wall Sherwood numbers caused by different Ro-numbers

Flow Field And Heat Transfer Calculations

The flow field in the first section, up to $x/d_h = 9$, is relatively uniform, showing only slight variations between the stationary and the rotating case. The effect of

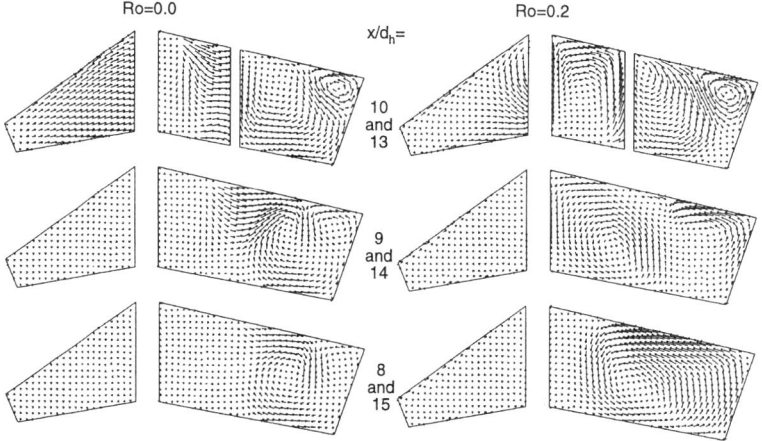

Figure 4: Secondary flow field with $Re = 25000$ at three cross sections, $Ro = 0$ and 0.2

rotation in the first section is much smaller than the effect of the turn and therefore, is almost invisible in Figure 4. The flow field after the bend reveals a high degree of three dimensionality mostly due to the forced turn. The bend and the guide vane induce a set of vortices that is much stronger than the vortices induced by the Coriolis forces. Nevertheless, rotation still has a strong effect on the development of the flow field after the bend, as can bee seen from Figure 4. The merging of various sets of counter rotating vortices is clearly different in the two cases investigated. To give an impression of the complexity of the flow field, Figure 5 shows the streamlines close to the wall. This picture reveals the three-dimensional nature of the separation zone that forms on top of the guide vane.

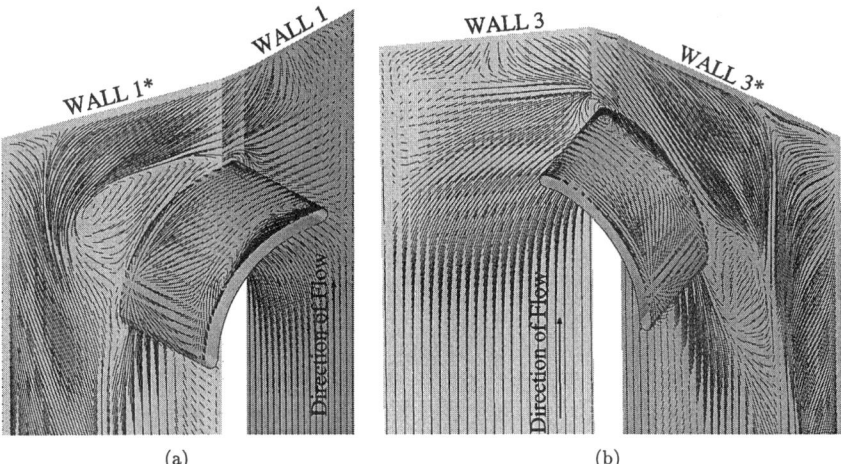

Figure 5: Streamlines close to the wall in the area of the guide vane ($Re = 25000$, $Ro = 0$)

The calculation of Nusselt numbers is very challenging because of the high sensitivity of the wall heat transfer to the speed and temperature gradients at the wall. In order to give the best possible description of these gradients, a two-layer approach with $k - \epsilon$ turbulence modeling in the core flow and a low Reynolds formulation close to the wall has been chosen. The resulting Nusselt number ratio distributions are presented in Figure 6. When comparing these results with the experiments presented in Figure 2, it is visible that some major features could be predicted. The increase in heat transfer at the end of the guide vane could be predicted, and even the different extent of that area on wall 1* and 3* was captured by the calculation. On the other hand, some predicted areas of low heat transfer in the corners of the bend or over the guide vane could not be observed in the experiment. However, the overall magnitude of Nusselt numbers is in good agreement with the experiment but local discrepancies still remain.

Regional, circumferential averaging of Sherwood/Nusselt numbers along x_u at each streamwise location x/d_h is a good way of describing the streamwise development of these values. Figure 7 shows circumferentially averaged Sherwood/Nusselt number ratios obtained from the experiments and from the CFD calculations. In the case

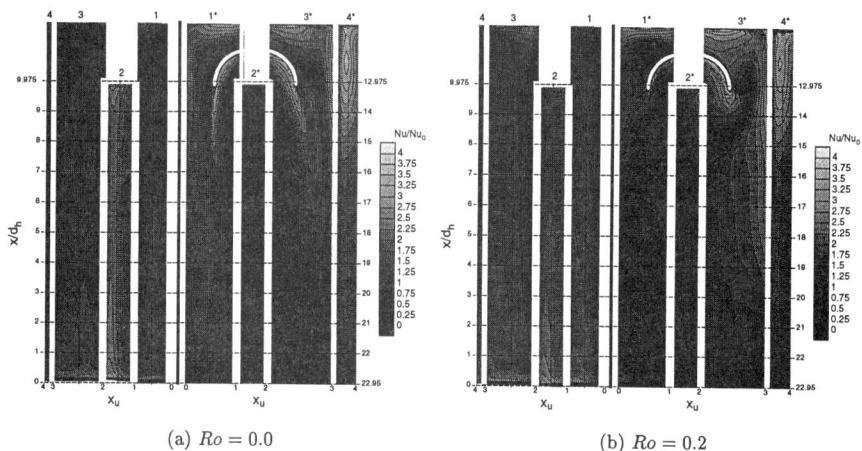

Figure 6: Local Nu/Nu_0 distribution for $Re = 25000$, $Ro = 0.0$ and $Ro = 0.2$

without rotation, the tendencies observed in the experiments could be matched fairly well over the whole domain. The differences in the gradients at the first pass can be explained by the uncertainty of the experimental inflow conditions into the test section. In addition, underprediction of the heat transfer in the order of 10% to 30% can be observed, but the overall shape of the curves is matched. This underprediction may be partly explained by the experimental shortcoming of measuring near the edges of the channel walls. In these regions, however, a low heat/mass transfer is present which would lower the experimental average if taken into account. Another striking difference is the predicted drop of Nusselt numbers right after the beginning of the bend region ($x/d_h = 10.5$). It is caused by separation in the corners of the bend, but could not be observed in the experiments.

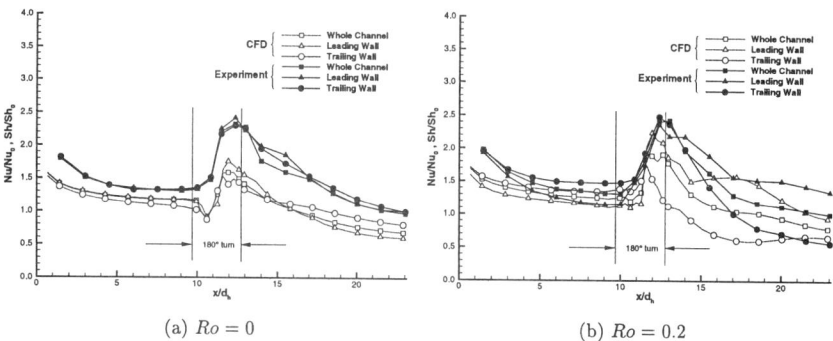

Figure 7: Averaged Nusselt/Sherwood number ratios for $Re = 25000$ plotted over x/d_h

For the case with rotation, fairly good agreement between calculation and experiment is observed up to the bend region. In the second pass, the calculation significantly underpredicts the heat transfer, but still shows some correct tendencies. The leading wall 3*, for example, shows the highest transfer rates, the trailing wall 1* the lowest. Nevertheless, strong differences between CFD and experiments remain in this region, raising questions about the correct prediction of the underlying flow field for this case. At the outlet of the domain, the experiment shows a strong difference between leading wall 3* and trailing wall 1* that is much less pronounced in the computed results. In the CFD computation, it seems that the turn-induced secondary motions are dominant much further downstream, whereas the experiments point to a quicker decay of these motions and therefore, a faster establishment of the new, Coriolis force driven flow field.

SUMMARY AND OUTLOOK

The results presented in this paper have shown the complex nature of heat/mass transfer in a rotating geometry with near engine shape. The naphthalene sublimation technique was used to gain detailed information on local mass transfer. These experiments show, that the effect of rotation is important to the heat/mass transfer in this type of flows, even though the induced secondary flows are small in comparison to the secondary flow that is induced by the sharp turn.

CFD is a tool that can be used to gain more insight into this kind of flows and to predict the overall range of Sherwood/Nusselt numbers. Nevertheless, currently available CFD codes still have some difficulties in precisely predicting local heat transfer distributions.

Future experiments should focus on the measurement of the flow field in order to find out if the difficulties that CFD encounters are mostly because of a poor prediction of the flow field or because of problems with the treatment of near wall regions. Meanwhile, other available turbulence models could be used to investigate their potential in predicting this kind of complex flows.

Acknowledgement: The work presented was supported by the German ministries BMBF and BMWi as part of the joint research programme "Engine 3E".

REFERENCES

1. K.-H. Presser, Experimentelle Überprüfung der Analogie zwischen konvektiver Wärme- und Stoffübertragung bei nichtabgelöster Strömung, *Wärme- und Stoffübertragung*, Bd. 1 (1967), pp 225-236
2. R. J. Goldstein, H. H. Cho, A Review of Mass Transfer Measurements Using Naphthalene Sublimation,*Exp. Thermal and Fluid Science*, 1995, pp. 416-434
3. L. Rathjen, D. K. Hennecke, M. Elfert, S. Bock, E. Henrich, Investigation of Fluid Flow, Heat Transfer and Pressure Loss in a Rotating Multi-Pass Coolant Channel With an Engine-Near Geometry, *Presented at the 14th ISABE Symposium, No. 99-7201, 1999, Florence, Italy*
4. R. J. Goldstein, M. Y. Jabbari, J. P. Brekke, The Near -Corner Mass Transfer Associated With Turbulent Flow in a Square Duct, *Wärme- und Stoffübertragung*, Bd.27 (1992), pp 265-272

Analyses of Heat Transfer in Stationary and Rotating Ribbed Blade Cooling Passages using Computational Fluid Dynamics

R.A. BREWSTER and S. JONNAVITHULA

adapco
60 Broadhollow Road
Melville, NY 11747 USA

ABSTRACT: Computational fluid dynamics (CFD) predictions of the flow patterns and heat transfer in simplified ribbed-wall turbine blade cooling passages were performed for representative stationary and rotating conditions. Analyses have been performed with different mesh densities and using different turbulence models to assess the sensitivity of predictions to these variables. Computed local heat transfer results are compared to measurements available in the literature to assess their accuracy. The results generally agree well with experiment, although the peak values of the heat transfer coefficients were under-predicted in the first leg of the channel. Some sensitivity to mesh density was seen, while the choice of near-wall turbulence model appeared to have little effect.

INTRODUCTION

The need for understanding and improving the cooling of gas turbine blades is well known. Testing and experimentation can often be very expensive and time-consuming. As a result, there has been an increasing desire for computational techniques that can accurately predict the heat transfer characteristics of blade cooling passage designs under realistic operating conditions. In this paper, a commercial computational fluid dynamics (CFD) code, STAR-CD[1], is used to analyze geometries having the essential characteristics of common turbine blade cooling passage designs.

Although a number of CFD analyses of representative cooling passage geometries have been performed in recent years[2-3], the advent of advanced computing technology and parallel processing now make calculations on very dense meshes feasible. Analyses of models containing millions of computational cells are now performed routinely in industry. In addition, continuing developments in the area of turbulence modeling are expected to lead to improvements in the accuracy of predictions. This paper presents results of CFD predictions using very high mesh densities. In addition, results are provided from analyses using several different turbulence models. Comparisons to measurements are used to assess the accuracy of the predictions.

TEST CONFIGURATION AND MEASUREMENTS

Figure 1 shows the geometry analyzed in this paper. This represents a portion of a geometry which had previously been investigated experimentally[4], and for which heat transfer coefficient data on the leading, trailing and lateral surfaces are available for both stationary and rotating conditions. The geometry has ribs skewed at a 45° angle and staggered on the leading and trailing surfaces. The lateral surfaces are smooth.

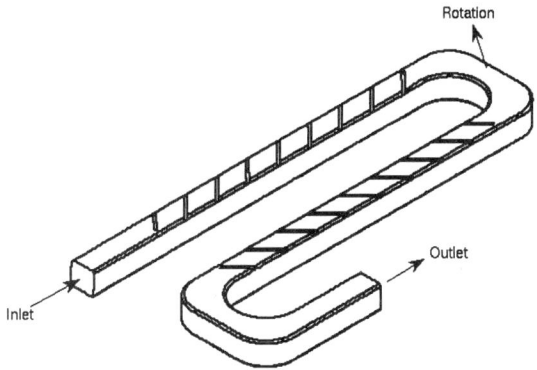

Figure 1: Schematic of the Cooling Passage Configuration

In the experiments, all walls of the channels were heated to a constant temperature, except for the inner surfaces of the bends, which were insulated. Steady-state heat flow and fluid temperature measurements were made over various portions of the geometry for both stationary and rotating (at 550 rpm) conditions

CFD ANALYSES

All numerical analyses were performed using the STAR-CD CFD package[1]. STAR-CD solves the steady or transient fully compressible and viscous mass, momentum and energy conservation equations on unstructured meshes. Analyses may be performed in serial (one processor) or parallel (multi-processor) mode. The models and methods employed in the CFD simulations are described in more detail in the following sections.

Computational Models

The computational models were constructed using the samm[5] (semi-automatic meshing methodology) software package. samm produces a mesh by trimming a hexahedral volume grid to produce a high quality mesh which consists primarily of hexahedral cells and some polyhedral (trimmed hexahedral) cells. In addition, layers can be extruded from the surface mesh to produce a high-quality near-wall mesh.

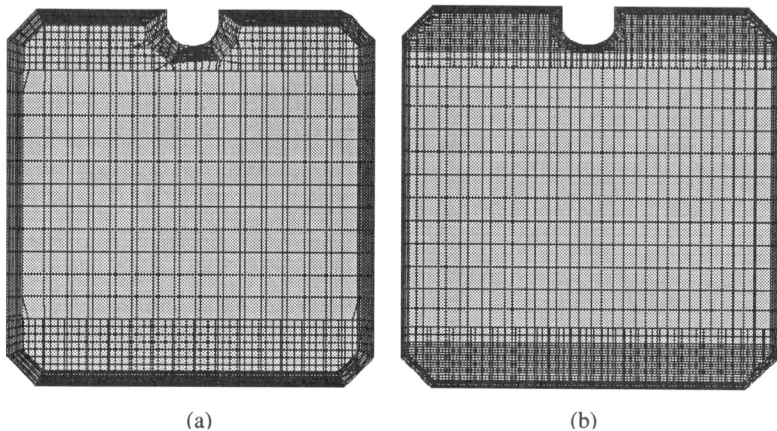

Figure 2: Section Plots Showing the Mesh Structure for the (a) Coarse (1.49 million cells) and (b) Fine (2.74 million cells) Models

Figure 2 shows the meshes used in the present study. The coarser of the two models, shown in Figure 2(a), consists of approximately 1.49 million cells, including 10 near-wall extruded cell layers.

The second of the two models consists of approximately 2.74 million cells, including 10 near-wall extruded cell layers, and is shown in Figure 2(b). For this mesh, special effort was made to refine the mesh behind the ribs, in order to capture the details of wakes behind the ribs. The very dense near-wall mesh was also required to accommodate the requirements of the two-layer wall turbulence model, as discussed below.

Not shown in Figures 1 and 2 were extensions of the mesh upstream of the inlet plane and downstream of the outlet plane. The inlet extension was added to simulate an assumed fully-developed flow condition at the inlet, while the outlet extension was necessary to provide a suitable outflow condition.

Turbulence Models and Thermophysical Properties

Turbulence in the flow was modeled using the Reynolds time averaging procedure. For the results presented here, the RNG variant of the k-ε model was used[1].

In conjunction with the RNG model, a two-layer model was used to model the turbulence in the boundary layer. In this approach, a one-equation algebraic model was used to compute the turbulence dissipation (ε) within the boundary layer, rather than using logarithmic wall functions. The two variants of the one-equation turbulence dissipation models used in the present study are those due to Norris & Reynolds[1], and Wolfstein[1].

The working fluid for the analyses was air. The air was assumed to be compressible, obeying the ideal gas law. The viscosity of the air was assumed to vary with temperature according to the Sutherland law. The specific heat and thermal conductivity of the air were assumed to be constant at 1006 J/kg K and 0.02637 W/m K, respectively.

Boundary Conditions

The inlet boundary conditions correspond to a Reynolds number of approximately 25,000, and are summarized in Table 1. The inlet turbulence conditions in Table 1 were assumed values based on prior experience. For the rotating cases, the rotational speed was 550 rpm about an axis lying in the plane of the cooling passage and located 548.64 mm (21.6 inches) upstream of and parallel to the inlet plane. In these cases, the equations of motion were solved in a rotating reference frame.

Table 1: Summary of Blade Passage Model Boundary Conditions

CONDITION	VALUE
Inlet Mass Flow Rate	5.918×10^{-3} kg/s
Inlet Temperature	299.82 K
Inlet Turbulence Intensity	5.0 %
Inlet Turbulence Length Scale	1.27 mm
Outlet Condition	Zero-gradient outflow
Wall Thermal Condition	344.26 K / Adiabatic

Computational Parameters

A bounded, gradient-based TVD (Total Variation Diminishing) scheme called MARS[1] was used for the spatial differencing of the convective terms. The MARS scheme is second-order accurate. The SIMPLE pressure correction scheme was used together with STAR-CD's conjugate gradient solvers.

The analyses were performed in parallel on four IBM SP2 200 MHz processors. For the coarse (1.49 million cells) model, each executable required 200 Mb of RAM. These analyses typically converged in approximately 900-1000 iterations, requiring approximately 80 CPU hours.

For the finer (2.74 million cells) model, each executable required approximately 445 Mb of RAM. Convergence for these analyses was achieved 900-1000 iterations, requiring approximately 170 CPU hours.

RESULTS AND DISCUSSION

For the purposes of presenting results and comparing with the experimental heat transfer data, the channel was divided into the twelve regions shown in Figure 3. Regions A and L are 38.1 mm (1.5 inches) long, while regions B-D and G-I are 50.8 mm (2.0 inches) long. Regions E, F, J and K each encompass one-half of the two 180° bends.

Results of the CFD calculations are shown in Figures 4-7. Figures 4 and 6 show a comparison between the computed and measured heat transfer enhancement factors on the leading and trailing surfaces under stationary and rotating conditions, respectively. The Nusselt number values were normalized by the fully-developed Nusselt number under non-rotating conditions to obtain the enhancement factors.

Values of the enhancement factor are provided for each of the twelve channel regions shown in Figure 3. The Nusselt number for each of the regions A-L was computed as:

$$Nu = Q_{i,j} D_h / [k A_{i,j} (T_w - T_{b,j})]$$

where $Q_{i,j}$ is the total heat transfer for wall i of region j (i.e. i is the leading or trailing wall and j is the region of the channel A-L), $D_h = 12.7$ mm (0.5 inch) is the hydraulic diameter of the channel, k is the thermal conductivity of the air, $A_{i,j}$ is the total area of wall i of region j, $T_w = 344.26$ K is the prescribed wall temperature, and $T_{b,j}$ is the bulk (average) temperature of region j. The bulk temperature for each region was computed as the mean of the mass-flux-weighted inlet and outlet temperatures for that region.

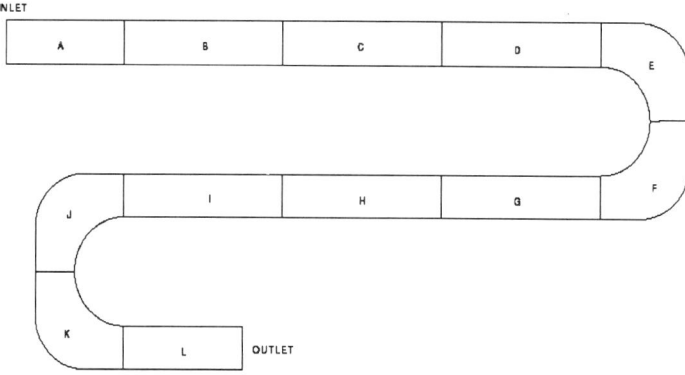

Figure 3: Definitions of Regions A-L of the Passage Geometry

The Nusselt numbers computed in this way were then normalized by the stationary, fully-developed Nusselt number to obtain the enhancement factor. For the conditions considered in this paper, the value of stationary, fully-developed Nusselt number was 58.

Figures 5 and 7 provide detailed in-plane velocity distributions at various sections through the fine model, under stationary and rotating conditions, respectively. In Figure 7 (rotating case), the velocity vectors are relative to the rotating (at 550 rpm) blade passage.

The results of the stationary analyses shown in Figure 4 show generally good qualitative and quantitative agreement with experiment, although in all cases, the peak enhancement factors are under-predicted by the CFD analyses for the first leg. This may be a function of the assumed geometric and turbulence conditions, since the agreement is very good downstream of the first leg.

The choice of near-wall turbulence model is seen to have little effect on the results, while the increased density of the fine model mesh does show some improvement. The first data point represents an unenhanced region of the channel (i.e. no ribs). Downstream of this region, until the first bend is reached, the ribs promote disruption of the boundary layer, resulting in significantly enhanced heat transfer. Note that the data for the "leading" and "trailing" surfaces are not identical because the ribs on these surfaces are staggered relative to each other.

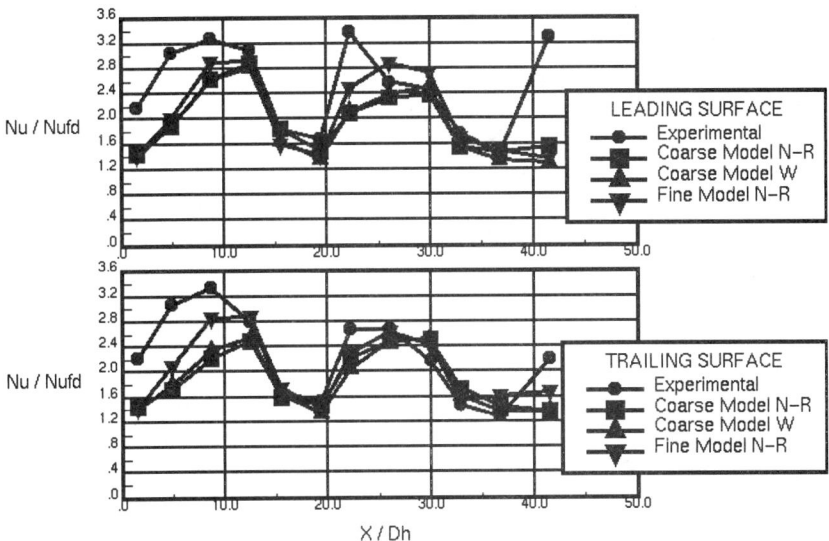

Figure 4: Comparison of Experimental and Computed Heat Transfer Enhancement Factors for Stationary Channels (N-R: Norris-Reynolds two layer model; W: Wolfstein two layer model)

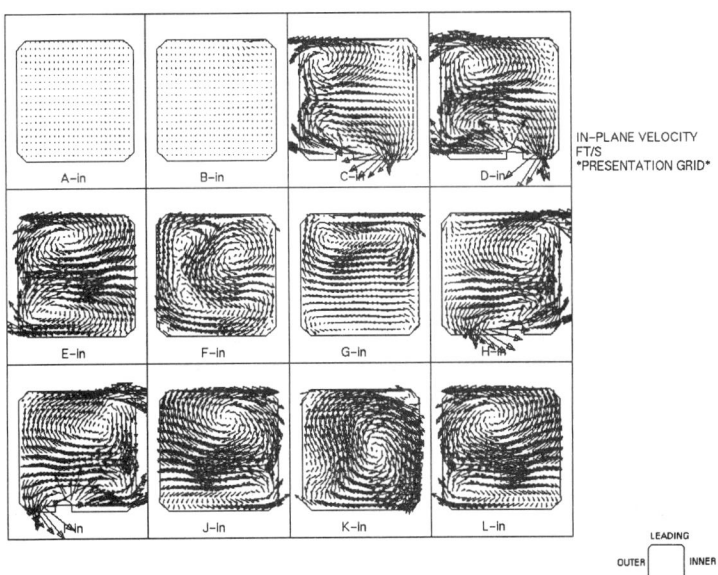

Figure 5: Plots of the In-Plane Velocity Field at Various Sections for the Stationary Analysis of the Fine (2.74 million cells) Model with the Norris-Reynolds Two-Layer Model (View is from Upstream)

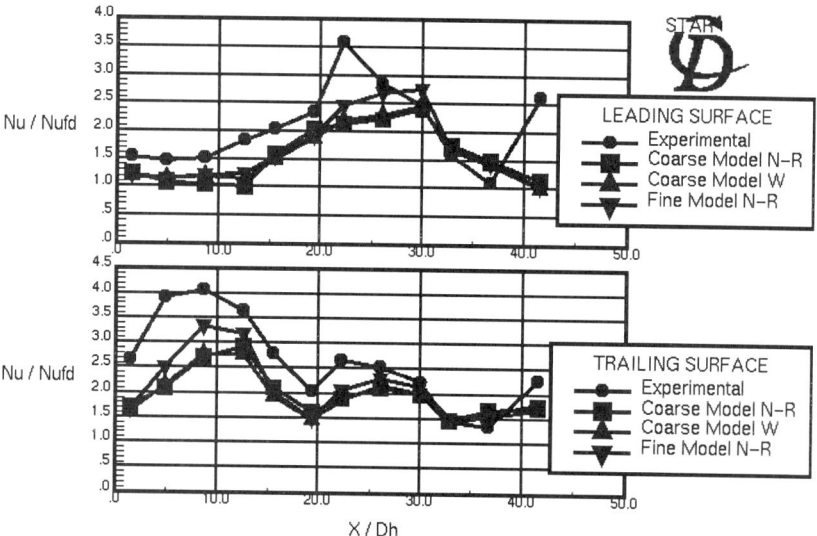

Figure 6: Comparison of Experimental and Computed Heat Transfer Enhancement Factors for Rotating Channels (N-R: Norris-Reynolds two layer model; W: Wolfstein two layer model)

Through the first turn, the heat transfer enhancement factors decrease due to the absence of the ribs in these regions. In the second leg of the passage, the enhancement factors increase once again, although not to the levels seen in the first leg. In the outlet section L, the CFD results significantly underpredict the enhancement factors because the experimental test section had ribs in this region, but the CFD model did not.

In Figure 5, the in-plane velocity field is shown at a number of locations. Sections A-in and B-in are before the flow has encountered the ribs. After the flow has entered region B, the flow near the upper and lower surfaces is directed toward the outer walls, setting up counter-rotating recirculating zones, as seen in sections C-in, D-in and E-in of Figure 5. Through the bend, the flow reverses and recirculation zones rotating in the opposite directions are seen in the second leg.

A comparison of measured and computed enhancement factors for the rotating cases is shown in Figure 6. Once again the CFD simulations follow the general trends of the experimental data, and the fine model results show the best agreement. Very little difference is seen between the Norris-Reynolds and Wolfstein two-layer models.

Lower enhancement factors (compared to the non-rotating case) are seen on the leading surface due to the weaker flow there. Conversely, the trailing surface shows somewhat higher enhancement factors.

Sections A-in and B-in of Figure 7 show the weak rotational flow upstream of the ribbed sections. Downstream of the beginning of the ribbed section of the first leg, the ribs direct the flow toward the outer wall, while the rotation sets up a flow from the trailing to the leading surfaces along the inner and outer walls. The resultant of these two effects is the flow shown in section E-in.

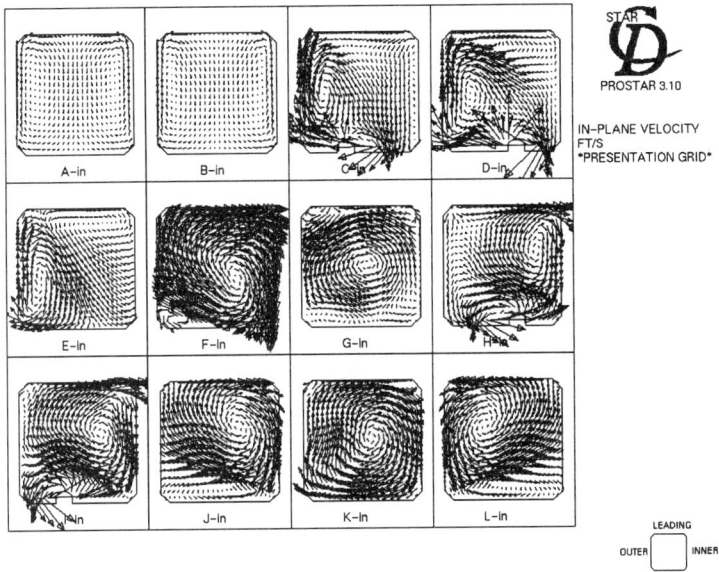

Figure 7: Plots of the In-Plane Relative Velocity Field at Various Sections for the Rotating Analysis of the Fine (2.74 million cells) Model with the Norris-Reynolds Two-Layer Model (View is from Upstream)

CONCLUSIONS

CFD simulations of flows in simplified ribbed turbine blade cooling passages were performed and the results compared to measurements under typical non-rotating and rotating conditions. The agreement was generally good, with the exception of the first leg where assumptions about the inlet turbulence and geometry may have played a role. The results also showed that the choice of near-wall turbulence model had little effect on the results, while the use of a denser mesh improved the agreement with experiment.

The primary effect on heat transfer enhancement was due to the presence of the ribs. The effect of the rotation was secondary. Therefore, it is concluded that the denser mesh better captured the disruption of the flow caused by the ribs. Furthermore, even the very dense meshes used in these analyses (on the order of millions of cells) were insufficient to capture all of the effects of the flow disruption caused by the ribs.

REFERENCES

1. STAR-CD V3.1 Methodology & User Manuals. 1999. Computational Dynamics, London, UK.
2. Bonhoff, B., U. Tomm, B. Johnson & I. Jennions. 1997. Heat Transfer Predictions for Rotating U-Shaped Coolant Channels with Skewed Ribs and Smooth Walls, ASME Paper 97-GT-162.
3. Abauf, N. & D.M. Kercher. 1994. Heat transfer and turbulence in a turbulated blade cooling circuit, ASME J. Turbomachinery, **116**:169-177.
4. Johnson, B.V., J.H. Wagner, G.D. Steuber & F.C. Yeh. 1994. Heat transfer in rotating serpentine passages with trips skewed to the flow, ASME J. Turbomachinery, **116**:113-123.
5. The samm v2.02.01 User Manual. 1999. adapco Software, Melville, NY, USA.

PREDICTION OF PRESSURE LOSS AND HEAT TRANSFER IN INTERNAL COOLING PASSAGES

K. HERMANSON, S. PARNEIX, J. VON WOLFERSDORF & K. SEMMLER

ALSTOM POWER Switzerland, CH-5401 BADEN, Switzerland

ABSTRACT: This paper reports CFD-simulations of the turbulent flow, pressure loss and heat transfer occurring in ribbed passages. The channel section is rectangular, with an aspect ratio of 2.04. Ribs are square cross-section, their height is 10% of the channel height, and their inclination is varied from 90° to 33°. Reynolds number is 30,000. Three turbulence models (k-ε wall functions and 2-layer, V2F) are used and compared to the experimental data of Cho et al.[1]. All three models accurately predict the pressure losses due to the ribs and the qualitative heat transfer distribution on the ribbed wall. However, only the V2F model can accurately reproduce the absolute heat transfer levels, this at all inclination angles. The correlation developed by Han and co-workers for smaller rib-heights under-predicts the friction factor and wall heat transfer level on the current configuration. This shows the danger of using a correlation outside of its application range.

NOMENCLATURE

Cp	Specific heat	S	Wall surface area
D_h	Channel hydraulic diameter	T_b	Channel bulk temperature
e	Rib height	T_w	Channel wall temperature
f	Friction factor	U_b	Channel bulk velocity
f_0	Smooth wall friction factor	W	Channel width
h	Heat transfer coefficient, $q/(T_w-T_b)$	y^+	Normalized distance to the wall
H	Channel height	α	Rib inclination
Nu	Nusselt number (h D_h / λ)	Δp	pressure loss across one rib pitch
Nu_0	Smooth walls Nusselt number	η	Heat transfer performance
P	Inter-rib axial pitch	λ	Heat conductivity
Pr	Prandtl number (μCp / λ)	μ	Molecular viscosity
q	Surface heat flux	ρ	Density
Re	Reynolds number ($\rho U_b D_h$ / μ)		

INTRODUCTION

Modern day demands for higher efficiencies in gas turbine engines require the gas turbine cycle to operate at higher pressure ratios and, of greatest significance to the turbine designer, at high turbine inlet temperatures. Turbine inlet temperatures can exceed the allowable blade metal alloy temperature by more than 500°C, so the blades can survive only with the use of effective cooling methods. Cooling reduces the mean temperature of the blade material and includes both external (film or transpiration cooling) and internal cooling techniques (convection and/or impingement cooling).

For both industrial and aero gas turbines, the ability of predicting the internal flow and heat transfer in blade cooling passages is essential if accurate blade metal temperature and blade lifetime are to be calculated quickly. Note that a decrease of 25°C in metal temperature can in some circumstances double the life of a high-pressure turbine blade. Also a better prediction capability would allow a better minimization of the flow taken from the compressor to cool the turbine blades, reducing the thermodynamic penalties, thus improving the overall gas turbine cycle efficiency, which helps reducing emissions.

The convective heat transfer within the internal passages of a turbine blade is usually augmented with the use of turbulator ribs. The ribs are designed to introduce additional flow mixing, through secondary flow generation and turbulence enhancement, increasing heat transfer locally. The optimum design is achieved only with a balance of heat transfer and friction factor augmentation. For a more efficient and cost-effective ribbed channel design, there is an impending need for improvement and validation of numerical heat transfer predictions, namely near-wall turbulence models. Computational Fluid Dynamics (CFD) codes must be benchmarked with experimental data to monitor both flowfield, pressure and heat transfer predictions on both a quantitative and qualitative level. Novel improvements in turbulence modeling [2-9] have recently allowed a jump in the confidence that one can have in CFD in accurately predicting turbulent heat transfer.

The objective of this work has been to compute the turbulent flow, the pressure loss and thermal fields in a ribbed passage while the rib inclination is varied from 90° to 33° (see Figure 1). To this end, the CFD-code Fluent [10], using unstructured hybrid 3D meshes (Centaur [11]), has been considered. Three turbulence models have been tested and compared against the recent pressure-loss and analogous mass-transfer (naphthalene sublimation technique) experimental data from Cho et al. [1]:

- Standard k-ε model with wall functions
- Standard k-ε model with the 2-layer near-wall approach
- V2F model

A comparison with a standard correlation developed by Han and co-workers [12,13,14] has also been performed. Additionally, some information, which was not available in the experiments, namely the flow structure and the heat transfer on the rib and on the side-walls, has been taken from the CFD-simulations and is also reported.

REFERENCE GEOMETRY AND DATA

The present configuration consists in a very long rectangular-section ribbed channel of aspect ratio W/H=2.04. Ribs are square-section, inline, their height is 10% of the channel height and inter-rib pitch is 10 rib heights. Inclination of the ribs is varied between 90° and 33°. The two side-walls have been denoted leading and trailing side-walls, depending on their location relatively to the coolant flow and the ribs (see Figure 1). Reynolds number, based on bulk velocity and channel hydraulic diameter is 30,000.

Mass-transfer measurements (naphthalene) have been taken in the fully-developed region along the ribbed wall [1]. The ribs were not coated. Heat-mass transfer analogy was invoked to deliver heat transfer enhancements, Nu/Nu_0. Wall pressure has been measured along the overall channel, and the fully-developed friction factor has then been derived. Since only the fully-developed region is of interest here, only one inter-rib pitch has been modeled (see Figure 1) with cyclic boundary conditions on the inlet/outlet boundaries, following Patankar's methodology [15]. Constant wall temperature is considered to be the most appropriate thermal boundary condition for mass transfer experiments.

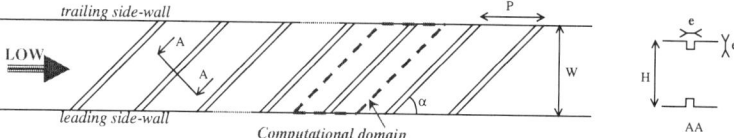

Figure 1 *Geometry of the ribbed channel configuration*

NUMERICAL METHODOLOGY

Numerical process

As stated in the previous section, only one rib-module has been considered. The full height of the channel has been modeled, although the current configuration is symmetric. This allowed checking the symmetry of the solution (even so the grid is not symmetric), which adds some confidence in the calculations. Also, rotation effects could then be modeled direclty. Unstructured hybrid meshing has been applied here with Centaur [11]. The surface mesh is covered with triangles. From this surface mesh, a set of prism layers develop and allow high grid quality in the near-wall regions where gradients are mostly normal to the boundary. Then tetrahedrals are used to fill in the rest of the domain, which allows meshing highly complex geometry. The unstructured finite-volume CFD-code Fluent 5 [10] is used for computing the fully-developed flow and heat transfer. The incompressible formulation with constant air-properties was utilized. Convergence was supposed to be reached when all normalized residuals were smaller than 10^{-4}, except energy smaller than 5.10^{-7}; 2nd order of discretization was used on all the equations.

Turbulence modeling

The wall function [16] is the most popular solution for near-wall turbulence modeling. It involves patching a prescribed semi-empirical boundary condition onto a basic model. The function (a log-law) attempts to specify how the boundary layer and associated turbulence vary near the wall. This algebraic function is then imposed on the first computational node close to the wall (see Figure 2). One major constrain is that this computational node should be located far enough from the solid wall, well into the turbulent part of the boundary layer ($y^+>30$), which is not always feasible. It is valid in configurations where the flow is parallel to the wall and no pressure gradients are present. The wall function is available in Fluent 5, for all turbulence models (k-ε, RNG, RSM).

The alternative to wall-functions is the so-called "low-Reynolds" turbulence models. These models are valid up to the solid wall and use the no-slip boundary condition directly at the wall. One of these models, the two-layer approach [17], attempts to model the boundary layer by patching two models, one for the outer turbulent flow, one for the flow in the wall layer (see Figure 3). The location where this patching occurs is dependent on the distance from the solid wall. The two-layer model offers the advantage of avoiding the wall-functions, which are not universal. However, patching has a tendency to be sensitive to the distribution of computational cells, an undesirable property. The 2-layer model is available in Fluent 5, for all turbulence models (k-ε, RNG, RSM).

The V2F model was developed at Stanford University [2,3]. It originated from the desire to eliminate the need to patch or damp models in order to predict phenomena like heat transfer or flow separation and to try introducing some of the physics of near-wall turbulence. It has some of the merits of Reynolds-stress models, while remaining computationally attractive. For instance, it takes into account the anisotropy of near-wall turbulence near solid walls, while it retains the robust eddy-viscosity assumption. As a result, V2F is a single model, valid throughout the whole flow domain, automatically becoming a wall model close to solid boundaries (see Figure 4). Several years of basic research at Stanford University have consistently proved that V2F was an effective solution to near-wall modeling. For instance, adverse-pressure gradient flows, separated flows, impinging flows, 3D flows including 3D boundary layers and horseshoe vortices were successfully modeled[4-8]. The V2F model has been made available by Cascade Inc.[18] as a module working in Fluent 5.

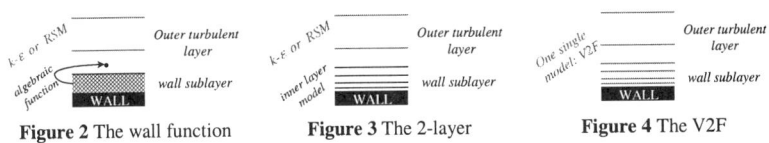

Figure 2 The wall function   **Figure 3** The 2-layer   **Figure 4** The V2F

Grid independence

Grid independence has been checked by using three different grids with 200 000, 450 000 and 1 million cells. Refinement has been done with smaller triangles and finer prism layers along the walls, and smaller tetrahedrals in the middle of the duct. The difference with the fine grid in friction factor and ribbed-wall Nusselt number was 4.8 and 1.1% (resp. 2.2 and 0.7) for the coarse mesh (resp. the middle-size mesh). The grid with 450 000 cells (see Figure 5) was thus considered appropriate. A similar study has been successfully performed for the wall-function model (leading to basic meshes of about 150 000 cells). However, note that for the 90°-ribs case, the y^+-constrain can not be fulfilled while keeping a fine enough resolution on the ribs (y^+=30 along the ribbed wall corresponding to approximately a third of the rib height).

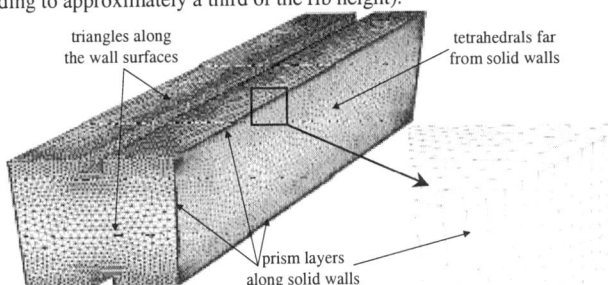

Figure 5 *Basic hybrid mesh (about 450 000 cells)*

COMPARISON WITH EXPERIMENTAL DATA

Global design variables

Two main variables are of primary importance for the design of internal cooling turbine blade: the pressure loss and the average wall heat transfer. The former will allow evaluating the coolant mass flow throughout the cooling system. The later, which also depends on the former, will evaluate the average cooling efficiency. A cooling ribbed channel can be better analyzed with the following non-dimensionalized parameters:
- The friction factor, $f = (D_h \Delta p) / (2 \rho U_b^2 P)$
- The heat transfer enhancement: $Nu / Nu_0 = Nu / (0.023 \, Re^{0.8} \, Pr^{0.4})$
- The heat transfer performance: $\eta = (f/f_0) / (Nu/Nu_0)^{1/3}$, with $f_0 = 0.046 \, Re^{-0.2}$

Figure 6 and Figure 7 show the results obtained for different rib inclinations. One can see that all turbulence models are well capable of reproducing the pressure loss induced by the ribs (with a slight better performance for the V2F model), except the wall-function formulation for the 90°-case, which can partially be explained by the non-adequacy of the mesh regarding the y^+-limitations. However, the average wall heat transfer results show a different picture. Only the V2F model is able to accurately predict the ribbed-wall Nusselt number for all inclination angles. The k-ε model, with both near-wall formulations, cannot reproduce the numbers, and even the qualitative trends. Indeed, this model would predict that the 33°-case is optimal for a heat transfer performance point of view (see Figure 8), which is not at all indicated by the experiment. For this channel aspect ratio, the heat transfer performance is actually about the same between 90° and 45°, and then decreases for α=33°. The V2F model reproduces this trend very well.

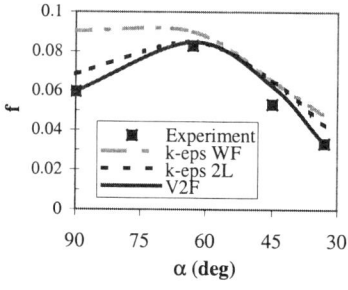

Figure 6 *Friction factor*

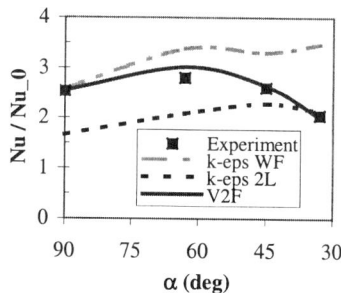

Figure 7 *Ribbed-wall averaged heat transfer*

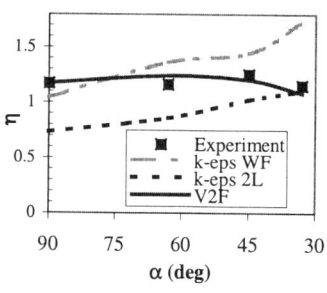

Figure 8 *Heat transfer performance*

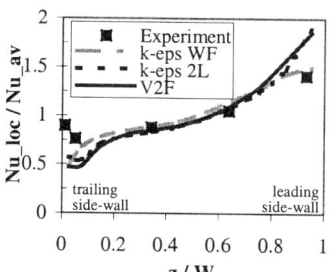

Figure 9 *Local spanwise heat transfer (α=45°)*

Local heat transfer

Besides the two average values studied previously, the local heat transfer distribution can be of importance. Note that this parameter plays a smaller role in the design because the material conductivity of blades tends to smooth out heat transfer gradients. However, if these gradients become big enough, they will significantly influence the blade wall local temperature. For blade cooling applications, the streamwise-averaged spanwise distribution (i.e. distribution along the airfoil chord for a radial cooling channel) is of major interest. It has been normalized by the average ribbed-wall Nusselt number taken for each individual simulation. This way, errors on the absolute heat transfer level are not visible and only the qualitative spanwise distribution is shown. Only the 90° and 45° cases have been studied experimentally [1]. The 90°-ribs do not create any significant spanwise heat transfer gradients (not shown here). However the secondary flow created by the inclined ribs (see Figure 13) strongly influences the local Nusselt distribution, with higher values close to the leading-sidewall (see Figure 9), where the secondary flow is impinging on the ribbed wall. One can see that all turbulence models reproduce this trend quite well. Only an over-prediction of the gradients is present close to the side-walls; however, note that experimental data at these locations might have a higher uncertainty.

Comparison with a standard correlation

Han et al.[12,13,14] have developed a standard correlation for ribbed-channels, based on a set of experimental data, in particular, with several rib heights, pitches, inclinations, this with different channel aspect-ratios. The heat transfer on the rib was included ("*the thin layer of glue ensured electrical isolation but thermal conduction from foil to brass ribs*"[14]), so the correlation has been corrected due to the added wetted-surface effect for a fair comparison with the experiment. Unfortunately, there is no data for this geometry, which gives the rib to ribbed-wall heat transfer ratio[19]. So, only an average Nusselt number is given (heat transfer on the rib equal to heat transfer between the ribs):

$Nu_{global} = Nu_{correlation} (S_{projected\ area} / S_{global})$,   *global area = rib + ribbed wall*
$Nu_{global} = (S_{ribbed\ wall} / S_{global}) Nu_{ribbed\ wall} + (S_{rib}/S_{global}) Nu_{rib}$, *with* $Nu_{rib} \approx Nu_{ribbed\ wall}$

The correlation under-predicts the pressure losses for the inclined ribs (about 15%, see Figure 10) and the heat transfer levels on the ribbed wall (up to 20%, see Figure 11). Some simulations have been performed with heat transfer, either only on the ribbed-wall, or on all the channel-walls. The latter case delivers smaller heat transfer levels, as expected (the coolant is heating up along the other walls), but the difference is only 5% and cannot explain the 20% under-prediction from the correlation. In fact, the latter has been developed with experiments using smaller ribs (rib height of 5% channel height). So, it seems to be dangerous to extrapolate the correlation outside of its experimental rib-height range. Note that this result is consistent with the finding of *Rau et al.*[20]

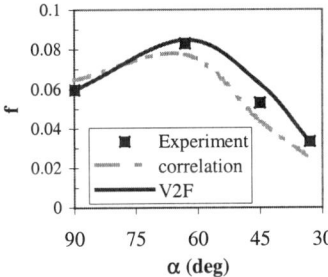 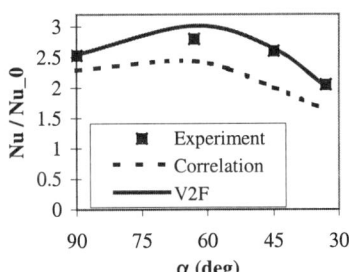

Figure 10 *Friction factor*   **Figure 11** *Ribbed-wall averaged heat transfer*

FURTHER ANALYSIS

Some information, which is important for the design, was missing in the experimental study, i.e. side-wall heat transfer, rib heat transfer, flow structure. However, the V2F model has been shown to agree well with the experimental data available for all rib inclination angles. So this single model will be used to analyze this missing information. Some qualitative comparison with some literature data will also be added.

Side wall heat transfer

Heat transfer levels as high as on the ribbed-wall can be obtained along the leading side-wall, on which cool air, entrained by the secondary flow, impinges (see Figure 12). On the trailing side-wall, small heat transfer occurs, but it is still up to 50% bigger than the smooth-wall heat transfer levels. This can be expected, because the presence of the ribs creates high turbulence levels around them, increasing heat transfer. This latter qualitative trend has been found in some experiments [12,21]. However, the existing data shows a big scatter in the quantitative levels, so one cannot conclude about the CFD-accuracy. In any case, the heat transfer levels along the webs is far from being equal to the smooth-wall heat transfer. Although the webs are not directly in contact with the hot gas, this information is important because an under-estimation of the heat transfer, locally on the web, will over-predict the local metal temperature on the webs, so it will under-predict the thermal stresses, and so will over-predict the life of the blade.

Rib heat transfer

Heat transfer levels on the ribs have been found to be 15 to 50% greater than on the ribbed wall (see Figure 12). Again, this trend qualitatively agrees very well with some experimental data, although performed on different rib configurations [19] (ratio between rib and ribbed-wall Nusselt number is between 1.3 and 2.0, depending on the rib height, but on a square-section channel). When high inclined ribs are used, the rib heat transfer area becomes a significant part of the total heat transfer area. So the heat transfer level on the rib itself becomes of importance and will influence the overall Nusselt numbers.

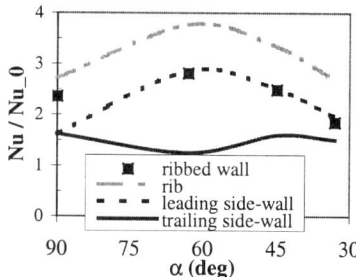

Figure 12 *Heat transfer on all walls (V2F)*

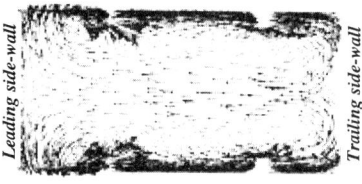

Figure 13 *Secondary flow structure in plane orthogonal to channel axis (45°-case)*

Flow structure

The secondary flow occurring in the present ribbed channels is shown on Figure 13. It is enhancing the mixing between the coolant in the middle of the duct and the hot walls. Also it allows a flow motion even downstream of the ribs eliminating dead zones, which are present for the 90°-case. Note that the confidence level in the prediction of the secondary flow is quite high [22]. For a square-section duct, two main secondary vortices are created. For the current aspect ratio of 2, four structures can be observed. This is mainly due to the fact that a slice orthogonal to the channel axis cuts more than 1 rib per ribbed-wall (see Figure 13). Also the intensity of the secondary motion decreases when channel aspect ratio increases (spanwise velocity equal to 30% of the bulk velocity in the middle of the duct for W/H=2.04, instead of 50% for W/H=1). This is consistent with the decrease of heat transfer when the duct aspect ratio is increasing (W/H>1).

CONCLUSIONS

CFD-simulations of pressure loss and heat transfer in ribbed passages have been performed. The modern V2F turbulence model agrees very well with the experimental data, for all rib-inclinations studied (90 to 33°). The standard k-ε model can only reproduce the pressure loss and spanwise relative heat transfer distribution. The k-ε prediction of heat transfer levels is poor, even qualitatively (e.g. it does not show the right trend when the rib-inclination is varied). Heat transfer levels on the ribs and on the sidewalls have also been computed (V2F). Values higher than the ones found respectively on the ribbed-wall and the smooth walls have been obtained. Accurate predictions of these parameters are also requested for a correct heat transfer analysis of blade multi-pass cooling systems. Some more experimental data is thus needed for extending the CFD-validation at this level. Finally, rotation effects (buoyancy and Coriolis) should be studied before CFD can be used with high confidence in turbine blade cooling design.

REFERENCES

1. Cho, H.H., Wu, S.J., and Kwon, H.J. 1999. Local Heat/Mass Transfer Measurements in a Rectangular Duct with Discrete Ribs. ASME Paper No. 99-GT-121.
2. Durbin, P.A. 1991. Near-wall turbulence closure modeling without "damping functions". Theoret. Comput. Fluid Dynamics. **3**: 1-13.
3. Durbin, P.A. 1993. Application of a near-wall turbulence model to boundary layers and heat transfer. Int. J. Heat and Fluid Flow. **14**: 316-323.
4. Durbin, P.A. 1995. Separated flow computations with the k-ε-v^2 model. AIAA J. **33**: 659-664.
5. Behnia, M., Parneix, S. & Durbin, P.A. 1998. Prediction of heat transfer in an axisymmetric turbulent jet impinging on a flat plate. Int. J. Heat and Mass Transfer. **41**: 1845-1855.
6. Parneix, S., Durbin, P.A. & Behnia, M. 1998. Computation of 3-D turbulent boundary layers using the V2F model. Flow Turbulence and Combustion Journal. **60**: 19-46.
7. Parneix, S., Behnia, M. & Durbin, P.A. 1999. Predictions of turbulent heat transfer in an axisymmetric jet impinging on a heated pedestal. J. Heat Transfer. **121**: 43-49.
8. Manceau, R. & Parneix, S. 2000. Turbulent heat transfer predictions using the v^2-f model in a finite element code. Int. J. Heat and Fluid Flow. **21**: 320-328.
9. Iacovides, H. 1998. Computation of flow and heat transfer through rotating ribbed passages. Int. J. Heat and Fluid Flow. **19**: 393-400.
10. FLUENT User's Guide, Version 5.1. 1999. **I-IV**, Fluent Inc.
11. Khawaja, A., Kallinderis, Y., Irmisch, S., Lloyd, J., Walker, D. & Benz, E. 1999. Adaptive hybrid grid generation for turbomachinery and aerospace applications. AIAA Paper No 99-0916.
12. Han, J.C. & Park, J.S. 1988. Developing heat transfer in rectangular channels with rib turbulators. Int. J. Heat Mass Transfer. **31**: 183-195.
13. Han, J.C. 1988. Heat transfer and friction characteristics in rectangular channels with rib turbulators. J. Heat Transfer. **110**: 321-328.
14. Han, J.C., Ou, S., Park, J.S. & Lei, C.K. 1989. Augmented heat transfer in rectangular channels of narrow aspect ratios with rib turbulators. Int. J. Heat Mass Transfer. **32**: 1619-1630.
15. Patankar, S.V., Liu, C.H. & Sparrow, E.M. 1977. Fully developed flow and heat transfer in ducts having streamwise periodic variations of cross-sectional area. J. Heat Transfer. **99**: 180-186.
16. Launder, B.E. & Spalding, D.B. 1974. The numerical computation of turbulent flows. Computer Methods in Applied Mechanics and Engineering. **3**: 269-289.
17. Chen, H.C. & Patel, V.C. 1988. Near-wall turbulence models for complex flows including separation. AIAA J. **26**: 641-648.
18. Cascade Inc. 1999. www.turbulentflow.com.
19. Taslim, M.E. & Wadsworth, C.M. 1997. An experimental investigation of the rib surface-averaged heat transfer coefficient in a rib-roughened square passage. J. Turbomachinery. **119**: 381-389.
20. Rau, G., Cakan, M., Moeller, D. & Arts, T. 1998. The effect of periodic ribs on the local aerodynamic and heat transfer performance of a straight cooling channel. J. Turbomachinery. **120**: 368-375.
21. Cakan, M. 2000. Aero-thermal investigation of fixed rib-roughened cooling passages. VKI Lecture Series 2000-03. 1-73.
22. Bonhoff, B., Parneix, S., Leusch, J., Johnson, B.V., Schabacker, J. & Bölcs, A. 1999. Experimental and numerical study of developed flow and heat transfer in coolant channels with 45 degree ribs. Int. J. Heat and Fluid Flow. **20**: 311-319.

Numerical Simulation of Local Heat Transfer in Rotating Two-Pass Square Channels

A.I. KIRILLOV[a], V.V. RIS[a], E.M. SMIRNOV[b], and D.K. ZAITSEV[b]

[a]*Department of Thermoengineering,* [b]*Department of Aerodynamics,*
St.-Petersburg State Technical University, St.-Petersburg, 195251, Russia

ABSRACT: 3D turbulent air flow and heat transfer developing in a two-pass square channel rotating in the orthogonal mode are simulated using the high-Re k-ε turbulence model and a recently developed modification of wall functions. Auxiliary problem for accurate definition of inlet boundary conditions formed by a long unheated upstream section is considered. Details of flow structure are presented. Local heat transfer results are compared with experimental data.

INTRODUCTION

A two-pass rotating channel is a representative model for simulations of the conditions typical for internal serpentine cooling passages in gas turbine rotor blades. Local heat transfer in two-pass channels has been studied experimentally by several research groups for the last decade[1-3]. Recently 3D simulation has been started to use for getting more knowledge about features of flow developing in the channel[4-7].

Real flow and heat transfer phenomena are made very complicated by the presence of a sharp U-bend, and also by Coriolis and rotational buoyancy forces. The Coriolis-influenced flow field and the heat transfer from the individual surfaces depend also on the channel orientation with respect to the rotation axis.

Three-dimensional numerical simulation is expected to be an effective tool for prediction of the flow and local heat transfer distributions at various operating conditions. However, additional efforts are needed to assess whether engineering turbulence models are able to predict phenomena of such complexity.

FLOW CONFIGURATION

Basically, conditions adopted at the experiments of Dutta & Han[2] are considered in the present work. The only difference consists in use of constant wall temperature conditions over the heated region instead of streamwisely-constant wall heat flux arranged in the measurements. Figure 1 shows the flow geometry for case when the turn is aligned with the rotation axis. The flow is determined by the Reynolds number, Re=$W_m D/\nu$, (W_m is bulk velocity) the rotation number, K=$\omega D/W_m$, the normalized mean radius, R_m/D=$(R_0 + L/2)/D$, and the temperature factor, ε_T=$(T_{wall} - T_{in})/T_{mean}$. Among the

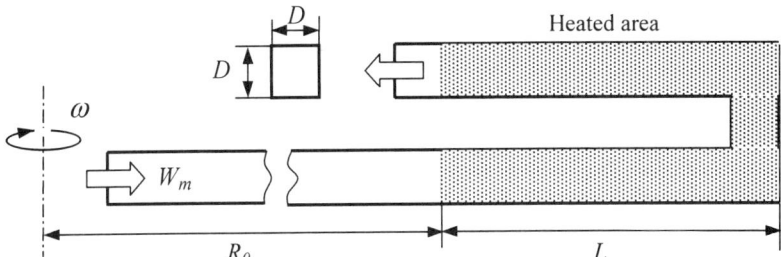

Fig. 1 Flow geometry for case of parallel orientation of the turn with respect to the rotation axis

measurement data provided[2] runs with the Reynolds number of 5000 are of the major interest for the present study since the Coriolis and the buoyancy effects are the most pronounced (K=0.15, ε_T=0.13, R_m/D=55). The heated area is 26D long.

MATHEMATICAL MODEL

Governing Equations

Governing equations are the full Reynolds-averaged Navier-Stokes equations written with the Boussinesq's approximation for incorporation of the centrifugal buoyancy. The gravitational inertia is neglected due to its small magnitude compared to the rotationally induced centrifugal force. The energy equation is written in the form of static enthalpy balance. The governing equations of the standard high-Re k-ε model[8] in combination with the wall function technique are used for turbulence modeling.

Wall Functions

For application of the standard k-ε turbulence model it is usually recommended to place the first near-wall grid point at a distance satisfying the following condition: $y^+ > 30$, where y^+ is the conventional normalized wall coordinate. In case of internal flows controlled by curvature and/or rotation, this condition usually contradicts the accuracy requirement for resolution of secondary flow itself and its effect on the primary flow. The most severe situation is associated with the Reynolds numbers lower than 10,000 that is just the case in the present study. To avoid this contradiction, one can use any low-Re turbulence model. However, for complex 3D problems of engineering orientation this way, requiring a grid strongly clustered near the wall, is still less attractive than the application of the wall function technique. An alternative way is to generalize/modify the wall functions to shift the limit of applicability of the technique up to $y^+ = 3...5$ that would be enough in terms of flow and heat transfer resolution.

The near-wall velocity distribution employed in the present study as a wall function is described by the following three-layer approximation[8]

$$u^+ = y^+, \quad y^+ < 5 \tag{1}$$

$$u^+ = 5 \cdot \ln y^+ - 3.04719, \quad 5 < y^+ < y_b^+ \tag{2}$$

$$u^+ = \kappa^{-1} \ln(y^+ E), \quad y^+ > y_b^+; \quad \kappa = 0.41, \quad E = 9.0 \tag{3}$$

where y_h^+ is defined by the equality of u^+-values given by (2) and (3). For the energy equation, the generalized heat transfer wall function[8,10] is used.

Standard wall functions for the k-ε governing equations are given by [8]

$$k = u_\tau^2 / C_\mu^{0.5}, \quad \varepsilon = u_\tau^3 / \kappa y \tag{4}$$

where u_τ is the friction velocity. For the problem under consideration the standard formulation (4) results in a considerable overprediction of heat transfer rate on any computational meshes sufficient for secondary flow resolution. To get more accurate results, modified wall functions suggested recently by one of the authors[11] are used for definition of turbulent parameters at the first computational point away of the wall. The modification consists in incorporation of a damping factor, D, into expressions (4) as

$$k = D^n u_\tau^2 / C_\mu^{0.5}, \quad \varepsilon = u_\tau^3 / \kappa y D^{1-2n} \tag{5}$$

where n is an empirical constant. Substituting (5) into the relation for the eddy viscosity

$$v_t = C_\mu k^2 / \varepsilon = \kappa u_\tau y D \tag{6}$$

shows that D may be treated as a conventional near-wall damping factor. To define D, the following expressions[12] are used

$$D = (y^+ / y_0^+)^2, \quad y^+ < y_0^+; \qquad D = 1, \quad y^+ \geq y_0^+; \qquad y_0^+ = 33 \tag{7}$$

Test computations of 2D channel/pipe flows at the Reynolds numbers ranging from 3,000 to 20,000 have shown that overall, for the damping factor adopted, the best agreement with experimental data for the friction coefficient and heat transfer rate is achieved at $n \cong 0.35$.

COMPUTATIONAL ASPECTS

The computational domain includes the heated region, an upstream section one hydraulic diameter long, and a downstream section of the same length. Adiabatic wall boundary conditions are imposed for both the upstream and downstream sections. The computational domain is covered by a 113x21x21 nonuniform mesh in case of the standard wall functions (4) and by a 113x33x33 mesh in case of the modified wall functions (5). It should be stressed that the further grid refinement in duct cross-sections is insensible when accounting for limitations associated with wall-function technique. Sure, the modified wall functions allow a closer distance of the first computational point to the wall.

The computations have been performed with a well-validated academic code (named SINF). This advanced 3D Navier-Stokes solver is based on the second-order finite-volume spatial discretization using the cell-centered variable arrangement and body-fitted block-structured grids. The artificial-compressibility method is used to link the velocity and pressure fields through the continuity equation. In case of an unsteady problem, the artificial-compressibility technique is applied at each physical time step. The pseudo-time stepping is performed with an effective implicit method. A QUICK-type upwind scheme is employed to compute convective fluxes on the stage of residual computation. The numerical dissipation introduced in the stabilizing operator of the left-hand side of the linearized governing equations is proportional to the spectral radius of the Jacobian matrices of the convective flux vectors. Additional details for the solver can be found elsewhere[13,14].

AUXILIARY PROBLEM FOR INLET CONDITION DEFINITION

In the experiments of Dutta & Han[2] an upstream non-heated channel of about $40D$ long was assembled to provide the conditions of fully-developed isothermal flow at the inlet section of the heated region. For numerical simulation purposes an obvious and attractive way for inlet boundary condition definition is to compute beforehand characteristics of the fully-developed flow on the base of the formulation neglecting streamwise flow variations. It is known that such a simplified problem admits a mirror-symmetrical solution, however it can be unstable with respect to non-symmetrical perturbations[15,16]. The problem formulated without assuming the mirror symmetry may have no steady state solutions. It just occurs for the chosen flow geometry and the Reynolds and the rotation numbers. In the present work the fully-developed isothermal turbulent flow has been computed on the base of the unsteady formulation[16,17] using the same grids for the channel cross section as for the main problem (21x21 or 33x33).

Figure 2 shows oscillations of normalized transversal velocity at a monitoring point placed in the middle plane at the distance of 0.08D from the trailing wall. Four snapshots given in Figure 3 illustrate oscillations of secondary flow. The velocity fields averaged over the period of developing self-oscillations (used then as inlet boundary conditions for the two-pass channel non-isothermal flow problem) are compared in Figure 4 with the steady-state solution computed on the base of the mirror-symmetry formulation used earlier by various authors[18,19]. One can see that the symmetry formulation overestimates the effect of the near-trailing-wall vortices.

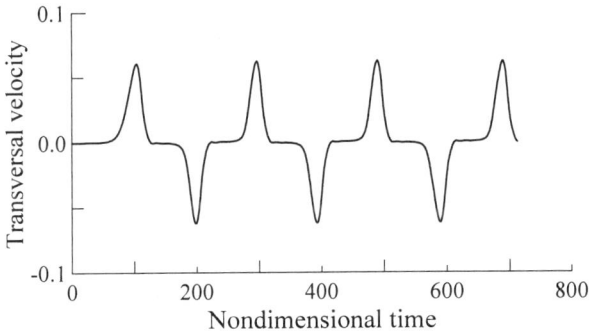

Fig. 2 Oscillations of velocity obtained with unsteady formulation for fully-developed flow

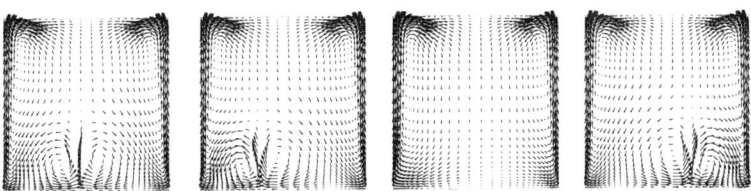

Fig. 3 Snapshots of unsteady cross circulation computed for fully-developed flow

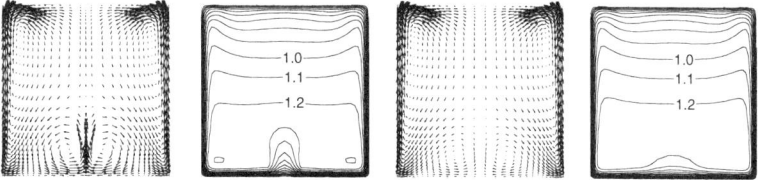

Fig. 4 Comparison of secondary flow patterns and streamwise velocity contours computed for fully-developed flow with (left) the symmetry and (right) the whole-section formulation

RESULTS FOR TWO-PASS CHANNEL

Figure 5 illustrates complicated secondary-flow patterns and normalized temperature distributions in the two-pass channel. The effect of the sharp 180 deg turn is well pronounced over the whole second pass. Details of flow in the turn region are shown in Figures 6 and 7. As one can conclude from Figure 7, the near-surface motion is significantly different for the trailing and the leading walls. Remarkable that the reversal flow developing in the trailing-wall/outer-wall corner penetrates deeply into the first pass.

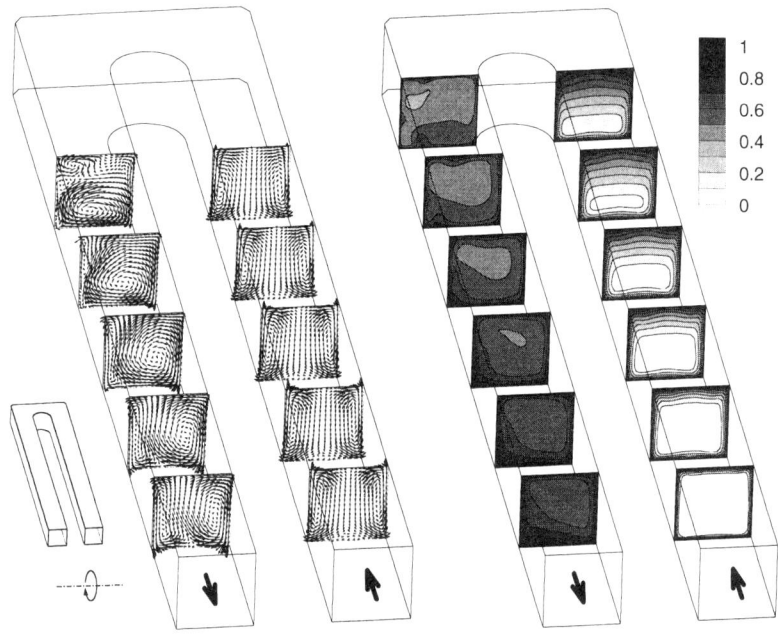

Fig. 5 Secondary flow patterns and normalized temperature contours

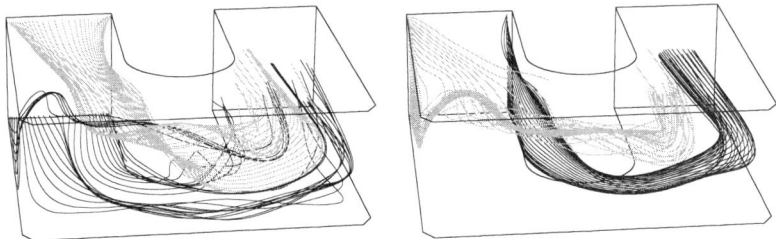

Fig. 6 Streamline patterns in the flow turn region

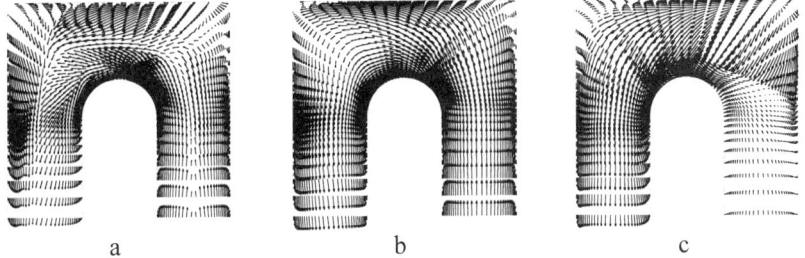

Fig. 7 Velocity vector patterns in the flow turn region: (a) plane at the distance of 0.02D from the trailing wall, (b) middle plane, (c) plane at the distance of 0.02D from the leading wall

Figure 8 compares side-averaged heat transfer results obtained in the present work with data of Dutta & Han[2]. One can see that the modified wall functions lead to a much better agreement with the measurement results. The computational model with the modified wall functions reproduces well spatial oscillations of local heat transfer in the flow turn region. The only notable drawback consists in some overprediction of heat transfer rate on the leading wall in the first pass and on the trailing wall in the second pass. It seems that it is due to neglecting the direct action of the Coriolis force on turbulent transport in the standard k-ε turbulence model. To improve the results one can test in the future a rotation-modified version of the turbulence model involving a correction based on the Richardson number[19].

To extract the centrifugal buoyancy effects, computations have been also performed with constant-density approximation. Figure 9 compares the results of buoyant flow and constant-density flow computations performed with the modified wall functions. For the set of determining parameters adopted, the influence of buoyancy on local heat transfer is pronounced for the first pass only. Here, the buoyancy action results in the augmentation of the heat transfer rate. This trend is similar to that obtained previously for trailing and leading walls with a low-Re algebraic second moment closure[20]. The buoyancy has only a minor influence on heat transfer on the inner and the outer walls.

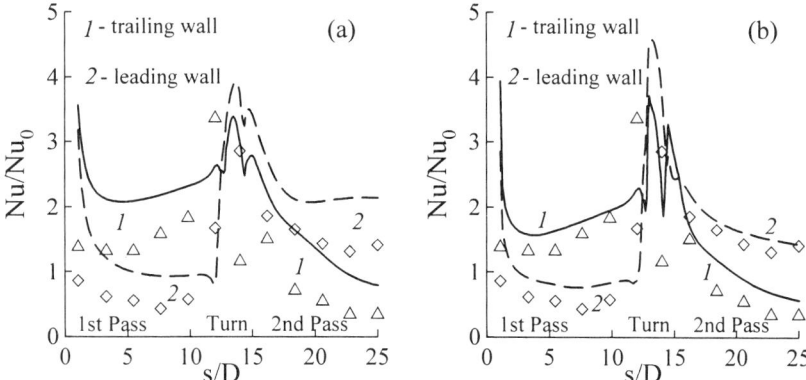

Fig. 8 Variation of side-averaged heat transfer rate along the channel. Curves show results of present computations using (a) standard and (b) modified wall functions for k-ε turbulence model in comparison with (symbols) experimental data given by Dutta & Han[2] for Re=5000, K=0.15

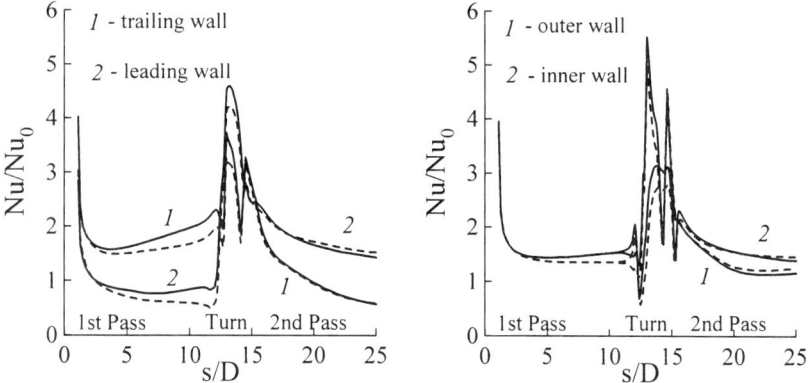

Fig. 9 Effect of buoyancy on side-averaged heat transfer rate: (solid lines) buoyant flow computations and (dashed lines) constant-density flow computations

CONCLUSIONS

3D turbulent air flow and heat transfer developing in a two-pass square channel rotating in the orthogonal mode have been simulated using the high-Re k-ε turbulence model with the standard and modified wall functions. Account of the molecular viscosity effect via incorporation of damping factor into the boundary conditions for turbulence parameters has resulted in a significant improvement of accuracy of local heat transfer prediction and secondary flow resolution.

An auxiliary problem implying time-dependent computations of fully-developed turbulent flow has been considered for accurate definition of inlet boundary conditions formed by a long unheated section upstream of the two-pass channel. More computations are needed to estimate allowable freedom in variations of inlet conditions.

The effect of the flow turn manifests itself over the whole second pass. There is also a significant influence of the turn on the upstream region.

For the problem considered, the influence of buoyancy on local heat transfer is pronounced for the first pass only. The buoyancy has a minor influence on heat transfer on the inner and the outer walls.

AKNOWLEDEMENTS

The study has been supported by the Russian Foundation of Basic Research, grant 98-02-18127.

REFERENCES

1. Wagner, J.H., B.V. Johnson & F.C. Kopper. 1991. Trans. ASME. J. Turbomachinery. **113**: 321-330.
2. Dutta, S. & J.-C. Han. 1996. Trans. ASME, J. Heat Transfer, **118**: 578-584.
3. Mochizuki, S., J. Takamura, S. Yamawaki & W.-J. Yang. 1994. Trans. ASME. J. Turbomachinery. **116**: 133-140.
4. Sathyamurthy, P.S., K.C. Karki & S.V. Patankar. 1994. ASME Paper. 94-GT-197: 1-9.
5. Iacovides H., B.E. Launder & H-Y. Li. 1996. Int. J. Heat and Fluid Flow. **17**: 22-33.
6. Hwang, J.J. & D.Y. Lai. 1998. Int J. Heat Mass Transfer. **41**: 979-991
7. Song, B., Amano, R.S. 2000. ASME TURBO-EXPO'2000, Munich, Germany, Rep. 2000-GT-0228.
8. Launder, B.E. & D.B. Spalding. 1974. Comput. Methods Appl. Mech. Eng. **3**: 269-289.
9. Taylor C., C.E. Thomas & K. Morgan. 1981. Int. J. Num. Meth. Fluids. **1**: 295-304.
10. Rosten, H.I. & J.K. Worrell. 1988. PHOENICS Journal. **1**: 81-109.
11. Smirnov, E.M. 1999. Private communication.
12. Zyabrikov, V.V. & L.G. Loitsanskii. 1987. Izvestiya AN SSSR. MZhG. No. 5: 45-53.
13. Agaphonov B.N., V.D. Goryachev, V.G. Kolyvanov, V.V. Ris, E.M. Smirnov & D.K. Zaitsev. 1999. *In* Finite Volumes for Complex Applications II. R. Vilsmieer et al., Eds.: 743-750. Hermes Sci. Publ., Paris.
14. Smirnov, E.M. 2000. Solving the full Navier-Stokes equations for very-long-duct flows using the artificial compressibility method. *In* Proceedings of the ECCOMAS 2000 Conference, Barcelona, Spain, 11-14 September 2000. In press.
15. Nandakumar, K., H. Raszillier & F. Durst. 1991. Phys. Fluids A. **3**: 770-781.
16. Kirillov, A.I., V.V Ris & E.M. Smirnov. 1998. *In* Proceedings of the 11th International Heat Transfer Conference, Kyongju, Korea, August 23-28, 1998, Vol. 3: 45-50.
17. Kirillov, A.I., V.V Ris & E.M. Smirnov. 1994. *In* Proceedings of the First Russian National Heat Transfer Conference, Moscow, November 1994, Vol. 1: 102-105. In Russian.
18. Iacovides, H. & B.E. Launder. 1991. Trans. ASME. J. Turbomachinery. **113**: 331-338.
19. Khodak, A.E., A.I. Kirillov, V.V. Ris V.V. & E.M. Smirnov. 1992. *In* Computational Fluid Dynamics'92. Ch. Hirsch et al. Eds. Vol.2: 597-604. Elsevier Science Publishers B.V.
20. Bo. T., H. Iacovides & B.E. Launder. 1995. Trans. ASME. J. Turbomachinery. **117**: 474-484.

EXPERIMENTAL DETERMINATION OF AVERAGE TURBULENT HEAT TRANSFER AND FRICTION FACTOR IN STATOR INTERNAL RIB- ROUGHENED COOLING CHANNELS

L.BATTISTI*, P. BAGGIO**

*Department of Mechanical and Structural Engineering
Turbomachinery Laboratory, Faculty of Material Engineering
**Department of Civil and Environmental Engineering
Faculty of Environmental and Land
University of Trento, 38100 Trento, Italy

ABSTRACT: in gas turbine cooling design, techniques for heat extraction from the surfaces exposed to the hot stream are based on the increase of the inner heat transfer areas and on the promotion of the turbulence of the cooling flow. This is currently obtained by casting periodic ribs on one or more sides of the serpentine passages into the core of the blade. Fluid dynamic and thermal behaviour of the cooling flow have been extensively investigated by means of experimental facilities and many papers dealing with this subject have appeared in the latest years. The evaluation of the average value of the heat transfer coefficient most of the time is inferred from local measurements obtained by various experimental techniques. Moreover the great majority of these studies are not concerned with the overall average heat transfer coefficient for the combined ribs and region between them, but do focus just on one of them.

This paper presents an attempt to collect information about the average Nusselt number inside a straight ribbed duct. Series of measurements have been performed in steady state eliminating the error sources inherently connected with transient methods. A low speed wind tunnel, operating in steady state flow, has been built to simulate the actual flow condition occurring in a rectilinear blade cooling channel. A straight square channel with 20 transverse ribs on two sides has been tested for Re of about $3 \cdot 10^4$, $4.5 \cdot 10^4$ and $6 \cdot 10^4$. The ribbed wall test section is electrically heated and the heat removed by a stationary flow of known thermal and fluid dynamic characteristics.

NOMENCLATURE

Alphanumeric symbols
A   area of the surface
c   blade chord

cp   specific heat
Dh   channel hydraulic diameter
e   rib height

f	friction factor	e	effective
h	enthalpy	f	bulk
h	adductive heat transfer coefficient	g	gas (hot stream)
k	thermal conductivity	h	hub
k_G	constant	conv	convective
L	channel length	joule	heating
$\dot{m}$	mass flow	in	test inlet section
n	number of channels	loss	loss in the surrounding
p	spacing between ribs	out	test outlet section
$\dot{q}$	thermal power	o	reference, smooth
S	channel perimeter	p	rib
T	temperature	w	wall
U	velocity main component		
x	blade radial coordinate		
y	Ainley exponent[15]		

Non-dimensional numbers

Bi — Biot number — $Bi = \dfrac{hD}{k}$

Pr — Prandtl number — $Pr = \dfrac{c_{p,c}\,\mu_c}{k_c}$

Nu — Nusselt number — $Nu = \dfrac{hD_h}{k_c}$

Re — Reynolds Number — $Re = \dfrac{\rho D_h U}{\mu_c}$

Greek symbols

ΔP — pressure drop
μ — air dynamic. viscosity
ρ — air density
σ — blade solidity

Subscripts
b — blade
c — coolant (air)

(dimensions of symbols are given according to S.I. standard)

INTRODUCTION

Component durability is of prime importance for advanced HP gas turbine stages. Current combustor outlet temperatures and stage loading levels require efficient cooling systems for blade and disks to ensure the design lifetime. Although an isothermal temperature distribution in the blade should result in the lowest thermal stresses, a variation of the cooling effectiveness not exceeding 20% actually represents a good design goal for the thermo-mechanical fatigue control. The requirements of effective air cooled turbine blades are high cooling effectiveness, low pressure drop, acceptable stress levels, light weight and ease of manufacture.

Advanced HPT vanes and blades internal cooling system design provide repeated rib-turbulators cast on two opposite walls of cooling passages whose typical features are depicted in the figure 1.

Pattern and configuration (aspect ratio, blockage ratio) of the rib-roughened channels depend on profile local thickness, taper ratio, and external stream heating load distribution. Internal cooling efficiency is

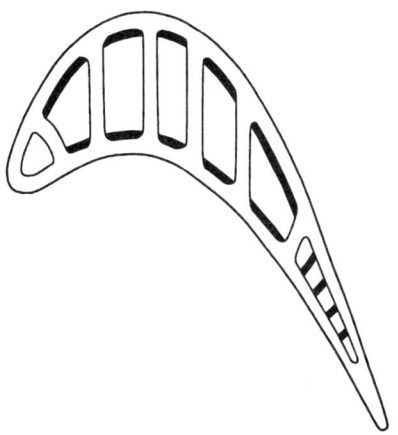

Fig.1 - Typical advanced gas turbine blade

given by the following equation:

$$\frac{T_{g,e} - \overline{T}_b}{T_{g,e} - \overline{T}_{c,e,h}} = \frac{\lambda}{1+\lambda} e^{-\left\{\frac{\lambda}{1+\lambda}\frac{\overline{h}_g S_g x}{\dot{m}_c c_{p,c}}\right\}} \quad \text{where} \quad \lambda = \frac{\overline{h}_c S_c}{\overline{h}_g S_g}$$

The higher is the coefficient λ, the smaller is the required cooling air flow for a given blade temperature. It can be shown that the ratio λ, for assigned gas flow conditions, depends essentially on the geometric factor G which can be expressed as[1]:

$$G = \left(\frac{L}{\sigma}\right)^{0.8}\left\{\frac{Nu}{Nu_o} + \left(1 + \frac{A_p}{A_c}\right)^{1.2}\right\}\frac{c^{0.8-y}}{k_c}\left\{\frac{S_c^{1.2}}{A_{c,o}}\right\}, \text{ where the parameter } \left\{S_c^{1.2}/A_{c,o}\right\}, \text{ after}$$

some manipulations, can be more conveniently written for design purposes as $k_G \, n^{0.2}/D_{h,a}^{0.8}$. This parameter actually controls the effectiveness of an air cooled blade.

Efficiency can be raised by increasing the parameter $\left\{Nu/Nu_o + \left(1 + A_p/A_c\right)^{1.2}\right\}$. The factors Nu/Nu_o and $\left(1 + A_p/A_c\right)^{1.2}$ represent respectively the fluid dynamic contribution (turbulence level in the core flow) and the geometrical (surface augmentation) contribution to the enhancement factor due to the presence of ribs. In rib-roughened channels high heat transfer regions are established whose extension has been found to be a function of geometrical features as rib height, aspect ratio, pitch and angle of attack according to

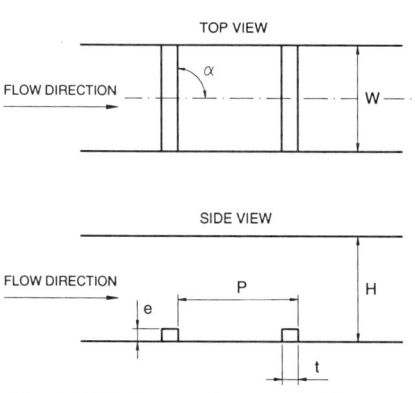

Fig. 2 - Turbulators main geometrical features

the definitions shown in figure 2. In figure 3, the contribution of the ribbed surface to the total heat exchange surface in the channel is depicted for two opposite 90° ribbed sides and current channel aspect ratios when the p/e ratio varies. More complex schemes of rib patterns have been investigated, such as staggered, inclined or broken ones. Obviously, increased complexity of geometrical features lead to a better heat transfer performance but also to increased friction losses. As a consequence, internal heat transfer cannot be considered separately from coolant flow rate consumption and channel chocking. Geometry optimisation is therefore always necessary. Pitch over rib height ratio from 8.5 to 10, and blockage ratios from 10% to 20% are currently used in HPT design.

Transient and steady state facilities are

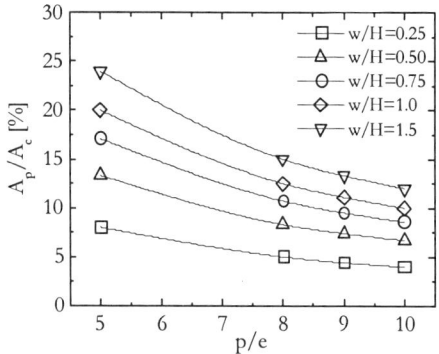

Fig. 3 - Surface augmentation ratio versus blockage ratio for different ribs spacing factors – two ribbed sides

used to simulate, according to similarity parameters, flow and heat transfer characteristics of internal cooling channels.

The evaluation of the averaged values of the heat transfer coefficient starting from measurements on a single wall is complicated and rather time consuming. Local heat transfer coefficient have to be independently averaged on each channel wall and then composed to get information on the average heat transfer coefficient. The main limitation of this kind of approach is the computation of the surface averaged heat transfer coefficient of the single wall. Its actual distribution is not easily predicted because of the strong three dimensional patterns of the flow on the surface. Large variations arise moving for instance from the centre line to lateral sides[2]. A large body of literature is available on turbulator equipped channels covering local thermohydraulic performance having different shapes and flow regimes[3,4,5,6].

The use of discrete sensors are useful when local conditions are investigated and this experimental approach is particularly suitable to compare the effect of geometrical features on the heat transfer and gradients in metal temperature. In transient facilities, steady state conditions in velocity and temperature profile of the stream are quickly established and the transient variation of the wall temperatures give information on the heat flow impinging locally on the surface[7,8,9]. General mapping techniques such as steady state liquid crystal thermography is increasingly used in convective heat transfer measurements[10,11].

An alternative approach is to evaluate the mean heat transfer coefficient in a significant portion of the channel wall in stationary conditions. The present work proposes a simple approach to obtain information on average heat transfer for ribbed channels This method allows not only to carry out accurate measurements of the average heat transfer flow rate on selected regions of ribbed channels without resorting to "artificial" compositions of local data, but also, when coupled with techniques giving local heat transfer data, to obtain information about the actual contribution to the total heat transfer given by different areas of the channel walls Results will be presented for model simulating typical mid chord region channel having square aspect ratio. Results are presented for two ribbed wall configuration and different Reynolds numbers

DESCRIPTION OF THE EXPERIMENTAL APPARATUS AND TEST PROCEDURE

In order to obtain information about attainable average heat transfer coefficients for gas turbine internal cooling channels design, a model was built using as similarity parameters the Reynolds number, the Prandtl number, and the temperature ratio. The Eckert number has been neglected. The averaged heat transfer coefficient of the test section is evaluated in steady state conditions.

The facility is an open flow loop that operates in aspiration mode sucking ambient air through a quadrant vane nozzle inlet. The main components are an unheated turbulent flow developing entrance, a heated test section, an orifice flow meter, a mass flow throttling valve, a plenum and a centrifugal blower. Stabilised electrical power supply, sensors and DAS complete the equipment.

A twin parallel square channel having a length to hydraulic diameter ratio (L/D) of 20, and wooden made, guide the air to the test section. The air extracts a measurable amount of heat from a uniformly electrically heated wall of the in the test section. A sketch of the test section is given in figure 4 and a picture in figure 5.

The latter is divided in three parts of approximately equal length, two of them having four Plexiglass walls, and one having a wall made of copper. This wall is common for the two channels to minimise heat losses.

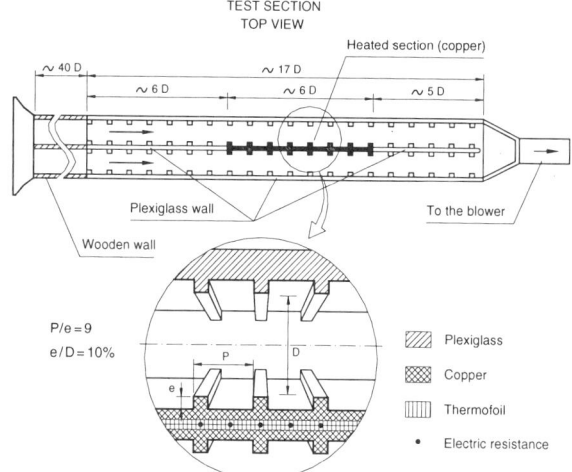

Fig.4 - Sketch of the test section

The heat source is an electrical resistance squeezed in a copper sandwich inserted between the two copper walls. This assembly allows to quickly exchange copper foils with different surface geometrical configuration. For the present investigation, seven ribs have been machined on the copper foil, the remaining are made of Plexiglass and glued on the preceding and following sections. Periodicity of the flow is therefore ensured.

The two channels are symmetrical with respect to the central plane, and sections can be exchanged from downstream to upstream. Conductive losses were minimised by placing outside a sheet of insulating foam. Heat losses measured through the test section wall amounted to about 2.5% of the total heat flow rate. Wall temperature for each heat transfer

Fig.5 – Picture of the facility

plate has been measured at three locations along the streamline by thermocouples embebbed on the copper close to the copper skin. Two thermocouples have also been placed inside two copper ribs.

Their temperature readings were found to be the same within a fraction of degree. The copper block has been therefore considered isothermal. (This was confirmed by the value of the Biot number which was equal to about 0.011 and then the thermal resistance is lumped on the surface). The averaged heat transfer coefficient was determined from bulk air flow rate and supplied electrical power measurements. The air inlet temperature has been sampled in the centre of the cross section assuming uniform temperature profile. This assumption was confirmed by tests. The outlet temperature has been obtained from 20 points mass flow weighted temperature measurements in a section far downstream of the test section. A check of the energy balance has been made comparing the electric power measurements with the enthalpy balance given by the equation:

$$\dot{q} = \dot{m}_c \left(h_{c,out} - h_{c,in} \right)$$

A close agreement (a few percent) has been found.
Supplied electrical power has been corrected accounting for heat losses. Fully turbulent flow is established in the entrance developing channel. Heat transfer measurements were performed in steady state. In this state we can write:

$$\dot{q}_{conv} = \dot{q}_{joule} - \dot{q}_{loss}$$

The radiation contribution to the heat transfer coefficient is small (less than 5 W/m²K) and as usual has been included in the adductive coefficient. This adduction heat flow rate results from the contribution of the whole channel wall (ribs and floor) and then can be considered the real average flow rate resulting from actual measurement and not from a regional weighted average of local values.

Linear increase of the flow temperature was assumed between inlet and outlet measurement section, and fluid properties evaluated at the average temperature and pressure. Average heat transfer coefficient have been compared with the smooth wall channel Dittus Boelter correlation and expressed as enhancement factor:

$$EF = \frac{Nu}{Nu_o} = \frac{\dfrac{D_h}{k_c} \dfrac{\dot{q}_{conv}}{(T_f - T_w)}}{0.023 \, Re^{0.8} \, Pr^{0.4}}$$

Four pressure tabs along the centreline of the ribbed side and four along the smooth wall were used for static pressure drop measurements. The pressure drop across the test channel was based on the isothermal

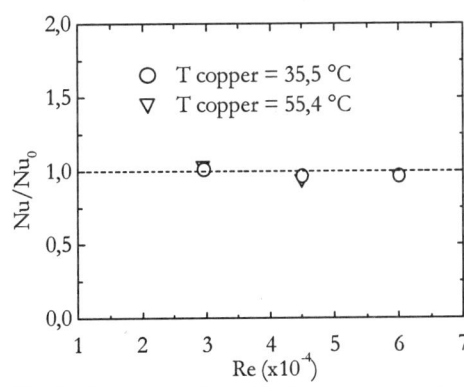

Fig. 6 – Enhancement factor for the smooth channel

conditions. The averaged friction factor was calculated as: $\bar{f} = \Delta P \Big/ 4 \dfrac{L}{D_h} \rho \dfrac{\overline{U}^2}{2}$

The friction factor was normalized by the friction factor for fully developed turbulent flow in smooth circular tubes ($10^4 \leq Re \leq 10^6$) proposed by Blasius as: $f_o = 0.046 \, Re^{-0.20}$.

Experimental uncertainties, after the method of Kline and Mc Clintock[12] has been determined to be ± 1.5% (15:1) for the Reynolds number using the flow meter during the heat transfer measurements. The uncertainty related to the friction factor was less than ± 3.5% (15:1). Accuracy of the method is mainly related to air temperature rise through the test section. Therefore the worst results were obtained for low copper plate temperature and high Reynolds numbers. In these conditions the Nusselt number uncertainty resulted in ± 4.5% (15:1).

RESULTS AND DISCUSSION

Results are presented as heat transfer enhancement factor: Nu/Nu_o versus Reynolds number. As preliminary tests, smooth copper foil walls have been installed and the enhancement factor computed for $Re = 3 \cdot 10^4$, $4.5 \cdot 10^4$ and $6 \cdot 10^4$. Electrical power has been adjusted in the range between 60 and 300 W. Temperature of the copper plate ranged from 30 to 80 °C. Negligible variation of the heat transfer coefficient was observed by varying the copper surface temperature. The good accuracy of the measurement system is evident from the results presented in figure 6. For the ribbed configuration, the geometrical configuration- 90° double side square ribs was considered. The spacing of the ribs was kept at 9 times the ribs height (p/e=9), and the blockage ratio (e/D) equal to 10%. Levels of electrical power and plate temperature are the same as for the smooth tests. The results have been compared with published data and shown in figure 7. Enhancement factor drops down as Reynolds number increases. Dispersion of data is also due to some difference in the examined geometrical features. The blockage ratio is for instance strongly affecting the flow structures and therefore the heat transfer coefficient. Rau[13] used liquid crystal thermograpy. Thermocouples have been used in the remaining analysed works. All considered authors compute the average heat transfer coefficient by means of surface averaging of discrete wall temperature measurements.

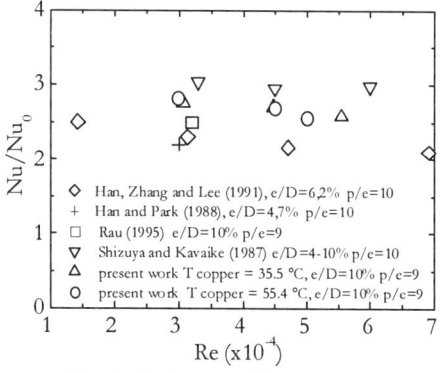

Fig. 7 – Enhancement factor versus Re

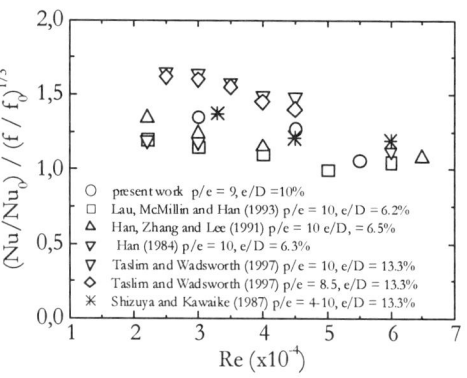

Fig. 8 – Thermal performance versus Re

The lower values reported from most of the literature can be explained considering that in such works measurements of heat transfer coefficient over the ribs are usually not performed. As a consequence a significant contribution to the overall heat transfer coefficient is not taken into account. Depending on the p/e ratio the term $(A \cdot h)_{ribs}$ can take values ranging from about 30 to 50 percent of $(A \cdot h)_{ribs+floor}$ [9,14]. The thermal performance based on the same pumping power is given[4] by $(Nu/Nu_o)/(f/f_o)^{\frac{1}{3}}$, and shown in figure 8. The highest thermal performance is obtained for low Reynolds numbers. The experimental data fit well with measurements of other authors performed with lower and higher pitch to chord and blockage ratios.

CONCLUSIONS

A model, simulating a straight blade internal cooling channel, suitable to perform steady state tests has been presented in this contribution. Similarity parameters as Reynolds number, Prandtl number and Temperature ratios have been properly selected to obtain flow conditions close to real conditions in HP gas turbine stages.

Preliminary wall heat transfer measurements and thermal performance evaluation in 90° square double side rib-roughened channel have been carried out. The resulting data have been compared with available literature data.

An alternative approach has been used to obtain information on average heat transfer for ribbed channels which has not already appeared on the acknowledged literature. This method allows to carry out accurate measurements of the average heat transfer flow rate on selected regions of ribbed channels without resorting to "artificial" compositions of local data.

REFERENCES

1. Battisti L. "Driving Concepts for Chord Size Choice in Air Cooled Blades" to be presented at ATI 2000 55° Congress, 15-20 Sept. 2000, Bari, Italy.
2. Han, J,C, Ou S., Park J.S. and Lei C.K., "Augmented Heat Transfer in Rectangular Channels of Narrow Aspect Ratio with Rib Turbulators", Int. Journal of Heat Mass Transfer, Vol. 32, N.9, pp. 1619-1630, 1989
3. Han J.C. "Heat Transfer and Friction Characteristics in Rectangular Channels with Rib Turbulator", ASME Journal of Heat Transfer, Vol. 110, pp. 321-328, 1988.
4. Han J.C. and Park J.S, "Developing Heat Transfer in Rectangular Channels with Rib Turbulator" Int. J. Heat and Mass Transfer Vol. 31 N.1 pp.183-195, 1988
5. Han J.C., Zhang Y.M. and Lee C.P., "Augmented Heat Transfer in Square Channels with Parallel, Crossed, and V-Shaped Angled Ribs" Journal of Heat Transfer Vol. 113 pp.590-596, 1991
6. Shizuya M. and Kawaike K., "Experimental Investigation of Blade Internal Cooling Methods Using Ribs and Fins 87 - Tokyo - IGTC - 65.
7. Battisti L., Arts. T., "Wall Heat Transfer Measurements in Rib-Roughened Cooling Channels by Means of a Transient Technique", Ati 51° Convegno ATI, Udine 16-20 Settembre 1996
8. Battisti L., Schmeer T., "Experimental Study of the Surface Heat Transfer Enhancement in a Rib Roughened Blade Coolant Channel By Means of double Layer Thin Films" Proceedings of the 55° Eurotherm Seminar, Santorini, Greece, Sept. 1997.
9. Battisti L., Determinazione sperimentale dello scambio termico in un modello di canale di raffreddamento di pala di turbina a gas con promotori di turbolenza per mezzo di sensori a film sottile in regime transitorio", Atti del Convegno MIS-MAC V Metodi di Sperimentazione nelle Macchine Roma, Febb. 1998.
10. Rau G., Çakan M., Moeller D., Arts T. "The Effect of Periodic Ribs on the Local Aerodynamic and Heat Transfer Performance of a Straight Cooling Channel, VKI Preprint 1996-11, Feb.1996
11. Ireland P.T., Jones T.V., "The Measurement of Local Heat Transfer Coefficients in Blade Cooling Geometries", AGARD CP-390.
12. Kline S.J., Mc Clintock F.A. , "Describing Uncertainties in Single-Sample Experiments", Mechanical Engineering, Jan, 1953
13. Rau G., "The blockage effect of turbulators in a rectilinear cooling channel". Presented in "Heat Transfer and Cooling in Gas Turbine" von Karman Institute Lecture series, May 1995

14. Taslim, M.E., and Wadsworth, C.M. " An experimental investigation of the rib surface averaged heat transfer coefficient in a rib-roughened square passage", ASME Journal of Turbomachinery, VOl.119, pp.381-389, April 1997
15. Ainley D. G. "Internal Air Cooling for Gas Turbines", Aeronautical Research Council, R & M 3013, 1957
16. Han J.C. "Heat Transfer and Friction in Channel With Two Opposite Rib-Roughened Walls" Int. J. Heat Transfer, Vol. 106 pp.774-781, 1984
17. Liou T.M., Hwang J.J. "Turbulent Heat Transfer Augmentation and Friction in Periodic Fully Developed Channel Flows" Int. J. Heat Transfer, Vol. 114 pp.56-64, 1992

Contribution of Heat Transfer to Turbine Blades and Vanes for High Temperature Industrial Gas Turbines Part 2 : Heat Transfer on Serpentine Flow Passage

KEN-ICHIRO TAKEISHI and SUNAO AOKI

Takasago Research and Development Center, Mitsubishi Heavy Industries Ltd.,
2-1-1, Shinhama, Arai-Cho, Takasago City, 676-8686, Japan
Tel. +81-794-45-9705, Fax. +81-794-45-6089, E-mail: takeishi@wl.trdc.mhi.co.jp

ABSTRACT: The improvement of the heat transfer coefficient of the 1st row blades in high temperature industrial gas turbines is one of the most important issues to ensure reliable performance of these components and to attain high thermal efficiency of the facility. This paper deals with the contribution of heat transfer to increase the turbine inlet temperature of such gas turbines in order to attain efficient and environmentally benign engines. Following the experiments described in Part 1, a set of trials was conducted to clarify the influence of the blade's rotating motion on the heat transfer coefficient for internal serpentine flow passages with turbulence promoters. Test results are shown and discussed in this second part of the contribution.

INTRODUCTION

The trend in the development of advanced industrial gas turbines for high thermal efficiency goes towards increasing the turbine inlet temperature (TIT) to a 1500°C level. Highly sophisticated turbine blade cooling is necessary to maintain performance and life of these components. Typical modern gas turbine blades are internally cooled by supplying compressed air through multi-pass rib-roughened serpentine flow passages. Figure 1 shows the cooling scheme of the 1st blade for Mitsubishi's 1500°C class G-type gas turbine. It is designed as a serpentine flow cooling structure with cooling air supply from the root section. Angled turbulence promoters are installed in the flow passages to increase the heat transfer coefficient in the internal cooling structure. In addition, the blade is equipped with full coverage film cooling.

Many investigators have studied the heat transfer coefficient of serpentine flow passages with turbulence promoters in stationary conditions. However, for the heat transfer in a rotating flow passage with and without turbulence promoters, there is only a limited amount of data available in the open literature. The rotating motion of turbine blades gives rise to Coriolis and centrifugal buoyancy forces which can affect coolant flow patterns and influence the local heat transfer coefficient in the serpentine flow passage. The present experimental work has been performed to study the influence of such rotating motion on the heat transfer coefficient of a serpentine flow passage with and without turbulence promoters.

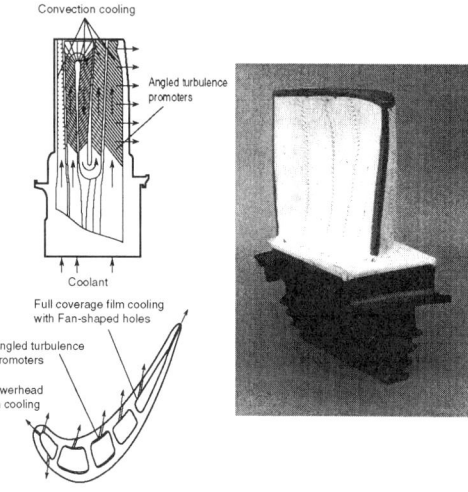

Figure 1 Cooling scheme of the first stage blade for 1500°C class industrial gas turbine

NOMENCLATURE

De = Equivalent hydraulic diameter
f = Friction factor
G = Flow rate
L = Test section length
M = Mass flux ratio ($=\rho_a U_a / \rho_g U_g$)
N = Rotational speed [rpm]
Nu = Nusselt number ($=\alpha De/\lambda$)
ΔP = Pressure drop
Re = Reynolds number ($=UDe/\nu$)
Ro = Rotation number ($=\omega De/U$)
T = Temperature
U = Velocity
x = Distance

α = Heat transfer coefficient
λ = Thermal conductivity
ν = Kinematic viscosity
ρ = Density
ω = Angular velocity

Subscripts:
a : Air
w : Wall

HEAT TRANSFER FOR SERPENTINE FLOW PASSAGE

Similarly to the first row vane, the first stage blade receives also the hot gas emanating from the combustion chamber. However, due to the rotating motion of the turbine blade, the gas temperature distribution is averaged in circumference direction. The movement of the blade also reduces the effective stagnation temperature for the blades by about 100°C compared to the average total temperature of the 1st stage vane. Complicated cooling schemes as used for the 1st stage vane may not be applicable for the rotating blade. The currently most successful cooling configuration is the one with a serpentine flow passage into which turbulence promoters have been added. In stationary conditions, many studies have been conducted to improve the heat transfer coefficient by means of turbulence promoters and to decrease the pressure loss, e.g., by Burggraf[1], Han et al.[2][3][4], Metzger et al.[5], Taslim and Spring[6][7][8], and Webb[9].

The effect of rotation on the heat transfer characteristics of a rotating coolant channel with smooth walls was the subject of studies by Morris and Ghavarni[10], Morris and Salemi[11], Wagner et al.[12][13], and Yang et al.[14].

Wagner et al.[15] and Johnson et al.[16] investigated the heat transfer coefficient in multi-pass square channels with ribs normal and skewed to the flow. Parsons et al.[17] and Han and Zhang[18] reported results on the effect of wall temperature variation on the heat transfer in a two-pass square channel with normal and angled ribs.

A rotational movement of the flow passage may alter the heat transfer characteristics due to the appearance of Coriolis and buoyancy forces. Since studies on this effect are scarcely reported in the open literature, the present work was conducted to clarify whether such an influence exists. A corresponding experimental study has been performed using an enlarged heat transfer model simulating a typical flow passage of the first stage turbine blade.

Experimental Apparatus

Figure 2 shows the schematic diagram of the heat transfer test model for a three-pass (two-turn) passage. The passage has a rectangular cross-section of 20mm × 32.6mm (aspect ratio of 1.63). The rib height-to-channel hydraulic diameter ratio is 0.0524 and the rib pitch-to-height ratio is 10. By varying the flow rate, Reynolds numbers between 20,000 and 150,000 can be achieved in this facility. The rib-roughened channel models the internal cooling passages in modern gas turbine airfoils. The rib height and pitches, and the Reynolds number range are typical of turbine airfoil cooling applications. The rectangular test section consists of two Bakelite sidewalls of dimension H and two Bakelite walls of dimension B located directly opposite of each other, respectively. Filtered and dried air is used as the test fluid. The local heat transfer coefficient is derived from the measured heat flux of an electrically heated nickel foil divided by the difference of the main stream temperature and the wall temperature, which was measured by embedded thermocouples in the Bakelite wall. Turbulence promoters are inserted on the leading and trailing walls on each flow passage whose detailed geometry and dimensions are shown in Figure 3. In order to measure the local and the averaged heat transfer coefficient, 35 K-type thermocouples are embedded in the trailing and leading side walls of each of the three flow passages within one pitch of turbulence promoters. In total, 35×2×3 = 210 thermocouples are installed. However, all these gauges can only be used in stationary conditions, while for a rotating conditions the total number of simultaneously used thermocouples is limited to nine out of 35 for each location (hence, the total number of thermocouples used in this case is 54). Since this is insufficient for an accurate determination of the spatial surface temperature distribution, this measurement was therefore conducted by means of

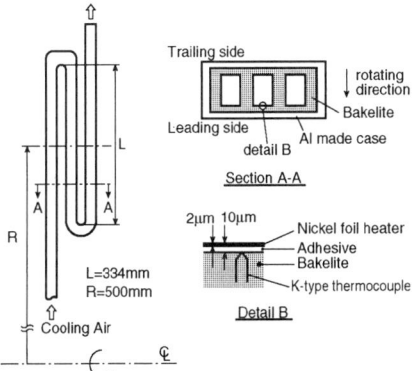

Figure 2   Schematics of test section

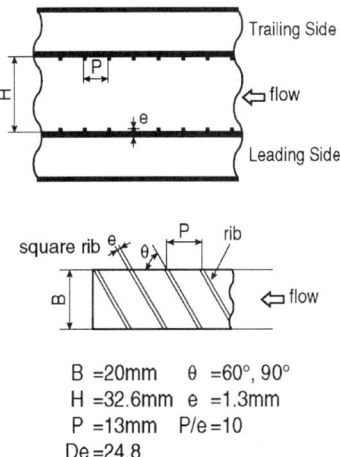

Figure 3   Cross section of the test channel

recording the color change of liquid crystals painted on the surface of the nickel foil heaters.

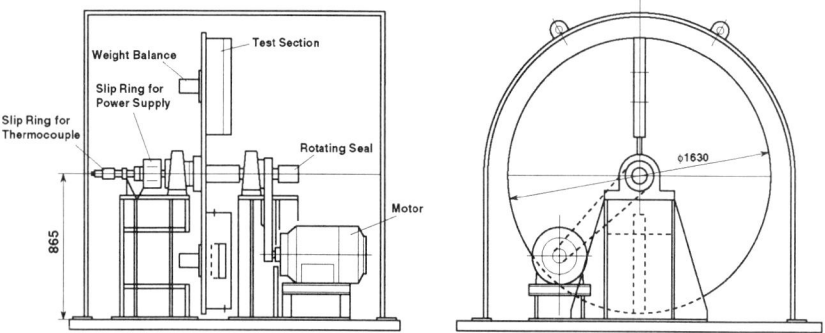

Figure 4   Rotating test rig

A schematic of the rotating test facility and relevant dimensions are shown in Figure 4. The heat transfer test model shown in Figure 2 is mounted on the test section. The filtered and dried air from a compressor flows through a rotary seal into the hollow rotating shaft. A standard JSME sharp edge orifice[19] is used to measure the flow rate. After passing the test section the air is released into the atmosphere. Electric power for the nickel heater is supplied through a slip ring mounted on the opposite side of the shaft and of the rotary seal. Wall temperature and cooling air temperature are measured by the thermocouples, whose signals are transmitted through another slip ring (see Figure 4).

The calculation of the local heat transfer coefficient uses the measured local net heat transfer rate q, the area A of the smooth wall exposed to the cooling air, the local wall temperature T_w at the position of the thin nickel foil and the local air temperature T_a. The latter value is determined from measured air temperatures at the starting point and the end of heated flow passage. The typical measured wall and air temperatures are 40~45°C and 10~15°C respectively. The density ratio ($\Delta\rho = (\rho_a - \rho_{wall})/\rho_a$) is about 0.11.

$$\alpha = q/A(T_w - T_a) \quad (1)$$

The local net heat flux q is derived from the electric power generated by the nickel foil heater minus the heat loss by heat conduction through the wall to backside.

In fully developed channel flow, the friction factor f can be determined by measuring the pressure drop ΔP across the flow channel and calculating the flow velocity U from measured mass flow rate G of the air. With these data, the friction factor f can be calculated from the following definition.

$$f = \Delta P / 4(L/De)(\rho U^2/2) \quad (2)$$

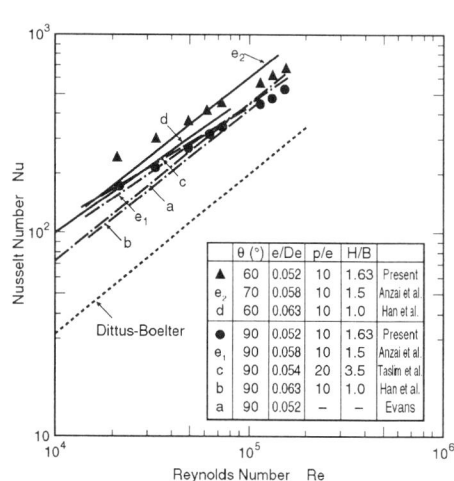

Figure 5   Heat transfer coefficients in rectangular ribbed ducts

Test Results and Discussion

Figures 5 to 8 show typical heat transfer and friction test results for the stationary, i.e., non-rotating condition. Results from the open literature are presented for comparison in order to confirm the accuracy of the test.

The Reynolds number is based on the hydraulic diameter De as shown in the nomenclature. The results shown in Figure 5 indicate a relationship of the form $Nu \sim Re^{n}$. For flow passages with traverse ribs ($\theta = 90°$), the present data agree very well with the results obtained by Taslim [8]. Within the range $1.5 \times 10^5 > Re > 2 \times 10^4$, the heat transfer coefficient of the flow passage with 60° angled ribs is about 20% higher than that of traverse ribs, while the exponent n in the Nu-Re relation has a slightly smaller value than in the case of traverse ribs.

Flow visualization tests with liquid crystals have been conducted to clarify the mechanism of the heat transfer increase in the flow passage with 60° angled ribs. Typical results are shown in Figure 6 and Figure 7. In the passage with traverse ribs, the local heat transfer coefficient has its maximum between the ribs where the flow re-attaches as is clearly apparent in Figure 6.

Figure 7 shows the heat flux distribution for a passage with 60° angled ribs, which is obviously different from the pattern seen for traversing ribs (Figure 6). The results indicate that very high heat flux region appear at the corners of the angled ribs, caused by a strong swirling flow that has been generated by the ribs in the passage.

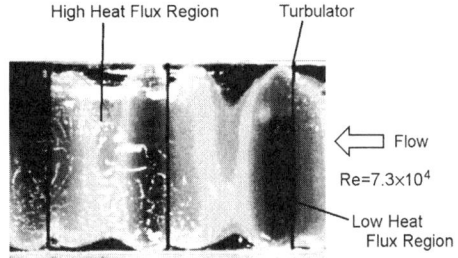

Figure 6 Flow visualization of the passage with traverse rib by liquid crystal

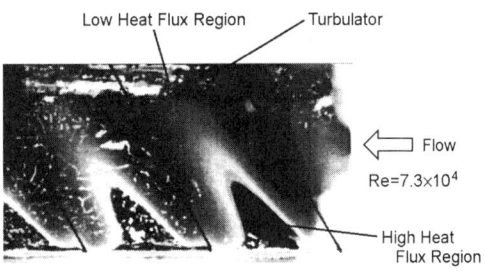

Figure 7 Flow visualization of the passage with angled rib by liquid crystal

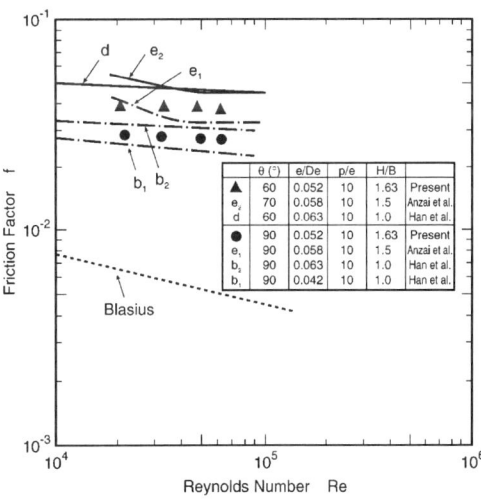

Figure 8 Average friction factor versus Reynolds number

The characteristics of the flow in the passage with angled ribs are a) separated flow at the top of the ribs, b) circulating flow behind the ribs, and c) re-attached flow and strong swirling flow along the angled ribs.

The increase in heat transfer coefficient by 20% compared to the case of a flow passage with traversing ribs is caused by this swirling flow, which dramatically enhances the heat transfer coefficient within the triangular region defined by the angled ribs and the side wall.

Figure 8 shows the experimentally obtained friction factor (following equation (2)) in comparison with results from the open literature. The results obtained here indicate an independence of the Reynolds number. The friction factor for the flow passage with traverse ribs fits very well with data the open literature but in the case of 60° angled ribs the values are slightly below those obtained by Anzai [20] and Han [3].

Figures 9(a) to 9(c) present typical test results that demonstrate the effect of the rotating motion on the heat transfer coefficient in passages with and without ribs. The heat transfer measuring point is located at the middle of the passage height at a non-dimensional distance of x/De=6.7 behind the 180° turn. Figures 9(a) to 9(c) indicate the variations of the average Nusselt number (derived as the mean value of nine local Nusselt

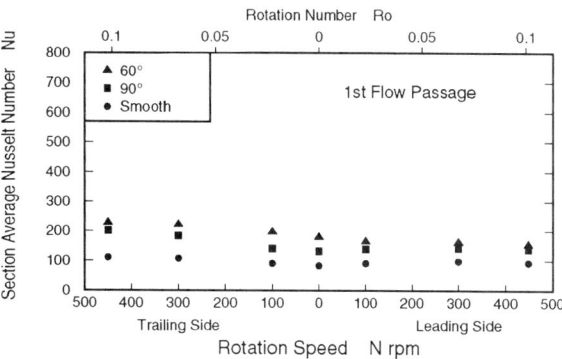

Figure 9(a) Effect of rotation on Nu in 1st passage with/without traverse and angled ribs

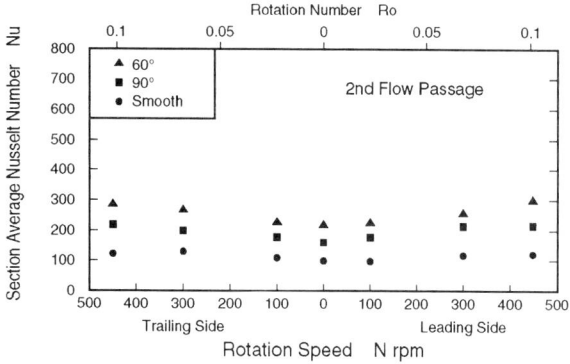

Figure 9(b) Effect of rotation on Nu in 2nd passage with/without traverse and angled ribs

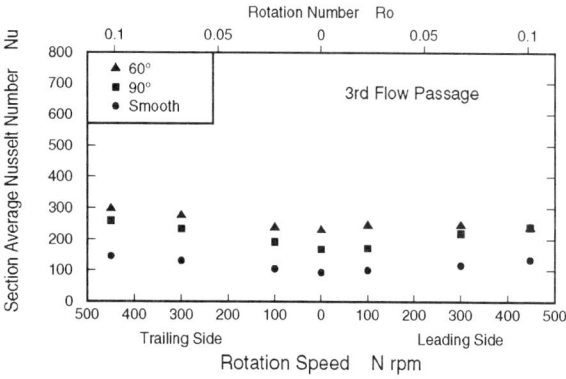

Figure 9(c) Effect of rotation on Nu in 3rd passage with/without traverse and angled ribs

numbers) with rotation speed. Figure 9(a) clearly shows that in the first passage the trailing surface Nusselt number increases with an increase of the rotation speed, while the leading surface Nusselt number is largely uniform and shows only a very weak dependence on the rotational speed. Contrary to the first flow passage, it appears that the Nusselt number of the leading side and trailing side in the second pass and of the trailing side in the third pass are more sensitive to the rotation speed compared to that in the first flow passage. Also, the heat transfer coefficients of the flow passage with turbulence promoter are more sensitive to the rotation speed compared to that of the flow passage without turbulence promoter.

These results differ from Wagner et al[15] who conducted the rotating heat transfer test using a model with square cross section, i.e., an aspect ratio of one. Wagner has conducted heat transfer test using a four-pass square duct. The first, second, and third passes were instrumented to measure an averaged heat transfer coefficient using copper blocks. However, the results presented were mostly for the first two passes. It was shown that increasing the rotation rate causes a significant increase in heat transfer on the trailing surface of the first pass and a comparatively weaker increase on the leading surface of the second pass. Wagner et al.[13] also showed that the heat transfer coefficients from the first pass leading and second pass trailing sides decrease with an increase in rotation number.

The discrepancy between Wagner's results and ours in the tendency of the heat transfer coefficient is attributed to the different geometries used: the flow pattern generated by the Coriolis force and hence the influence of the rotation speed on the heat transfer characteristics may be weakened for high aspect ratios (as in the present case) and by high turbulence mixing at the 180° turns.

The enhancement on the heat transfer coefficient by the turbulence promoter is dominant. It is estimated that the reason why the Nusselt number of the second pass increases with increasing rotation number can be found in an enhancement of the heat transfer by inward main flow against centrifugal buoyancy flow. Another reason why the Nusselt number keeps uniform on the leading side of the first pass and increases on the trailing side of the second pass may be the difference of measuring location. Considering the position of the heat transfer measurement point, the heat transfer coefficient of our test may be influenced more strongly by the effect of higher turbulent mixing generated at the 180° turns, which is dominating over the rotational effects at middle height.

CONCLUSIONS

The following conclusions were obtained through the measurements of heat transfer coefficient on rotating ducts with and without turbulence promoters.

1) The heat transfer characteristics of the turbulence promoters showed good agreement with previous experiments. Among the tests, the angled turbulence promoter showed the best performance.
2) The strong swirling flow along the angled ribs enhanced the heat transfer coefficient of the flow passage with 60° angled ribs at the corner of the angled ribs and increased the averaged heat transfer coefficient.
3) For the serpentine cooling passage, the effect of rotation on the heat transfer is established by a decreasing Nusselt number on the trailing side and a largely uniform or increasing Nu on the leading side for the 1st flow passage. Heat transfer is enhanced by the rotation on both leading and trailing side for the 2nd and 3rd flow

passage. It is estimated that the difference to the results from Wagner et al. [13] is caused by the passage aspect ratio (i.e., secondary flow effects) and high turbulence mixing at the 180° turns.

ACKNOWLEDGMENTS

The authors wish to express their gratitude to Mitsubishi Heavy Industries, Ltd. for permission to publish this paper and also would like to thank Prof. Dr. H. Kleine of Tohoku University for valuable comments on the first draft.

REFERENCES

1. Burggraf, F. 1970. Augmentation of Convective Heat and Mass Transfer, A. E. Bergles and R. L. Webb. **Eds.**: 70-79. ASME, New York.
2. Han, J.C., Glicksman, L.R. and Rohsenow, W.N. 1978. Int. J. Heat Mass Transfer. **21**: 1143-1156.
3. Han, J.C., Park, J.S. and Lei, C.K. 1985. J. of Eng. for Gas Turbine and Power, **107**: 628-635.
4. Han, J.C., Zhang, Y.M., and Lee, C.P. 1992. J. of Turbomachinery, **114**: 872-880.
5. Metzger, D.E., Fan, C.S. and Yu, Y. 1990. Compact Heat Exchanger : A Festschrift for A.L., London, Hemisphere Publishing Co. 151-167.
6. Taslim, M.E. and Spring, S.D. 1987. AIAA Paper No. AIAA-87-2009.
7. Taslim, M.E. and Spring, S.D. 1988. AIAA Paper No. AIAA-88-3014.
8. Taslim, M.E. and Spring, S.D. 1991. AIAA Paper No. AIAA-91-2033.
9. Webb, R.L., Eckert, E.R.G. and Goldstein, R.J. 1971. Int. J. Heat Mass Transfer, **14**: 281-289
10. Morris, W.D. and Ghavarni-Nasr, G., 1990. ASME Paper No. 90-GT-138.
11. Morris, W.D. and Salemi, R., 1991. ASME Paper No. 91-GT-17.
12. Wagner, J.H., Johnson, B.V. and Hajek, T.J., 1989. ASME Paper No. 89-GT-272.
13. Wagner, J.H., Johnson, B.V. and Kopper, F.C., 1990. ASME Paper no. 90-GT-331.
14. Yang, W.J., Zhang, N. and Chioi, J., 1992. J. of Heat Transfer, 114: 354-361.
15. Wagner, J.H., Johnson, B.V., Graziani, R.A. and Yeh, F.C. 1992. J. of Turbomachinery, **114**: 847-857.
16. Johnson, B.V., Wagner, J.H., Steuber, G.D. and Yeh, F.C. 1994. J. of Turbomachinery, **116**: 113-123.
17. Parsons, J.A., Han, J.C. and Zhang, Y.M. 1994. Int. J. of Heat and Mass Transfer, **37**: 1411-1420.
18. Han, J. C. and Zhang, Y. M. 1992. J. Heat Transfer, **114**: 850-858
19. JIS Z8762-1995
20. Anzai, S. et al. 1991. J. of the Gas Turbine Society of Japan, Vol. 19, **75**: 65-73. (in Japanese)
21. Evans, D. M. and Noble, M. L. 1978. ASME Paper 78-GT-33

SECONDARY FLOW EFFECT TO HEAT TRANSFER OF A DUCT WITH DISCRETE RIB TURBULATORS

K. TATSUMI*, H. IWAI*, K. INAOKA** and K. SUZUKI*

* Department of Mechanical Engineering
Kyoto University, Kyoto 606-8501, Japan
** Department of Mechanical and Systems Engineering
Doshisha University, Kyoto 610-0321, Japan

ABSTRACT: In the present study, three-dimensional numerical simulation is carried out for discrete rib arrays to investigate the local fluid and heat transfer characteristics. Rib pitch and gap width were varied in several steps. Furthermore, two kinds of rib arrays were studied, i.e. an array of ribs having the gaps at both spanwise ends and a staggered array of ribs with gaps at both spanwise ends and with gaps in the middle. Eventually, the staggered arrayed ribs is suggested to be best in performance of heat transfer.

INTRODUCTION

Because of its high power generation efficiency and low emission rate of pollutants, gas turbine is widely spread for industrial use, especially as a power generator. In order to raise the efficiency of a gas turbine cycle, as is well known, it is necessary to raise the gas turbine inlet temperature. On the other hand, a serious problem of fracture can occur to the material elements at high temperature, especially of the turbine blades. So a lot of studies have been done for cooling techniques. As an effective convective cooling device, rib turbulator is popularly used for gas turbine cooling.

Many investigators, in the past few decades, have carried out numerous numerical and experimental studies of the flows and the related heat transfer around the rib turbulator. A standard full-span rib was experimentally[1] and numerically[2] investigated in various conditions. By these meritorious studies, basic characteristics have been studied in detail for this case. To increase the efficiency of the rib turbulator, further researches have been done on changing the rib geometry covering inclined rib, discrete ribs and V-shaped ribs based on experimental works[3,4,5]. These ribs produce a highly three-dimensional flow and thermal fields inside the cooling passage. Among these, Hu et al.[6] reported that discrete ribs produce several types of secondary flows, especially longitudinal vortices downstream the rib. However, no detail measurement has been done experimentally and its structure has not been fully investigated. Tatsumi et al.[7] carried out a three-dimensional numerical simulation over a single rib mounted in a duct flow and its overall features of flow and its related heat transfer have been studied. In the practical turbine blade, turbulence promoting ribs are arranged in a spatially periodical manner in the form of a row through the cooling passage. It is reasonable and essential to systematically analyze the fluid and heat transfer characteristics of such discrete ribs periodically attached to the duct wall.

In the present study, three-dimensional numerical simulation is performed for the ribs periodically attached on to the bottom wall of a duct. Investigation will be made for the effect of three dimensional unsteady flow fields including the effect of secondary flow and its related heat transfer. Two-different kinds of discrete ribs are investigated, i.e. an array of ribs having the gaps at both spanwise ends and a staggered array of ribs having gaps at both spanwise ends, and the following one having a gap in the middle of the span of the channel.

COMPUTATIONAL PROCEDURE

Three-dimensional, time dependent continuity, momentum and energy equations were solved in the numerical computations. For the turbulence model, a non-linear eddy viscosity model proposed by the UMIST group[8] was applied in the computation. Turbulent heat flux, on the other hand, was assumed to be proportional to the temperature gradient. Turbulent Prandtl number, Pr_t was assumed to be constant and equal to 0.9.

Fully implicit forms of finite difference equivalents of the governing equations were solved numerically along the time axis. For the finite-differencing of the governing equations, a forth-order central-difference scheme was employed for the diffusion terms while the fifth-order upwind scheme was employed for the convection terms in the case of momentum equation. In the energy equation and in the transport equations of k and $\tilde{\varepsilon}$, a first-order upwind scheme was adopted for the convection terms and a central-difference scheme for the diffusion terms. For the solution algorithm of the pressure correction equation, SIMPLE algorithm was employed. In order to relax the solutions of these elliptic differential equations, iterative procedure was applied at each time step. In this procedure, alternating direction implicit (ADI) method was also combined.

The computational domain in the present study is illustrated in Fig. 1 together with the coordinate system to be used. Rib height to duct height ratio, H/H_d and Reynolds number based on the hydraulic diameter of the duct were set respectively to be H/H_d=0.15 and Re=10,000. Rib pitch to rib height ratio, p/H and gap width to duct width ratio, W_G/W_d were varied in several steps, i.e. p/H=6, 10 and 15; W_G/W_d=0.15, 0.20 and 0.25. The aspect ratio of the duct (AR=H_d/W_d) was set equal to unity, i.e. a square shaped duct.

For boundary conditions at walls, all variables except temperature were set equal to zero. For the thermal boundary conditions, heat flux was set constant at all the duct and rib wall surfaces in all the computations.

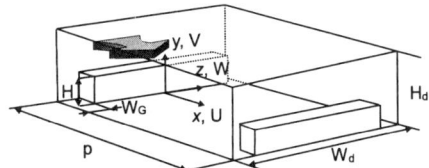

Figure 1: Computational domain.

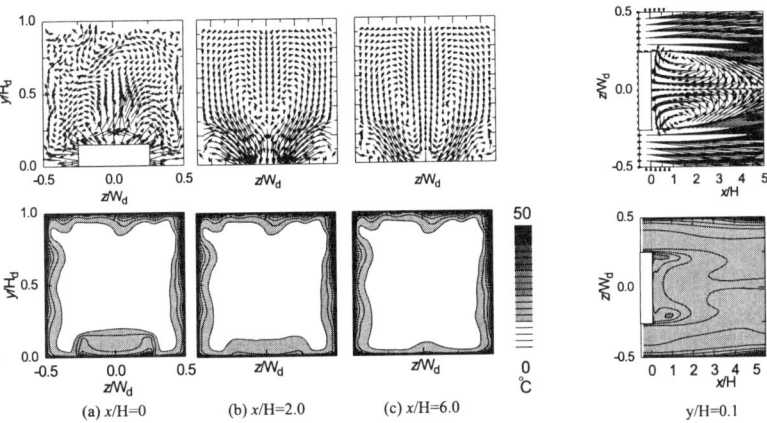

Figure 2: Velocity vectors and isotherms on y-z plane(left) and x-z plane(right). (p/H=10, W_G/W_d=0.25)

RESULTS AND DISCUSSION

Normal arrayed discrete ribs

Discrete ribs are reported to generate secondary flow downstream the rib. This flow structure largely differs from that of a standard full-span rib[4]. Especially, longitudinal vortices are generated from the rib leading edge and its corner edges[6,7]. These secondary flows enhance the local heat transfer from the duct bottom- and sidewalls by including washing effect at each wall surface. On the contrary, the friction factor decreases due to of the gap existence, by which flow resistance is reduced noticeably[7].

Figure 2 illustrates the secondary flow velocity vectors in the cross-sectional planes(left) at the locations of x/H=0, 2 and 6 and the one in the horizontal plane(right) at y/H=0.1. Isothermal contours in the same plane are shown together at the lower column. Through the computations, flow largely fluctuated accompanying numerous vortices. The results were taken in time mean value, however, for the present case, not enough time length was taken and, thus, the distributions are not fully symmetric.

Due to the pressure gradient existing behind the rib, the fluid passing the gaps moves inwards to the duct centerline. By this flow pattern, the flow passing the leading edge of the rib also tends to turn towards the centerline as it moves downstream

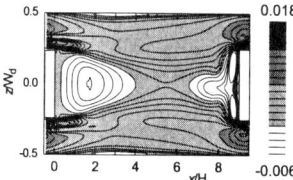
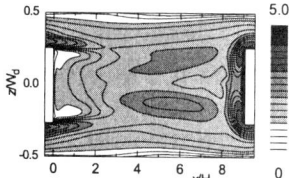

Figure 3: C_f on bottom wall. ($p/H=10$, $W_G/W_d=0.25$)

Figure 4: Nu/Nu_s on bottom wall. ($p/H=10$, $W_G/W_d=0.25$)

near the reattachment point. This is indicated in Fig. 2-(left)-(b). Downstream the reattachment point, this flow, then, generates longitudinal vortices. These vortices are observed in Fig. 2-(left)-(c). Cooler fluids are carried toward the duct surfaces by these vortices, which is indicated by the isotherm curvature. On the other hand, another rib wake is characterized by a pair of tornado type vortices generated downstream the rib, which is shown in Fig. 2-(right). These vortices are mainly caused by the high speed fluid passing the rib gaps and they are stirred up to the level of the rib height. By these tornado type vortices, flow stagnation points are observed at the center of each vortices and affect the local heat transfer, which will be discussed in the next paragraph.

Figures 3 and 4 show the spatial distributions on the duct bottom wall of the local skin friction coefficient, C_f and Nusselt number, Nu. For the C_f distribution, the white parts correspond to the area where it takes a negative value. The reattachment point location, which is defined at the location where C_f changes its sign from negative to positive, is $x/H=4.1$ if measured at the centerline where it takes the maximum value. The reattachment length from the rib is shortened as it gets near the duct sidewalls. This is the effect of the flow passing the rib gaps. The flow passing each gaps are accelerated and this causes the nearby local pressure decrease, which makes the earlier reattachment.

For Nu, the value is normalized by its spatially averaged value of a smooth duct. Two areas of large Nu are observed symmetrically on both sides of the centerline downstream the reattachment point. At these locations, longitudinal vortices exist and the downward flow of them cools the duct bottom surface, which was mentioned above. On the contrary, small Nu are shown near the rib rear surface and near the duct sidewalls. The former region corresponds to the center of the tornado type vortices where stagnation points exist. The latter region corresponds to the location of the reverse flow shown in Fig. 3.

Rib pitch to rib height ratio effect

Rib pitch to rib height ratio effect has been reported in the previous works[1]. For a full-span rib, the ratio of 10 is reported to be effective considering heat transfer. However, for the case of the present discrete ribs, the rib pitch to rib height ratio has not been fully investigated in detail. In the present section, the ratio is varied in three steps, i.e. $p/H=6$, 10 and 15 under the condition of $Re=10,000$ and $H/H_d=0.15$.

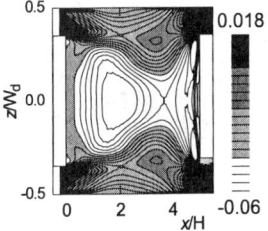
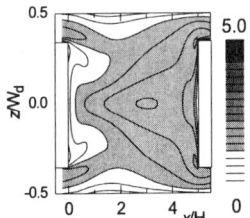

Figure 5: C_f on bottom wall. (p/H=6, W_G/W_d=0.15)

Figure 6: Nu/Nu_s on bottom wall. (p/H=6, W_G/W_d=0.15)

Table 1 : Spatial average of Nusselt number, Nu_m/Nu_s

	W_G/W_d=0.25	W_G/W_d=0.20	W_G/W_d=0.15
p/H=6	1.15	1.29	1.40
p/H=10	1.99	1.83	1.86
p/H=15	-	1.97	2.18

Figures 5 and 6 illustrate the spatial distribution of C_f and Nu on the duct bottom wall of the case of p/H=6. In Fig. 5, here also, white parts correspond to the negative value of C_f as in the previous section. For the case of p/H=6, white part covers the overall area between the ribs, which indicates the merge of the two recirculation zone downstream and upstream the rib. Here, no reattachment point appears. Comparing the Nu distributions of different rib pitch cases to each other, we can see that for the case of p/H=6, heat transfer is not effectively enhanced as in the case of p/H=10. This strongly relates to the disappearance of the flow reattachment, which enhances the local heat transfer.

Table 1 shows the spatially averaged Nusselt number, Nu_m over one streamwise pitch of the ribs in a form of normalized with counterpart of a smooth duct, Nu_s. The case of p/H=15 takes the largest value among the three cases while p/H=6 the worst result of heat transfer. As mentioned above, for the case of p/H=6, no reattachment was observed near the duct centerline and local heat transfer enhancement reflects on the lower spatially averaged value. On the other hand, for the case of p/H=15, a highly effective heat transfer is obtained because of the flow reattachment and boundary layer modification by the longitudinal vortices.

Staggered arrayed ribs

Further computations were carried out for a condition of a cooling passage having two different kinds of discrete ribs arrayed in a staggered formation: one rib having gaps in both side ends (here after DR-1) and the other having a gap in the middle (here after DR-2) mounted alternately in the streamwise direction at the pitch of $10H$. The rib height to duct height ratio was set equal to H/H_d=0.15 and the gap width to duct width ratio W_G/W_d=0.25 and W_G/W_d=0.5, respectively. Reynolds number was set equal to the one in the previous computation, i.e. Re=10,000.

Figures 7 and 8 show the C_f and Nu distributions on the duct bottom wall. Reverse flows, which correspond to the white part in Fig. 8, appear alternately behind

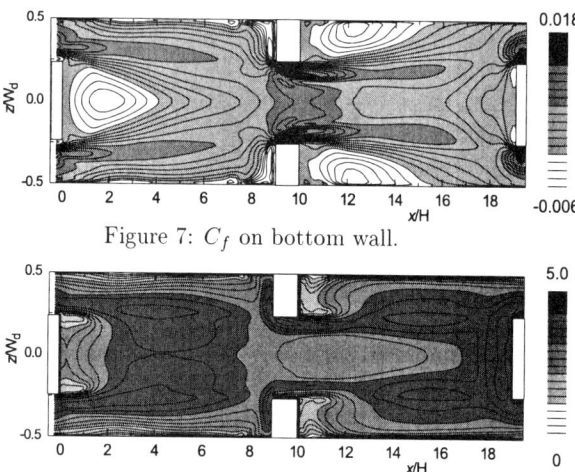

Figure 7: C_f on bottom wall.

Figure 8: Nu/Nu_s on bottom wall.

both, the DR-1 rib and the DR-2 ribs. However, comparing with Fig. 3, the flow recirculation zone behind DR-1 rib takes a triangle shape while flow recirculation zone is hardly observed in front of the rib. These might be the effect of the existence of the DR-2 rib. As the gap locates in the middle of the duct, the flow passing the DR-1 rib is strongly led toward it. By this, the recirculation zone is suppressed and form a triangle shape. On the other hand, the flow passing over the leading edge of the DR-2 ribs, moves outwards to the duct sidewalls as it gets near the reattachment point. Related to this flow pattern, the pressure near the front wall of DR-1 rib is decreased and the size of the wall near circulation zone is decreased.

Large Nu is observed in a vast area of the duct bottom wall. Comparing with Figures 4 and 6, it is suggested that the present staggered arrayed case enhance the local heat tranfer than in the previously discussed normal arrayed case. For the spatially averaged value also, the value of the staggered arrayed case is Nu_m/Nu_s=3.1, which is larger than the normal arrayed case.

Figures 9 and 10 illustrate the Reynolds stresses, $-\overline{uv}/U_{ref}^2$ and $-\overline{vw}/U_{ref}^2$ in the cross-sectional planes respectively for the case of staggered arrayed ribs and normal arrayed ribs. $-\overline{uv}/U_{ref}^2$, takes a slightly larger value in the case of staggered arrayed ribs comparing with the normal arrayed ribs. Furthermore, large value of $-\overline{uv}/U_{ref}^2$ is observed in a wider area in the spanwise direction for the staggered case at the location of x/H=2.0 than the normal arrayed one. This might be one of the reasons of why large Nu is obtained in a large area on the bottom wall. Furthermore, for $-\overline{vw}/U_{ref}^2$ also, larger value can be seen for staggered case than the other case.

Figure 11 shows the turbulence kinetic energy, k/U_{ref}^2 near the duct bottom wall for each cases; staggered arrayed case and normal arrayed case. Here also, the kinetic energy takes a larger value for the staggered case especially downstream each rib corners. These suggest that the momentum exchange is very active by having two alternative ribs in a row than a unit arrayed case, thus enhance the local heat transfer on the duct bottom wall.

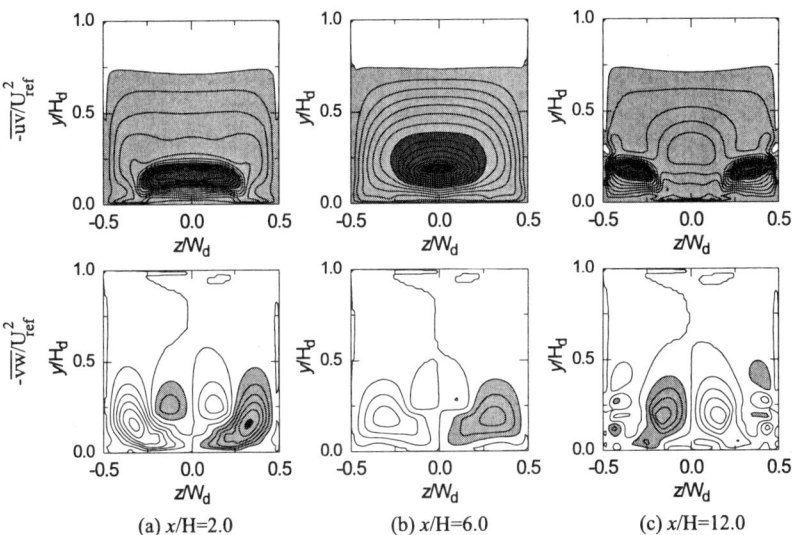

Figure 9: Reynolds stress contours. (Staggered arrayed case)

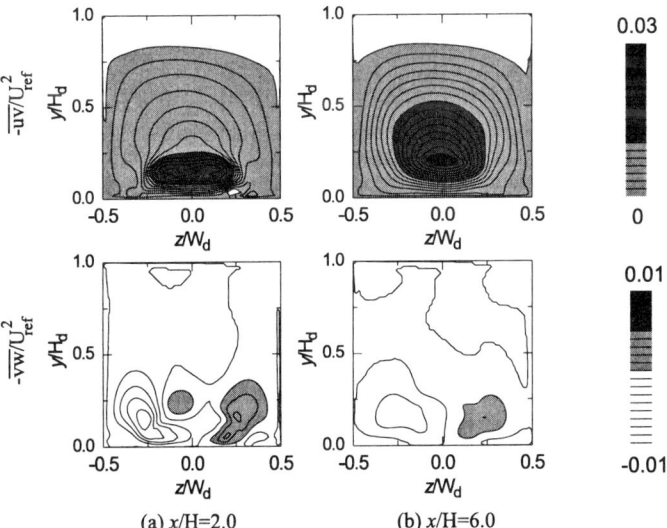

Figure 10: Reynolds stress contours. (Normal arrayed case)

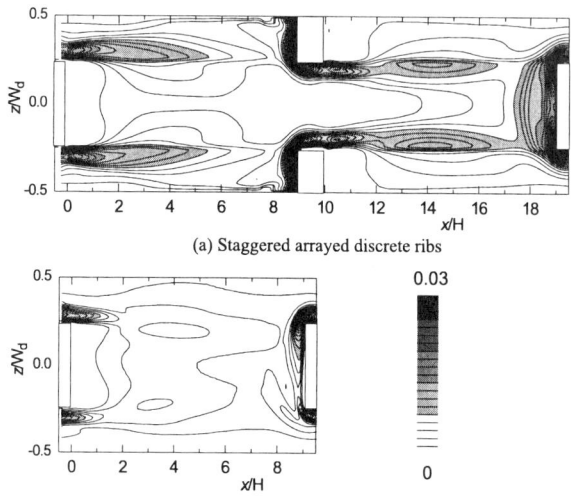

Figure 11: Turbulence kinetic energy, k/U_{ref}^2 near bottom wall.

CONCLUSIONS

Numerical simulation was carried out to investigate the fluid and heat transfer characteristics for discrete rib arrays. The wakes of discrete ribs were characterized by two main vortices. One was the longitudinal vortices appearing downstream the reattachment point, and the other was the tornado type vortices located behind the rib. These vortices largely affected the local heat transfer. For the comparison of normal arrayed ribs and staggered arrayed ribs, the staggered case showed a fair increase in heat transfer.

REFERENCES

1. Han, J. C. 1988. *In* ASME, J. of Heat Transfer, Vol. 110, pp. 321-328.
2. Tatsumi, K., Iwai, H., Inaoka, K. and Suzuki, K., 2000. *In* Proceedings of the 3rd Int. Sympo. on Turbulence, Heat and Mass Transfer, Nagoya, Japan, April, pp. 207-214.
3. Lau, S. C., Kukreja, R. T. and McMillin, R. D., 1991. *In* Int. J. of Heat and Mass Transfer, Vol. 34, No. 7, pp. 1605-1616.
4. Lau, S. C., McMillin, R. D. and Han, J. C., 1991.*In* ASME J. of Turbomachinery Vol. 113.
5. Balatka, K., Mochizuki, S. and Murata, A., 1997.*In* Proceedings of the 1st Pacific Sympo. on Flow Visualization and Image Processing, Honolulu, U.S.A., February, pp. 219-244.
6. Hu, Z. and Shen, J., 1996.*In* ASME Paper 96-GT-313.
7. Tatsumi, K., Iwai, H., Inaoka, K. and Suzuki, K., 2000.*In* Proceedings of the 8th Int. Sympo. on Transport Phenomena and Dynamics of Rotating Machinery (ISROMAC-8), Honolulu, U.S.A., March, Vol. 2, pp. 801-806.
8. Suga, K., 1995. *In* UMIST, Dept of Mechanical Engineering, Report TFD/95/11.

Development of Non-destructive Inspection Method for the Performance of Thermal Barrier Coating

M.MORINAGA, T.TAKAHASHI

Thermal Engineering department Yokosuka Research Laboratory
Central Research Institute of Electric power Industry
2-6-1 Nagasaka, Yokosuka-shi, Kanagawa-ken 240-0196 JAPAN

ABSTRACT: This paper shows that our proprietary non-destructive inspection method can be used to effectively measure the thermal barrier performance of the thermal barrier coating used to coat gas turbine hot parts by the results of numerical analysis and laboratory experiments.

INTRODUCTION

A gas turbine will be operated at ever higher temperatures to obtain high efficiency. On this account, thermal barrier technology to protect gas turbine hot parts from high temperature gas becomes increasingly important. Accordingly, the evaluation of the thermal barrier performance of the thermal barrier coating (TBC) becomes important, too. TBC thermal barrier performance is decided by thermal resistance of TBC (in other words, TBC thickness and TBC thermal conductivity). Both decrease of TBC thickness by erosion and sintering of TBC bring TBC thermal barrier performance degradation. To date, however, no effective inspection method for the thermal barrier performance of TBC has been developed. In this paper, we describe an effective nondestructive inspection method to evaluate the thermal barrier performance of TBC[1,2].

MEASUREMENT PRINCIPLES

In our proprietary non-destructive inspection method, a laser is used to heat a measurement object point, and then a radiation thermometer or an infrared camera measures the surface temperature at that point. The measured temperature are then used to evaluate the thermal resistance of the TBC. In this non-destructive inspection method, it is important to separate the heating wavelength and the measuring wavelength to remove any effects of reflected light. Heating and measuring the target must also be performed at a constant rate. Fig.1 shows a trial application of this non-destructive inspection method (following, TBC transient heating method) to a gas turbine combustor.

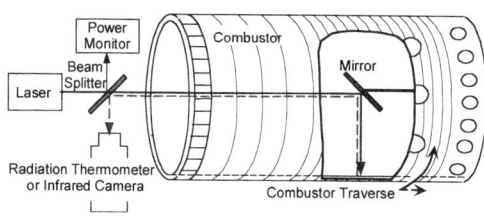

Fig.1 Conceptional Figure of Our Proprietary Non-destructive Inspection Method applied to Gas Turbine Combustor

Table 1 Numerical Analysis Conditions

parameters	Conditions
Thermal Conductivity	virgin TBC, age TBC(15%up)
Heat Flux	0.4, 0.6, 0.8 × 10⁶W/m²
Heating Area	φ2, φ4 mm
Traverse Speed	10, 20mm/s
TBC Thickness	0, 0.1, 0.15, 0.2, 0.3mm
Base Metal Thickness	2.0, 2.5, 3.0mm
Surface Direction	horizontal, vertical
Bond Coating	exist, none
Peeling	exist, none

RESULTS OF NUMERICAL ANALYSIS

To make clear the characteristics of the TBC transient heating method and to evaluate its validity, this chapter shows the results of unsteady state conduction analysis using a finite element method program.

Calculation Conditions

TBC thickness and TBC thermal conductivity are used as thermal barrier performance parameters for measurement conditions, and measurement surface direction was selected according to base metal thickness, heat strength, heating area and heat-source traverse speed(Table 1). Underlined values in the table indicate standard conditions. Time integration utilizes backward difference (Eulerian implicit analysis). Calculations were made in steps of 0.01s.

The calculation model consists of a three-dimensional plate model using a two-layer construction of the TBC and the base metal. Since the thermal conductivity of the bond coating layer is similar to the base metal and the layer is thin, this conductivity is omitted to simplify the calculations. We have confirmed that the omission of the bond coating layer does not affect the results of the analysis.

The material used for the TBC is ZrO_2-$8Y_2O_3$, and the base metal is Hastelloy X. The thermal diffusivity, specific heat, and density measured in our laboratory are used as data for the thermal properties of both the TBC and the base metal[3,4]. In addition, for thermal properties of aged materials, data measured from actual gas turbine aged materials is used, as well as data measured from simulated materials examined in a state of high-temperature oxidation.

Calculation Results

Fig.2 shows temperature change of each section for elapsed time in standard conditions, and Fig.3 shows an example of temperature distribution for the model overall. For the elapsed time using a TBC transient heating method, the time that the laser heating area reached to the center of the measurement point was set as 0s, and the following study utilizes the elapsed time 0s value for the calculated temperature. The

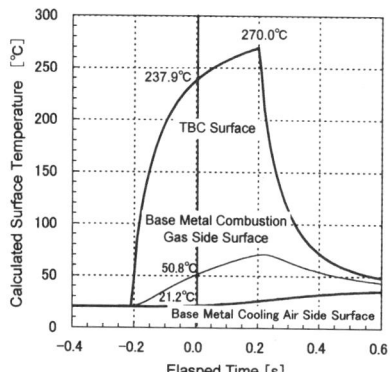

Fig.2 An Example of Numerical Analysis Results of Calculated Temperature on Combustor

following section will discuss the effects of calculated temperature on each parameter.

(1) Effects of Heating Conditions

Heat strength, heating area and heat source traverse speed can be offered as heating condition parameters. The effects of each parameter are shown below.

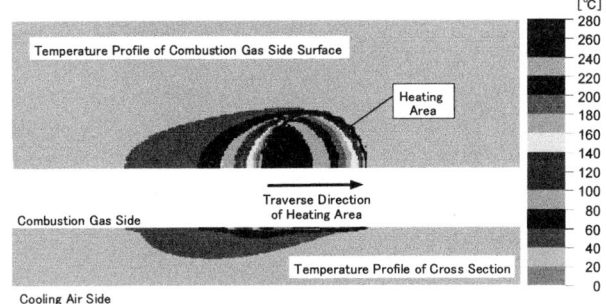

Fig.3 An Example of Temperature Distribution for the model in standard conditions

a) Effects of Heat Strength

Increasing heat strength causes a linear increase in calculated surface temperature (Fig.4). The temperature difference between calculated surface temperatures of virgin TBC and aged TBC can be increased by increasing the heat strength.

b) Effects of Heat-source Traverse Speed

Increasing the speed of heat-source traverse can reduce the overall inspection time, but this also reduces the calculated temperature. Because of this, we can reduce the temperature difference between the calculated temperatures of virgin TBC and aged TBC approximately 25 percent by doubling the heat-source traverse speed from 10 mm/s to 20 mm/s.

c) Effects of Elapsed Time

When organizing the calculated temperature difference between virgin TBC and aged TBC using elapsed time, the calculated temperature difference begins to rise from an elapsed time of –0.28s, attaining a maximum value at an elapsed time of 0.13s. In addition, including the center of measuring temperature area in the heating area results in an elapsed time between –0.2 to 0.2s. At the point showing the maximum value, the calculated temperature difference rises approximately 17 percent and the measuring temperature area is shifted behind the heating area. As a countermeasure, we suggest using an infrared camera and making a wide-area temperature measurement to obtain a greater calculated temperature difference.

(2) Effects of TBC Thickness

Reducing TBC thickness results in a linear drop in calculated temperature (Fig.5). For a variation of 1 μm of TBC thickness, the calculated temperature varies –0.70degC/μm.

(3) Effects of TBC Thermal Conductivity

The difference between the calculated temperature when measuring virgin TBC and aged TBC (thermal conductivity rises about 15 percent) is 33.2degC, and there is a calculated

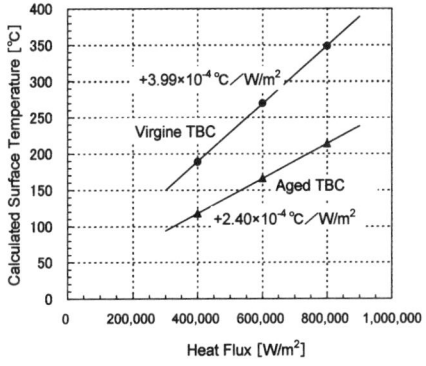

Fig.4 Effect of Heat Flux for Calculated Surface Temperature

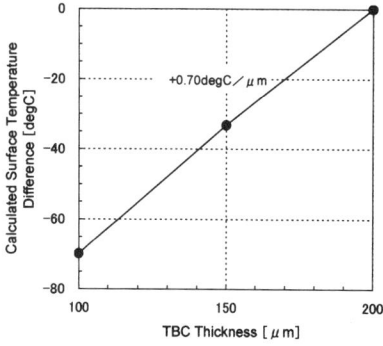
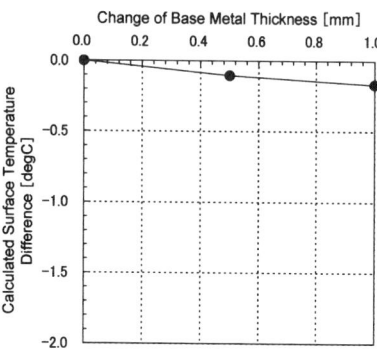

Fig. 5 Effect of TBC Thickness for Calculated Surface Temperature

Fig. 6 Effect of Change of Base Metal Thickness for Calculated Surface Temperature

temperature variation of –2.2degC/% for each 1 percent variation in the TBC thermal conductivity.

(4) Effects of Base Metal Thickness

When using TBC transient heating method, there is almost no rise in the surface temperature of the rear surface of the measurement section (Fig.6), and we can ignore the effects of changes in the base metal thickness on the calculated temperature. This is a very important consideration when dealing with complex shapes of metal parts on actual gas turbine hot parts.

(5) Detecting Peeled Sections

When peeling occurs in TBC, it obstructs thermal conduction from the TBC to the base metal, and a high temperature is detected at the surface of the section with peeling. When peeling occurs with standard conditions of TBC transient heating method, the maximum calculated temperature of the section with peeling is 377°C, which is clearly higher than when peeling has not occurred. The high-temperature section follows the shape of the peeling, and so by indicating temperature distribution based on the infrared camera measurements and radiation thermometer data, it is possible to detect the shape of the peeling.

Evaluating Suggested Measurement Methods

If we consider detecting TBC thermal resistance with a precision of ±5% (corresponding to approximately ±2degC of base metal surface temperature under operation), a variation of ±10 μm in TBC thickness results in a detected temperature variation of ±7degC, and a variation of ±5% in TBC thermal conductivity corresponds to a ±11degC variation in detected temperature. However, these temperature variations can easily be captured with current equipment performance, and we have determined that when TBC transient heating method are applied to actual gas turbine hot parts to evaluate the thermal barrier performance of TBC, the methods have sufficient measurement precision.

Verification With Laboratory Tests

This chapter will discuss the test results with different parameter variations using test apparatus manufactured based on TBC transient heating method.

Overview of Test Apparatus

The test apparatus used was a radiation thermometer based on TBC transient heating method and using a CO_2 laser as a heat source. Objects measured were plate test piece. These tests used a fixed heat source, and the test piece were moved. The speed at which the test piece were moved involved computer control and adjustment of the motorized X axis translation stage. The heating laser beam was shined from the normal direction of the test piece. Fig.7 shows an overview of the test apparatus, and Table 2 shows the apparatus specifications.

Looking at the wavelength used for heating and measuring temperature, we find that the CO_2 heating laser has a wavelength of 10.6 μm, and the radiation thermometer has a detection wavelength of 5 to 6 μm. These were selected to keep the heating and the measuring wavelengths separate and so avoid any influence on the measurement from the laser light reflected from the surface of the test piece.

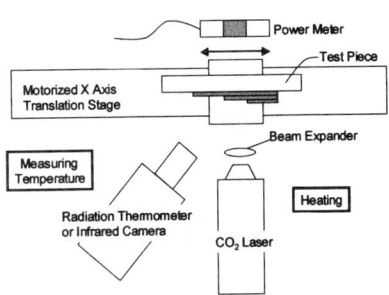

Fig.7　Overview of the Test Apparatus

Table 2　Equipment Specifications

Equipment	Specifications	
CO_2 Laser	Wavelength	10.6 μm
	Average Power	12.5W
	Polarization	Linear
	Transverse Mode	TEM00
	Power Stability	±5%
	Beam Diameter	ϕ 2.7mm
	Beam Divergence	7.6mrad
Radiation Thermometer	Wavelength	5~6µm
	Detected Range	~200°C
	Cooling System	Forced Air
Motionrized X Axis Translation Stage	Max Speed	120mm/s
	Traverse Range	±250mm

The heat strength was controlled by adjusting the output of a CO_2 laser and adjusting the traverse speed of the test piece using the motorized X axis translation stage. The beam profile was adjusted using a beam expander.

Heat strength was measured using a laser power meter placed behind the motorized X axis translation stage, before and after the test piece was heated by a CO_2 laser. Data was recorded every 100 ms for two items: the test piece surface temperature measured with a radiation thermometer and the laser output measured with a laser power meter.

Test Piece

In consideration of thermal properties, SUS304 (2 mm thick) was selected as an equivalent material to the base metal of actual gas turbine hot parts, and atmosphere plasma spray coated TBC was used in commercial TBC powder with thickness varied in stages of 100 μm, 200 μm and 300 μm. When performing the tests, test piece measurement positions were changed to suit the test parameters. Fig.8 shows the shapes of the test piece, and Table 3 shows the composition of the commercial TBC powder. For the TBC thermal values, the thermal diffusivity was measured with the laser flash method, and the specific heat was measured with a differential scanning calorimeter (DSC) by Central Research Institute of Electric Power Industry.

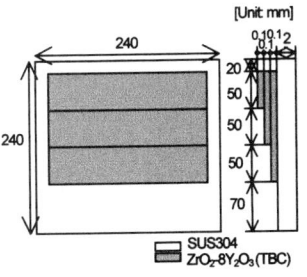

Fig.8　The Shape of Test Piece

Table3　Contents of TBC

Contents [wt %]	ZrO_2	Y_2O_3	HfO_2	SiO_2	Al_2O_3	TiO_2	Fe_2O_3
	90.41	7.51	1.74	0.15	0.09	0.08	0.02
Thermal Conductivity　$\lambda = 7.2086 \times 10^{-5} t + 0.59132$ [W/mK] (room temperature~500°C)							

Test Results

Fig.9 shows an example of test results. As the graph shows, increasing the TBC thickness causes a rise in the measured temperature. The next section will discuss the effects of each parameter on the detected temperature.

(1) Effects of Heating Conditions

Fig.10 shows the detected temperature for each heat-source traverse speed adjusted for heat strength. A positive straight-line correlation can be seen between the heat strength and the detected temperature. By increasing the heat strength, we can increase the

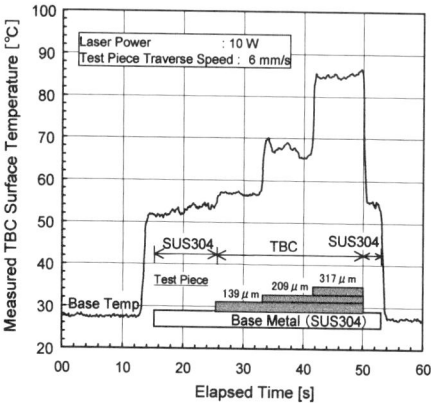

Fig.9 An Example of Test Results

detected temperature. Both increasing the heat strength and reducing the heat-source traverse speed have been determined to increase the detected temperature. However, we can imagine such restricting factors as TBC surface damage and a relative reduction in detected temperature differential accompanying increased heat flow input to the base metal.

Fig.11 shows detected temperatures for each laser output adjusted for heat-source traverse speeds. A negative straight-line correlation can be seen between the heat-source traverse speed and the detected temperature.

(2) Effects of TBC Thermal Resistance

Fig.12 shows the detected temperature for each heat-source traverse speed adjusted for TBC thermal resistance. A positive straight-line correlation can be seen between TBC thermal resistance and detected temperature.

(3) Effects of Base Metal Thickness

Effects of variations in base metal thickness were evaluated using aluminum plates mounted on the rear surface of the test pieces. The aluminum plate mountings displayed a tendency for the detected temperature to drop, and the more the heat-source traverse speed was reduced, the greater the width of the drop in temperature. In the test conditions in these tests, when the heat-source

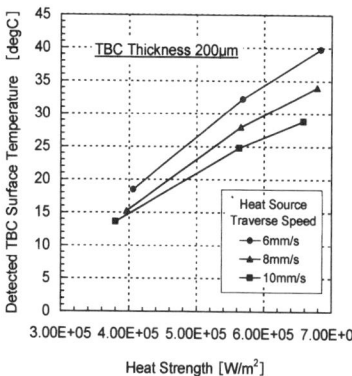

Fig.10 Effect of Heat Strength for Detected Temperature

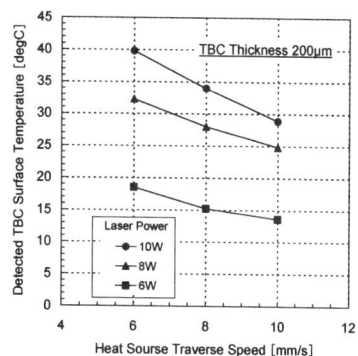

Fig.11 Effect of Heat Source Traverse Speed for Detected Temperature

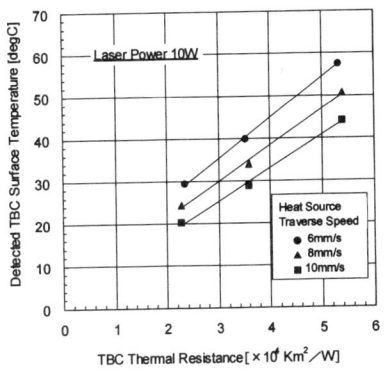
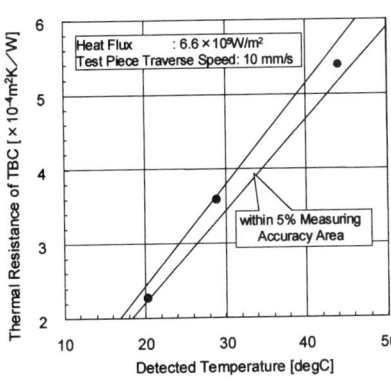

Fig.12 Effect of TBC Thermal Resistance

Fig.13 The Relation between Detected Temperature and Thermal Resistance of TBC

traverse speeds exceeded 6 mm/s, the amount of temperature drop in detected temperature was around 3 percent, and this value corresponds to less than 1degC base metal temperature conversion, and so we can safely ignore it.

(4) Effects of Heating Patterns

The beam profile changes from an acute conical shape (gauss distribution) to a flat shape according to the beam expander setting. Depending on the beam expander settings, the fluctuations thought to occur in the measured temperature due to reflections at the surface of the test piece do not occur. Also, since the heat strength distribution is reduced within the heating area, the shift in the field of measured temperature reduces the margin of error in measurement, and so the beam expander setting is required.

Evaluation Results at the Laboratory Level

A Comparison of Test Results with Varied Parameters and Numerical Analysis Results

We confirmed that the test results with varied parameters showed the same trends as the numerical analysis results. This verified the validity of the measurement principles of TBC transient heating method and the numerical analysis results. However, in the test results with varied parameters, the detected temperatures were lower than with the numerical analysis results, but this is thought to be caused by a large measured temperature field in relation to the heating area, and by a low TBC laser light absorptivity.

A Study of Measurement Precision

When measuring on actual gas turbine hot parts, a calibration line is first created indicating the relationship between previously detected temperature and thermal resistance, and then measurement is made on the actual gas turbine hot parts. When creating the calibration line, evaluation includes the measurement margin of error of the parts, and so the problem that occurs in measurement precision when measuring on actual gas turbine hot parts is the dispersion of the measurement data. Dispersion has a tendency to decrease in inverse proportion to the heat strength and the heat-source traverse speed. Within these test conditions, we were able to detect TBC insulation performance with a measurement precision of ±5% using heating output conditions of 10 W. Fig.13 shows an example with the allowable range of dispersion of detected temperature collectively entered in the measurement data. In this measuring method, at the laboratory level, we were able to confirm that measurement of TBC thermal barrier performance was possible at ±5% by selecting measurement conditions. However, when measuring on actual gas turbine hot parts, it will be necessary to consider the effects of the surface condition (emissivity, roughness) of the measurement object. Because of this, when applying to actual gas turbine hot parts, it should be necessary to make temperature correction using such methods as simultaneously measuring the temperature and the emissivity.

CONCLUSION

We have demonstrated the validity of our proprietary non-destructive inspection method of TBC thermal barrier performance.

We have manufactured test equipment based on TBC transient heating method devised by this laboratory, and with actual combustor simulation materials, we have performed tests with variations in measuring parameters such as heat strength, heat-source traverse speed, TBC thickness and base metal thickness. Those results have shown a good conformity with the results of numerical analysis performed in advance of the actual gas turbine hot parts, and have confirmed the validity of TBC transient heating method.

We have confirmed the possibility of using TBC transient heating method to evaluate TBC thermal resistance of actual combustor simulation materials according to selected measuring parameters, and we have evaluated the method as being valid at the laboratory level.

ACKNOWLEDGEMENT

Based on the success we have had to date, we are planning to design and manufacture hypothesized test apparatus for TBC transient heating methods, and conduct a variety of studies using test measurements of actual gas turbine hot parts TBC.

Finally, we would like to take this opportunity to express our gratitude for the understanding and assistance we have received from the electrical power companies, and we would like to continue to ask for their very valuable assistance in the future.

REFERENCES

1. M.Morinaga, T.Takahashi and T.Fujii. 1998. Development of Non-destructive Inspection for the Performance of Thermal Barrier Coating – No.1 Numerical Analysis for TBC Thermal Resistance Detecting Method with Transient Heating -. CRIEPI Report W97021
2. M.Morinaga, T.Takahashi. 1999. Development of Non-destructive Inspection for the Performance of Thermal Barrier Coating – Part.2 Experimental Study for TBC Thermal Resistance Detecting Method with Transient Heating -. CRIEPI Report W99005
3. T.Fujii, T.Takahashi. 1998. Estimation of Thermophysical Properties of Coating Layers for Gas Turbine Hot Parts - Part 1: Measurement of Thermophysical Properties of Coating Layer and Superalloy, and Comparison between Virgin and Aged Material -. CRIEPI Report W97017
4. T.Fujii, T.Takahashi. 1999. Estimation of Thermophysical Properties of Coating Layers for Gas Turbine Hot Parts - Part 2: Comparison between Virgin and Aged Thermal Barrier Coatings in terms of Plasma Spray Powder Size and Shape -. CRIEPI Report W98017

Numerical Modelling of Flow and Heat Transfer in the Rotating Disc Cavities of a Turboprop Engine

JOHN FARAGHER[1] AND ANDREW OOI[2]

[1]*Airframes & Engines Division, Defence Science and Technology Organisation, 506 Lorimer Street, Fishermans Bend, Victoria 3207, Australia.*

[2]*Mechanical and Manufacturing Engineering Department, University of Melbourne, Parkville, Victoria 3052, Australia.*

ABSTRACT

A numerical analysis of the flow and heat transfer in the cavity between two co-rotating discs with axial inlet and radial outflow of fluid, a configuration common in gas turbine engines, is described. The results are compared with the experimental data of Northrop and Owen[1]. The effectiveness of the k - ε turbulence model with the two-layer zonal model for near-wall treatment of Chen and Patel[2] is tested for this type of flow. Using three-dimensional models it is shown that modelling discrete holes at the outlet as opposed to a continuous slot, which is the approximation inherent in the two-dimensional axisymmetric model, has little effect on the predicted Nusselt number distribution along the disc surface. Results of a conjugate heat transfer analysis of a spacer in the turbine section of a turboprop engine are then presented.

INTRODUCTION

The creep life of turbine discs depends strongly on temperature. The low-cycle fatigue life of turbine and compressor discs depends on the mechanical and thermal stresses they experience and on the disc material properties, which vary with temperature. The mechanical stresses depend on engine speed; and in general they are larger than the thermal stresses, which depend on the temperature gradients. Even though the thermal stresses are smaller than the mechanical stresses, they do significantly affect the life of the discs. To determine temperature gradients in the discs requires accurate predictions of both the flow in the cavities between the discs, and also the heat transfer in the complex boundary layers that are formed.

This paper describes a CFD analysis of the flow and heat transfer in the cavity between two co-rotating discs, with axial inflow and radial outflow of fluid. The CFD method is then applied to the conjugate heat transfer analysis of a spacer in the turbine section of a turboprop engine. All CFD predictions employed the commercial CFD software FLUENT to solve the Navier-Stokes equations, using primitive variables and finite volume discretisation with collocated arrangement of dependent variables. The SIMPLE scheme was used to couple the velocity and pressure fields.

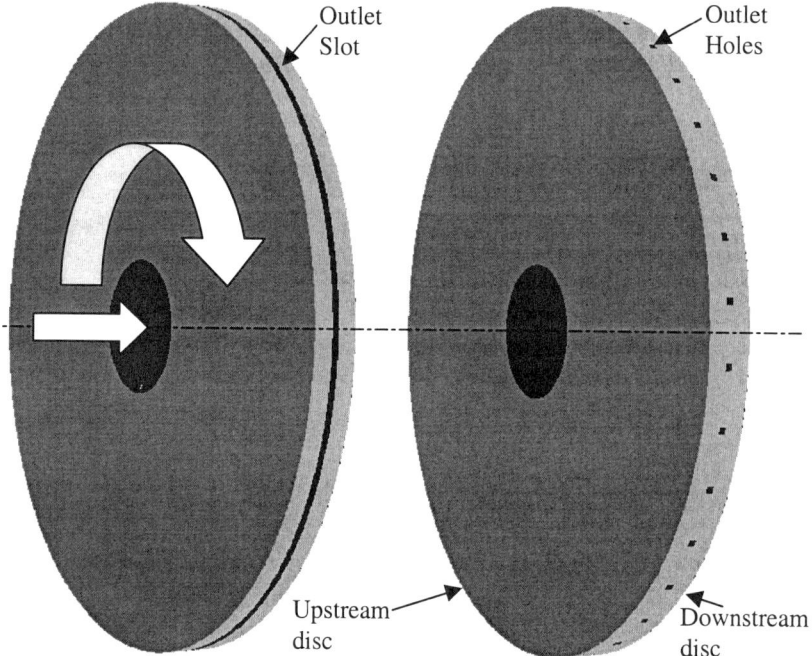

Figure 1. The two configurations of the system that were modelled.

HEAT TRANSFER IN A ROTATING DISC CAVITY

Northrop and Owen[1] measured the turbulent flow and heat transfer in a cavity bounded by two flat co-rotating discs. One disc had a central hole through which fluid entered the cavity. The flow left the cavity through a perforated shroud joining the two discs at their rims. The cavity dimensions were: outer radius b = 428 mm, inner radius (central hole) a = 89 mm, cavity width s = 59 mm. They presented distributions of Nusselt number as a function of radial coordinate. The geometry is based on the turbine discs in a gas turbine engine where air is supplied to the bore of the discs to cool them, and flows outward to rejoin the mainstream gas flow at the rim of the discs. The two configurations of the system that are modelled in this paper are shown in Figure 1. The standard k - ε turbulence model with the two-layer zonal model for near-wall treatment (Chen and Patel[2]) has been tested. CFD predictions of the heat transfer in this cavity using a two-dimensional axisymmetric model are compared with the experimental results of Northrop and Owen. In their experiments the shroud had 32 holes equally spaced around the circumference. In the two-dimensional axisymmetric model these holes must be approximated by a slot. Three-dimensional models of both the hole and slot configurations are then used to determine whether this difference in the outlet geometry significantly affects the heat transfer from the discs.

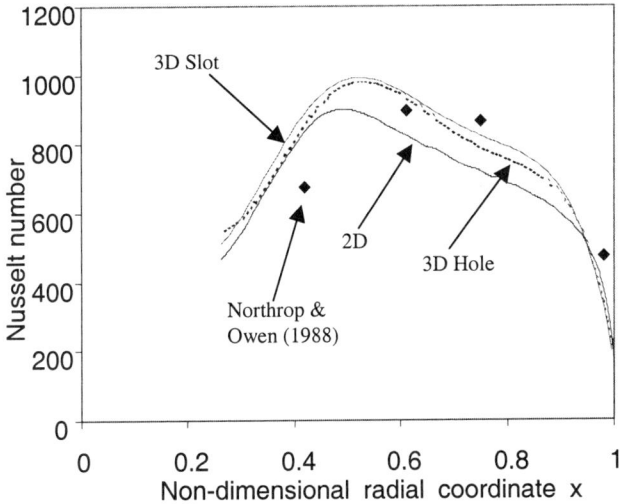

Figure 2. Nusselt number distributions for the upstream disc for Re = 3.2e6.

Two-dimensional axisymmetric model

The cavity was modelled using a two-dimensional axisymmetric solver with swirl. The grid had 209 x 950 grid points. To simulate the experiment the boundary conditions were set as follows. The inlet boundary condition was specified with the swirl and radial components of velocity relative to the absolute frame of reference set to zero, so that the fluid at the central inlet was not rotating with the discs. The axial component of the inlet velocity was set to 3.55 m/s. Velocity outlet, pressure outlet and outflow boundary conditions were tried, with negligible difference between the predicted Nusselt number distributions on the discs. The turbulence model was the standard $k - \varepsilon$ model with the two-layer zonal model near the walls. The rotational speed of the discs was 255.7 rad/s (2442 rpm), and $Re_\phi = 3.2 \times 10^6$. Figure 2 shows a plot of Nusselt number against radial coordinate for the upstream disc. The results from the two-dimensional axisymmetric model are within 20 percent of the experimental data. This agreement is not as close as that reported in previous CFD predictions for the same geometry using the Launder and Sharma turbulence model[3] and the Morse turbulence model[4] where the Nusselt number distributions were within 9 percent of the experimental data. The Launder and Sharma turbulence model in FLUENT was tested by the current authors and the agreement with the experimental data was slightly worse than for the standard $k - \varepsilon$ model.

Three-dimensional turbulent flow with heat transfer

For the three-dimensional models it is sufficient to model an 11.25° sector of the disc assembly of Northrop and Owen[1]. Two models were created: one with an outlet slot and the other with the sector centred on an outlet hole and periodic symmetry planes halfway between this outlet hole and the adjacent outlet holes as shown in Figure 3. The grid comprised approximately 200,000 hexahedral cells. The turbulence model and boundary conditions were the same as for the two-dimensional model.

The results in Figure 2 show that the three-dimensional models, with the hole and the slot, produced very similar Nusselt number distributions to each other. This demonstrates that this change in the outlet boundary conditions has very little affect on the flow in the rest of the cavity. Both of the three-dimensional models produced a trend similar to the two-dimensional model results, with a higher peak and higher values of Nusselt number in the outer half of the cavity. The fact that the three-dimensional grid is coarser than the two-dimensional grid is probably the cause of the difference between the Nusselt number distributions predicted by these models. The two-dimensional grid was refined until grid-independent results were achieved but this has not been done with the three-dimensional grid.

For the three-dimensional model with an outlet hole there was also negligible variation of the Nusselt number distributions in the tangential direction, i.e. there was no significant difference between the results at the middle of the sector and those at the symmetry plane.

These results show that the approximation of the holes by a slot is not the cause of the difference between the computed and experimental results, and indicate that the use of a two-dimensional axisymmetric model is sufficient. Further work is required to determine the reason for the difference between the computed and experimental Nusselt number distributions. Possible areas for improvement are the turbulence model and refining the grid in the boundary layers on the discs.

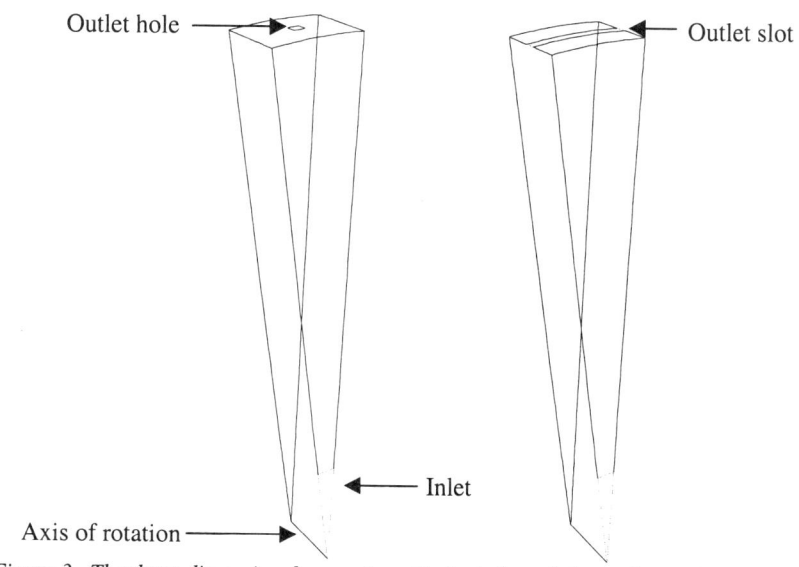

Figure 3. The three-dimensional geometry with the hole and slot outlet configurations.

APPLICATION TO A TURBOPROP ENGINE

A heat transfer analysis is required in order to support an investigation of the creep life of turbine components in a turboprop engine. The primary component of interest is the 2nd-3rd stage spacer as shown in Figure 4. A conjugate flow and heat transfer analysis has been used to predict the convection heat transfer between the turbine spacer and the cooling air in the rotating cavities on either side of this spacer, and to predict the conduction heat transfer within the spacer.

The grid comprised two separate fluid zones with a total of 80,000 cells and three solid zones with a total of 55,000 cells. The standard k-ε turbulence model with wall functions was used. The discs and the spacer have an angular speed of 14,239 rpm (1491 rad/s). The inlet and outlet slots through which the air enters and leaves the cavities on either side of the spacer are labelled in Figure 4. The inlet velocity was set to 73 m/s to give a mass flow of 0.04 kg/s at each inlet. An outflow boundary condition was used at each outlet.

The spacer is made of an austenitic iron-base superalloy material. The material properties used in the analysis were as follows:
Density = 7900 kg/m^3
Specific heat = 460 J/kgK
Thermal conductivity = 21 W/mK.

The predicted streamlines are shown in Figure 4. The flow is clearly divided into four distinct regions as described by Owen and Rogers[5]. These are: a source region where the flow is being entrained into the boundary layers, non-entraining Ekman-type boundary layers, an interior core where there is no axial or radial velocity, and a sink region at the outlet. It is noteworthy that although the fluid enters the cavities with a radial velocity component and zero axial velocity it behaves more like the flow in a cavity with an axial inlet. Rather than the flow simply dividing equally between the two disc surface boundary layers, as is typical for a rotating cavity with a radial inlet, all of the fluid forms a wall jet on the right side of the cavity, and then half of it breaks away from that wall and crosses the cavity to be entrained in the boundary layer on the other wall. The fact that the flow pattern shown with the complex engine geometry in Figure 4 is similar to that observed with the simpler geometry used above gives confidence in applying the CFD analysis to the real engine.

The thermal boundary conditions were based on thermocouple measurements from engine tests. On the two discs the temperature was specified on the surfaces not in contact with the fluid. Measured values were only available for three points on each disc, so linear interpolation was used between these points. These temperatures are marked on Figure 4. For the 2nd stage disc the temperature was 739 K at the bore, 770 K at the midpoint and 842 K at the rim. For the 3rd stage disc the temperature was 730 K at the bore, 772 K at the midpoint and 871 K at the rim. A constant temperature of 947 K was specified on the top surface of the spacer. The temperature of the cooling air at the inlets was set to 740 K. There was no ingress of fluid at the outlets. The predicted temperature contours for both the fluid and solid regions are shown in Figure 5. They show a very steep decrease in the temperature radially inward from the rim of the spacer. This steep gradient would cause high thermal stresses in this region. These temperatures will be used as one of the inputs to a finite element analysis of the creep life of the spacer.

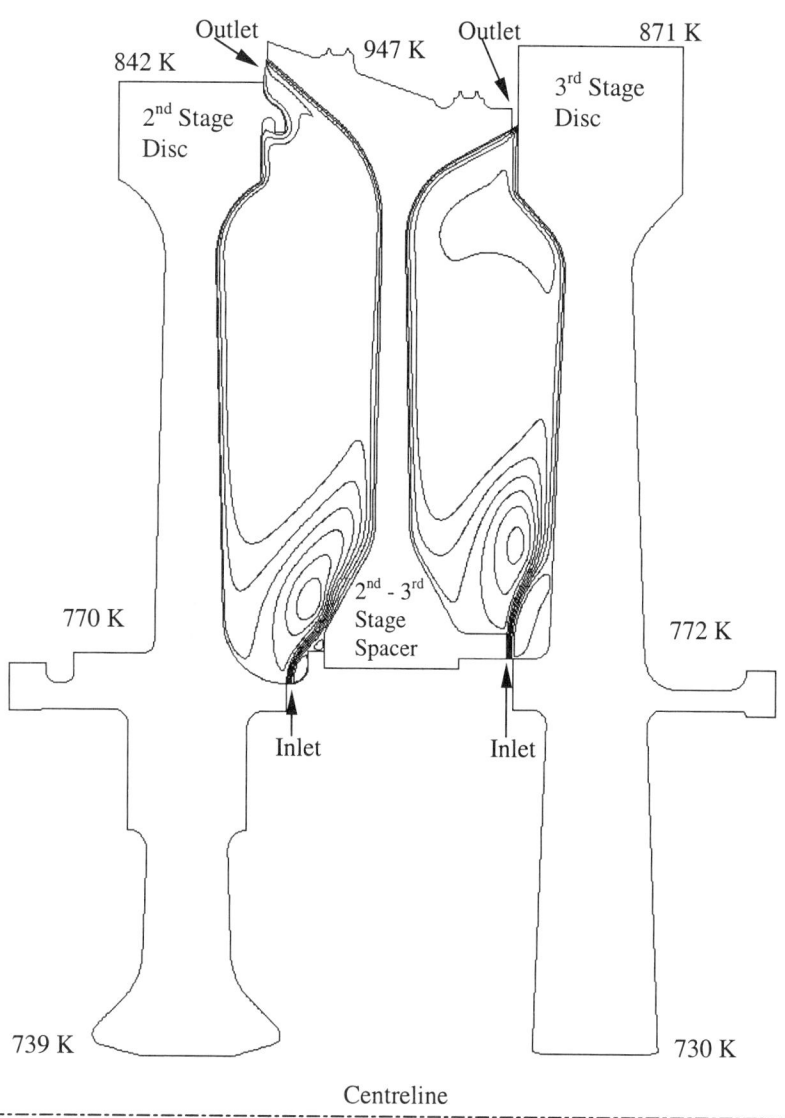

Figure 4. The predicted streamlines and specified thermal boundary conditions for turbine disc cavity region of a turboprop engine.

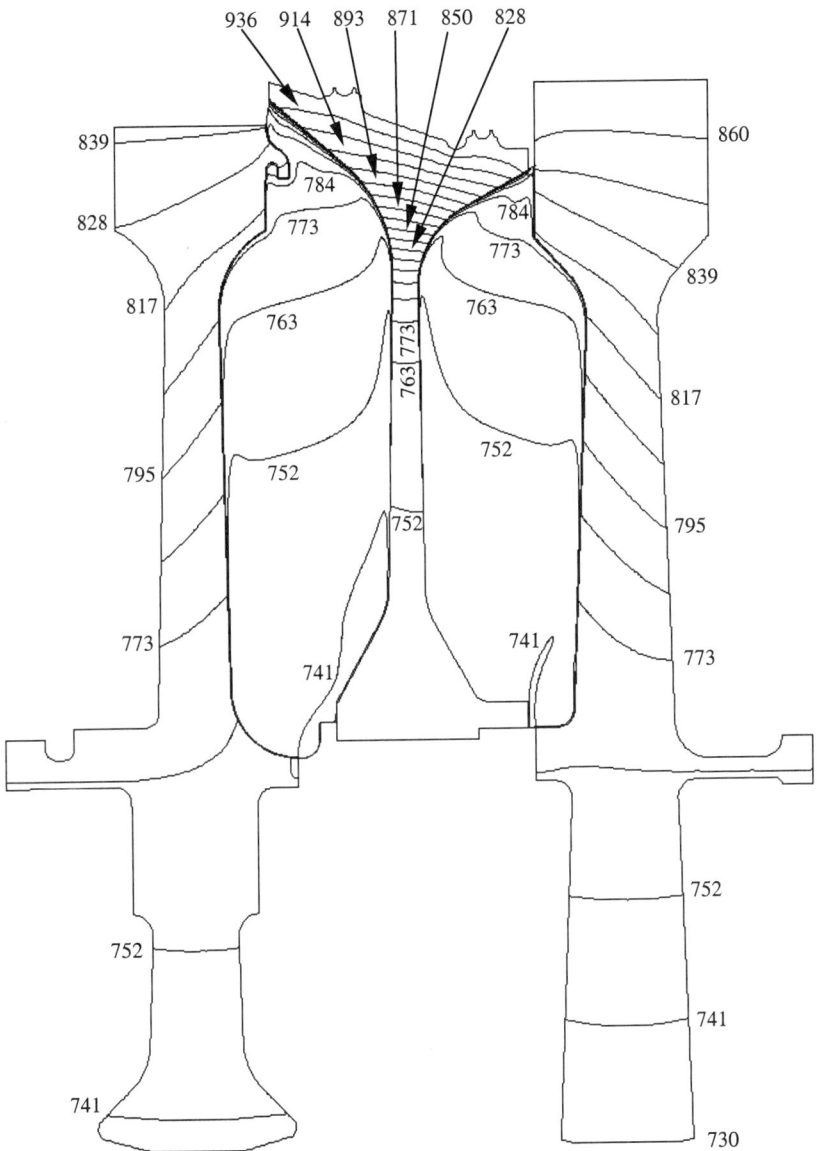

Figure 5. The predicted temperatures for both the fluid and solid zones of the turbine disc cavity region for a turboprop engine (temperatures are in degree Kelvin).

CONCLUSIONS

A numerical analysis of the turbulent flow and heat transfer in the cavity between two co-rotating discs with axial inlet and radial outflow of fluid has been performed. The predicted Nusselt number distributions are within 20 percent of the experimental data of Northrop and Owen[1]. By using three-dimensional models it has been shown that non-axisymmetric outlet geometry, which cannot be incorporated in the two-dimensional axisymmetric model, has little effect on the predicted Nusselt number distribution along the disc surface. Further work testing different turbulence models is required to improve the results.

A conjugate heat transfer analysis of the convective and conductive heat transfer in the 2nd - 3rd stage spacer of a turboprop engine has been performed. The analysis predicts a steep temperature gradient near the rim of the spacer. These data will enable a creep life analysis to be performed. The similarity between the flow pattern predicted with the complex engine geometry and that predicted with the simpler geometry lends support to the applicability of the CFD method to the real engine.

REFERENCES

1. Northrop, A. & Owen, J.M., 1988, Int. J. Heat & Fluid Flow, 9(1), pp27-36.
2. Chen, H.C. & Patel, V.C., 1988, AIAA Journal, 26(6), pp641-648.
3. Wilson, M., Chen, J.X. & Owen, J.M., 1996, Proc. Computers in Reciprocating Engines and Gas Turbines, IMechE HQ, London, 9-10 January.
4. Morse, A.P. & Ong, C.L., 1992, J. Turbomachinery, 114, 247-255.
5. Owen, J.M. and Rogers, R.H., 1995, Flow and Heat Transfer in Rotating Disc Systems: Vol. 2, Rotating Cavities, Research Studies Press, Taunton, U.K.

Verifying Heat Transfer Analysis of High Pressure Cooled Turbine Blades and Disk

SHIGEMICHI YAMAWAKI

Advanced Technology Department, Aero-Engine & Space Operations
Ishikawajima-Harima Heavy Industries Co.,Ltd
229, Tonogaya, Mizuho-machi, Nishitama-gun, Tokyo, 190-1297, JAPAN

ABSTRACT: **To demonstrate cooling and heat transfer technology, a core engine test was conducted with a turbine inlet temperature 1700°C. Measurement data were compared with predictions for a vane, a blade, and a disk. Measured cooling effectiveness of the blade and the vane agreed well with predictions. CFD analysis was carried out for verification of the heat transfer coefficient which was adopted from a heat conduction analysis over the disk. The CFD model including bolt heads showed better results than an axisymmetric model.**

INTRODUCTION

To improve the performance of gas turbines, the specification for turbine inlet temperature continues to increase. One example is the super/hypersonic transport propulsion system research project in Japan which began in 1989 and has recently been completed[1]. A turbine inlet temperature was planned in the 1700°C class. To demonstrate cooling and heat transfer technology, a core engine test was conducted.

Heat transfer around a turbine blade and vane and a disk is very complicated in an actual engine. To maintain sufficient life time of these hot parts, verifying heat transfer analysis is very important.

Generally, various empirical correlations are applied to thermal boundary condition in a heat transfer analysis and are modified to meet measurement data[2]. But it is difficult to develop general correlations for satisfying geometrically different engines, because original correlations are based on experimental data in a simple configuration.

CFD analysis is used increasingly for estimating flow field and heat transfer distribution around blades and disks and for applying heat transfer analysis instead of using empirical methods. Although CFD analysis is a very useful tool, it is necessary to improve prediction accuracy and to reduce calculating time, while paying attention to analysis model and grids.

In this paper, measurement data were compared with predictions for a vane, a blade, and a disk. CFD analysis was carried out for verification of the heat transfer coefficient which was adopted from a heat conduction analysis over the disk.

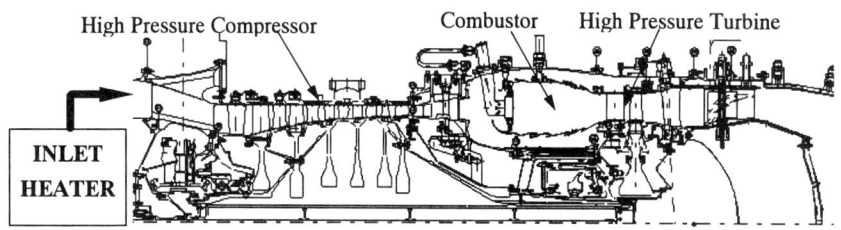

Fig. 1 High Temperature Core Engine

EXPERIMENTS

A single stage high pressure turbine, installed in a high temperature core engine[3](Fig. 1), was instrumented at many measurement locations for metal and secondary air temperatures, and for static pressures. The inlet temperature of the core engine was controlled by a inlet heater. The maximum turbine inlet temperature(TIT) was 1700°C. The maximum discharge temperature of the high pressure compressor was about 560°C. The turbine inlet pressure was about 0.25MPa. Although this pressure means lower heat flux than in actual engines, it also causes lower measurement error for thermocouples. Steady state and a transient conditions were measured. During transient operation, the TIT was lower because the inlet heater was not operated.

The turbine blades and vanes were fully film-cooled by shaped holes. The inside of

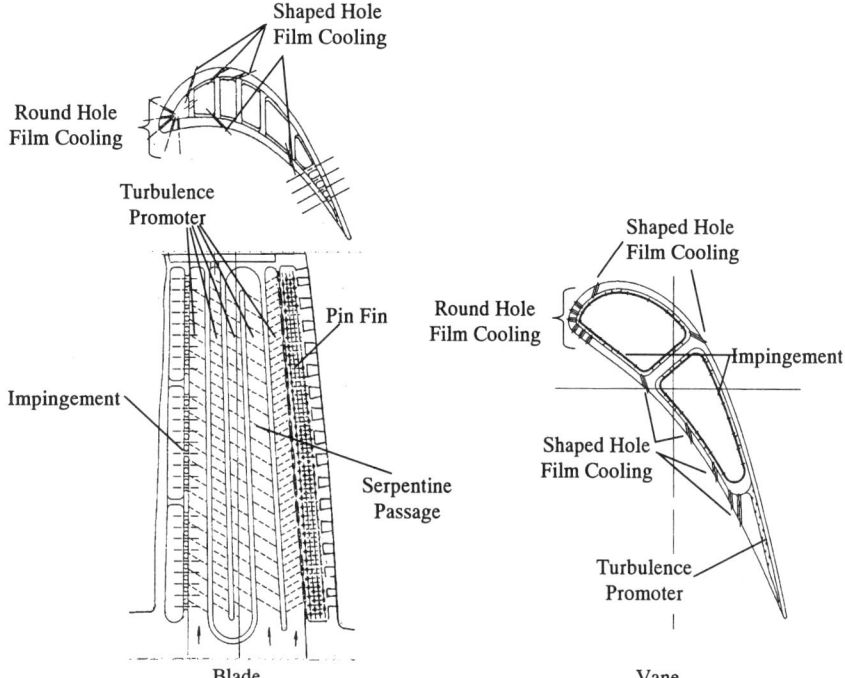

Fig.2 Cooling Configuration of Turbine Blade and Vane

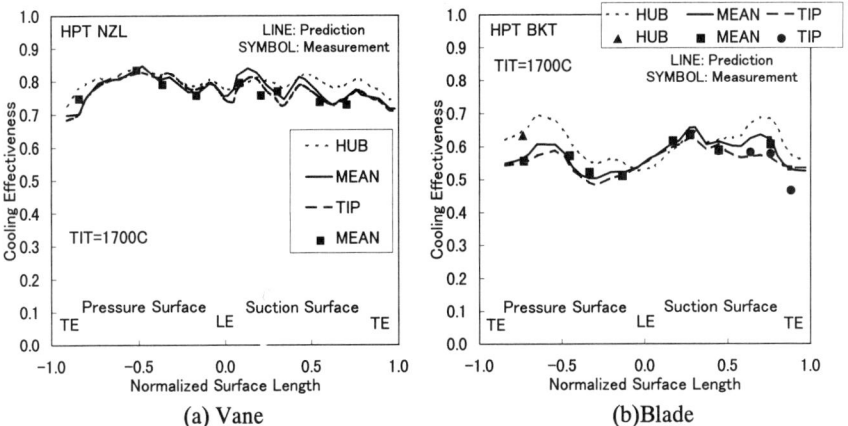

Fig. 3 Comparison of measured and predicted cooling effectiveness

blades and vanes were cooled by impingement and turbulence promoters as seen in conventional turbine blades.(Fig.2) A film hole configuration was selected after a few basic tests at low temperature[4,5]. Prior to the engine test, scaled up model tests were conducted in a hot wind tunnel for verification of cooling design. A thermal barrier coating was not used in order to measure a net cooling performance in the engine test. The surface temperature on the vanes, blades, and disks were measured by embedded thermocouples. For the vane, small diameter stainless sheath thermocouples were set in a groove on the vane. For the blade, alumel and chromel lines were set on the surface with alumina coating. For the disk, the spot welded junctions of the sheathed thermocouple edges were covered with Inconel thin plate. Heat conduction errors at the junction of thermocouples resulted in errors of about 20K for the vane and blade, and about 3K for the disk.

RESULTS AND DISCUSSIONS

Turbine Blades

The original cooling design showed the predicted high cooling effectiveness in the scaled up model test. After manufacturing of actual turbine blades and vane, the core engine tests were carried out.

Fig. 3 shows distributions of cooling effectiveness, defined as

$$\eta = (T_g - T_w)/(T_g - T_c),$$

where T_g is the assumed gas temperature calculated from verified cycle analysis and measured temperature profiles in combustor rig tests; T_w is the measured surface temperatures; T_c is the measured compressor discharge air temperature.

Results showed a high cooling effectiveness and flat distribution for the vane and the blade. A prediction was carried out using an in-house 2D heat conduction analysis for three sections: 20%, 48% and 80% span from the hub. Measured and predicted cooling effectiveness agreed well for both blade and vane. One exception is the blade tip near the trailing edge where a lower effectiveness was measured than was predicted due to tip leakage flow.

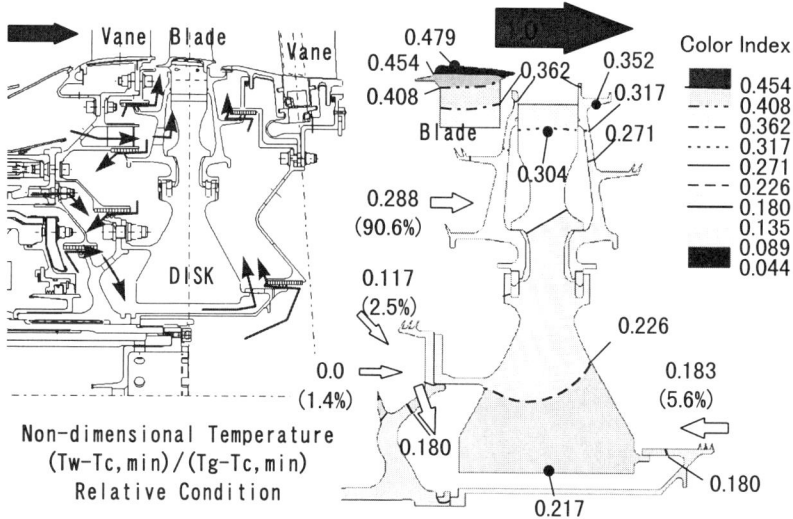

Fig.4 Steady State Temperature Distribution (Analysis Result)

Turbine Disk

The temperature prediction was carried out on an axi-symmetric model with 3D corrections provided by commercial heat conduction analysis software. Fig.4 shows predicted steady-state temperature distribution on the disk. The temperature is normalized by the highest and the lowest temperature according to the following equation.

$$\Phi = (T_w - T_{c,min})/(T_g - T_{c,min})$$

where T_w is disk surface temperature, T_g is main flow gas temperature, $T_{c,min}$ is the lowest secondary air temperature supplied near disk bore. All temperatures are relative. Main cooling air for the disk is the high pressure compressor discharge air which is supplied by the swirler located on the inner side of the vane.

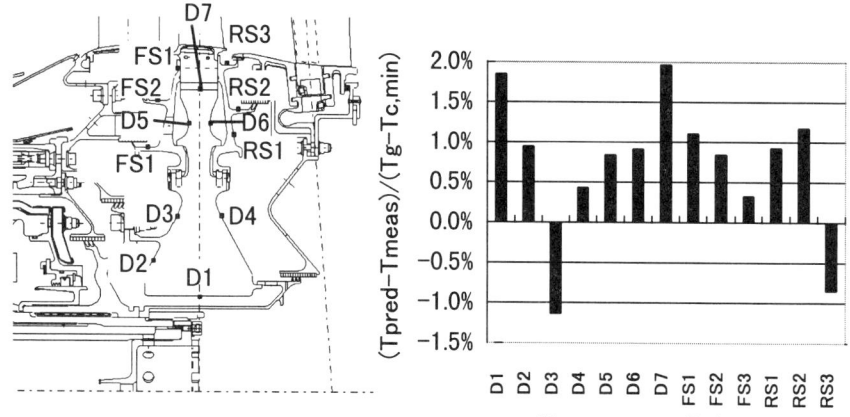

Fig.5 Prediction Error in Steady State Condition

The rotational Reynolds number at steady state condition is given by

$Re_\Phi = \rho\Omega b^2/\mu = 3.4\times 10^6$

where Ω is rotational speed, b is the disk rim radius, ρ and μ are density and viscosity at the main secondary air supply conditions. Nondimensional mass flow rate of main secondary air at steady state condition is

$Cw = w/(\mu b) = 5.1\times 10^4$

where w is mass flow rate.

Prediction error, the difference between the predicted and measured temperature, is shown in Fig.5. The error is defined by the following equation,

error = (Tpred−Tmeas)/(Tg−Tc,min)

where Tpred is predicted temperature, Tmeas is measured temperature. At most measurement points, predicted temperatures are higher than measured. All prediction errors are lower than 2% of the temperature range between gas temperature and the lowest secondary air temperature; 1% error means about 11K. These results are based on the prediction applied with accurate secondary air supply temperatures which is estimated by the measured air temperature at each supply location. In the steady state condition, temperature differences between secondary air and disk surface are generally small. We found that to reduce prediction error, it is important to estimate correct secondary air supply temperature

Comparisons between measured and predicted transient temperature for a typical location of the disk are shown in Fig.6 and Fig.7. These profiles are normalized with measured temperature at the engine start, Tstart, and the highest measured temperature in a steady state condition, Tmax. In these figures, measured and predicted temperature profiles agree well when the engine is in the start to idle condition, the idle to steady state condition, and in the engine decelerate. In the transient measurement, the inlet heater was not operated. Steady state conditions are not the same temperature level as previous figures, because the turbine inlet temperature is low.

The measured transient temperature change showed faster response than predicted on the front face of the disk surface as shown in Fig.8. This means that the heat transfer coefficient applied in the predictions is too low. To produce the same temperature

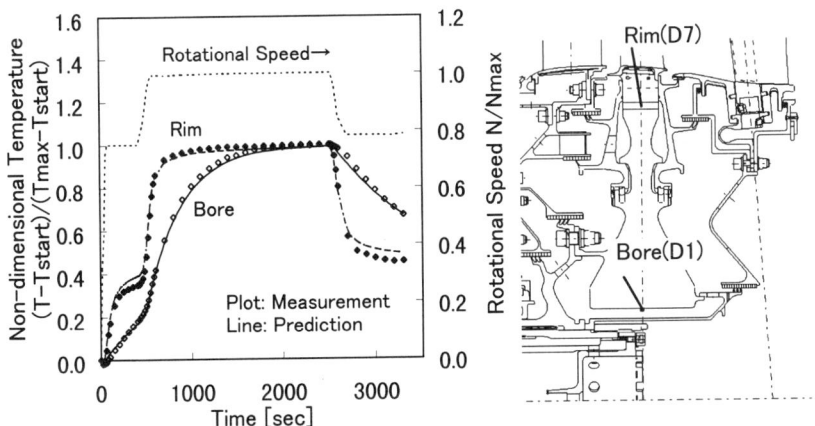

Fig.6 Comparisons of measured and predicted transient temperature for the disk rim and bore

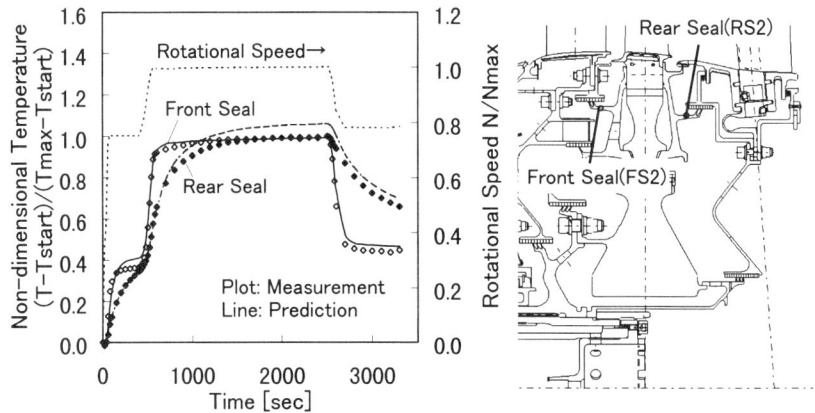

Fig.7 Comparisons of measured and predicted
transient temperature for the disk front and rear seal

response, the heat transfer coefficient should be 4 times larger than the original base value. The rotor-stator cavity for this measurement location is not geometrically simple and has bolt heads on the rotor side. Initially, the relative velocity on the disk was assumed to be very low, because the lower side of this cavity is an enclosed rotor surface with bolt heads on one surface. To understand the possibility of the higher heat transfer, CFD analysis was applied to the disk cavity. Our in-house developed CFD code[6] using unstructured grid with a standard high Reynolds number k-ε model with wall function (Launder and Spalding) was used in the analysis.

This cavity is a radial inflow type disk cavity. In the CFD analysis, 0.84 swirl ratio was assumed at the secondary air inlet boundary. The swirl ratio is defined as,

Ratio = V_θ / U,

where V_θ is circumferential velocity component; U is local rotating speed. Fig.9 shows

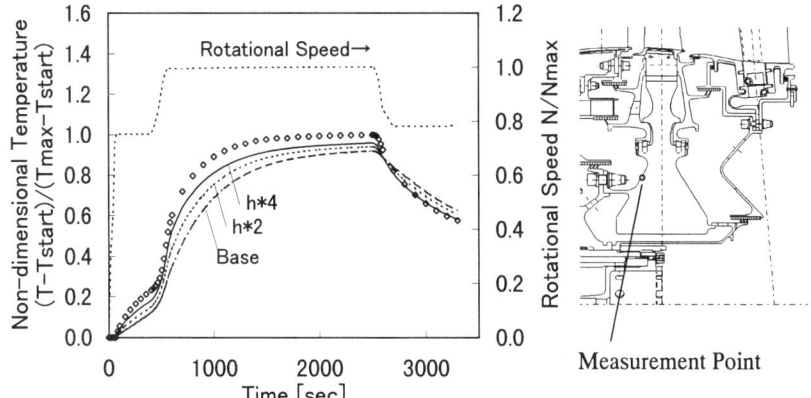

Fig.8 Comparisons of measured and predicted
transient temperature for the disk front web

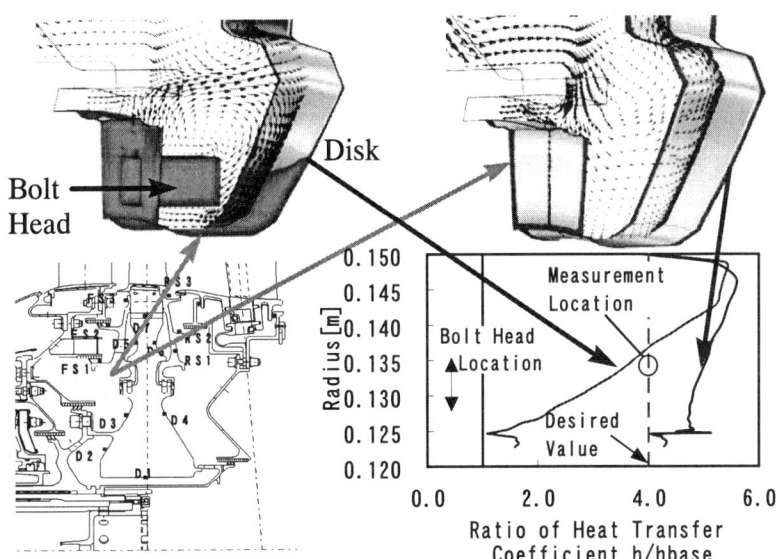

Fig. 9 CFD results for the disk

two flow vectors and the distributions of the heat transfer coefficient. The difference between the two analyses is that one includes bolt heads, and the other is an axisymmetric model. The bolt heads produce different secondary flow near the disk surface. Both results of analysis show higher heat transfer coefficient compared with original h_{base}, which was estimated by a correlation for a simple rotor-stator cavity with assumed low relative velocity. Although low relative circumferential velocity, which means high swirl ratio, was observed in the result of the CFD analysis, unexpected high radial flow component appeared in the analysis. This relative flow may produces high heat transfer on the disk. The lower curve of the two, (closer to zero) with the bolt head model, produces a better agreement with a desired heat transfer coefficient and consequently better temperature response.

CONCLUSIONS

The core engine test was completed successfully at a turbine inlet temperature of 1700°C.
Conclusions of this paper are as follows:
1. The turbine blade and vane with shaped holes showed high cooling effectiveness in the core engine test. Measured cooling effectiveness agrees well with predictions.
2. For the geometrically complicated disk cavity, the results of CFD analysis showed the higher heat transfer coefficient which is expected from the measured temperature response on the disk. The CFD model including bolt heads showed better results than an axi-symmetric model.

ACKNOWLEDGMENTS

This work was conducted under the entrustment contract with NEDO (New Energy and Industrial Technology Development Organization) as a part of the National Research and Development Program (Industrial Science and Technology Frontier Program) of Agency of Industrial Science and Technology (AIST), Ministry of International Trade and Industry (MITI).

REFERENCES

1. Ishizawa, K. 1999. Overall Review of Engineering Research for Super/Hyper-sonic Transport Propulsion System (HYPR). *In* Proceedings of the International Gas Turbine Congress 1999 Kobe, Japan, November 14-19: 211-218.
2. Dixon, J. A. 1999. Gas Turbine Critical Component Temperature Predictions for Fatigue Life and Integrity Considerations, 14th International Symposium on Air Breathing Engines, Florence, Italy, September 5-10, paper, ISABE 99-7117
3. Fujimura, T. et al. 1998. Reaseach and Development of High Temperature Core Engine for HST, 34th AIAA/ASME/SAE/ASEE Joint Propulsion Conference & Exhibit, Cleveland, OH, July 13-15,paper AIAA 98-3279, July 13-15
4. Yamawaki, S., et al. 1996. Study of Film Cooling with Shaped Holes for a Hydrogen Combustion Turbine Blade, *In* Proceedings of the 11th World Hydrogen Energy Conference, Stuttgart, Germany, June 23-28, Veziroglu, T. N. et al. Eds: 1905-1908.
5. Haven, B.A. et al. 1997. Anti-Kidney Pair of Vortices in Shaped Holes and Their Influence on Film Cooling Effectiveness, The International Gas Turbine & Aeroengine Congress & Exhibition, ASME Paper, 97-GT-45, June 2-5
6. Ohkita, Y., et al. 1997. Numerical Simulation of Flow and Heat Transfer in 3D Complicated Geometries Using Unstructured Grids, 13th AIAA Computational Fluid Dynamics Conference, June 29-July 2, AIAA-97-1948

COMPUTATION OF FLOW AND HEAT TRANSFER IN ROTATING CAVITIES WITH PERIPHERAL FLOW OF COOLING AIR

Muhsin KILIÇ

*Uludağ University, Faculty of Engineering and Architecture,
Department of Mechanical Engineering, Görükle Campus, TR-16059 Bursa, TÜRKİYE*

ABSTRACT: Numerical solutions of the Navier-Stokes equations have been used to model the flow and the heat transfer that occurs in the internal cooling-air systems of gas turbines. Computations are performed to study the effect of gap ratio, Reynolds number and the mass flow rate on the flow and the heat transfer structure inside isothermal and heated rotating cavities with peripheral flow of cooling air. Computations are compared with some of the recent experimental work on flow and heat transfer in rotating-cavities. The agreement between the computed and the available experimental data is reasonably good.

INTRODUCTION

Rotating-disc systems can be used to model the flow and the heat transfer that occurs in the internal cooling-air systems of gas turbines. The flow and the heat transfer between corotating compressor or turbine discs are modelled by a rotating cavity. The main parameters that affect the distributions of the cavity local heat transfer coefficient are coolant flow rate, disc temperature, rotation speed, and cavity configuration.

The geometries in real engines are very complicated. In order to understand the flow over these complicated surfaces, it is usual to simulate the geometries by plane rotating-disc systems, as shown in Figure 1. The free disc, Figure 1.(a), provides a base for all rotating-disc systems. As mentioned above, a turbine disc usually rotates next to either a stationary or another rotating disc, and these can be simulated with a rotor-stator system or a rotating cavity as shown in Figures 1.(b) and 1.(c-f), respectively.

As shown schematicaly in Figure 1(f), in some gas turbine configurations, the cooling air for the rotating turbine discs is supplied through a stationary casing at the periphery of the disks. Cooling air enters the annular rotating cavity between the disks through nozzles in the stationary casing and leaves through the small clearances between the disks and the casing.

It is convenient to define some of the nondimensional variables that specify the system. The nondimensional radii, x and x_a, are defined as

$$x = r/b \quad , \quad x_a = a/b \tag{1}$$

the gap ratio, G, as
$$G = s/b \tag{2}$$
the rotational Reynolds number, Re_ϕ, as
$$Re_\phi = \frac{\rho \Omega b^2}{\mu} \tag{3}$$
and the non dimensional flow rate (for a superposed inflow or outflow), C_w, as
$$C_w = \frac{\dot{m}}{\mu b} \tag{4}$$
The local Nusselt number, Nu, is defined as
$$Nu = \frac{rq}{k(T_s - T_{ref})} \tag{5}$$
where T_{ref} is a suitable reference temperature (the temperature of the air at inlet to the system was used in this work).

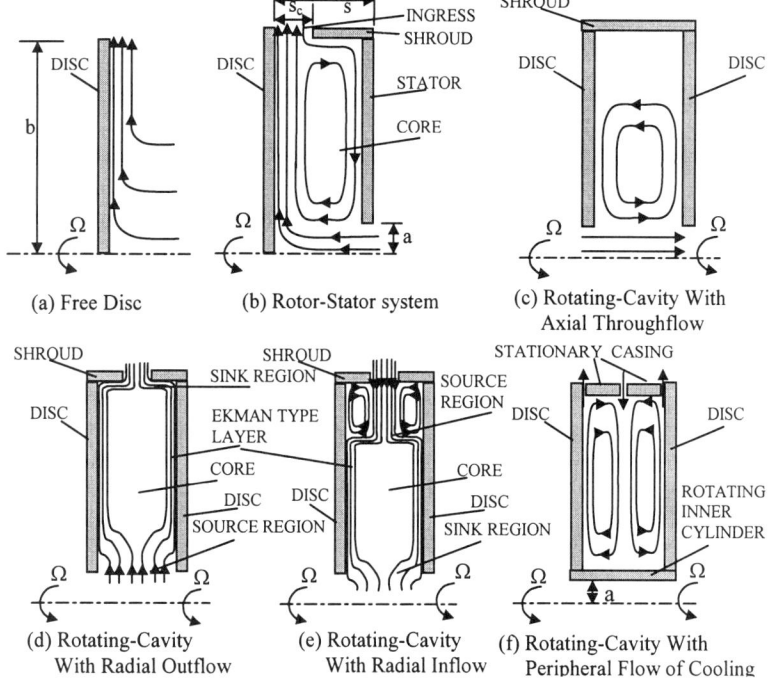

Figure 1. Schematic diagram of rotating-disc systems.

Mirzaee et al.[1] reported a combined computational and experimental study of the heat transfer in a rotating cavity with a peripheral inflow and outflow of cooling air for rotational Reynolds numbers up to $Re_\phi = 1.5 \times 10^6$ and flow rates up to $|C_w| = 3000$. In the core, the tangential component of velocity is invariant with z and the measured values V_ϕ conform to a combined free and forced vortex, or Rankine vortex (see Owen and Rogers[2]), where $V_\phi / (\Omega r) = A\, x^{-2} + B$ and the coefficients A and B depend on C_w and Re_ϕ. Although, the measured values of $V_\phi/(\Omega r)$ show a Rankine-vortex behaviour that is not accurately captured by the computations in their study. The use of a "Richardson correction" in the Launder-Sharma low Reynolds number k-ε turbulence model improves the agreement between the computed and measured values of $V_\phi /(\Omega r)$. The measured and computed Nusselt numbers show that Nu increases as Re_ϕ and $|C_w|$ increase.

In this paper, an extension of the study of Mirzaee et al.[1] is presented. A computational study of the flow and heat transfer is presented for rotating cavities with peripheral inflow and outflow of cooling air for rotational Reynolds numbers up to Re_ϕ = 1.5×10^6 and flow rates up to $|C_w|$ = 3000. In order to investigate the effect of gap ratio on the flow and heat transfer in the cavity, two different gap ratios, G = 0.2 and 0.3, are considered. Comparisons with the available flow and heat transfer measurements are performed. Satisfactory agreement between the numerical and experimental results are obtained.

COMPUTATIONAL PROCEDURE

The computer code used was a finite-volume, axisymmetric elliptic multigrid solver, employing the Launder-Sharma low-Reynolds-number k-ε turbulence model, details could be found in the references Kılıç[3], Kılıç et al.[4], Gan et al.[5]. In cylindrical-polar coordinate system, the time averaged, axisymmetric, steady-state conservation equations for mass, momentum, energy, and the turbulence quantities turbulence kinetic energy, k, and its dissipation rate, ε, together with the low-Reynolds number k-ε turbulence model, were solved to predict the flow and the heat transfer in the cavity. A finite difference mesh of 169x113 (axial-radial) nodes was employed, with a combination of geometrically expanding/contracting grid spacings to cluster nodes near the discs and the casings. For all the cases considered, the grid point closest to the disk was set to ensure that y^+ < 0.5 and geometric expansion/contraction factors did not exceed 1.20. A Richardson number based correction to the k-ε turbulence model was employed[1,6].

The computational geometry matched the experimental one described by Mirzaee et al.[1] :the discs radii were b=382 mm, and inner cylinder radius was a=191 mm (a/b = 0.5), and the two discs were spaced an axial distance of s=113 mm (G=s/b=0.3). The cooling air inlet was modelled by a circumferential slot which is 11.3 mm wide and located midway between the discs. The air left the system through the two symmetric clearances between the stationary casings and the discs; each clearance was 1 mm wide.

For the solid boundaries (i.e. discs and casings surfaces), no-slip boundary conditions were used. For the inlet slot, the air was assumed to enter the cavity radially with uniform radial velocity and without swirl at r = b, so that $V_r = \dot{m}/(2\pi\rho b s_c)$, $V_\phi = 0$, $V_z = 0$. The mass flow rates out of the systems, at the clearances adjacent to each disc, were equal and a uniform radial velocity was imposed to ensure the mass balance between the inflow and outflows. For non-isothermal computations, the fluid was assumed to enter the cavity with uniform temperature, and the temperature distribution along the heated disc wall was obtained from a polynomial fit to the experimental data of Mirzaee et al.[1], the other disc and the casings were assumed to be adiabatic. At the outlet clearances zero normal derivative conditions were applied for the other variables.

RESULTS AND DISCUSSION

Computed flow structure:

Figure 2 shows the computed streamlines in the r-z plane for $Re_\phi = 3.75 \times 10^5$, flow rates $|C_w|$ = 0, 1500 and 3000 and G = 0.2. There is recirculating flow, symmetric about

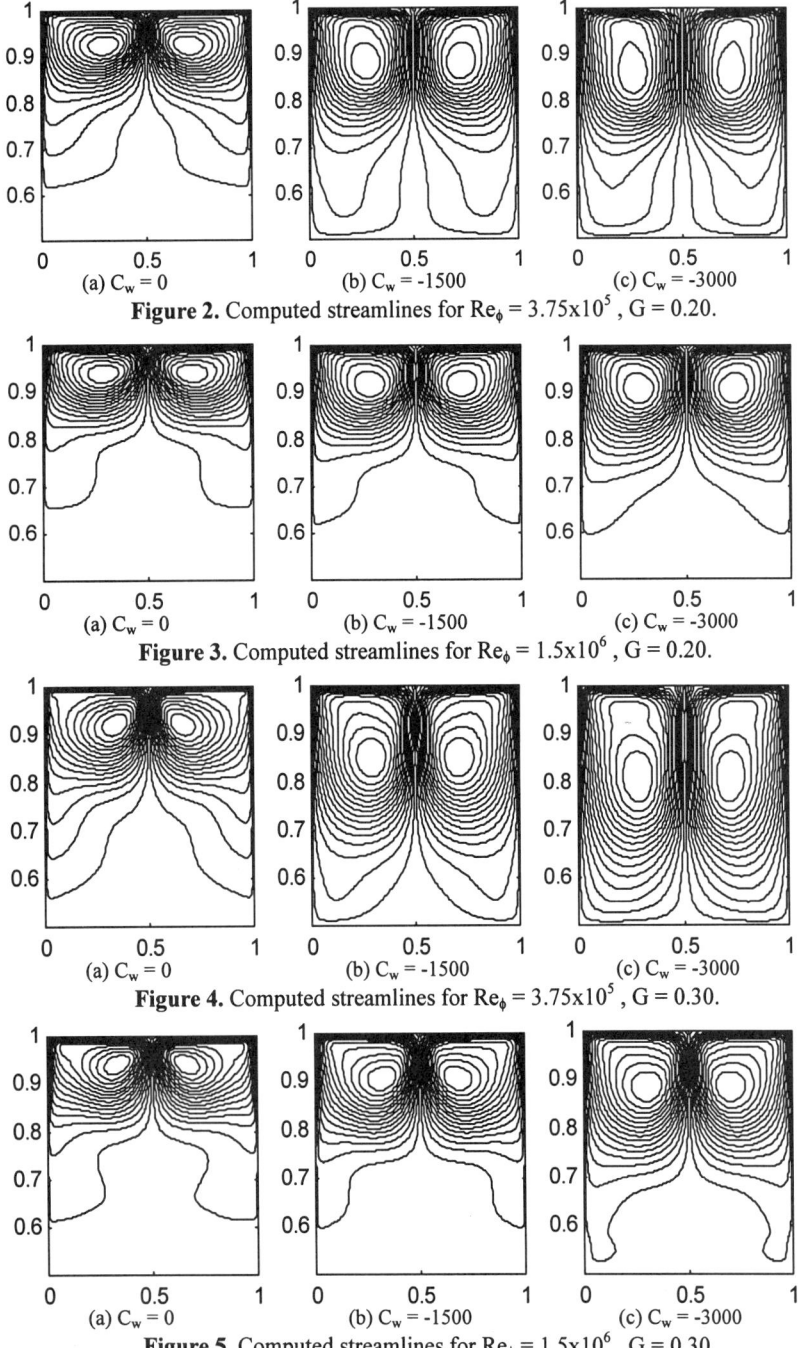

Figure 2. Computed streamlines for $Re_\phi = 3.75 \times 10^5$, $G = 0.20$.

Figure 3. Computed streamlines for $Re_\phi = 1.5 \times 10^6$, $G = 0.20$.

Figure 4. Computed streamlines for $Re_\phi = 3.75 \times 10^5$, $G = 0.30$.

Figure 5. Computed streamlines for $Re_\phi = 1.5 \times 10^6$, $G = 0.30$.

the midplane. The radial extent of the recirculation zone increases with the nondimensional mass flow rate. The location of the recirculation zone core moves towards the center of the cavity with increasing nondimensional mass flow rate. This is caused by the exchange of angular momentum between the incoming air and the rotating air. In Figure 3, the computed streamlines for $Re_\phi = 1.5 \times 10^6$, flow rates $|C_w| = 0$, 1500 and 3000 and $G = 0.2$, are given. Similar conclusions can be drawn from the streamlines in Figure 3. Comparing Figure 2 and Figure 3, it can be said that increasing Re_ϕ causes a decrease in the radial extent of the recirculating zone. This is probably caused by the increasing pumping power of the discs. For $G = 0.30$, the computed streamlines are given in Figure 4 and 5. Flow structures are similar to the cases presented in Figure 2 and 3. However, the radial extents of the recirculating zone are larger than the cases for $G = 0.2$. The location of the recirculation zone core moves towards the center of the cavity with increasing gap ratio.

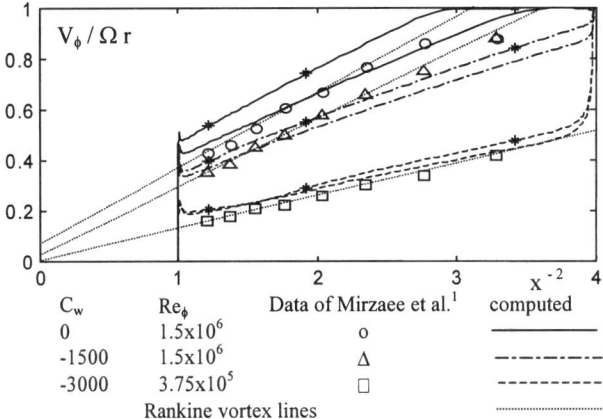

Figure 6. Variation of $V_\phi /\Omega r$ with x^{-2} for $z/s = 0.8$.(Lines with * for $G = 0.20$, others for $G = 0.30$).

Figure 6 shows the variation of $V_\phi /(\Omega r)$ with x^{-2}. Although there are some differences between the measured and computed results, Rankine-vortex behaviour is captured by the computations. It can be seen from Figure 6 that decreasing G increases the solid body rotation in the cavity.

For $Re_\phi = 3.75 \times 10^5$ and 1.5×10^6, the variations of $V_r /(\Omega r)$ with z/s are presented in Figures 7 and 8 respectively. The results presented for $|C_w| = 0$, 1500 and 3000, $G = 0.2$ and 0.3. Computations of the radial velocity showed that there is radial outflow in thin boundary layers on the disk and radial inflow in the core between the boundary layers. The magnitude of the radial velocity in the core increases with the mass flow rate. For $G = 0.3$ at large x values the computed velocity distributions show a peaky distribution, but this is not present at the cases for $G = 0.2$.

Heat transfer results:

Figures 9 and 10 show comparisons between the computed and the measured (Mirzaee et al.[1]) distribution of temperature and Nusselt numbers. The results are presented for $Re_\phi = 3.75 \times 10^5$ and 1.5×10^6, flow rates $|C_w| = 0$, 1500 and 3000 and $G = 0.2$ and $G=0.3$. The variations of temperature with x are shown in Figure 9 (a) and (b),

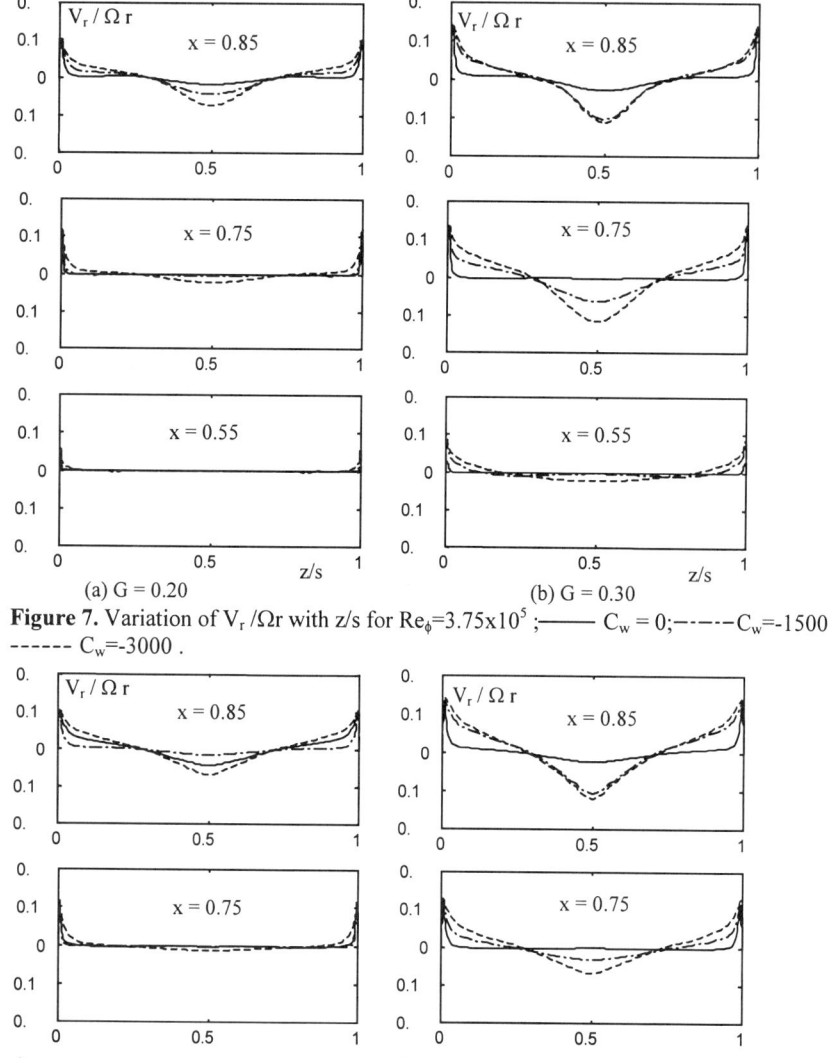

Figure 7. Variation of $V_r/\Omega r$ with z/s for $Re_\phi=3.75\times10^5$;——— $C_w = 0$;— - — - C_w=-1500 ; — — — — C_w=-3000 .

Figure 8. Variation of $V_r/\Omega r$ with z/s for $Re_\phi=1.50\times10^6$;——— $C_w = 0$;— - — - C_w=-1500 ; — — — — C_w=-3000 .

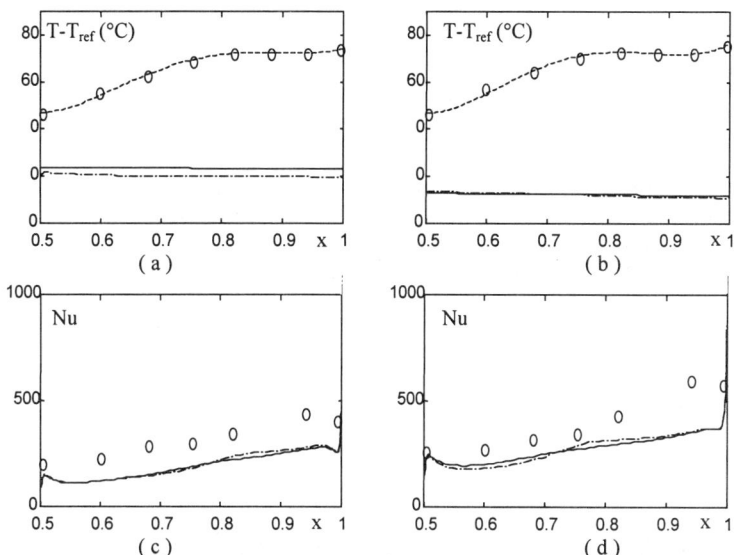

Figure 9. Variation of temperatures and Nusselt numbers with x for $Re_\phi = 3.75 \times 10^5$: (a) and (c) for $C_w = -1500$, (b) and (d) for $C_w = -3000$, o Measurements, G=0.30, Mirzaee et al.[1] ; ----- fitted curve (heated disc); —·—·— computed (adiabatic disc), G = 0.20; ——— computed (adiabatic disc), G = 0.30.

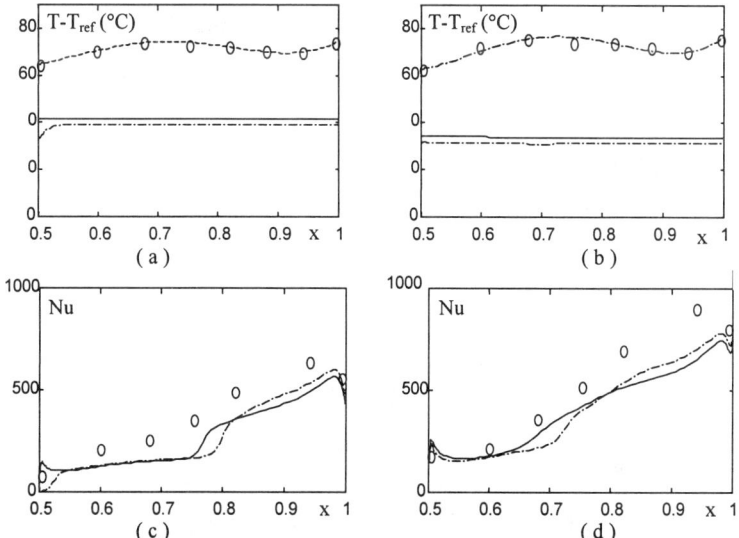

Figure 10. Variation of temperatures and Nusselt numbers with x for $Re_\phi = 1.5 \times 10^6$: (a) and (c) for $C_w = -1500$, (b) and (d) for $C_w = -3000$, o Measurements, G=0.30, Mirzaee et al.[1] ; ----- fitted curve (heated disc); —·—·— computed (adiabatic disc), G = 0.20; ——— computed (adiabatic disc), G = 0.30.

Figure 10 (a) and (b). Referring to these Figures, the fitted temperatures on the heated disk were used as boundary conditions for the solution of the energy equation. Computed temperature distributions on the adiabatic discs can be seen in Figure 9 and 10. In general, the temperatures on the adiabatic discs for G = 0.2 are smaller than the cases for G = 0.3.

Considering the Nusselt number distributions given in Figure 9 and 10, it can be seen that the computed values of Nu exhibit the experimental trends but the experimental values of Nu are underestimated especially at larger values of x. Comparing Figure 9 and 10, it can be seen that the magnitude of the Nusselt numbers increases with increasing Re_ϕ and $|C_w|$. It can be also seen that the existence of the recirculation zone increases the values of the Nusselt numbers. For x > 0.8, the values of the Nusselt numbers for G=0.3 are smaller than the cases for G=0.2. For x < 0.8, however, the values of the Nusselt numbers for G=0.3 are greater than the cases for G=0.2. This is probably related to the existence of the recirculation zone and the magnitude of the recirculation velocities as well as the magnitude of the angular velocity of the core.

CONCLUSION

Computation of flow and convective heat transfer in rotating cavities with peripheral flow of cooling air has been performed. The effects of rotational Reynolds number, mass flow rate, gap ratio are investigated. Computations are compared with some of the recent experimental work on flow and heat transfer in rotating-cavities. The agreement between the computed and the available experimental data is reasonably good. The following conclusions can be drawn from the results.

Rankine-vortex behaviour is captured by the computations. The location of the recirculation zone core moves towards the center of the cavity with increasing gap ratio and nondimensional mass flow rate. Computations of the radial velocity showed that there is radial outflow in thin boundary layers on the disk and radial inflow in the core between the boundary layers. The magnitude of the radial velocity in the core increases with the mass flow rate.

For the same conditions, the temperatures on the adiabatic discs increases with increasing gap ratio. The existence of the recirculation zone increases the values of the Nusselt numbers. The magnitude of the Nusselt numbers increases with increasing Re_ϕ and $|C_w|$.

ACKNOWLEDGMENTS

The author wish to thank the Research Fund of the Uludağ University for providing financial support for this work under the project number 99/30.

REFERENCES

1. Mirzaee, I., Gan, X., Wilson, M. and Owen, J.M. 1998. J. Turbomachinery, Vol. 120: 818-823.
2. Owen, J.M. and Rogers, R.H. 1995. Flow and Heat transfer in rotating systems. Vol 2: Rotating-Cavities, Research Studies Press, Taunton, U.K..
3. Kılıç, M. 1993. Flow Between Contra-Rotating Discs, *Ph.D. Thesis*, University of Bath, Bath, UK.
4. Kılıç, M., Gan, X. and Owen, J.M. 1994. J. Fluid Mech., Vol. 281:119-135.
5. Gan, X., Kılıç, M. and Owen, J.M. 1994. Int. J. Heat Fluid Flow, Vol.15:438-446.
6. Sloan, D.G., Smith, P.J., and Douglas Smoot, L., 1986. Prog. Energy Combust. Sci., Vol.12:163-250.

Index of Contributors

Abdon, A., 417–423
Acharya, S., 110–125, 424–431
Aoki, S., 305–312, 473–480
Arts, T., 126–134

Bae, J.C., 233–240
Baggio, P., 464–472
Bario, F., 313–320
Barthet, S., 313–320, 369–376
Bataille, F., 385–392
Battisti, L., 464–472
Bazdidi-Tehrani, F., 393–400
Bock, S., 432–439
Boelcs, A., 297–304
Bousgarbiès, J.L., 337–344
Brereton, G., 52–63
Brevet, P., 409–416
Brewster, R.A., 440–447
Bunker, R.S., 64–79

Chen, P-H., 353–360
Cho, H.H., 233–240, 281–288
Choi, J.H., 281–288
Chyu, M.K., 27–36
Croce, G., 273–280

Ding, P-P., 353–360
Dittmar, J., 321–328
Dorignac, E., 337–344, 409–416
Dutta, S., 162–178

Eriksson, L-E., 241–248

Faragher, J., 497–504
Foucault, E., 337–344
Frank, S.L.F., 257–264

Ginibre, P., 377–384
Glezer, B., 222–232
Goldstein, R.J., xi, 1–10, 147–161
Gritsch, M., 401–408
Guo, S.M., 361–368

Han, B., 147–161
Han, J-C., 162–178
Hennecke, D.K., 432–439
Hermanson, K., 448–455
Hoda, A., 110–125
Holmer, M-L., 241–248
Hung, M-S., 353–360

Ijichi, N., 289–296
Inaoka, K., 481–488
Ishida, H., 345–352
Iwai, H., 481–488
Iyer, G.R., 265–272

Jonnavithula, S., 440–447
Jung, I.S., 321–328

Kaszeta, R.W., 37–51
Kiliç, M., 513–520
Kim, W.S., 233–240
Kimoto, H., 345–352
Kirillov, A.I., 456–463
Kleinstück, R., 432–439
Kulisa, P., 369–376
Kumada, M., 289–296

Lallemand, A., 385–392
Langston, L.S., 11–26
Leboeuf, F., 95–109
Lee, H-W., 329–336
Lee, J.S., 321–328, 329–336
Lefebvre, M., 377–384
Liamis, N., 377–384

Mahmoodi, A.A., 393–400
Martelli, F., 80–94
Mathelin, L., 385–392
Morinaga, M., 489–496

Noirault, P., 337–344
Nuntadusit, C., 345–352

Obata, M., 289–296
Oldfield, M.L.G., 361–368
Ooi, A., 497–504
Owen, J.M., 206–221

Parneix, S., 448–455

Rathjen, L., 432–439
Rawlinson, A.J., 361–368
Rhee, D.H., 281–288
Ris, V.V., 456–463

Sargison, J.E., 361–368
Saumweber, C., 401–408
Schulz, A., 135–146, 249–256, 321–328, 401–408
Semmler, K., 179–193, 448–455
Sgarzi, O., 95–109
Shih, T.I-P., 52–63
Simon, T.W., 1–10, 37–51
Smirnov, E.M., 456–463
Srinivasan, V., 1–10
Sunden, B., 241–248, 417–423
Suslov, D., 249–256
Suzuki, K., 481–488

Takahashi, H., 345–352
Takahashi, T., 489–496

Takeishi, K.-I., 305–312, 345–352, 473–480
Tatsumi, K., 481–488
Tyagi, M., 110–125

Ukai, T., 345–352

Vogel, G., 297–304
Von Wolfersdorf, J., 179–193, 448–455
Vullierme, J.J., 337–344, 409–416

Weigand, B., 179–193
Wilson, M., 206–221
Wittig, S., 249–256, 321–328, 401–408

Yamawaki, S., 505–512
Yavuzkurt, S., 265–272
Yoshida, T., 194–205
Yu, M.S., 233–240

Zaitsev, D.K., 456–463
Zhou, F., 424–431